PLANT TAXONOMY

Second Edition

AF782860

ABOUT THE AUTHOR

O P Sharma, with his 35 research articles published in national and international journals, 29 books written for university students, and 40 years experience of teaching, is an able researcher, established Indian author, and an experienced teacher. His areas of research include pollen morphology, angiosperm's anatomy and mycology, with special focus on Indian Cyperaceae (with particular interest on *Cyperus)*. Over a dozen of Dr Sharma's books have been published through internationally known publishers like Tata McGraw-Hill Education Pvt. Ltd., and Macmillan India Ltd. He has also revised *Economic Botany*, an internationally renowned text by late Professor Albert F Hill (Harvard University, USA), a publication of McGraw-Hill, New York. Encouraging reviews of his books have been published in reputed scientific journals, and his books on *Practical Botany* have received appreciations from some eminent botanists including J D Dodge (England), T Christensen (Denmark), J M Herr (USA) and C R Metcalfe (England).

Immediately after passing his MSc Botany in first division from C.C.S. University, Meerut and thereafter PhD from the same university, Dr Sharma started his teaching career in 1967 as a faculty member of Botany Department, Meerut College, Meerut, and retired from his active services as a Reader from the same department in 2007. Besides attending several national and international workshops, symposia and conferences during his four decades of teaching career, Dr Sharma is still enjoying his post-retirement inning as an active author.

PLANT TAXONOMY

Second Edition

O P Sharma
Retired Reader
Department of Botany,
Meerut College, Meerut

McGraw Hill Education (India) Private Limited

Published by McGraw Hill Education (India) Private Limited,
Registered Office: Anjana Complex No: 5/90 A, Butt Road, St. Thomas Mount, Chennai – 600016, Tamil Nadu, India.

Copyright © 2009, by McGraw Hill Education (India) Private Limited. All Rights Reserved.

21st reprint 2025

This book is protected under copyright law. Any unauthorized copying, reproduction, distribution, adaptation, conversion, transmission, storage or use of this publication or any part or extract of it, through any means or any medium, including electronic, mechanical, photocopied, recorded, or otherwise, whether for any direct/indirect commercial gain or not, and without the prior written consent of the publisher, is explicitly prohibited and will result in severe legal consequences. Such restrictions apply to social media channels/groups, YouTube videos, audiobooks, AI applications, and any other means of copying, reproducing, distributing, adapting, converting, transmitting, storing the content of this book, using current or future technologies.

This edition can be exported from India only by the publisher,
McGraw Hill Education (India) Private Limited

ISBN-13: 978-0-07-014159-9
ISBN-10: 0-07-014159-2

22 23 24 25 26 MAG 29 28 27 26 25

Information contained in this work has been obtained by McGraw Hill Education (India), from sources believed to be reliable. However, neither McGraw Hill Education (India) nor its authors guarantee the accuracy or completeness of any information published herein, and neither McGraw Hill Education (India) nor its authors shall be responsible for any errors, omissions, or damages arising out of use of this information. This work is published with the understanding that McGraw Hill Education (India) and its authors are supplying information but are not attempting to render professional services. If such services are required, the assistance of an appropriate professional should be sought.

Typeset at Bharati Composers, D-6/159 A, Sector – VI, Rohini, Delhi 110085, and
Printed and bound in India at Magic International Pvt. Ltd., Sector 31 (Industrial), Greater Noida 201306.

Visit us at: www.mheducation.co.in
Toll free in India: 1800 103 5875
Write to us at: support.india@mheducation.com
CIN: U80302TN2010PTC111532

CONTENTS

PREFACE TO THE SECOND EDITION

Plant Taxonomy has been revised to reflect global perspectives with particular emphasis on the current syllabi requirements of the Indian Universities. Published first in 1993 and reprinted 19 times in subsequent years upto 2008, speaks volumes about its overwhelming reception by the readers. The developments in the field of plant systematics, coupled with new and exciting discoveries contributing to the progress in taxonomical fields providing new information in classification of plants, have given impetus to bring out the revised edition. The new edition was also needed because of several changes made in the syllabi of Indian Universities, and also because of variations in the market trends and examination patterns.

The second edition of ***Plant Taxonomy*** is designed to present the current principles, practices and techniques of plant taxonomy and contemporary classifications, and also to describe some of the other important angiospermic families and groups. Those who are familiar with the first edition will be aware that there has been change of format to a larger page size, and also a completely different layout of the text. A considerable number of illustrations and chapter-wise revision questions have also been added in the revised text.

The major aim of this book is to provide a general overview of plant taxonomy in the most effective and positive manner to the students. This revised edition of ***Plant Taxonomy*** is a broad, up-to-date synthesis of this active and fascinating field of botany. Besides retaining all chapters of the first edition, this revised edition now also includes some new chapters, *viz.* Taxonomic Structure, Examination of Plant Specimen, Molecular Taxonomy, and Table of Major Characters of Discussed Families. Also added in this new edition are the detailed discussions of 16 more families of dicotyledons (*viz.* Saxifragaceae, Begoniaceae, Passifloraceae, Araliaceae, Caprifoliaceae, Campanulaceae, Ericaceae, Plumbaginaceae, Loganiaceae, Polemoniaceae, Pedaliaceae, Plantaginaceae, Aristolochiaceae, Piperaceae, Cannabinaceae, and Salicaceae) and 4 families of monocotyledons (*viz.* Bromeliaceae, Cannaceae, Juncaceae, and Typhaceae), thus increasing the number of discussed families from 69 to 89 out of total number of 200 families of flowering plants discussed by Bentham and Hooker. This number of discussed families in this book is now larger than all other available books on plant taxonomy in the Indian market. The revised edition also discusses some new subtopics in certain existing chapters as suggested by reviewers of the book. These subtopics include (i) Deep-freezing methods, (ii) Restructuring of ICBN with the addition of the details of St. Louis Code (1999) and

Vienna Code (2005), (iii) Phylocode, a new system of nomenclature, and (iv) Classification and major characters of subclasses and orders of dicotyledons and monocotyledons. In Comparative Tables of Selected Families, 13 new tables have been added, thus increasing the total number from 20 to 33 comparative tables. Another new feature is an index-listing of all relevant medicinal usages of discussed plants given at the end of the book. These additions will prove to be of great help to the students while preparing for their examination.

The major highlights of this book are as follows:

- Complete coverage of all important topics in Plant Taxonomy
  - Plant Classification
  - Plant Collection and Specimen Preparation
  - Identification and Botanical Nomenclature
  - Herbarium and Botanical Gardens
  - Phylogeny
  - Classification System and Description of 89 Families
- Four recent application-based Chapters
  - Numerical Taxonomy
  - Chemotaxonomy
  - Serotaxonomy
  - Molecular Taxonomy
- Examination Preparation Tools to aid students memorise basics and prepare efficiently for their examinations
  - All 89 Families summarised in a Table
  - 33 questions on 'Differentiate between the families'
  - 164 Chapter-end Examination Review Questions
- Rich pedagogy
  - More than 900 rich plant organs sketches and floral diagrams
  - Around 40 tables
- Updated as per International Botanical Congress Guidelines, (St. Louis, Missouri, July–August 1999) organised by IAPT: St. Louis Code, The International Code of Botanical Nomenclature.

Plant Taxonomy, in its present form, should now meet the needs of undergraduate students of all universities of the country taking botany as their major subject. It should also cater to the complete requirements of postgraduate students of Botany, Agriculture and Forestry in majority of the Indian Universities. As has already been in the past, the book should also be useful for and liked by students preparing for AIPMT, CPMT, NET, SLET, IAS, IFS, PCS and several other major competitive examinations.

A major highlight of this revised edition is that all chapters are largely independent, so that the teacher may choose the desired sequence of topics according to his/her syllabus requirements. The level of presentation is primarily for undergraduate students and based mainly on the assumption that the students will have had an introductory Botany course. The author hopes that the book in its present form will prove stimulating to serious amateurs, teachers as well as professionals who are specialists in other fields but use classifications and other taxonomic information about flowering plants.

Before going to press, the revised manuscript was critically reviewed by some experts in the field, including Dr Tariq Husain, Angiosperm Taxonomy and Herbarium Division, National Botanical Research Institute, Lucknow; Dr N K Soni, Department of Botany, Dr H S Gour University, Sagar; and Dr P J Handique, Department of Biotechnology, Gauhati University, Guwahati. Besides several positive comments and suggestions of these reviewers, one of them even commented that, "*This book is above all the books available on Plant Taxonomy in the country*". For the pains taken by these reviewers, the author acknowledges their help and thank them all by heart.

Without the help of my dear student, Dr Mayank Uday Charaya, Professor of Botany, C.C.S. University, Meerut, this book would have not come in its present form. I express my heartfelt feelings to Professor Charaya and his team for searching latest information through the Internet. For several types of suggestions, technical clarifications and encouragements, I also express my gratitude to Dr R Shiam (my teacher from BSc to PhD), Dr N P Saxena (Meerut), Dr Lokendra Singh (Meerut) and Dr H P Pandey (University of Allahabad).

At this stage, I cannot ignore the help, support, cooperation and total dedication that my wife Dr (Mrs) Kanti D Sharma, PhD extended to me during the entire period of the preparation of this book. She deserves all my love and appreciation. Throughout this period, I had no option but to ignore my grandchildren (Kuhu and Karan), though deeply unwillingly. To them, I can simply say that I love both of you to my fullest.

I would be grateful if users of the book notify me of any errors or omissions that come to their notice. Comments and suggestions for improvements may be sent to tmh.sciencemathsfeedback@gmail.com (*kindly mention the title and author name in the subject line*).

Jun 2009

O P Sharma
+91-9837566555

PREFACE TO THE FIRST EDITION

As opposed to the various other branches of Botany, Plant Taxonomy, has for a long time lacked an appropriate text, especially at the undergraduate level. Currently a student can choose between the voluminous reference books meant for advanced and the much abbreviated college texts. I have tried an approach that would interest and inspire all those who take up Plant Taxonomy. The aim of this book is to provide a general overview of Plant Taxonomy in an effective manner—a simplified but factual account.

My teaching experience of over 25 years has convinced me that a study of the basic principles of Plant Taxonomy and of selected families of angiosperms provide sufficient material for a beginning course on this subject at the undergraduate level. Consequently, I have discussed the basic principles of the discipline and then moved on to some selected families of dicotyledons and monocotyledons.

Botanists agree that presently there are about 2,45,000 species of flowering plants on earth and these have been grouped into 197–430 families, depending on the authority concerned. However, a basic knowledge of angiosperms can be acquired by a study of a selected group of families. I have selected 69 families of flowering plants, prominent in India. The criteria for selection are easy availability of their several 'representatives' unusual characteristics, economic importance, typical representation of a particular taxa.

In limiting my discussion to 69 families, I have had to deliberately overlook many which are important and commonly available in several parts of the world except the Indian subcontinent. While arranging the families I have followed Bentham and Hooker's system since (i) It is followed in almost all Indian herbaria, (ii) Majority of the Indian flora are based on this system, (iii) It is the most appropriate system for studying angiosperms in laboratories.

The subject matter is divided into twenty chapters which are by-and-large independent so that the teacher may choose the desired sequence of topics. Chapters 1–3 discuss the traditional aspects of Plant Taxonomy including the history and classification. Chapters 4–6 discuss the methods, techniques and principles of plant collection, Plant identification and nomenclature. Chapter 7 discusses the role of various Botanical disciplines—External Morphology, Vegetative Anatomy, Floral Anatomy, Cytology, Palynology, Embryology, Chemistry, Ecology, Palaeobotany, Electron Microscopy—in solving various taxonomic problems.

Chapters 8–10 deal with Numerical Taxonomy, Chemotaxonomy and Serotaxonomy, respectively. Latest information on phylogeny of angiosperms, the various theories regarding their origin and evolution, is presented in Chapter 11. Chapters 12–14 expand on various botanical institutions—botanical libraries, herbaria and botanical gardens.

Floral formulae and floral diagrams, from which it is generally possible to identify a family, are explained in Chapter 15. In Chapter 16, position is assigned to 72 commonly available families in some well-known systems of classification. Chapter 17 discusses over 550 common botanical terms used in taxonomic description.

Chapters 18 and 19 provide descriptions, illustrations and related information on 69 families of flowering plants—57 dicotyledons and 12 monocotyledons. Description of each family includes its systematic position, filed recognition, distribution, general characteristics, pollination and dispersal, general floral formula, economic importance, systematics and phylogeny, and also the descriptions of some common plants chosen as typical representatives of the family. Chapter 20 contains twenty tables of comparison between different families.

Although this book is primarily for undergraduate students of universities in the Indian subcontinent, I am also hopeful that it will benefit postgraduates, teachers and professionals.

I would be grateful if the reader notifies me of any errors or omission. I also welcome comments and suggestions from students and fellow teachers.

May, 1993 **O P Sharma**

INTRODUCTION

1.1 TAXONOMY AND SYSTEMATICS: SYNONYMS OR INDEPENDENT BRANCHES

The "science of the classification of organisms according to their resemblances and differences" is the definition of *taxonomy* as stated by Abercrombie *et al.* (1980) in the 7th edition of *The Penguin Dictionary of Biology.* According to them the word *systematics* is "often used synonymously with taxonomy, but sometimes interpreted more widely to include also the identification, practice of classification and nomenclature". However, Andrew Sugden in the 1986 edition of *Longman's Illustrated Dictionary of Botany* defined *taxonomy* as "the science of classification and relationships of organisms", and *systematics* as "the part of classification that involves the arrangement of organisms into related groups".

Jones and Luchsinger (1987) believe that although there is no agreement among the botanists "for the distinctions between *systematics* and *taxonomy*", some botanists treat them as two separate branches. According to these botanists, *systematics* is "the study of the diversity of plants and their identification, naming, classification and evolution" while *taxonomy* is "restricted to the study of classification" (Jones and Luchsinger, 1987).

Solbrig (1966) maintained that the two fields of inquiry (viz. taxonomy and systematics) have absolutely different concern. But Heywood (1967) and Ross (1974) treated systematics as a field covering the scientific study of the diversity and differentiation of organisms and the relationships that exist between them, and consider *taxonomy* as a part of systematics. Morse (1974), while presenting a system of computer programming to help in plant identification, treated taxonomy and systematics as two separate branches. Stace (1980) also considered these two as separate disciplines in his book entitled *Plant Taxonomy and Biosystematics*.

However, the terms *taxonomy* and *systematics* have been so loosely and interchangeably used in the past that to establish a proper delineation between the two is extremely difficult. In actual practice, and also in the present text, the two terms are used synonymously and deal with the study of classification, its principles, procedures and rules. Lam (1959) and Turrill (1964) also used them as synonyms.

The Greek meaning of taxonomy is "arrangement by rules" and of systematics is "to put together". The term taxonomy was first coined and used by the famous French botanist A.P. de Candolle in 1813 in his book *Theorie Elementaire de la Botanique.*

1.2 SYSTEMATICS AND THE SYSTEMATIST

1.2.1 Systematics

Radford (1986) defined systematics as "The study of phenotypic, genetic and phylogenetic relationship among taxa".

If seen from the *fundamental point of view* systematics is the study of the nature, causes, patterns, and trends in variation among various taxa.

If seen from the *structural point of view* systematics is the study of the basic taxonomic components and evolution of various taxa.

If seen from the *functional point of view* systematics is the study of characters from different fields of evidence for establishing relationships among various taxa.

If seen from the *developmental point of view* systematics is the study of variations among various taxa for the determination of character correlations as well as their relationships.

If seen from the *theoretical point of view* systematics is based on the idea that because of great diversity in the biological world there exist some 'discontinuous units' that can be classified, described, identified and named on the basis of evolution of these texa.

If seen from the *philosophical point of view* systematics is an ever-evolving as well as unending field for understanding taxonomic and evolutionary concepts, processes and principles.

1.2.2 Systematist

The systematist is a student, researcher, and scholar who studies, classifies, identifies, describes, names, observes, synthesizes or analyses the variations within the populations, species and higher taxa.

From the functional standpoint, a taxonomist has the following activities or duties to perform:

1. As a *classifier*, a taxonomist determines the position and rank for new taxa.
2. As an *identifier*, a taxonomist distinguishes or identifies new taxa and establishes the diagnostic characters for old taxa.
3. As a *describer*, a taxonomist determines the circumscription for a specimen according to an established system of classification.
4. As a *nomenclaturist*, a taxonomist assigns names to new taxa and determines the correct names for old taxa according to the rules laid down by the latest International Code of Botanical Nomenclature.
5. As a *coiner*, a taxonomist selects the proper Greek or Latin prefixes or suffixes for scientific names.

6. As an *observer*, a taxonomist observes the characteristics of organisms, population and taxa for a better understanding of their taxonomy.
7. As an *analyzer*, a taxonomist decides which characters of the individual, population or taxa are to be analyzed.
8. As a *synthesizer*, a taxonomist collects information from various aspects (morphology, cytology, anatomy, etc.) of the organisms for the development of new systems of classification and identification.
9. As a *theorizer*, a taxonomist develops new concepts on the basis of recent findings and reinterprets old concepts.
10. As a *historian*, a taxonomist provides better understanding of the old classical concepts in the light of the new researches in the field.
11. As a *generator*, a taxonomist generates effective methods for collection, analysis and presentation of data.
12. As a *conservator*, a taxonomist suggests new approaches for the preservation of threatened and endangered species.

1.2.3 Every Human Being is a Taxonomist

Every human being always behaves as a taxonomist. It is so because each one of us identifies, classifies, describes, and names while taking decisions about the food we eat, the beverages we drink, the homes we purchase, the clothes we wear, the friends we like, the games we play, the religion we believe in, and the politics we exercise. While talking about sex, sickness, sin, and sorrow we talk differently to children, old people, specialists, and parents, and thus behave as a taxonomist in the daily routine life. By smelling we identify garlic, by tasting we recognise sugar, and by seeing we name the animal as elephant, and thus behave as taxonomists. We describe that road is long, grass is green and ball is round, and thus work as a taxonomist.

1.3 OBJECTIVES, GOALS AND AIMS OF PLANT SYSTEMATICS

1.3.1 Objectives

1. To prepare a scheme of classification that provides phenetic, natural or phylogenetic relationships among plants.
2. To establish a suitable method for identification, nomenclature and description of plant taxa.
3. To provide an inventory of plant taxa that suits local, regional, and continental needs.
4. To create an understanding of the evolutionary processes.
5. To train the students of plant sciences in regard to the diversity of organisms and their relationship with other biological branches.

1.3.2 Goals

1. To acquire the fundamental values of plants systematics.
2. To know about the basic concepts and principles of plant systematics.

3. To be aware of the importance of taxonomic relationships in plant systematics.
4. To develop the knowledge of applicability of plant systematic studies.

1.3.3 Aims for the Study of Regional Flora

1. To know how to collect specimens.
2. To know how to prepare specimens for future preservation.
3. To know how a manual should be used.
4. To know how to use identification keys.
5. To recognize divisions, classes, orders, families, genera, and species.
6. To know how the plants are described.
7. To know how the diversity in species may be related with the regional habitat diversity.
8. To become familiar with the basic taxonomic principles, and with at least one system of plant classification.

1.4 HIERARCHICAL STAGES OF A SYSTEMATIST

The following may be the hierarchical grades or stages of a systematist:

A *natural scientist is a systematist* who provides or gathers information about the relationships between phenetic, genetic, and ecological variations among the organisms.

A *systematist is a taxonomist* who develops a profound knowledge of the specific aspects of systematics.

A *taxonomist is a classifier* who names, describes, classifies and identifies the organisms.

A *classifier is a student* who establishes the named, delimited, and identified taxa for establishing the diversity among organisms.

A *student* of diversity among organisms is *a classifier, taxonomist, systematist and natural scientist* who provides information to the society about the phenetic, genetic, and ecological variations among the organisms for the upliftment of mankind.

1.5 HIERARCHICAL CATEGORIES OF TAXONOMY

The International Code of Botanical Nomenclature (Voss *et al.* 1983) has recognized the following categories or ranks of taxa along with their endings, as mentioned in Table 1.1.

1.6 BASIC COMPONENTS OF TAXONOMY

Classification, identification, description and nomenclature are the four basic components of taxonomy.

Classification is the arrangement of botanical groups with definite circumscriptions by position and rank according to artificial criteria, phenetic similarities, or phylogenetic relationships.

Identification is the determination of similarities or dissimilarities between the two elements. Under identification we make a direct comparison of the characteristic features of a specimen with those present in the already existing keys for identification.

Table 1.1 Series of ranks and endings provided by the International Code of Botanical Nomenclature (1983)

Ranks of Taxa	Endings of ranks above genus	Examples
Division	phyta	Pterophyta, Magnoliophyta
Subdivision	phytina	Pterophytina
Class	opsida	Pteropsida, Magnoliopsida
Subclass	opsidae	Pteropsidae
Order	ales	Rosales, Asterales
Suborder	ineae	Rosineae
Family	aceae	Rosaceae, Asteraceae
Subfamily	oideae	Rosoideae
Tribe	eae	Roseae
Subtribe	inae	Rosinae
Genus	us, a, um, on, es, etc.	*Rosa*, *Pinus*
Subgenus		
Section		
Subsection		
Series		
Subseries		
Species		
Subspecies		
Variety		
Subvariety		
Form		
Subform		

Description is the orderly recording of maximum possible characters of a taxon, individual plant, plant part, or object.

Nomenclature is a simple system under which the individual taxonomic groups of plants are scientifically named.

1.7 FUTURE OF PLANT TAXONOMY

The evidence gathered from written records indicates that man has been categorizing or classifying organisms for more than 2000 years. Incorporation of new researches, evidences, methods and techniques have developed the science of systematics to its present status. But a lot more is still to be done. Introduction of mathematical data, statistical details and the use of computers in the recent past have brought tremendous changes in the field of taxonomy. The use of electron microscopy in several cases has entirely changed the old concepts. Chromatography and spectrochemistry have also brought unexpected changes in several existing ideas.

In the future, several new aspects in the field of systematics, will be explored. Some of them are undermentioned:

1. According to Stace (1980) about 3,00,000 species of green plants, over 1,50,000 fungi and a few thousand bacteria are known to the biologists. But in future several thousands of new species of plants still await discovery, description, and naming in different parts of the world.
2. Monographs of many plant groups are to be prepared.
3. Many groups, treated more than a century ago, badly need a revision in the light of new discoveries and techniques.
4. Explorations of several unknown floristic regions are highly essential.
5. Several old floras are to be revised.
6. Speciation understanding in several groups is yet to be done.
7. Many old basic fields and taxonomic principles need re-evaluation.
8. Biology of a majority of the plant species has not been studied so far.
9. Several mathematical and statistical schemes for the determination of relationships have been developed. These schemes are yet to be used in many cases. More mathematical logic is to be used in taxonomic investigations.
10. Several 'anomalous groups' are still waiting for their proper placing in the systems of classification.
11. In future, taxonomists will have to use the expertise and knowledge of genetic diversity. This would help in the introduction of new crops and also in the improvement of old crops.
12. New and improved methods of studying nucleic acids will have to be used by the taxonomists in future for the basic understanding of the relationships among the groups of organisms.

Test Your Understanding

1. Taxonomy and Systematics: Are they synonyms or two independent branches?
2. Enlist any six functions, a taxonomist has to perform.
3. Write a brief note on the fact that every human being is a taxonomist.
4. International Code of Botanical Nomenclature has recognised some basic categories or ranks of taxa. Name any five of them along with their endings and an example of each of them.
5. Describe the future of plant taxonomy in about 100 words.

Suggested Reading

Bell, C.R., 1969, *Plant Variation and Classification*, Macmillan, London.

Davis, P.H. and V.H. Heywood, 1963, *Principles of Angiosperm Taxonomy*, Oliver & Boyd, Edinburgh.

Heywood, V.H., 1967, *Plant Taxonomy*, Edward Arnold Publishers, London.

Jones, S.B. and A.E. Luchsinger, 1987, *Plant Systematics* (2nd Ed.), McGraw-Hill, New York.

Lam, H.J., 1959, 'Taxonomy: General principles and angiosperms,' *Vistas in Botany* **2**: 4–75.

Morse, L.E., 1974, 'Computer assisted storage and retrieval of the data of taxonomy and systematics,' *Taxon* **23**: 29–43.

Radford, A.E., 1986, *Fundamentals of Plant Systematics*, Harper and Row, New York.

Stare, C.A., 1980, *Plant Taxonomy and Biosystematics*, University Park Press, Baltimore.

Voss, E.G., 1983, *International Code of Botanical Nomenclature*, Bohn, Scheltema and Holkema, Utrecht, Netherlands.

HISTORY OF PLANT TAXONOMY

2.1 HISTORY AND DEVELOPMENT OF PLANT CLASSIFICATION

2.1.1 Earlier than Man Could Read and Write

Not much is known today about the botanical knowledge of our preliterate ancestors. But it is clear that they knew by experience the plants that were edible and others which were not. The preliterate mankind also possessed some kinds of linguistic mechanisms for correct distinctions among different kinds of plants.

2.1.2 Age of Theophrastus, Secundus, Dioscorides and Parasara

The advent of writing and printing changed the entire picture of the ancient preliterate mankind. Theophrastus (370–285 B.C.), the "*grandfather of the modern botany*", was the greatest botanical writer of the distant past. He was a student of Plato and Aristotle, and studied botany under their philosophic guidance at Athens. He classified the plants into four groups: herbs, subshrubs, shrubs, and trees. He also distinguished between the nonflowering plants (Cryptogams) and flowering plants (Phanerogams). He suggested that calyx and corolla are the modified leaves. He described nearly 500 plants in details, and certain names (e.g. *Asparagus, Daucas,* and *Narcissus)* are used even today in the same sense. The details of his works are available to the world in the form of books entitled "*Enquiry into Plants*" and "*The Causes of Plants*".

Caius Plinius Secundus (23–79 A.D.), known to the botanical world as Pliny the Elder, was a lawyer and also served in the Roman army till his death caused by the heart failure. Voluminous and important works of Pliny are available in the form of 37 volumes of *Natural History.* He described the biological, medicinal and agricultural aspects of the plants known to the world up to his time in these volumes, and Pliny's *Natural History* was among the first books to be printed by the movable type in the late 15th century. The word 'stamen' in its modern sense was first used by Pliny.

Pedanios Dioscorides (62–128 A.D.) was a contemporary of Pliny the Elder. He was actually a Physician in the Roman army and belonged to a Roman province, Cilicia. His monumental work is compiled in the form of *Materia Medica.* It described the botany, mainly the medicinal aspects of about 600 species of plants. *Materia Medica* was considered as a highly esteemed source book for those practicing medicine in those days. The importance and fame of *Materia Medica* of Dioscorides may be guessed from the fact that the Emperor Flavius Anicius Olybrius (500 A.D.) presented a beautifully illustrated copy of the book to his daughter (Princess Juliana Anicia) as a precious gift at the time of her marriage. Juliana's copy of *Materia Medica* remained for several years in Constantinople (Istanbul) and was later transferred to Vienna, where it still exists under the name of *Codex Juliana.* Several plant names (e.g. *Aloe, Aristolochia, Anemone, Phaseolus)* as suggested by Dioscorides are still in use in the present botanical literature.

The book titled *Vrikshayurveda,* written by an Indian, Parasara, is one of the earliest Indian works describing plants in a scientific manner. Though it was written before the beginning of the Christian era, the scientifically described plant classification and distribution in this book led the famous systematist Albert E. Radford (1986) of USA to state that Parasara *had some kind of hand lens or microscope.* There exist several other references which show that the early civilizations of India, Egypt and China did have a definite knowledge of the plant taxonomy.

2.1.3 Taxonomy in Middle Ages

Little is known about the development of taxonomy during the early one thousand years of the Christian era. Then came the Middle Ages or Medieval Ages (about A.D. 1100–1500). Albertus Magnus (A.D. 1200–1280) has been the most famous plant taxonomy worker of this period. Commonly called "Doctor Universalis" or "Aristotle of the Middle Ages" by his contemporaries and historians, Magnus employed a scheme of classification of plants that recognised monocots and dicots, and separated nonvascular plants from vascular plants. Worthmentioning are the names of two Muslim scholars of twelfth century. Ibn-Sina who authored *Canon of Medicine,* and Ibn-al-Awwam of Spain who described over 600 species of plants and interpreted accurately the sexuality in plants.

2.1.4 Herbalists

After the Medieval Ages (A.D. 1500) the history of plant taxonomy was influenced tremendously by two things, the invention of printing and the development of the science of navigation. The printing technology lowered the cost of books and increased literacy. During the early years of printing, the medically oriented books on plants became quite popular. Printed forms of ancient texts had many superfluous and irrelevant writings and this actually prompted several interested persons to write and publish their own *botanical medical books.* These books were called *herbals* and their authors were called *herbalists.*

The science of navigation prompted sailors to go on long voyages. This resulted in the exploration of several new areas of the world, and, in turn, increased man's practical knowledge of plant taxonomy.

At the advent of 16th century the first herbals published were "*Gart der Gesundheit*" and "*Hortus Sanitalis*". These herbals had crude illustrations of plants and were published without an attribution of authorship. However, the 16th century is considered as the "time of great herbalists". The best known among the herbalists belong to Germany. Among them were Otto Brunfels (1464–1534) known for his herbal *Herbarum Vivae Eicones,* Jerome Bock (1489–1554) for his herbal *Neu Kreuterbuch,* and Leonhart Fuchs (1501–1566) for his herbal *De Historia Stirpium.* All these herbalists are considered as the "German Fathers of Botany" and their herbals exhibit some excellent illustrations and detailed taxonomic descriptions of several available plants. However, they did not emphasize on any system of classification of plants.

2.1.5 Taxonomy during Seventeenth Century

Andrea Caesalpino (1519–1603), an Italian, was the first scientist who worked for achieving a rational scheme of classification of plants. Another name of repute among the taxonomists of the 17th century is that of Gaspar Bauhin (1560–1624), a Swiss botanist. He compiled all the different kinds of plants known to the science till then in a register called *Pinax Theatri Botanici.* His register, which had an account of different names used by various workers for each plant, is considered an authoritative discussion of synonymy in systematic botany. He, first, established the distinction between the concept of genus and species and also initiated the use of the binomial nomenclature, in some cases though not consistently.

John Ray (1627–1705), an English taxonomist, is another great contributor of the seventeenth century. His works are published in two books entitled *Methodus Plantarum Nova* (1682) and *Historia Plantarum* (1686). He also suggested a scheme of classification of plants.

Joseph Pitton de Tournefort (1656–1708) is another significant name in the field of taxonomy, mainly for his publication titled *Institutiones Rei Herbariae.* He arranged over 9,000 kinds of plants in about 700 genera grouping them in classes.

2.1.6 Period of Linnaeus

The eighteenth century belongs clearly to Carl Linnaeus (1707–1778) from the point of view of history of taxonomy. He is the creator of the modern system of nomenclature.

A son of a Swedish clergyman, Linnaeus was educated at the universities of Lund and Uppasala, and obtained the degree of M.D. in Netherlands. After practicing medicine for a few years, he became a Professor of natural history at the University of Uppasala, where he spent the rest of his life.

Linnaeus is considered as the first taxonomist who showed that the reproductive features were of paramount importance in taxonomy. He was the originator of the sexual system of classification, in which he recognized 24 classes, mainly on the basis of number, length, union and certain other characters of stamens. He was the first to use consistently the "*binomial system of nomenclature*", in which each organism is represented first by its generic name followed by the name of its species. This scheme of nomenclature provided by Linnaeus is used throughout the world till today. The plant taxonomy findings of Linnaeus were published in May 1753 in the form of his work *Species Plantarum.*

The name of Linnaeus is commemorated today in the form of a well-established scientific society, *The Linnaean Society of London,* and journals, such as *Linnaea.* Contribution of Linnaeus has been so great that it may well be stated that so far the 'plant taxonomy' would survive in the world, Linnaeus would be remembered.

2.1.7 Natural System Approach

Though Linnaeus (1753) was the first one to establish a system of classification of plants based on reproductive parts, several totally unrelated plants were classified together (e.g. *Prunus* was classified along with *Cactus* because of the same number of stamens) in his system of classification. This fact compelled the botanists after Linnaeus to realise that no single character is intrinsically or naturally more important than any other character. An approach to a natural system of classification first took seed in France, where Michel Adanson (1727–1806) emphasised the fact that in several cases natural characters are more useful than the others. This theory of Adanson was later recognised as *phenetic taxonomy*.

The first scheme of classification based on natural characters was presented in 1789 by Antoine-Laurent de Jussieu (1748–1836), a great botanist of France. All the four members of A.L. de]ussieu's family (Antoine, Bernard, Joseph, and Antoine-Laurent) made notable contributions to the science of plant taxonomy. The plants resembling each other in a set of characters were grouped together in A.L. de Jussieu's scheme of classification, and therefore, it was purely natural in its approach. He presented his scheme of classification in his *Genera Plantarum Secundum Ordines Naturales Disposita*.

Another family of botanists, contemporary to A.L. de Jussieu (1748–1836) was that of Augustin Pyramus de Candolle (1778–1841). A.P. de Candolle presented a new classification of plants in his book *Theorie elementaire*. He followed the approach of the natural system in his scheme and put all alike plants together. An attempt to prepare an account of all available higher plants of the world was undertaken by A.P. de Candolle in his *Prodromus Systematis Naturalis Regni Vegetabilis* in 1816, and he continued with this unfinished project till his death in 1841. The same project was continued by his son, Alphonse (1806–1893) until 1873. After 1873, Alphonse and his son, Anne Casimir de Candolle (1836–1918) published the details of this project in the form of monographs, but this 'great project' could never be completed.

The latest, the best and a highly recognised natural system of classification was proposed by George Bentham (1800–1884) and Joseph Dalton Hooker (1817–1911). They classified plants strictly on the basis of a natural scheme. They made thorough observations of the material from the herbaria, took very little help from the literature existing at that time, and presented their well-known scheme of classification in their book titled *Genera Plantarum*. The Bentham and Hooker's system of classification is still supposed to be the best classification system, especially from the practical laboratory point of view.

2.1.8 The Phylogenetic Approach

Publication of Darwin's theory of organic evolution in 1859 in the form of his well-known book, *On the Origin of Species*, provided the real base for the development of modern systematics. Scientists started working on the line of thought that the life of plants as well as animals has continually changed on the earth over a period of time. Botanists started working on the concept of evolution regarding the development of a classification system of plants.

S. Endlicher (1805–1849) and A.W. Eichler (1839–1889), the two German botanists, were the first to start working along this line of thought. They proposed the phylogenetic systems of the classification of plants, and their schemes were later modified and developed by A. Engler (1844–1930) and K. Prantl (1849–1893)[1]. Engler and Prantl's monumental work was published in several volumes in the form of *Die Naturlichen Pflanzenfamilien* between 1887 and 1915. Published in the 1915 volume

[1] Details are mentioned in Chapter 3 (Classification).

of *Pflanzenfamilien* their scheme indicates that they started from simplest plants (e.g. Salicaceae in case of dicotyledons because the members of this family possess simplest floral structures) and ended with plants of complex floral structures (e.g. Compositae or Asteraceae in case of dicotyledons because members of this family possess the height of floral complexity). Majority of the present day botanical institutions and publications follow Engler and Prantl's scheme of classification.

Another important work, contemporary to Engler and Prantl, was that of C.E. Bessey[1] (1845–1915), a Professor of botany at the University of Nebraska. Bessey grouped the flowering plants on the basis of their evolutionary relationships. He categorised them on the basis of characters of primitiveness and advanceness.[1] These guiding characters or principles were named as 'dicta' by Bessey. Richard von Wettstein (1862–1931), an Austrian botanist, and Hans Hallier (1868–1938), a German botanist, are two other plant taxonomists who suggested phylogenetic systems of classification.

John Hutchinson (1884–1972), a British botanist also suggested a widely recognized phylogenetic system of plant classification which he published in two volumes of his well-known book *The Families of Flowering Plants.* First published in 1926, the 3rd edition of this great book appeared in 1973, shortly after his death on 2nd September 1972 at the age of 88. He classified the flowering plants on the basis of 24 general principles.

2.1.9 Some Current Contributions

Four major systems of plant classification have been published during the last few decades, by Armen Takhtajan (1910–) of Leningrad (Russia), Arthur Cronquist (1919–) of New York (USA), Rolf Dahlgren (1919–) of Copenhagen (Denmark), and Robert F. Thorne (1920–) of Claremont (USA).

Armen Takhtajan (1910–) of Botanical Institute of Academy of Sciences, Leningrad, proposed a system of classification of flowering plants in Russian in 1954. Its English version was first published in 1969 in *Flowering Plants: Origin and Dispersal.* A new version of his classification[1] is published in *Botanical Review* of 1980. According to him the angiosperms are monophyletic and evolved from some very ancient group of gymnosperms. He divided the angiosperms (Division: Magnoliophyta) into two classes, i.e. Magnoliopsida (=Dicotyledons) and Liliopsida (= Monocotyledons).

Arthur Cronquist (1919–) of New York Botanical Garden proposed a comprehensive system of classification of angiosperms.[2] First appeared in 1968 in *The Evolution and Classification of Flowering Plants,* the latest version of Cronquist's classification is published in 1981 in *"An Integrated System of Classification of Flowering Plants".*[3] Cronquist considered the seed ferns (Pteridosperms) as the probable ancestors of angiosperms. He divided angiosperms into Magnoliopsida and Liliopsida. The Magnoliopsida, equivalent to Dicotyledons, consists of 64 orders, 318 families and about 1,65,000 species, whereas the Liliopsida consists of 19 orders, 65 families and about 50,000 species.

Rolf Dahlgren (1919–) of Botanical Museum of University of Copenhagen proposed a new scheme[4] of classification of angiosperms. First published in 1975 in *Botanische Notiser,* the revised and improved versions of his scheme appeared in 1980 in *Botantcal Journal of Linnean Society,* in 1981

[1] Details are discussed in Chapter 3, Article No. 3.6.1.
[2] Details are discussed in Chapter 3, Article No. 3.6.2.
[3] Published by Columbia University Press, New York.
[4] Details are mentioned in Chapter 3, Article No. 3.6.3.

in *Phytochemistry and Angiosperm Phylogeny,* and in 1983 in *Nordiac Journal of Botany.* He also considered angiosperms to be monophyletic, and divided the class Magnoliopsida (= Angiospermae) into two subclasses *viz.* Magnoliidae (= Dicotyledoneae) and Liliidae (= Monocotyledoneae).

Robert F. Thorne (1920–) of Rancho Santa Ana Botanic Garden, Claremont, California, published a phylogenetic system of classification[1] of plants. First published in 1968 in *Aliso,* the revised and enlarged outlines of Thorne's scheme were published in 1981 in *Phytochemistry and Phylogeny* and in 1983 in *Nordiac Journal of Botany.* Thorne has tried to establish the phylogenetic relationships among the higher taxa of flowering plants. He divided the class Annonopsida (= Angiospermae) into subclasses Annonidae (= Dicotyledoneae) and Liliidae (= Monocotyledoneae). Thome divided Annonidae (= Dicotyledoneae) into 19 superorders, 41 orders, 56 suborders, 297 families, 350 subfamilies, 9,640 genera and 1,73,370 species, and Liliidae (= Monocotyledoneae) into 9 superorders, 12 orders, 17 suborders, 53 families, 102 subfamilies, 2,615 genera and 52,120 species.

2.2 TAXONOMY—OUR CONTEMPORARY

The plant kingdom is very large and diversified. We are faced with an estimated total of about 3,00,000, or if fungi included, 4,50,000 plants. This large number includes many different kinds of plants of a vast range in size and structure. This entire living world of plants forms the subject matter of the plant taxonomy. The raw material of the taxonomic botany is the diversity of plant life in all its aspects. One is, therefore, faced with exceedingly diverse taxonomical activities and techniques. The algologist collecting algae in the arctic region, the plant collector in the tropical forest, the botanist describing plants of a herbarium, the biochemist studying amino acids by chromatography, the cytogeneticist studying breeding systems, the cytologist finding the chromosome number, the electron microscopist going down to molecular details, the botanical artist depicting plants with his pen, brush and ink, all contribute to the growth and development of plant taxonomy. Taxonomy is the only branch of botany that involves the interests and abilities of so many practitioners. From a plant collector to a computer programmer, from a biochemist to an electron microscopist, and from a mycologist to a taxonomist, every one has room to show his talents in this branch. Thus, we can say that taxonomy is truly our contemporary.

2.3 CHRONOLOGICAL DEVELOPMENT OF TAXONOMY IN INDIA

Indians of very old times had a satisfactory knowledge of plant taxonomy is evident from the studies of the earliest Indian civilization as well as from the treatises such as *Vrikshayurveda* of Parasara, *Charaka Samhita, Sushrut Samhita* and voluminous Indian classics such as *Vedas, Ramayana, Mahabharata,* etc. However, a complete chronology of the development of plant taxonomy in India is beyond the scope of this book. The major attempts of the botanical explorations of India during the past few centuries and some of the important published works are undermentioned in Table 2.1.

2.4 SOME INDIAN JOURNALS RELATED TO TAXONOMY

1. *Annals of the Royal Botanic Garden,* Calcutta (now Kolkata)
2. *Bulletin of the Botanical Society of Bengal*

[1] Details are mentioned in Chapter 3, Article No. 3.6.4.

3. *Bulletin of the Botanical Survey of India*
4. *Indian Forest Records*
5. *Indian Journal of Forestry*
6. *Journal of Bombay Natural History Society*
7. *Journal of Economic and Taxonomic Botany*
8. *The Indian Forester*
9. *The Indian Journal of Agricultural Science*
10. *The Journal of the Asiatic Society of Bengal*
11. *The Records of the Botanical Survey of India*

Table 2.1 Chronology of the major published taxonomic works of India

Year A.D.	Researcher/Explorer	Major published/botanical work
1565	Garcia d'Orta	Published *Os Colquios in Goa*; described common Indian medicinal plants.
1578	C. Acosta	Published *Tractado de las Drogas*
1670	Heinrich van Rheede	Published *Hortus Malabaricus*
1728–1785	John Gerard Koenig	Formed society called "*The United Brothers*" in order to promote and study Indian plants
1746–1793	Robert Kyd	Founded Botanical Gardens of Calcutta
1751–1815	William Roxburgh	Published *Flora Indica* and *Plantae Coromandelianus*; prepared 2382 coloured drawings of Indian plants; popularly called *Linnaeus of India*
1817	Nathaniel Wallich	Published *Plantae Asiaticae Rariores*
1824	J.F. Watson	Published *Flora of Kumaon*
1833	J.F. Royle	*Flora of Kashmir*
1839	J. Grahm	*Flora of Bombay*
1840	J.W. Masters	*Flora of Calcutta*
1844	W. Munro	*Flora of Agra*
1857	J. Long	*Flora of Bengal*
1859	W. Elliot	*Flora of Andhra*
1859	T. Anderson	*Flora of Lucknow*
1869	J.L. Stewart	*Flora of Punjab*
1872–1897	J.D. Hooker	*Flora of British India* published in the form of seven volumes
1887	George King	Started the publication of the journal *The Annals of the Royal Botanic Garden*
1890	George King	Established Botanical Survey of India
1901–1908	T. Cooke	*Flora of Presidency of Bombay*
1902	H. Collett	*Flora Simlensis*

(*Contd.*)

1903	D. Prain	*Bengal Plants*
1903–1922	J.F. Duthie	*Flora of Upper Gangetic Plains and the Adjacent Siwaliks and Sub-Himalayan Tract*
1908	T.F. Bourdillon	*The Forest Trees of Travancore*
1910	H.H. Haines	*A Forest Flora of Chota Nagpur*
1915–1936	J.S. Gamble and C.E.C. Fisher	*Flora of Presidency of Madras*
1921	K. Rangacharier	*A Handbook of South Indian Grasses*
1921–1925	H.H. Haines	*The Botany of Bihar and Orissa*
1923	C.E. Parkinson	*Forest Flora of Andaman Islands*
1929	P.V. Mayurnathan	*The Flowering Plants of Madras City and its Immediate Neighbourhood*
1929	E. Blatter and W.S. Millard	*Beautiful Indian Trees*
1932	P.F. Fyson	*The Flora of South Indian Hill Stations*
1932–1940	V. Kanjilal	*Flora of Assam*
1940	N.L. Bor	*Common Grasses of United Provinces*
1949	R.L. Bahadur and S. Ghosh	*Poisonous Plants of India*
1953	S.P. Agahakar	*Medicinal Plants*
1960	B.P. Pal	*Beautiful Climbers of India*
1962	H. Santapau	*The Flora of Saurastra*
1962	K. Subramanian	*Aquatic Angiosperms*
1964	P. Maheshwari and U. Singh	*Dictionary of Economic Plants in India*
1964	G.S. Puri *et al.*	*Flora of Rajasthan*
1965	M.S. Randhawa	*Flowering Trees*
1965	J.K. Maheshwari	*Illustrated Flora of Delhi*
1966	B.P. Pal	*The Roses in India*
1966	V.D. Vartak	*Enumeration of Plants from Gomantak* (*Goa*)
1968	R.K. Gupta	*Flora Nainitalensis*
1968	R. N. Chopra and S.L. Nayyar	*Glossary of Indian Medicinal Plants*
1968	S.L. Jindal	*Ornamental Bulbous Plants*
1973	S.V. Ramaswamy and B.A. Raji	*Flora of Bangalore District*
1973	Santapau and Henry	*Dictionary of Flowering Plants in India*
1974	B.P. Pal and Vishnuswarup	*Bougainvilleas*
1975	M.B. Raizada	*Supplement to the Duthie's Flora of Upper Gangetic Plains*
1975	M.A. Rau	*High Altitude Flowering Plants of Western Himalayas*
1976	T.N. Srivastava	*Flora Gorakhporensis*
1976	C.J. Saldanah and D.H. Nicolson	*Flora of Hasan District, Karnataka*
1976	R. Rao	*Flowering Plants of Travancore*
1977	M.M. Bhandari	*Flora of Rajasthan Desert*

(*Contd.*)

1977	M. Oomachan	*The Flora of Bhopal*
1978	M.B. Raizada and H.O. Saxena	*Flora of Mussoorie*
1978	N.C. Nair	*Flora of Punjab Plains*
1979	A.S. Rao	*Orchids of India*
1979	V.N. Naik	*The Flora of Osmanabad*
1979	B.L. Sapru *et al.*	*Flora of Ladakh*
1983	U. Singh, A.M. Wadhwani and B.M. Johri	*Dictionary of Economic Plants in India*
1984	B.D. Naithani	*Flora of Chamoli*
1984	H.J. Chowdhary and B.M. Wadhwa	*Flora of Himachal Pradesh*
1984	O. Polunin and A. Stainton	*Flowers of Himalayas*
1984	C.J. Saldanha	*Flora of Karnataka*
1986	M.K. Kaul	*Weed Flora of Kashmir Valley*
1986	CSIR Publication	*The Useful Plants of India*
1987	Som Deva and H.B. Naithani	The *Orchid Flora of N.W. Himalayas*
1988	R.K. Gupta	*The Living Himalaya Plant Exploration and Phytogeography*
1988	S.P. Vij (ed.)	*Biology, Conservation and Culture of Orchids*
1992	H.K. Bakhu	*Herbs That Heal*
1994	S.K. Jain	*Medicinal Plants* (5th ed.)
1995	K. Haridarshan, G.P. Shukla and B.S. Beniwal	*Medicinal Plants of Arunanchal Pradesh*
1997	T.K. Chatterjee	*Herbal Options*
1998	B. Bhattacharyya and B.M. Johri	*Flowering Plants: Taxonomy and Phylogeny*
2001	S.K. Bhattacharjee	*Handbook of Medicinal Plants*
2004	K.M. Nadkarni	*Medicinal Plants of India*
2005	S.K. Bhattacharjee and L.C. Dey	*Medicinal Herbs and Flowers*
2006	T.C. Narendran	*An Introduction to Taxonomy*
2007	A.K. Singh	*Sedges and Grasses of Eastern Uttar Pradesh*

2.5 A NOTE ON BOTANICAL SURVEY OF INDIA

George King, the Chief Executive Officer of the Royal Botanic Gardens, established the Botanical Survey of India in 1890 with the main objective to coordinate the botanical work, being done in different parts of India at that time. Regional offices of BSI were established at Madras[1] with M.A. Lawson as the Director for southern India, at Saharanpur with J.F. Duthie as the Director for northern India, and at Pune with T. Cooke as the Director for the western India. The publication of a journal *The Records of the Botanical Survey of India* was also started by George King. The Industrial Section of the Indian Museum at Kolkata and the office of the Reporter of Economic Products were brought under BSI in 1911. But the coming few decades were a period of inactivity

[1] Renamed now Chennai.

for BSI mainly because of lack of funds and manpower. In 1939 the conditions deteriorated so that the post of Director at Kolkata remained suspended, and BSI was surviving only in the form of a Curator, Industrial Section, and a Systematic Assistant in the herbarium at Sibpur (Howrah).

The Government of independent India then came to the rescue of BSI. E.K. Janaki Ammal was appointed as the Officer-on-Special Duty in 1952 to reorganise BSI. The first unit was established in the form of The Central Botanical Laboratory at Allahabad with Janaki Ammal as the Director. Four Regional Circles were established with their headquarters at Dehradun of Northern Circle, Coimbatore of Southern Circle, Shillong of Eastern Circle, and at Pune of Western Circle. In 1962 the Central Botanical Laboratory was shifted from Allahabad to Kolkata, and at Allahabad a new regional station for Central India was established. In 1972 two new circles (Andaman and Nicobar Circle, and Arid Zone Circle) were established. Arunachal Pradesh Circle with its headquarter at Itanagar was created in 1977, and Sikkim-Himalaya Circle with its head office at Gangtok in 1979. The Deccan Circle at Hyderabad, and High Altitude Circle at Solan (Himachal Pradesh) were established in 1984.

BSI headquarter office at Kolkata coordinates the research and other activities of all units, mainly on the basis of the scientific policies of the Central Government. It also maintains links of BSI with other major research institutions of the country such as CSIR, ICAR, ICMR, etc. A senior scientist of BSI is also posted at the Royal Botanic Gardens, Kew (England) to maintain a link between the two organizations as well as for many technical inquiries and clarifications regarding nomenclature, etc.

Some of the major publications of BSI include the *Bulletin of Botanical Survey of India, The Records of the Botanical Survey of India, Annual Reports of BSI,* and *Newsletters.* During the last 50 years, over 4000 research papers have been published by the scientists of BSI, and the herbaria of the Survey hold over 2 million plant specimens.

Test Your Understanding

1. Who is known as the "grandfather of modern Botany"?
2. The monumental work of which of the great physician of Roman army is compiled in the form of *Materia Medica*?
3. Write a note on "herbalists" in about 100 words.
4. Give brief details of some discoveries made by Linnaeus.
5. Linnaeus obtained degree of M.D. and practiced ________ for few years and then became a Professor of ________ at the University of ________ in Netherlands.
6. Armen Takhtajan, Arthur Cronquist, Rolf Dahlgren and Robert F. Thorne belonged to which countries?
7. J.F. Duthie has been the author of the flora of which region of India?
8. Name three Indian journals related to plant taxonomy.
9. Write a detailed note on Botanical Survey of India.

Suggested Reading

Arber, A., 1938, *Herbals: Their Origin and Evolution,* Cambridge University Press, England.

Core, E.L., 1955, *Plant Taxonomy,* Prentice-Hall, Englewood Cliffs, N.J.

Cronquist, A., 1968, *The Evolution and Classification of Flowering Plants,* Thomas Nelson & Sons, Ltd., London.

Ewan, J.A., 1969, *A Short History of Botany,* Hafner Publishing Comp., New York.

Gibbs, R.D., 1963, *History of Chemical Taxonomy,* Academic Press, London.

Green, J.R., 1909, *History of Botany 1860–1900,* Clarendon Press, Oxford.

Maheshwari, P. and R.N. Kapil, 1963, *Fifty Years of Science in India: Progress of Botany,* Indian Sci. Congr. Assoc., Calcutta.

Mayr, E., 1982, *The Growth of Biological Thought,* Harward Univ. Press, Cambridge.

Rao, R.S., 1973, Angiosperm Taxonomy, (In) *A Decade* (1963–72) *of Science in India: Progress of Botany,* pp. 32–36. *Indian Sci. Congr. Assoc.* Calcutta.

Santapau, H., 1958, *History of Botanical Research in India, Burma and Ceylon,* Part II, *Systematic Botany of Angiosperms,* The Bangalore Press, Bangalore.

Sneath, P.H.A., 1957, Application of Computers in Taxonomy, *J. Gen. Microbiol.* **17**: 201–226.

Steere, W.C., 1958, *Fifty Years of Botany,* McGraw-Hill Book Co., New York.

Turrill, W.B., 1938, The Expansion of Taxonomy, *Biol. Rev.* **13**: 342–373.

Wilmott, A.J., 1950, Systematic Botany from Linnaeus to Darwin, (In) *Lectures on the Development of Taxonomy,* Linnean Society, London.

CHAPTER 3

CLASSIFICATION

3.1 WHAT IS CLASSIFICATION?

Andrew Sugden (1984) defined the word "classification" in *Longman Illustrated Dictionary of Botany* as "the naming of species and their grouping into families, orders, divisions, etc." Radford (1986) stated that "classification is the arrangement of groups of plants with particular circumscriptions by rank and position according to artificial criteria, phenetic similarities, or phylogenetic relationships."

In its simplest form, classification is the placement of plants, animals and objects into groups and categories for a clear understanding, proper study and effective organization.

3.2 RANKS OF PLANT CLASSIFICATION

Species, genus, family, order, class, and division are the six main ranks of plant classification in an ascending order. Each rank has its subcategories, i.e. towards the higher ranks, subform, form, subvarieties, varieties, and subspecies are the subcategories of *species*; subsection, section, and subgenus are the subcategories of *genus*; subtribe, tribe, and subfamily are the subcategories of *family*; suborder is the subcategory of order; subclass is the subcategory of *class*; and subdivision is the subcategory of *division*.

Several species are included within a genus, several genera within a family, several families within an order, several orders within a class, and several classes are included within a division.

3.3 TYPES OF SYSTEMS OF CLASSIFICATION

The past taxonomic literature described three basic categories of systems of classification viz. *artificial systems, natural systems* and *phylogenetic systems.* But Radford (1986) has described following four types of the systems of classification:

1. *Artificial Classifications* These systems use the habit and importance to man as the taxonomic characters. Some advocates of artificial systems of classifications were Theophrastus (370–285 B.C.), Secundus (23–79 A.D.), Dioscorides (62–128 A.D.), Magnus (1200–1280

A.D.), Brunfels (1464–1534), Bock (1489–1554) and Fuchs (1501–1566). Details of the works of these workers are mentioned in Chapter 2 (Articles No. 2.1.2–2.1.4).

2. *Mechanical Classifications* These systems used one or a few selected taxonomic characters to group taxa. Some mechanical classifications were given by Caesalpino (1519–1603), Bauhin (1560–1624), Ray (1627–1705), Tournefort (1656–1708) and Linnaeus (1707–1778). Details of the works of these workers are mentioned in Chapter 2 (Articles No. 2.1.5 and 2.1.6).
3. *Natural Classifications* These systems of classifications used as many taxonomic characters as possible to group taxa. Some of the natural systems of classification were given by Adanson (1727–1806), A.L. de Jussieu (1748–1836) and his three family members (Antoine, Bernard, and Joseph), A.P. de Candolle (1778–1841) and his son Alphonse (1806–1893), and Bentham (1800–1884) and Hooker (1817–1911). Some details of the works of these workers are mentioned in Chapter 2 (Article No. 2.1.7).
4. *Phylogenetic Classifications* These systems of classification used as many taxonomic characters as possible in addition to the phylogenetic (evolutionary) interpretations. Some of the phylogenetic systems of classification were proposed by Eichler (1839–1889), Engler (1844–1930) and Prantl (1849–1893), Bessey (1845–1915), Wettstein (1862–1931), Hallier (1868–1938), Hutchinson (1884–1972), Takhtajan (1980), Cronquist (1981), Dahlgren (1983) and Thorne (1983). Some historical details of the works of these workers are mentioned in Chapter 2 (Articles No. 2.1.8 and 2.1.9).

3.4 SOME IMPORTANT SYSTEMS OF CLASSIFICATION

3.4.1 John Ray (1627–1725)

An English biologist, John Ray[1] treated over 18,000 species in the last edition of his book *Methodus Plantarum Nova* published in 1703. He developed a classification system on the basis of the form relationships by grouping together the plants that resembled one another. Ray was the first to divide herbs, shrubs and trees into Dicotyledons and Monocotyledons on the basis of the presence of two or one cotyledons. Broadly, he divided the plants as under:

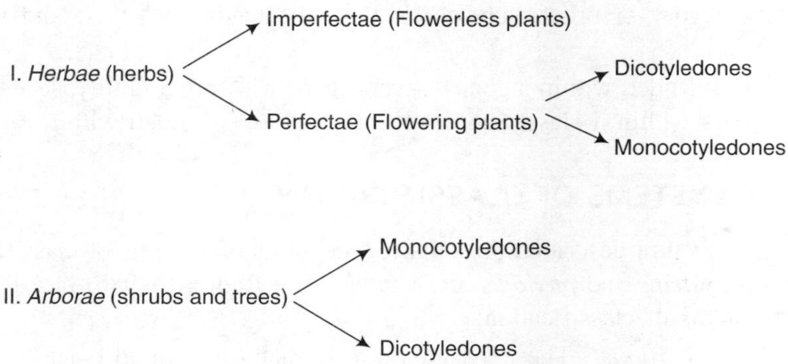

[1] For historical details, see Chapter 2 (Article No. 2.1.5).

3.4.2 Carl Linnaeus[1] (1707–1778)

Linnaeus,[2] the *father of taxonomy* and creator of binomial system of nomenclature is considered as "the most prodigious systematist of all times" (Lawrence, 1951). He was born on May 23, 1707 in Rashult, southern Sweden.

In 1727, at the age of twenty, Linnaeus entered the University of Lund to study medicine. In 1728 he was transferred to the University of Uppasala, and published his first research paper on sexuality of plants in 1729. In 1737 he received M.D. degree from the University of Harderwijk of Netherlands. In Netherlands, he became the personal physician of a wealthy banker George Clifford. Clifford was interested in plants. Linnaeus remained for three years in Netherlands, travelled extensively through Europe, and collected plants. These three years were the most creative years of his life. In 1741, he returned to Sweden, became the Professor of botany and medicine at the University of Uppasala, and continued in the same position till his death on January 10, 1778.

Linnaeus was so popular as a teacher that his classes on field trips were attended often by 200 or more students. He was buried in the Cathedral at Uppasala "unshaven, unwashed, unclad, enveloped with a sheet" as per his own wish and request.

Three universally known works of Linnaeus are the *Systema Naturae* published in 1735, which contains the outline of his system of classification; *Genera Plantarum* published in 1737 that contains the description of several genera; and *Species Plantarum* published in 1753 in two volumes, which contains his detailed views on plant identification and description of species.

Plant collections of Linnaeus were sold by his wife in 1783 to J.E. Smith, a British botanist, who was one of the founders of Linnean Society of London. The entire plant collection is now housed in the office of the Linnean Society in Burlington House, Piccadilly, London.

The Linnaeus's system of classification is an artificial sexual system. However, Radford (1986) described it as a mechanical system of classification. Linnaeus divided plants into 24 classes, mainly on the bases of number, union, and length of stamens. The classes were divided into orders. On the basis of the number of stamens in each flower Linnaeus placed all algae, fungi, mosses and ferns under one class Cryptogamia.

All the 24 classes of the classification of Linnaeus are undermentioned:

1. *Monandria* (with 1 stamen), e.g. *Scirpus.*
2. *Diandria* (with 2 stamens), e.g. *Veronica.*
3. *Triandria* (with 3 stamens), e.g. *Iris.*
4. *Tetrandria* (with 4 stamens), e.g. *Ulmus, Mentha.*
5. *Pentandria* (with 5 stamens), e.g. *Primula.*
6. *Hexandria* (with 6 stamens), e.g. *Berberis, Rumex.*
7. *Heptandria* (with 7 stamens), e.g. *Aesculus.*

---

[1]He was also known as Carl Linné or Carl von Linné. Lawrence (1951) and most of the other taxonomists mentioned the spelling of his name as "Carolus Linnaeus". But some recent taxonomists (Radford, 1986; Jones & Luchsinger, 1987) spelled his name as Carl Linnaeus.

[2]For more details of the history of Linnaeus, consult Chapter 2 (Article No. 2.1.6).

8. *Octandria* (with 8 stamens), e.g. *Fagopyrum.*
9. *Enneandria* (with 9 stamens), e.g. *Ranunculus.*
10. *Decandria* (with 10 stamens), e.g. *Acer.*
11. *Dodecandria* (with 11–19 stamens), e.g. *Euphorbia.*
12. *Icasandria* (with 20 or more stamens, episepalous), e.g. *Rosa.*
13. *Polyandria* (with 20 or more stamens, attached to axis), e.g. *Papaver.*
14. *Didynamia* (stamens didynamous), e.g. *Linnaea.*
15. *Tetradynamia* (stamens tetradynamous), e.g. Cruciferae.
16. *Monadelphia* (stamens monadelphous), e.g. Malvaceae.
17. *Diadelphia* (stamens diadelphous), e.g. *Trifolia.*
18. *Polyadelphia* (stamens polyadelphous), e.g. *Hypericum.*
19. *Syngenesia* (syngenesious condition), e.g. Compositae.
20. *Gynandria* (stamens adnate to gynoecium), e.g. Orchidaceae.
21. *Monoecia* (plants monoecious), e.g. *Typha.*
22. *Dioecia* (plants dioecious), e.g. *Salix, Urtica.*
23. *Polygamia* (plants polygamous), e.g. *Empetrum.*
24. *Cryptogamia* (flowers concealed), e.g. algae, fungi, mosses, ferns.

The classification system of Linnaeus dominated the botanical scene for over seven decades and was replaced by natural classification systems of A.L. de]ussieu and A.P. de Candolle based on form relationships.

3.4.3 Antoine-Laurent de Jussieu (1748–1836)[1]

A natural system of classification was proposed by A.L. de Jussieu in 1789 in his *Genera Plantarum Secundus Ordines Naturales Disposita.* He recognized one hundred *orders* of the plants which are now called *families.* He divided the plants into three main groups, i.e. Acotyledones, Monocotyledones and Dicotyledones. He mainly emphasized on the number of cotyledons and their presence or absence, number of petals and their presence or absence, and position of stamens. This system firmly established the philosophy of the natural system among the botanical community. All the hundred *orders* (now called *families*) were arranged by A.L. de Jussieu into following 15 classes:

Groups	*Class*
1. *Acotyledones* (plants without cotyledons, i.e. algae, fungi, mosses, etc.)	I
2. *Monocotyledones* (plants with one cotyledon)	
Stamina hypogyna (stamens hypogynous)	II
Stamina perigyna (stamens perigynous)	III
Stamina epigyna (stamens epigynous)	IV
3. *Dicotyledones* (plants with two cotyledons)	
Stamina hypogyna (stamens hypogynous)	V

[1] For historical details see Chapter 2 (Article No. 2.1.7).

Stamina perigyna (stamens perigynous)..VI
Stamina epigyna (stamens epigynous)..VII
Monopetalae
Corolla hypogyna (corolla hypogynous)..VIII
Corolla perigyna (corolla perigynous)..IX
Corolla epigyna (corolla epigynous)
Antheris connatis (anthers connate) X
Antheris distinctis (anthers free) XI
Polypetalae
Stamina hypogyna (stamens hypogynous) XII
Stamina perigyna (stamens perigynous)..XIII
Stamina epigyna (stamens epigynous)..XIV
Declines irregularis (corolla generally absent; male and female flowers on different plants) XV

3.4.4 Augustin Pyramus de Candolle (1778–1841)[1]

One more natural system of classification was proposed by A.P. de Candolle in 1813 in his book *Theorie Elementaire de la Botanique*. He divided plants into two major groups i.e. Cellulares (non-vascular plants), and Vasculares (vascular plants). In his another work, *Prodromus,* he described all the species of the vascular plants known to the world then. In this work he described 58,000 species of dicotyledons belonging to 161 families.

A brief outline of his plan is undermentioned:

1. *Vasculares* (plants with vascular bundles)
   Class 1. Exogenae (Dicotyledoneae: vascular bundles in ring; 2 cotyledons)
   (A) Diplochlamydeae (both calyx and corolla present)
   (a) Thalamiflorae (polypetalous, hypogynous)—Orders 1–46.
   (b) Calyciflorae (perigynous or epigynous)—Orders 47–84.
   (c) Corolliflorae (gamopetalous and hypogynous)—Orders 85–108.
   (B) Monochlamydeae (only calyx present)—Orders 109–128.
   Class 2. Endogenae (Monocotyledoneae; vascular bundles scattered; cotyledon one)
   (A) Phanerogamae (flowers present)—Orders 129–150.
   (B) Cryptogamae (flowers absent or hidden)—Orders 151–155.
2. *Cellulares* (Plants without vascular bundles or cotyledons)
   Class 1. Foliaceae (leafy and sexual)—Orders 156–157.
   Class 2. Aphyllae (nonleafy and without known sexes)—Orders 158–161.

[1]For historical details see Chapter 2 (Article No. 2.1.7).

The system of A.P. de Candolle was easy and simple, and surpassed all other systems. But its major drawback was the inclusion of vascular cryptogams among the monocots.

3.4.5 George Bentham (1800–1884) and Joseph Dalton Hooker (1817–1911)

The most accepted natural system of classification was proposed by Bentham and Hooker in their *Genera Plantarum* published during July 1862 and April 1883. Bentham, a self-trained British botanist, and Hooker, the Director of the Royal Botanic Gardens, Kew (England), described all known genera of seed plants in three volumes of their *Genera Plantarum,* published in Latin.

Bentham and Hooker's system of classification is still used and followed in several herbaria of the world. In most of the Indian herbaria too, the plants are arranged according to this system of classification. It is supposed to be the best system for the students to identify plants in the laboratories. This is so because Bentham and Hooker prepared the generic descriptions of the plants from their own observations and not by copying from the available literature. Large genera are divided into sections and sub-sections, and the description of genera is complete and accurate.

In all, they described 97,205 species belonging to 7,569 genera of 200 families of flowering plants in three volumes of *Genera Plantarum* as shown in Table 3.1.

Table 3.1 Number of orders, genera and families described by Bentham and Hooker

Groups	Orders (families)	Genera	Species
Dicotyledons			
(a) Polypetalae	82	2,610	31,874
(b) Gamopetalae	45	2,619	34,556
(c) Monochlamydeae	36	801	11,784
Gymnosperms	3	44	415
Monocotyledons	34	1,495	18,576
Total	200	7,569	97,205

The Bentham and Hooker's system of classification is clearly derived from the systems of de Jussieu and de Candolle. Bentham and Hooker divided all Phanerogams or seed plants into Dicotyledons, Gymnosperms and Monocotyledons. Ranales were placed in the beginning and grasses at the end in this classification. A summary outline of their classification is mentioned below:

(A) *Dicotyledons* (Reticulate venation; two cotyledons; pentamerous flowers).

1. Polypetalae (Corolla of separate petals)

Series I. Thalamiflorae (stamens many; hypogynous; disc absent).

Order 1. *Ranales:* Ranunculaceae, Magnoliaceae, Annonaceae, Nymphaeaceae and four more families.

2. *Parietales:* Papaveraceae, Capparidaceae, Cruciferae, Violaceae and five more families.

3. *Polygalineae:* Polygaleae and three more families.

4. *Caryophyllineae:* Caryophyllaceae, Portulacaceae and two more families.
5. *Guttiferales:* Guttiferae and five more families.
6. *Malvales:* Malvaceae, Tiliaceae and Sterculiaceae.

Series II. Disciflorae (stamens hypogynous; disc present).

Order 1. *Geraniales:* Geraniaceae, Rutaceae, Meliaceae, and eight more families.
2. *Olacales:* Olacineae, and two more families.
3. *Celastrales:* Rhamnaceae, and three more families.
4. *Sapindales:* Sapindaceae, Anacardiaceae and Sabiaceae.

Series III. Calyciflorae (stamens perigynous or epigynous; ovary generally inferior).

Order 1. *Rosales:* Leguminosae, Rosaceae, and seven more families.
2. *Myrtales:* Combretaceae, Myrtaceae, Lythraceae, and three more families.
3. *Passiflorales:* Cucurbitaceae, Begoniaceae, and five more families.
4. *Ficoidales:* Cactaceae, Ficoideae.
5. *Umbellales:* Umbelliferae, and two more families.

2. Gamopetalae (petals of corolla are partially or completely fused).

Series I. Inferae (inferior ovary).

Order 1. *Rubiales:* Rubiaceae and Caprifoliaceae.
2. *Asterales:* Compositae, and three more families.
3. *Campanales:* Campanulaceae, and three more families.

Series II. Heteromerae (ovary superior; androecium of one or two series; carpels more than two).

Order 1. *Ericales:* Ericaceae, and five more families.
2. *Primulales:* Primulaceae, and two more families.
3. *Ebenales:* Sapotaceae, and two more families.

Series III. Bicarpellatae (ovary superior; androecium of one series; carpels two).

Order 1. *Gentianales:* Oleaceae, Apocynaceae, Asclepiadaceae, and three more families.
2. *Polemoniales:* Convolvulaceae, Solanaceae, and three more families.
3. *Personales:* Scrophulariaceae, Pedaliaceae, Bignoniaceae, Acanthaceae, and four more families.
4. *Lamiales:* Labiatae, Verbenaceae, and two more families.

3. Monochlamydeae (Petals absent).

Series I. *Curvembryeae* (embryo coiled, ovule generally one): Amaranthaceae, Chenopodiaceae, Polygonaceae, and four more families.

Series II. *Multiovulatae aquaticae* (ovules many; immersed aquatics): Podostemaceae.

Series III. *Multiovulatae terrestris* (ovules many; plants terrestrial): Nepenthaceae, and two more families.

Series IV. *Microembryeae* (embryo very minute): Piperaceae, and three more families.

Series V. *Daphnales* (ovary with one carpel and one ovule): Proteaceae, and three more families.

Series VI. *Achlamydosporeae* (usually inferior ovary; one locule, with 1–3 ovules): Loranthaceae, Santalaceae and Balanophoreae.

Series VII. *Unisexuales* (flowers unisexual): Euphorbiaceae, Urticaceae, and seven more families.

Series VIII. *Ordines anomali* (families of uncertain relationship): Ceratophyllaceae, and three more families.

(B) *Gymnospermae* (naked-seeded plants): Gnetaceae, Coniferae, Cycadaceae.

(C) *Monocotyledons* (parallel venation; one cotyledon; trimerous flowers).

Series I. *Microspermae* (inferior ovary; minute seeds): Orchidaceae, and two more families.

Series II. *Epigynae* (inferior ovary; large seeds): Iridaceae, Amaryllidaceae, and five more families.

Series III. *Coronarieae* (superior ovary; coloured perianth): Liliaceae, Commelinaceae, and six more families.

Series IV. *Calycineae* (superior ovary; green perianth): Juncaceae, Palmae, Flagellariaceae.

Series V. *Nudiflorae* (perianth usually absent; superior ovary): Typhaceae, Araceae, and three more families.

Series VI. *Apocarpae* (carpels free): Alismaceae, and two more families.

Series VII. *Glumaceae* (reduced perianth; bracts large, scaly): Cyperaceae, Gramineae, and three more families.

Merits of the System of Bentham and Hooker

1. It is the first great natural system of classification.
2. It is very easy to follow for all practical purposes, and that is why Kew Herbarium and several other herbaria of the world, including India, are arranged according to this system.
3. This system was never planned by Bentham and Hooker on the basis of phylogeny, although the theory of organic evolution was already announced by Darwin and Wallace in 1859. So this system should not be criticized on the basis of phylogeny.
4. Ranales have been given a primitive position in this system. Recent taxonomic findings also indicate that Ranales are the most primitive living angiosperms.
5. In this system the monocots are derived from dicots. Several recent taxonomic findings support this view.

Demerits of the System of Bentham and Hooker

1. The position of gymnosperms in between dicots and monocots in this system is its foremost demerit. This arrangement is made without considering the affinities among these groups.
2. Several important floral characters have been neglected in this system.

3. In this system some of the closely related families have been separated and placed under different orders (*cohorts*). In the same way, a number of unrelated families have been grouped nearer. Few examples are undermentioned:
   (i) All the families of Series Curvembrae are related to Caryophyllaceae of series Thalamiflorae of Polypetalae.
   (ii) Podostemaceae of Series Multiovulatae aquaticae of Monochlamydeae deserves a placing under Rosales of Series Calyciflorae of Polypetalae.
   (iii) Nepenthaceae (of Series Multiovulatae-terrestris of Monochlamydeae) is related more to family Saracenniaceae (of orders Parietales of Thalamiflorae).
   (iv) Laurineae of Series Daphnales is related closely with Magnoliaceae of Ranales.
4. Advanced families, such as Orchidaceae, have been considered primitive in this system by placing them in the beginning.
5. The entire arrangement of monocots is unnatural and unphylogenetic in this system.

3.4.6 August Wilhelm Eichler (1839–1889)

Eichler (1883), a German botanist, put forth a widely accepted system of classification of plants. This system may be considered as a transitional phylogenetic system. The schematic representation of Eichler's classification is undermentioned:

Plant Kingdom

1. *Cryptogamae* (Nonseed plants)
   Divisions 1. *Thallophyta:* Algae, Fungi, Lichens
   2. *Bryophyta:* Liverworts and Mosses
   3. *Pteridophyta:* Equisetineae, Lycopodineae and Filicineae
2. *Phanerogamae* (Seed plants)
   Divisions 1. *Gymnospermae*
   2. *Angiospermae*
   Class 1. Monocotyledones (7 orders)
   2. Dicotyledones
   Subclass 1. Choripetalae (20 orders)
   2. Sympetalae (9 orders)

3.4.7 Adolf Engler (1844–1930) and Karl Prantl (1849–1893)

Engler and Prantl's system of classification has also been described as a "transitional phylogenetic system" by Jones and Luchsinger (1987). However, Radford (1986) and most of the other taxonomists describe it as a phylogenetic system. Engler, a Professor at the University of Berlin, proposed a system of classification based on that of Eichler. Between 1887 to 1915, he published it along with his associate, Prantl, in several volumes of *Die Naturlichen Pflanzenfamilien.* This, profusely-illustrated work of Engler and Prantl, provides keys and description of all the plant families known to them at that time. They classified all the plants from algae to angiosperms. This system is used in most of the non-British herbaria of the world.

The followers of Engler and Prantl published revised classification in several successive editions of *Syllabus der Pflanzenfamilien.* The twelfth edition of *Syllabus*, dealing angiosperms, was edited by Melchior in 1964.

The most noteworthy features of the Engler and Prantl's system of classification are that they (i) placed monocots before dicots, (ii) considered orchids to be more evolved than grasses, and (iii) considered apetalous and catkin-bearing dicots primitive to the dicots bearing petals and simple unisexual flowers.

Engler and Prantl's system divided the plant kingdom into following *14 divisions:*

1. *Schizophyta* (classes Schizomycetes and Schizophyceae), 2. *Myxothallophyta* (class Myxomycetes), 3. *Flagellatae,* 4. *Dinoflagellatae,* 5. *Bacillariophyta,* 6. *Conjugatae,* 7. *Heterocontae.* 8. *Chlorophyceae,* 9. *Charophyta,* 10. *Phaeophyceae,* 11. *Rhodophyceae,* 12. *Eumycetes* (Fungi), 13. *Archegoniatae* or *Embryophyta-Asiphonogama* (subdivisions Bryophyta and Pteridophyta), 14. *Embryophyta-Siphonogama* (subdivisions Gymnospermae and Angiospermae).

Subdivision—A*ngiospermae*

Class 1. *Monocotyledoneae*

Orders: 1. Pandanales (*Typhaceae,* etc.)
2. Helobiae (*Alismataceae,* and 6 more families)
3. Triuridales (*Triuridaceae*)
4. Glumiflorae (*Cyperaceae, Gramineae*)
5. Principes (*Palmae*)
6. Synanthae (*Cyclanthaceae*)
7. Spathiflorae (*Araceae, Lemnaceae*)
8. Farinosae (*Commelinaceae*, and 12 more families).
9. Liliiflorae (*Juncaceae, Liliaceae, Amaryllidaceae, Iridaceae,* and 5 more families)
10. Scitamineae (*Musaceae,* and 3 more families)
11. Microspermae (*Orchidaceae* and *Burmanniaceae*)

Class 2. *Dicotyledoneae*

Subclass 1. *Archichlamydeae*

Orders: 1. Verticillatae (*Casuarinaceae*)
2. Piperales (*Piperaceae*, and 2 more families)
3. Hydrostachyales (*Hydrostachyaceae*)
4. Salicales (*Salicaceae*)
5. Garryales (*Garryaceae*)
6. Myricales (*Myricaceae*)
7. Balanopsidales (*Balanopsidaceae*)

8. Leitneriales (*Leitneriaceae*)
9. Juglandales (*Juglandaceae*)
10. Julianiales (*Julianiaceae*)
11. Batidales (*Batidaceae*)
12. Fagales (*Fagaceae, Butolaceae*)
13. Urticales (*Moraceae, Urticaceae, Ulmaceae*)
14. Podostemonales (*Podostemonaceae*)
15. Proteales (*Proteaceae*)
16. Santanales (*Santalaceae, Loranthaceae*, and 5 more families)
17. Aristolochiales (*Aristolochiaceae*, and 2 more families)
18. Balanophorales (*Balanophoraceae*)
19. Polygonales (*Polygonaceae*)
20. Centrospermae (*Chenopodiaceae, Amaranthaceae, Nyctaginaceae, Caryophyllaceae,* and 6 more families)
21. Ranales (*Ranunculaceae, Nymphaeaceae, Magnoliaceae, Annonaceae*, and 15 more families)
22. Rhoeadales (*Papaveraceae, Capparidaceae, Cruciferae,* and 4 more families)
23. Sarraceniales (3 families)
24. Rosales (*Rosaceae, Leguminosae,* and 15 more families)
25. Pandanales (*Pandanaceae*)
26. Geraniales (*Geraniaceae, Rutaceae, Meliaceae, Euphorbiaceae,* and 17 more families)
27. Sapindales (*Anacardiaceae,* and 22 more families)
28. Rhamnales (*Rhamnaceae, Vitaceae*)
29. Malvales (*Malvaceae, Tiliaceae, Bombacaceae, Sterculiaceae,* and 3 more families)
30. Parietales (*Violaceae,* and 30 more families)
31. Opuntiales (*Cactaceae*)
32. Myrtiflorae (*Myrtaceae, Combretaceae,* and 21 more families)
33. Umbelliflorae (*Umbelliferae*, and 2 more families)

Subclass 2. Metachlamydeae (*Sympetalae*)

Orders
1. Diapensiales (*Diapensiaceae*)
2. Ericales (*Ericaceae*, and 3 more families)
3. Primulales (*Primulaceae*, and 2 more families)
4. Plumbaginales (*Plumbaginaceae*)
5. Ebenales (*Sapotaceae*, and 6 more families)
6. Contortae (*Apocynaceae, Asclepiadaceae, Oleaceae,* and 3 more families)

7. Tubiflorae (*Convolvulaceae, Boraginaceae, Verbenaceae, Labiatae, Solanaceae, Scrophulariaceae, Bignoniaceae, Pedaliaceae, Acanthaceae,* and 13 more families)
8. Plantaginales (*Plantaginaceae*)
9. Rubiales (*Rubiaceae*, and 4 more families)
10. Cucurbitales (*Cucurbitaceae*)
11. Campanulatae (*Campanulaceae, Compositae*, and 4 more families)

Merits of the System of Engler and Prantl

1. This is a convenient and well-known filing system of several herbaria of the world.
2. Polypetalae and Monochlamydae of Bentham and Hooker were merged by Engler and Prantl into one single subclass Archichlamydae.
3. This system treated families such as Orchidaceae and Compositae as advanced families.
4. In this system several closely related families (e.g. Liliaceae, Juncaceae, Iridaceae, and Amaryllidaceae) are treated close to one another.
5. Abundant illustrations are provided along with the description of the families.
6. The system is provided with exhaustive keys of families and orders.
7. The description of each family also contains a summary of its embryology, morphology, anatomy, and geographical distribution.

Demerits of the System of Engler and Prantl

1. Monocots have been placed before dicots in this system. But this system has been reversed in the 1964 edition of *Syllabus der Pflanzenfamilien.*
2. Naked flowers of Amentiferae have been treated as primitive in this system.
3. Helobiae, consisting of primitive forms, have been placed between two advanced orders Glumiflorae and Pandanales.
4. Araceae are derived from Liliaceae, but Engler and Prantl placed Araceae before Liliaceae.
5. This system fails to recognize the significance of reduction, and because of this "simple" were equated with "primitive" according to Cronquist (1965).

3.4.8 Charles Edwin Bessey (1845–1915)

A system of classification that tries to reflect evolution is called phylogenetic system. Bessey, a student of Asa Gray, and a Professor at the University of Nebraska, contributed a pure phylogenetic system of plant classification in 1894 after being highly impressed by the evolutionary ideas of Darwin. It could be published in the final form in 1915, shortly after his death. Bessey closely followed Bentham and Hooker's system of classification. His system was based on certain guiding principles of primitive and advanced characters called *dicta*[1]. He considered angiosperms monophyletic and derived from a cycadeoid ancestor having bisexual strobili.

[1] Some of the Besseyan principles or "dicta" are mentioned in Chapter 13 (Article 13.3).

According to Bessey (1915), Ranales were the primitive angiosperms. One branch of Ranales developed into monocots, and the other into dicots. This system is often called *Ranalian concept of evolution*. On being diagramed, this system took the form of a cactus plant, and came to be called *Bessey's cactus* (Fig. 3.1).

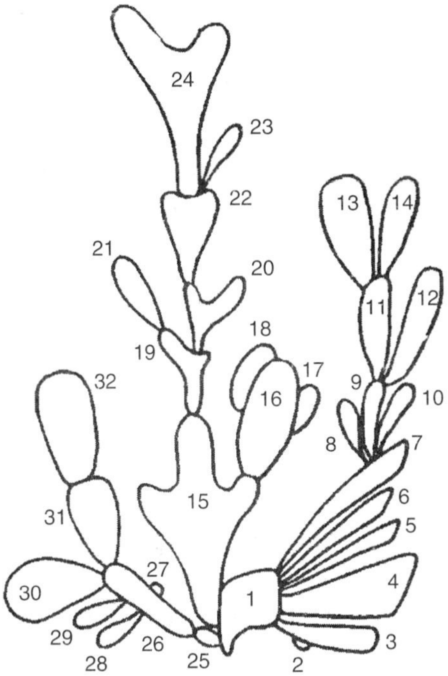

1. Ranales
2. Serraceniales
3. Malvales
4. Geraniales
5. Guttiferales
6. Rhoeadales
7. Caryophyllales
8. Ebenales
9. Primulales
10. Ericales
11. Geraniales
12. Polemoniales
13. Scrophulariales
14. Lamiales
15. Rosales
16. Myrtales
17. Cactales
18. Lossales
19. Celastrales
20. Umbellales
21. Sapindales
22. Rubiales
23. Campanulales
24. Asterales
25. Alismatales
26. Liliales
27. Hydrales
28. Arales
29. Palmales
30. Graminales
31. Iridales
32. Orchidales

Fig. 3.1 Bessey's cactus—Ranalian concept of evolution.

In this system of classification, the *angiosperms* are divided as under:

Class—ALTERNIFOLIAE (Monocotyledoneae)

Subclass—*Strobiloideae*

Orders—Alismatales (Families 1–9), Liliales (Fam. 10–22), Arales (Fam. 23–25), Palmales (Fam. 26), Graminales (Fam. 27–31).

Subclass—*Cotyloideae*

Orders—Hydrales (Fam. 32), Iridales (Fam. 33–43), Orchidales (Fam. 44–45).

Class—OPPOSITIFOLIAE (Dicotyledoneae)

Subclass—*Strobiloideae*

Superorder—*Apopetalae—Polycarpellatae*

Orders—Ranales (Fam. 46–69). Malvales (Fam. 70–81), Sarraceniales (Fam. 82–83), Geraniales (Fam. 84–105), Guttiferales (Fam. 106–125), Rhoeadales (Fam. 126–132), Caryophyllales (Fam. 133–149).

Superorder—*Sympetalae—Polycarpellatae*

Orders—Ebenales (Fam. 150–154), Ericales (Fam. 155–160), Primulales (Fam. 161–165).

Superorder—*Sympetalae—Dicarpellatae*

Orders—Gentianales (Fam. 166–171), Polemoniales (Fam. 172–177), Scrophulariales (Fam. 178–187), Lamiales (188–191).

Subclass—*Cotyloideae*

Superorder—*Apopetalae*

Orders—Rosales (Fam. 192–214), Myrtales (Fam. 215–229), Loasales (Fam. 230–234), Cactales (Fam. 235), Celastrales (Fam. 236–259), Sapindales (Fam. 260–274), Umbellales (Fam. 275–277).

Superorder—*Sympetalae*

Order—Rubiales (Fam. 278–282), Campanulales (Fam. 283–286), Asterales (Fam. 287–300).

Merits of the System of Bessey

1. Bentham and Hooker placed Gymnosperms in between Dicotyledons and Monocotyledons. Such is not the case in Bessey's system of classification.
2. Monochlamydeae has been completely abolished by Bessey. Families of Monochlamydeae have been distributed near their allies in Dicotyledons (Oppositifoliae).
3. Ranales are the starting point among Dicotyledons.
4. In Bessey's system the families with an inferior ovary follow the families possessing a superior ovary.

Demerits of the System of Bessey

1. Monocots have been assigned a position prior to dicots, which is not proper.
2. Hypogyny, perigyny and epigyny have been emphasized too much in this system.

Comparison of Englerian and Besseyan Concepts

1. Englerian school believed the *primitive flowers* to be apetalous and unisexual, while Besseyan school believed them polypetalous and bisexual.
2. Englerian school believed the wind pollination primitive while it was the insect pollination according to Besseyan school.
3. According to Englerian concept, *dicots began with* Amentiferae while according to Besseyan concept they began with Ranales.
4. Englerian school believed that *monocots were derived from a gymnospermous* stock while according to Besseyan school they derived from some primitive dicots.
5. The main philosophy of Englerian school is that simple flowers are primitive while Besseyan school believed that flowers with complex structures are more so.

3.4.9 John Hutchinson (1884–1972)

An intentional phylogenetic system was produced by John Hutchinson, a British botanist, associated with the Royal Botanic Gardens, Kew. His system is close to that of Bessey, and is published in his *Genera of Flowering Plants* (1964–1967) and *Families of Flowering Plants* (1973).

Hutchinson based his system of classification on the *twenty four principles* mentioned below in a condensed form:

1. Evolution is both downwards and upwards.
2. Evolution does not necessarily involve all organs at the same time.
3. Generally, the evolution has been consistent.
4. Broadly, shrubs and trees are more primitive than herbs in any one genus or family.
5. In comparison with climbers, shrubs and trees are older in any one genus or family.
6. Perennials are older than annuals and biennials.
7. Aquatic flowering plants are derived from terrestrial ancestors.
8. Dicots are primitive compared to monocots.
9. Spiral arrangement of vegetative and floral leaves is primitive to cyclic arrangement.
10. Usually, simple leaves are more primitive than compound leaves.
11. Bisexual flowers are less advanced than unisexual, and dioecious plants are more recent than monoecious.
12. Inflorescence is more advanced than the solitary flowers.
13. Types of aestivation are evolved from contorted to imbricate to valvate.
14. Polymerous flowers (many-parted) precede oligomerous (few-parted) flowers.
15. Apetalous flowers are derived from flowers with petals.
16. Polypetaly is more primitive than gamopetaly.
17. In comparison to zygomorphy, actinomorphy is more primitive.
18. Hypogyny is more primitive than perigyny. The most advanced condition is epigyny.
19. Apocarpy is primitive than syncarpy.
20. Polycarpellary condition precede the condition of a few carpels.
21. Endospermic seeds with a small embryo are primitive to non-endospermic ones with a large embryo.
22. Flowers with many stamens are primitive to flowers with few stamens.
23. Plants with separate anthers are primitive compared to those with fused anthers or filaments.
24. Aggregate fruits are more highly evolved than single fruits.

An outline of the Hutchinson's system of classification, up to the level of orders, is mentioned below. Common families of some of the orders are mentioned in parenthesis.

Phylum *ANGIOSPERMAE*

Subphylum 1. DICOTYLEDONES

Division I. *Lignosae*

Orders: 1. Magnoliales (*Magnoliaceae*); 2. Annonales (*Annonaceae*); 3. Laurales; 4. Dilleniales; 5. Coriariales; 6. Rosales (*Rosaceae*); 7. Leguminales; 8. Cunoniales; 9. Styracales; 10. Araliales; 11. Hamamelidales; 12. Salicales; 13. Leitneriales; 14. Myricales; 15. Balanopsidales; 16. Fagales (*Fagaceae*); 17. Juglandales; 18. Casuarinales (*Casuarinaceae*); 19. Urticales (*Moraceae, Urticaceae*); 20. Bixales; 21. Thymeleales (*Nyctaginaceae*); 22. Proteales (*Proteaceae*); 23. Pittosporales; 24. Capparales (*Capparidaceae*); 25. Tamaricales; 26. Violales (*Violaceae*); 27. Polygalales; 28. Loasales; 29. Passiflorales; 30. Cucurbitales (*Cucurbitaceae, Begoniaceae*); 31. Cactales (*Cactaceae*); 32. Tiliales (*Tiliaceae, Sterculiaceae, Bombacaceae*); 33. Malvales (*Malvaceae*); 34. Malpighiales; 35. Euphorbiales (*Euphorbiaceae*); 36. Theales; 37. Ochnales; 38. Ericales; 39. Guttiferales; 40. Myrtales (*Myrtaceae*); 41. Celastrales (*Salvadoraceae*); 42. Olacales; 43. Santalales (*Loranthaceae, Santalaceae*); 44. Rhamnales (*Vitaceae*); 45. Myrsinales; 46. Ebenales (*Sapotaceae*); 47. Rutales (*Rutaceae*); 48. Meliales (*Meliaceae*); 49. Sapindales; 50. Loganiales (*Oleaceae*); 51. Apocynales (*Apocynaceae, Asclepiadaceae*); 52. Rubiales (*Rubiaceae*); 53. Bignoniales (*Bignoniaceae, Pedaliaceae*); 54. Verbenales (*Verbenaceae*).

Division II. *Herbaceae*

Orders: 55. Ranales (*Ranunculaceae*); 56. Berberidales; 57. Aristolochiales (*Aristolochiaceae*); 58. Piperales (*Piperaceae*); 59. Rhoeadales (*Papaveraceae, Fumariaceae*); 60. Brassicales (*Brassicaceae*); 61. Resedales; 62. Caryophyllales (*Caryophyllaceae*); 63. Polygonales (*Polygonaceae*); 64. Chenopodiales (*Chenopodiaceae, Amaranthaceae*); 65. Onagrales; 66. Gentianales; 67. Primulales; 68. Plantaginales; 69. Saxifragales; 70. Sarraceniales; 71. Podostemales; 72. Umbellales (*Umbelliferae*); 73. Valerianales; 74. Campanulales; 75. Goodeniales; 76. Asterales (*Compositae*); 77. Solanales (*Solanaceae, Convolvulaceae*); 78. Personales (*Acanthaceae, Scrophulariaceae*); 79. Geraniales (*Geraniaceae*); 80. Polemoniales; 81. Boraginales (*Boraginaceae*); 82. Lamiales (*Labiatae*).

Subphylum 2. MONOCOTYLEDONES

Division. 1. *Calyciferae*

Orders: 83. Butomales (*Butomaceae*); 84. Alismatales; 85. Triuridales; 86. Juncaginales; 87. Aponogetonales; 88. Potamogetonales; 89. Najadales; 90. Commelinales (*Commelinaceae*); 91. Xyridales; 92. Eriocaulales; 93. Bromeliales; 94. Zingiberales (*Musaceae, Zingiberaceae*).

Division. 2. *Corolliferae*

Orders: 95. Liliales (*Liliaceae*); 96. Alstromeriales; 97. Arales (*Araceae*); 98. Typhales (*Typhaceae*); 99. Amaryllidales (*Amaryllidaceae*); 100. Tridales;

101. Dioscoreales; 102. Agavales; 103. Palmales (*Palmae*); 104. Pandanales; 105. Cyclanthales; 106. Haemodorales; 107. Burmanniales; 108. Orchidales (*Orchidaceae*).

Division 3. *Glumiflorae*

Orders: 109. Juncales (*Juncaceae*); 110. Cyperales (*Cyperaceae*); 111. Graminales (*Gramineae*).

Merits of the System of Hutchinson

1. Families and orders in this system are of small size and comprise only very closely related taxa.
2. This system is in confirmity with the modern views of the phylogeny of angiosperms as it considers the Ranales and Magnoliales as the starting points among dicots.
3. Monocots are discussed after dicots in this system.

Demerits of the System of Hutchinson

1. From the point of view of plant identification this system is not of much utility.
2. Dicots have been divided into two major groups (Lignosae and Herbaceae) in this system. This kind of classification is outdated as the habit used to be the main basis of classification in the past era of Aristotle. Secondly, it also separates the families that have close affinities.

3.5 COMPARISON OF SYSTEMS OF CLASSIFICATION

Table 3.2 provides points of comparison between three main systems of classification proposed by Bentham and Hooker, Engler and Prantl, and Hutchinson.

3.6 CURRENT SYSTEMS OF CLASSIFICATION

Armen Takhtajan (1980) of Russia (formerly U.S.S.R.), Arthur Cronquist (1981) of U.S.A., Rolf Dahlgren (1983) of Denmark, and Robert F. Thorne (1983) of U.S.A. published systems[1] of plant classification during last few decades. These workers used the recent data and information of palaeobotany, biochemical systematics, and the ultrastructural details unveiled by the electron microscope, along with the information of the traditional sources such as morphology and anatomy of plants.

3.6.1 Armen Takhtajan (1980)[1]

The latest version of Takhtajan's classification is published in 1980 in *Botanical Review*. According to him, the angiosperms are monophyletic and developed from some very ancient group of gymnosperms. He is an adherent of the Besseyan philosophy of taxonomy.

Some of the criteria, used by Takhtajan in determining the relative degree of advancement of flowering plants are undermentioned: (i) Woody plants are primitive compared to herbaceous plants; (ii) Deciduous woody plants are evolved from evergreen plants; (iii) Parallel venation is most advanced; (iv) Most primitive leaf arrangement is alternate; (v) Stomata with subsidiary cells are

[1] For the historical details of these systems, refer Chapter 2 (Article No 2.1.9).

Table 3.2 Comparison of systems of classification

S. No.	Bentham and Hooker	Engler and Prantl	Hutchinson
1.	It is a natural system of classification.	It is a transitional phylogenetic system.	It is a phylogenetic system.
2.	It is based on de Candolle's system of classification.	It is based on Eichler's system of classification.	It is based on Bessey's system of classification.
3.	Fixity of species is the main idea behind this system.	Darwinian theory of descent is the main idea behind this system.	Considerable knowledge of phylogeny is the basis of this classification.
4.	Dicots ar divided into (i) Polypetalae, (ii) Gamopetalae, and (iii) Monochlamydae.	Dicots are divided into (i) Archichlamydae, and (ii) Metachlamydae.	Dicots are divided into (i) Lignosae, and (ii) Herbaceae in this system.
5.	Gymnosperms are placed in between dicots and monocots.	Gymnosperms are placed before dicots.	Gymnosperms are placed before dicots.
6.	Monocots are treated after dicots.	Monocots are treated before dicots.	Monocots are treated after dicots.
7.	Orchidaceae and some other advanced families are regarded as primitive, and treated in the beginning.	Orchidaceae are treated as advanced.	Orchidaceae are treated as advanced.
8.	Families with free petals are placed in Polypetalae, with fused petals in Gamopetalae, and without petals in Monochlamydeae.	Polypetalae and Monochlamydeae are placed together under a single group Archichlamydeae. Families with fused petals are placed under Metachlamydeae.	Emphasis is on resemblances in place of differences, and so the Gamopetalae and Monochlamydeae are distributed according to their relationship in Polypetalae.
9.	Monocotyledonous families are arranged in 7 series.	Monocotyledons are arranged in 11 orders.	Monocotyledons are arranged in 29 orders.
10.	All flowering plants are placed in 200 families.	280 families of flowering plants have been recognized.	Number of recognised angio-spermic families is 411.
11.	Gramineae is the last family of monocots.	Orchidaceae is the last family of monocots.	Gramineae is the last family of monocots.

primitive while those lacking subsidiary cells are advanced; (vi) Unilacunar nodes are derived from trilacunar or pentalacunar nodes; (vii) Xylem fibres evolved from tracheids to libriform fibres, through fibre tracheids; (viii) Cymose inflorescence is primitive while racemose is derived; (ix) Flowers with an indefinite or a variable number of their floral parts are primitive; (x) Pollen grains with their exine lacking any external sculpturing are primitive while those having various types of sculptures are advanced; (xi) Apocarpous gynoecium is the characteristic of primitive taxa; (xii) Unitegmic ovules developed from bitegmic ovules; (xiii) Basic type of ovule is anatropous type; all others are derived ones; (xiv) Basic and most primitive type of female gametophyte is 8-nucleate *Polygonum*-type; (xv) Primitive condition is porogamy, and the derived conditions are mesogamy and chalazogamy;

(xvi) Most primitive and basic type of fruit is a many-seeded follicle which develops from a multi-carpellary apocarpous gynoecium.

Takhtajan divided Magnoliophyta (= Angiospermae) into two classes—Magnoliopsida (Dicots) and Liliopsida (Monocots). He further divided Magnoliopsida into 7 subclasses, 20 superorders and 71 orders, and Liliopsida into 3 subclasses, 8 superorders, and 21 orders.

Takhtajan (1980) recognizes a total of 92 orders and 410 families among angiosperms. An outline of his classification up to the level of orders is undermentioned:

DIVISION: MAGNOLIOPHYTA (ANGIOSPERMAE)

Class: *Magnoliopsida* (Dicotyledones)

Subclass 1. Magnoliidae

Superorder I. Magnolianae

Orders: 1. Magnoliales (Annonales), 2. Illiciales, 3. Laurales, 4. Piperales, 5. Aristolochiales.

Superorder II. Rafflesinae

Order: 6. Rafflesiales.

Superorder III. Nymphaeanae

Orders: 7. Nymphaeales, 8. Nelumbonales.

Subclass 2. Ranunculidae

Superorder IV. Ranunculanae

Orders: 9. Ranunculales, 10. Papaverales, 11. Sarraceniales.

Subclass 3. Hamamelididae

Superorder V. Hamamelidanae

Orders: 12. Trochodendrales, 13. Circidiphyllales, 14. Eupteleales, 15. Didymelales, 16. Hamamelidales, 17. Eucommiales, 18 Urticales, 19. Barbeyales, 20. Casuarinales, 21. Fagales, 22 Balanopales, 23. Leitneriales.

Superorder VI. Juglandanae

Orders: 24. Myricales, 25. Juglandales.

Subclass 4. Caryophylladae

Superorder VII. Caryophyllanae

Orders: 26. Caryophyllales, 27. Polygonales.

Superorder VIII. Plumbaginanae

Order: 28. Plumbaginales.

Subclass 5. Dilleniidae

Superorder IX. Dillenianae

Orders: 29. Dilleniales, 30. Paeoniales, 31. Theales, 32. Violales, 33. Begoniales, 34. Capparales, 35. Tamaricales, 36. Salicales.

Superorder X. Ericanae

Orders: 37. Ericales, 38. Ebenales, 39. Primulales.

Superorder XI. Malvanae

Orders: 40. Malvales, 41. Euphorbiales, 42. Thymeleales.

Subclass 6. Rosidae

Superorder XII. Rosanae

Orders: 43. Saxifragales, 44. Rosales, 45. Fabales, 46. Connariales, 47. Podostemales, 48. Nepenthales.

Superorder XIII. Myrtanae

Order: 49. Myrtales.

Superorder XIV. Rutanae

Orders: 50. Rutales, 51. Sapindales, 52. Geraniales, 53. Polygalales.

Superorder XV. Aralianae

Orders: 54. Cornales, 55. Apiales.

Superorder XVI. Celastranae

Orders: 56. Celastrales, 57. Santalales. 58. Balanophorales, 59. Rhamnales, 60. Elaeagnales.

Superorder XVII. Proteanae

Order: 61. Proteales.

Subclass 7. Asteridae

Superorder XVIII. Gentiananae

Orders: 62. Gentianales, 63. Oleales, 64. Dipsacales, 65. Loasales.

Superorder XIX. Lamianae

Orders: 66. Polemoniales, 67. Lamiales, 68. Scrophulariales.

Superorder XX. Asteranae

Orders: 69. Campanuales, 70. Calycerales, 71. Asterales.

Class: *Liliopsida* (*Monocotyledones*)

Subclass 1. Alismalidae

Superorder I. Alismatanae

Orders: 1. Alismatales, 2. Najadales.

Subclass 2. Liliidae

Superorder II. Triuridanae

Order: 3. Triuridales.

Superorder III. Lilianae

Orders: 4. Liliales, 5. Smilacales, 6. Burmanniales, 7. Orchidales, 8. Bromeliales.

Superorder IV. Juncanae

Orders: 9. Juncales, 10. Cyperales.

Superorder V. Commelinanae

Orders: 11. Commelinales, 12. Eriocaulales, 13. Restionales, 14. Hydatellales, 15. Poales.

Superorder VI. Zingiberanae

Order: 16. Zingiberales.

Subclass 3. Arecidae

Superorder VII. Arecanae

Orders: 17. Arecales, 18. Cyclanthales, 19. Pandanales, 20. Typhales.

Superorder VIII. Aranae

Order: 21. Arales.

Merits of the System of Takhtajan

1. Dicots (Magnoliopsida) are discussed prior to monocots (Liliopsida).
2. Dicots begin with the Magnoliales which are universally considered to be the most primitive living angiosperms.
3. In this system the families are small homogeneous units made up of closely related genera.
4. Division of Dicots into two traditional groups of Engler and Prantl, i.e. Archichlamydae and Metachlamydae, has been abolished in this system.
5. Alismatales, which are considered to be the most primitive living monocots these days, are the starting point of monocots in this system.

Demerit of the System of Takhtajan

One strong demerit is the extremely narrowly defined taxa in this system that has resulted in the unwarranted splitting of the related groups.

3.6.2 Arthur Cronquist (1981)

On the basis of a wide range of taxonomic characters of phylogenetic importance, Cronquist (1981) presented a phylogenetic classification. With the help of synoptic keys and detailed charts, he grouped the angiospermic families into orders and subclasses on a worldwide basis. Pteridosperms (*seed ferns*) were considered by him as the probable ancestors of angiosperms.

Class Magnoliopsida (= Dicots) in this system has been divided into six subclasses, i.e. Magnoliidae, Hamamelidae, Caryophyllidae, Dilleniidae, Rosidae, and Asteridae. Of them, the Magnoliidae were thought to be the basal complex, and all the remaining five subclasses were derived from it separately. Asteridae formed the most advanced group of dicots.

Class Liliopsida (= Monocots) in this system has been divided into five subclasses, i.e. Alismatidae, Arecidae, Commelinidae, Zingiberidae, and Liliidae. It is proposed by Cronquist (1981) that Liliopsida have arisen from aquatic ancestors.

Cronquist (1981) recognises a total of 2 classes, 11 subclasses, 83 orders and 383 families and about 2,19,300 species among the angiosperms. A brief outline of Cronquist's system of classification up to the level of orders is mentioned below. *The number mentioned in the parenthesis of each order signifies the number of the families in that order.*

Division—MAGNOLIOPHYTA

Class—1. *Magnoliopsida* (Dicots)

Subclass 1. *Magnoliidae*

Orders: 1. Magnoliales (10), 2. Laurales (8), 3. Piperales (3), 4. Aristolochiales (1), 5. Illiciales (2), 6. Nymphaeales (5), 7. Ranunculales (8), 8. Papaverales (2).

Subclass 2. *Hamamelidae*

Orders: 1. Trochodendrales (2), 2. Hamamelidales (5), 3. Daphniphyllales (1), 4. Didymelales (1), 5. Eucommiales (1), 6. Urticales (6), 7. Leitneriales (1), 8. Juglandales (2), 9. Myricales (1), 10. Fagales (3), 11. Casuarinales (1).

Subclass 3. *Caryophyllidae*

Orders: 1. Caryophyllales (12), 2. Polygonales (1), 3. Plumbaginales (1).

Subclass 4. *Dilleniidae*

Orders: 1. Dilleniales (2), 2. Theales (18), 3. Malvales (5), 4. Lecythiadales (1), 5. Nepenthales (3), 6. Violales (24), 7. Salicales (1), 8. Capparales (5), 9. Batales (2), 10. Ericales (8), 11. Diapensiales (1), 12. Ebenales (5), 13. Primulales (3).

Subclass 5. *Rosidae*

Orders: 1. Rosales (24), 2. Fabales (3), 3. Proteales (2), 4. Podostemales (1), 5. Haloragales (2), 6. Myrtales (12), 7. Rhizophorales (1), 8. Cornales (4), 9. Santalales (10), 10. Rafflesiales (3), 11. Celastrales (11), 12. Euphorbiales (4), 13. Rhamnales (3), 14. Linales (5), 15. Polygalales (7), 16. Sapindales (15), 17. Geraniales (5), 18. Apiales (2).

Subclass 6. *Asteridae*

Orders: 1. Gentianales (6), 2. Solanales (8), 3. Lamiales (4), 4. Callitrichales (2), 5. Plantaginales (1), 6. Scrophulariales (12), 7. Campanulales (7), 8. Rubiales (2), 9. Dipsacales (4), 10. Calycerales (1), 11. Asterales (1).

Class 2. *Liliopsida* (Monocots)

Subclass 1. *Alismatidae*

Orders: 1. Alismatales (3), 2. Hydrocharitales (1), 3. Najadales (10), 4. Triuridales (2).

Subclass 2. *Arecidae*

Orders: 1. Arecales (1), 2. Cyclanthales (1), 3. Pandanales (1), 4. Arales (2).

Subclass 3. *Commelinidae*

Orders: 1. Commelinales (4), 2. Eriocaulales (1), 3. Restionales (4), 4. Juncales (2), 5. Cyperales (2), 6. Hydatellales (1), 7. Typhales (2).

Subclass 4. *Zingiberidae*

Orders: 1. Bromeliales (1), 2. Zingiberales (8).

Subclass 5. *Liliidae*

Orders: 1. Liliales (15), 2. Orchidales (4).

3.6.3 Rolf M.T. Dahlgren (1983)

Rolf Dahlgren (1980, 1981, 1983), Professor at the Botanical Museum of the University of Copenhagen, Denmark, proposed a new classification on the basis of the characters drawn from several branches of botany including anatomy, embryology, and chemistry. He believes that "none of the extant groups of flowering plants is ancestral to any other present-day group. Thus, the Magnoliaceae-Ranunculaceae assemblage is not ancestral to other flowering plants but has simply retained a large number of primitive features" (Jones and Luchsinger, 1987). Dahlgren has been of the view that angiosperms evolved in one particular line of gymnosperms, and are thus monophyletic. Magnoliiflorae have been considered as most primitive dicotyledons while Alismatiflorae have been regarded as most primitive monocots.

Dicots (Magnoliidae) have been divided into 25 superorders and 85 orders while monocots (Liliidae) have been divided into 8 superorders and 25 orders. An outline of Dahlgren's classification up to the level of orders is mentioned below:

Class: *MAGNOLIOPSIDA* (= Angiospermae)

Subclass (1) MAGNOLIIDAE (Dicotyledoneae)

Superorder 1. *Magnoliiflorae:* Orders—1. Annonales, 2. Aristolochiales, 3. Rafflesiales, 4. Magnoliales, 5. Lactoridales, 6. Chloranthales, 7. Illiciales, 8. Laurales, 9. Nelumbonales.

Superorder 2. *Nymphaeiflorae:* Orders—10. Piperales, 11. Nymphaeales.

Superorder 3. *Ranunculiflorae:* Orders—12. Ranunculales, 13. Papaverales.

Superorder 4. *Caryophylliflorae:* Order—14. Caryophyllales.

Superorder 5. *Polygoniflorae:* Order—15. Polygonales.

Superorder 6. *Plumbaginiflorae:* Order—16. Pumbaginales.

Superorder 7. *Malviflorae:* Orders—17. Malvales, 18. Urticales, 19. Euphorbiales, 20. Thymeleales, 21. Rhamnales, 22. Elaeagnales.

Superorder 8. *Violiflorae:* Orders—23. Violales, 24. Cucurbitales, 25. Salicales, 26. Tamaricales, 27. Capparales, 28. Salvadorales.

Superorder 9. *Theiflorae:* Orders—29. Dilleniales, 30. Paeoniales, 31. Theales.

Superorder 10. *Primuliflorae:* Orders—32. Primulales, 33. Ebenales.

Superorder 11. *Rosiflorae:* Orders—34. Trochodendrales, 35. Cercidiphyllales, 36. Hamamelidales, 37. Geissolomomatales, 38. Balanopales, 39. Fagales, 40. Juglandales, 41. Myricales, 42. Casuarinales, 43. Buxales, 44. Cunoniales, 45. Saxifragales, 46. Droserales, 47. Gunnerales, 48. Rosales.

Superorder 12. *Podostemiflorae:* Order—49. Podostemales.

Superorder 13. *Proteiflorae:* Order—50. Proteales.

Superorder 14. *Myrtiflorae:* Orders—51. Haloragales, 52. Rhizophorales, 53. Myrtales, 54. Chrysobalanales.

Superorder 15. *Fabiflorae:* Orders—55. Fabales.

Superorder 16. *Rutiflorae:* Orders—56. Sapindales, 57. Rutales, 58. Polygalales, 59. Geraniales, 60. Balsaminales, 61. Tropaeolales.

Superorder 17. *Santaliflorae:* Orders—62. Celastrales, 63. Vitales, 64. Santalales.

Superorder 18. *Balanophoriflorae:* Order—65. Balanophorales.

Superorder 19. *Araliflorae:* Orders—66. Pittosporales, 67. Araliales.

Superorder 20. *Asteriflorae:* Orders—68. Campanulales, 69. Asterales.

Superorder 21. *Solaniflorae:* Orders—70. Solanales. 71. Boraginales.

Superorder 22. *Corniflorae:* Orders—72. Fouquieriales, 73. Ericales, 74. Eucommiales, 75. Sarraceniales, 76. Cornales, 77. Dipsacales.

Superorder 23. *Loasaliflorae:* Order—78. Loasales.

Superorder 24. *Gentianiflorae:* Orders—79. Goodeniales, 80. Oleales, 81. Gentianales.

Superorder 25. *Lamiiflorae:* Orders—82. Scrophulariales, 83. Hippuridales, 84. Lamiales, 85. Hydrostachyales.

Subclass (2) LILIIDAE (Monocotyledoneae)

Superorder 1. *Alismatiflorae:* Orders—1. Hydrocharitales, 2. Alismatales 3. Zosterales.

Superorder 2. *Triuridiflorae:* Order—4. Triuridales.

Superorder 3. *Ariflorae:* Order—5. Arales.

Superorder 4. *Liliiflorae:* Orders—6. Dioscoreales, 7. Asparagales, 8. Liliales, 9. Burmanniales, 10. Orchidales.

Superorder 5. *Bromelliflorae:* Orders—11. Velloziales, 12. Bromeliales 13. Haemodorales, 14. Philydrales, 15. Pontederiales, 16. Typhales.

Superorder 6. *Zingiberiflorae:* Order—17. Zingiberales.

Superorder 7. *Commeliniflorae:* Orders—18. Commelinales, 19. Hydatellales, 20. Juncales, 21. Cyperales, 22. Poales.

Superorder 8. *Areciflorae:* Orders—23. Arecales, 24. Cyclanthales, 25. Pandanales.

3.6.4 Robert F. Thorne (1983)

On the basis of characters of comparative morphology, ultrastructure, embryology, palaeobotany, cytology, plant geography, pollen and seed morphology, and host-parasite relationships, Thorne (1983) published a pure phylogenetic system. He also believed in the monophyletic origin of angiosperms.

Thorne divided class Angiospermae (Annonopsida) into 2 subclasses, 28 superorders (with the ending of *iflorae*), 54 orders, 73 suborders, 350 families, 12,255 genera and 2,25,490 species. A brief outline of his classification up to the level of orders is mentioned below:

Class: ANGIOSPERMAE (Annonopsida)

Subclass 1. DICOTYLEDONEAE (Annonidae)

Superorder 1. *Annoniflorae:* Orders—1. Annonales, 2. Nelumbonales, 3. Paeoniales, 4. Berberidales.

Superoder 2. *Nymphaeiflorae:* Order—5. Nymphaeales.

Superorder 3. *Rafflesiflorae:* Order—6. Rafflesiales.

Superorder 4. *Theiflorae:* Orders—7. Theales, 8. Ericales, 9. Fouquieriales, 10. Ebenales, 11. Primulales, 12. Polygonales.

Superorder 5. *Chenopodiiflorae:* Orders—13. Chenopodiales, 14. Geraniales.

Superorder 6. *Celastriflorae:* Order—15. Celastrales.

Superorder 7. *Santaliflorae:* Orders—16. Santalales, 17. Balanophorales.

Superorder 8. *Violiflorae:* Orders—18. Violales, 19. Capparales.

Superorder 9. *Malviflorae:* Orders—20. Malvales, 21. Urticales, 22. Rhamnales, 23. Euphorbiales.

Superorder 10. *Rutiflorae:* Order—24. Rutales.

Superorder 11. *Proteiflorae:* Order—25. Proteales.

Superorder 12. *Hamamelidiflorae:* Orders—26. Hamamelidales, 27. Casuarinales, 28. Fagales.

Superorder 13. *Rosiflorae:* Orders—29. Rosales, 30. Pittosporales.

Superorder 14. *Loasiflorae:* Order—31. Loasales.

Superorder 15. *Myrtiflorae:* Order—32. Myrtales.

Superorder 16. *Gentianiflorae:* Orders—33. Oleales, 34. Gentianales, 35. Bignoniales, 36. Lamiales.

Superorder 17. *Solaniflorae:* Orders—37. Solanales, 38. Campanulales.

Superorder 18. *Corniflorae:* Order—39. Cornales, 40. Araliales, 41. Dipsacales.

Superorder 19. *Asteriflorae:* Order—42. Asterales.

Subclass 2. MONOCOTYLEDONEAE (Liliidae)

Superorder 1. *Liliiflorae:* Order—1. Liliales.

Superorder 2. *Triuridiflorae:* Order—2. Triuridales.

Superorder 3. *Alismatiflorae:* Orders—3. Alismatales, 4. Zosterales, 5. Najadales.

Superorder 4. *Ariflorae:* Order—6. Arales.

Superorder 5. *Cyclanthiflorae:* Order—7. Cyclanthales.

Superorder 6. *Pandaniflorae:* Order—8. Pandanales.

Superorder 7. *Areciflorae:* Order—9. Arecales.

Superorder 8. *Typhiflorae:* Order—10. Typhales.

Superorder 9. *Commeliniflorae:* Orders—11. Commelinales, 12. Zingiberales.

Merits of Thorne's System

1. Annonales are universally accepted as the most primitive living angiosperms, and they form the starting point of this system.
2. Traditional division of dicots into Archichlamydeae and Metachlamydeae has been abolished in this system.
3. In this system, by and large, the closely related taxa are placed nearer to one another.
4. The families of "Amentiferae" (a group of unrelated families) have been distributed in different orders, and the group Amentiferae has been abolished.
5. Inclusion of orders Malvales, Urticales, Rhamnales and Euphorbiales under one superorder Malviflorae is also a commendable aspect of this system.

6. Placing of the related orders Cornales and Dipsacales in one superorder Corniflorae is also the merit of this system.

Demerits of Thorne's System

1. This system is not of much practical utility in the identification of plants.
2. The view of Thorne that angiosperms might have originated from some Pteridospermous members in early Cretaceous times is also not accepted by several taxonomists.

Test Your Understanding

1. What are the basic categories of systems of classification? Dioscorides, A.P. de Candolle, and Bentham and Hooker proposed which type of system of classification?
2. Who is called the "father of taxonomy"?
3. The classical works entitled *Systema Naturae*, *Genera Plantarum*, and *Species Plantarum*, have been authored by which taxonomist/taxonomists?
4. Linnaeus divided plants into 24 classes. Name at least 10 of the classes proposed by him.
5. Bentham and Hooker described ________ species belonging to ________ genera of ________ families of flowering plants in his *Genera Plantarum*.
6. Give an outline of the system of classification proposed by Bentham and Hooker.
7. Write some merits and demerits of the system of Bentham and Hooker.
8. Who has authored "*Die Naturlichen Pflanzenfamilien*"?
9. Hutchinson based his system of classification on how many principles? Write at least 10 of these principles.
10. Give a comparison of the systems of Bentham and Hooker and Hutchinson.

Suggested Reading

Bentham, G. and J.D. Hooker, 1862–1883, *Genera Plantarum,* 3 Vols., London.

Bessey, C.E., 1915, Phylogenetic taxonomy of flowering plants, *Ann. Mo. Bot. Gard.* **2**: 109–164.

Candolle, A.P. de., 1813, *Theorie elementaire de Ia botanique,* Paris.

_______ 1824–1873, *Prodromus systematics naturalis regni Vegetabilis,* 17 Vols., Paris.

Cronquist, A., 1981, *An Integrated System of Classification of Flowering Plants,* Columbia University Press, New York.

Dahlgren, R., 1980, A revised system of classification of angiosperms, *Bot.J. Linn. Soc.* **80**: 91–124.

_______ S. Rusendal-Jensen and B.J. Nielsen, 1981, A revised classification of the angiosperms with comments on the correlation between chemical and other characters. In D.A. Young and D.S. Seigler (eds.), *Phytochemistry and Angiosperm Phylogeny,* Praeger, New York.

_______ 1983, General aspects of angiosperm evolution and microsystematics, *Nord. J. Bot.* **3**: 119–149.

Eichler, A.W., 1875–1878, *Bluthendiagramme construirt und erlautert,* 2 Vols. Leipzig.

Engler, A. and K. Prantl, 1887–1915. *Die Naturlichen Pflanzenfamilien,* 23 Vols., Leipzig.

Hutchinson, J., 1964–1967, *The Genera of Flowering Plants,* 2 Vols., Clarendon, Oxford.

_______ 1969, *Evolution and Phylogeny of Flowering Plants,* Academic Press, London.

_______ 1973, *The Families of Flowering Plants* (3rd ed.), Clarendon, Oxford.

Jussieu, A.L. de., 1789, *Genera Plantarum Secundum Ordines naturales disposita,* Paris.

Linnaeus, C., 1735, *Systema Naturae,* Lugduni Batavorum.

_______ 1737, *Genera Plantarum,* Lugduni Batavorum.

_______ 1753, *Species Plantarum,* 2 Vols., Stockholm.

Takhtajan, A., 1980, Outline of the classification of flowering plants (Magnoliophyta), *Bot. Rev.,* **46**: 225–359.

Thome, R.F., 1976, A phylogenetic classification of the angiospermae, *Evol. Biol.,* **9**: 35–106.

_______ 1983, Proposed new realignments in the angiosperms. *Nord. J. Bot.* **3**: 85–117.

CHAPTER 4

TAXONOMIC STRUCTURE

Subordination of organisms in groups under groups is a fact of nature, and it results from evolutionary divergence. When an ancestral form splits into subgroups which undergo divergence during course of time, the resultant descendants are modified forms but resemble each other in decreasing degrees. Members of a species resemble each other because they are genetically more close to each other. This resemblance, however, decreases progressively in different species of the same genus, different genera of the same family, different families of the same order, different orders of the same class and so on. And, to establish such relationships is one of the main concerns of plant taxonomy. In any biological classification, all this is accomplished by taxonomic hierarchy. Different hierarchical categories of plant taxonomy have been mentioned under Article 1.5 of Chapter 1 (Table 1.1).

4.1 CONCEPT OF TAXA

Taxon (pl. taxa) may be defined as "a named taxonomic group of any rank". Thus at family level, taxa may be represented by the Ranunculaceae and Rosaceae, while *Ranunculus* and *Rosa* are examples of generic taxa. The term "*taxon*" was actually coined to replace clumsy phrases such as taxonomic entity and taxonomic unit. Furthermore, the organisms contained within a rank (e.g., species, genus, or order) can also be referred to as *taxa*.

4.1.1 Taxonomic Characteristics and Taxonomic Characters

Characters by which an organism or group of organisms can be recognised is called a *characteristic*. For example, flowers are characteristic of angiosperms and wood is characteristic of trees. On the other hand, any part or shape of an organism that makes it possible to classify the organism is called a *character*. For example, characters used in classification include the shape of the leaves, arrangement of the reproductive organs, etc.

Taxonomic characters are actually features, such as form, structure, behaviour, and physiology, that are assessed in isolation from the rest of the plant by taxonomists, for making comparisons and interpretations. It is important to differentiate between *characters* and *character states*. For example, width of the leaf may be a character, while leaves 6 mm wide are an expression of that character, i.e., its character state. Often, characters are referred to as "good" or "bad" but this is strictly relative.

Thus, a good diagnostic character, such as compound leaves in a group of plants that mainly possess entire leaves, would be a bad character for separating taxa in a group in which leaf divisions were either variable or often compound.

4.1.2 Problems of Hierarchy in Taxonomic Structures

It has been established that taxonomic structure is a hierarchical system. For the effective working of this system, there should exist clear concepts regarding various taxa in plant taxonomy. Differing from non-living things, a biologist faces peculiar type of problems in circumscribing the different categories or taxa in biology. It is so because they show dynamic, evolving and variable populations. To solve this problem, any concept of taxa or taxonomic category has to be flexible to some extent. It should bc able to meet the special requirements of biology. Variability within and also in between the groups has caused many subjective judgements. At higher levels of hierarchy, this has also resulted in various types of controversies. For example, traditionally, there were two kingdoms, viz., Animalia and Plantae. Such a classification has been challenged by several modern workers. Copeland (1947) suggested four kingdoms, viz. Monera, Protista, Metaphyta and Metazoa. Whittaker (1969) gave a five-kingdom classification of organisms, namely Monera, Protista, Plantae, Fungi and Animalia. Edwards (1976) recognized seven kingdoms under two superkingdoms (Procaryota and Eucaryota).

It has, however, been observed that the degree of arbitrariness in classification decreases as we proceed down in the hierarchy, and it reaches to its minimum at a particular level i.e., the species.

4.2 CONCEPT OF SPECIES

A species is usually the smallest unit of classification. It includes individuals which are alike and can breed with each other. Species are sometimes divided into *subspecies* and *varieties* on the basis of small differences between populations. Species are given Latin binomial names.

Species is, thus, the fundamental unit of study in taxonomy, comprising all the populations of one breeding group that normally are permanently separated from other such groups by marked discontinuities. In case, the crossing between species does not occur then the resulting hybrids are normally sterile, thus maintaining the reproductive barrier between species. This broadly genetic definition of a species does not hold good for species that reproduce by self-fertilisation or by asexual means or to extinct species. In all these cases, a species is delimited by observation of the similarities between its members and dissimilarities between it and the other species.

With reference to botanical nomenclature, the *specific epithet* forms the second part of the binomial (the first part being the generic name) and it is always written in lower case. In plants, the ending of the specific epithet always agrees with the gender of the generic name.

There exist several infraspecific categories in the taxonomic hierarchy, but *subspecies* and then the *varieties* are probably the most widely used. Groups of similar species are placed in genera.

4.2.1 Subspecies

A subspecies is a taxon within a species. All subspecies of a species differ in small ways. Although they can breed with each other, they are usually found in different places or in different populations.

While naming a subspecies, a third Latin name is put after the binomial. *Binomial* is the Latin name of a species consisting of two words, of which first one is the name of the genus to which the species belongs and the second one is the name which distinguishes the species from other species in the same genus.

Subspecies is, thus, the rank subordinate to species in the taxonomic hierarchy. The term 'subspecies' is used when two or more populations are separated in some way (e.g. morphologically or ecologically) throughout their range. However, they are not usually genetically isolated. Generally, if 90% or more of a group of infraspecific individuals are recognizably distinct from another similar group, then each may be ranked as subspecies. Often, this is referred to as "*90% rule*".

The abbreviation "ssp." or "subsp." is used to indicate a subspecies, e.g. *Portulaca pilosa* ssp. *pilosa* and *Daucus carota* subsp. *gummifer.*

4.2.2 Variety

The variety is a rank subordinate to species but above the category "form" in the taxonomic hierarchy (Table 1.1, Chapter 1). Varieties are actually morphological variants, which may or may not have a clear geographical distribution. Sometimes they represent only habit phase or colour. A variety designated by one author may be designated as a subspecies or form by another author. A variety produced by agricultural or horticultural techniques and not normally found in natural populations is called a *cultivated variety* or *cultivar.* In a variety, there may also be recognised *subvarieties.*

4.2.3 Form

"Form" is the lowest rank (see Table 1.1, Chapter 1) normally used by taxonomists for sporadic distinct variants that sometimes occur in populations. Forms may be relatively minor genetic variants of a variety or subvariety but their effects can be conspicuous. Commonly observed forms are those in which flower colour is modified, e.g., occurrence of albino individuals in a population of purple-flowered plants. Although, a form is more commonly used to distinguish variants of subspecies and varieties, it may also be related directly to a species. Sometimes, "subforms" may also be recognised within a form.

Above-mentioned details show that a genus is a group of species and a subspecies or a variety or a form are parts of a species. Workers like Levin (1969), Cock (1977) and Wiley (1980) have defined "species" in their own ways but still the definition of the term "species" is a matter of great discussion and controversy. Various views have been put forward by the biologists during last few centuries. For the sake of convenience these can all be classified into four major concepts, viz., (i) nominalistic concept, (ii) taxonomic concept, (iii) biological concept, and (iv) alternative concept of species. All these are discussed below, in brief.

(a) Nominalistic Concept of Species

According to Slobodchikoff (1976), the nominalistic concept of species suggests that the "nature produces the individuals and nothing more". Species is nothing but the brainchild of man and are not objectively real. Different species can be explained only in terms of formal relationship and not on the characteristics of organisms. According to workers like Burma (1954) and Spurway (1955), species have been invented as a device to refer to large number of individuals collectively. Evolutionists like

Haldane (1956) believe that species do not exist as taxa and that it is simply a unit of convenience. About "species", nominalists think that individuals are related and hence included in a species due to similarity whereas the evolutionary point of view of workers like Mayr (1969, 1976) is that similarity is due to evolutionary relationship, i.e., it is a common descent.

(b) Taxonomic or Typological Concept of Species

The name 'species' comes from the Latin root "*specere*" which means "to look at". It, therefore, refers to appearance. Aristotle, the great philosopher believed that biological species are highly variable and also opined that hybridization between different species resulted into new species. Such examples have also been given that different breeds of dogs have come into existence from breeding of bitches with tigers, lions, wolves, foxes and even goats. Theophrastus, the Father of Modern Botany, wrote a book "*Enquiry into Plants*" on transmutation of species. He opined that species changed when they are transplanted in different soil and living in different climate. Several ancient workers reported that orange was a new species created by grafting lemon on pomegranate. Similarly, banana originated from the seed of date-palm (Zirkle, 1935) inserted into the corm of *Colocasia.*

(1) Aristotelian Concept of Essentialism Aristotle suggested that for every natural group, there is an inner "*essence*" that makes them what they actually are and that this is real. This concept is also known as the "typological concept of species". According to this concept, as mentioned by Simpson (1961), "every natural group of organisms, hence every natural taxon in classification has an invariant, generalised or idealised pattern shared by all members of the group". Species were taken to be fixed units in nature, their number determinate and they were characterised by universal types or "essence". The essence is like floral diagram of a family, and all the genera are supposed to fit in this and anything less or more is inconsequential. According to this concept, each species is supposed to have a basic plan which is unchangeable and defined also by the universal characteristics possessed by all members of the group. Several evolutionists refused to accept the taxonomic concept of species because "they do not necessarily reflect actual species in nature."

But, what exactly constitute the "essence" is a basic question. Supporters of the "concept of essence" believe that for all practical purposes, the "essence" is deduced by observing a number of individuals. The constant and invariable ones should constitute the "essence".

(2) John Ray's Concept of Species Botanists, such as John Ray (1686), put forward the theory of "breeding relationship as the basis of species". He also suggested that there exist variability within species. In his *Historia Plantarum* published in 1686, Ray suggested "great care should be taken in deciding what constitutes a species and what sort of characters are insufficient for species delimitations".

(3) View of Linnaeus Linnaeus (1707–1778) recognised two different kinds of variations among organisms, the true differences created by the Creator and constituting the "essence" and the intraspecific variations which were considered to be "sport of nature" or "accidents". Linnaeus (1737) warned that "sport" should not be neglected. "If neglected, these elusive ghosts glide away and are gone". Linnaeus, however, realised in the later years that species can arise by hybridisation.

(4) Darwin's view Darwin (1859), however, provided a sound basis by explaining evolution of species by natural selection. According to him, the organisms are characterised by variation. The geometrical increase in the number of individuals takes place due to fertility of the organisms. But,

it is subject to nature check i.e. the natural selection. Amongst these individuals only those survive which possess an inherent advantage over the others in the population. These inherent advantages are inheritable. For thousands of generations, the selection continues and new variants take the place of the original ones in a rapidly changing environment. Mechanism of heredity was, however, not known much during the times of Darwin.

(5) Neo-Darwinism Rediscovery of Mendel's work in 1900 brought forward a new theory of evolution in the 20th century, after the death of Darwin. It is called *neo-Darwinism.* It includes Darwin's theory of natural selection and the more recent knowledge of genetics and inheritance through chromosomes. Due to this theory, phylogenetic relationship came to be accepted as the basis of organismal similarities. It has been suggested that groups of organisms are not related because they are similar, but they are similar because they have a common descent.

(c) Biological Concept of Species

Concept of species has altogether changed during last century due to considerable amount of work done on various aspects, such as (i) genetic basis of variations, (ii) reproductive mechanisms, (iii) population structure, (iv) breeding behaviour, etc. All these have led to the formulation of *biological concept of species* by Mayr (1969), which has been supported by workers like Stebbins (1970), Heywood (1974), and others. Mayr (1969) defined the biological concept of species by genetic kinship as evidenced by breeding behaviour. Grant (1959) defined such a species as "a community of cross-fertilising individuals linked together by bonds of mating and isolated reproductively from other species". This shows that biological species is formed by "groups of interbreeding natural populations that are reproductively isolated from such other groups". Two points mainly considered in biological concept of species are (i) interbreeding between the members of the same species, and (ii) reproductive isolation between the members of different species. The biological concept of species has received wide acceptance among the modern biologists and it has inspired several studies on various aspects of biology, especially genetics and breeding behaviour.

One of the greatest shortcomings of biological species concept is that it is not applicable to the non-sexual organisms.

(d) Alternative Concepts of Species

(i) Concept of Evolutionary Species Grant (1971) proposed the concept of evolutionary species which is equally applicable both sexual and asexual organisms. This concept as defined by Grant represents "a spatio-temporal lineage of populations that evolve separately from other lineages and has its own ecological niche". In its present form, however, this concept is too vague to be applied in all forms.

(ii) Concept of Sibling Species In several plant and animal species, there exist many instances where "morphologically similar or identical populations are reproductively isolated". Such species have been named as *sibling species* by Mayr (1963). According to Amadon and Short (1976), the term "sibling species" is used as a "purely descriptive term having no genetic or taxonomic implication". Stayskal (1972) criticized the use of the term "sibling species" because it exactly means of "same parentage". Instead, he used the alternative term "*aphanic species*" for "sibling species".

(iii) Concept of Ecological Species According to Van Velen (1976), "species are maintained for the most part ecologically and not reproductively", and this is called concept of ecological species. This

concept suggests that (i) genes are of minor importance in evolution, (ii) the control of evolution is mainly by ecology and the constraints of individual development, and (iii) selection works primarily on phenotypes which are the main building blocks of community. Van-Velen defined species as "a lineage which occupies an adaptive zone minimally different from that of any other lineages outside its range".

(iv) Concept of Selection Species Slobodchikoff (1976) viewed species as a unit of selection and this formed the basis of the *concept of selection species*. According to him, "species is a system of genetically similar individuals and populations maintained as a cohesive unit by a set of selection pressures that balance the disruptive forces imposed by environmental factor, mutations or genetic recombinations". The entire concept of Slobodchikoff turns round selection pressures.

Conclusion: All the above-mentioned concepts clearly indicate that there is no universally applicable definition for species. In biology, the rank of species covers a variety of situations, and therefore, the concept or definition of species should be excercised by considerable flexibility. *To conclude*, it may be mentioned that species is a group of individuals that (1) actually or potentially interbreed with each other but not with other such groups, (2) show continuous morphological variation within the group but which is distinct from other such groups. Taxonomically, species are grouped into genera and divided into subspecies and varieties or, horticulturally, into cultivars.

4.3 CONCEPT OF GENUS

Most widely accepted definition of the word "*genus*" (pl. genera) is that "it is a group of related species". In the taxonomic hierarchy, "*genus*" is an important rank which is subordinate to family, but above the rank of species. It is a group of obviously homogeneous species. The generic name forms the first part of the binomial (the second part being the *specific epithet*), e.g. *Cicer arietinum*.

Genus is usually a singular noun and is written in Latin with a capital initial letter. Collections of similar genera are grouped into families. Davis and Heywood (1963) have suggested three main parameters that should provide guidelines for the determination of the generic status, viz. (i) naturalness, (ii) delimitation of closely related genera, and (iii) practicability of keeping them distinct or including them in other genera.

The concept of monophyletic taxa has proved quite satisfactory in determining the naturalness and demarkation between different genera. Monophyletic group means that its members must have descended from a common ancestor and it must include all the descendants of that ancestor. Legendre and Vaillancourt (1969) defined genus as a monophyletic group of species which occupies a definite adaptive zone. According to them, all the species descending from a common ancestor have to be included in the same genus. All non-monophyletic genera have to be abandoned because they do not represent natural groups.

In the light of recent developments the circumscription and delimitation of the genera may also be redefined. For example, the "Eugenia complex" of Myrtaceae may include all the new world species in the genus *Eugenia* Linn. and all the old world species in the genus *Syzygium* Gaertn.

However, while revising the generic limits, it should be done mainly on the basis of a worldwide study because it needs in-depth knowledge of the range and patterns of variations. Instead of taking a single arbitrarily selected character for revising a genus, taxonomists should better be guided by what is called a synthetic approach. More emphasis should be given on natural grouping.

4.4 CONCEPT OF FAMILY

A taxon consisting of related genera is called a family. The Latin names of families usually end with "–aceae" e.g. Ranunculaceae, Malvaceae, etc. Eight exceptions to such an ending, however, are Compositae, Cruciferae, Gramineae, Guttiferae, Labiatae, Leguminosae, Palmae and Umbelliferae. International Code of Botanical Nomenclature has, however, proposed alternative names for these families as Asteraceae for Compositae, Brassicaceae for Cruciferae, Poaceae for Gramineae, Hypericaceae for Guttiferae, Lamiaceae for Labiatae, Fabaceae for Leguminosae, Arecaceae for Palmae, and Apiaceae for Umbelliferae. Many workers, however, still use older names.

A family is a major category in taxonomic hierarchy comprising groups of similar genera. Some taxonomists believe that "families" represent the highest natural grouping. Groups of similar families are placed in orders.

Larger families may be split into *tribes*. The *tribe* is, therefore, a rank subordinate to family but superior to genus in the taxonomic hierarchy (see Table 1.1, Chapter 1). The term '*tribe*' is applied to assemblages of similar genera within large families. The Latin names of tribes have the ending "-eae", e.g. tribe Saniculeae in the family Umbelliferae or Apiaceae. Similar tribes may be grouped together in subfamilies. Tribes may also be split into subtribes. The Latin names of subtribes have the ending "-inae".

Ideally, families have to be natural and monophyletic. The characters, by which families are delimited vary with the groups, but they are of more obvious nature than that of genera and species. Both vegetative and reproductive characters are used in delimiting families. For example, Cactaceae is recognised by the cactoid habit of the members while insectivorous families (e.g. Nepenthaceae and Droseraceae) are characterised by their distinct insect-trapping mechanisms. Compositae and Umbelliferae are recognised by their characteristic inflorescences while Poaceae is recognised by their fruits.

Two different types of families are *definable* and *indefinable* (Walters, 1961). The families which are very homogenous and natural groups are called *definable*, e.g. Brassicaceae and Apiaceae. These can be easily recognised from such other familier because they are homogenous natural groups and their component genera are ill-defined. On the other hand, indefinable families include great diversity of structure and are not as distinctive as the definable families. Most of the present day plant families are indefinable. Their member genera differ among themselves so much that they can be easily separated, as in Ranunculaceae. Genera of indefinable families are easily definable due to clear morphological discontinuities.

Sometimes, the question of inclusion of a particular genus in a given family becomes largely a matter of choice of taxonomist. For example, *Nyctanthes* is included by some taxonomists in Oleaceae, by others in Verbenaceae and by still others in Nyctanthaceae.

4.5 TAXA ABOVE FAMILY LEVEL

Order, Class, Division, and Kingdom are the categories above the level of family used in taxonomic hierarchy. They are also sometimes divided into their lower categories like suborder, subclass and subdivision. One more category "superorder" has been included by Takhtajan (1969) to accommodate one more evolutionary node. Ideally, all these groups should be monophyletic. Regarding evolution of these higher taxa (Kubitzki, 1977), much information is, however, not available.

Order is a taxon consisting of families. The Latin names of orders usually end with "-ales", e.g. Rosales. However, some orders, which were erected prior to the compilation of the International Code of Botanical Nomenclature, end in "-ae", e.g. Tubiflorae, Glumiflorae. Groups of families, thought to possess a degree of phylogenetic unity, are placed in an order.

Class is a taxon consisting of orders. In Botany, class is a taxonomic rank below division and above order. The names of classes end in "-phyceae" in algae, -mycetes in fungi, or -opsida in other plants.

Division is a major taxon, which is made up of classes. Three main divisions of land plants are bryophytes, pteridophtes and spermatophytes. A division is the second highest category in the taxonomic hierarchy, placed above the classes and below the kingdom. The Latin names of division terminate in "-phyta", e.g. Tracheophyta. In place of division, several botanists now use the term phylum.

Kingdom, the largest of all the taxa, is actually the highest level in the hierarchy of taxonomic ranks. In older systems, of classification, there are only two kingdoms, viz., plant kingdom and animal kingdom. In some modern systems of classifications, fungi are considered in a separate kingdom—Mycota. Similarly, some taxonomists prefer to place unicellular organisms in their own kingdom, the Protista. Whittaker (1969) proposed a five-kingdom system of classification of living organisms, namely, Monera, Protista, Plantae, Fungi and Animalia.

Test Your Understanding

1. Give a brief account of concept of species.
2. Explain briefly the concept of taxa.
3. Give an account of problems of hierarchy in taxonomic structures.
4. Describe briefly the biological concept of species.
5. Differentiate between: (a) subspecies and variety, and (b) variety and form.

Suggested Reading

Amadon, D. and L.L. Short, 1976, Treatment of subspecies approaching species status. *Syst. Zool.* **25**: 161–167.

Cock, A.G. 1977, Bernard's symposium – the species concept. *Bio. J. Linn. Soc.* **9**: 1–30.

Edwards, P. 1976, A classification of plants into higher taxa based on cytological and biochemical criteria. *Taxon* **25**: 529–54.

Grant, V. 1971, *Plant speciation.* New York.

Haldane, J.B.S. 1956, Can a species concept be justified? *Syst. Assoc. Publ.* **2**: 95–96.

Heywood, V.H. 1974, Principles and concepts in the classification of higher taxa. *Pl. Syst. Evol. Suppl.* **1**: 1–12.

Legendre, P. and P. Vaillancourt 1969, A mathematical model for the entities, species and genus, *Taxon* **18**: 234–252.

Levin, D.A. 1969, The nature of plant species, *Science* **204**: 381–384.

Mayr, E. 1969, The biological meaning of species. *Biol. J. Linn. Soc.* **1**: 311–320.

_______ 1976, Is the species a class or an individual? *Syst. Zool.* **25**: 19.

Slobodchikoff, C.N. 1976. *Concepts of species.* Vol. III. Pennsylvania.

Van Valen L. 1976. Ecological species, multispecies and oaks. *Taxon* **25**: 233–239.

Walters, S.M. 1961, The shaping of angiosperm taxonomy, *New Phytol.* **60**: 74–84.

Wiley, E.O. 1980, Is the evolutionary species fiction? *Syst. Zool.* **29**: 76.

PLANT COLLECTION AND SPECIMEN PREPARATION

Plant collection and field preparation of specimens are the fundamental aspects of study, training and research in plant systematics. Herbarium[1] specimens are the permanent records of plant species of a particular place at a given time. Therefore, the plants should be carefully collected, selected, and the herbarium specimens should be properly prepared and preserved.

5.1 WHICH TYPE OF SPECIMENS SHOULD BE COLLECTED?

The points to be kept in mind during a plant collection are:

1. Collect entire, vigorously growing typical specimens.
2. Select such individuals that represent almost all phases of the natural population.
3. Avoid collecting insect-damaged specimens.
4. Collect underground parts (e.g. roots, bulbs, rhizomes, tuber, etc.) of herbaceous perennials.
5. Collect those specimens of flowering plants that contain flowers, fruits and seeds, because keys are prepared mainly on the basis of these characters.
6. Specimens larger than the size of a single sheet, should be divided and pressed on a series of sheets.
7. Collect plants with the leaves intact as different kinds of foliage prove helpful in identification.
8. Collect the bark and wood samples of the woody plants.
9. Avoid collecting rare or uncommon plants. Never collect the only plant of a species at a locality.

5.2 FIELD EQUIPMENT AND SUPPLIES

Some of the commonly used field equipment and supplies needed for plant collection and preparation of herbarium specimens are:

[1]For details of some national and international herbaria and their upkeeping, refer Chapter 15.

1. *Field Press* It is made up of a pair of hardwood, metal or strong plywood frames of 12 by 18 inch.
2. *Driers or Blotters* These are the sheets of heavy blotting papers or of other moisture absorbing material of 11 by 16 inch dimensions. Old newspapers are also used as driers.
3. *Straps or Ropes* Rope, sash cord, or a pair of strong web straps of about 4–5 feet length are used to tighten the press.
4. *Corrugate Ventilators* Sheets of corrugated cardboard of 12 by 18 inch are used in between the driers (= blotters) as ventilators when plants are dried by artificial heat. They provide space for the passage of air through the press to remove moisture.
5. *Field Notebook* An indispensable item of a plant collector is a permanently bound, small field notebook with ruled pages. It is used to record the full data (such as date, place, locality, habitat, elevation, local name, collector number,[1] etc.) of the collected plant. The data is recorded in the field notebook in the field itself.
6. *Digging and Clipping Tools* Trowel, diggers, hammer, pruning shears, garden clippers, geological pick, and heavy sheath knife, etc. are all used for digging or clipping the plants.
7. *String Tags* These are made of waterproof material, and are used for labelling plants that are not immediately pressed.
8. *Vasculum* Vasculum is a container made of tin or aluminium sheet and contains a hinged light lid. Plants, which are not pressed in the field, are placed in a vasculum to preserve their freshness for sometime.
9. *Collecting Bags* These are plastic bags, used as containers for fresh specimens.
10. *Collection Bottles* These are glass or plastic bottles with leakproof screw caps. These are used for collecting small-sized material, to be preserved in liquid preservatives.
11. *Liquid Preservative* Formalin-aceticacid-alcohol (FAA)[2] is the most common liquid preservative used for anatomy materials. 6:3:1 mixture of chloroform, 95% ethyl alcohol and glacial acetic acid is often used for cytology materials.
12. *Hand Lens* A 5 ∞ or 10 ∞ lens is used in the field for observation and identification.
13. *Waxed Paper* Sheets of waxed paper are used for pressing viscid or weak-looking plants.
14. *Cardboard Storage Boxes* These are the boxes or containers used to store dry materials.
15. *Maps* These are useful in the field for determining the localities for particular species.
16. *Colour Charts* These are used for determining the actual colour of the flower parts in the field itself.
17. *Camera and Film* These are used for taking photographs of important plants in the field.
18. *Other Field Equipment and Supplies* Altimeter, compass, pocket knife, soft lead pencils, insect repellant, portable plant drier, and seed envelopes are some other useful equipment for plant collection.

[1] A *collector number* is a numerical series starting with 1 and continuing throughout the lifetime of the collector.
[2] Composition of FAA should be 70% ethyl alcohol (90 cc), formalin (5 cc) and glacial acetic acid (5 cc).

5.3 ORGANISATION OF THE FIELD PRESS

Following should be the organisation of the field press:

Hardwood or plywood press

Corrugate ventilator

Drier or blotter

10 sheets of torn newspapers (each sheet will contain one specimen)

Drier or blotter

Corrugate ventilator

Drier or blotter

10 sheets of torn newspapers (each sheet will contain one specimen)

Drier or blotter

Corrugate ventilator (continue the sequence until 100 specimens i.e. 10 groups of 10 newspapers, are included)

Hardwood or plywood press.

5.4 HOW TO PRESS PLANT SPECIMENS?

Specimens, after being cut or dug, should be pressed as soon as possible. The specimen can be placed carefully on a pressing sheet, i.e. a newsprint sheet or a blotter, taking care so that there is no folding or overlapping of parts. The extra leaves or branches are removed, if necessary. Plants too large to fit in the 11 ∞ 16 inch fold of a newspaper or a blotter may be bent into a V, N, or M figure. The specimens should not protrude from the fold of the paper.

Specimens should be arranged in such a way so that some upper and some lower surfaces of leaves are exposed. Flowers or inflorescences should be spread out thoroughly for a complete view. A few flowers can be cut longitudinally and pressed, if possible. Each sheet should contain a collector's number which refers to the notes in the collector's field notebook.

In the press, the specimens in the specimen paper are placed in between two driers or blotters. A corrugate ventilator is often inserted before the next specimen papers and driers are added. The usual sequence, as shown above under Section 5.3 (*Organisation of the Field Press*), is corrugate ventilator, drier, sheets of specimens in newspaper, drier, corrugate ventilator, drier, and so on.

All the specimens are placed for pressing in such a way so that the entire bundle is ultimately of almost uniform thickness in the middle and sideways. Once the plant specimens are arranged, the press is tightly bound with ropes or straps to prevent wrinkling of the specimens. The press is now ready for drying.

5.5 DRYING OF SPECIMENS

Specimens should be dried as rapidly as possible to get the best results. In the usual process of drying, the press containing the specimens is placed in the sun. After about 24 hours, the press is opened, and the specimens are placed in fresh blotters. Any last rearrangement of the plant parts may be made at this time. The press, along with the specimens in fresh blotters, is now again bound tightly. The wet blotters are dried in the sun for reuse. For three-four days the wet blotters or driers are changed daily, until the specimens are completely dried. The press becomes loose when the plants are completely dried.

Artificial heat may also be used for the drying process. But the specimens are never to be dried in an oven. In humid regions, or the rainy season, the plants may be dried in a drier. *A drier* is made in the form of a wooden box, 3 feet in length and 18 inches in breadth, made up of 1/10 inch thick boards. Five light bulbs of 60 watts are fitted inside for producing heat. Small openings are made at the bottom of the box for the entry of air which gets heated and will dry the plants. In this way, a press containing 100 specimens can be dried in 8–12 hours.

An excellent *drying chamber* may also be made by using a wooden or metal box with an open top that will accommodate a press. The box is equipped with an electric heater with a fan.

5.6 MOUNTING OF SPECIMENS

The process, by which a dry specimen is attached to a mounting paper or a herbarium sheet, and a label is affixed at the lower right corner of the sheet, is called *mounting*. The standard size of a herbarium sheet is 28.75 by 41.25 cm. Heavy, hand-made cardsheets are used as herbarium sheets.

Before mounting, the dry specimens are dipped in a saturated solution of mercuric chloride in ethyl alcohol to prevent infection of fungi, insects, etc. In place of mercuric chloride, lauryl pentachlorophenate is also used, because the former is a virulent poison.

A good quality glue or paste is applied to the back of the specimens for affixing them onto the herbarium sheets. Common animal glue is used in India for this purpose. Some of the good quality glues, used in developed countries, include Wilhold 128, Elmer's Glue-All, Swifts Z-5032, or Nicobon B. Strips of gummed paper are used as an additional aid to hold heavy and woody specimens. Loose parts (e.g. seeds, fruits, flowers, etc.) and dissected parts are placed in a paper packet and pasted on the same herbarium sheet. Use of cellulose tapes or stiching of the plant parts with thread should be avoided.

5.7 DEEP-FREEZING METHODS

These days, deep-freezing methods are being adapted in some herbaria to prevent infection of fungi, insects, etc. instead of mercuric chloride and other harmful chemicals.

5.7.1 Portable Freezer Unit

Recently, Davis and Gauthier (2008) of University of Idaho, Moscow, evaluated and found that a portable refrigerator/freezer unit, powered by the 12 v DC connection in a vehicle, maintained a constant and colder temperature than a similar-sized cooler filled with ice. They found it useful for preserving the quality of seeds and cuttings of plants collected in the field.

Portable freezer units offer two main benefits over traditional ice coolers. First, "refrigeration control offers temperature stability independent of exterior temperature without the worries about ice-melting". Second, such units "can function well in remote locations where replenishment of ice may not be feasible". These units help very effectively in collecting seeds which require the essential maintenance of appropriate temperature. *Engel MT 35* is a portable freezer developed by Engel USA, Jupiter, Florida (USA). The main drawback of Engel MT 35 is "its need to remain inside the vehicle when sampling, although longer cables could resolve this concern" (Davis and Gauthier, 2008).

5.7.2 Deep-freeze Arctic Vault

Recently, in the early months of 2008, thousands of seeds, plants and crop varieties from several countries belonging to all continents, have been shipped and preserved in a vault in the Arctic Circle. According to the organisers of the Arctic Circle, all these would be preserved here in deep-freezers for thousands of years for future study and use. In this Norwegian-built Svalbard Global Seed Vault (SGSV), more than 2 lakh crop varieties will be stored in the deep-freeze conditions in the coming years. The seeds of several crops including rice, wheat, beans, sorghum, sweet potatoes, lentils, etc. have already been shipped to the vault's built quite deep under the Arctic permafrost in this Norway's Svalbard archipelago. A very large number of seeds and crop varieties have been given by Consultative Group on International Agricultural Research (CGIAR), which maintains seeds of over 6 lakh plant varieties in its various centres located in different parts of the world. A major centre of this organisation is in Mexico City which maintains a bank of over 1.5 lakh unique samples of wheat and its relatives from more than 100 countries and also a very vast collection of maize. Seed duplicates from CGIAR shipped to this deep-freeze Arctic Vault were provided by various agricultural research centres located in Benin, Colombia, Ethiopia, India, Kenya, Mexico, Nigeria, Peru, Philippines and Syria. These CGIAR collections are referred as "*crown jewels*" of international agriculture. They include the world's largest and most diverse collections of rice, wheat, maize and beans.

5.8 LABELLING OF SPECIMENS

A label, usually of 6.5 by 10.5 cm (Jones and Luchsinger, 1987) dimension, is pasted on the lower right side of a herbarium sheet. It contains definite information about the specimen pasted on the herbarium sheet. Ideally, it contains a miniature essay on the plant and its habitat. The paper of the label should be of high rag quality. .

Following information should be incorporated in a label:

1. *Heading,* mentioning the name of the institution or individual, as well as the names of the state and country.
2. *Family name.*
3. *Scientific name,* i.e. genus and species names, as well as authority.
4. *Locality,* identifying the location of the collection with latitude and longitude, as well as distance from a well-known location.
5. *Habitat,* mentioning moisture, vegetation type, soil, slope, etc.
6. *Date of collection.*
7. *Field notes* from the notebook of the collector.
8. *Collector,* indicating the name of the collector.
9. *Collection number,* which is a numerical sequence that begins with 1 and continues throughout the collector's lifetime.

10. *Local or vernacular name.*
11. *Additional information,* such as flower and leaf colour, related plants, bark, economic usage, and so on.

5.9 IDENTIFICATION OF SPECIMENS

Finally, pressed, dried and labelled specimens are identified with the help of appropriate floras, manuals, and monographs. A duplicate specimen is to be retained before sending a specimen to the specialist for identification.

For more details of identification of plants, refer to Chapter 7.

Test Your Understanding

1. Make a list of at least 10 commonly used field equipment needed for plant collection and preparation of herbarium.
2. What is the most commonly used liquid preservative used for plant materials?
3. What should be the organisation of the "field press"?
4. Write a note on pressing and drying of plant specimens.
5. What information should be incorporated in labelling of specimens?
6. Give an account of some deep-freezing methods.

Suggested Reading

Archer, W.A., 1950, New plastic aid in mounting herbarium sheets, *Rhodora* **52**: 298–299.

Croat, T.B., 1978, Survey of herbarium problems, *Taxon* **27**: 203–218.

Davis, A.S. and M.M. Gauthier, 2008, Portable refrigerator freezer provides stable temperature for plant material collection, *Native Plants Journal* **9**(1): 41–43.

Davis, P.H., 1961, Hints for hard pressed collectors, *Watsonia* **4**: 283–289.

Lee, W.L., B.M. Bell and J.F. Sutton, 1982, *Guidelines for Acquisition and Management of Biological Specimens,* Assoc. Syst. Collections, Lawrence, Kansas.

Savile, D.B.O., 1962, *Collection and Care of Botanical Specimens,* Canada Deptt. Agricul. Publication No. 11, 13.

Smith, C.E., 1971, *Preparing Herbarium Specimens of Vascular Plants,* Agricultural Information Bulletin No. 348, U.S. Govt. Printing Press, Washington, D.C.

EXAMINATION OF A PLANT SPECIMEN

Since, taxonomy is essentially a practical activity, examination of a plant specimen is one of the most important work of a taxonomist. Without this, nothing can proceed in plant taxonomy. This may be done in field, garden and/or in the laboratory with the help of specimens of living plants or of well-mounted plants.

6.1 ❁ EQUIPMENT NEEDED FOR EXAMINATION OF PLANT SPECIMEN

Very simple type of equipment is needed for examining plant specimen. Some of the commonly required articles include a hand lens (10X and 15X), a sharp rajor, blades, a pair of dissecting needles, a pair of forceps, collection bottles, polythene bags, dissecting microscope, compound microscope, a good drawing pencil, eraser, notebook, some old newspapers, and a good plant identification manual.

6.2 ❁ INSTRUCTIONS TO BE FOLLOWED WHILE EXAMINING A PLANT SPECIMEN

While examining a plant specimen, certain guidelines should be followed for getting best possible results. Some of such guidelines are mentioned below:

1. Plants should be examined in upwards succession, beginning with the roots and ending with the flowers and fruits.
2. Examine the given plant first with the naked eye and then with the help of the hand-lens.
3. Study the flowers with great care, again in ascending succession, i.e., starting from flower-stalk through the calyx, corolla and androecium upto the tips of stigmas.
4. Make a record of all possible characters before the flower is cut in any way.
5. Use the hand-lens for recording the details like the way the anthers open, the aestivation, the placentation, etc.
6. At least two flowers should be sectioned with the help of rajor or sharp blade, one horizontally and the other down the middle. The former is required to know the placentation and

also to complete the floral diagram. The latter is necessary to see if the ovary is superior or inferior and also to enable a half-flower to be drawn.

7. In half-flower drawings, cut surfaces should be drawn first, indicated by double lines, and the background filled in later.
8. Floral diagram and other drawings should be drawn quite large in size.
9. The parts of the drawings should be labelled as much as possible.
10. If the given specimen has flowers at different stages, all stages should be examined and drawn as much as possible.
11. It is often easier to see the aestivation in horizontal sections of buds. But better placentation can be seen in young fruiting stage.
12. Make the maximum use of available material in an intelligent way as much as possible.
13. Develop a habit of making keen and precise observation of the given plant specimen.
14. For practical taxonomic work, accurate recording of the information obtained and making their neat and correct diagrams is also highly essential.

6.3 GUIDE TO THE EXAMINATION OF PLANT SPECIMEN

Undermentioned is the scheme which should be used for the examination, description and identification of a given plant specimen.

6.3.1 Habit

(1) Is the plant a *herb* or *woody*?

If it is a *herb*, then:

is it annual, biennial or perennial?

is it erect, prostrate, trailing, twining or climbing?

is it succulent?

is it aquatic?

If it is *woody*, then:

is it a shrub or tree?

is it evergreen or deciduous?

(2) Any other peculiar habit such as epiphyte, saprophyte, parasite, etc.

6.3.2 Root

(1) Is it tap root or adventitious or fibrous root?

(2) Branched or unbranched.

(3) Whether it shows any other special form such as fusiform[1], napiform, conical, tuberous, nodulose, annulated or moniliform, etc.?

[1] For explanation of hundreds of technical terms used in this chapter, refer to Chapter 19 "550 Terms of Plant Description".

6.3.3 Stem

(1) Is it herbaceous or woody?

(2) Is it erect, prostrate, climbing or twining? In case of a climber, note the means of climbing like tendril, hook, spine or any other outgrowth?

(3) Is it cylindrical or angular? If angular, then note the number of angles.

(4) Is it unbranched or branched? If branched, then note the mode of branching.

(5) Is it hollow or solid, jointed or unjointed?

(6) Is it smooth, waxy, hairy or spiny?

(7) Is it green or of any other colour?

(8) Also note any other special modifications such as:

 (i) rhizome, tuber, bulb, corm.

 (ii) Whether it is a runner, sucker, stolon or offset?

 (iii) Is it a phylloclade or cladode?

(9) Also note features of bud and bark, if present.

6.3.4 Leaf

(1) Note *arrangement*, i.e., whether the leaf is alternate, opposite or whorled.

(2) Note *attachment to stem*, i.e., whether it is sessile or petiolate.

(3) Note *stipules*, i.e., whether stipulate or exstipulate.

 (i) If stipulate, then note the type of stipules, i.e., scaly, free-lateral, adnate, interpetiolar, ochreate, foliaceous, etc.

(4) Note whether leaf is simple or compound.

 (i) If compound, then mention whether pinnately compound or palmately compound.

 In case it is pinnately compound, then note whether it is unipinnate, bipinnate, tripinnate or decompound.

 In case the leaf is palmately compound, then note whether it is unifoliate, bi-, tri-, quadri-, or multi-foliate.

(5) Note *shape* of leaf, i.e., whether it is acicular, linear, lanceolate, elliptical, oval, ovate, obovate, oblong, rotund, cordate, reniform, oblique, spathulate, sagittate, hastate, cuneate, deltoid, falcate, pedate or lyrate.

(6) Note *margin* of the leaf, i.e., whether it is entire, repand, sinuate, serrate, dentate, runcinate, crenate, ciliate or spinous.

(7) Note *apex* of leaf, i.e., whether the apex is acute, acuminate, obtuse, cuspidate, truncate, retuse, emarginate, mucronate or cirrhose.

(8) Note *surface* of leaf, i.e., whether it is rough, glabrous, glaucous, glutinose, spiny or hairy.

(9) Note *venation* of leaf, i.e., whether it is reticulate or parallel.

(i) If it is *reticulate*, then note whether it is unicostate or multicostate. If multicostate, then also note whether it is convergent or divergent type.

(ii) If it is *parallel*, then note whether it is unicostate or multicostate. If it is multicostate, then also note whether it is convergent or divergent type.

6.3.5 Inflorescence

(1) Note whether the inflorescence is *racemose* or *cymose*, or special types, or flowers are solitary.

(2) If *racemose*, then note whether it is a raceme, spike, spikelet, panicle, catkin, spadix, corymb, umbel or capitulum

(3) If *cymose*, then note whether it is a monochasial, dichasial or polychasial cyme.

(4) If *special* type, then note whether it is a cyathium, or verticillaster, or hypanthodium.

6.3.6 Flower

Note whether the flower is:

(i) bracteate or ebracteate, (ii) bracteolate or without bracteoles, (iii) sessile or pedicellate, (iv) complete or incomplete, (v) unisexual or bisexual, (vi) dimerous, tri-, tetra- or pentamerous, (vii) hypogynous, perigynous or epigynous, and (viii) colour.

6.3.7 Perianth

A collective term used together for calyx and corolla when there is no distinction between the two, as in monocotyledons. Note the number of tepals, i.e., parts of perianth, and whether they are free (polyphyllous) or fused together (gamophyllous). Also note the colour of tepals and their aestivation.

6.3.8 Calyx

(1) Of how many sepals, is it composed?

(2) Are the sepals free (polysepalous) or united above the base (gamosepalous)?

(3) Whether aestivation is valvate, twisted, imbricate, quincuncial or vexillary?

(4) Colour of sepals, generally green or of any other colour.

(5) Caducous or persistent.

6.3.9 Corolla

(1) Number of petals.

(2) Whether free (polypetalous) or united (gamopetalous).

(3) Aestivation, as noted in case of calyx.

(4) Shape of corolla, i.e., whether cruciform, caryophyllaceous, rosaceous, tubular, campanulate, rotate, funnel-shaped, papilionaceous, bilabiate, ligulate or personate.

(5) Appendages, if any, present on corolla such as nectary, corona, spur, etc.

6.3.10 Androecium

(1) Number of stamens (write 'indefinite' if more than ten).
(2) Count the number of whorls of stamens and note whether or not some stamens are reduced into staminodes.
(3) Note whether stamens are free or united.
(4) Notc the nature of cohesion, i.e., whether monadelphous, diadelphous, polyadelphous, syngenesious or synandrous.
(5) Nature of adhesion, i.e., whether epipetalous, gynandrous or possess any other special character.
(6) Whether stamens are alternipetalous or obdiplostemonous.
(7) Inserted or exerted.
(8) Also note whether filament is long, short or flattened. Whether they show didynamous or tetradynamous condition?
(9) Monothecous or dithecous.
(10) Attachment of the anthers, i.e., whether basifixed, adnate, dorsifixed or versatile.
(11) Whether introrse or extrorse or laterorse?
(12) Also note the presence of appendages, if any, e.g., hair, scales, staminal corona, etc.
(13) Whether any disc is present outside or inside the stamens?

6.3.11 Gynoecium

Note the following:

(1) Number of carpels.
(2) Whether carpels are free (apocarpous) or fused (syncarpous).
(3) Superior, inferior or semi-inferior nature of ovary.
(4) Number of locules, i.e., unilocular, bilocular or multilocular.
(5) Number of ovules in each locule.
(6) Placentation type, whether marginal, axile, parietal, free-central, basal or superficial.
(7) Number of styles.
(8) Number, shape and any type of modification of stigma.
(9) Presence or absence of disc below the ovary.

Difference Between Gynoecium, Pistil and Carpel

The *gynoecium* is the female part of a flower consisting of one or more pistils. A *pistil* is the female reproductive organ of a flower, consisting of ovary, style and stigma. *Carpel* is the female reproductive unit of a flower, consisting of the ovary with ovules. Most angiosperms have several carpels, which are joined together at their margins to form the ovary.

How to Calculate the Number of Carpels in a Pistil?

For calculating the number of carpels in a pistil, proceed as follows:

(1) Observe Carefully the Outline of the Ovary

(i) If it is cleary asymmetrical in cross-section, the pistil is likely to be made up of only one carpel.

(ii) If it is symmetrical, the pistil is likely to be made up of two or more carpels. In case it is symmetrically lobed into two or more lobes, the number of lobes may indicate the number of carpels.

(2) Now Observe Carefully the Styles

(i) If the number of styles is more than one, then the pistil is made up of two or more carpels, indicating that number of styles indicate the number of carpels.

(ii) In case there is only one style, the pistil may be made up of one or more than one carpels. In such a condition, proceed as mentioned below in (3).

(3) Observe Carefully the Stigmas

(i) If there are present more than one stigmas, the pistil is made up of two or more carpels. If there is only one stigma, then pistil may be made up of one or more than one carpel.

(ii) If the stigma is asymmetrical or horse-shoe shaped, it is made up of only one carpel.

(iii) If the stigma is symmetrical, the pistil is more likely to be made up of two or more carpels.

(iv) If the stigma is lobed symmetrically into two or more lobes, the pistil is more likely to be made up of two or more carpels. This indicates that the number of stigma lobes may indicate the number of carpels.

(4) If the Stigma is Unlobed, then Proceed as Under

Cut a transverse section of the ovary across the middle and note whether the ovary is divided into two or more compartments or loculi? If so, then the pistil is made up of two or more carpels, indicating that the number of loculi indicate the number of carpels. If not, then pistil may be made up of one or even more than one carpels.

(5) Observe the Number of Placentae Visible Clearly in a Cross-section

(a) If placentae are more than one, then the pistil is made up of two or more carpels, indicating that the number of placentae indicates the number of carpels.

(b) If only one placentum is present, then pistil is more likely to be made up of only one carpel.

By using a combination of the above-mentioned five methods, it is possible to ascertain if the pistil is *simple* or *compound*.

(i) If the pistil is *simple*, then gynoecium is *apocarpous* and made up of a single free carpel.

(ii) If the pistil is *compound*, then gynoecium is *syncarpous* and made up of two or more *fused carpels*.

In the syncarpous condition, also note the numbers of styles, stigmas and loculi in the ovary, which together decide about the number of carpels making up the pistil.

6.3.12 Fruit

Record the type of fruit, if present, i.e., whether it is simple, aggregate or multiple type.

(1) If *simple*, then note whether it is dry or fleshy.

(a) If *dry and simple*, then note whether it is *indehiscent* (e.g., achene, utricle, caryopsis, cypsella, or nut) or *dehiscent* (e.g., follicle, legume, siliqua, silicula or capsule) or *schizocarpic* (e.g., cremocarp, cacervulus, regma, lomentum or samara).

(b) If *fleshy and simple*, then note whether it is berry, hesperidium, pepo, drupe or pome.

(2) If *aggregate*, then note whether it is etaerio of achenes, or etaerio of follicles, etaerio of drupes or etaerio of berries.

(3) If it is a *multiple fruit*, then note whether it is a sorosis or syconus.

6.3.13 Seed

Record the following:

(1) Number of seeds in a fruit.

(2) Whether it is endospermic or non-endospermic.

(3) Number of cotyledons.

(4) Shape, size and surface ornamentations of the seed.

6.3.14 Floral Formula

Write the floral formula of the given specimen according to the details given in Chapter 17 and Table 17.1.

6.3.15 Floral Diagram

Make a floral diagram of the given specimen according to the guidelines discussed in Chapter 17.

Test Your Understanding

1. What major guidelines one should follow while examining a plant specimen?
2. What should you note in a leaf while examining a plant specimen?
3. Make a systematic list of characters you should note in the male parts of a flower while examining a plant.
4. How can you calculate the number of carpels in the pistil of a flower?
5. From the point of view of examining a plant, how should you study its fruit?

Suggested Reading

Featherly, H.I. 1954, *Taxonomic Terminology of the Higher Plants*, Iowa State College Press, Ames.

Lawrence, G.H.M. 1951, *Taxonomy of Vascular Plants*, Macmillan Company, New York.

Sharma, O.P. 2007, *A Manual of Practical Botany Vol. II* (9th ed.), Pragati Prakashan, Meerut.

Stearn, W.T. 1983, *Botanical Latin* (3rd ed.), Davis and Charles, Newton Abbot, England.

PLANT IDENTIFICATION

7.1 WHAT IS IDENTIFICATION?

Identification is the basic process of classification. But it is not the same as naming of an individual. In the biological sense, *identification* is the determination of the group to which a specimen belongs. The process of identification usually includes a direct comparison of an unknown specimen with the already classified, circumscribed, and named taxa. This process also includes the use of *keys* of various types, *computer-based methods* and variety of *polyclave-type* devices.

In brief, *identification* may variously be defined as (i) the determination of similarities or differences between two specimens; (ii) the direct comparison of the features of a specimen in hand with those in keys in order to arrive at a name; (iii) the assignment of an unidentified taxon to the correct class in an established system of classification; (iv) the determination of a name for a particular specimen in relation to an already established system of identification.

7.2 CHARACTERS OFTEN CONSIDERED BEFORE PLANT IDENTIFICATION

1. Determine herbaceous or woody, and annual or perennial nature of the specimen.
2. Note the leaf type, phyllotaxy, and venation.
3. Note the presence or absence and type of stipule on young shoots.
4. Whether or not milky or coloured sap is present in the leaf, stem or other plant parts.
5. Note the distribution and kinds of surface coverings (i.e. hairs, trichomes, spines, etc.).
6. Observe the flower and name its parts.
7. Count the number of sepals and petals, and also note their arrangement, i.e. aestivation.
8. Note whether perianth is present in one series, more series, or absent.
9. Observe whether the sepals and petals are separate or fused.
10. Note whether pappus (e.g. Asteraceae) or epicalyx (e.g. Malvaceae) or similar structures are present.
11. Note whether a nectar-secreting disc is present in the flowers (e.g. Rutaceae).

12. Determine whether the flowers are actinomorphic or zygomorphic.
13. Observe the number and attachment of stamens. Also note if there is any fusion of anthers or filaments.
14. Note whether the stamens are antipetalous (e.g. Chenopodiaceae and Amaranthaceae), alternipetalous or obdiplostemonous (e.g. Caryophyllaceae).
15. Count the number of pistils, styles and stigmas of the gynoecium.
16. Cut a transverse section of the ovary, count the number of locules, number of ovules per locule, and also observe the placentation.
17. Cut a longitudinal section of the entire flower through its centre. Note the position of the ovary and fusion of the perianth.
18. Determine the number of carpels.

After carefully examining all the above characters, the next step in identification is keying.

7.3 IDENTIFICATION WITH KEYS

A *key* is an artificial arrangement or analytical device whereby a choice is provided between two contradictory statements resulting in the acceptance of one and the rejection of the other. A single pair of contradictory statements in a key is called a *couplet.* Each statement of a couplet is termed a *lead.* Leads are usually the best contrasting characters. Characters following the lead are called *secondary key characters.*

Keys are of two types: punched cards keys and dichotomous keys.

7.3.1 Punched Cards Keys

These consist of cards having names of all the taxa (i.e. all species, genera or families for which the key is meant) printed on all of them. On one of the corners of each card is printed any one character and a definite number. All the taxa showing this particular character are indicated by a perforation in front of their names, and the taxa lacking this character are without any perforation. The number of the cards and the characters chosen for the purpose are the same. For identifying a plant, only those cards showing characters possessed by this plant, are selected. Combination of characters shown by this plant will allow only one perforation in the selected set of cards. The plant is then referred to that particular family to which the card shows this perforation.

7.3.2 Dichotomous Keys

These keys consist of pairs of contrasting characters or *couplets,* each statement of which is a *lead.* Both the leads are numbered, and begin with the same word as far as possible.

Dichotomous keys are of two general types i.e. indented key and bracket key.

(a) In the *indented* or *yoked key,* each of the couplets is indented a fixed distance from the left margin of the page. An example of the indented key is given below in the form of identification of 5 common genera of Ranunculaceae (Buttercup family), i.e, *Ranunculus, Clematis, Anemone, Aquilegia* and *Delphinium*:

Fruit a group of achenes; unspurred flowers.

Petals absent

Sepals usually 4; involucre absent.......... *Clematis*

Sepals usually 5; involucre present.......... *Anemone*

Petals present.......... *Ranunculus*

Fruit a group of follicles; spurred flowers.

Spurs 5; flowers regular.......... *Aquilegia*

Spur 1; flowers irregular.......... *Delphinium*

(b) In the *bracket* or *parallel key,* the two couplets are always next to each other in consecutive lines on the page. At the end of each line in the key, there is either a number or a name referring to a couplet. An example of the bracket key is given below, in which all the same five genera of Ranunculaceae (i.e. *Ranunculus, Clematis, Anemone, Aquilegia* and *Delphinium*) are identified:

(1) Fruit a group of achenes; unspurred flowers.......... (2)

(1) Fruit a group of follicles; spurred flowers..........(4)

(2) Petals absent..........(3)

(2) Petals present.......... *Ranunculus*

(3) Sepals usually 4; involucre absent.......... *Clematis*

(3) Sepals usually 5; involucre present.......... *Anemone*

(4) Spurs 5; flowers regular.......... *Aquilegia*

(4) Spur 1; flowers irregular.......... *Delphinium*

7.4 SOME UNCONVENTIONAL IDENTIFICATION METHODS

Keys are the conventional or traditional methods of identification. Some highly improved, recent but unconventional methods of identification include *polyclave identification, computerised identification* and other identification methods.

7.4.1 Polyclave Identification

Polyclave method is a system of identification in which a choice of several characteristics can be used. A *polyclave* is actually a multientry, order-free key implemented in several different formats. In a polyclave method of identification the user of the system is free to choose any character, in any order or sequence, and thus avoid the rigid format of traditional dichotomous keys.

One form of the polyclave is a diagnostic key in which cards are utilised. The cards are placed on top of one another to eliminate taxa which disagree with the plant to be identified. The second form of the polyclave is a computer-stored multientry key, and the third form is a printed table which gives the status of different taxa and characters useful for separating the taxa.

7.4.2 Computerised Identification

Computers are machines planned only for the evaluation of a computer programme or algorithm. An algorithm is a series of logical steps or instructions by which an identification can be made.

Therefore, the computer can be programmed to identify only if an initial algorithm has been prepared successfully by an expert systematist.

In the computerised identification of plants, the major research efforts are grouped into four major approaches: (1) computer-constructed keys, (2) computer-stored dichotomous keys, (3) automated pattern-recognition systems, and (4) simultaneous character-set methods.

A system of computer programme to help in the plant identification and other related aspects of plant taxonomy has been presented by Morse (1974). This programme of Morse includes routines for plant identification, key construction, description preparation, taxa comparison and also the production of punched-card field keys.

7.4.3 Other Identification Methods

Some other unconventional, but sometimes quite effective, methods of plant identification include the use of techniques of *chromatography, spectroscopy* and *optical scanners,* etc. *Statistical methods* are also in the current use, in some developed countries, for plant identification.

7.5 ARTIFICIAL KEYS FOR THE IDENTIFICATION OF 51 COMMON FAMILIES OF ANGIOSPERMS

7.5.1 KEY-1 (Major Groups)

1. Plants with seeds; ovules enclosed within the ovary .... *Angiosperms*
  2. Flowers mostly tetra- to pentamerous; two cotyledons .... *Dicotyledons*
    3. Both sepals and petals present.
      4. Petals are free; flowers with calyx and corolla .... *Polypetalae*
        5. Polysepalous; petals hypogynous.
          6. Dome-shaped, unexpanded thalamus .... *Thalamiflorae*
          6. Thalamus expanded into a disc .... *Disciflorae*
        5. Gamosepalous; inferior ovary .... *Calyciflorae*
      4. Petals are fused; flowers with calyx and corolla .... *Gamopetalae*
        7. Ovary inferior .... *Inferae*
        7. Ovary superior
          8. More than two carpels .... *Heteromerae*
          8. Only two carpels .... *Bicarpellatae*
    3. Either sepals or petals are present; rarely both absent .... *Monochlamydae*
  2. Flowers mostly trimerous; one cotyledon .... *Monocotyledons*
1. Plants with seeds; ovules exposed, i.e. not enclosed within the ovary .... *Gymnosperms*

7.5.2 KEY-2 (Thalamiflorae)

DICOTYLEDONS

Polypetalae

Thalamiflorae

Dome-shaped unexpanded thalamus; polysepalous; hypogynous; ovary superior.

1. Carpels free (apocarpous).
   2. Herbs or soft-wooded climbing shrubs .... *Ranunculaceae*
   2. Woody shrubs or trees; rarely climbing
      3. Leaves stipulate .... *Magnoliaceae*
      3. Leaves exstipulate........ *Annonaceae*
1. Carpels fused (syncarpous)
   4. Parietal placentation.
      5. Flowers tetramerous.
         6. Androgynophore present .... *Capparidaceae*
         6. Androgynophore absent; anthers tetradynamous........ *Cruciferae*
      5. Flowers pentamerous.
         7. Stamens indefinite; flowers actinomorphic .... *Papaveraceae*
         7. Stamens 5; flowers zygomorphic .... *Violaceae*
   4. Axile placentation; calyx valvate; stamens monadelphous........ *Malvaceae*
   4. Axile placentation; calyx valvate; stamens in different whorls .... *Tiliaceae*
   4. Free-central placentation; stamens 10; obdiplostemonous .... *Caryophyllaceae*

7.5.3 KEY-3 (Disciflorae)

DICOTYLEDONS

Polypetalae

Disciflorae

Thalamus expanded into a disc; ovary superior.

1. Leaves alternate; pinnately compound, gland-dotted; stamens not monadelphous... *Rutaceae*
1. Leaves are not gland-dotted; stamens monadelphous .... *Meliaceae*

7.5.4 KEY-4 (Calyciflorae)

DICOTYLEDONS

Polypetalae

Calyciflorae

Calyx gamosepalous; thalamus cup-shaped; disc thin, often lining the calyx tube.

1. Calyx inferior; generally hypogynous to perigynous; leaves stipulate.
   2. Leaves pinnately compound; calyx with odd sepal anterior; fruit a pod .....*Leguminosae*

3. Corolla descending-imbricate; flowers zygomorphic.........*Fabaceae or Papilionaceae*
3. Corolla ascending-imbricate; flowers zygomorphic............................*Caesalpinoideae*
3. Corolla valvate; flowers actinomorphic ....*Mimosoideae*

2. Calyx with odd sepal posterior; fruit of various types but not a pod...............*Rosaceae*

1. Calyx superior; flowers epigynous; leaves exstipulate.
4. Shrubs or trees.
5. Leaves gland-dotted and exstipulate ....*Myrtaceae*
4. Herbs.
6. Mostly climbers; flowers unisexual .... *Cucurbitaceae*
7. Mostly herbs and not climbers; flowers bisexual ....*Umbelliferae*

7.5.5 KEY-5 (Inferae)

DICOTYLEDONS

Gamopetalae

Inferae

Petals fused; ovary inferior.

1. Leaves opposite or whorled; stamens epipetalous; dichasial cyme inflorescence.................. *Rubiaceae*
1. Leaves alternate or opposite; stamens syngenesious; head or capitulum inflorescence.......... *Asteraceae or Compositae*

7.5.6 KEY-6 (Bicarpellatae)

DICOTYLEDONS

Gamopetalae

Bicarpellatae

Petals fused; flowers with calyx and corolla; ovary superior; two carpels.

1. Flowers actinomorphic.
2. Leaves opposite; milky sap present.
3. Stamens 5, epipetalous; axile placentation; gynostegium absent...........*Apocynaceae*
3. Stamens 5, connected with stigma to form gynostegium; marginal placentation........ *Asclepiadaceae*
2. Leaves alternate; milky sap generally absent.
4. Plants erect.
5. Carpels obliquely placed; swollen placenta .... *Solanaceae*
5. Carpels not obliquely placed; placenta not swollen ....*Boraginaceae*
4. Plants climbing or prostrate; carpels medianly placed....................... *Convolvulaceae*
1. Flowers zygomorphic.
6. Bracts large, leafy; stem with swollen joints; seeds with jaculators ....*Acanthaceae*

6. Bracts minute or absent; stem not with swollen joints; seeds not with jaculators.
   7. One or two ovules in each locule.
      8. Ovary entire, style terminal, stigma simple.. *Verbenaceae*
      8. Ovary entire, style long, stigma bifid .... *Pedaliaceae*
      8. Ovary tetralocular, style gynobasic ....*Labiatae*
   7. Many ovules in each locule.. *Scrophulariaceae*

7.5.7 KEY-7 (Monochlamydae)

DICOTYLEDONS

Monochlamydae

Flowers incomplete; either calyx or corolla or sometimes both absent.

1. Flowers bisexual.
   2. Leaves stipulate, stipules ochreate ....*Polygonaceae*
   2. Leaves exstipulate.
      3. Leaves opposite, bracts coloured.. *Nyctaginaceae*
      3. Leaves alternate.
         4. Flowers with dry, scarious bracts and perianth............................ *Amaranthaceae*
         4. Perianth sepalloid or green .... *Chenopodiaceae*
1. Flowcrs uniscxual.
   5. Inflorescence cymose or raceme; ovary unilocular... *Urticaceae*
   5. Inflorescence cyathium; ovary trilocular.. *Euphorbiaceae*
   5. Inflorescence catkin; fruit capsule; 2–4 stigmas ....*Salicaceae*
   5. Inflorescence catkin; fruit drupe or sorosis; 2 stigmas.....................................*Moraceae*

7.5.8 KEY-8 (Monocotyledons)

Monocotyledons

1. Arborescent; stem with prominent scars of leaf bases; flowers in fleshy spikes ....
   Arecaceae or *Palmaceae*
1. Not arborescent; stems with no prominent leaf bases.
   2. Herbaceous, aquatic; apocarpous ....*Alismaceae*
   2. Plants generally terrestrial; usually syncarpous.
      3. Inferior ovary.
         4. Inflorescence spadix...*Musaceae*
         4. Inflorescence not spadix.
            5. Parietal placentation .... *Orchidaceae*
            5. Placentation usually axile, not parietal.
               6. Six fertile stamens... *Amaryllidaceae*

6. Only one fertile stamen; other stamens transform into staminodes .... *Zingiberaceae*

3. Superior ovary.
   7. Well-developed perianth.
      8. Leaves with sheathing base .... *Commelinaceae*
      8. Leaf bases not sheathing .... *Liliaceae*
   7. Perianth ill-developed or reduced to hairs or bristles.
      9. Flowers arranged on spadix, fruit berry .... *Araceae*
      9. Flowers arranged in spikelets; fruit indehiscent.
         10. Phyllotaxy 1/3; culm triangular; perianth reduced to bristles or hairs; fruit nut .... *Cyperaceae*
         10. Phyllotaxy 1/2; culm cylindrical; perianth reduced to lodicules; fruit caryopsis .... *Poaceae* or *Graminae*

Test Your Understanding

1. Differentiate between punched card keys and dichotomous keys.
2. Give an account of dichotomous keys and their types with reference to plant identification.
3. Make an artificial key to differentiate major groups of dicotyledons.
4. How can you differentiate between Rubiaceae and Asteraceae making an artificial key?
5. Using an artificial key, differentiate between any four families of your course belonging to Thalamiflorae.

Suggested Reading

Bossert, W., 1962, "Computer techniques in systematics", In *Systematic Biology,* National Academy of Science Publication, Washington.

Hansen, B. and K. Rahn, 1969, Determination of angiosperm families by means of a punched-card system, *Dan. Bot. Ark.* **26**: 1–46.

Harrington, H.D. and L.W. Durrel, 1957, *How to Identify Plants?* The Shallow Press, Chicago.

Hutchinson, J., 1967, *Key to the Families of Flowering Plants of the World,* Clarendon Press, Oxford.

Morse, L.E., 1971, Specimen identification and key construction with the time-sharing computer, *Taxon* **20**: 269–282.

_______ 1974, Computer-assisted storage and retrieval of the data of taxonomy and systematics, *Taxon* **23**: 29–43.

Osborne, D.V., 1963, Some aspects of the theory of dichotomous keys, *New Phytol.* **62**: 144–160.

Pankhurst, R.S.. 1970, A computer programme for generating diagnostic keys, *Computer J.* **13**: 145–151.

_______ 1971, Botanical keys generated by computer, *Watsonia* **8**: 357–368.

_______ 1974, Automated identification in systematics, *Taxon* **23**: 45–51.

_______ 1975, *Biological identification with Computers*, Academic Press, London.

Saldanha, C.J. and C.K. Rao, 1975, *A punched-card key to the Dicot Families of South India,* Amarind Publishers, Bangalore.

PLANT NOMENCLATURE, BOTANICAL NAMES AND PHYLOCODE

(A) PLANT NOMENCLATURE

8.1 FUNDAMENTALS OF PLANT NOMENCLATURE

8.1.1 Definitions

Assignment of definite names to plants is called *plant nomenclature.*

In the present botanical world, the nomenclature involves the principles governed by rules formulated and adopted by International Botanical Congresses. The rules developed by IBC are listed formally in a *code* called International Code of Botanical Nomenclature (Voss *et al.,* 1983), abbreviated as ICBN. The major goal of ICBN is to provide one correct name for each *taxon.*

Taxa (singular, *taxon*) are the taxonomic groups of any rank. The ascending hierarchy of taxa include species, genus, family, order, class and division.

8.1.2 Who is a Nomenclaturist?

A nomenclaturist is a taxonomist who assigns names to new taxa, determines the right names for old taxa according to the rules of International Code of Botanical Nomenclature, and finalises the right name for a specimen according to an established system of classification.

8.1.3 Why is Nomenclature Needed?

Imagine for a few moments about the state of affairs if there are no *names* for all the things we see, make or handle. How strange and chaotic the life would be in such a condition? The entire business affairs of the world would stop, and practically there will be no "give and take" of knowledge. It would also be impossible for all of us to communicate our feelings to others. Practically, the life would stop. Therefore, nomenclature for everything is needed.

8.2 COMMON NAMES AND SCIENTIFIC NAMES

Vernacular or *common names* are made up of words from the native language of the country or the region. They may vary in different countries as well as in different regions of the same country. They may be Portuguese words in Portugal; Spanish in Spain; English in England, U.S.A. and Australia; and Hindi, Bangla, Oriya or Tamil, etc. in U.P., West Bengal, Orissa and Tamilnadu, respectively. 'Papita', 'Kela', 'Gazar' and 'Tamatar', of India are known in U.S.A. as papaya, banana, carrot and tomato, and by other names in Spain. There is therefore no international uniformity in common names.

On the other hand, *scientific names* which are based mainly on Latin language have international uniformity. By the name *Saccharum officinarum*, all botanists of the world would understand that it is sugarcane.

Jones and Luchsinger (1987) have stated that common names present five major problems mentioned in the following table:

S. No.	Common Names	Scientific Names
1.	These are not universal. They vary in different languages.	These are universal and are recognised throughout the world.
2.	They do not provide information indicating generic and family relationships.	They provide information regarding these relationships.
3.	A well-known plant may have hundreds of common names.	A well-known plant has only one scientific name.
4.	Sometimes, two or more plants have the same common name.	Two or more plants always have different scientific names.
5.	Many species do not have any common names.	All known plants have a scientific name.

The scientific name of a plant consists of two separate words. The first word designates the *genus* of the organism and the second word designates the *species*. Details of this two-name system were first given by Linnaeus (1753) in his famous *binomial system of nomenclature.*

8.3 BINOMIAL NOMENCLATURE

8.3.1 What is Binomial Nomenclature?

In the earlier days plant names were long and descriptive e.g. in the herbal of Clusius (1583) a species of willow is named *Salix pumila angustifolia altera*. Then, in 1623, Gaspar Bauhin (1560–1624) devised a plan of adopting two names for each plant in his *Pinax Theatri Botanici*. But it was Carolus Linnaeus (1707–1778), the great Swedish naturalist, to whom the actual credit goes for devising and methodically employing the binomial system of nomenclature. Linnaeus employed this system in the first edition of his *Species Plantarum* in 1753. According to this system the scientific name of a plant consists of two Latin or latinized words: (1) The first is the name of the genus, i.e. generic name or *generic epithet*, and (2) the second is the name of the species, i.e. *specific epithet*. For example, the botanical name of sugarcane is *Saccharum officinarum*. The first word (*Saccharum*) designates the *genus* of the plant and the second word (*officinarum*) designates the *species* of this genus.

8.3.2 Generic Name

It is always a noun, and written with a capitalised initial letter and the remainder small. The generic name is also always singular in number. It may have following types of origin:

1. Several generic names may be in honour of the names of well-known persons, e.g. *Theophrasta* in honour of Theophrastus and *Candollea* in honour of A.P. de Candolle.
2. They may be descriptive, with reference to some common characteristics of the included species, e.g. *Cercocarpus* (coiled fruit), *Xanthoxylum* (yellow wood), etc.
3. They may be of poetic or mythological origin, e.g. *Theobroma* (god's food).
4. They may also be the aboriginal name of the plants, e.g. *Betula* and *Quercus* which were the old Greek names for Birch and Oak.

8.3.3 Specific Epithet

According to the 1983 recommendations of the International Code of Botanical Nomenclature (Voss, *et al.*, 1983), all specific epithets be written with a small initial letter. But if the specific epithet is derived from common names, or from former generic names, or from a person's name, the initial letter of the epithet may be a capital letter. In the typed or handwritten matter, both the specific epithet and generic names should be underlined. They should be printed in *italics* or *boldface*. The authority, written after the specific epithet, is never underlined.

The specific epithet is often an adjective. According to Article 23 of ICBN, a specific epithet (i) may be a name in honour of a person, or (ii) may be derived from a geographical location, or (iii) may originate from an old common name, or (iv) may be derived from some characteristics of the plant, or (v) it may also be named arbitrarily.

8.3.4 Authority

The name of the species is incomplete if it is not followed by full or abbreviated name(s) of the author(s). For example, *Pyrus malus* is incomplete. The complete name is *Pyrus malus* L. where L. is abbreviated for Linnaeus. Citation of the full or abbreviated form of the author is necessary because this will verify the date or time of the first valid publication of the name of a particular taxon.

8.4 INTERNATIONAL CODE OF BOTANICAL NOMENCLATURE (ICBN)

8.4.1 What is ICBN?

The *International Code of Botanical Nomenclature* (ICBN) is the set of rules and recommendations dealing with the formal *botanical names* that are given to plants. Its main aim is that each taxon or taxonomic group of plants has only one *correct botanical name* and that is accepted throughout the world. Two main principles of ICBN are listed below:

(1) *Priority* is the guiding principle in botanical nomenclature. The ICBN sets the formal starting date of plant nomenclature at 1 May, 1753, which is the publication date of *Species Plantarum* by Linnaeus.

(2) Each botanical name is fixed to a taxon by a *type*, which is almost invariably dried plant material usually deposited and preserved in a herbarium.

Very few hard rules of ICBN apply above the taxonomic rank of family. Each new edition of ICBN supersedes the earlier editions and is retroactive back to 1753.

8.4.2 Who can Change ICBN?

The ICBN can only be changed by an International Botanical Congress (IBC), with the International Association for Plant Taxonomy providing the supporting infrastructure.

8.4.3 To which Organisms does ICBN Apply?

The ICBN applies not only to plants, as they are now defined, but it also applies to other organisms studied traditionally by botanists, e.g., Cyanobacteria (blue-green algae), fungi, photosynthetic protists and taxonomically related non-photosynthetic groups. For fossils, there are special provisions in ICBN.

8.4.4 Whether ICBN is Applicable to Cultivated Plants?

No. For naming of cultivated plants, there is a separate code called the International Code of Nomenclature for Cultivated Plants (ICNCP).

8.4.5 History of ICBN

Linnaeus in 1737 and again in 1751 proposed the elementary rules of naming plants in his *Philosophia Botanica*. Then in 1813, A.P. de Candolle set forth a detailed set of rules regarding plant nomenclature in his *Theorie elementaire de la botanique*. The same rules of Linnaeus, A.P. de Candolle and his son Alphonse de Candolle were later evolved into our present *International Code of Botanical Nomenclature* (ICBN).

Alphonse de Candolle convened the *First International Botanical Congress* in 1867 in Paris. It was attended by the botanists of several countries. They adopted a set of rules of plant nomenclature, most of which were proposed by A. de Candolle. These excellent rules of plant nomenclature are known as *de Candolle rules* or *Paris Code* of 1867. Subsequent meetings of the International Botanical Congress were held in 1892 (*Rochester Code*), 1905 (*Vienna Code*), 1907 (*American Code*) and 1910, but a general agreement, regarding the internationally acceptable rules of plant nomenclature, was reached in the meeting of the IBC at Cambridge in 1930. Lawrence (1951) has discussed the detailed history of the Code and may be quoted that in 1930 at the Cambridge Congress "for the first time in botanical history, a code of nomenclature came into being that was international in function as well as in name". This code is called the *International Code of Botanical Nomenclature*. Scientists in the International Botanical Congresses suggest the modifications or amendments which are incorporated in the ICBN on a regular basis.

The International Code of Botanical Nomenclature, 1983, was adopted by the Thirteenth International Botanical Congress, Sydney (Australia) in August 1981, and the Chairman of the editorial committee was E.G. Voss. ICBN is divided into three parts, i.e. *Principles, Rules* and *Recommendations.*

8.4.6 Principles of ICBN

The philosophical basis of the Code is formed by the following six principles:

"1. Botanical nomenclature is independent of zoological nomenclature.
2. The application of names of taxonomic groups is determined by means of nomenclatural types.
3. The nomenclature of a taxonomic group is based upon priority of publication.
4. Each taxonomic group with a particular circumscription, position, and rank can bear only one correct name, the earliest that is in accordance with the Rules, except in specific cases.
5. Scientific names of taxonomic groups are treated as Latin regardless of their derivation.
6. The Rules of nomenclature are retroactive unless expressly limited."

8.4.7 Rules and Recommendations of ICBN

According to the ICBN (1983) the detailed provisions of the Code "are divided into *Rules,* set out in Articles, and *Recommendations".* It is stated under point No. 4 of the *Preamble* of ICBN (1983) that the main object of the *rules* is to put the nomenclature of the past into order and to provide for that of the future. It also states that the names contrary to a rule cannot be maintained.

As mentioned under point No. 5 of the *Preamble* of ICBN (1983) the *Recommendations* deal with subsidiary points, their object being to bring about greater uniformity and clearness, especially in future nomenclature. It also states that names contrary to a recommendations cannot, on that account, be rejected. They are, however, not examples to be followed.

ICBN *Rules* and *Recommendations* apply to all organisms treated as plants (including fungi but excluding bacteria) as well as fossils or non-fossils. Nomenclature of bacteria is governed by *International Code of Nomenclature of Bacteria* (ICNB) and not by ICBN.

Point No. 10 of ICBN (1983) states that the latest edition of the Code supersedes all previous editions of future studies.

8.4.8 Some Later Codes

St. Louis Code (1999)

The 16th International Botanical Congress was held at St. Louis, Missouri, USA in August 1999, and the International Code of Botanical Nomenclature adopted in this Congress is called *St. Louis Code.* Botanists of 85 nations attended this Congress. Hectic discussions were made mainly in view of the facts that "as many as two-thirds of the world's 3,00,000 plant species are in danger of extinction in nature during course of the 21st century", and almost every aspect of the life of human beings depends on plants. Botanists made some *resolutions* and *recommendations,* of which some major ones are undermentioned:

(A) Resolutions Six resolutions were passed in this International Botanical Congress at St. Louis.

(1) Resolution I Botanists agreed on 10 major aspects in this resolution, of which some are mentioned below. They call for:

(i) "the establishment of a new coordinating body associated with United Nations to monitor the status of plants throughout the world, detect those in most danger, and take steps to conserve them in nature, in botanic gardens, or in gene banks";
(ii) "securing additional funds for study of plants throughout the world";
(iii) "making all the information about plants generally available on the Internet";
(iv) placing additional emphasis on the importance of the "survival of biodiversity throughout the world";
(v) maintaining "an active census of the status of each country's plants" at national level;
(vi) "actively developing and implementing plans to conserve the world's economic plants ....";
(vii) "devoting special attention to the conservation of medicinal plants ....";
(viii) "funding internationally an ongoing programme of research on plant population biology...."

(2) Resolution II Botanists of the 85 nations attending this Congress call on governments and policy makers to:

(i) "recognize the importance of developing and maintaining scientific expertise, provide resources for the education and training of scientists, and maintain career opportunities".... especially in biological sciences;
(ii) "actively develop floras and detailed accounts of plants of all regions ....";
(iii) "support collaborative programmes between and among developed and developing countries";
(iv) "ensure high priority be given to the maintenance of botanical museums, herbaria, libraries, gardens, living plant collections and gene banks ....";

(3) Resolution III Botanists in this resolution resolved to:

(i) "increase our knowledge of diversity and relationships of plants ...." and "to make that knowledge accessible to all";
(ii) "advocate to policy makers the relevance of plant sciences ....", and thus "maintain the quality of human life on earth";

(4) Resolutions IV to VI These three resolutions are regarding the (i) importance and programmes of biodiversity (Resolution IV); acceptance of the decisions of Nomenclature Section of the Congress (Resolution V); and selection of sites for future Congresses to facilitate the "attendance of botanists from all regions of the world" (Resolution VI).

(B) Recommendations Several recommendations have been made in ICBN in the *St. Louis Code* by botanists attending the XVI International Botanical Congress, of which only some are undermentioned:

1. The scientific names under the jurisdiction of the *Code*, irrespective of rank, "are consistently printed in *italic type*. The *Code* sets no binding standard in this respect, as typography is a matter of editorial style and tradition, not of nomenclature".

2. To set off scientific names even better, the use of *italics* for technical terms and other words in Latin, "has now been abandoned".
3. For style of bibliography, the titles of the books are "abbreviated in conformity with *Taxonomic Literature*, ed. 2 by Stafleu & Cowan (1976–1988; with supplements by Stafleu & Mennega, 1999–2000)".
4. For style of bibliography the *journal titles* are abbreviated in conformity with *Botanico-periodicum-huntianum* (1968) and its supplement (1991).
5. Author citations of scientific names are standardized in conformity with *Authors of Plant Names* by Brummitt and Powell (1992).
6. The single largest area of change in *St. Louis Code* concerns typification, where many proposals have been made on *Lectotypification* (*for details refer to original code*).
7. All reference to registration of new botanical names, to become "mandatory from a future date, be deleted from the Code ....".
8. Fossil plant nomenclature underwent profound changes in St. Louis Code (*for details refer to original Code*).
9. Fungal nomenclature was only affected in a marginal way by decisions of the St. Louis Congress.
10. One new term "*isonym*" has been introduced into the Code, "defined to mean the same name used independently by different authors at different times ....".
11. In the field of terminology, the terms "homotypic synonym", "heterotypic synonym" and "replacement names" were accepted as optional equivalents of the earlier "nomenclatural synonym", "taxonomic synonym" and "avowed substitute".
12. The terminations *-viridae*, *-virales*, *-virinae*, and *-virus* were "outlawed for names of subclasses, orders, subtribes and genera, respectively.

Vienna Code (2005)

The latest XVII International Botanical Congress was held in Vienna in 2005 (XVI being at St. Louis, Missouri, USA), and the International Code of Botanical Nomenclature accepted in this Congress is called *Viena Code*. It is written entirely in English and has been translated in several other languages. One of the reasons invoked for the choice of Vienna as the site of XVII Congress was that the second International Botanical Congress had been held there exactly 100 years earlier, i.e., in 1905. It was in this IInd Congress in 1905 that the first internationally developed rules governing nomenclature of plants were accepted, and these rules were recognised as *Vienna Rules*. The XVII IBC held on 12–16 July, 2005 at Vienna was attended by 198 registered members carrying 402 institutional votes. On the other hand, St. Louis Congress at Missouri, held six years ago in 1999 was attended by 297 members carrying 494 institutional votes.

The *Vienna Code* does not differ substantially in overall presentation and arrangements from the *St. Louis Code*, and the number of Articles remains the same. Some selected additions and recommendations of *Vienna Code* are listed below:

1. The most notable feature of *Vienna Code* "is the inclusion for the first time a Glossary, which appears in Appendix VII. This Glossary is very tightly linked to the wording of the Code, and "only nomenclatural terms defined in the Code can be included".

2. The scientific names under the jurisdiction of *Code*, irrespective of rank, are consistently printed in *italic type*. The "*Code* sets no binding standard in this respect, as typography is a matter of editorial style and tradition not of nomenclature".
3. The titles of the *books* in bibliographic citations are abbreviated in conformity with "*Taxonomic Literature*. ed. 2, by Stafleu and Cowan (1976–1988; with 6 supplements by Stafleu and Mennega, 1992–2000)", but with capital initial letters.
4. For titles of the *journals* in bibliographic citations, the abbreviations follow the *Botanico-periodicum-huntianum*, ed. 2 (2004).
5. Author citations of scientific names appearing in the Code are "standardized in conformity with *Authors of Plant Names*, by Brummitt & Powell (1992)"; these are "also adopted and updated by the *International Plant Names Index*".
6. "Perhaps the most important single decision incorporated into the *Vienna Code* was to deal with what many have recognized as a *bomb waiting to explode*, the publication status of *theses submitted for a higher degree*". The *Code* decided that "no independent non-serial publication stated to be a thesis submitted for a higher degree on or after 1 January 1953 would be considered an effectively published work without a statement to that effect or other internal evidence" (*for details, readers should consult the original Code*).
7. Regarding valid publication of names, *Vienna Code* made it clear that "names be composed only of letters of Latin alphabet, except as otherwise provided in the Code."
8. St. Louis Code accepted that all fossil taxa should be treated as *morphotaxa*. In *Vienna Code*, this has not been considered appropriate. A distinction between a morphotaxon and a regular fossil taxon is now established in *Vienna Code*.
9. In *Vienna Code*, it was agreed that "the starting date for valid publication of suprageneric names of spermatophytes, pteridophytes and bryophytes be 4 August 1789, the date of publication of Jussieu's *Genera Plantarum*".
10. *Vienna Code* also established that parenthetic author citation is not permitted at suprageneric ranks.
11. Four family names, previously overlooked in Berchtold and Presl's rare, later, multivolume work of the same name (1823–1825) have been updated: *Aquifoliaceae*, *Cornaceae*, *Potamogetonaceae* and *Punicaceae*.
12. Under the concept of "minimum invalidity" (Art. 33.10), the "rules determining when a rank is denoted by a misplaced term (and hence not validly published) were clarified and made more practical".
13. From 1 January, 2007, a new combination, a new generic name with a basionym, or an allowed substitute "is not validly published unless its basionym or replaced synonym is cited".
14. *Vienna Code* establishes that only if validly published does a name have any status; indeed, unless otherwise indicated, the word "name" in the Code means a name that has been validly published".

8.4.9 Future Botanical Congress and Legality of ICBN

The International Code of Botanical Nomenclature (ICBN) is published under the authority of the International Botanical Congresses (IBC). The next meeting of the IBC will be held in Melbourne, Australia, from 23–30 July, 2011. Similar to other international codes of nomenclature, ICBN also has no legal status and is dependent on the voluntary acceptance of its rules by scientists in general and botanists in particular.

8.5 SOME IMPORTANT RULES OF NOMENCLATURE

1. *Ranks and Endings of Taxa* In an accepted system of classification, each individual plant is treated as belonging to a number of taxa of different ranks. Generally, the species is considered as the basic unit of classification. Other main ranks in the flowering plants in an ascending order are genus, family, order, subclass and class. However, ICBN (Voss *et al.*, 1983) has mentioned 22 different ranks and some standardized grammatical endings *(suffixes)* for the ranks from division down to the level of genus. All these ranks are mentioned under Article No. 1.5 (Table 1.1, Chapter 1).
2. *Principle of Priority* One plant might have been described under different botanical names by various plant nomenclaturists in different parts of the world. But, according to the "principle of priority" "each taxon is to be known by its earliest name". For example, *Cleome gynandra* Linn. was first described and named by Linnaeus in 1753. Then he himself changed its name as *Cleome pentaphylla* Linn. In 1824 de Candolle recognized three separate genera (*Cleome, Polanisia* and *Gynandropsis*) and named the Linnean genus as *Gynandropsis pentaphylla* (Linn.) DC. Iltis in 1960 merged *Gynandropsis* and *Cleome* into one genus, i.e. *Cleome.* So according to the "Principle of priority", the oldest name *(Cleome gynandra* Linn.) is the correct valid name.

   Because of Principle of Priority, names of several plants have now been changed.
3. *Type Method* The type method is a legal device to provide the correct name for a taxon. A *type specimen* is a herbarium sheet (or rarely a drawing or a photograph) of a specimen which was used by the author to provide its authentic description. According to Article 9 of ICBN the type of a genus is a species (e.g. the type of genus *Vernonia* is *V. noveboracensis* (L.) Michx.), and the type of a family is a genus (e.g. *Aster* is the type genus of Asteraceae).

   Several *kinds of types* designated by ICBN are undermentioned:

   i. *Holotype* It is a specimen used by the author in the original publication as the nomenclatural type.

   ii. *Isotype* It is a duplicate specimen of the holotype, i.e. from the same collection, with the same locality, date and number as the holotype.

   iii. *Lectotype* It is a specimen selected by a competent worker from the original material studied by the author of the species, when no holotype was designated or when the holotype has been destroyed or lost.

   iv. *Neotype* It is a specimen selected to serve as a substitute for the holotype when all material, on which the name of the taxon was based, is missing.

v. *Nomenclatural Type* It is that element with which the name of a taxon is permanently associated.

vi. *Syntype* It is one of the two or more specimens cited by an author of a species when no holotype was designated; or, a syntype is any one of the two or more specimens originally designated as types.

vii. *Paratype or "Co-types"* It is a specimen other than isotype or holotype. If two or more specimens have been cited as types by the author, the remaining cited specimens are called *"Co-types"* or *paratypes.*

viii. *Topotype* It is a specimen collected from the same locality from where the holotype was collected.

4. *Synonyms and Related Definitions* A name rejected due to misuse or difference in taxonomic judgement is called *synonym.*

   A specific or intraspecific name which has priority and is retained when transferred to a new taxon, is called a *basionym.*

   A case in which two or more identical names are based on different types, of which only one can be a legitimate name, is called a *homonym.*

   An illegitimate binomial, in which the name of the genus and the name of the species is the same, is called a *tautonym,* e.g. *Armoracia armoracia* (L.) Britton.

   An automatically created legitimate tautonym for infraspecific or infrageneric taxa is called an *autonym,* e.g. *Hypericum* subgenus *Hypericum* section *Hypericum.*

5. *Citation of Author* Some of the rules related to author citation are undermentioned:

   i. *Original Author* The name of a taxon is complete and accurate only when it is followed by a full or abbreviated form of the author(s) who first validly published the concerned name, e.g. *Liliaceae* and *Lilium superbum* are incomplete; the complete names are *Liliaceae* Adans. and *Lilium superbum* Linn.

   ii. *Joint Author* If two authors have jointly published the name of a taxon, the names of both the authors should be cited and linked by the words *et* or & e.g. *Illicium griffithi* Hook & Thoms (or Hook et Thoms).

   iii. *Rank Alteration* When a taxon of a lower rank is upgraded in a higher rank but retains its name, the author's name who published it first should be cited in the bracket It is to be followed by the name of the author who made the alteration, e.g. Allioni raised the rank of variety *Medicago polymorpha* var. *orbicularis* L. to the species rank. Therefore, it becomes *Medicago orbicularis* (L.) All.

   iv. *Name Proposal* When the name of a taxon is proposed but not validly published by one author, and is later on validly published by another, the word *ex* should be used as a connecting link between the name of the former author and the name of the subsequent author, e.g. *Gossypium tomentosum* Nutt ex Seem.

6. *Names of Cultivated Plants* Wild plants brought under cultivation retain their original names.

7. *Latin Diagnosis* The diagnosis, details and description of new taxa published before 1st January, 1955 were accepted by ICBN as valid, irrespective of the language. After this date the description of any new taxa would be considered valid only if accompanied by Latin diagnosis.
8. *Effecttve and Valid Publication* The publication of new names and description are effective and considered valid only when they are distributed in a printed form to the general public or to at least ten well-established botanical institutions.

   A *validly published names*, as specified in the ICBN, is the one in which the basic provisions are (1) effective publication, (2) publication in the form specified for the name of each category of taxa, (3) publication with a description, or a reference to a previously published description, of the taxon to which the name applies, (4) accompanied by a Latin description or by a reference to a previously and effectively published Latin description of the taxon, and (5) an indication of the nomenclatural type. On the other hand, an *effectively published name* is the one published in printed matter generally available to botanists.
9. *Choice of Names when the Taxon Rank is Changed* When the rank of a taxon is changed (*viz.* a species becomes a genus or *vice-versa*) the earliest legitimate name in its new rank is its correct name.
10. *Choice of Names when same-rank Taxa are United* When two or more taxa of the same rank (*viz.* two or more genera, two or more species, etc.) are united into one, the oldest legitimate name of these taxa would be retained as the name of the united taxon.
11. *Retention of Names of Divided Taxa* When a genus or a species is divided into two or more genera or species, respectively, the original name of the genus or species must be retained.
12. *Retention of Names of Taxa on Transference* When a subdivision of a genus is transferred to another genus, or a species is transferred to another genus without the change of the rank, the original legitimate name must be retained. For example, *Hydrocotyle asiatica* L., on transference to genus *Centella*, must be named as *Centella asiatica* (L.) Urban.
13. *Rejection of Names* Names of taxa must be rejected in the following conditions:
    i. When the names are illegitimate (i.e. if it is a tautonym, later homonym, rejectable generic name, nomenclaturally superfluous, etc.).
    ii. When the names give different meanings, and become a permanent source of confusion.
    iii. When the characters of the name of the taxon are derived from two or more different elements.
    iv. When the generic names coincide with a morphological term, or are unitary designations to species, or are words not intended as names.
    v. When the specific names are tautonyms, or are published without any consideration of binomial system, or are words not intended as names.
14. *Names of Different Taxa* According to the International Code of Botanical Nomenclature (1983) the following should be the pattern of naming different taxa:
    i. Genera and higher ranks should be *monomials,* e.g. *Aesculus* L., *Rosa* L., Rosaceae Juss.

ii. Species should be *binomials,* e.g. *Gossypium tomentosum* Nutt.
iii. Subspecies should be *trinomials,* e.g. *Hibiscus moscheutos* ssp. *palustris* (L.) Clausen.
iv. Varieties should be *quadrinomials,* e.g. *Lilium catesbaei* Walter ssp. *catesbaei* var. *longii* Fernald.

(B) BOTANICAL NAMES

Botanical names are either Latin words or words that have been latinized from Greek or some other language.

The name of a species is a *binomial* and consists of a generic name and a specific epithet. A specific epithet is the second part of the binomial. However, one must be quite clear about the fact that it is wrong to use the specific epithet alone to designate a particular species; it must always be used with a generic name to form the binary combination for that species. The specific epithets are formed from nouns, adjectives, etc. and may join these words with a large number of prefixes and suffixes.

8.6 COMMON PREFIXES USED IN SPECIFIC EPITHETS

Some Latin prefixes of *numbers* are as follows:

1. uni-(L.): *uniflorus* (one-flowered)
2. bi-(L.): *bifoliatus* (two-leaved)
3. tri-(L.): *triangularis* (with 3 angles)
4. quadri-(L.): *quadrangularis* (with 4 angles)
5. quinque-(L.): *quinquefolius* (5-leaved)
6. sex-(L.): *sexangularis* (6-angled)
7. septem-(L.): *septemlobus* (7-lobed)
8. octo-(L.): *octoflorus* (8-flowered)
9. noveme-(L.): *novemneris* (with 9 nerves)
10. decum-(L.): *decumlobus* (with 10 lobes)

Some of the Greek prefixes of numbers include *mon-* (for one, e.g. *monandra*), *di-* (for 2), *tri-* (for 3), *tetra-* (for 4), *penta-* (for 5), *hex-* (for 6), *hepta-* (for 7), *octo-* (for 8), *ennea* (for 9) and *deca-* (for 10, e.g. *decapetalus*).

Some of the other common prefixes of numbers are *amphi-* (Gr.): for two kinds; *diplo-* (Gr.): for double; *haplo-* (Gr.): for single; *multi-* (L.): for many; *poly-(Gr.)*: for many; *a-* or *ab-* (L.): away from; *ecto-* (Gr.): outside; *endo-* (Gr.): inside; *inter-* (L.): between, and; *intra-* (L.): within.

8.7 COMMON SUFFIXES USED IN SPECIFIC EPITHETS

Some of the common suffixes along with one example of each of them are undermentioned:

-aceus: *crustaceus*; -alis: *digitalis*; -aris: *angularis*; -arium: *aquarium*; -aticus: *aquaticus*; -estris: *campestris*; -eus: *roseus*; -ilis: *sexitilis*; *-osus: foliosus.*

8.8 PLANT PARTS USED AS EPITHETS

Almost all plant parts, including root, stem, leaf, bud, flower, fruit and seeds, have been used as names of specific epithets. Some of the specific epithets, along with the related plant parts in parenthesis, are undermentioned:

receptaculum (for receptacle), *sepalum* (for sepals), *carpellum* (for carpel), *loculus* (for locules), *stylus* (for style), *petalum* (for petals), *discus* (for disc), *ovarium* (for ovary), *ovulum* (for ovule), *pedicellus* (for pedicel), *pistillum* (for pistil), *fructus* (for fruit), *sperma* (for seed), *folium* and *phyllon* (for leaf), *rhiza* (for root), *caulos* (for stem), etc.

8.9 SPECIFIC EPITHETS LINKED WITH COLOUR

Some of the specific epithets linked with colour, along with the name of the colour in the parenthesis, are undermentioned alphabetically:

albicans (whitish), *albus* (white), *atrovirens* (dark green), *aureus* (golden yellow), *calcareus* (chalky white), *candidus* (shining white), *croceus* (saffron yellow), *flavidus* (slightly yellow), *flavus* (pale yellow), *fulvus* (dull yellow), *glaucus* (grey-green), *luteolus* (pale yellow), *niger* (black), *niveus* (snow white), *purpureus* (purple), *roseus* (rosy), *violaceous* (violet), *virens* (green) and, *viridis* (green).

8.10 SPECIFIC EPITHETS LINKED WITH GEOGRAPHY

Some of the geography-linked specific epithets, along with their related geographical places in parenthesis, are undermentioned alphabetically:

africanus (of Africa), *americanus* (of America), *arabicus* (of Arabia), *argentinus* (of Argentina), *asiaticus* (of Asia), *australiensis* (of Australia), *austriacus* (of Austria), *brasiliensis* (of Brazil), *canadensis* (of Canada), *chinensis* (of China), *cubensis* (of Cuba), *europaeus* (of Europe), *germanicus* (of Germany), *helveticus* (of Switzerland), *indicus* (of India), *italicus* (of Italy), *mexicanus* (of Mexico), *sinensis* (of China), *virgnicus* (of Virginia) and, *zeylanicus* (of Ceylon).

8.11 SPECIFIC EPITHETS LINKED WITH SIZE

Some of size-linked epithets are mentioned below alphabetically:

altus (altitude or tall), *exaltatus* (very tall), *giganteus* (gigantic or very large), *grandis* (large), *humilis* (dwarf), *major* (greater), *minor* (less), *minutus* (very small), *procerus* (very tall), *pumilus* (dwarf) and, *robustus* (stout or robust).

8.12 SPECIFIC EPITHETS LINKED WITH HABIT

Epithets, along with the name of the related habit in parenthesis, are undermentioned alphabetically:

arborescens (arborescent), *dichotomus* (dichotomous), *erectus* (erect), *furcatus* (forked), *prostratus* (prostrate), *ramosus* (branched), *repens* (creeping) and, *stoloniferus* (stoloniferous).

8.13 SPECIFIC EPITHETS LINKED WITH HABITATS

Some of such epithets, along with the name of the related habitat in the parenthesis, are listed below alphabetically:

amphibius (living both on land and in water), *aquaticus* (living in water), *arvensis* (growing in ploughed field), *campestris* (of field), *hypogeus* (underground), *lacustris* (of ponds or lakes), *littoralis* (growing on seashores), *maritimus* (of sea), *montanus* (of mountains), *palustris* (of swamps), *rupestris* (on rocks), *sativus* (cultivated), *sylvaticus* (of woods), *sylvestris* (growing in woods) and, *terrestris* (growing in dry soil).

(C) PHYLOCODE: A NEW SYSTEM OF NOMENCLATURE

8.14 WHAT IS PHYLOCODE?

Recently, Cantino and de Queiroz (2001) proposed a new code for naming organisms by clear "reference to phylogeny (ancestry and descent), rather than on the basis of the Linnaean hierarchy of taxonomic categories (species, genus, family, and so on)", as stated by Robinson and Kommedahl (2002), and this "new code" has been named by then as "*Phylocode*". They also proposed a new term "*clades*" in this code. According to them, "clades" is a "group(s) of species comprising a common ancestor and all its descendants". Clades are the constituents of the "*tree of life*". All clades should have explicit and unambiguous names that do not change with time. The current systems of nomenclature are based on Linnaean hierarchy, and they do not "boast such immutability".

The International Code of Phylogenetic Nomenclature is known in short as "*Phylocode*". It is a developing "draft for a formal set of rules governing phylogenetic nomenclature". The current version of Phylocode is specifically designed "to regulate the naming of clades, leaving the governance of *species* names up to the rank-based codes".

8.14.1 How has Phylocode Come into Existence?

Phylocode is not a new idea. It is actually based on ideas presented in the literature since the late 1980s and, more formally, on the outcome of a workshop held at the Harvard University Herbaria, Cambridge, Massachusetts, USA - on 7–9 August 1998. In this workshop, a draft proposal was put forward, and this draft covers only the naming of clades. It was also decided in this workshop that rules governing species names will be added later. However, as a temporary measure, "Linnaean binomial nomenclature is used in the draft Phylocode where species names are needed".

8.14.2 How are the Species Named in Phylocode?

As mentioned above under Article 8.14.1, the final rules governing species names have not been decided yet. According to Robinson and Kommedahl (2002), "the form that species names should take in the Phylocode" is still controversial. Cantino *et al.* (1999) earlier presented 13 possibilities for naming species phylogenetically and compared these methods with each other and also with the Linnaean system. It is, however, for the scientific community to finally decide whether or not the "Phylocode should become the sole code governing the names of the taxa" as mentioned by Robinson and Kommendahl (2002).

8.14.3 What is the Basic Difference between Phylocode and Linnaean Binomial Nomenclature?

The fundamental characteristic "that distinguishes the Phylocode from the conventional hierarchic nomenclatural systems is its *ranklessness*" according to Robinson and Kommendahl (2002). The proposed Phylocode will cover the naming of clades and species, but in this Code, these terms will refer not to ranks but to different kinds of biological entities. It is so because in Phylocode both (clades and species) are the products of evolution that are discovered, rather than created, by systematists, and both have an objective existence according to Cantino and de Queiroz (2001).

8.14.4 What would be the Starting Date for Phylocode?

Regarding the starting date for Phylocode it has been mentioned by Robinson and Kommedahl (2002) stated that the starting date for the new Code, "which has not yet been decided", will coincide with the publication of a companion volume providing definitions of clade names, which are widely used.

8.14.5 What does Phylocode Advisory Group now want from the Scientific Community?

There is an advisory group which is now working on formulation and implementation of Phylocode. This group is now coordinating work and seeking comments and ideas concerning this proposal from as many people as possible. Anybody interested is welcome to review the current draft of Phylocode and communicate it to the Phylocode Advisory Committee.

8.15 PRINCIPLES OF THE PHYLOCODE

The proposed Phylocode is based on certain principles. It allows freedom of taxonomic opinion with reference to hypotheses about relationships. It only deals how the names are to be applied with reference to a given phylogenetic hypothesis. As published in Division 1 (Principles) of Phylocode, the five basic principles of Phylocode are reference, clarity, uniqueness, stability, and phylogenetic context.

(1) Reference The most basic purpose of names of taxon is to "provide a means of referring to taxa, as opposed to indicating their characters, relationships, or membership".

(2) Clarity The names of taxon should be "unambiguous in their designation of particular taxa. Clarity in nomenclature is achieved through explicit definitions.

(3) Uniqueness For promoting clarity, "each taxon should have only one accepted name, and each accepted name should refer to only one taxon".

(4) Stability During course of time, the names of taxa should not change. "As a corollary, it must be possible to name newly discovered taxa without changing the names of previously discovered taxa". There should, therefore, be a complete stability in the names of taxa.

(5) Phylogenetic context The concern of Phylocode is only with the "naming of taxa and the application of taxon within a phylogenetic context".

8.16 PHYLOCODE: AN OVERVIEW

The phylogenetic nomenclature will be regulated by the Phylocode because the latter will provide rules for the following:

(1) How to decide which combinations of names and definitions will be considered validly published (*for details, refer to Chapter II of Phylocode*)?

(2) Which of the names and definitions will be considered homonyms or synonyms (*for details, refer to Articles 13 and 14*)?

(3) Which one of a set of homonyms or synonyms will be finally considered vaild?

Furthermore, the Phylocode will only allow:

(i) the naming of clades (*for details, refer to Article 1.1*), and

(ii) the use of specimens, species, and amorphies as specifiers (*for details, refer to Article 11*).

8.17 REGISTRATION DATABASE OF PHYLOCODE

8.17.1 RegNum

The future of the Phylocode is still undecided. If and when it is finally implemented, the "Phylocode will be associated with a registration database called *RegNum*. The RegNum will store the names of all clades and definitions that will be considered potentially valid".

The utility of RegNum will be to "provide a publicly-usable tool for associating clade names with definitions". In the latter stages, this will be "associated with sets of subtaxa or specimens through phylogenetic tree databases (such as *TreeBASE*)". RegNum will also be an important tool in taking a decision that "which one of a number of synonyms or homonyms will be considered valid".

8.18 MAIN EVENTS IN THE HISTORY OF PHYLOCODE

1. As mentioned earlier under Article 8.14.1, the idea of Phylocode came as an outcome of a workshop of systematists held at Harvard University in August 1998. An advisory group of leading systematists was formed in this workshop.
2. In April 2000, these systematists prepared a draft of Phylocode and made it public on the *web*. They also invited comments from the systematists globally.
3. In July 2002, a second workshop on Phylocode was held at Yale University, USA. Several modifications were made in the rules and recommendations of Phylocode in this workshop.
4. From July 6, 2004 to July 9, 2004, the First International Phylogenetic Nomenclature Meeting took place in Paris (France). It was attended by about 70 systematists and evolutionary biologists from 11 nations. A society was formed, namely *International Society for Phylogenetic Nomenclature* (ISPN). An advisory group was formed to have a complete control on the development of Phylocode.

5. From June 28, 2006 to July 2, 2006 the Second International Phylogenetic Nomenclature meeting took place at Yale University, USA.
6. From July 21, 2008 to July 23, 2008, the Third International Phylogenetic Nomenclature meeting was held at Dalhousie University, Halifax, Nova Scotia, Canada.

8.19 FUTURE OF PHYLOCODE

Still there exists some definite criticism of the Phylocode, and therefore, the Code as such is still controversial. Till the late 2007, the "number of supporters for the official adoption of Phylocode is still small", and it is still uncertain when, actually, "the Code will be implemented, and how widely it will be followed". Majority of its supporters believe that the registration database or RegNum should be made popularised as much as possible among the scientific community, as a first major step in this direction. A lot, however, depends on following the recommendations of the Third International Phylogenetic Nomeclature meeting held at Dalhousie University, Canada between July 21, 2008 and July 23, 2008.

Test Your Understanding

1. Instead of common names, why do we need scientific names for all plants? Give three differences between common names and scientific names.
2. What is binomial nomenclature? Who proposed it and when?
3. What is International Code of Botanical Nomenclature? It applies to which organisms? Who can make changes in ICBN?
4. Write any four principles of ICBN.
5. Give an account of resolutions and recommendations of St. Louis Code.
6. Write at least six major recommendations made by Vienna Code.
7. The XVIII meeting of International Botanical Congress is scheduled to be held in July, 2011. Name the host city and country of this meeting.
8. Define: (a) holotype, (b) isotype, (c) lectotype, and (d) neotype.
9. What is the difference between a validly published name and an effectively published name?
10. Give a brief account of phylocode: a new system of nomenclature.

Suggested Reading

Cantino, P.D. and de Queiroz K., 2001, *Phylocode: a Phylogenetic Code of Biological Nomenclature* (draft document): www.ohiou.edu/phylocode/printable.html.

________, Bryant, H.N. and Lee M.S.Y., 1999, *Species names in phylogenetic nomenclature. Syst. Biol.* **48**: 790–807.

Clausen, R.T., 1938, On the citation of 'authorities' for botanical names, *Science* **88**: 299–300.

_______ 1941, On the use of the terms "subspecies" and "variety", *Rhodora* **43**: 157–167.

Fosberg, F.R., 1942, Subspecies and variety, *Rhodora* **44**: 153–157.

Greuter, W. (ed.), 2000, *International Code of Botanical Nomenclature*. Regnum Vegetabile. **138**: Konigstein (Germany), 16th Int. Bot. Cong. (St. Louis Code).

Heller, J.C., 1964, The early history of "binomial nomenclature", *Huntia* **1**: 33–70.

Jeffrey, C., 1973, *Biological Nomenclature,* Edward Arnold Ltd., London.

Johnson, A.T., 1971, *Plant Names Simplified,* W.H. & L. Collingeridge Ltd., London.

Lawrence, G.H.M., 1951, *Taxonomy of Vascular Plants,* Macmillan, New York.

McVaugh, R., R. Ross and F.A. Stafleu, 1968, *An Annotated Glossary of Botanical Nomenclature,* Regnum Vegetabile, Bohn.

Nicolson, D.H., 1974, *Orthography of names and epithets: Latinization of personal names, Taxon* **23**: 549–561.

_______ and R.A. Brooks, 1974, Orthography of names and epithets: Stems and compound words, *Taxon* **23**: 163–177.

Radford, A. E., 1986, *Fundamentals of Plant Systematics,* Harper & Row, New York.

Robinson, P. and T. Kommedahl, 2002. *Phylocode: A New System of Nomenclature, Sci. Editor* **25**(2): 52.

St. John, H., 1958, *Nomenclature of Plants,* Ronald, New York.

Smith, A.C., 1957, Fifty years of botanical nomenclature, *Brittonia.* **9**: 2–8.

Voss, E.G. (ed.), 1983, *International Code of Botanical Nomenclature (ICBN),* Regnum Vegetabile, Vol. 97. Bohn, Scheltema & Holkema, Utrecht.

MODERN TRENDS IN PLANT TAXONOMY

It is true that morphological characters of the plants have provided the foundation and framework for taxonomy. These characters have been used extensively in the preparation of classification systems, diagnostic keys, etc., and are still indispensable to the systematists. But in the light of the modern developments, the complete knowledge of taxonomy is possible only when a synthetic approach, in the light of the principles of various disciplines, is applied. These disciplines (viz. plant anatomy, cytology, palynology, embryology, physiology, ecology, genetics, plant geography, etc.) have played a significant role in plant taxonomy. The old ideas of classification and phylogeny of the plants have changed enormously because of the newer aspects of taxonomy such as biosystematics, cytotaxonomy, chemotaxonomy, molecular taxonomy, etc.

Applicability of the data of some of these branches (anatomy, cytology, palynology, etc.) of botany to the modern taxonomic procedures are discussed briefly in this chapter.

9.1 EXTERNAL MORPHOLOGY IN RELATION TO TAXONOMY

According to Radford (1986) the morphological data of plants are easily observable and obtainable, and are "thus used most frequently in taxonomic studies". The evidence from the external morphology provides the "basic language for plant characterization, identification, classification and relationships" (Radford, 1986).

Angiosperms are identified by morphological characters. Natural groups are defined mainly with the help of floral characters. The basic similarity in the morphology of flowers, fruits and seeds within different species, genera, families and orders provides a sound base in characterising taxonomic groups. Morphology provides majority of the characters used in the construction of systems of taxonomy. Therefore, morphology is closely related to taxonomy and would continue to reign over it for many more years to come (Ogura, 1964; Holttum, 1968; Cronquist, 1975).

Some of the *vegetative characters* that play a major role in plant taxonomy and in deducing phylogeny include growth habit, phenological characters, underground organs, stem, leaves, petiole and stipules.

The *floral characters* which are used commonly in plant taxonomy include types of inflorescence and flower, perianth structure, floral symmetry, union of floral leaves in each whorl, types of androecium and stamens, gynoecium and carpels, ovules, and also the characters of bracts, bracteoles and pedicels. Types of fruit and seed also provide good diagnostic features useful at various levels of classification.

The specific examples showing such a relationship are mentioned below:

1. *Growth habit* (herbs, shrubs, and trees) may be variable or constant within a genus or a family, e.g. all Brassicaceae members are herbaceous. Members of Asteraceae are both herbaceous as well as woody.
2. *Growth patterns* have been used in defining taxonomic groups above the generic level in some families, e.g Zingiberaceae.
3. *Seedling characters,* such as germination, cotyledonary characters, leaf dimorphism, etc. have been of systematic value in Convolvulaceae (Sampathkumar, 1982), Brassicaceae (Gomez–Campo and Maria, 1974), etc.
4. *Underground parts,* such as roots, tubers, etc. are of some taxonomic value in plants. Tubers are helpful in the taxonomy of *Dioscorea* and Cyperaceae, whereas in *Ranunculus* and *Aristolochia* the species are delimited on the basis of the shape of the root.
5. *Leaf characters,* such as arrangement, type, form, duration and venation are widely used in both the classification and identification. In *Ulmus* and *Betula*, the species are delimited only on the basis of leaf characters. In *Trifolium,* the species are separated on the basis of stipule morphology. *Dalbergia* species are distinguished on the basis of their leaflet size, shape and arrangement on the rachis. Various venation patterns have been described in the leaves of *Glossopteris* by Pant (1958).
6. *Inflorescence and flower* are the two main organs on which the classification systems and several other principles of taxonomy are based. Species are distinguished on the basis of calyx types in *Anthyllis*, and shape and degree of inflation of bracts in *Calystegia*. Nectaries and floral discs are of great diagnostic value in Brassicaceae. Species are distinguished on the basis of staminodes in *Scrophularia,* anthers in *Eucalyptus,* staminal appendages in *Alyssum* and degree of branching of inflorescence in *Nepetia.*

   Pollen wall ornamentations have yielded very useful taxonomic information.

   A detailed account of the floral characters and their impact on taxonomy has been given by Lawrence (1951) and Davis and Heywood (1963).
7. *Fruit characters* are used in distinguishing different families of Rhoeadales. Genera are delimited in Moraceae on the basis of fruit characters.
8. *Seed characters* have been used widely in the construction of diagnostic keys and in distinguishing species. The generic separation of *Glinus* Linn. and *Mollugo* Linn. is based mainly on seed characters, the former bears appendaged seeds while the latter bears non-appendaged seeds. Features of seeds such as colour, shape and sculpturing prove helpful in distinguishing species in *Drymaria* of Caryophyllaceae. The number and shape of the seeds is the only reliable distinction between *Anthericum* and *Chlorophytum* of Liliaceae. Two genera of Cruciferae (*Sisymbrella* and *Nasturtium*) are distinguished on the basis of the

presence or absence of mucilage on testa of seeds. Several species and genera of Acanthaceae, Asclepiadaceae, Convolvulaceae and Malvaceae are distinguished on the basis of their seed characters.

9.2 VEGETATIVE ANATOMY IN RELATION TO TAXONOMY

In determining relationship between different genera, families, orders and other taxonomic categories, the anatomical characters are most useful. Anatomical data have also solved several phylogenetic problems. Metcalfe (1968) identified several herbarium specimens using vegetative anatomy.

According to Radford (1986) some of the basic evidentiary anatomical characters of well-established taxonomic value are the type, size, shape, wall sculpture and pattern of wood cells, stelar patterns, types of vascular bundles, rays, ground tissue and parenchyma, epidermal and mesophyll tissue, stomata, trichomes, sclereids, nodes, phloem cells, etc. A few of them are discussed below.

9.2.1 Epidermis

Shape, wall thickness, wall sculpturing and inclusions in the epidermal cells are some of the epidermal characters of taxonomic importance in different families and genera. Occlusion of stomata by an unidentified substance in Winteraceae, papillate epidermal cells in Graminae, presence and distribution of silica bodies in Cyperaceae, sclerification of the wall of the epidermal cells in some genera of Compositae, and presence of very narrow epidermal cells in Stylidiaceae are some of the characters of useful taxonomic importance.

Sharma and Shiam (1984) described the taxonomic importance of silica bodies in 22 Indian species of *Cyperus.* They also reported the presence of cuticular papillae as a character of taxonomic significance in *Cyperus pilosus*, and formation of crater-like pores in *C. flabelliformis* (Sharma and Shiam, 1981 a, b).

9.2.2 Stomata

Studies on the morphology and development of different stomatal types are expected to provide clues to various evolutionary trends among families of angiosperms. They may also prove to be helpful in assigning taxa of uncertain affinities to proper positions (Paliwal and Anand, 1978). Some of the stomatal characters of taxonomic significance are their morphology and ontogeny, number and arrangement of subsidiary cells, and their relationship with other epidermal cells.

About 31 different patterns of stomata and subsidiary cells have so far been reported in vascular plants by different workers, including Pant (1965), Stace (1965) and Van Cotthem (1970). These stomatal patterns appear "to be most valuable at higher taxonomic levels" (Jones and Luchsinger, 1987), e.g. these patterns have been used to characterise the subclasses of monocotyledons.

Significance of stomata has been confirmed in the taxonomy of Gramineae (Prat, 1932), Epacridaceae (Watson, 1962), Combretaceae (Stace, 1965), Rubiaceae (Pant and Mehra, 1965), Acanthaceae (Paliwal, 1966), Umbelliferae (Guyot, 1971), Papilionaceae (Kothari and Shah, 1975), Araliaceae and Umbelliferae (Kannabiran and Krishnamurthy, 1979), Myrtaceae (Vauwyk *et al.* 1982), Cyperaceae (Sharma and Shiam, 1984), and several other angiospermic families.

9.2.3 Trichomes

Epidermal hairs or trichomes are of much taxonomic significance because they exhibit great diversity of form, size, structure and function. Certain species of *Vernonia* are differentiated on the basis of their trichomes. They are also of great value in analysing the suspected hybrids of Compositae and several other families.

Trichome types and their distribution are useful characters in distinguishing various genera of Fabaceae and Icacinaceae. Position of *Nyctanthes* in Oleaceae has been confirmed by Inamdar (1967) on the basis of the structure and ontogeny of trichomes. Presence of sessile glandular hairs in both Typhaceae and Sparganiaceae (Solereder and Meyr, 1933) confirms their close affinity with each other. Cruciferae has been divided into tribes and genera mainly on the basis of types of trichomes (Schulz, 1936). Species of *Digitalis* have been divided into two groups mainly on the basis of presence or absence of glandular hairs. Trichomes in *Parthenium argentatum* (Compositae) are T-shaped while in *P. incanum* they are whiplike with a long thread. Species in various genera of Labiatae are also separated on the basis of presence or absence of glandular hairs. Trichomes have been the main basis of the formation of generic key for the Indian members of Compositae (Ramayya, 1969).

9.2.4 Stem Anatomy

Anatomy of vegetative structures has been of importance in separating higher categories, such as gymnosperms from angiosperms, and monocots from dicots, and not commonly so at lower levels. But, on the basis of stem anatomy, Oleaceous genus *Nyctanthes* is considered to have a closer affinity to Verbenaceae (Stant, 1952). In the subgenus *Genuini* of *Juncus,* Stace (1970) has shown that anatomy of the stem can be used to distinguish the majority of its British species. *Dioscorea* species are also distinguished on the basis of stem anatomy.

Transformation of cortex into transfusion tissue in *Casuarina,* structure of stem endodermis in families such as Asteraceae and Piperaceae, presence of bicollateral vascular bundles in two alternate rings in Cucurbitaceae, and occurrence of cortical and medullary bundles in some families such as Amaranthaceae, Chenopodiaceae and Nyctaginaceae, are some of the features of taxonomic importance. For a detailed study, readers may refer to the profusely illustrated two volumes of Metcalfe and Chalk (1950).

9.2.5 Nodal Anatomy

According to Dickison (1975) correlations of nodal anatomy with some other features might help significantly in tracing the phylogeny of angiosperms. Philipson and Philipson (1968) also attempted to use nodal anatomy as an aid to taxonomy.

Sinnott (1914) considered the trilacunar node as primitive, and unilacunar and multilacunar nodes as advanced. In all members of Centrospermales, the node is unilacunar. However, there may be variations in the nodal anatomy even within a single plant. Subfamily Icacinoideae of family Icacinaceae may be divided into two sections, one with unilacunar nodes and the other characterized by trilacunar nodes.

According to Paliwal and Anand (1978) majority of dicotyledons possess trilacunar nodes. Unilacunar nodes are found in Laurales, Caryophyllales, Ericales, Ebenales, Primulales, Myrtales and a few Tubiflorae and Asteridae members. According to them, multilacunar nodes are found in Magnoliales, Piperales, Trochodendrales, Umbellales and Asterales.

9.2.6 Petiole Anatomy

Metcalfe and Chalk (1950) and Howard (1963) have suggested that the petiole anatomy might also be of taxonomic significance. According to Howard (1963) families, genera and even species in some cases may be identified by petiole characters, such as its position on stem, presence or absence of stipules, its vascularization, nodal structure, number of traces, etc.

Petiole anatomy of 64 species of *Baphia* of Leguminosae (Soladoye, 1982) and some species of *Phlomis* and *Eremostachys* of Labiatae (Azizian and Cutler, 1982) provide clear support of its use in the taxonomy of these genera. Petiolar vascularization has also been helpful in the classification of *Rhododendron* and some other genera.

9.2.7 Leaf Anatomy

Leaf anatomy provides various characters of taxonomic importance. Leaf anatomical studies associated with C_{4-} photosynthetic pathway have resulted (Brown, 1975) in a revised classification of several genera of the grass family. A prominent chlorenchymatous bundle sheath is present in C_{4-} plants. Characters of taxonomic significance in leaf anatomy include nature and thickness of epidermis, mesophyll types, pattern of sclerenchyma, venation patterns, crystals, etc.

Koyama (1967) and Govindrajalu (1966, 1968, 1974, 1975 and 1976) studied the leaf anatomy of several species of Cyperaceae and formulated keys to identify various species of *Cyperus, Fuirena,* etc.

Leaf anatomy has been used widely in several taxonomically different groups such as Euphorbiaceae, Cyperaceae and Gramineae of Angiosperms and Coniferae of Gymnosperms. Benson (1962) in *Ranunculus,* Webster (1967) in *Euphorbia* and *Chamaesyce,* Metcalfe (1968) in several genera of Cyperaceae, and Vidakovie (1957) in *Pinus* have used several characters of leaf anatomy in differentiating species.

Patterns of the distribution of sclerenchyma in *Carex* and *Festuca* have been used in distinguishing species. Sclerenchyma is also used in differentiating two genera of Velloziaceae *viz. Vellozia* and *Barbacenia.* Taxonomic implication of leaf anatomy of several genera of Musaceae, Zingiberaceae, Xanthorrhoeaceae and Ericaceae has also been established by several workers.

9.2.8 Sclereids

Sclereids are the cells with very thick lignified walls. They have been used as the diagnostic tools in several taxa. They are extremely rare in monocots, except in certain genera of Araceae, Agavaceae, Arecaceae and a few other families. In dicots, they are more common in woody forms than in herbaceous ones. Rao and Das (1981) have shown their taxonomic value in about 30 species of *Limonium.*

The systematic value of sclereids has been specifically ascertained in Limoniaceae, Nymphaeaceae, Theaceae, Oleaceae, Connaraceae, and a few genera of Araceae, Acanthaceae, Ericaceae and

Melastomaceae. The forms of sclereids may be characteristic of a species or genus, and may also be of taxonomic value.

9.2.9 Specialised Cells and Cell Contents

Specialised cells of peculiar morphology and function are also important diagnostic tools in many taxa. Motor cells are characteristic of grass leaf epidermis while the bulliform cells occur in the epidermis of many monocotyledonous families. Silica cells are common in sedges while the presence of latex-containing cells is the diagnostic feature of succulent plants.

Microscopic characters of cell contents such as starch grains (e.g. *Solanum tuberosum*), protein bodies (some Cactaceae), albuminoids (e.g. *Laportea*), large silica bodies (Musaceae, Arecaceae and Zingiberaceae), calcium oxalate crystals (e.g. *Eichhornia, Allium*), cystoliths (e.g. Moraceae and Urticaceae), and tanniniferous cells (e.g. Xyridaceae), etc. are important diagnostic tools, and at times prove extremely helpful in delineating species, genera and families.

According to Nair *et al.* (1977) the epidermal cells of the male plants of *Myristicta fragrans* contain simple crystals while those of the female plants contain compound crystals or druses.

9.2.10 Wood Anatomy

The characters of secondary wood are the most important anatomical features that have been used in resolving problems of taxonomy and phylogeny. "The evolutionary series of vessel elements has been used in combination with other morphological features to develop hypotheses about phylogeny of angiosperms" (Jones and Luchsinger, 1987). Bailey (1957), Swamy (1958), Stern (1978), Gregory (1980), Purkayastha (1980) and Fahn (1987) have clearly established the role of wood anatomy in taxonomy.

The secondary xylem of different species varies in the size, shape, arrangement, and relative amounts of its constituents. These variations have been the major source of development of keys by which plant species can be identified. Botanical species have also been identified from a piece of timber by observing its anatomical features. Such an identification is of commercial and industrial importance. Clarke (1938), Phillips (1948) and Fahn (1987) have prepared keys for differentiating several genera and species of gymnosperms and angiosperms by using characters of wood anatomy. Keys have also been prepared for separating species of *Frankenia, Phyllyrea* and *Marrubium* by using wood anatomy characters.

Wood anatomy has prompted the allocation of *Amborella, Tetracentron* and *Trochodendron* to their respective independent families. It has also played its role in the institution of a new taxon, the Degeneriaceae (Baily and Smith, 1942); retention of *Sarcandra* in Chloranthaceae (Swamy, 1953), and transfer of *Idenburgia* from Monimiaceae and its inclusion in *Nouhuysia.*

Wood anatomy has been used at almost all taxonomic levels. It has helped in deciding the systematic position of primitive vesselless families such as Amborellaceae, Tetracentraceae, Trochodendraceae and Winteraceae, all of Magnoliales of Angiosperms. Phylogenists have suggested that because of the presence of specialised wood, Amentiferae cannot be considered primitive. Amentiferae includes families, such as Fagaceae, Juglandaceae and Betulaceae.

Wood anatomy has also provided evidence for assigning a definite position to taxa of uncertain affinity. Placement of *Myristicta* close to Lauraceae, delimitation of different subgenera of *Quercus* of Fagaceae, and non-inclusion of Calycanthaceae in Rosales or Myrtales are all supported by the wood anatomy.

The type and arrangement of vascular bundles, type of secondary growth, and different characters of xylem and phloem elements have been used in solving phylogenetic problems and delineating different taxa.

Some important wood elements of taxonomic importance are undermentioned:

1. *Vessel elements* with pittings on their lateral walls, vessel abundance, presence of solitary or aggregate vessel groupings, sculpturing on vessel walls, and distribution of vessels are characters of systematic value.
2. *Vascular rays,* their width, dimensions, abundance and degree of their wall thickness are useful taxonomic criteria.
3. *Axial parenchyma,* its distribution, arrangement, length and width of its cells, thickness and lignification of its wall, and its absence are all characters of taxonomic significance.
4. Presence or absence of *storied* wood is an important taxonomic character.
5. Presence or absence of *latex vessels*, *resins*, *gums*, *crystals,* etc. in the wood are also the characters of taxonomic importance.

9.3 FLORAL ANATOMY IN RELATION TO TAXONOMY

The importance of floral anatomy in relation to taxonomy has been described by several workers, including Eames (1953), Puri (1952, 1958, 1962), and Murty and Puri (1980). In majority of these studies, main emphasis is laid on the path and distribution of vascular bundles within the receptacle and floral parts.

Puri (1954) has shown strong affinities between Passifloraceae and Moringaceae on the basis of the orientation of placental strands with reference to floral axis. On the basis of floral anatomy, Kale and Pai (1979) have supported the removal of genus *Trichopus* from family Dioscoreaceae and creation of a new family Trichopodiaceae, with *Trichopus* as its sole representative.

Floral anatomical studies have been useful in solving some fundamental questions like the nature of flower, carpel, inferior ovary, and also several problems related with homologies, phylogeny and taxonomy. Using clearing techniques, Melville (1960) investigated the floral vasculature of large number of angiosperms and proposed the *gonophyll theory* of the origin of angiosperm flowers.

According to Puri (1958) the structure of the gynoecium is essentially alike in Capparidaceae, Cruciferae, Moringaceae and Papaveraceae. The placentation in these families is parietal, but the placental strands are inverted and occur on the inner side of the secondary marginal bundles possessing normal orientation. This study has given the support to the view that parietal placentation has been derived from axile placentation, and Moringaceae should be included in order Rhoeadales. On the basis of floral anatomical studies, Roy (1949) proposed the removal of *Trapa* from Onagraceae and its inclusion in a separate family Trapaceae. Eyde (1975) interpreted the vestigial organs and modified parts of flowers by studying the floral anatomy of several taxa.

Contribution of floral anatomy in resolving the taxonomic position of some disputed taxa is discussed below:

1. Floral anatomy of Annonaceae, Calycanthaceae and Menispermaceae confirms that all these families originated from Ranunculaceae.
2. Floral anatomy of Polemoniaceae and Caryophyllaceae suggest that the former have been derived from a caryophyllaceous stock.
3. Uniformity in floral vasculature of Solanaceae and Scrophulariaceae suggests that both should be included in one single order, Scrophulariales.
4. Formerly, Cyperaceae and Gramineae were treated together in one single order. But Hutchinson (1973) treated them separately in Cyperales and Graminales. Floral anatomical studies of both the families support their separation as suggested by Hutchinson.
5. Separation of *Paeonia* from Ranunculaceae and its inclusion under a separate family Paeoniaceae was supported by the floral anatomical studies.
6. Transfer of *Hydrocotyle asiatica* L. to the genus *Centella* in the form of *Centella asiatica* L. was confirmed by studies of floral anatomy.
7. Systematists have suggested that there is no close affinity between *Cyrtandromoea* and other members of Gesneriaceae, and it should be transferred to Scrophulariaceae. Floral anatomical studies also show a close relationship between *Cyrtandromoea* and members of Scrophulariaceae because of the presence of several lateral traces in carpels, a bilocular ovary, and absence of a disc in both.
8. *Lilaea*, earlier included under family Scheuchzeriaceae, was later separated in an independent family Lilaeaceae by systematists. Floral anatomy supported the removal of *Lilaea* from Scheuchzeriaceae because both differ in their vascular supply of flower and number of ovules.

9.4 CYTOLOGY IN RELATION TO TAXONOMY, OR CYTOTAXONOMY

Study of cell is referred as *cytology*. But according to Solbrig (1968) and Stebbins (1971), only the details of the *chromosomes* have been used in resolving many taxonomic problems. Utilization of the characters and phenomena of cytology for the explanation of taxonomic problems is referred as *cytotaxonomy*.

The term *karyotype* is used for the phenotypic appearance of the somatic chromosomes. The diagrammatic representation of the karyotype is termed as *idiogram*.

The characteristics of the chromosomes, which have proved to be of taxonomic value, include (1) chromosome number, (2) chromosome size, (3) chromosome morphology, and (4) chromosome behaviour during meiosis.

9.4.1 Chromosome Number

Usually, all individuals within a species possess the same chromosome number. In angiosperms, the haploid chromosome number varies between n = 2 (*Haplopappus gracilis* of Compositae) and n = 132 (*Poa littorosa* of Gramineae). However, the majority of them show a range between n = 7 and n = 12. The highest chromosome number recorded for vascular plants is found in a pteridophyte, *Ophioglossum reticulatum* (2n = 1260).

Variation or constancy in the chromosome number, within taxa of different categories, prove to be important characters for taxonomic groupings.

The chromosome number in some plants remains constant in all species, e.g. all species of *Pinus* and *Quercus* possess n = 12 chromosomes. Such numbers are called *constant chromosome numbers.* The species bearing constant chromosome numbers are called *homoploids.* These numbers are, of course, of limited importance, but prove useful in knowing a particular genus. In several genera of vascular plants polyploid series are present. *Polyploids* are the plants which possess higher chromosome numbers because of the multiplication of genomes or chromosome sets, e.g. different species of *Aster* have n = 9 or n = 18 or n = 27, etc. Such a series of polyploidy, in which the chromosome numbers of a taxon are in the proportion of its exact multiples, is called *euploidy.* On the other hand, if the chromosome numbers of a group bear no simple numerical relationships with each other, the series is called *aneuploidy,* e.g. different species of *Brassica* bear n = 6, 7, 8, 9, or 10.

9.4.2 Chromosome Size

The chromosome size has been very useful in understanding relationships in several taxa. In most plants, the length of a chromosome varies from 0.5 to 30 μ. Among monocots, the members of Zingiberaceae possess small chromosomes, of Iridaceae small to medium-sized, of Amaryllidaceae large-sized, while those of Liliaceae possess chromosomes of varying sizes. Except *Eleocharis* and *Fimbristylis,* all the members of Cyperaceae possess extremely small chromosomes. According to Stebbins (1938) the chromosome size is characteristic of only certain groups and families, and not related to the phylogeny of angiosperms.

9.4.3 Chromosome Morphology

Usually the chromosomes are characterized as *median, submedian, subterminal* or *terminal* with reference to their length and the position of the *centromere.* The centromere location marks the position of the *primary constriction.* Additional constrictions are called *secondary constrictions.* Occasionally, a secondary constriction may be present near the terminal end of a chromosome, separating its small segment called *satellite.* The chromosomes may be *symmetrical* or *asymmetrical.* Symmetrical ones possess two equal arms and a median centromere. Asymmetrical ones possess unequal arms and subterminal centromeres.

Relative length of the arms of the chromosomes, position of the centromere, presence of satellites, etc. are some characters of taxonomic significance.

9.4.4 Chromosome Behaviour at Meiosis

Significant taxonomic information may be obtained by the study of the behaviour of the chromosomes during meiosis. Degree of sterility and occurrence of hybridisation are determined by the behaviour of chromosomes during meiosis. Abnormalities in meiosis, such as non-pairing, crossing over, unequal interchanges or translocations, bridge formation, lagging chromosomes, etc. have all proved to be of systematic value.

9.4.5 Systematic Value of Cytological Studies

According to Jackson (1971) chromosome characteristics have been found to be useful at all taxonomic levels. A few of such common characteristics of systematic value are undermentioned:

1. Members of *Cyperaceae* and *Juncaceae* possess chromosomes with diffuse or non-localized centromeres, and also show inverted meiosis. This reflects a close association between these two families.
2. In *Ranunculaceae,* reshuffling of the genera has been done in the light of cytological data. All the genera with n = 7, 8 and 9, as well as genera with long chromosomes and short chromosomes are accommodated in tribes Anemoneae and Helleboreae. In *Thalictrum* and *Anemonella* of Anemoneae, and *Isopyrum* and *Aquilegia* of Helleboreae, the base number is n = 7 and chromosomes are small-sized. Because of such a karyological similarity, all these genera have now been grouped in tribe Thalictreae. In the same way, genera *Coptis* and *Zanthorhiza* of Helleboreae have n = 9 and very small chromosomes and therefore grouped under Coptideae.
3. *Yucca* had long been treated as a member of Liliaceae because of the superior ovary, and *Agave* of Amaryllidaceae because of the inferior ovary. Hutchinson shifted both *Yucca* and *Agave* to *Agavaceae.* The presence of bimodal karyotype with 25 small and 5 large chromosomes in both of them, justified their placement in one single family, as done by Hutchinson.
4. In the subfamily Bambusoideae of Gramineae n = 12, and in the subfamily Poideae n = 7. This indicates that the chromosome numbers have proved to be of taxonomic utility also at the subfamily level. Stebbins (1958) provided information on the evolution of grasses on the basis of cytogenetics.
5. The hexaploid wheat (*Triticum aestivum*) has a genome constitution of AABBDD. In this case the genome A has been contributed by *Triticum monococcum,* genome B by *Aegilops speltoides,* and genome D by *Aegilops squarrosa.* This shows that by such a study of meiosis the parentage of many polyploid taxa may be located.
6. The basic chromosome number in Loranthaceae is n = 9 while in Viscaceae there is a series of aneuploid numbers ranging between 10 and 14. Wiens (1975), therefore, suggested their separation from each other on the basis of the cytological evidence.
7. Delimitation of the tribes in Compositae has been done on the basis of chromosome numbers.
8. On the basis of cytological studies, Lewis (1951, 1953) submerged the genus *Godetia* in *Clarkia* (Onagraceae).
9. Naik (1977) differentiated three species of *Chlorophytum* of Liliaceae on the basis of cytological data. According to him *C. bharuchae* has 2n = 16 chromosomes while *C. glaucum* and *C. glaucoides* have 2n = 42. Both the latter species, having 2n = 42, differ in their karyomorphology.
10. Warburg (1938) studied taxonomy of Geraniales on the basis of cytological studies.
11. Manton (1932) confirmed the formation of subdivisions of Brassicaceae on the basis of cytological studies. All the subfamilies have different base chromosome numbers.

12. Genus *Cistus* (Cistaceae), formerly included in *Helianthemum,* has chromosome number 8 while *Helianthemum* has base chromosome number 9. So, *Cistus* should be recognized as a separate genus.
13. Genera *Physaria* and *Lesquerella* of Brassicaceae were treated as single genus by several taxonomists. Their different cytological details suggest them to be treated as two different genera.
14. A new classification of the genus *Narcissus* of Amaryllidaceae has been proposed by Fernandes (1951) on the basis of cytological studies.
15. Stebbins (1971) distinguished grasses on the basis of size and number of chromosomes. According to him bambusoid grasses have small and many chromosomes, chloridoid grasses have small and few chromosomes, festucoid grasses have large chromosomes, and panicoid grasses have medium-sized chromosomes.
16. According to Mahabale and Cheenaveeraiah (1953) the palm species with n = 16 chromosomes have pinnate leaves while the palm species with n = 18 have palmate leaves. However, there exist exceptions in both the cases.
17. Cheenaveeraiah (1962) suggested on the basis of karyotypic studies of species of *Aegilops* that its section *Sitopsis* should either be shifted from *Aegilops* to *Triticum* or it should be given the rank of a new genus.
18. Sharma (1956), on the basis of his studies of Araceae, Amaryllidaceae and Dioscoreaceae, proposed that the changes in karyotypes of somatic tissue play a distinct role in evolution. He further proposed (Sharma, 1964) that large chromosomes, low chromosome number and symmetrical karyotype represent a primitive status, while small chromosomes, high chromosome number and extreme asymmetry of karyotype represent the advance status. These principles provided interesting results in taxonomy of Alismataceae, Liliaceae, Amaryllidaceae and Dioscoreaceae.

9.5 PALYNOLOGY IN RELATION TO TAXONOMY

9.5.1 Pollen Characters and Related Studies

Study of pollen and spores is called *palynology.* The shape and symmetry of pollen grain, the architecture of its wall, exine stratification, sculpture and structure; and type, number, position, shape and structure of its aperture, are some of the basic characters, which prove useful at all taxonomic levels.

Palynological characters have been used in solving several taxonomic problems, including the repositioning of several disputed taxa, and interpretation of problems relating to the origin and evolution of different groups (Nair, 1980). Cronquist (1981) and several other workers have made the exclusive use of pollen characters in providing classification of angiosperms. Erdtman (1963) used the pollen characters in discussing and solving the taxonomic problems of 105 families. Pollen morphology has proved to be of great value in the classification of Compositae (Stix, 1960), Acanthaceae (Raj, 1961), Euphorbiaceae (Kohler, 1965), Gentianaceae (Nilsson, 1967), Saxifragaceae (Nair, 1970), Asclepiadaceae (Radford, *et al.,* 1974), Caryophyllidae (Cronquist, 1981), etc. Heywood (1967) has gone up to the extent of stating that exine details of pollen are such that they can be used in plant identifications much in the way that fingerprints are used for the identification of criminals.

9.5.2 Monosulcate and Tricolpate Pollen

Jones and Luchsinger (1987) mentioned that angiosperms contain two basic kinds of pollen grains: monosulcate and tricolpate. *Monosulcate pollen grains* are characteristic of primitive dicotyledons, several monocotyledons, pteridosperms, and cycads. Such pollen grains are boat-shaped in outline and possess one long germinal furrow and a germinal aperture. *Tricolpate pollen grains* are characteristic of advanced dicotyledons. Such pollen grains have three germinal apertures and are globose in shape.

9.5.3 Stenopalynous and Eurypalynous

Such a taxon, in which the type of the pollen is constant and characteristic, is called *stenopalynous* or *unipalynous,* e.g. Gramineae is unipalynous, having monoporate aperture and somewhat smooth exine surface. On the other hand, such taxa, in which pollen types vary in size, aperture, exine stratification, etc., are called *eurypalynous* or *multipalynous,* e.g. Euphorbiaceae is multipalynous with grains of various aperture forms.

Cruciferae, Labiatae and Asclepiadaceae are unipalynous families while Acanthaceae, Rubiaceae, Compositae and Verbenaceae are highly multipalynous families.

9.5.4 NPC-System

The classification of pollen is based on the *number-position-character analysis,* called NPC-system. Palynological studies suggest that the taxa with the same general NPC-formula be grouped together, and those showing different NPC, separately. NPC-system helps in providing a three-dimensional classification, and also in the preparation of diagnostic keys below the family level. Cronquist (1981) also emphasized on the importance of the NPC-system in solving several taxonomic problems. For example, the pollen grains of Parietales (as investigated by means of "palynological compass needle") are 3-tremes (N_3), zonotreme (P_4) and colporate (C_5). Therefore, the NPC-formula for Parietales is 345.

9.5.5 Some Other Pollen Characters of Systematic Value

1. *Size and shape* of pollen are important diagnostic characters. The shape may be circular, triangular, etc. in polar view, whereas it may be prolate, perprolate, spheroidal, subpolar, etc. in equatorial view.
2. *Apertures* may be proximal (gymnosperms and monocotyledons), distal (pteridophytes), zonal (dicotyledons) or global (dicotyledons).
3. *Exine sculpturing* may be of depression type (e.g. reticulate, foveolate, striate, fossulate, areolate, scrobiculate, etc.) or excrescence type (e.g. granulose, verrucate, spinulose, tuberculate, spinose, gemmate, clavate, etc.).
4. *Pollen associations,* such as their arrangement in tetrads, presence of pollinia, etc. are characteristic of several angiospermic families.

5. *Number of nuclei* in the pollen grains at the time of their release is a valuable taxonomic and phylogenetic character.

9.5.6 Some Examples of Role of Palynology in Taxonomy

1. Author (Sharma, 1987) investigated the pollen morphology of 16 Indian species of *Cyperus,* and prepared a key to differentiate all of them on the basis of pollen characters. Of the 16 investigated species, the pollen grains of six (*Cyperus iria, C. difformis, C. squarosus, C. triceps, C. flabelliformis* and *C. paniceus* var. *roxburghianus*) are predominantly l-colpate, two (*C. exaltatus* and *C. pumilus*) are predominatly 2-colpate, four (*C. rotundus, C. laevigatus, C. alulatus* and *C. bulbosus*) are predominantly 4 aperturate with 3 colpi and 1 pore, three (*C. compactus, C. kyllingia* and *C. globosus)* are predominantly pantoaperturate, and one (*C. digitatus*) is predominantly 1-porate. Individual species are also differentiated on the basis of pollen characters.
2. On the basis of apertural morphoforms of pollen, Nair (1974) proposed a palynological classification. He divided the entire plant kingdom into (i) *Amorphosporophyta,* including thallophyta with no apertural markings, (ii) *Trimorphosporophyta,* including Archegoniatae with monolete, trilete, or alete forms, and (iii) *Polymorphosporophyta,* including angiosperms with different forms of apertural types.
3. According to Meyer (1975) gymnosperms contain alveolar or granular ektexine and laminated endexine, whereas angiosperms contain columnar or granular ektexine and nonlaminated endexine.
4. A massive exine and thin intine is present in angiosperm pollens. But in certain taxa among monocots (e.g. Musaceae, Amaryllidaceae, Cannaceae, Zingiberaceae, Costaceae and Heliconiaceae), the exine is highly reduced and the intine is well-developed.
5. Several angiospermic taxa have distinctive pollen types. Gramineae have smooth sulcate pollen, Malvaceae and Compositae have typically spinulose exine, and Plumbaginaceae have verrucate exine. Pollens are binucleate in Magnoliidae and trinucleate in Caryophyllidae.
6. The exine pattern is useful in recognising different species of a genus. According to Nair (1974) pollens are pilate in *Bauhinia acuminata,* striate in *B. krugii,* verrucate in *B. retusa,* reticulate in *B. racemosa,* spinulate in *B. malabarica,* and reticulate tuberculate in *B. purpurea.*
7. Phylogenetic relationships are determined using pollen characters in several cases. Genus *Cicer* is usually placed under the tribe Viceae of Papilionaceae. Palynological studies of *Cicer* suggest that it is more close to the tribe Ononideae and should be transferred from Viceae to Ononideae (Clarke and Kupicha, 1976).
8. Pollen grains are associated in tetrads in several families of dicots (Annonaceae, Winteraceae, Mimosaceae, Droseraceae, etc.) and monocots (Cyperaceae and Juncaceae). In Asclepiadaceae and Orchidaceae, they form pollinia.
9. *Elastic threads* join massulae in the pollinia in Orchidaceae (Vijayraghavan and Shukla, 1980). *Viscin threads* of similar nature are found in Onagraceae.

10. According to Kooiman (1972) 2-nucleate pollen grains of Lamiaceae are tricolpate while 3-nucleate pollen grains are 6-colpate.
11. Palynological evidences have supported the separation of Paeoniaceae from Ranunculaceae, Nelumbonaceae from Nymphaeaceae, Fumariaceae from Papaveraceae, and Bombaceae from Malvaceae.
12. Removal of *Podophyllum* from Berberidaceae and its inclusion in a separate family Podophyllaceae is effected on palynological grounds.
13. Pollen characters prove useful in distinguishing the genera *Salix* and *Populus* of Salicaceae, *Phytolacca* and *Rivinia* of Phytolaccaceae, and several genera of Betulaceae, Primulaceae and Acanthaceae.
14. According to Nair (1980), three well-defined evolutionary stocks among angiosperms (i.e. Monocot stock, Magnolian-dicot stock, and Ranalian-dicot stock) form the palynological basis of the triphyletic theory of angiosperms.
15. The genus *Krameria,* variously placed under Leguminosae, Polygalales, etc. was recommended to be placed in a separate monotypic family Krameriaceae by Simpson and Skvarla (1981) on palynological grounds.
16. Palynological studies suggest that Thunbergioideae of Acanthaceae should be given a family status.

9.6 EMBRYOLOGY IN RELATION TO TAXONOMY

Embryology is the study of micro- and megasporogenesis, gametophyte development, fertilization, and development of endosperm, embryo, and seed coats. Embryological evidences have been used in solving the taxonomical problems at almost all levels. These evidences have resolved the doubtful systematic positions of several taxa. However, the role of embryology in solving taxonomic problems was first brought into prominence by a German embryologist, Schnarf in 1931. And in the later years, excellent detailed treatments of the role of embryology were published by Johansen (1950), Maheshwari (1950, 1964), Kapil (1962), Johri (1963, 1967), Davis (1966), Sporne (1969), Philipson (1974), Dickison (1974), Palser (1975), Bhojwani and Bhatnagar (1979), Kapil and Bhatnagar (1980), etc. According to Jones and Luchsinger (1987), the embryological characters have proved to be of significant help "in determining relationships within families, genera and species", and have proved to be less useful "at the rank of order, subclass, or class".

9.6.1 Basic Embryological Evidentiary Characters

According to Maheshwari (1950, 1964), Bhojwani and Bhatnagar (1979) and Radford (1986), some basic embryological characters which have proved to be of special importance in taxonomic considerations include (i) presence and type of anther tapetum; (ii) number and arrangement of anther loculi; (iii) type of anther endothecium; (iv) quadripartition of microspore mother cell; (v) mature pollen grains; (vi) development, structure, position, vasculation, and orientation of ovule; (vii) origin of sporogenous tissue in ovule; (viii) megasporogenesis and development of embryo sac; (ix) presence of aril; (x) form of embryo sac; (xi) fertilization; (xii) type of embryo; (xiii) type of embryogeny; (xiv) endosperm formation; (xv) type of haustorium formation; (xvi) seed-coat; (xvii) cotyledons, etc.

9.6.2 Some Examples of Role of Embryology in Taxonomy

A lot of literature has been published on the role of embryology in solving taxonomic problems during the last 40 years. It is not possible to review it entirely. Some examples are undermentioned:

1. *Dicots and Monocots* Angiosperms are universally divided into dicotyledons and monocotyledons. This primary classification of angiosperms is based on one major embryological character, i.e. number of cotyledons.
2. *Caryophyllales* Embryological characters, such as trinucleate pollen, bitegmic crassinucellate ovules which are campylotropous or amphitropous, seed with peripheral embryo, and perisperm with little or no endosperm, are the characters which are found only in Caryophyllales, more widely known as Centrospermae (Cronquist, 1968).
3. *Helobiae* This monocotyledonous order, treated as a subclass in some recent systems of classification, is characterised by the presence of helobial type of endosperm.
4. *Orchidales* The distinguishing embryological character of the members of this order is the presence of undifferentiated embryo and very little or no endosperm.
5. *Podostemaceae* Members of this family are recognised because of the formation of pseudoembryo sac which is formed by the disintegration of nucellar cells below the embryo sac. The presence of paired pollen grains, tenuinucellate ovules and prominent suspensor haustoria, and absence of antipodals and triple fusion are other characteristics of the family.
6. *Onagraceae* The family is recognized by the presence of *Onagrad*-type of embryo-sac. Such an embryo sac is derived from the micropylar megaspore of the tetrad. The lower three chalazal megaspores do not disintegrate. In this type of embryo sac, antipodals are absent.
7. *Cyperaceae* In flowering plants, four functional microspores develop from each microspore mother cell. But in Cyperaceae, each microspore mother cell gives rise to only one pollen grain. Out of the four nuclei, formed by the meiosis of a microspore mother cell, three are cut off on one side and do not form pollen grains. The remaining fourth nucleus divides mitotically to give rise to a vegetative cell and a generative cell. Mitosis in the generative cell results in the formation of two sperm cells.
8. *Lemnaceae* Phylogenetic studies indicate that Lemnaceae have been derived either from the Helobiales or from the Araceae. But on the basis of the embryological studies, Maheshwari (1954, 1958) suggested that Lemnaceae have been evolved from Araceae and not from Helobiales.
9. *Crassulaceae* Embryological studies of Crassulaceae suggest that it should be placed in the order Rosales close to the family Saxifragaceae.
10. *Loranthoideae and Viscoideae* Members of the subfamily Loranthoideae, with triradiate pollen, *Polygonum*-type embryo sac and composite endosperm show the presence of suspensor and polyembryony. On the other hand, members of the subfamily Viscoideae contain spherical pollen, *Allium*-type embryo sac, non-composite endosperm and show the general absence of suspensor and polyembryony. These differences clearly suggest that both the subfamilies should be elevated to the rank of families, i.e. Loranthaceae and Viscaceae.

11. *Paeonia* It has been treated by most of the taxonomists as a member of the monogeneric tribe Paeonieae of the family Ranunculaceae. But *Paeonia* differs from Ranunculaceae in its chromosome number, vascular anatomy, floral anatomy as well as in embryological details. It shows a unique embryogeny. Its young zygote first becomes coenocytic and then the cell formation starts from its periphery. From some of the peripheral cells develop the embryo initials. *Paeonia* bears arillate seeds and a follicle fruit while in Ranunculaceae the seeds are non-arillate and the fruit is an achene. All these details suggest that *Paeonia* belongs to an independent family Paeoniaceae.
12. *Exocarpus* Because of the presence of a naked ovule and pollen chamber, *Exocarpus* was removed from Santalaceae of angiosperms and was treated as a member of the family Exocarpaceae near Taxaceae in gymnosperms. But the presence of a typical angiospermic flower, *Polygonum-type* of embryo sac, cellular endosperm and some other embryological characters, led Ram (1956) to confirm that *Exocarpus* belongs to the family Santalaceae of angiosperms, and not with gymnosperms.
13. *Trapa* Majority of the taxonomists treat *Trapa* as a genus of Onagraceae while others consider it to belong to the family Hydrocaryaceae. But its embryological details (*Polygonum-type* embryo sac, absence of endosperm, well-developed suspensor haustorium, extremely reduced one cotyledon, etc.) suggest that *Trapa* should be treated under an independent family Trapaceae.
14. *Pentaphragma* It has been treated under the families Campanulaceae, Boraginaceae and Pentaphragmataceae by different workers. But its similarities with Campanulaceae in its embryological details (i.e. ovule, endothecium, embryo sac, endosperm and embryogeny) deserve its final inclusion in this family.
15. *Butomus* The presence of the *Polygonum*-type of embryo sac in *Butomus* and the *Allium*-type of embryo sac in the other genera of the family Butomaceae, suggest that only *Butomus* should be retained in Butomaceae. The other genera of this family should be transferred either to Alismataceae or to Limnocharitaceae.
16. *Parnassia* It is generally treated as a member of the family Saxifragaceae. But the total dissimilarity of its embryological details with the other genera of Saxifragaceae suggests its removal from Saxifragaceae and inclusion in a separate family Parnassiaceae.
17. *Peganum* It has been variously treated as a member of Rutaceae or Zygophyllaceae. But its embryological details resemble those of Linaceae. The recent embryological findings, therefore, suggest its inclusion in an independent family, Peganaceae near Linaceae.

All the abovementioned examples confirm that the embryology plays a definite and significant role in solving taxonomic problems.

9.7 CHEMISTRY IN RELATION TO TAXONOMY[1]

Application of chemistry to taxonomy is called *chemical taxonomy* or *Chemotaxonomy*[1]. Chemical evidences are used in determining the relationship among taxa of different categories. Some of the

[1] For details of chemotaxonomy, see Chapter 11.

major classes of the chemical evidence include flavonoids, alkaloids, amino acids, fatty acids, aromatic compounds, terpenoids, polysaccharides, carotenoids, etc.

Cronquist (1981) cited following examples to indicate the use of chemistry in solving taxonomic problems:

1. Caryophyllales produce betalains and not anthocyanins.
2. Polygonales produce anthocyanins and not betalains.
3. Juglandales are aromatic plants while Fagales are non-aromatic.
4. Highly aromatic compounds are found in Lamiaceae.
5. Alkaloids are very common in Solanaceae.
6. Sapindaceae have plenty of tannins.

9.8 ECOLOGY IN RELATION TO TAXONOMY

Ecological evidences also contribute to our understanding of several taxonomic problems. Some of the basic ecological evidentiary characters include habitat (water, sand, mud, rock, marsh, desert, etc.), abiotic relations (light, moisture, climate, soil, etc.), biotic relations (symbiotic, nutritional, reproductive, etc.), spatial relations (areal, distributional, etc.), and temporal relations (succession, phenology, periodicity, etc.)

Ecological evidences prove helpful in understanding the distribution of taxa, variation within taxa, and adaptations of plants. These evidences have also been used for providing an explanation about the origin of the flowering plants, and also about the development of classification of several taxa, mainly below the level of the genus. Ecological studies also show that many morphological features of plants have a direct correlation with the environmental factors, such as light, humidity, water, etc. Ecologists make a substantial contribution to the systematics by examining ecotype variation, soil specializations, seedling establishment ecology, seed dispersal mechanism, etc.

9.9 PALAEOBOTANY IN RELATION TO TAXONOMY

Information regarding the correct evolutionary history of the present-day plants is gathered from the fossil records or *palaeobotany*. Any theory of evolution of plants is incomplete without any palaeobotanic evidence. Both, microfossils (e.g. pollen) and macrofossils (e.g. stems, leaves, etc.) are used as the source of the systematic data. Palaeobotanic studies prove helpful in determining the evolution of the floras of the past, in the comparative morphology, as well as in determining the ecological conditions of the past.

According to Beck (1976) and Taylor (1981) the palaeobotanic studies are helpful in the phylogeny of angiosperms. Dilcher (1979) is of a firm opinion that the palaeobotany provides important information about several aspects of the early history of angiosperms, as well as their origin and diversification. Number of new concepts have been developed about the primitive flower types by Dilcher (1979) on the basis of his palaeobotanic studies. These studies also suggest that diversified representatives of the modern *Amentiferae* were commonly present in Middle Eocene. Palaeobotanists are also now of a firm opinion that the Middle Eocene flora may provide definite information about the origin of plants in general and the evolution of flower in particular.

9.10 ELECTRON MICROSCOPY IN RELATION TO TAXONOMY

Electron microscopy has brought revolution in all biological fields, and so it has its implications on taxonomy too. Ultrastructural studies, which include scanning electron microscopy (SEM) and transmission electron microscopy (TEM), have provided definite clues to the phylogeny of angiosperms. Pollen grains, trichomes, spores, seeds and plant surface have been investigated in detail using electron microscopy.

The taxonomic value of sieve-element plastids in the classification of flowering plants was demonstrated by Behnke (1972) using transmission electron microscopy. Two classes of plastids were recognized, i.e. s-type and p-type. The s-type plastids accumulate starch while p-type accumulate only proteins or starch and protein both. Valuable taxonomic information is provided by TEM studies in certain taxa. Different types of wall structure in some members of Compositae, dilated cisternae of endoplasmic reticulum in some genera of Cruciferae and Capparidaceae, and presence of crystalloids in the cotyledon mesophyll cells in Cucurbitaceae are some characters of systematic value.

Test Your Understanding

1. Explain in detail the role of external morphology in relation to taxonomy.
2. Give a brief account of some modern trends in plant taxonomy.
3. In determining relationship between different genera, families, orders and other taxa, anatomical characters are most useful. Explain with examples.
4. Describe the importance of floral anatomy and cytology in relation to taxonomy.
5. Explain in brief the role of embryology in relation to taxonomy.

Suggested Reading

Bailey, I.W., 1957, Wood anatomy in the study of phylogeny and classification angiosperms, *J. Arnold Arbor. Harv. Univ.* **38**: 243–54.

Behnke, H. D., 1972, Sieve tube plastids in relation to angiosperm systematics: An attempt towards a classification by ultrastructural analysis, *Bot. Rev.* **38**: 155–97.

_______ 1977, Transmission electron microscopy and systematics of flowering plants, *Pl. Syst. Evol. Suppl.* **1**: 155–78.

Bendz, G. and J. Santesson, 1973, Chemistry in botanical classification, *Proc. 25th Nobel Symp. Acad.*, New York.

Cole, G. T. and H.D. Behnke, 1975, Electron microscopy and plant systematics, *Taxon* **24**: 3–15.

Davis, G.L., 1966, *Systematic Embryology of the Angiosperms*, Wiley, New York.

Davis, P.H. and V.H. Heywood, 1965, *Principles of Angiosperm Taxonomy*, Van Nostrand, Princeton, N.J.

Dickison, W.C., 1975, The bases of angiosperm phylogeny: Vegetative anatomy, *Ann. Mo. Bot. Gard.* **62**: 59–62.

Erdtman, G., 1966, *Pollen Morphology and Plant Taxonomy,* Vol. I. *Angiosperms,* Hafner, New York.

Eyde, R.H., 1975, The bases of angiospenn phylogeny: Floral anatomy, *Ann. Mo. Bot. Gard.* **62**: 521–37.

Gibbs, R.D., 1974, *Chemotaxonomy of Flowering Plants,* 4 Vols. McGill-Queens Univ. Press, Montreal.

Harborne, J.B. and B.L. Turner, 1984, *Plant Chemosystematics,* Academic Press, New York.

Howard, R.A., 1963, The vascular structure of the petiole as a taxonomic character, *Adv. Hort. Sci.* **3**: 7–13.

Jackson, R.C., 1971, The karyotype in systematics, *Ecol. Syst.* **2**: 327–68.

Johri, B.M., 1963, Embryology and taxonomy (In) *Recent Advances in the Embryology of Angiosperms,* Maheshwari, P. (ed.). International Soc. Pl. Morphology, New Delhi.

_______ 1967, Angiosperm's embryology and taxonomy, *Bull. Natn. Inst. Sci. India* **34**: 263–68.

Jones, S.B. Jr. and A.E. Luchsinger, 1987, *Plant Systematics* (2nd ed.), McGraw-Hill, Singapore.

Kapil, R.N., 1962, Some recent examples of the value of embryology in relation to taxonomy, *Bull. Bot. Surv. Ind.* **4**: 57–66.

Kruckeberg, A.R., 1969, The implications of ecology for plant systematics, *Taxon* **18**: 92–120.

Maheshwari, P., 1964, Embryology in relation to taxonomy, (In) W.B. Turrill (ed.) *Vistas in Botany* IV. Macmillan, New York.

Metcalfe, C.R., 1964, An anatomist's view on angiosperm classification, *Kew Bull.* **9**: 427–40.

_______ and L. Chalk, 1950, *Anatomy of the Dicotyledons,* Vols. I. & II, Oxford Univ. Press, Oxford.

Murty, Y.S. and V. Puri, 1980, Floral anatomy in relation to taxonomy, *Glimp. Pl. Res.* **5**: 222–243.

Nair, P.K.K., 1980, *Glimpses in Plant Research* V. *Modern Trends in Plant Taxonomy,* Vikas Publ., New Delhi.

Paliwal, G.S. and S.K. Anand, 1978, Anatomy in relation to taxonomy, *Acta Bot. Indica* **6**: 1–20.

Philipson, W.R., 1974, Ovular morphology and the major classification of dicotyledons, *Bot. J. Linn. Soc.* **68**: 89–108.

Puri, V., 1952, Floral anatomy in relation to taxonomy, *Agra Univ. J. Res.* (*Sci*) **1**: 15–35.

_______ 1958, Floral anatomy and taxonomy, *Ind. Bot. Ser. Mem.* **1**: 15–18.

_______ 1962, Floral anatomy in relation to taxonomy, *Bull. Bot. Surv. India.* **4**: 161–65.

Radford, A.E., 1986, *Fundamentals of Plant Systematics,* Harper and Row, New York.

Sharma, A.K., 1954, Cytology as an aid in taxonomy, *Bull. Bot. Soc. Bengal* **18**: 1–4.

Sharma, O.P., 1987, Pollen morphology of *Cyperus, Bangladesh J. Bot.* **16**(2): 141–50.

_______ and R. Shiam, 1981a, Occurrence of cuticular papillae in *Cyperus, Curr. Sci.* **50**(5): 236.

_______ and R. Shiam, 1981b, Pore formation in Cyperaceae: A new report, *Curr. Sci.* **50**(2): 452–53.

_______ 1984, Epidermal structures of culm in *Cyperus* with a discussion of silica bodies in Cyperaceae, *Bangladesh J. Bot.* **13**(1): 16–24.

Smith, P.M., 1976, *The Chemotaxonomy of Plants,* Edward Arnold, London.

Stace, C.A., 1965, Cuticular studies as an aid to plant taxonomy, *Bull. Br. Mus.* (*Nat. Hist.*) *Bot.* **4**: 1–78.

NUMERICAL TAXONOMY

10.1 WHAT IS NUMERICAL TAXONOMY?

The analysis of various types of taxonomic data by mathematical or computerized methods is called *numerical taxonomy* or *taximetrics*. This approach of systematics involves the numerical evaluation of the similarities or affinities between taxonomic units, and then arrangement of these units into taxa on the basis of their affinities.

Adanson (1763), a French botanist, was the first to put forward a plan for assigning numerical values to the similarity between organisms. He tried to use as many characters as possible for the classification, and such classifications were recognised as *Adansonian classifications*. Adansonian principles have developed several new methods in taxonomy during recent years, particularly after 1960, and these methods are all included under numerical taxonomy.

The use of modern electronic data-processing techniques and several other methods of the numerical taxonomy have helped in the evolution of several new classifications of plants during the past 40 years. Several objectives and qualitative methods are employed in their evolution.

According to Heywood (1967) the numerical taxonomy may be defined as "the numerical evaluation of the similarity between groups of organisms and the ordering of these groups into higher ranking taxa on the basis of these similarities."

10.2 PRINCIPLES OF NUMERICAL TAXONOMY

The principles on which Numerical Taxonomy is based are also called *neo-Adansonian principles*. Sneath and Sokal (1973) have enumerated following seven principles of numerical taxonomy:

1. The greater the content of information in the taxa of a system of classification and more the characters on which it is based, the better will be the classification system.
2. Each character has equal weightage in creating a new taxa.
3. The overall similarity between any two entities is a function of "their individual similarities in each of the many characters in which they are being compared" (Sneath and Sokal, 1973).

4. Many different taxa can be recognized because correlations of characters differ in the groups of organisms which are studied.
5. By assuming some evolutionary mechanisms and pathways, the phylogenetic conclusions may be drawn from the taxonomic structure of a group and from character correlations.
6. Under numerical taxomony, the science of taxonomy is viewed and practiced as an empirical science.
7. Numerical taxonomy suggests that phenetic similarity is the base of classifications.

10.3 LOGICAL STEPS OF NUMERICAL TAXONOMY

Following are some of the successive steps used in the construction of taxonomic groups:

10.3.1 Operational Taxonomic Units (OTUs)

The lowest ranking taxa or the basic units in any specific study are called *operational taxonomic units* or *OTUs.* An OTU may be an individual, a species, a genus, or a taxon of other higher rank. The rank of OTUs may, therefore, differ from study to study. Usually, the species is treated as an OTU under numerical taxomony.

10.3.2 Unit Taxonomic Characters or Attributes

A *unit character* is a taxonomic character of two or more states, which, within the study at hand, cannot be subdivided logically, except for the subdivision brought about by the changes in the method of coding as mentioned by Sneath and Sokal (1973). Usually the phenotypic characters are used as unit characters, e.g. the presence or absence of an awn in a spikelet of Poaceae.

Maximum number of unit characters, and certainly not less than 50 in any case, should be used to obtain a fairly stable and reliable classification. Too few characters impart unreliability. A new item of information should be contributed by each unit character. The characters which are disqualified or not useful in numerical taxonomy are called *inadmissible characters.*

Proper selection of the unit characters should be based on the following: (i) they should come from all the parts of the organisms, (ii) they should belong to all the stages of the life cycle of the organism under study, (iii) they should contain all variable characters within the group; and (iv) due weightage should be given to characters related to distribution of the organisms, their ecology, morphology and physiology.

10.3.3 Types of Basic Characters or Attributes

Following are the major types:

1. *Binary Characters* These characters contain two contrasting states, such as the presence or absence of some features, e.g. presence or absence of stipules.
2. *Multistate Characters* These may be of following two types:
   (a) *Qualitative multistate characters* These characters contain three or more contrasting forms, and each form is ranked on equal footing, e.g. flower colour blue, orange or red.

(b) *Quantitative multistate characters* These characters represent measures of the size on a continuous scale such as length, breadth, height, or weight, etc., e.g. length of a fungal spore, or amount of mucilage produced by an alga.

10.3.4 Coding of Characters

The selected unit characters are assigned a symbol or mark, e.g. 1, 2, 3 …, or + and –, or 1 +, 2 +, 3 +, etc. This is called coding of characters. The symbol NC is used for "no comparison". Coding of characters may be of following types:

1. *Two-state Coding* The characters in this type of coding are divided as 1 and 0, or as + and –. The characters recorded as 1 or + are the positive characters while those recorded as 0 or – are negative. The word NC is used if an organism is not possessing a given character present in other organisms.
2. *Multistate Coding* Each quantitative multistate character may be expressed by a single numerical value, such as 1, 2, 3, 4 … corresponding to the range of the variation of the character. On the other hand, the qualitative multistate characters cannot be arranged in a reliable sequence or definite order. Therefore, the qualitative characters are converted conveniently into new characters, and coded as some alphabetic symbols depending on the computer programme.

10.3.5 Estimation of Resemblance

Between any two operational taxonomic units, the resemblance is estimated or measured in terms of similarity, i.e. percentage of characters in which they agree, or in terms of dissimilarity, i.e. percentage of characters in which they do not agree. In numerical taxonomy, the dissimilarity is specially useful in the construction of taxonomic models or maps because it is the taxonomic distance between the positions of the operational taxonomic units in the phenetic appearance.

Phenetic resemblance between taxonomic groups is usually estimated by three methods, viz. (a) coefficients of association, (b) coefficients of correlation, and (c) measurement of taxonomic distance between OTUs.

Coefficients of association are measured by the following formula:

$$S = \frac{Ns}{Ns + Nd}$$

where S is the numerical index; Ns = the number of positive features shared by any two OTUs; and Nd = the number of positive features in one OTU and number of negative features in the other OTU.

Coefficients of correlation (r) are measured by the following formula:

$$rjk = \frac{\sum^{n}[(Xij - \bar{X}i)(Xik - \bar{X}k)]}{\sqrt{\left\{\left[\sum^{n}(Xij - \bar{X}j)^2\right]\left[\sum^{n}(Xik - Xk)^2\right]\right\}}}$$

where j and k stand for two units under comparison; Xij stands for the value of the character i in unit j; Xik stands for the value of the character i in unit k; $\overline{Xj}$ *and* $\overline{Xk}$ stand for the mean for all the characters in units j and k; n stands for number of characters.

Taxonomic distance (d) is measured by the following formula:

$$djk = \sqrt{\left\{\sum^{n}(Xij - Xik)\frac{2}{n}\right\}}$$

where Xij is the character state of unit j for the character i; Xik is the character state of unit k; symbol $\sum^{n}$ stands for the sum over n characters; the value of the taxonomic distance (d) is the distance in a phenetic space divided by $\sqrt{n}$.

10.3.6 Cluster Analysis

Clusters are the groups of OTUs. With the help of estimation of resemblances, the affinities of different OTUs are determined. OTUs of similar affinities are grouped together in different taxa. A taxonomic system is constructed on the basis of resemblances and differences between groups of OTUs or clusters. The main features of the cluster analysis are arranged in the form of a dendrogram.

10.3.7 Phenons and Ranks

Groups of similar organisms recognised by numerical methods are called *phenons.* Phenons are equivalent to various taxonomic groups. However. the term *phenon* is not a synonym of the term *taxon.* A phenon may or may not be equivalent to the ranks of the classical taxonomy, such as species, genus, tribe, family, etc.

Nair (1984) has stated that "the delimitation of phenons is done by drawing a horizontal line (Fig. 10.1) across the dendrogram at a similarity value. A line at 75%, for example, creates five 75-phenons 1; 7; 3, 5, 6; 4, 9, 10; and 2, 8; while that at 80% creates six phenons".

10.4 ADVANTAGES OF NUMERICAL TAXONOMY

Some of the major advantages of numerical taxonomy are undermentioned:

1. If compared with the conventional taxonomy, the data in numerical taxonomy are collected from more variety of sources, such as morphology, physiology, chemistry, amino acid sequences, proteins, genetical and cytological studies.
2. Much taxonomic work can be done by less highly skilled workers by using numerical taxonomy.
3. Numerically coded data may be used easily for the creation of keys, maps, descriptions, catalogues, etc. with the help of existing electronic data processing systems in herbaria and other taxonomic institutions such as botanical gardens and libraries.
4. Since numerical methods are more sensitive in delimiting taxa they provide better keys, maps and classification systems in comparison to the conventional taxonomic methods.
5. The conventional taxonomy has improved by using numerical taxonomy because more and better-described characters are used in the latter.

6. Numerical taxonomy has provided several fundamental improvements in the conventional principles of taxonomy in general and classification systems in particular.
7. In the light of numerical taxonomy several existing biological concepts have also been reinterpreted.

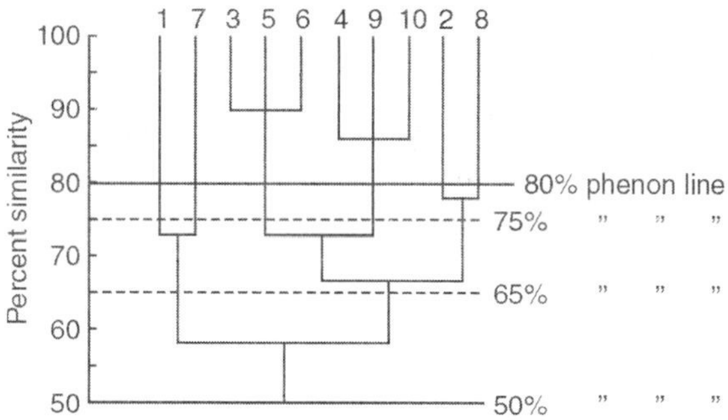

Fig. 10.1 Dendrogram to show formation of phenons. (From Naik, V.N. 1984, *Taxonomy of Angiosperms.* Used with permission of Tata McGraw-Hill, New Delhi).

10.5 APPLICATIONS OF NUMERICAL TAXONOMY

Undermentioned are a few examples of the applications of numerical taxonomy:

1. Numerical methods are used considerably to study the similarities and dissimilarities in bacteria, and other microorganisms.
2. Numerical taxonomy is used for delimitation of several angiospermic genera, including *Solanum* (Soria and Heiser, 1961), *Oryza* (Morishima and Oka, 1960) and *Onosis* (Cook, 1969).
3. Several angiospermic taxa are reclassified using numerical methods.
4. Sneath and Sokal (1973) have mentioned several examples of the application of numerical taxonomy in several angiospermic genera, including *Apocynum, Crotalaria, Cucurbita, Chenopodium, Oenothera, Salix, Zinnia*, barley cultivars, maize cultivars, wheat cultivars, etc.

Test Your Understanding

1. What is numerical taxonomy? Is it different from taximetrics?
2. Explain in brief the successive steps/logical steps of numerical taxonomy.
3. What do you mean by basic characters in numerical taxonomy? What are their main types? Explain the meaning of "coding of characters".
4. In relation to numerical taxonomy, write down the formulae of measuring (a) coefficients of association, (b) coefficicent of correlation, and (c) taxonomic distance.
5. Write in brief some applications and advantages of numerical taxonomy.

Suggested Reading

Cole, A.J., 1969, *Numerical Taxonomy,* Academic Press, London.

Cullen, J., 1968, Botanical problems of numerical taxonomy, (*In*) V.H. Heywood (ed.) *Modern Methods in Plant Taxonomy,* Academic Press, London, pp. 175–184.

Rohlf, F.J. and R.R. Sokal, 1965, Coefficients of correlations and distance in numerical taxonomy, *Univ. Kansas Sci. Bull.* **45**: 3–27.

Sneath, P.H.A. and R.R. Sokal, 1962, Numerical taxonomy, *Nature.* **193**: 855–860

________ and ________ 1973, *Numerical Taxonomy,* W.H. Freeman and Company, San Francisco.

Sokal, R.R. and P.H.A. Sneath, 1963, *Principles of Numerical Taxonomy,* W.H. Freeman and Company, San Francisco and London.

CHAPTER 11

CHEMOTAXONOMY

11.1 WHAT IS CHEMOTAXONOMY?

The approach of taxonomy in which chemical features of plants are used in developing classifications or in solving taxonomic problems is called *Chemotaxonomy, Chemosystematics, Chemical Taxonomy, Chemical Plant Taxonomy,* or *Plant Chemotaxonomy.* Some of its basic features or evidentiary characters include alkaloids, flavonoids, carotenoids, polysaccharides, terpenoids, fatty acids, amino acids, aromatic compounds, etc.

The chemical evidences are useful in establishing relationships among taxa is exemplified (Cronquist, 1981) by the (i) presence of aromatic plants in Juglandales, (ii) production of betalains and not anthocyanins by the members of Caryophyllales, (iii) presence of highly aromatic compounds in the members of Lamiaceae, (iv) presence of alkaloids in Solanaceae, and (v) presence of tanniferous plants in Sapindaceae.

11.2 PURPOSE OF CHEMOTAXONOMY

Chemotaxonomy has been used at all levels of the hierarchy of classification starting from the rank of 'variety' up to the rank of 'division' in plants. Moreover, chemical evidences have been used in all the groups of the plant kingdom starting from simple organisms, such as fungi and bacteria, up to the most highly advanced and specialised groups of angiosperms. These evidences are used in the classification of plants with two main purposes, (i) to develop such taxonomic characters which may improve the existing systems of plant classification, and (ii) to develop the present-day knowledge of phylogeny or evolutionary relationships of plants.

11.3 A BRIEF HISTORY

The history of the knowledge or use of chemical characters of plants and their utility in plant classification is very old. Socrates, the great Greek philosopher, was forced to take his own life by drinking a broth of poison hemlock, i.e. *Conium maculatum* of family Apiaceae or Umbelliferae. In his *Materia Medica*, Dioscorides in the 1st century A.D. described several aromatic mints. But some of the pioneer botanists, who employed chemical evidences in plant classification, are Nahemia Grew (1673), William Withering (1785), A.P. de Candolle (1804), Abbott (1886) and Geshoff (1891). However, the modern phase of chemotaxonomy began with the publication of the work of Abbott (1886) in *Botanical Gazzette.* McNair published a series of papers on the use of chemical evidences in solving taxonomical problems during 1917 and 1945.

The major research centre in the field of chemotaxonomy in the world is *The Institut fur Pharmakognosie of the University of Kiel, Germany.*

Some of the books providing details of chemotaxonomy are by Hawkes (1968), Swain (1973), Gibbs (1974), Bisby *et al.* (1980), Young and Seigler (1981), and Harborne and Turner (1984).

11.4 CHEMICAL CHARACTERS AND THEIR USE IN TAXONOMY

No suitable classification of the chemical characters and their use in taxonomy is developed so far. Naik (1984) divided these characters into three categories, i.e. (i) directly visible characters, e.g. starch grains, raphides, silica, gypsum, etc., (ii) chemical test characters, e.g. phenolics, betalains, oils, fats, waxes, alkaloids, etc., and (iii) proteins. On the other hand, Jones and Luchsinger (1987) divided the natural chemical plant products useful in taxonomy, on the basis of their molecular weight, into two major groups, i.e. (i) low molecular weight compounds with a molecular weight of 1000 or less, called *micromolecules,* e.g. amino acids, alkaloids, fatty acids, terpenoids, flavonoids, etc., and (ii) high molecular weight compounds with a molecular weight of over 1,000, called *macromolecules,* e.g. proteins, DNA, RNA, complex polysaccharides, etc. Sometimes, the term *semantides* is used for the information-carrying proteins, such as DNA, RNA, etc.

In the absence of a suitable classification, a discussion of some taxonomically important chemical compounds, along with a few examples of their systematic value, is undermentioned:

11.4.1 Flavonoids

Most widely and most effectively used compounds in chemotaxonomy are the flavonoids. These are the phenolic glycosides consisting of two benzene rings linked together through a heterocyclic pyrane ring. The reasons for their wide use in chemotaxonomy include their widespread distribution, chemical complexity, the relative physiological stability, quick and easy identification, and great structural variation.

Some of the common *classes of flavonoids* include anthocyanidins (e.g. cyanidin, delphinidin, etc.), flavones (e.g. apigenin, levtolin), biflavonyls (e.g. amentoflavone), flavonols (e.g. kaempferol, quercetin), flavonones (e.g. naringenin), isoflavones (e.g. orobol), isoflavonoids (e.g. ferreirin), chalcone (e.g. butein) and aurone (e.g. sulphuretin).

Some of the specific examples of the use of flavonoids in chemotaxonomy are undermentioned:

1. Alston (1967) documented hybridization between the species of *Baptisia* by using chromatographic patterns of flavonoids.
2. Dahlgren (1980, 1981) used flavonoids in clarifying certain relationship while suggesting his famous system of angiosperm classification.
3. *Psilotum* contains biflavonyls, and this brings it closer to lycopods than leptosporangiate ferns, as proposed by some workers.

4. Ellagic acid, a polyphenol, is confined to woody dicots and not found at all in monocots.
5. Analysis of leaf flavonoids of Liliaceae, Juncaceae, Cyperaceae and Gramineae (Williams, 1976) suggest that all these families have arisen from Liliaceous ancestors.
6. Williams *et at.* (1983) identified six ornamental species of *Fuchsia* on the basis of their leaf flavonoids.
7. Belzer and Ownbey (1971) studied the flavonoid pattern of 5 species of *Tragopogon* of Compositae and their F_1 and F_2 hybrids, and noted that all the species and their population show different chromatographic results.
8. Isoflavone iridin is found only in the section Pogonivis of genus *Iris*.
9. South American species of *Eucryphia* of the family Eucryphiaceae may easily be distinguished on the basis of their flavonoid glycosides.
10. Species of *Malus* are classified on the basis of their flavonoids by Williams (1982).
11. Presence of five leaf flavonoids (tricin, luteolin, glycosyl flavones, bisulphate complex and 5-glucoside) in both Gramineae and Palmae, link them together.
12. Species of *Spirodella,* unidentifiable on morphological grounds, may be distinguished on the basis of their flavonoid chemistry.

11.4.2 Betalains

Betalains differ from flavonoids and other phenolic compounds owing to the presence of nitrogen in them. However, these are functionally equivalent to phenolics. Betalains are popularly known as nitrogenous anthocyanins. They comprise of red to violet betacyanins and yellow betaxanthins. Some of the aspects of taxonomic value of betalains are undermentioned:

1. All the ten families of the order Centrospermae (i.e. Aizoaceae, Amaranthaceae, Basellaceae, Caryophyllaceae, Chenopodiaceae, Didieraceae, Molluginaceae, Nyctaginaceae, Phytolaccaceae and Portulacaceae) produce betalains and are placed together in a single order Caryophyllales in modern schemes of classification.
2. Betalains are also reported in the family Cactaceae. This family, once placed elsewhere (i.e. in the order Cactales) in the classification schemes, should be placed under the betalains-containing order Caryophyllales or in the older order Centrospermae.

11.4.3 Terpenoids

These are a biogenic group of diverse compounds with mevolonic acid as their precursor (Goodwin, 1970). Some regard them as components of essential oils. According to the number of isoprene units present in a terpenoid molecule, they may be of five categories: (i) *hemiterpenes* (C_5H_8) e.g. tiglic acid, (ii) *monoterpenes* ($C_{10}H_{16}$). e.g. myrcene, (iii) *sesquiterpenoids* ($C_{15}H_{24}$), e.g. parnesol, (iv) *diterpenoids* ($C_{20}H_{32}$), e.g. phytol and, (v) *triterpenoids* ($C_{30}H_{48}$), e.g. squalene.

Terpenoids have been used extensively in the chemotaxonomy of mints, umbellifers, *Citrus* plants, and gymnosperms. Some of their chemotaxonomic applications are mentioned below:

1. Origin of certain *Citrus* cultivars was determined by studying their rind and leaf terpenoid pattern.

2. Geographical races of *Pseudotsuga menziesii* were distinguished by studying the terpenoids of their cortical oleoresins.
3. In the taxonomy of Compositae and tribe Genisteae of Fabaceae, the petal carotenoids, which form a prominent group of terpenoids, prove very useful.
4. Mirov (1961) used the gum terpentines in the taxonomy of *Pinus*.
5. Some specific terpenoids are found in certain families, e.g. cucurbitacins in Cucurbitaceae, asperuloside in Rubiaceae, and sesquiterpene lactones in Compositae.

11.4.4 Steroids

True steroids possess two methyl groups and are mostly alcohols or esters. They are located in plant cutins and perform the function of water proofing. Their distribution proved helpful in the taxonomy of the genera of tribe Veratreae of family Liliaceae. These genera contain the steroid veratum according to Kupchan *et al.* (1961).

11.4.5 Iridoid Compounds

Monoterpenoid cyclopentanoid lectones represent a separate class of compounds of taxonomic significance, called *iridoids.* Some of the aspects of their taxonomic value are mentioned below:

1. Because of the presence of an iridoid compound, aucubin, genus *Buddleia* was transferred from Loganiaceae to Buddleiaceae, near Scrophulariaceae.
2. The iridoid compound, arsperuloside, is the characteristic of Rubiaceae whereas aucubin is found in Orobanchaceae, Scrophulariaceae and Cornaceae.
3. On the basis of the presence of iridoids, the interfamilial relationships were clarified between Cornales, Scrophulariales, Caprifoliales, and Rubiales.
4. Several iridoid-families (e.g. Rubiaceae, Scrophulariaceae, Cornaceae, Dipsacaceae and Caprifoliaceae) are considered the putative progenitor of Compositae.

11.4.6 Alkaloids

These are the basic nitrogen-containing organic compounds, usually with a heterocyclic ring. Alkaloids show significant effects on the central nervous system of animals, and are well-known for their medicinal, chemical, poisonous, and systematic viewpoints. Some of their taxonomic applications are undermentioned:

1. Jones and Luchsinger (1987) stated that alkaloids are useful in taxonomic studies in *Veratrum* (Liliaceae), *Papaver* and *Argemone* of Papaveraceae, as well as in *Lycopodium, Lupinus,* and Caryophyllales.
2. Morphine is produced only by *Papaver somniferum,* and strychnine is found only in *Strychnos.*
3. Because of the accumulation of isoquinoline alkaloids containing protopine, in both Fumariaceae and Papaveraceae, the two are considered closely related.
4. Lupin alkaloids are the characteristic of the tribes Sophoreae, Genisteae and Podalyrieae of subfamily Lotoideae of Leguminosae.

5. Distribution patterns of alkaloids in two genera (*Genista* and *Adenocarpus*) of the tribe Genisteae of Fabaceae and the genus *Ammodendron* of the tribe Sophereae of the same family suggest the transfer of *Ammodendron* from Sophereae to the tribe Genisteae.
6. Several modern taxonomists, including Takhtajan, Cronquist, Thorne and Dahlgren, rearranged the families with apocarpous gynoecia in Magnoliidae on the basis of the distribution of benzylisoquinoline alkaloids.

11.4.7 Glucosinolates

Glucosinolates is a group of about 85 thioglucosides which on hydrolysis form glucose and a corresponding isothiocyanate. They are also called mustard oil glucosides because they are confined to the taxa of the mustard family (Cruciferae) and some related families of Capparales.

Formerly, Cruciferae, Capparaceae, Fumariaceae and Papaveraceae were all included under one order Rhoeadales. But the recent findings support the view that glucosinolate-producing families (Cruciferae and Capparaceae) should be placed separately under Capparales, and the alkaloid-containing families (Fumariaceae and Papaveraceae) under Papaverales. Members of Bataceae and Gyrostemonaceae also contain glucosinolates. Glucosinate patterns are also used to document hybridization in some genera of Cruciferae, e.g. *Cakile*.

11.4.8 Amino Acids

These are the building blocks of proteins and are universally distributed in plant tissues. Several workers have studied the taxonomic potential of amino acids. A few examples of their taxonomic significance are mentioned below:

1. *Lathyrus* species could be grouped under seven infrageneric groups on the basis of the association of amino acids within the seeds.
2. According to Turner and Harborne (1967), canavanine is found only in the taxa of the subfamily Lotoideae of Leguminosae.
3. Lathyrine is found only in the genus *Lathyrus* while azetidine-2-carboxylic acid is found only in Agavaceae, Liliaceae and Amaryllidaceae.
4. Acetyl ornithine is identified as the main amino acid of Fumarioideae.
5. Species of the section Gummiferae of the genus *Acacia* can be recognised easily on the basis of the amino acid contents of their seeds.

11.4.9 Lipids and Waxes

Lipids are the esters of fatty acids with glycerol, mostly made up of carbon, hydrogen and oxygen. Plant waxes are esters of long-chain alcohols with long-chain fatty acids. They may contain free alcohols, free fatty acids, aldehydes, ketones or hydrocarbons. Some of the taxonomic applications of lipids and waxes are undermentioned:

1. Restricted distribution of some fatty acids is characteristic of certain families, e.g. petroselinic acid of Umbelliferae, erucic acid of Cruciferae, chaulmoogric acid of some taxa of Flacourtiaceae, and xymenyric acid of Santalaceae.

2. Proportions of fatty acids in lipids are constant in both genera and species in Palmae (Hilditch, 1952).
3. Malvalic acid exists in the members of Malvaceae while capric acid is characteristic of Lauraceae and Lythraceae.
4. Fatty acids of bacterial lipids are not found in other plants.

11.4.10 Nucleic Acids

Unavailability of the necessary expertise and equipment, required for the nucleic acid technique, to most of the taxonomists, hindered the progress in this direction. However, eight species of *Atriplex* of Chenopodiaceae were compared on the basis of their DNA nucleotide sequences. DNA of wheat, rye and barley of Triticeae, compared with that of oat of Aveneae, showed marked differences. Moreover, wheat DNA is more similar to the DNA of rye than that of barley. According to Goldberg *et al.* (1972), all the 13 species of *Cucurbita* contain mainband and satellite DNA whose densities are constant throughout in all species of this genus.

11.4.11 Proteins

Proteins are used in chemotaxonomy owing to their large complex molecules, universal distribution, presence in large quantity in some organisms, and also because of the knowledge of several simple and rapid methods of protein analysis, such as electrophoresis. Phylogenetic affinities in Triticinae were demonstrated by Johnson and Hall (1965) by the process of protein electrophoresis. This process also helped in establishing a close relationship between *Vicia* and *Lathyrus.* By exploiting peroxidase, esterase and acid phosphatase pattern, Symenoidis and Tsekos (1984) suggested that the genus *Taeniantherum,* formerly considered as a part of the genus *Hordeum,* should be treated as an independent genus.

As the proteins represent the direct products of the DNA code, their comparative analysis may be used as fundamental taxonomic characters. Protein analysis by electrophoresis and other methods may be used to determine whether differences actually exist among plants that are treated as distinct species.

11.4.12 Cyanogenic Compounds

The ability of some plants to release poisonous compounds, such as hydrocyanic acid, amygdalin, etc., after injury to their cells, is called *cyanogenesis*, and these poisonous compounds are called *cyanogenic compounds.* Plants containing these compounds are called *cyanophoric plants.* Over two thousand vascular plants have so far been established as cyanophoric. Taxa belonging to Araceae, Gramineae, Juncaceae, Juncaginaceae and Scheuchzeriaceae are commonly cyanogenic. Several dicotyledonous families belonging to Asteridae, Rosidae and Dilleniidae are also cyanogenic.

Test Your Understanding

1. What do you mean by chemotaxonomy? Explain its purpose and history in brief.
2. With the help of suitable examples, explain the use of chemical characters in solving taxonomical problems.

3. Explain some aspects of taxonomic value of flavonoids, terpenoids, alkaloids and amino acids.
4. Write an essay on chemotaxonomy.

Suggested Reading

Alston, R.E. and B.L. Turner, 1963, *Biochemical Systematics,* Prentice-Hall, Englewood Cliffs, N.J.

Bisby, F.A., J.G. Vaughan and C.A. Wright (eds.), 1980, *Chemosystematics: Principles and Practices,* Academic Press, London.

Crawford, D.J. and D.E. Giannasi, 1982, Plant chemosystematics, *Bioscience* **32**: 114–118.

Cronquist, A., 1980, Chemistry in plant taxonomy: An assessment of where we stand, (In) F.A. Bisby *et al. Chemosystematics: Principles and Practices,* Academic Press, London.

Gibbs, R.D., 1974, *Chemotaxonomy of Flowering Plants,* 4 Vols. McGill-Queens Univ. Press, Montreal.

Harborne, J.B., 1964, *Biochemistry of Phenolic Compounds,* Academic Press, London.

_______ 1967, *Comparative Biochemistry of the Flavonoids,* Academic Press, London.

_______ and B.L. Turner, 1984, *Plant Chemosystematics,* Academic Press, New York.

Hawkes, J.G. (ed), 1968, *Chemotaxonomy and Serotaxonomy,* Academic Press, London.

Jones, S.B. and A.E. Luchsinger, 1987, *Plant Systematics,* McGraw-Hill, New York.

Smith, P.M., 1976, *The Chemotaxonomy of Plants,* Edward Arnold, London.

Swain, T. (ed.), 1973. *Chemistry in Evolution and Systematics,* Butterworth, London.

Young, D.A. and D.J. Seigler, 1981, *Phytochemistry and Angiosperm Phylogeny,* Praeger, New York.

CHAPTER 12

MOLECULAR TAXONOMY

12.1 WHAT IS MOLECULAR TAXONOMY?

Molecular taxonomy, also known as *molecular systematics*, is the use of the structure of *molecules* to obtain information on an organism's evolutionary relationships. (As is well-known, the *molecule* is the smallest unit of an element or compound which occurs naturally. Molecules consist of more than one *atom*, e.g., a molecule of hydrogen consists of two hydrogen atoms (H_2)). Some prefer to call molecular systematics as *molecular phylogenetics*. The results of a molecular phylogenetic analysis are expressed in the form of a tree called *phylogenetic tree* (Hillis and Moritz, 1996, Wikstro, *et al*., 2001).

An International Workshop on Molecular Systematics was organised at Bhagalpur University, Bhagalpur (India) during 11–16 February 2007. In this workshop Pandey (2007) stated that Molecular systematics includes a series of approaches in which phylogenetic relationships are inferred using information from macromolecules of the organisms. The type of molecular data acquired include those from DNA sequences, DNA restriction sites, microsatellites, RAPDs (Random Amplified Polymorphic DNA) and AFLPs (Amplified Fragment Length Polymorphism) in this study.

12.2 SOME BASICS USED IN THE TECHNIQUES AND APPLICATIONS OF MOLECULAR TAXONOMY

DNA, RNA, and proteins are present in all living organisms. It has been proved genetically that closely related organisms generally have a high degree of agreement in the molecular structure of these substances. Contrary to this, molecules of distantly related organisms usually show a pattern of dissimilarity. Molecular taxonomists use such data to build a "*relationship tree*" that provides a picture of probable evolution of various organisms. This is actually the basic aspect which is used in the techniques and applications of molecular taxonomy.

The most common approach in majority of the molecular taxonomical studies is the comparison of sequences for genes using techniques of sequence alignment to identify similarity among various organisms.

In recent times, molecular phylogeny is also applied in *DNA barcoding*. In this technique the species of an individual organism is identified using small sections of mitochondrial DNA.

Molecular phylogeny techniques are also used in the field of human genetics specially in: (i) *genetic testing* to determine a child's paternity, and also (ii) in the branch of *criminal forensics* focussed on evidence known as "*genetic fingerprinting*". These days, this is commonly used globally.

12.3 SOME TECHNIQUES USED IN MOLECULAR TAXONOMY

Commonly used techniques in molecular taxonomy are (i) extraction, (ii) amplification, (iii) sequencing of genes, (iv) sequence alignment, and (v) phylogenetic analysis. Young taxonomists should be trained in these techniques by the experts and it can be achieved by organising workshops, time to time, in different parts of the globe.

An old technique to determine the divergences between the genotypes of individuals has been *DNA-DNA hybridisation* but this has now been superseded by the above-mentioned recent techniques. This old technique was based on the use of entire genotype rather than focussing on particular sections of DNA.

12.4 HOW ARE CHEMOTAXONOMY AND MOLECULAR TAXONOMY RELATED?

Application of chemical evidence in taxonomy is called *chemotaxonomy*. It makes use of proteins, harmones, enzymes, carbohydrates and other molecules, which get separated using many techniques including *chromatography*. During last four decades these techniques have been largely replaced by *DNA-sequencing* under *molecular taxonomy*. The DNA-sequencing technique produces the exact sequences of *nucleotides* or *bases* in either DNA or RNA segments which are extracted using different techniques. (A *nucleotide* is a molecule with a pentose sugar, a phosphate group and a purine or pyrimidine base containing nitrogen. Nucleotides are the units which form the long chain polymers, nucleic acids i.e. DNA and RNA. A *base* is a either purine or pyrimidine unit).

Techniques of molecular taxonomy such as DNA-sequencing are generally considered superior for evolutionary studies because the actions of evolution are finally reflected in genetic sequences.

12.5 WHERE DO WE STAND TODAY IN KNOWING DNA-SEQUENCING TECHNIQUES AND WHY?

DNA – sequencing technique is a long, very technical and expensive process to sequence the entire DNA of an organism, i.e., its *genome*. (The *genome* is the genetic material on the sets of chromosomes in a cell. The smallest genome consists of all genes on a haploid set of chromosomes). Due to all these complicate techniques, the process of DNA-sequencing has so far been done for only a few species.

12.6 PRESENT STATUS OF MOLECULAR TAXONOMY IN INDIA

Majority of the so-far globally completed molecular systematic studies have been performed in the developed countries, mainly because they are long and quite expensive. However, with the rich biological resources in many developing countries, molecular taxonomy will encouragingly advance the field of systematics and several related bilological branches "if more scientists and students in these countries are involved in molecular systematic studies". India is one of the richest countries of the world in biodiversity. In spie of this, not many molecular studies have been carried out due to lack of proper training in the field of molecular taxonomy. The first workshop, aiming to provide training to young scientists in molecular taxonomy of flowering plants was organised under the auspices of Indian Association for Angiosperm Taxonomy (IAAT) at Bhagalpur University in 2007.

12.7 WHAT DOES A TYPICAL MOLECULAR SYSTEMATIC ANALYSIS REQUIRE?

A typical molecular systematic analysis requires the sequencing of approximately 1000 base pairs. At any location within such a sequence, the bases found in a given position may vary between different organisms. The particular sequence present in a given organism is referred to as its haplotype. (A *haplotype* is actually a particular set of alleles at several very closely linked loci). Technically, since there are four base types, with 1000 base pairs, we could have 4^{1000} distinct haplotypes. In a molecular systematic analysis, the haplotypes are actually determined for a well-defined area of genetic material. At the end of the process, the haplotypes from a smaller number of individuals from a definitely different taxon are determined. These are technically called *out group.* Then the base sequences for the haplotypes are compared. Initially, the difference between two haplotypes is determined by counting the number of locations where they have different bases. This is technically called as the number of *substitutions.* The difference between organisms is mentioned as the *percentage divergence.* It is calculated by dividing the number of substitutions by the number of analysed base pairs.

After the determination of the divergences between all pairs of samples, a *triangular matrix* of differences is resulted. It is submitted to some definite statistical calculations, and finally a *dendrogram* is resulted. (A *dendrogram* is a branching diagram after the style of a family tree reflecting similarities or affinities of some sort). From the dendrograms, it is calculated whether or not they provide any clues about the current ideas about taxonomy of the group. *Evolutionary trees* are prepared using all these studies.

12.8 ASSUMPTIONS AND USES OF MOLECULAR TAXONOMY

Some of the characteristics of molecular taxonomy and the assumptions involved there in are listed below:

1. Being essentially a *cladistic* approach, the molecular taxonomy assumes that classification must correspond to phylogenetic descent, and that all valid taxa must be monophyletic. (The word *cladistics* refers to the method of classifying organisms into groups or taxa based on recency of common descent, as judged by the possession of shared-derived characteristics. A branching diagram showing the relationships between groups of organisms determined by the methods of cladistics is called a *cladogram.*)
2. Molecular taxonomy often uses the molecular clock assumption which emphasizes that quantitative similarity of genotype is a sufficient measure of the recency of genetic divergence. But this assumption does not always hold good in relation to *speciation* (evolutionary process in which new *species* are produced).

3. Mitochondrial DNA is very conveniently used globally for molecular systematic analysis in animals.
4. Molecular systematics helps in establishing the relationship of different plant groups at DNA level.
5. It has also unlocked the treasure chest of information on evolutionary history of organisms.
6. By the use of conserved molecular sequences, it is now possible to define and identify a very large number of species of plants and animals.
7. Evolutionary patterns of biodiversity are also now investigated using DNA data.
8. DNA taxonomy has its definite role in genome mapping and biodiversity conservation.
9. Molecular systematics is now highly useful not only in plant identification but also in knowing genealogical history of plants as well as animals.
10. It also helps in understanding the phytogeography, which can ultimately help in conservation of biodiversity.
11. DNA-based *molecular markers*, used for designing DNA-based molecular probes, have also been developed under the branch of molecular systematics (*details discussed elsewhere in this chapter under Article 12.10*).
12. Modern tools and techniques of DNA analysis can find out differences at single-nucleotide level and sequences with these differences are then said to possess *single nucleotide polymorphism* (*SNP*). SNP studies, if carefully performed, can differentiate between two similar-looking organisms.
13. Amplified Fragment Length Polymorphism (AFLP) technique produces large sets of *polymorphic markers* that are used to analyse closely related taxa. AFLP data are also used in studying polyploidy evolution.
14. Molecular systematics is now used in *Gene Bank search* using definite methodology. With the help of computers, sequence alignment is now done using different softwares.
15. By the techniques of PAUP (Phylogenetic Analysis Using Parsimony) cladistic analysis is now done using both morphological and molecular data.

12.9 CRITICISM AND FUTURE OF MOLECULAR SYSTEMATICS

Above-mentioned assumptions and uses of molecular systematics are not totally uncontroversial among biological systematists. Being a cladistic method, molecular systematics is criticised by systematists as the science of "*cladistics*" in general. Some even go upto the extent of calling it a "mistake to replace a classification based on visible and ecologically relevant characteristics by one based on genetic details that may not even be expressed in the phenotype".

Inspite of all this, molecular systematics is gaining increasing acceptance. As studies are proceeding making gene sequencing easier and cheaper, the science of molecular systematics is being used and applied for more and more groups. Considering the overall scenario, some molecular systematic studies have already lead to radical revisions of many accepted taxonomic groups.

12.10 A NOTE ON MOLECULAR MARKERS

Molecular markers are DNA-based markers used widely in various fields including DNA-fingerprinting and also as forensic tools. These have revolutionised the entire scenario of biological sciences and occupied their definite position in various fields, including genetic engineering, taxonomy, physiology, embryology, etc. They are widely used globally to bring about automation in the process of *genome analysis*. The discovery of polymerase chain reaction (PCR) has brought about a new class of DNA profiling markers which has resulted in the development of (i) marker-based gene tags, (ii) maping-based cloning of agronomically important genes, (iii) variability studies, (iv) phylogenetic analysis, (v) synteny mapping, (vi) marker-assisted selection of desirable genotypes, etc. (Joshi, *et al.*, 1999). DNA-markers can now provide data that can be analysed objectively.

Molecular markers actually include biochemical constituents (e.g., secondary metabolites in plants) and macromolecules, viz., proteins and DNA. DNA-based molecular markers are widely used in biological fields, and in this regard *DNA-fingerprinting* was introduced for the first time by Jeffrey *et al.,* in 1985. According to these workers, DNA-fingerprinting describes the "bar-code-like DNA fragment patterns generated by multilocus probes after electrophoretic separation of genomic DNA fragments" (Joshi, *et al.*, 1999). This technique is now widely used in analysis of genome evolution, taxonomy, plant breeding, population genetics, and diagnostics. Joshi *et al.* (1999) listed some properties desirable for ideal DNA-markers. These include (i) highly polymorphic nature, (ii) frequent occurrence in genome, (iii) easy access or easy availability, (iv) easy and fast assay, and (v) high reproducibility, etc.

Various types of molecular markers used to estimate DNA polymorphism are generally categorised as (i) hybridisation-based markers, and (ii) PCR (polymerate chain reaction)-based markers. The versatile technique of PCR was invented by Saki *et al.*, (1985). Joshi *et al.*, (1999) described in detail various types of DNA-markers and their application in plant genome analysis and breeding. One of the most recent applications of molecular markers has been shown in sex identification of dioecious plants.

12.11 SOME SPECIFIC EXAMPLES SHOWING ROLE OF MOLECULAR SYSTEMATICS IN PLANTS

(1) Endress (2002) reviewed the details of the morphology and angiosperm taxonomy in molecular era and concluded that the goals of comparative morphology have shifted in the present molecular era. According to him, the "morphology no longer plays the primary role in phylogenetic studies". New opportunities for morphology are now opening up "that were not present in the premolecular era", e.g., (i) phylogenetic studies with combined molecular and morphological analysis, (ii) reconstruction of evolution of morphological features based on molecularly derived cladograms, (iii) redefined analysis of morphological features induced by inconsistencies of previous molecular and molecular phylogenetic analyses, and (iv) exploration of the evolution of morphological traits by integration of comparative structural and molecular developmental genetic aspects. He has also mentioned that the field of angiosperm taxonomy in the molecular era "is still in its infancy in botany; its advancement is one of the major goals of evolutionary botany" (Endress, 2002).

(2) Hussain *et al.* (2008) studied molecular analysis of dicot-monocot split and relationship among major angiosperm groups. Using RAPD (random amplified polymorphic DNA)-technique, they established relationship among major angiospermic groups based on cladistic analyses. "The phylogenetic trees of relationship derived from molecular data confirm dicots as the ancestral class of monocots, there seems no dicot-monocot split. Dicots form an ancestral class of magnoliids and the monocot lineage was derived from one of the basal magnoliids. Their studies also "pinpoint dicots as the possible close relatives of the monocots" (Hussain *et al.*, 2008).

Test Your Understanding

1. Define molecular taxonomy. What are the basics used in the techniques of molecular taxonomy?
2. What is DNA-barcoding?
3. What is the relation of chemotaxonomy and molecular taxonomy?
4. Write a note on the status of molecular taxonomy in India.
5. Define the terms (a) DNA-sequencing, (b) dendrogram, (c) cladistics.
6. Describe the typical requirements for a molecular systematic analysis.
7. Give an account of some assumptions and uses of molecular taxonomy.
8. Write a detailed note on molecular markers.

Suggested Reading

Endress, P.K., 2002, *Morphology and angiosperm systematics in the molecular era. Bot. Rev.* **68**: 545–570.

Hillis, D.M. and C. Moritz, 1996, *Molecular Systematics* (2nd ed.), Sinauer Associates Incorporated (ISBN 0–87893–282–8).

Hussain, M.B., V. Verma and Z.A. Malik, 2008, Molecular analysis of dicot-monocot split and relationship among major angiosperm groups. *Afr. Jour of Pl. Sci.* **2**(1): 1–4.

Jeffrey, A.J., V. Wilson and S.L. Thein, 1985. *DNA-fingerprinting. Nature* **314**: 67–73.

Joshi, S.P., P.K. Ranjekar and V.S. Gupta, 1999, *Molecular markers in plant genome analysis. Curr. Sci.* **77**: 230–240.

Pandey, A.K., 2007, *Molecular systematics* (Report of International Workshop), *Curr. Sci.* **92**(7): 881–882.

Saki, R.K., S. Scharf, F. Faloona, K.B. Mullis, G.T. Horn, H.A. Erlich and N. Arnheim, 1985. *PCR-technique. Science* **230**: 1350–1354.

Wikstro, M.N., V. Savolainen and M. Chase, 2001, *Evolution of the angiosperms: Calibrating the family tree. Proc. Roy. Soc. Lond.* B. **268**: 2211–2220.

CHAPTER 13

SEROTAXONOMY

13.1 WHAT IS SEROTAXONOMY?

The application or utility of serology in solving taxonomic problems is called *serotaxonomy.* The study of antigen-antibody reactions is called *serology.* The substance capable of stimulating the formation of an antibody is called *antigen.* A highly specific protein molecule produced by plasma cells in the immune system is called *antibody.* The antibodies combine chemically with specific kinds of antigens. Proteins are the useful carriers of taxonomic information and are most widely used antigens in serotaxonomy.

Serology is useful in the following ways:

1. It expresses similarities and dissimilarities amongst different taxa, and these data are helpful in taxonomy. It determines the degree of similarity between species, genera, families, etc. by comparing the reactions of antigens from various plant taxa with antibodies raised against the antigen of a given taxon.
2. It helps in comparing non-morphological characteristics, the knowledge of which is useful in taxonomy.
3. Single proteins from different plant taxa are also compared by serological techniques.

13.2 SOME OTHER RELATED TERMS

1. *Serology* According to Boyden (1964) it is the branch of biology which deals with "the nature and interactions of antigenic material and antibodies". But according to Smith (1976) serology is the "study of origin and properties of *antisera.*"
2. *Agglutinogens* Antigens are also called agglutinogens.
3. *Agglutinins* Antibodies that participate in agglutination process.
4. *Agglutination* A type of antigen-antibody reaction that results in visible clumps of organisms or other materials.

5. *Precipitins* These are the antibodies "capable of combining with and reacting upon antigenic material because of certain determinant groups" (Fairbrothers, 1968). Precipitins are the antibodies causing precipitation.
6. *Precipitin Reaction* It is the *in vitro* reaction between solutions or suspensions of antigen and antibody, with the formation of a precipitate.
7. *Reference Reaction* It is the "reaction between an antiserum and the antigenic material used to stimulate its formation" (Fairbrothers, 1968).
8. *Haptens* These are the substances which are unable to induce antibody formation by themselves, but are able to react specifically with antibodies.
9. *Serological Reactions* Antigen-antibody reactions *in vitro* are known as serological reactions.

13.3 GENERAL FEATURES OF SEROLOGICAL REACTIONS

1. The reaction is specific, an antigen combining only with its homologous antibody and *vice versa*.
2. Entire molecules react and not the fragments.
3. Antigens or antibodies do not pass through denaturation during these reactions.
4. Combination occurs at the surface, and it is firm but reversible.
5. Both antigens and antibodies take part in the formation of precipitates or agglutinates.
6. Antigens and antibodies can combine in varying proportions.

13.4 BRIEF HISTORY OF SEROTAXONOMY

Nuttal (1901) was the first biologist to compare the immunochemical specificity of serum proteins for systematic purposes. Other notable serologists of that time were Kowarski (1901), Bertarelli (1902) and Magnus (1908). It was Dunbar (1910) who showed that proteins from pollen, seeds and leaves of rice were serologically distinct. A school of serology was founded in 1914 by Gohlke at Koenigsberg in Germany, and in the later years Germany became the centre of serological studies. Mez (1926) and Moritz (1934) of Germany and Boyden (1942) of America developed the technique of serology in the later years. Notable contributions on the application of serology to taxonomy are those of Rives (1923), Hawkes and Lester (1966), Smith (1968, 1969, 1972), Kloz (1971), Fairbrothers (1967, 1975, 1977, 1978, 1983) and Crawford and Giannasi (1982).

13.5 GENERAL PROCESS OF SEROTAXONOMY

In serotaxonomical studies, the protein extract of plant or animal origin, i.e. antigen, is injected into the blood stream of a rabbit or an experimental animal, to form the antibodies. In response to a specific antigen a specific antibody is produced. The serum, called antiserum, is then made to react *in vitro* with the antigenic protein as well as with proteins of other taxa, the affinities of which are to be determined. The amount of the precipitation shows the degree of the protein homology. The whole process may be explained as under:

Suppose, we are to ascertain the closeness of a taxon P with taxa R, S and T. The proteins extracted from P are injected into a rabbit, in which the antibodies are produced. The antibodies are then extracted from the blood of the rabbit in the form of an antiserum. When the antiserum is allowed to react with the original protein extract from P, a total coagulation takes place. When this antiserum is allowed to react with the protein extracts from the taxa R, S and T, the degree of coagulation varies. This degree of coagulation is related directly to the closeness between the taxa being compared.

13.6 EXAMPLES OF SEROTAXONOMIC IMPORTANCE

Out of the several cases of the implications of serological data in the classification of angiosperms, some are mentioned below:

1. According to Fairbrothers (1983) serological data have been used in the classification of orders and the assignment of families in Apiales, Caryophyllales, Capparales, Fagales, Cornales, Magnoliales, Juglandales, Papaverales, Rubiales, Ranunculales, Scrophulariales, Typhales, Primulales, etc.
2. Fairbrothers and Johnson (1959) separated six species of *Bromus* on the basis of serological studies.
3. According to Jensen (1967), serological characteristics within Ranunculaceae show a close similarity between *Aconitum-Delphinium, Actaea-Cimicifuga, Anemone-Clematis* and *Ranunculus-Myosurus,* and suggest a common ancestry for *Aquilegia, Leptopyrum* and *Thalictrum.*
4. While studying the serological assessment of relationships within Solanaceae, Hawkes and Tucker (1968) observed a strong serological relationship between *Solanum, Nicotiana, Hyoscyamus, Datura* and *Salpiglossis.*
5. Hillebrand and Fairbrothers (1970) made a phytoserological systematic survey of Caprifoliaceae and suggested the monophyletic origin of the family. According to these workers there is a close relationship between Caprifoliaceae and Cornaceae on serological grounds.
6. On the basis of serological studies, Johnson and Fairbrothers (1965) suggested that the genera *Magnolia* and *Michelia* show the closest affinity within Magnoliaceae.
7. Simon (1971) showed a close relationship between Nymphaeaceae and Nelumbonaceae on the serological ground.
8. Serotaxonomic findings of Pickering and Fairbrothers (1970) in Umbelliferae support the classification of the family into Apioideae, Saniculoideae and Hydrocotyloideae, and also suggest that Apioideae is more close to Saniculoideae than to Hydrocotyloideae.
9. Gartner (1978) suggested some serological evidences for assigning *Phaseolus aureus* and *P. mungo* to the genus *Vigna.*
10. *Hydrastis* of Berberidaceae has more serological similarities with Ranunculaceae than Berberidaceae.
11. On the basis of serological results, Mez and Ziegenspeck (1926) prepared a *stammbaum* or *family tree* for the entire plant kingdom.

12. According to Lee and Fairbrothers (1978), two genera (*Galium* and *Asperula*) of Rubiaceae show similar serological characters, and both of them differ from other genera of this family.
13. Kloz (1971) applied the serological techniques in the classification of Leguminosae.
14. Hawkes and Lester (1968) showed a relationship amongst the tuber-bearing taxa of *Solanum* by serological techniques.
15. Fairbrothers *et al.* (1975) supported the separation of *Illicium* from Magnoliaceae to Illiciaceae, and of *Schizandra* from Magnoliaceae to Schizandraceae, on the basis of serological studies.
16. Piechura and Fairbrothers (1983), on the basis of their serotaxonomic studies, supported the view that taxa of Oleoideae evolved from those of Jasminoideae.

Test Your Understanding

1. Differentiate between serology and serotaxonomy.
2. Define the following terms in one sentence only.
   (a) Agglutinogens, (b) Agglutinins, (c) Agglutination
3. Describe some ways of usefulness of serology.
4. Discuss, in brief, the general process of serotaxonomy.
5. Give at least five examples which show the importance of serotaxonomy.

Suggested Reading

Boyden, A., 1964, Perspectives in systematic serology, In C.A. Leone (Ed.) *Taxonomic Biochemistry and Serology,* Ronald Press, New York, pp. 75–99.

Fairbrothers, D.E., 1977, Perspectives in plant serotaxonomy, *Ann. Mo. Bot. Gard.* **64**: 147–160.

_________ 1983, Evidence from nucleic acid and protein chemistry, in particular serology, in angiosperm classification, *Nord. J. Bot.* **3**: 35–41.

Hawkes, J.G. (ed.), 1968, *Chemotaxonomy and Serotaxonomy,* Academic Press, London.

CHAPTER 14

PHYLOGENY: ORIGIN AND EVOLUTION OF ANGIOSPERMS

14.1 WHAT IS PHYLOGENY?

The history of development of a race, or simply evolutionary history, is called *phylogeny.*

Radford (1986) defined the word 'phylogeny' as "the study of the genealogy and the evolutionary history of groups of organisms", and the word *genealogy* he defined as "the study of ancestral relationships and lineages". However, according to Jones and Luchsinger (1987) "the evolutionary development or lineage of a taxon is its phylogeny".

The phylogeny of angiosperms, in the light of the above-mentioned definitions, therefore, includes the knowledge of the ancestors of angiosperms, place and time of their origin, the primitive angiosperms and the interrelationships among their major groups.

14.2 WHAT ARE ANGIOSPERMS?

Angiosperms, generally known as *flowering plants,* constitute that subdivision of seed plants whose members have a stem with xylem tissue containing vessels and bear their seeds within one or more closed carpels. Another subdivision of seed plants is *gymnosperms.* Seeds are enclosed within the carpel or carpels in angiosperms while they are borne exposed on the carpels in gymnosperms.

One remarkable feature of the angiosperms is the possession of a complex reproductive structure called *flower.* A flower is a structure usually containing sepals, petals, stamens and carpels. However, there exist flowers which consist of only a single stamen or a single ovary.

Some of the major *characteristics* of the angiosperms are (1) vessels in the xylem, (2) phloem contains sieve elements and companion cells, (3) embryo sac is generally *Polygonum*-type or eight-nucleate (1 egg, 2 synergids, 2 polar nuclei, and 3 antipodals), (4) double fertilisation, (5) closed carpels, (6) after fertilization the ovary becomes a fruit, enclosing one or more seeds, and (7) a reduced gametophytic generation. However, there exist numerous exceptions to all these characteristics.

Angiosperms have the greatest number of plant species (over 300,000 species) inhabiting the earth. They include annuals, perennials, herbs, shrubs, and trees, ranging from very small to gigantic sizes.

There exist parasitic, epiphytic, saprophytic and even insectivorous species among angiosperms. They include two classes, Monocotyledoneae and Dicotyledoneae. Some of the major *differences between the monocots and dicots* are undermentioned in Table 14.1:

Table 14.1 Major differences between dicots and monocots

S.No.	Dicots	Monocots
1.	Cotyledons 2 (rarely 1, 3, or 4).	Cotyledon 1.
2.	Leaves mostly with reticulate venation.	Leaves mostly with parallel venation.
3.	Intrafascicular cambium usually present.	Intrafascicular cambium absent.
4.	Vascular bundles usually in rings.	Vascular bundles usually scattered.
5.	Floral parts usually in sets of 5.	Floral parts usually in sets of 3.
6.	Pollen typically triaperturate.	Pollen uniaperturate.

14.3 GENERAL PRINCIPLES OF ANGIOSPERM PHYLOGENY

According to Thorne (1976), the botanists working on angiosperm phylogeny face three major problems, viz. (1) incomplete and inadequate fossil records, (2) prevalence of convergent evolution, and (3) vast modifications in the structure of these plants. However, in spite of the meagre fossil records of angiosperms, their available fossil data and some other characters indicate definite trends in their phylogeny. The problem of the convergent evolution has been tackled, to some extent, by assembling and evaluating the information from all possible sources.

On the basis of the studies of the comparative morphology and palaeobotany, Bessey (1915) presented some guiding principles of the angiosperm phylogeny. These principles were called by Bessey as *dicta*. In the light of modern discoveries, these *Besseyan Principles* were modified by several modern phylogenists, including Cronquist (1968), Takhtajan (1969), Hutchinson (1973), Thorne (1976), etc. Undermentioned are some of the modified "*Besseyan Principles*", which are commonly used in the angiosperm phylogeny:

1. Usually the woody plants have preceded the herbs, climbers and vines.
2. Annuals have been derived from both biennials and perennials, and biennials have developed from perennials.
3. Terrestrial habit preceded the aquatic or epiphytic habit.
4. Dicotyledons are more primitive than monocotyledons.
5. Alternate leaves are primitive while opposite or whorled leaves are advanced.
6. Simple leaves preceded compound leaves.
7. Bisexual flowers are primitive while unisexual flowers are advanced.
8. Multi-membered flowers are primitive while those with a few members are advanced.
9. Actinomorphic flowers preceded zygomorphic flowers.
10. Free perianth parts are primitive compared to fused perianth parts.
11. Flowers with petals preceded apetalous flowers.
12. Many free stamens are primitive compared to a few or fused stamens.

13. Hypogynous condition is primitive while epigynous and perigynous conditions are derived.
14. Apocarpous condition is primitive to the syncarpous condition.
15. Pollens having one pore preceded those with 2 or more pores.
16. Arillate seeds with a small embryo and well-developed endosperm are primitive than non-arillate seeds with a large embryo and ill-developed endosperm.
17. Axile placentation is primitive while free-central placentation is derived.
18. Single fruits preceded aggregate fruits.
19. Capsule fruits preceded drupe or berry fruits.

14.4 ANGIOSPERM PHYLOGENY: A GENERALLY ACCEPTED PICTURE

The available fossil evidences show that the angiosperms originated roughly 165 million years ago in the Mesozoic era, probably in the Jurassic period or perhaps even earlier than that, and appeared in great outburst in Upper Cretaceous time i.e., about 110 million years ago. But the actual origin or true relationships of this large group of plants, dominating the earth today, is still not known.

Earliest known fossil angiosperms are the members of Ranales as well as of Amentiferae. Because of this, some phylogenists believe that Ranales is the basal group in the evolution of angiosperms while others believe that Amentiferae should be considered as most primitive. But till today, all these are mere assumptions about the possible ancestors of angiosperms. Some assumptions consider the seed ferns to be the ancient while others consider some group of the Cycads, etc. The sum total of all we know about it is that suddenly, in Upper Cretaceous time, there appeared a large outburst of already highly evolved and advanced angiosperms.

14.5 MONOPHYLETIC OR POLYPHYLETIC ORIGIN

The question of phyly of the angiosperms is still not solved, mainly because of their inadequate fossil records. But this much is sure that the angiosperms are a natural group, and contain such characters which make them unique from all other vascular plants (Takhtajan, 1969; Stebbins, 1974).

Faegri (1980), Dahlgren (1983) and several other phylogenists believe that angiosperms are *monophyletic,* i.e. a group originated from a single ancestor at a single time in the past. They believe so because present-day angiosperms show remarkable consistency in their characters, i.e. uniform staminal structure, characteristic endothecial layer of the anther wall, double fertilization, formation of triploid endosperm, and presence of sieve tubes in all. Hickey and Doyle (1977) also support the *monophyletic* origin of angiosperms on the basis of their studies of monosulcate pollens. Dahlgren (1983) believes that the ancestor of the present-day angiosperms was a gymnospermous member. However, no definite fossil evidences are available in favour of the monophyletic origin.

Eames (1961), Cronquist (1965), Meeuse (1970, 1975), Krassilov (1977) and several other phylogenists have been of the view that angiosperms are *polyphyletic,* i.e. dicots and monocots originated at different times from different primitive stocks, and attained their present status through parallel or convergent evolution. Fossil records, variety in perianth and the nature of carpel in both dicots and monocots also support the theory of *polyphylesis.* The primitive orders of both the monocots and dicots do not show any close relationship in their characters, and this also favours the polyphyletic origin.

However, in the light of the available fossil records and other characters, it appears that angiosperms, as a group, are monophyletic, and their families or groups of families are polyphyletic. To find out the exact ancestors of angiosperms, more fossil records should be investigated.

14.6 FOSSIL RECORDS OR TIME OF ANGIOSPERMS ORIGIN

The available fossil records indicate that the angiosperms spread over the entire earth from the equator to the polar regions about 100 million years ago and during a period of only about 10 million years. During this period, the vegetation was dominated by the gymnosperms and ferns. But some phylogenists, such as Wolf *et al.* (1975), believe that they originated in the early Mesozoic or even in the late Palaeozoic, i.e. about 250 million or more years ago. However, the majority of the fossil records point out their origin to be about 130 to 135 million years ago during the early Cretaceous, and the same view is supported by Taylor (1981) on the basis of his studies of fossil pollens.

The angiosperms dominated the earth during the later parts of the Cretaceous period. Fossils of the monosulcate pollen grains are reported from the lower Cretaceous rocks. Such type of pollen grains are characteristic of the primary monocotyledons and dicotyledons, both. In the early years of Cretaceous, the angiosperms split into monocots and dicots and both of them became prevalent in the terrestrial flora along with ferns and gymnosperms in the later years.

Professor David Dilcher of Indiana University and his associates (Dilcher, 1979; Retallack and Dilcher, 1981) provided sufficient evidence of the importance of fossil records in the early history of angiosperms. They discovered a 95 million years old fossil bisexual flower, thereby indicating the coexistence of unisexual and bisexual flowers in the early history of the flowering plants.

Fossil pollens and leaves of several Magnoliales, Ranunculales, Theales, Hammamelidales and some monocots have been reported from the Maestrichtian stage of the Upper Cretaceous period, i.e. about 70 million years ago. Muller (1970) reported pollen of members of Compositae from the Oligocene of Tertiary period of the Cenozoic era, i.e. only about 25 to 30 million years ago.

14.7 PROBABLE ANCESTORS OF ANGIOSPERMS: SOME THEORIES

To identify the ancestors of the flowering plants is not easy and this aspect has been discussed in detail by phylogenists. It is difficult to find a group of plants, that was not at sometime or the other considered the ancestor of angiosperms. Various orders of gymnosperms (Bennettitales, Cordaitales, Cycadofilicales, Cycadales, and Gnetales), pteridophytes (Psilophytales, Lycopodiales and Equisetales), and even algae have been considered as the probable ancestors of angiosperms by different workers. Charles Darwin called the origin of angiosperms an "abominable mystery" in the absence of definite evidence which could help in tracing their probable ancestors. In spite of the accumulation of huge literature the problem still remains unresolved.

Following are the views of various workers:

1. *Isoetes-Monocotyledon Theory* Campbell (1928) advocated that monocotyledons have been derived from *Isoetes* via Ophioglossaceae. Linear leaves and cormous habit of *Isoetes* were correlated with some simple aquatic plants of monocots. Details of the embryo of *Isoetes*

and the anatomy of its older sporophyte show similarity with some lower aquatic monocots, such as *Najas flexilis.* But neither any fossil evidence nor similarity between the angiosperm flower and sporophyll of *Isoetes* exist to support this theory.

2. *Pteridosperm Theory* Majority of the phylogenists (Andrew, 1947; Arnold, 1949; Thomas, 1955; Cronquist, 1968) believe that seed ferns (Pteridosperms or Cycadofilicales) were the ancestors of angiosperms. Vascular histology, stelar structure and available fossil records also favour this theory. The most interesting evidence in favour of the pteridosperm theory is the absence of vessels in the secondary wood of some primitive angiosperms and in seed ferns. In both, the pteridosperms and angiosperms, the sporangial development is eusporangiate, and the amphiphloic stele is also common. But it is not possible to explain the complex pteridosperm seed in terms of the simple angiosperm ovule.

   Cronquist (1968), however, mentioned that although "most students of phylogeny provisionally accept the seed ferns as the probable ancestors of the angiosperms", yet it should be accepted only after a definite fossil evidence is traced.

3. *Caytonialean Theory* Caytoniales, a Middle Jurassic order of fossil gymnosperms, seem to have a few angiosperm-like features, and are looked upon as their possible ancestors. The ovules in this group of seed ferns were semi-enclosed in small pouches, their leaves had reticulate venation, and their sporangia resembled superficially with the anthers of angiosperms. However, the caytonialian order was strongly pinnately veined and the angiosperm carpel is palmate. Most angiospermic ovules have two integuments while Caytoniales had single integument.

   This theory was first proposed by Thomas (1925, 1936) and was supported by Stebbins (1974) on the basis of the homology of ovules of angiosperms with those of Caytoniales.

4. *Glossopteridalean Theory* Dahlgren (1983) has considered the seed fern order Glossopteridales as the probable ancestors of angiosperms. He suggested that the "Cupules" in the members of this order correspond to the outer integument of an angiospermic ovule. Magasporophylls of glossopterids are identical with a typical leaf of angiosperms. However, there is no similarity between the pollen of the two.

5. *Bennettitalean Theory* On the basis of the nature and organization of the reproductive structures of the members of Bennettitales and their similarity with the flower of angiosperms, Bennettitales were considered as the ancestors of angiosperms. This theory was first proposed by Saporta and Marion (1885) followed by Arber and Parkin (1907).

   The strobili of the Mesozoic genus *Cycadeoidea* resemble the flowers of *Magnolia.* Both these structures are bisexual and contain an elongated axis having bracts, microsporophylls and megasporophylls. However, the stamens of *Magnolia* are free while the microsporophylls of Bennettitales are connate. The seeds of *Magnolia* and other primitive angiosperms are with copious endosperm and small embryo while those of Bennettitales were non-endospermic with a large embryo. The angiospermic stem has a small pith, thick vascular cylinder and thin cortex while that of Bennettitales had a large pith, thin vascular cylinder and thick cortex.

6. *Coniferales-Amentiferae Theory* Coniferales are treated as the probable ancestors of angiosperms by several phylogenists including Engler (1892) and Rendle (1904). They pointed out several resemblances between conifers and angiosperms, and treated Amentiferae as the

most primitive dicots. Flowers in amentifers (Casuarinaceae, Salicaceae, Fagaceae, etc.) are simple and naked like those of conifers. Fertilization in conifers is similar to that of angiosperms (Doyle, 1945). The seed-scale complex of *Juniperus* also resembles the gynoecia of Amentiferae. However, definite differences prevail between the ovuliferous scales of conifers and the angiosperm carpel.

7. *Gnetales-Angiosperm Theory* Gnetales, the transitional group between angiosperms and gymnosperms, are also considered as the ancestors of angiosperms. *Gnetum* bears angiosperm-like leaves, female gametophyte and vessel-bearing wood. Gnetales have two cotyledons like dicots. Their ovules bear two integuments as is the case with most of the angiosperms. Stamens of all the three genera of Gnetales (*Gnetum, Ephedra* and *Welwitschia*) are similar to those of angiosperms. Gametophytes of *Gnetum* and *Welwitschia* are highly reduced like those of angiosperms. However, the vessels of Gnetales originated in an entirely different manner from those of angiosperms (Cheadle, 1953). Moreover, the details of the vascular anatomy strongly refutes the suggestion of the existence of any similarity between the two groups.
8. *Durian Theory* Corner (1949) believed that *Durio zibethinus,* a member of Bombacaceae found in Burmese and Malayan forests, is the only "surviving member of primitive angiosperms", and after the name of the genus, the theory is named as "*durian-theory*". *D. zibethinus* is a cauliflorous tree with large spiny capsules and fleshy arillate seeds. Arillate genera are found in about 45 families of angiosperms according to Corner (1949), mostly distributed in the tropics, like *Durio.* According to him within these arillate families, a reduction series can be traced from genera with arillate fruits to genera having fruits of other types. Parkin (1953) and Eames (1961) criticised this theory.
9. *Pentoxylon Theory* Meeuse (1961) compared *Pandanus,* a monocot, with Pentoxylales, a group of fossil pteridophytes, and listed a number of common features between the two. He proposed that *Pandanus* "is a direct descendent of a group of plants practically identical" with Pentoxylales. *Pandanus* and Pentoxylales, according to Meeuse (1961), resemble each other in their stem and leaf anatomy, male and female inflorescence, and pollen. Both bear terminal tufts of strap-like leaves in spiral arrangement, copious endosperm and minute embryo. However, Pant and Kidwai (1971) attributed these similarities, between *Pandanus* and Pentoxylales, to the parallel evolution.
10. *Czekanowskiales Theory* Jones and Luchsinger (1987) stated that "the Czekanowskiales, an order of Jurassic seed plants belonging to the class Ginkgopsida, are regarded by some as early angiosperms". They further mentioned that similar to Caytoniales, the ovule in this order also "was surrounded by a cupule having a flange that somewhat resembled a stigmatic surface".

14.8 PRIMITIVE ANGIOSPERMS

Undermentioned are some of the views regarding the type of primitive living flowering plants:

1. Some phylogenists believe that the first flowering plants were upland plants that diversified in tropical or arid regions about 250 million years ago in Permian. Absence of fossils from this period is explained by pointing out that upland plants are usually not preserved as fossils.

2. A few other phylogenists are of the view that the primitive flowering plants were tropical mesophytic trees with pinnate leaves and the fruits in the form of clusters of large arillate follicles.
3. Another view is that the early flowering plants were insect-pollinated, *Magnolia*-like, evergreen trees of tropical upland regions. Fossils, resembling the present-day Magnoliaceae are common among the fossil remains of Lower Cretaceous period. Phylogenists believe that the primitive angiosperms had a minute embryo embedded in the copious oily endosperm, and the same is true of the embryo and endosperm of *Magnolia.* All these characters led some botanists to call Magnoliales as *Living Fossils.*
4. According to Stebbins (1974), primitive angiosperms were small woody plants with small leaves and moderate-sized flowers.
5. According to the Takhtajan (1980), the primitive angiosperms were woody plants with simple, entire, pinnately-veined leaves; moderate-sized, solitary, axilary or terminal flower having perianth of modified bracts; leaf-like stamens; monosulcate pollen; and conduplicate carpel.
6. Hutchinson (1973) believed that the primitive angiospermic flowers were medium-sized and grouped together in lateral clusters, much like those of *Drimys winteri* of the family Winteraceae.
7. Engler and Prantl (1925) and Wettstein (1935) regarded Amentiferae as the primitive angiosperms because their flowers do not possess petals, and inflorescences resemble the cones of gymnosperms.
8. Bailey (1956) regarded Ranales as the most primitive order of living dicotyledons. This is supported by the evidences from floral anatomy, wood anatomy and pollen grain structure.

14.9 PLACE OF THE ORIGIN OF ANGIOSPERMS

Where the angiosperms might have first originated, is also not yet finally decided. Three divergent views regarding the place of their origin are undermentioned:

1. Heer (1868) suggested that the angiosperms originated at high latitudes in the north polar regions. Hooker, Asa Gray, Arnold, etc. supported his view. From the polar Arctic and Antarctic regions the angiosperms spread in successive waves across the whole earth. However, Croizat (1952) and Takhtajan (1969) rejected firmly the Arctic or Antarctic origin of flowering plants.
2. Hallier (1912) proposed that the angiosperms originated first in the *basin of the Pacific Ocean,* at places such as New Zealand, New Caledonia, Andes and the Hawaiian Islands. According to Takhtajan (1969), the western part of the basin of Pacific Ocean is particularly rich in the primitive angiospermic families and covers the areas such as Assam, Burma, China, Japan to Australia, New Zealand, New Caledonia and Fiji, and therefore the angiosperms probably originated in this part of the eastern and south-eastern Asia and Australia.
3. According to Dutoit (1947), the discovery of some angiosperm-like inflorescences from the Triassic beds of South Africa indicates that this must be the place of the origin of angiosperms.

14.10 ORIGIN OF MONOCOTS: SOME VIEWS

1. *Hegelmaier* (*1878*) suggested that the embryo of monocots originated as one of the two cotyledons of a typical dicot embryo that failed to develop. This hypothesis was named as *abortion hypothesis*. This view was suppored by workers such as Metcalfe (1936) and Eames (1961).
2. *Engler* (*1892*) proposed that monocots originated from Isoetales, Ophioglossales and a few other pteridophytes.
3. *Sargant* (*1903, 1908*) proposed that monocots originated as a result of the adaptation of the aquatic habit to the geophyllous or marshy habit, and called them as *aquatic geophytes*.
4. *Hallier* (*1914*) concluded that monocots originated from a dicotyledonous family Lardizabalaceae.
5. *Cheadle* (*1953*) has shown that vessels in dicots and monocots originated independently, and therefore the ancestors of monocots must have been some vesselless dicots such as Nymphaeales or Magnoliales.
6. *Hutchinson* (*1959*) opined that monocots originated from Ranunculaceae.
7. *Takhtajan* (*1961*) believed that monocots might have originated from some extinct vesselless herbs having apocarpous gynoecium and monocolpate pollen similar to the modern Nymphaeales.
8. *Stebbins* (*1974*) was of the opinion that monocots probably originated from some *Drymis*-like ancestor having vesselless wood and short internodes.
9. *Burger* (*1977*) opined that monocots were derived from some dicot ancestors similar to Chloranthaceae of Piperales. He believed that Piperales were related more closely to monocots than any other living dicots. He suggested that simple flowers, like those of Chloranthaceae, came together by loosing their internodes and formed trimerous flowers of several monocots and Piperales.

14.11 LINES OF EVOLUTION IN ANGIOSPERMS

Ranales are considered as the most primitive angiosperms because of (i) absence of vessels in over 100 living representatives of families (Magnoliaceae, Winteraceae, Alismaceae, etc.) of this order, (ii) presence of a triradiate ridge on the pollen grains of the genus *Schizandra* of Magnoliaceae, (iii) binucleate pollen at the time of its release from a microsporangium in several Magnoliaceae, Winteraceae, Ranunculaceae, etc., and (iv) peculiar floral morphology and anatomy, etc.

The general opinion of phylogenists is that at least *three* main lines of general evolution are observed from the Ranalian plexus. (1) The *first line* consists of Rhoeadales, Sarraceniales, Parietales, Malvales, Geraniales, Sapindales, Rhamnales and some Tubiferales, (2) the *second line* attains the climax of Campanulales through Rosales, Myrtales, Umbellales and Rubiales, and (3) the *third line* included Liliales and Orchidales.

In the *first line* the hypogynous condition remained unchanged. Rhoeadales and Sarraceniales evolved parallaly from Ranales. Development of the gynophore in Capparidaceae shows a definite relation between Parietales and Rhoeadales. Malvales evolved through Bombacaceae. There exist definite relationships between Malvales and Geraniales. Geraniales are related to Sapindales, and Sapindales to Rhamnales. Fusion of sepals, petals and carpels, as well as the epipetalous condition is seen in Solanaceae. Labiatae is the most advanced family of this line. Besides the fusion of sepals, petals and carpels and epipetalous condition, the number of floral parts in Labiatae are reduced to four or two.

In the *second line* of evolution a change is observed from hypogyny to epigyny through perigyny. The change from hypogyny to epigyny is seen clearly in Rosales. The formation of hypanthodium shows a change of perigyny into epigyny in Myrtales. Epigyny and syncarpy in Umbellales show an advancement over Myrtales. Umbellales are related to Rubiales through the characters such as umbel-like inflorescence, suppression of calyx, epigyny, etc. Campanulales top this line with the presence of pappus, irregular flowers, and fusion of floral parts.

The *third line* shows a few resemblances between monocotyledons (e.g. Alismataceae) and dicotyledons (e.g. Ranunculaceae). Liliales and Ranales differ in the number of floral parts and fusion of carpels. Epigynous condition of Iridaceae and Amaryllidaceae is a stage of advancement over Liliales. There exist evidences of the origin of Gramineae from Liliaceae. But the reduction in number of stamens and carpels and also the loss of perianth parts indicate that Gramineae are more advanced than Liliaceae. Orchidaceae possess insect-pollinated, irregular and showy flowers, and is thus the most advanced family of monocotyledons.

Test Your Understanding

1. What do you mean by the word 'phylogeny'? Explain some general principles of angiosperm phylogeny.
2. Write brief notes on:
   (a) Monophyletic and polypyhyletic origin of angiosperms.
   (b) Primitive angiosperms.
3. Describe at least five theories, put forward by phylogenists, regarding the "probable ancestors of angiosperms".
4. Explain some views of scientists about the place of origin of angiosperms.
5. Phylogenists have opined at least three different lines of evolution of angiosperms. Explain them briefly.
6. Discuss some basic aspects of the origin and evolution of angiosperms.

Suggested Reading

Axelrod, D.I., 1952, A theory of angiosperm evolution, *Evolution.* **6**: 29–60.

Bessey, C.E., 1915, Phylogenetic taxonomy of flowering plants, *Ann. Mo. Bot. Gard.* **2**: 109–164.

Burger, W.C., 1981, The monocot theory of angiosperm evolution, *Evol. Theory* **5**: 189–225.

Cheadle, V.I., 1953, Independent origin of vessels in monocotyledons and dicotyledons, *Phytomorphology* **3**: 23–44.

Daghlian, C.P., 1981, A review of the fossil record of monocotyledons, *Bot. Rev.* **47**: 517–555.

Doyle, J.A., 1978, Origin of angiosperms, *Ann. Rev. Ecol. Syst.* **9**: 365–392.

Krassilov, V.A., 1977, The origin of angiosperms, *Bot. Rev.* **43**: 13–176.

Meeuse, A.D.J., 1975, Origin of the Angiosperms—Problem of inaptitude, *Phytomorphology* **25**: 373–379.

Pant, D.D. and P.F. Kidwai, 1971, The origin and evolution of flowering plants, *J. Indian Bot. Soc.* **50A**: 242–274.

Puri, V., 1967, The origin and evolution of angiosperms, *J. Indian Bot. Soc.* **46**: 1–14.

Scott, R.A., E.S. Berghoorn and E.B. Leopold, 1960. How old are the angiosperms? *Am. J. Sci.* **258A**: 284–299.

Stebbins, G.L., 1974, *Flowering Plants: Evolution Above the Species Level,* The Belknap Press, Cambridge.

Takhtajan, A., 1969, *Flowering Plants: Origin & Evolution,* Smithsonian Inst. Press, Washington.

Tutin, T.G., 1952, Phylogeny of flowering plants: Facts or fiction? *Nature* (London) **169**: 126–127.

Wolfe, J.A., J.A. Doyle and V.M. Page, 1975, The basis of angiosperm phylogeny: Palaeobotany, *Ann. Mo. Bot. Gard.* **62**: 801–824.

CHAPTER 15

BOTANICAL LIBRARY

15.1 BOTANICAL LIBRARY AND SYSTEMATICS

A botanical library is a storehouse of botanical books, journals, periodicals, magazines, photographs, movies, newspapers, dissertations, etc. All printed materials, drawings, films, paintings, and other similar articles belonging to the plant sciences are also stored there in a proper system.

It is also a documentation centre of botanical activities, and stores the data related to all the branches of plant sciences. It may also be called as a service facility to the botanists as it provides the required literature.

A botanical library exists for the benefit of the entire society in general and for the persons interested in botany in particular. However, it is such a huge repository of the world's botanical literature that no single individual can utilize this system in its entirety.

The important taxonomic publications, available in the taxonomy section of a botanical library include floras, monographs, literature indexes, and name indexes. The information regarding the plants of a given geographical region are available in a *flora,* while a *monograph* contains information on the systematics of a given taxonomic group of plants on a worldwide basis. The binomial system of nomenclature is used in both the floras and monographs. Various systems of indexing and abstracting the taxonomic literature are available in *Biological Abstracts* and *Bibliography of Agriculture.* For the name indexes of plants, the world botanists rely on books such as *Index Kewensis, Gray Herbarium Index* and J.C. Willis's *A Dictionary of the Flowering Plants and Ferns.*

15.2 COMMON TERMS USED IN A BOTANICAL LIBRARY

1. *Document* A written item such as a letter, a manuscript, notes, an article or a book is called a document.
2. *Documentation* An introduction to the written documents, specimens, etc. used as an evidence or a source of information in taxonomic studies is called documentation.
3. *Literature Retrieval* It is the process of locating and identifying by referring to the books, manuals, articles, and the exact information within the entire published literature that is of interest to the user.

4. *Document Retrieval* It is the process of locating and identifying both the published as well as unpublished documents such as specimens, manuscripts, illustrations, notes, films, etc.
5. *Voucher* It is a specimen or written document used for the verification of information.
6. *Citation* A quotation from a book or author is called a citation.
7. *Reference* A source of information with a definite direction is called a reference.
8. *Reference Book* An authoritative book consulted for facts or the background information is called a reference book.
9. *Bibliography* The list of books, research articles, or other writings, usually of one author or on one subject or topic is called bibliography.

15.3 USE OF BOTANICAL LIBRARY

Radford (1986) mentioned that the botanical library is useful to the systematists as an information source, a documentation centre, a data storehouse, and also as a service facility.

1. From the botanical library, taxonomists procure information to verify facts about the available literature, procedure, citation, classification, and all other related aspects of his/her project, and in that way the library is used *as an information source.*
2. The library is used *as a documentation centre* as it is a repository of rare publications, first editions, original manuscripts, illustrations, photographs, etc.
3. The library is also used *as a data storehouse* for research because it stores all the published data related to the systematic projects.
4. The big libraries are used *as a service facility* for providing the technical know-how of their available resources. A reference service of these libraries helps in procuring the requisite material, photocopies, microfilms, necessary for the projects in systematics. Most of the big libraries also have an interlibrary loan service for supply of materials not available locally.

15.4 BOTANICAL LIBRARY AS A TRAINING CENTRE

Several big libraries train the students for using the library in a proper and systematic manner, and thus function as training centres. The trained library staff trains the young students by (i) introducing them the references of common use, such as manuals, dictionaries, atlases, maps, encyclopedias, indexes, etc., (ii) familiarising them with the literature of the current developments such as periodicals, journals, monographs, recent bibliographies, etc., (iii) informing them of the proper rules for making catalogues, bibliography, etc., and (iv) providing them the way of analyzing the literature for scientific writings, etc.

15.5 CLASSIFICATION USED IN A BOTANICAL LIBRARY

In the botanical libraries, two systems of classification are used, i.e. the Dewey Decimal (DD) system and the Library of Congress (LC) system. The Dewey Decimal system is an old system of classification and is now being replaced by the LC-system in majority of the libraries of the world. For details of these systems, readers may refer Swift (1970).

15.6 MAJOR BOTANICAL LIBRARIES OF THE WORLD

A huge taxonomic literature is available in five great national libraries of the world, *viz.* Library of Congress in Washington, U.S.A.; Library of the Academy of Sciences in Leningrad, Russia; The National Library of Beijing, China; Library of the British Museum in London, England; and The Bibliotheque Nationale in Paris, France.

The other big American university libraries are at California (Berkeley), Cornell, Harvard, Iowa State, Minnesota, Illinois, Columbia, Wisconsin, Chicago, Michigan, Yale and Texas.

Several large botanical libraries are associated with the botanical gardens, museums and botanical societies. Chief among them are the libraries associated with the British Museum, the New York Botanical Garden, Missouri Botanical Garden, Gray Herbarium, Massachusetts Horticultural Society at Boston, and Academy of Natural Sciences at Philadelphia.

Some *major Indian botanical libraries* are associated with The Botanical Survey of India, Kolkata; National Botanical Garden, Lucknow; Forest Research Institute, Dehradun; Indian Agricultural Research Institute, New Delhi; and universities of Delhi, Chennai, Meerut, Allahabad, and Kolkata. A big botanical library was established at Meerut College, Meerut by Professor V. Puri during 1948 and 1967. Till 1970, it used to subscribe 166 research journals. It also contains over 8,000 books.

15.7 FUTURE INFORMATION SYSTEMS

The botanical libraries have been the main information systems during the last 500 years, i.e. since the invention of the movable type. But, the introduction of the high-speed electronic computers during the last few decades has brought about a colossus change in the entire scene. Instead of the printed pages, a large amount of the botanical information is now stored in computers in all parts of the world. With the help of the data stored in computers, the plants may be identified more easily and correctly. One may say it with confidence that the computers are now the main information tools. Detailed taxonomic data are now available on the pages of a computer memory, and these data are now easily available to the entire world via *internet.* The taxonomists are now able to get the latest information on a topic of his/her choice through computers. But this does not mean that botanical libraries are soon going to be replaced by computers. However, the present days are now of internet, and detailed taxonomic information is now easily available in computers.

Test Your Understanding

1. How will you define a botanical library?
2. Write a note on "botanical library as a training centre".
3. Give one major difference between Dewey Decimal (DD) system and Library of Congress (LC) system with reference to a botanical library.
4. Give a brief account of major botanical libraries of the world.
5. With reference to a botanical library, explain the following terms.
   (a) Document
   (b) Literature retrieval
   (c) Reference book
   (d) Bibliography

Suggested Reading

Radford, A.E., 1986, *Fundamentals of Plant Systematics,* Harper and Row, New York.

Swift, L.H., 1970, *Botanical Bibliographies: A Guide to the Bibliographic Materials Applicable to Botany,* Burgess Publish. Comp., Minnesota.

CHAPTER 16

HERBARIUM

16.1 HERBARIUM AND ITS LIMITS

A collection of dried and pressed plants arranged according to a classification system and available for study or reference is known as *herbarium* (plural, *herbaria*), a name first applied by Linnaeus.

Plant specimens are usually mounted on a sheet of high quality paper. Properly dried, pressed, and identified plant specimens are placed in thin paper folders (*species covers*), which are kept together in thicker paper folders (*genus covers*), and finally they are incorporated into the herbarium cupboards in their proper position. Some fleshy members (e.g. Cactaceae) are preserved in liquid preservatives instead of being pressed and dried. Bulky plant parts (e.g. dry fruits, cones of several gymnosperms, etc.) are dried without pressing and are stored in special boxes.

A herbarium may contain a few hundred locally collected plant specimens kept in a small place, or it may contain millions of them collected from different parts of the world and housed in a very big building. Herbaria in different countries remain associated with colleges, universities, scientific societies, research institutes, botanical gardens, or well-funded government organizations. They may contain a local collection, or flora of a district, state, country, continent or several continents. There are also private herbaria.

The *world's largest herbarium* is at Royal Botanic Garden, Kew, Richmond, Surrey, England, containing over 5 million specimens. But findings of Holmgren *et al.* (1981) suggest that the Herbarium of the Museum of Natural History, Paris contains over 6.5 million plant specimens. The *biggest Indian herbarium* is of Botanical Survey of India at Kolkata holding over 1.3 million specimens (Holmgren *et al.* 1981).

Over 1600 world's most important herbaria are listed in *Index Herbariorum* of Holmgren *et al.* (1981). Missouri Botanical Garden publishes a valuable monthly newsletter, entitled *Herbarium News*, covering the news of different herbaria of the world.

According to Shelter (1969), the herbarium resources of the world include as many as 250 million specimens. However, the total number actually reported from different institutions of the world was only 148 million specimens.

16.2 HISTORY OF HERBARIUM DEVELOPMENT

Luca Ghini (1490–1556) of Italy has been the initiator of the art of herbarium. He collected plants, dried and affixed them on paper in the form of herbarium specimens. Gherards Cibo, a student of Ghini, started collecting and preserving plant specimens in 1532. John Falconer, an Englishman, learned the art of preparing a herbarium, perhaps from Luca Ghini, and prepared one as early as 1553. Holmgren *et al.* (1981) mentioned that the great herbarium of the Museum of Natural History, Paris was founded in 1653.

Up to the time of Linnaeus (1707–1778), the usual practice of preparing a herbarium was mounting the specimens and binding them into volumes. The plants were usually sewn by thread. Linnaeus started the current practice of mounting specimens on single sheets of paper and storing them horizontally. By such a simple beginning of Linnaeus, herbaria have developed into facilities of housing millions of specimens, usually in steel cases. The Herbarium of the Indian Botanical Gardens, Kolkata (now known as Central National Herbarium) was established in 1832.

In U.S.A., the oldest herbarium was started at Salem College in 1772. The herbarium of Academy of Sciences, Philadelphia was started in 1812, University of Michigan in 1838, Missouri Botanical Garden in 1857, and U.S. National Herbarium in 1868. The large herbaria of U.S.A. are those of the Harvard University, the New York Botanical Garden, and the Missouri Botanical Garden.

In the earlier days the herbaria existed only in the form of a few pressed and dry specimens of local or regional significance. But now, many of them are the centres of advanced research in the field of taxonomy. Herbaria of Paris, Kew, Leningrad, Cambridge, Geneva and Kolkata are big centres of taxonomic research.

16.3 A MODERN HERBARIUM

According to Fosberg (1946), a modern herbarium is a "great filling system for information about plants, both primary in the form of actual specimens of the plants and secondary in the form of published information, pictures and recorded notes". However, according to Radford (1986), a modern herbarium "is a research, training, and service institution that serves as a reference centre, documentation facility, and data storehouse". It includes "diverse collections of flowering plants, gymnosperms, ferns, mosses, liverworts, lichens, fungi, algae, and fossils." It also contains "microscope slides, photographs, photomicrographs, wood specimens, camera lucida drawings, field notebooks, diaries, letters, unpublished reports, manuscripts, reprints, and botanical illustrations" according to Radford (1986).

Modern herbaria are utilized *as reference centres* for identification of plants by the botanists doing research work in taxonomy, ecology, agriculture, pharmacy, etc.

Modern herbaria are the *documentation centres* because they contain the specimens of new taxa, plants belonging to new discoveries, plants of economic importance, voucher specimens of cytological and cytogenetic studies, photographs of important plants, and all documents related with systematic research.

Geneticists, chemists, pharmacists, etc. use the modern herbaria as *a data storehouse.* They store data on ecology, habitat, distribution of plants, etc.

Modern herbaria provide training for young undergraduates, solve queries of scientists, environmentalists, doctors, public school students, etc., and thus work as *service institutions.*

16.4 TWENTY MAJOR HERBARIA OF THE WORLD

Undermentioned is the list of 20 major herbaria of the world, along with the number of their specimen holdings as worked out by Holmgren *et al.* (1981):

1. Museum of Natural History, Paris .... 6.5 million
2. Royal Botanic Gardens, Kew .... over 5 million
3. Komarov Botanical Institute, Leningrad .... over 5 million
4. Conservatory and Botanical Garden, Geneva .... 5 million
5. Combined Herbaria, Harvard University, Cambridge .... 4.5 million
6. New York Botanical Garden, Bronx .... 4.3 million
7. U.S. National Herbarium, Washington .... 4.1 million
8. British Museum of Natural History, London .... 4 million
9. University of Lyon, Lyon .... 3.8 million
10. Natural History Museum, Vienna .... 3.5 million
11. Missouri Botanical Garden, Saint Louis .... 2.9 million
12. Field Museum of Natural History, Chicago .... 2.4 million
13. University of Uppasala, Uppasala .... 2.2 million
14. National Botanical Garden of Belgium, Brussels .... 2.1 million
15. Botanical Garden and Botanical Museum, Berlin .... 2 million
16. Academy of Natural Sciences, Philadelphia .... 2 million
17. Royal Botanic Garden, Edinburgh .... 1.7 million
18. University of California, Berkeley .... 1.5 million
19. University of Michigan, Ann Arbor .... 1.4 million
20. Botanical Survey of India, Kolkata .... 1.3 million

However, the latest estimates show that in India, there exist two more herbaria having more than a million specimens. According to Tiagi and Kshetrapal (1988) these are (1) The Central National Herbarium, Indian Botanical Gardens, Kolkata (25,00,000 specimens), and (2) Herbarium of the Forest Research Institute, Dehradun (15,00,000 specimens). But in majority of the other Indian books, it is mentioned that herbarium of F.R.I. Dehradun has only about 300,000 specimens and not 15,00,000.

16.5 MAJOR INDIAN HERBARIA

Undermentioned is the list of some major Indian herbaria, along with the approximate number of their specimen holdings as mentioned by Naik (1984), and Tiagi and Kshetrapal (1988):

1. The Central National Herbarium, Kolkata .... 25,00,000 specimens
2. Herbarium of Forest Research Institute, Dehradun .... 3,00,000 specimens
(15,00,000 according to Tiagi and Kshetrapal, 1988)
3. Botanical Survey of India
   (i) Eastern Circle Herbarium, Shillong 10,00,000
   (ii) Southern Circle Herbarium, Coimbatore 1,75,000
   (iii) Western Circle Herbarium, Pune 50,000
   (iv) Northern Circle Herbarium, Dehradun 42,000
   (v) Central Circle Herbarium, Allahabad 45,000
4. Blatter Herbarium, Mumbai 1,00,000
5. National Botanical Garden Herbarium, Lucknow 80,000
6. Herbarium of Industrial Section, Indian Museum, Kolkata 50,000
7. Herbarium of Rajasthan University, Jaipur 30,000
8. Herbarium of School of Plant Morphology, Meerut College, Meerut 25,000
9. Herbarium of Delhi University, Delhi 15,000
10. Herbarium of Jiwaji University, Gwalior 15,000

16.6 FUNCTIONS OF HERBARIA

Some of the general functions of herbaria are mentioned below:

1. Plant specimens are permanently stored in herbaria, and therefore they are the major sources of information about plants and vegetation.
2. Preserved specimens of herbaria are used in almost all types of taxonomic research.
3. A picture of all species of a genus, or all the genera of a family may be gathered only in the herbarium.
4. The classification of the world's flora is based mainly on the herbarium material.
5. List of the endangered species of any region may be prepared only by herbarium specimens.
6. Limited individual collections are identified only with the help of herbaria.
7. Monographs of genera or families are prepared only by the herbarium specimens.
8. Our knowledge of the distribution of plants, evolution, and several taxonomic problems etc. is based mainly on the herbarium specimens.
9. Big herbaria provide training to young students in herbarium practices.
10. Exact area, region or location of the occurrence of important plants may be gathered from the herbarium specimens.
11. Herbaria preserve type specimens, and thus serve as a repository of chromosomes, chemotaxonomy, and experimental voucher specimens.

12. Herbarium material is used in studying the palynology, anatomy, and chemical aspects of desired plants.
13. Herbaria provide loan of specimens for study at other institutions.
14. Herbaria preserve the national plant wealth, and provide scientific information to the public regarding the plants.

16.7 PRECAUTIONS FOR USING HERBARIUM

1. Handle the plant specimens with extreme care because they are of scientific value and generally irreplaceable.
2. Dry specimens are brittle and easily damaged, and so keep the specimen sheets flat.
3. Store the specimens always in herbarium cases.
4. Keep the materials in folders when not in use.
5. Never keep heavy books or other heavy objects on the specimens.
6. Support the specimens with a ventilator when carrying them even for short distances.
7. If you notice that specimens are being damaged by insects, etc. inform the Curator immediately.
8. Do not write anything on sheets of herbarium, unless permitted to do so.
9. Do not reshelve specimens in the herbarium cases.
10. If you have taken specimens on loan from another institution, return them before the loan period is over.
11. The specimens, borrowed from other institutions, should be properly housed in fire-proof, dust-proof and pest-free cases.

Test Your Understanding

1. What do you mean by the word "herbarium"? Name the largest herbarium of the world. Where is it located and in which country?
2. Write a brief note on "a modern herbarium".
3. Make a list of 5 major herbaria of the world and also 5 major Indian herbaria.
4. Write six general functions of herbaria.
5. Comment on 'Modern herbaria work as service institutions'. What precautions does a herbarium require?

Suggested Reading

Croat, T.B., 1978, Survey of herbarium problems, *Taxon* **27**: 203–218.

DeWolf, G.P., 1968. Notes on making an herbarium. *Arnoldia* **28**: 69–111.

Fosberg, F.R. and M. Sachet, 1965, Manual for tropical herbaria, *Regnum Vegetabile* **39**: 5–132.

Holmgren, P.K., W. Keuken and E.K. Schofield, 1981, *Index Herbariorum,*, Part I. *The Herbaria of the World*, 7th ed. Bohn, Scheltema and Holkema, Utrecht.

Shelter, S.G., 1969, The herbarium: Past, present and future, *Proc. Biol. Soc. Washington* **86**: 687–758.

BOTANICAL GARDENS

17.1 WHAT ARE BOTANICAL GARDENS?

Botanical gardens are the institutions that maintain the living plant collections of different varieties of plants, including the ornamental and cultivated ones, wild, medicinal, of economic importance, of various geographical regions, of special interest, etc. They are of value not only to the botanists, horticulturists and foresters but also to the millions of tourists.

A big botanical garden contains plant species from several corners of the globe. It also includes greenhouses, a library, a herbarium, research laboratories, and several miscellaneous resources including photographs, paintings, illustrations, reprints, note-books and specimens of several types. It is, therefore, not merely a garden but a *botanical institution.*

Modern botanical gardens serve as centres for documentation, research, reference, data storage, education, conservation, and several other biological facilities to mankind.

At present there are over 600 botanical gardens in the world.

17.2 45 MAJOR BOTANICAL GARDENS OF THE WORLD: A CHRONOLOGICAL DIRECTORY

The gardens were maintained in all parts of the world since the early times also, as is evident from the old Indian classics such as Mahabharat, Ramayan and Vedas. For food, drugs, and beautiful flowers, man had begun to cultivate plants in gardens long before the dawn of history. The *Hanging Gardens* of Babylon are considered among the wonders of the ancient world. Several kinds of fruit, vegetable and drug plants were cultivated during the Middle Ages. Romans maintained small gardens as sources of medicine and as aids to medical studies.

Undermentioned are the historical details of some major botanical gardens of the world in a chronological order.

17.2.1 Botanical Gardens Established During 16th Century

1. The credit of establishing the first botanical garden of the world goes to *Luca Ghini* (A.D. 1490–1556), an Italian Professor of Botany. He established a *botanic garden at Pisa, Italy*

in 1543. He was its first director followed by Andrea Cesalpino (1554–1558). This garden does not exist today.

2. *Otto Botanico,* the *botanical garden of University of Padua, Italy,* was established by Professor Francis Bonafede on *June 29, 1545.* A large herbarium and a library are attached to this garden. More than 600 portraits of botanists are displayed in this garden.
3. *University Botanical Garden at Florence, Italy* was established in *1545* by Luca Ghini.
4. *Vatican Garden at Rome* was established in *1556* by Michele Mercati, a student of Andrea Cesalpino.
5. *Bologna Botanical Garden at Bologna, Italy* was founded in *1567,* and its first director was Ulisse Aldrovandi (1567–1605).
6. *Leiden University Botanical Garden* was established in *1587* at Leiden, Netherlands. Carolus Clusius, the Flemish-Austrian botanist, who was the director of this garden, was the first to establish the first greenhouse of the world in this garden in *1599.*
7. *Montpellier University Botanical Garden at Montpellier,* is one of the earliest botanical gardens in *France.* It was founded in *1593.*
8. *Botanical Garden, Heidelberg, Germany* was also established in *1593.*

17.2.2 Botanical Gardens Established During 17th Century

1. *Botanical Garden of Strasbourg, France* comes next in the chronological order. It was established in *1619.*
2. *Oxford University Botanic Garden, Oxford, England* was established in *1621.* This was the first botanical garden of Great Britain, and Earl of Danby was its initiator. In 1734, the first greenhouse of England was established in this garden.
3. *Jardin des Plantes,* a botanical garden of *Paris,* was established in *1640* under the name "Jardin du Roy". King Louis XIII of France ordered in 1626 to establish this royal botanic garden for the purpose of growing all kinds of medicinal herbs. It was actually an independent scientific institution, founded by the king to promote the teaching of pharmaceutical botany. Its present herbarium and palaeobotanical departments are among the world's best. It is actually the oldest and most important non-university botanical garden, existing today in the world.
4. *Botanical Garden of Groningen, Netherlands* was established in *1642.*
5. *Botanischer Garten,* one of the great botanical gardens of the world, was established in *1646* in *Berlin, Germany.* A.W. Eichler, Adolf Engler and L. Diels were the directors of this garden.
6. *Botanical Garden of Uppasala, Sweden* was founded in *1655* by O.J. Rudbeck. *Linnaeus* worked as the director of this garden from 1741 to 1777, and the original part of this garden is now maintained as a memorial to Linnaeus. With the efforts of Linnaeus, a collection of more 300 species of this garden grew to over 3000 species within seven years.
7. *Royal Botanic Garden of Edinburg, Scotland* was established in *1670.* Its first director was J. Sutherland.
8. *The Chelsea Physic Garden* was founded in London, England in *1673,* as the garden of the Society of Apothecaries.

9. *Amsterdam Botanical Garden, Netherlands* was founded in *1682.*
10. *Botanical Gardens of Tokyo Imperial University were* founded in *1684.*

17.2.3 Botanical Gardens Established During 18th Century

1. *Botanic Garden of University of Moscow,* Russia, was founded in 1707.
2. *Druggist's Garden of St. Petersburg,* Russia, was established in *1713.* St. Petersburg is the Leningrad of today. The reorganized form of this garden is now V.L. Komarov Botanical Institute of the Academy of Sciences, Leningrad, Russia. The herbarium attached to this garden is one of the largest in the world, holding over 5 million specimens.
3. First botanical garden in America was established by John Bartram in *1731* near *Philadelphia.*
4. *Botanical garden of Vienna, Austria,* was founded in *1754.*
5. *Jardin Botanico de Madrid, Spain,* was established in *1755.*
6. *Royal Botanical Garden,*[1] *Kew, England,* owes its establishment in *1760* (Radford, 1986). However, Core (1955) stated that these gardens "were officially opened in 1841, with Sir William J. Hooker as the first director". This is the largest botanical garden of the world.
7. *Botanic Garden of Cambridge University, Cambridge, England,* was founded in *1762.*
8. First economic gardens, for the cultivation of spices, etc., were established on the island of *St. Vincent,* British West Indies, in *1764.*
9. *Hortus Botanicus* of University of Budapest, was established in *1771.*
10. *Botanical Garden of Coimbra, Portugal,* was founded in *1773.*
11. *Indian Botanical Garden,*[2] *Kolkata,* formerly known as Royal Botanic Gardens or Kolkata Botanic Garden, was founded in 1787 by Lt. Colonel Robert Kyd. It is "one of the great botanical gardens of the world" (Core, 1955).

17.2.4 Botanical Gardens Established During 19th Century

1. *The Jardin Botanico de Rio de Janeiro, Brazil,* was established in *1808.*
2. *Universitets Botaniske Have* is the oldest botanical garden of Oslo, Norway, founded in *1814.*
3. First botanical gardens of Australia are the *Botanic Gardens of New South Wales, Sydney.* These were founded in *1816.*
4. *Botanic Garden at Buitenzorg, Java,* was founded in *1817.* It is well known for the cultivation of coffee and rubber.
5. *Conservatoire et Jardin Botaniques de Geneva* were also founded in *1817* in Geneva. A.P. de Candolle was its first director.
6. *Botanical Garden of Cape Town* was founded in *1848.*
7. *Government Botanical Garden, Kingston, Jamaica* was laid out in *1857.*

---

[1] For details see Artide No. 17.5.

[2] For details see Article No. 17.4.

8. *Singapore Botanic Gardens* were established in *1859*.
9. *Missouri Botanical Garden, St. Louis,* is the first great garden of USA. It was established by Henry Shaw in *1859*, and named as "*Shaw's Gardens*".
10. *Arnold Arboretum* at Harvard University, U.S.A. was founded in *1872* by James Arnold. Its first director was C.S. Sargent (1872–1927).
11. *New York Botanical Garden* in *New York* city was established in 1891. It is one of the greatest gardens of the world. Its first director was N.L. Britton.
12. *Jardin Botanico Municipal* of *Buenos Aires, Argentina* is the last well-known botanical garden of the world established during 19th century. It was founded in *1892*.

17.2.5 Botanical Gardens Established During 20th Century

Almost every university with a Botany department, now has a botanical garden. Some of the major gardens established during the early part of the century are undermentioned:

1. *Botanical Garden, National Museum of Natural History, Beijing (formerly Peking), China,* was founded in 1930.
2. *The Jardin Botanique de Montreal* of Canada was established in 1936.
3. *Botanical Garden, Forest Research Institute, Dehradun,* India was established in 1934 under the leadership of C.E. Parkinson.
4. *National Botanic Garden, Lucknow,* India was established in 1946.

17.3 MAJOR BOTANICAL GARDENS OF INDIA

17.3.1 Lalbagh or The Mysore State Botanical Garden, Bangalore

This famous botanical garden is considered to be the best in South India for its layout, maintenance, scenic beauty and scientific interest. Because of roses and other red-coloured flowers it was named as "Lalbagh" by Hyder Ali in 1760. Major Waugh was its director during 1799–1819. He introduced a number of foreign exotic plants in this garden. However, Dr. Cleghorn made it a real botanic garden in 1856. A tropical nursery was established in the garden in 1908. Rao Bahadur H.C. Jayaraja was the first Indian director of this garden. The garden is now a big centre of horticultural activities. It now has well-equipped laboratories for seed-testing and soil-testing, and also a grape orchard, tree nursery, fruit nursery, pot garden, economic garden, and a herbal garden.

17.3.2 Lloyd Botanic Garden, Darjeeling

It was started as a branch of the Royal Botanic Garden, Kolkata on a 40 acre land piece in Darjeeling. The land was donated by Mr. William Lloyd. Mr. A.G. Jeffrey was the first curator of this garden. Since 1910, this garden has become a major institution for the distribution of seeds, bulbs, and plants of temperate Himalayas to different parts of the world. It has a vast collection of plants from Myanmar, China and Japan. It has separate sections of coniferous and indigenous plants. A Rock Garden, Orchidarium, Bulbous section, Succulent section, Seed section, Herbarium of over 30,000 specimens, and Rosary are its major attractions. Its coniferous section has 45 species including Australian *Callitris*.

17.3.3 National Botanic Garden, Lucknow

This famous Indian botanical institution, established in its new form in 1946, is now known as National Botanical Research Institute, Lucknow. It is popularly known as Sikander Bagh, a name given by Nawab Wajid Ali Shah after his beloved Begum Sikander Mahal. Sikander Bagh, originally laid out by Nawab Sadat Ali Khan (1789–1814), was converted into a botanical garden in 1946 by Professor K.N. Kaul, its first director. The present garden and its laboratories are spread over 27 acres of land on the bank of river Gomti. Popular attractions of this garden are its Rosarium, Palm house, Cactus house, Fern house, Orchid house, and Orchards of mango, *Citrus* and guava. It has well-equipped laboratories of Plant Morphology, Aromatics, Cytogenetics, Plant breeding, Tissue culture, Virology, Palynology, Plant Physiology, Entomology, etc. The garden bears an added experimental research station at Banthra, about 20 km from Lucknow.

17.3.4 Botanical Garden of Forest Research Institute, Dehradun

Established in 1934 under the leadership of C.E. Parkinson, F.R.I. botanical garden is now among the famous gardens of the world. It covers an area of about 20 acres in New Forest Estate, Dehradun, and is the main Indian centre of research in problems related with plant introduction. There are about 700 species of plants belonging to about 400 genera and about 100 families in this garden. Over half of these 700 species have been introduced from different parts of the world. The garden has a greenhouse, a cactus house and a Plant Introductory Nursery. Its biggest attraction is a big herbarium holding over 30,00,00 plant specimens from all over the world.

17.4 LARGEST BOTANICAL GARDEN OF INDIA

Indian Botanical Garden, Kolkata, is the largest and oldest botanical garden of India. Formerly named as Royal Botanic Garden or Kolkata Botanic Garden, it was established in 1787 by Lt. Col. Robert Kyd. He founded it "not for the purpose of collecting rare plants as things of mere curiosity or furnishing articles for the gratification of luxury, but for establishing a stock for disseminating such articles as may prove beneficial to the inhabitants". The garden covers an area of about 273 acres of land. William Roxburgh, *The Father of Indian Botany,* was its second director and founded the world famous herbarium of this garden. The garden is now under the control of Botanical Survey of India. Dr. K Biswas was the first Indian to be appointed Superintendent of this garden in 1937.

The garden is now noted for potato cultivation and introduction of jute, sugarcane, tea, and quinine-yielding *Cinchona*. Cultivation of *Aloe,* coffee, India-rubber, cardamon, and Henbane are some of the special achievements of this garden.

The Great Banyan tree (*Ficus benghalensis*; Moraceae; Fig. 17.1), which is one of the largest trees in size in the world and mentioned also in the Guinees Book of World Records, is the main centre of attraction of this garden. It appears like a miniature forest in itself. Over 2880 of its prop roots are actually rooted in the ground. The circumference of the canopy of this single tree is more than 404 metre. It is considered to be over 250 years of age.

There are over 15000 species of plants in this garden from several countries. Some main attractions of the garden are its Palm-house, Orchid-house, Pinetum, Fernary, Cacti-collection, the giant water lily, *Victoria regia,* and the section of medicinal plants.

Fig. 17.1 "Great Banyan Tree", the largest tree in the world, mentioned also in Guinees Book of World Records.

17.5 LARGEST BOTANICAL GARDEN OF THE WORLD

Royal Botanical Garden, Kew, officially opened in 1841, and originally established in 1760 because of the interest of Princess Augusta of Wales, is the largest botanical garden of world. William Aiton was the first Curator of this garden. After the death of Princess Augusta in 1772, her Kew garden was united with the gardens of the Palace of Richmond, and the extensive area of the two gardens was named as Royal Botanic Garden, Kew, Richmond, England. At present, the herbarium of this garden contains over 5 million specimens, its arboretum has over 7000 species, and its glasshouses have over 13000 species. Its Jodrell Laboratory provides best research facilities in the world for cytologists, anatomists, geneticists and physiologists. A beautiful Alpine house, Rose garden, Bamboo garden, and a Lily pond are also attached to this garden. Because of all these facilities and vast developments, Kew Gardens are called the *botanical capital of the world.*

Some other remarkable features of Kew are Chelsea Physic Garden and world's best rock gardens. It is also well-known for some of its famous publications such as *Flora of British India, Index Kewensis* and *Kew Bulletin.*

After 1838, this garden was transferred from British Crown to British Parliament, and Sir William Jackson Hooker took over as its first director in 1841.

17.6 ROLE OF BOTANICAL GARDENS

1. Botanical gardens provide information of local flora and are used in the preparation of monographs.
2. Several gardens supply seeds and material for botanical investigations.

3. They provide information on food plants, ornamental plants, medicinal plants, etc.
4. Modern gardens supply living plant resources for research in systematics, horticulture, ecology, genetics, etc.
5. They also provide information on the protection of endangered species, and propagation of rare plants.
6. Big botanical gardens provide training facility to younger students by providing them laboratory, classroom, greenhouse, nursery, etc.
7. They supply facility for courses in local flora, horticulture, hybridization, plant propagation, etc. Their educational programmes include workshops and training sessions for teachers, students, naturalists, etc.
8. Gardens practice and provide training for the conservation of nature.
9. They provide instructions for home gardening, and care of plants in winter and summer.
10. Botanical gardens provide aesthetically pleasing environment and thus play a major role in providing sound mental health.

Test Your Understanding

1. "A botanical garden is not merely a garden but a botanical institution". Explain.
2. Approximately how many botanical gardens exist today in the world?
3. Write, at least, 3 names each of the botanical gardens of the world established during the 16th century and the 17th century, along with their years of establishment.
4. Who established the first botanical garden of the world at Pisa, Italy in 1543?
5. Describe briefly any three major botanical gardens of India.
6. The largest and oldest botanical garden of India is located in the eastern part of our country. Where is it located? Write a detailed note on this garden.
7. Describe in brief the largest botanical garden of the world.
8. Mention any 'five' major roles that botanical gardens play in our life.

Suggested Reading

Avery, G.S. Jr., 1957, Botanic gardens—What role today?, *Amer. J. Bot.* **44**(3): 268–271.

Hill, A.W., 1915, The history and functions of botanic gardens, *Annl. Miss. Bot. Gard.* **2**: 185–223.

Hyams, E. and W. Macquitty, 1969, *Great Botanic Gardens of the World,* Nelson, New York.

Stafleu, F.A., 1969, Botanical gardens before 1818, *Boissiera* **14**: 31–46.

Thompson, P.A., 1972, The role of botanic garden, *Taxon* **21**(1): 115–119.

Wyman, D., 1970, How to establish an arboretum or botanic garden? *Arbort. Bot. Gard. Bull.* **4**(52): 52–60.

FLORAL FORMULA AND FLORAL DIAGRAM

18.1 WHAT IS A FLORAL FORMULA?

The floral formula is the numerical representation of various parts of a flower. It also depicts the symmetry, interrelationship of various floral parts, and unisexuality as well as bisexuality in the flower.

Actually, the floral formula is a method of summarizing the characters of a flower by a kind of shorthand notation that exposes the essential facts. It describes the floral morphology of angiosperms with the aid of certain standard symbols. These symbols, when combined, constitute the so called *floral formula* for a particular flower.

18.2 SYMBOLS EMPLOYED IN FLORAL FORMULAE

There has been no attempt to standardize the symbols used in floral formulae on a universal basis. Different authors have used different symbols for the same part, e.g. calyx is variously designated as K, C and CA, and corolla as C, CO and COZ. However, the generally accepted symbols, used in the floral formula presentation by the taxonomists of Indian subcontinent, are presented in Table 18.1, and the same usage has been followed in the present book.

Each letter (i.e. K, C, P, A and G) is followed by the number of parts in its particular group, e.g. K_5 means 5 free sepals in the calyx whereas K(5) means 5 fused or united sepals in the calyx. If more than one whorl is present, each is shown separately, e.g. K_{2+4} means the presence of two whorls of calyx, of which one whorl contains 2 free sepals while the other whorl contains 4 free sepals. The same symbols are used in the case of corolla. P_{3+3} means the presence of two whorls of perianth, each of 3 free tepals, e.g. Liliaceae.

A_{2+4} means a tetradynamous condition of stamens, i.e. 2 stamens in one whorl and 4 stamens in the other whorl, e.g. Cruciferae. A_{2+2} means a didynamous condition, i.e. out of the 4 stamens, 2 are long and two short, e.g. Labiatae. $A(\propto)$ stands for a monadelphous condition i.e. many stamens united in one group, e.g. Malvaceae. $A_{(9)+1}$ indicates a diadelphous condition, i.e. stamens are united in two bundles (9 stamens in one bundle and 1 separate), e.g. Papilionaceae.

Table 18.1 Floral Symbols

Br	Bracteate	$\widehat{C\ A}$	Epipetalous stamens
Ebr	Ebracteate	$\widehat{P\ A}$	Epiphyllous stamens
Brl	Bracteolate	Std	Staminodes
Epik	Epicalyx	G_4	Tetracarpellary, free carpel
0 (zero)	Absence of a particular whorl	$G_{(\underline{4})}$	Tetracarpellary, syncarpous, ovary superior (hypogynous)
$\propto$	Indefinite number		
$\oplus$	Actinomorphic	G(4)–	Tetracarpellary, syncarpous, ovary semi-inferior (perigynous)
⦶	Zygomorphic		
♂	Male flower	$G(\bar{4})$	Tetracarpellary, syncarpous, ovary inferior (epigynous)
♀	Female flower		
⚥	Bisexual flower, or hermaphrodite condition	$G(\bar{\underline{4}})$	Tetracarpellary, syncarpous, ovary either superior or inferior
K	Calyx, sepals	Pistd.	Pistillode
C	Corolla, petals	$\widehat{A\ G}$	Androecium and gynoecium united
P	Perianth	2+4	2 in one set and 4 in another
A	Androecium	2–4	2 or 4 or 2 to 4
G	Gynoecium	X	Variable
( )	Cohesion of floral parts in a whorl		

An arc between two floral parts marked above them indicates their adhesion, e.g. $\widehat{C_5, A_5}$ indicates the epipetalous conditions of stamens; $\widehat{P_5, A_5}$ indicates the epiphyllous condition of stamens and; $\widehat{A_5, G}_{(2)}$ indicates the adhesion of androecium with gynoecium.

A line above the number of carpels in the gynoecium, i.e. $G(\bar{5})$ indicates an epigynous condition or inferior ovary, whereas a line below the number of carpels in the gynoecium, i.e. $G(\underline{5})$ indicates a hypogynous condition or superior ovary. A line at the side of the number of carpels in the gynoecium, i.e. G(5)—indicates a perigynous condition or semi-inferior ovary.

The floral formula of *Ranunculus* (buttercup) i.e. Br, $\oplus$, ⚥, K5, C5, $A\propto$, $G\underline{\propto}$ indicates a bracteate (Br), actinomorphic ($\oplus$), hermaphrodite (⚥) flower with a calyx (K) of 5 free sepals, corolla (C) of 5 free petals, androecium (A) of an indefinite number of free stamens and a gynoecium (G) of an indefinite number of free carpels with a superior ovary. On the other hand, the floral formula of *Campanula,* i.e. Br, $\oplus$, ⚥, K_5, C(5), A_5, $G(\bar{5})$ shows a bracteate, actinomorphic, hermaphrodite flower with 5 free sepals, 5 united petals to form a gamopetalous corolla, 5 free stamens, and 5 fused carpels to form a syncarpous gynoecium with an inferior ovary.

18.3 WHAT IS A FLORAL DIAGRAM?

A diagram, illustrating the relative position and number of parts in each of the sets of organs comprising a flower, is called the *floral diagram* (Fig. 18.1).

A floral diagram visually depicts the essential features of a flower in cross section. The parts of the flower are drawn by semidiagrammatic symbols or ideographs. The number of the whorls and

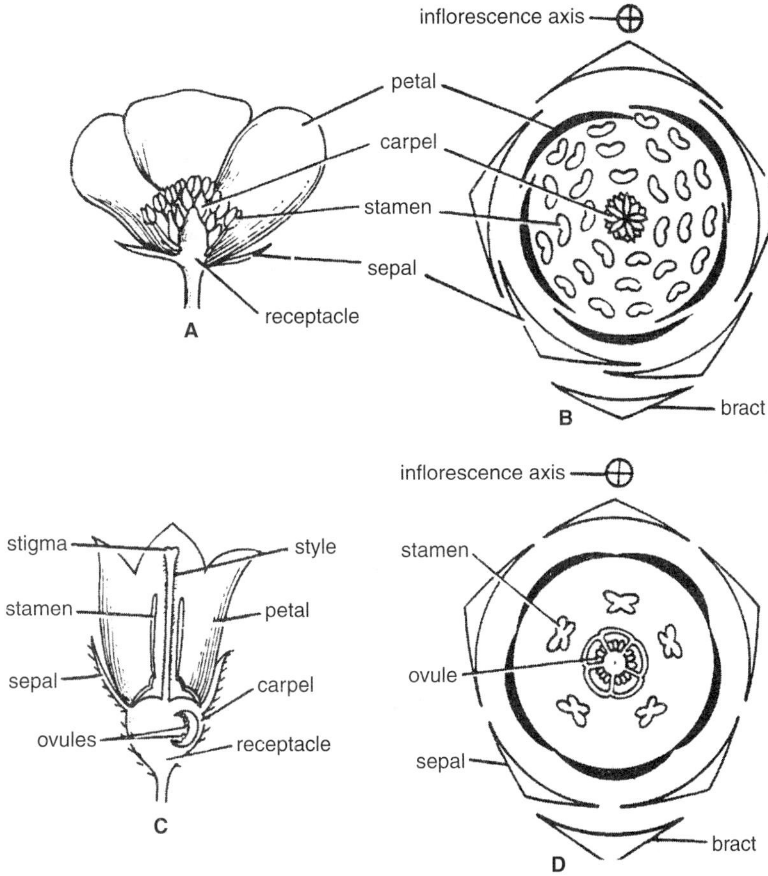

Fig. 18.1 A-B, Median vertical section of flower (A) and floral diagram (B) of *Ranunculus;* C-D, Median vertical section of flower (C) and floral diagram (D) of *Campanula.*

floral parts are both illustrated in a floral diagram. This also illustrates the fusion of the floral parts as well as the flower symmetry.

A floral diagram is thus the diagrammatic representation of the ground plan of a flower depicting the arrangement of various floral parts as projected in the transverse plane. In brief, a floral diagram is an ideal ground plan of a flower.

18.4 WHAT DOES A FLORAL DIAGRAM INFORM?

A floral diagram furnishes the following information:

1. Sex and symmetry of a flower.
2. Bracteate or ebracteate, and bracteolate or ebracteolate nature of a flower.
3. Number of floral cycles of a flower, i.e. its tetracyclic or pentacyclic nature.
4. Number of a floral parts in each cycle of a flower.

5. If the number of sepals or petals is odd, the floral diagram informs about the anterior or posterior position of the odd sepal or petal.
6. Type of aestivation of sepals and petals.
7. Number of whorls of stamens, their position in relation to the petals, attachment with the petals or their free nature, monothecous or dithecous condition, and introrse or extrorse nature, are all shown in a floral diagram.
8. Carpel number, their free or fused nature, number of locules, number of ovules in each locule, and the type of placentation are all shown by a floral diagram.
9. Position of all floral parts in relation with the mother axis.

18.5 HOW TO DRAW A FLORAL DIAGRAM?

The floral diagram is always drawn circular in outline. The different floral whorls are represented in concentric circles, the sepals on the outermost circle, then the petals, the stamens and carpels towards the inner side (Fig. 18.2).

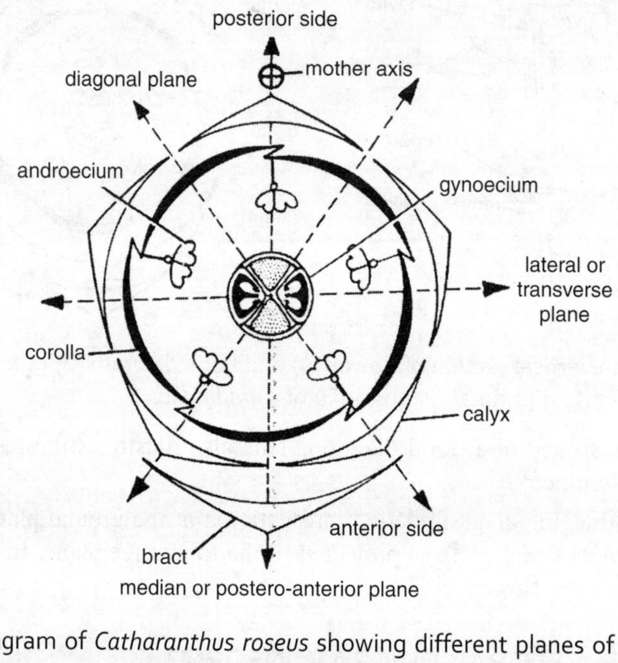

Fig. 18.2 Floral diagram of *Catharanthus roseus* showing different planes of flower.

The first step to draw a floral diagram is to examine mature floral buds which are due to open shortly but have not yet opened. Pluck the floral bud from the mother axis only after you have noted down the anterior and posterior sides. Floral parts are drawn in a floral diagram as they would be seen in their transverse sections below the mother axis. Make the floral diagram in the following sequential stages:

1. A very small circle is drawn above the floral diagram. This circle represents the *mother axis*. In actinomorphic flowers the mother axis circle is denoted as ⊕, but in zygomorphic flowers it is drawn as ⦶. If the flowers are terminal, the mother axis is not drawn.
2. In bracteate flowers, a section of *bract* is drawn below the floral diagram. In flowers without any bract, such a section is not drawn.
3. In bracteolate flowers, *bracteoles* are drawn in section on the left and right sides of the diagram.
4. Note the number of *sepals,* their arrangement in relation to the mother axis and their aestivation. Draw transverse sections of sepals between the mother axis and the bract, keeping all these points in view. In case of odd number of sepals, the odd sepal would be drawn either posterior or anterior to the flower, i.e. opposite the mother axis or opposite the bract, respectively.
5. The same procedure is repeated for *petals* as for sepals mentioned above. However, petals should be drawn alternate with the sepals.
6. If the flower is *zygomorphic,* petals are drawn of unequal sizes. Same may be the case with the sepals also in zygomorphic flowers.
7. If any sepal or petal is spurred, it is shown by drawing a loop at the back of that particular part in the floral diagram.
8. If parts of sepals or of petals are fused, draw lines to connect their edges together in the floral diagram.
9. In the *epipetalous* condition (i.e. when stamens are joined with petals), link the stamens and petals with small radial lines.
10. In case of bilabiate calyx or corolla, the two lips are joined by bulging lines.
11. Count the number of *stamens,* the number of whorls in which they are arranged, their cohesion and adnation to other floral parts, their position in relation to petals, their introrse or extrorse position, and draw them inside the petals in the floral diagram. Stamens are represented through transverse sections of the anthers. In the *obdiplostemonous* condition, the stamens of the outer whorl are drawn opposite to the petals. *Introrse* stamens face towards the centre whereas the extrorse towards the petals. *Staminodes* are represented either by an asterisk (*) or by a cross (×).
12. The *gynoecium* is represented by a transverse section of the ovary. Also draw the number of locules and the number of ovules in each locule, in the ovary. Type of placentation is also drawn,

Test Your Understanding

1. What is a floral formula? How does it differ from floral diagram?
2. Which symbols are used in a floral formula for the following?
   (a) Calyx (b) Corolla
   (c) Absence of a particular whorl (d) Zygomorphic

(e) Actinomorphic
(f) Bisexual flower
(g) Epipetalous stamens
(h) Female flower
(i) Tetracarpellary, syncarpous, superior
(j) Androecium and gynoecium united.

3. Explain various sequential stages of drawing a floral diagram. Illustrate your answer by drawing a floral diagram.

Suggested Reading

Featherly, H.I., 1954, *Taxonomic Terminology of the Higher Plants,* Iowa State College Press, Ames.
Rangaswamy, N.G., 1966, Floral diagrams and formulae: A reappraisal, *Sci. & Cult.* **31**(1): 33–34.
Schaffner, J.H., 1916, A general system of floral diagrams, *Ohio Jour. Sci.* **16**: 300–360.
Steam, W.T., 1983. *Botanical Latins* (3rd ed.), David and Charles, Newton Abbot, England.

CHAPTER 19

POSITION OF SOME SELECTED FAMILIES IN CLASSIFICATION SYSTEMS PROPOSED BY BENTHAM AND HOOKER, ENGLER AND PRANTL, HUTCHINSON, TAKHTAJAN, CRONQUIST, AND THORNE

Position of some selected families in some selected and widely accepted systems of classification is presented in Table 19.1. Majority of these families are discussed in the present book. Classification systems proposed by Bentham and Hooker (1862–1883) in their *Genera Plantarum,* Engler and Prantl (1887–1899) in 20 volumes of their *Die Naturlichen Pflanzenfamilien* revised by H. Melchior and E. Werdermann (1954–1964) in the form of *Syllabus der Pflanzenfamilien,* Hutchinson (1973) in his treatment entitled *The Evolution of Flowering Plants,* Takhtajan (1980) in *Botanical Review,* Cronquist (1981) in his book entitled *An Integrated System of Classification of Flowering Plants,* and Thorne (1983) in *Nordic Journal of Botany,* were taken into consideration in the present comparison.

Table 19.1 Position of selected families in various systems of classification

	Family	Bentham and Hooker (1862–1883)	Engler and Prantl (1954–1964)	Hutchinson (1973)	Takhtajan (1980)	Cronquist (1981)	Thorne (1983)
		Dicotyledons	***Dicotyledoneae***	***Dicotyledones***	***Magnoliopsida*** (Dicots)	***Magnoliopsida*** (Dicots)	***Annonidae*** (Dicotyledoneae)
1.	*Magnoliaceae*	Ranales	Ranales	Magnoliales	Magnoliales	Magnoliales	Annonales
2.	*Annonaceae*	Ranales	Ranales	Annonales	Annonales	Magnoliales	Annonales
3.	*Ranunculaceae*	Ranales	Ranales	Ranales	Ranunculales	Ranunculales	Berberidales
4.	*Nymphaeaceae*	Ranales	Ranales	Ranales	Nymphaeales	Nymphaeales	Nymphaeales
5.	*Papaveraceae*	Parietales	Rhoeadales	Rhoeadales	Papaverales	Papaverales	Berberidales
6.	*Fumariaceae*	(Treated under Papaveraceae)	(Treated under Papaveraceae)	Rhoeadales	Papaverales	Papaverales	(Treated under Papaveraceae)
7.	*Capparidaceae*	Parietales	Rhoeadales	Capparales	Capparales	Capparales	Capparales
8.	*Brassicaceae (Cruciferae)*	Parietales	Rhoeadales	Brassicales	Capparales	Capparales	Capparales
9.	*Violaceae*	Parietales	Parietales	Violales	Violales	Violales	Violales
10.	*Caryophyllaceae*	Caryophyllinae	Centrospermae	Caryophyllales	Caryophyllales	Caryophyllales	Chenopodiales
11.	*Portulacaceae*	Caryophyllinae	Centrospermae	Caryophyllales	Caryophyllales	Caryophyllales	Chenopodiales
12.	*Malvaceae*	Malvales	Malvales	Malvales	Malvales	Malvales	Malvales
13.	*Sterculiaceae*	Malvales	Malvales	Tiliales	Malvales	Malvales	Malvales
14.	*Tiliaceae*	Malvales	Malvales	Tiliales	Malvales	Malvales	Malvales
15.	*Bombacaceae*	Malvales	Malvales	Malvales	Malvales	Malvales	Malvales
16.	*Oxalidaceae*	(Treated under Geraniaceae)	Geraniales	Geraniales	Geraniales	Geraniales	Geraniales
17.	*Geraniaceae*	Geraniales	Geraniales	Geraniales	Geraniales	Geraniales	Geraniales
18.	*Rutaceae*	Geraniales	Geraniales	Rutales	Rutales	Sapindales	Rutales
19.	*Meliaceae*	Geraniales	Geraniales	Meliales	Rutales	Sapindales	Rutales
20.	*Rhamnaceae*	Celastrales	Rhamnales	Rhamnales	Rhamnales	Rhamnales	Rhamnales
21.	*Vitaceae*	Celastrales	Rhamnales	Rhamnales	Rhamnales	Rhamnales	Cornales
22.	*Sapindaceae*	Sapindales	Sapindales	Sapindales	Sapindales (Acerales)	Sapindales	Rutales
23.	*Anacardiaceae*	Sapindales	Sapindales	Sapindales	Rutales	Sapindales	Rutales

(*Contd.*)

24.	*Leguminosae* or *Fabaceae* (including *Papilionaceae, Caesalpiniaceae* and *Mimosaceae*)	Rosales	Rosales	Leguminales	Fabales	Fabales	Rutales
25.	*Rosaceae*	Rosales	Rosales	Rosales	Rosales	Rosales	Rosales
26.	*Combretaceae*	Myrtales	Myrtiflorae	Myrtales	Myrtales	Myrtales	Myrtales
27.	*Myrtaceae*	Myrtales	Myrtiflorae	Myrtales	Myrtales	Myrtales	Myrtales
28.	*Lythraceae*	Myrtales	Myrtiflorae	Myrtales	Myrtales	Myrtales	Myrtales
29.	*Cucurbitaceae*	Passiflorales	Cucurbitales	Cucurbitales	Cucurbitales	Violales	Violales
30.	*Cactaceae*	Ficoidales	Opuntiales	Cactales	Caryophyllales	Caryophyllales	Chenopodiales
31.	*Umbelliferae* (*Apiaceae*)	Umbellales	Umbelliflorae	Umbellales	Cornales (Apiales)	Apiales	Araliales (Treated under Araliaceae)
32.	*Rubiaceae*	Rubiales	Rubiales	Rubiales	Gentianales	Rubiales	Gentianales
33.	*Asteraceae* (*Compositae*)	Asterales	Campanulatae	Asterales	Asterales	Asterales	Asterales
34.	*Sapotaceae*	Ebenales	Ebenales	Ebenales	Ebenales	Ebenales	Ebenales
35.	*Primulaceae*	Primulales	Primulales	Primulales	Primulales	Primulales	Primulales
36.	*Oleaceae*	Gentianales	Contortae	Gentianales	Oleales	Scrophulariales	Oleales
37.	*Asclepiadaceae*	Gentianales	Contortae	Gentianales	Gentianales	Gentianales	Gentianales (Treated under *Apocynaceae*)
38.	*Apocynaceae*	Gentianales	Contortae	Apocynales	Gentianales	Gentianales	Gentianales
39.	*Boraginaceae*	Polemoniales	Tubiflorae	Boraginales	Polemoniales	Lamiales	Solanales
40.	*Convolvulaceae*	Polemoniales	Tubiflorae	Solanales	Polemoniales	Solanales	Solanales
41.	*Solanaceae*	Polemoniales	Tubiflorae	Solanales	Scrophulariales	Solanales	Solanales
42.	*Scrophulariaceae*	Personales	Tubiflorae	Personales	Scrophulariales	Scrophulariales	Bignoniales
43.	*Bignoniaceae*	Personales	Tubiflorae	Bignoniales	Scrophulariales	Scrophulariales	Bignoniales
44.	*Pedaliaceae*	Personales	Tubiflorae	Bignoniales	Scrophulariales	Scrophulariales	Bignoniales
45.	*Acanthaceae*	Personales	Tubiflorae	Personales	Scrophulariales	Scrophulariales	Bignoniales
46.	*Lamiaceae* (*Labiatae*)	Lamiales	Tubiflorae	Lamiales	Lamiales	Lamiales	Lamiales

(*Contd.*)

47.	*Verbenaceae*	Lamiales	Tubiflorae	Verbenales	Lamiales	Lamiales	Lamiales
48.	*Chenopodiaceae*	Curvembryae	Centrospermae	Chenopodiales	Caryophyllales	Caryophyllales	Chenopodiales
49.	*Amaranthaceae*	Curvembryae	Centrospermae	Chenopodiales	Caryophyllales	Caryophyllales	Chenopodiales
50.	*Polygonaceae*	Curvembryae	Polygonales	Polygonales	Polygonales	Polygonales	Polygonales
51.	*Nyctaginaceae*	Curvembryae	Centrospermae	Thymeleales	Caryophyllales	Caryophyllales	Chenopodiales
52.	*Loranthaceae*	Achlamydosporae	Santalales	Santalales	Santalales	Santalales	Santalales
53.	*Euphorbiaceae*	Unisexuales	Geraniales	Euphorbiales	Euphorbiales	Euphorbiales	Euphorbiales
54.	*Urticaceae*	Unisexuales	Urticales	Urticales	Urticales	Urticales	Urticales
55.	*Cannabinaceae*	Unisexuales	Urticales	Urticales	Urticales	Urticales	Urticales
56.	*Moraceae*	(Treated under Urticaceae)	Urticales	Urticales	Urticales	Urticales	(Treated under Urticaceae)
57.	*Casuarinaceae*	Unisexuales	Verticillatae	Casuarinales	Casuarinales	Casuarinales	Casuarinales
58.	*Salicaceae*	Ordines Anomali	Salicales	Salicales	Salicales	Salicales	Violales
		Monocotyledons	***Monocotyledoneae***	***Monocotyledones***	***Liliopsida***	***Liliopsida*** (Monocots)	***Liliidae***
59.	*Orchidaceae*	Microspermae	Microspermae	Orchidales	Orchidales	Orchidales	Liliales
60.	*Iridaceae*	Epigynae	Liliflorae	Iridales	Iridales	Liliales	Liliales
61.	*Amaryllidaceae*	Epigynae	Liliflorae	Amaryllidales	Liliales	(Treated under Liliaceae)	(Treated under Liliaceae)
62.	*Musaceae*	Epigynae (in Scitamineae)	Scitamineae	Zingiberales	Zingiberales	Zingiberales	Zingiberales
63.	*Zingiberaceae*	Epigynae (in Scitamineae)	Scitamineae	Zingiberales	Zingiberales	Zingiberales	Zingiberales
64.	*Cannaceae*	Epigynae (in Scitamineae)	Scitamineae	Zingiberales	Zingiberales	Zingiberales	Zingiberales
65.	*Liliaceae*	Coronarieae	Liliflorae	Liliales	Liliales	Liliales	Liliales
66.	*Commelinaceae*	Coronarieae	Farinosae	Commelinales	Commelinales	Commelinales	Commelinales
67.	*Juncaceae*	Calycinae	Liliflorae	Juncales	Juncales	Juncales	Commelinales
68.	*Arecaceae (Palmae)*	Calycinae	Principes	Palmales	Arecales	Arecales	Arecales
69.	*Typhaceae*	Nudiflorae	Pandanales	Typhales	Typhales	Typhales	Typhales
70.	*Araceae*	Nudiflorae	Spathiflorae	Arales	Arales	Arales	Arales
71.	*Cyperaceae*	Glumaceae	Glumiflorae	Cyperales	Cyperales	Cyperales	Commelinales
72.	*Poaceae (Gramineae)*	Glumaceae	Glumiflorae	Graminales	Poales	Cyperales	Commelinales

Test Your Understanding

1. Under which order has Magnoliaceae been placed in the systems of classification proposed by Bentham and Hooker, Hutchinson, and Thorne?
2. Brassicaceae (Cruciferae) has been placed under order Parietales by Bentham and Hooker, Rhoeadales by Engler and Prantl, Brassicales by Hutchinson, and ________ by Takhtajan (1980) and Cronquist (1981).
3. Bombacaceae has been placed under same order in almost all systems of classification. What is the name of that order?
4. Under which order has Asteraceae been treated in almost all systems of classification?
5. What are the new names of the following families?
   (i) Gramineae
   (ii) Palmae
   (iii) Labiatae
   (iv) Umbelliferae
   (v) Compositae

Suggested Reading

Bentham, G. and J.D. Hooker, 1862–1883, *Genera Plantarum,* 3 Vols., London.

Cronquist, A. 1981., *An Integrated System of Classification of Flowering Plants,* Columbia Univ. Press, New York.

Engler, A. and K. Prantl, 1954–1964, *Syllabus der Pflanzenfamilien,* 12th ed. 2 Vols., H. Melchior and E. Werdermann (eds.), Gebruder Borntraeger, Berlin.

Hutchinson, J., 1973, *The Families of Flowering Plants,* 3rd ed., Clarendon, Oxford.

Takhtajan, A., 1980, Outline of the classification of flowering plants (Magnoliophyta), *Bot. Rev.* **46**: 225–359.

Thorne, R.F., 1983, The classification of angiosperms (Annonopsida), *Nordic J. Bot.,* **3**: 102–111.

CHAPTER 20

550 TERMS OF PLANT DESCRIPTION

20.1 PLANT PARTS[1]

1. *Bud* An undeveloped, vegetative or floral shoot, covered with protective scales, or consisting of a short axis bearing primordia of leaves or floral parts.
2. *Flower* Reproductive structure of angiosperms, consisting usually of sepals, petals, stamens, and carpels. After fertilization the ovules of flower develop into seeds.
3. *Fruit* Mature ovary of flowering plants containing the seeds.
4. *Leaf* Produced from the buds on the stem, the leaves are photosynthetic and transpiring organs of the plant. They are usually green and expanded, and have a wide range of forms.
5. *Root* Organ of a plant that develops initially from the radicle, grows down into the soil, and functions for absorption and anchorage.
6. *Seed* Fertilized ripe ovule of flowering plants.
7. *Stem* Organ of a plant that develops initially from the epicotyl, grows mostly above the ground, and functions for support and conduction.

20.2 PLANT TYPES[1] (HABIT AND HABITAT)

1. *Annual* A plant that completes its entire life cycle, from seed to reproduction to death, in one single year.
2. *Aquatic* Organisms growing in water.
3. *Biennial* A plant that completes its entire life cycle in two years.
4. *Climber* A weak-stemmed plant with roots in the ground and climbing the other plants, etc. to support itself with the help of its tendrils or adventitious roots, etc. Usually climbers twist around their support.
5. *Creeper* A plant which is unable to support itself, and spreads along the ground.

[1] Arranged alphabetically.

6. *Epiphyte* A plant, growing on stems or branches of other plants, with no roots in the ground, but not taking nutrients on which it is growing.
7. *Habit* Appearance of an organism, e.g. herb, shrub, etc.
8. *Habitat* The place in which an organism or a community is found.
9. *Herb* A small, usually annual plant, with no wood in its stems or roots.
10. *Liana* An annual or perennial, elongate plant, with weak stem, which is often climbing.
11. *Parasite* An organism which takes all its nutrients from the tissues of another organism, usually with harmful effects.
12. *Perennial* A plant that grows and reproduces for many years.
13. *Shrub* A much branched, small, woody perennial plant with several branches from ground level upwards.
14. *Tree* A tall, woody, perennial plant with a single trunk which usually bears branches.
15. *Vine* A creeper of family Vitaceae.
16. *Xerophyte* A plant that lives in dry habitats or in a desert.

20.3 ROOT

20.3.1 Root Types[1]

1. *Adventitious Root* Root arising from any part of the plant body other than the radicle. It arises from an organ other than the root.
2. *Tap Root* The main, persistent, primary root of the plant which shows apical dominance and develops from the radicle.

20.3.2 Modified Roots[1] (Fig. 20.1)

1. *Aerial* A root arising from a part of a plant which is above the ground, e.g. betel.
2. *Annulated* Root having a series of ring-like swellings, e.g. *Psychotria.*
3. *Assimilatory* Green, chlorophyll-containing roots, e.g. *Trapa natans.*
4. *Buttress* Roots having plank-like or board-like growth on the upper side. These are supportive structures, e.g. *Terminalia bellirica.*
5. *Clasping or Climbing* These arise on the nodes of weak-stemmed plants, and help them to climb or fix on the support, e.g. *Pothos.*
6. *Conical* Cone like, i.e. broad at the base and tapers gradually towards apex, e.g. *Daucus carota.*
7. *Contractile* Roots capable of shortening and containing a wrinkled surface, e.g. *Crocus.*
8. *Fasciculated* Fleshy or tuberous roots in a cluster, e.g. *Asparagus.*
9. *Fibrous* Fine, thread-like roots, e.g. monocotyledons.
10. *Fleshy* Succulent roots, e.g. *Dahlia.*

[1]Arranged alphabetically.

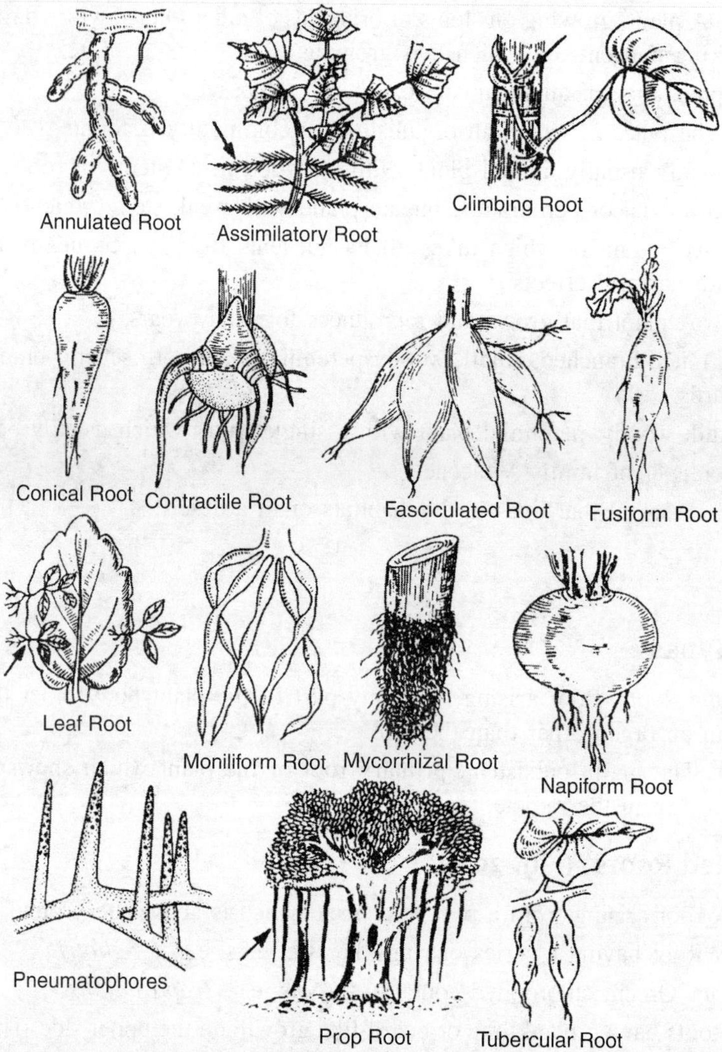

Fig. 20.1 Root modifications.

11. *Fusiform* Root thickened in the middle and tapering at the ends, e.g. *Raphanus sativus.*
12. *Haustorial* These are absorbing roots, present within the host of some parasitic plants, e.g. *Balanophora.*
13. *Knee or Aerating* These are vertical or horizontal, aboveground roots, e.g. *Heritiera.*
14. *Leaf Roots* Roots arising on the margin of leaf from adventitious buds, e.g. *Bryophyllum.*
15. *Moniliform* Root thickened at certain intervals giving a beaded appearance, e.g. *Momordica.*
16. *Mycorrhizal* Root in symbiotic relationship with fungal hyphae, e.g. *Monotropa.*

17. *Napiform* Root almost spherical at one end and tapering sharply at the lower part, e.g. turnip, *Beta vulgaris.*
18. *Nodulose* Root bearing a small knot at or near the apex, e.g. *Curcuma domestica.*
19. *Pneumatophorous or Respiratory* Spongy, aerating roots of marshy plants, e.g. *Rhizophora.*
20. *Prop or Stilt* Supportive roots which grow out from the bottom of the trunk into the ground, e.g. *Ficus benghalensis.*
21. *Tuberous or Tubercular* Fleshy roots appearing like stem tubers, e.g. *Ipomoea batatas* (Sweet Potato), *Mirabilis jalapa.*

20.4 STEM (Types, Surface, Forms and Modifications; Fig. 20.2)[1]

1. *Acaulescent* Stemless or with inconspicuous stem.
2. *Aerial* Stem which remains above the ground, e.g. *Sesbania.*
3. *Angular* Stem showing many angles in a transverse section, e.g. *Asparagus.*
4. *Arborescent* Woody and treelike, e.g. *Mangifera.*
5. *Ascending* Inclined upward.
6. *Branched* Stem with many branches, e.g. *Ranunculus.*
7. *Bud* Small embryonic stem tip bearing leaves or flowers, or both.
8. *Bulb* An underground, short, erect stem covered by fleshy leaves, e.g. *Allium cepa.*
9. *Bulbel* A young small bulb developed from the base of a large bulb.
10. *Bulbil* A bulblike body developed on the aboveground parts, e.g. *Agave.*
11. *Caudex* A short, hard, overwintering base of a perennial herb, e.g. palms.
12. *Caulescent* Possessing a distinct stem.
13. *Cespitose* Short, much branched, cushion-like plant.
14. *Cladode* A phylloclade in which a branch of a single internode becomes flat or leaf-like, e.g. *Asparagus, Ruscus.*
15. *Cladophyll* A leaflike, flattened, green stem.
16. *Climbing* Growing towards upper side by means of tendrils, petioles, adventitious roots, etc., e.g. pea, betel.
17. *Columnar* Erect with a strong trunk.
18. *Corm* Solid, round, bulblike, fleshy, underground stem, usually surrounded by membranous scales, e.g. *Gladiolus.*
19. *Cormel* Young small corm developing at the base of parent corm, e.g. *Amorphophallus.*
20. *Creeping* Trailing stem having roots throughout its length, e.g. *Oxalis.*
21. *Culm* Flowering and fruiting stem of sedges (Cyperaceae) and grasses (Gramineae), e.g. *Cyperus, Cynodon.*

[1] Arranged alphabetically.

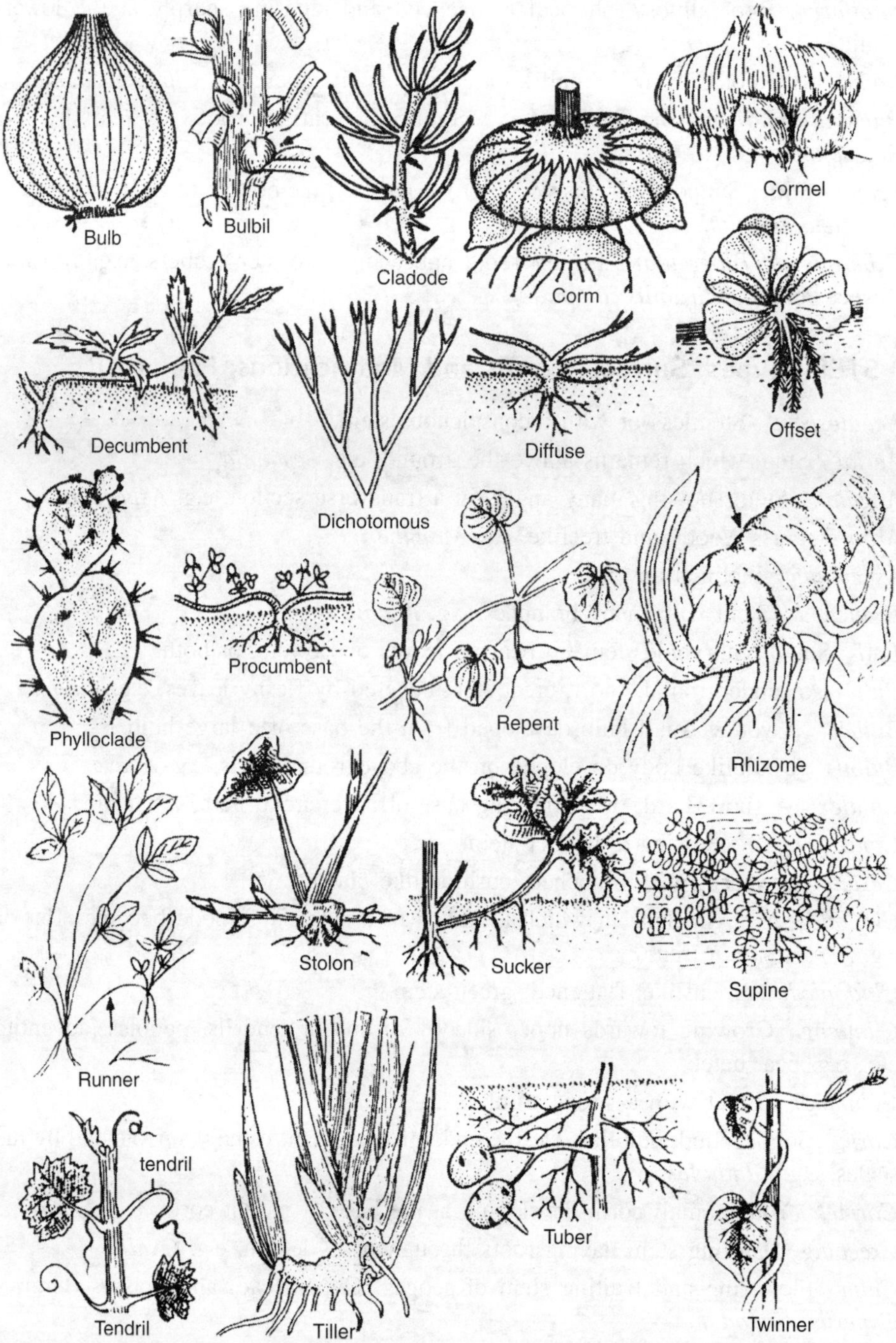

Fig. 20.2 Stem types and modifications

22. *Cylindrical* Stem showing circular outline in transverse section.
23. *Decumbent* Stem bending in one direction, e.g. *Tridax.*
24. *Dichotomous* Stem dividing equally into two, e.g. *Pandanus.*
25. *Diffuse* Branches of stem spreading in all directions, e.g. *Boerhaavia.*
26. *Eramous* With unbranched stems.
27. *Erect* Growing upright, e.g. *Abutilon.*
28. *Fastigiate* Strictly erect and parallel.
29. *Fistular* Stem hollow from inner side, e.g. wheat, bamboo.
30. *Fruticose* Woody and shrublike.
31. *Geniculate* Zigzag stem, or bending abruptly at node, e.g. *Vitis.*
32. *Glabrous* Smooth, or without hairs, e.g. Lemon.
33. *Glaucous* Shining and smooth.
34. *Hairy* Covered with hairs, e.g. *Calotropis.*
35. *Herbaceous* Soft, non-woody, dying to the ground at the end of the growing season, e.g. *Eclipta.*
36. *Internode* Region of stem between any two nodes, e.g. *Cynodon.*
37. *Laticiferous* Containing latex, e.g. *Calotropis.*
38. *Leaf Scar* A mark on the stem showing former place of attachment of leaf base or petiole.
39. *Lenticel* Lens-shaped or wartlike pore in the bark.
40. *Node* The area of the stem from which arise branches, leaves, or a leaf, e.g. sugarcane.
41. *Offset* A short, thick, creeping stem bearing a tuft of leaves above and a cluster of leaves below at the apex, e.g. *Eichhornia, Pistia.*
42. *Pachycauly* Thick, short, frequently succulent stems, e.g. cacti.
43. *Phylloclade* A flat or round, green, succulent stem with leaves either ill-developed or modified into spines, e.g. *Opuntia, Coccoloba.*
44. *Phyllode* A flat petiole which has the appearance of a leaf, e.g. Australian Acacia.
45. *Prickle* A sharp, pointed extension of epidermis or cortex, e.g. *Pisonia.*
46. *Primocane* The first-year, nonflowering stem.
47. *Procumbent or Prostrate* Growing flat or parallel on the ground, e.g. *Portulaca.*
48. *Pterocaulous* Winged stem.
49. *Pubescent* Covered with fine silky hairs.
50. *Ramose* Branched stem.
51. *Repent* Creeping stem.
52. *Rhizome* A horizontal, prostrate, or underground stem bearing scale-like leaves, e.g. ginger, *Canna.*
53. *Rootstock* A caudex, or a rhizome, or any underground portion of a plant, e.g. *Alocasia.*

54. *Runner* A horizontal stem, creeping aboveground, usually rooting and producing plants at the nodes, e.g. *Cynodon*.
55. *Sarcocauly* Fleshy stems.
56. *Scandent* Climbing stem.
57. *Scape* A leafless flowering stem arising from an underground stem, e.g. *Canna*.
58. *Scapose* Possessing a scape.
59. *Scars* Remains of a point of attachment of leaf, stipule, scale, bud, etc.
60. *Sclerocauly* Hard, woody stems.
61. *Spine* A stiff, pointed outgrowth.
62. *Stolon* A runner, or a horizontal stem rooting at the nodes, e.g. *Fragaria indica*.
63. *Stoloniferous* Bearing stolons.
64. *Subterranean* Underground.
65. *Sucker* A subterranean creeping stem, usually fast-growing and adventitious, e.g. *Mentha*, banana.
66. *Suffrutescent* Woody at the base.
67. *Supine* Prostrate stem with parts oriented upward.
68. *Tendril* Long, twisting appendage adapted for climbing, e.g. *Vitis*.
69. *Thorn* A sharp, pointed, reduced branch, e.g. *Duranta*.
70. *Tiller* A grass shoot developed from the base of the stem.
71. *Trailing* Stem sprawling on the ground with the help of adventitious roots.
72. *Tuber* Thick, enlarged fleshy tip of an underground stem, e.g. *Solanum tuberosum*.
73. *Turion* An overwintering bud.
74. *Twig* A short lateral branch of a woody stem.
75. *Twinner* Stem ascending by coiling on the support without any special device, e.g. *Abrus*.
76. *Virgate* Long, straight, slender or thin, stick-like.
77. *Woody* Hard in texture and possessing secondary xylem, e.g. *Mangifera indica*.

20.5 LEAF

20.5.1 Apex of Leaf (Fig. 20.3 A–M)

1. *Acuminate* Drawn out in the form of a long slender tail, or tapers to a protracted point (Fig. 20.3 A).
2. *Acute* Ending into a sharp point in the form of an acute angle but not drawn out (Fig. 20.3 B).
3. *Apiculate* Ending into a short, sharp, flexible point, or an apicula (Fig. 20.3 C).
4. *Aristate* Tapering to a very narrow, much elongated apex, or bearing a stiff awn (Fig. 20.3 D).
5. *Caudate* Containing a tail-like appendage (Fig. 20.3 E).

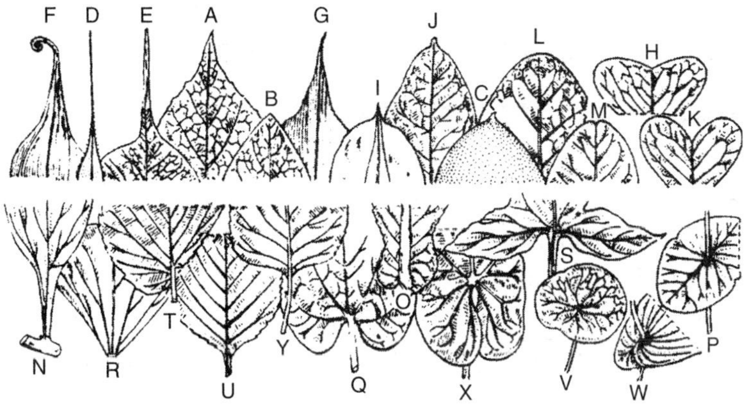

Fig. 20.3 A-M, Leaf apices; N-Y, Leaf bases.

6. *Cirrhose* Apex ending into a tendril like structure (Fig. 20.3 F).
7. *Cuspidate* Abruptly and sharply concavely constricted into a sharp and elongated pointed tip (Fig. 20.3 G).
8. *Emarginate* Containing a deep or shallow notch at the apex (Fig. 20.3 H)
9. *Mucronate* Terminating abruptly into a short and sharp point (Fig. 20.3 I).
10. *Mucronulate* Diminutive or smaller form of mucronate (Fig. 20.3 J).
11. *Obcordate* Deeply lobed at the apex (Fig. 20.3 K).
12. *Obtuse or Rounded* Blunt or rounded (Fig. 20.3 L).
13. *Retuse* Slightly notched obtuse apex (Fig. 20.3 M).

20.5.2 Base of Leaf[1] (Fig. 20.3 N–Y)

1. *Attenuate* Base showing a long gradual taper (Fig. 20.3 N).
2. *Auriculate* Bearing an appendage or ear-shaped part at the base (Fig. 20.3 O).
3. *Connate-perfoliate* Base of opposite sessile leaves, the stem appears passing through them (Fig. 20.3 P).
4. *Cordate* Heart-shaped (Fig. 20.3 Q).
5. *Cuneate* Wedge-shaped; triangular with the narrow end at the attaching point (Fig. 20.3 R).
6. *Hastate* Resembling with an arrowhead; halberd-shaped (Fig. 20.3 S).
7. *Oblique* Two sides of base being unequal-sized (Fig. 20.3 T).
8. *Obtuse* Rounded or blunt (Fig. 20.3 U).
9. *Peltate* Attached to its stalk inside the margin (Fig. 20.3 V).
10. *Perfoliate* Sessile leaf whose base completely surrounds the stem (Fig. 20.3 W).

---
[1] Arranged alphabetically.

11. *Sagittate* Base like an arrowhead, or triangular (Fig. 20.3 X).
12. *Truncate* Blunt, or appearing as if cut off at the end (Fig. 20.3 Y).

20.5.3 Kinds, Modifications and Parts of Leaf[1]

1. *Blade* Expanded, flat part of leaf.
2. *Bract* Reduced and modified leaf found in inflorescence.
3. *Bracteole* Small leaf found on the pedicel of flower.
4. *Cauline* Leaf arising only on the main axis.
5. *Cauline and Ramal* Leaves arising on the main axis as well as on side branches.
6. *Chaff* Bract present at the base of tubular flower, as in Asteraceae.
7. *Cotyledon* Embryonic leaf.
8. *Epetiolate* Sessile leaves.
9. *Epetiolulate* Sessile leaflets.
10. *Epicalyx* Leaves present below the true calyx, e.g. Malvaceae.
11. *Evergreen* Persistent.
12. *Exstipulate* Without stipules.
13. *Floral leaves* Calyx, corolla, stamens and carpels.
14. *Fly trap* Hinged leaves of insectivorous plants.
15. *Fugacious* Falling soon.
16. *Glume* Bract of the spikelets of sedges and grasses.
17. *Incomplete* Leaf without one or more parts, e.g. without petiole.
18. *Involucral bracts* Bracts present below the inflorescence in Asteraceae.
19. *Leaflet* A distinct or separate segment of leaf.
20. *Lemma* Outer bract subtending the grass floret.
21. *Ligule* Finger-like small part present at the upper end of leaf sheath.
22. *Pale* Chaff.
23. *Palea* Inner bract subtending the grass floret.
24. *Parts of leaf* Leaf base, stipule, petiole, lamina, midrib, veins, veinlets.
25. *Petiolate* Leaf with a distinct stalk or petiole.
26. *Petiolulate* Leaflets with a stalk.
27. *Phyllary* A bract subtending the inflorescence in Asteraceae.
28. *Pitcher* Tubular or pitcher-shaped leaf of insectivorous plants.
29. *Prophyll* Bracteole.
30. *Pulvinus* Swollen base of petiole or petiolule.
31. *Rachilla* Secondary axis of a compound leaf.

[1] Arranged alphabetically.

32. *Rachis* Main axis of a pinnately compound leaf.
33. *Radical* Leaves arising from bulb or other underground stem.
34. *Scaly leaf* Small, nongreen leaf.
35. *Seed leaves* Cotyledons.
36. *Sessile* Without petiole.
37. *Spathe* An enlarged bract enclosing an inflorescence.
38. *Sporophylls* Spore-bearing leaves.
39. *Stipel* Scale or gland present at the base of petiolule.
40. *Stipule* Scale, gland or blade-like structure at the base of petiole of leaf.
41. *Storage leaf* Fleshy, succulent leaf.
42. *Tendril* Coiled, modified leaflet.

20.5.4 Margin of Leaf[1] (Fig. 20.4)

1. *Biserrate* When each tooth of the serrated margin is again serrated.
2. *Ciliate* Margin bearing hairs.
3. *Crenate* With blunt, low rounded teeth.
4. *Crenulate* Finely crenate.
5. *Crispate* Curled and extremely undulate.
6. *Dentate* With sharp, marginal teeth pointing outward.
7. *Dissected* Cut or deeply divided into many segments.
8. *Divided* Cut into distinct segments or sections, extending to the midrib or base of the leaf.
9. *Entire* With a continuous smooth margin; lacking any teeth, lobes or indentations.
10. *Incised* Provided with sharp and irregular incisions.
11. *Lacerate* Torn or irregularly cut.
12. *Lobed* Provided with many lobes extending one third to one-half the distance between the midrib and the margin.
13. *Palmatifid* Cut about halfway down in a palmate form.
14. *Parted* Dissected or cut almost to the midrib.
15. *Pinnatifid* Parted in a pinnate way, or divided almost to the midrib.
16. *Repand* Wavy in appearance.
17. *Revolute* With the margin rolling inward i.e. towards the underside of the leaf.
18. *Runcinate* When the teeth of the serrated margin are pointing backward.
19. *Serrate* With the marginal teeth pointing towards the apex.
20. *Serrulate* When the teeth in the serrated margin are very minute.

[1] Arranged alphabetically.

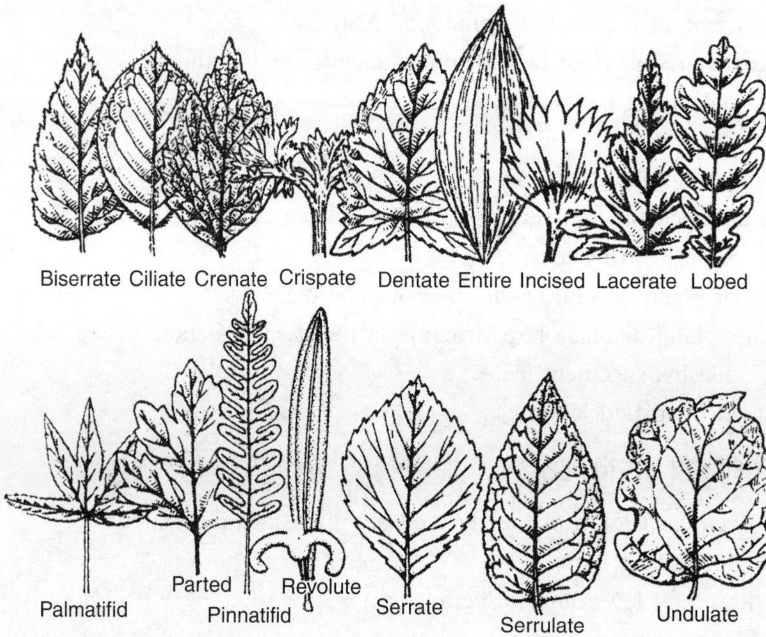

Fig. 20.4 Leaf margins.

21. *Sinuate* With a deeply wavy margin.
22. *Spinous* Provided with projecting spines.
23. *Undulate* With a slightly wavy margin.

20.5.5 Phyllotaxy[1]

1. *Alternate* Bearing one leaf at each node.
2. *Distichous* When alternate leaves appear on just two sides of the stem.
3. *Opposite* Bearing leaves paired at each node on opposite sides.
4. *Opposite decussate* When two successive opposite pairs of leaves occur at right angle to each other.
5. *Opposite superposed* When all the successive opposite pairs of leaves occur at the same plane.
6. *Radical or basal* When leaves often form a cluster at the ground level.
7. *Tristichous, pentastichous and octostichous* In the alternate arrangement of leaves, if the fourth leaf comes over the first one, the arrangement is called *tristichous*; if the sixth leaf comes over the first one after completing two revolutions of spiral, it is called *pentastichous*; and if the ninth leaf comes over the first one after completing three revolutions of the spiral, it is called *octostichous.*
8. *Whorled or verticillate* Bearing three or more leaves at each node.

[1]Arranged alphabetically.

20.5.6 Shape[1] of Leaf (Fig. 20.5)

1. *Acicular* Needle shaped.
2. *Auriculate* Ear shaped.
3. *Cordate* Heart shaped.
4. *Cuneate* Wedge-shaped with lower end narrow.
5. *Deltoid* Delta-like or triangular.
6. *Elliptical* Oval in outline, being narrowed to form rounded ends and widest at the middle.
7. *Falcate* Sickle-shaped.
8. *Filiform* Long, slender and thread-like.
9. *Hastate* Like an arrowhead.
10. *Lanceolate* Much longer than broad; or lance-shaped; or widening above the base and tapering towards the tip.
11. *Linear* Long, flat, and narrow with almost parallel sides.
12. *Lorate* Strap-shaped, or like a narrow strip of leather.
13. *Lyrate* Like a lyre, i.e. having a big terminal lobe and several smaller lateral lobes.
14. *Obcordate* Inversely cordate.
15. *Oblanceolate* Broad at the middle and tapering towards both the ends.
16. *Oblique* When two halves of the lamina are unequal.
17. *Oblong* Long, wide, with the parallel margins.
18. *Obovate* Upper terminal half broader than the lower basal half; opposite to ovate.
19. *Ovate* Egg shaped; broad at the base and narrowing towards apex.
20. *Pandurate* Resembling obovate with a distinct concavity along each basal side; fiddle-shaped.
21. *Pedate* Like the claw of a bird; or palmately lobed.
22. *Reniform* Kidney-shaped.
23. *Rhomboid* Rhombic-shaped.
24. *Rotund* Nearly circular.
25. *Runcinate* Irregularly serrate or sharply incised with the teeth pointing towards the base.
26. *Sagittate* Like an arrowhead; or triangular.
27. *Spathulate* Spoon-shaped.
28. *Subulate* Tapering from base to apex; awl-shaped.

20.5.7 Types of Stipule (Fig. 20.6)

1. *Adnate* Attached fully to petiole as wings.
2. *Basal* Stipules attached near the base of the petiole.

[1]Arranged alphabetically.

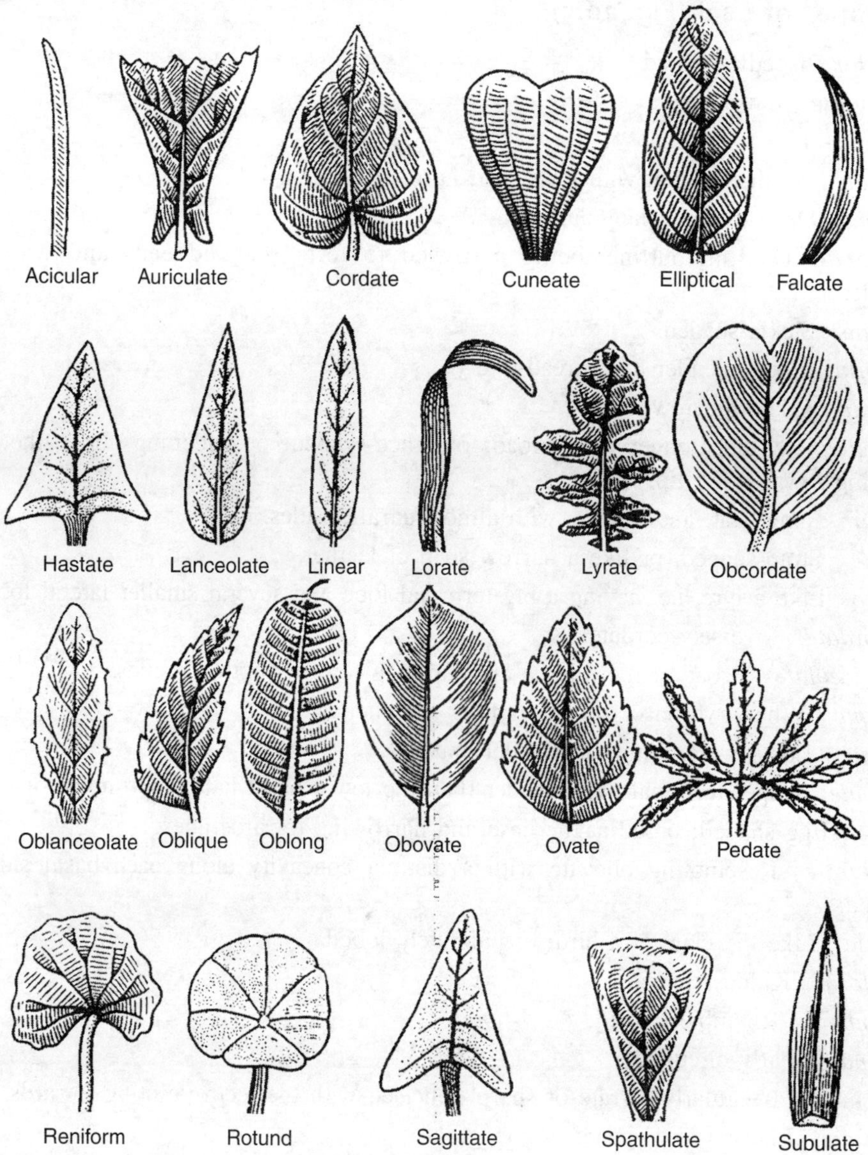

Fig. 20.5 Leaf shapes.

3. *Foliaceous* Stipule resembling with a large leaf.
4. *Interpetiolar* Present in between the petioles of opposite leaves.
5. *Intrapetiolar* Situated in the axil of leaves, and their margins fuse above the petiole.
6. *Lateral* Present laterally on the petiole.

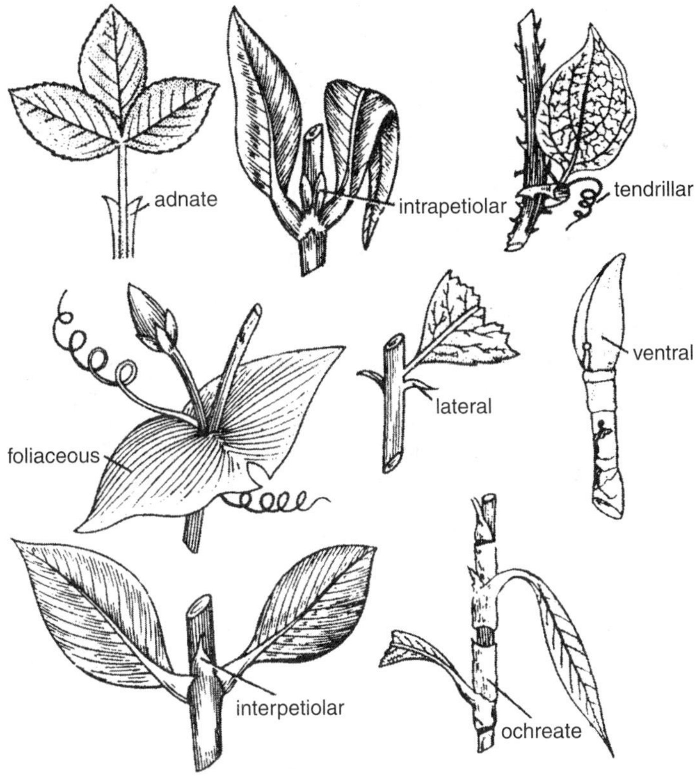

Fig. 20.6 Types of stipule.

7. *Ochreate* Fused to form an ochrea or a fused tubular nodal sheath around the internode.
8. *Photosynthetic* Large, green and leaflike or foliaceous.
9. *Scaly* Dry, small and membranous stipule.
10. *Sheathing or Protective* Enclosing a bud or flower like a sheath.
11. *Spinous* Stipule in the form of a hard spine.
12. *Tendrillar* Stipule modifying into a tendril.
13. *Ventral* Present on the ventral side of the petiole.
14. *Vestigial* Very small, minute, or remnant.

20.5.8 Types of Leaf (Fig. 20.7)

1. *Compound Leaf* A leaf, in which the leaf blade or lamina remains divided into smaller, bladelike parts or leaflets is called a compound leaf. A compound leaf may be palmately compound or pinnately compound.
2. *Palmately Compound Leaf* If the leaflets diverge from a common point at the end of the petiole, in the same way as the fingers from the palm of the hand, the leaf is called *palmately compound*. In such leaves, if a single leaflet is articulated to the petiole, it is called *unifoliate*;

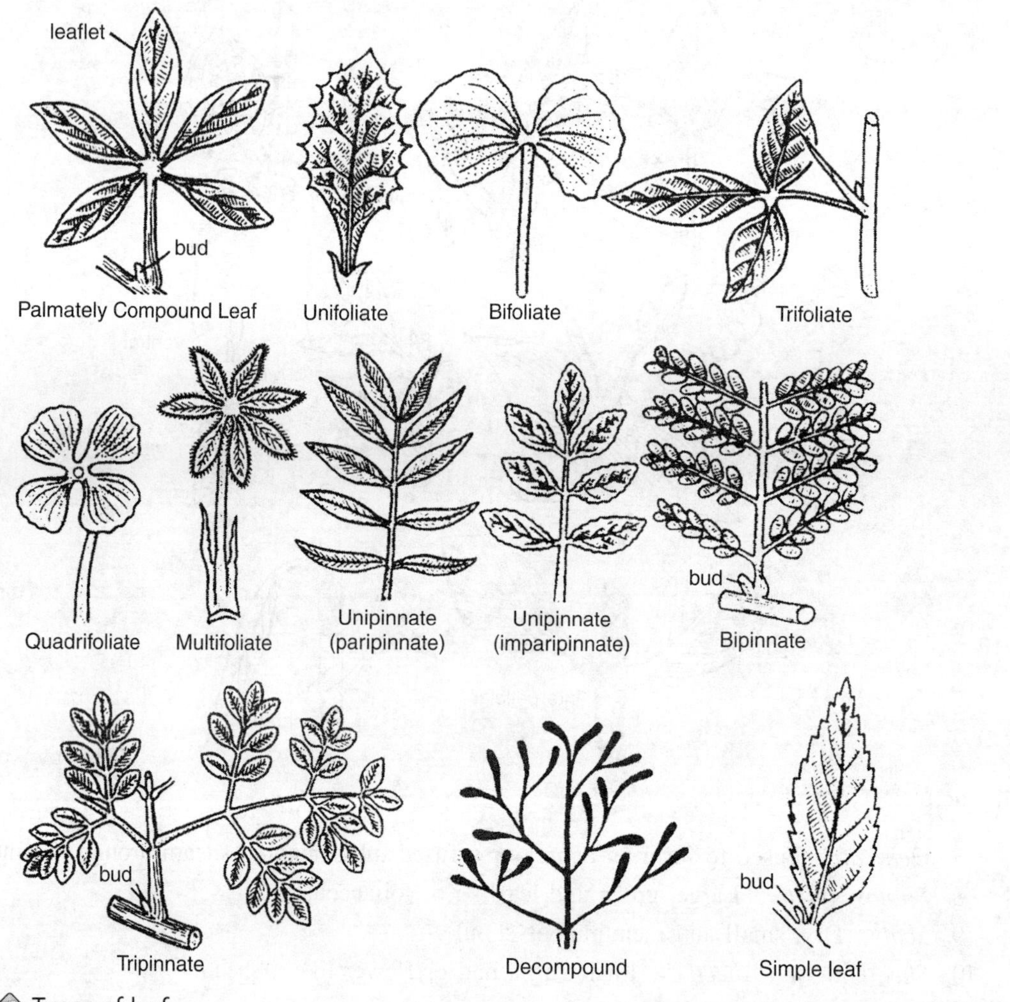

Fig. 20.7 Types of leaf.

if two, three or four leaflets are articulated to the petiole, it is called *bifoliate, trifoliate* or *quadrifoliate*, respectively; and if five or more leaflets are articulated to the petiole, it is called *multifoliate*.

3. *Pinnately Compound Leaf* If the leaflets are attached on both sides of one central rachis, the leaf is called *pinnately compound*. In such leaves, if the leaflets are attached directly on the midrib, the leaf is called *unipinnate. A* unipinnate leaf having even number of paired leaflets is called *paripinnate,* while that which contains an odd terminal leaflet is called *imparipinnate*. Such a pinnately compound leaf, in which the midrib produces secondary axis, and on the latter are present the leaflets, is called *bipinnate*; if the midrib of pinnately compound leaf produces secondary axis, and the latter produces the tertiary axis which

bears the leaflets, it is called tripinnate. If the leaf is more than thrice pinnate, it is called *decompound.*

4. *Simple Leaf* A leaf with the blade in a single part is called simple leaf. The single blade may, however, be variously divided.

20.5.9 Venation of Leaf (Fig. 20.8)

1. *Venation* The pattern of veins on the surfacc of a leaf.
2. *Parallel* When the veins run parallel to each other in the lamina of the leaf, the venation is called parallel, e.g. monocots. In this type of venation, if only one principal vein is present, it is called *unicostate,* and if several principal veins are present, it is called *multicostate.*
3. *Reticulate* When the pattern of the veins in the lamina of the leaf is like a network, e.g. dicots.
4. *Unicostate and Multicostate* In both parallel and reticulate venations, if only one principal vein is present, it is called *unicostate,* and if several principal veins are present, it is called *multicostate.* In the multicostate types, if the veins run in a curved manner from the base of the blade to its apex, it is called *convergent,* and if the veins arise at the base of the leaf blade and then diverge from one another towards the leaf margin, it is called *divergent.*

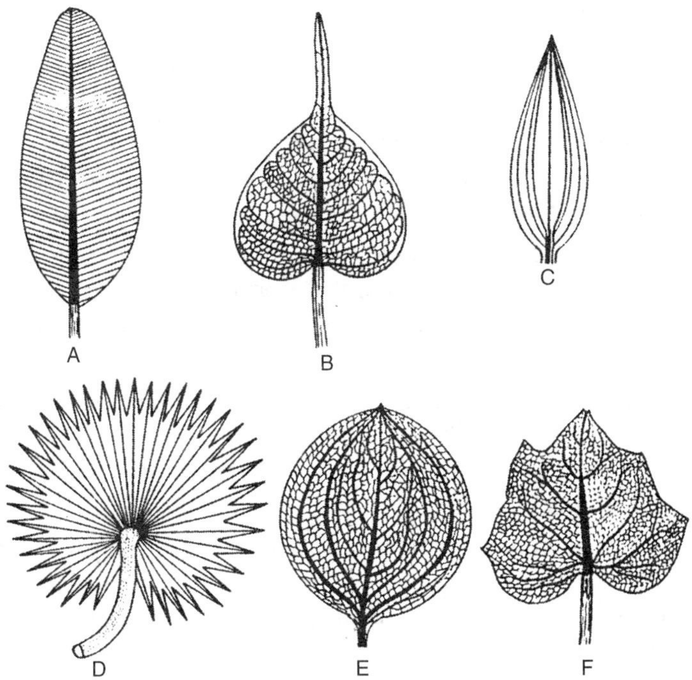

Fig. 20.8 Leaf venation. A: Unicostate parallel; B: Unicostate reticulate; C: Multicostate parallel convergent; D: Multicostate parallel divergent; E: Multicostate reticulate convergent; F: Multicostate reticulate divergent.

20.6 INFLORESCENCE

20.6.1 Basic Categories

Cymose, racemose, and special types are three basic categories of inflorescences (mode of the arrangement of the flowers on the floral shoot) of angiosperms.

1. *Cymose* In this type the growth of the main axis is checked soon by the development of a flower at the apex, and the lateral axis below the terminal flower also ends in a flower, and thus its growth is also checked. In cymose inflorescence, the terminal flower is the oldest and the young flowers are present on the lower side. Helicoid, circinnus, rhipidium, dichasium and polychasium are some of the examples of cymose.
2. *Racemose* In this type, the main axis does not terminate into a flower, but it keeps on growing continually and gives off flowers laterally in acropetal succession. Here the youngest flower is present at the apex and the older flowers towards the base. Raceme, spike, spikelet, panicle, catkin, spadix, corymb, umbel and capitulum or head are some of the examples of racemose.
3. *Special Types* Cyathium, verticillaster, and hypanthodium are some special types of inflorescences.

20.6.2 Types of Inflorescence[1] (Fig. 20.9)

1. *Catkin* A spike-like, deciduous, elongate, inflorescence, with scaly bracts, and unisexual, apetalous and sessile flowers.
2. *Cincinnus* A modified helicoid cyme having the short pedicels on the developed side.
3. *Compound Corymb* A branched corymb having pedicellate flowers.
4. *Compound Cyme* A branched cyme having pedicellate flowers.
5. *Compound Umbel* A branched umbel having primary rays arising at a common point with a secondary umbel arising from the tip of the primary rays.
6. *Corymb* A raceme whose lower stalks are longer than the upper ones, so the inflorescence has a flat top.
7. *Cyathium* A cup-shaped involucre containing nectar-secreting glands, a centrally-placed large female flower, and many male flowers.
8. *Dichasium or Simple Cyme* Determinate, dichotomous inflorescence of pedicellate flowers having pedicels of equal length.
9. *Glomerule* An indeterminate inflorescence having dense cluster of sessile or subsessile flowers.
10. *Head or Capitulum* A dense cluster of several sessile or subsessile flowers on a compound receptacle or torus.
11. *Helicoid Cyme* Curved and unbranched inflorescence of pedicellate flowers having the branches only on one side.

[1] Terms arranged alphabetically.

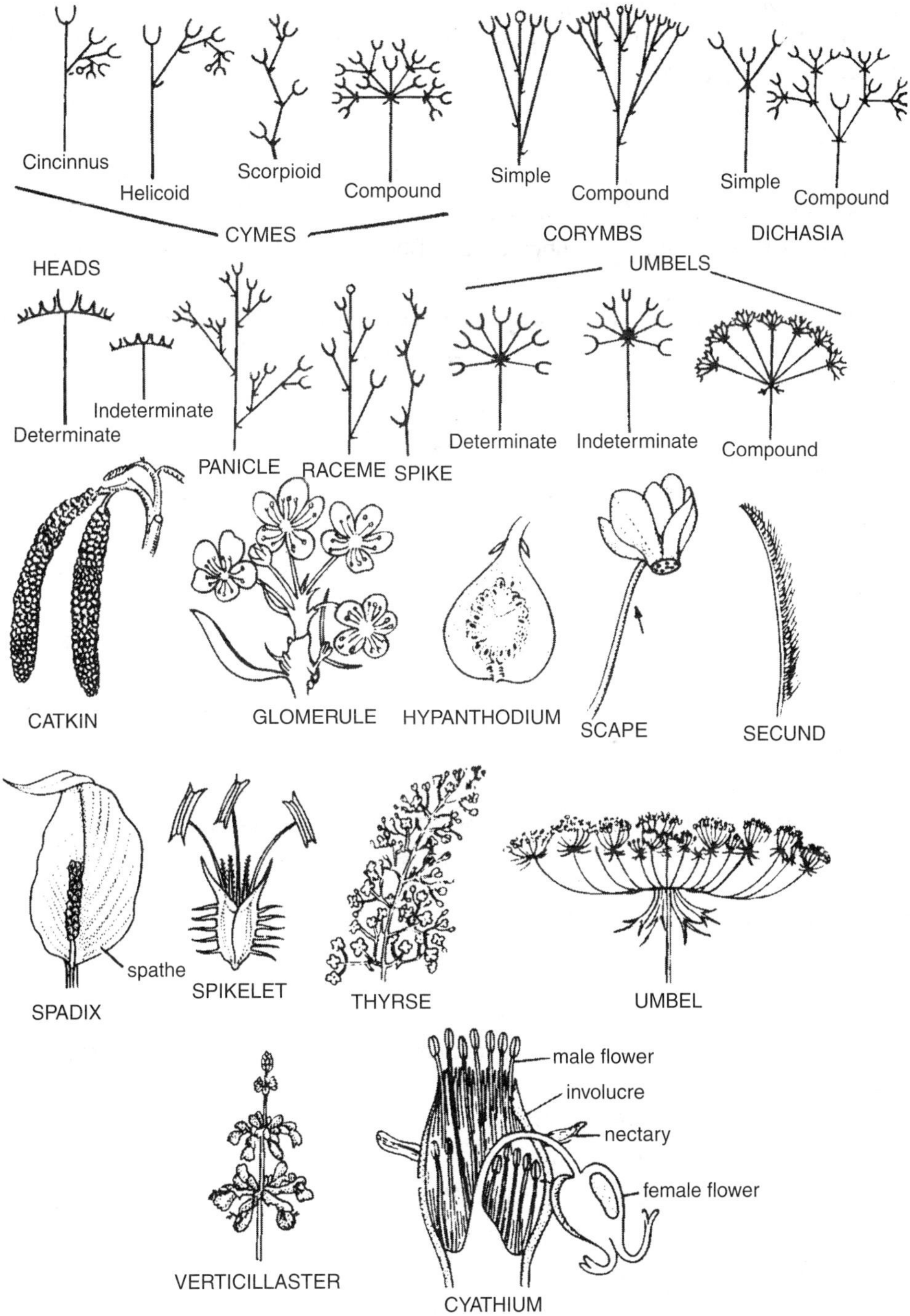

Fig. 20.9 ◈ Types of inflorescence.

12. *Hypanthodium* An inflorescence having sessile flowers on the wall of a concave capitulum, opening by a small ostiole. Male flowers are situated near the periphery and female flowers in the centre, e.g. *Ficus.*
13. *Panicle* Branched inflorescence with pedicellate flowers arranged in the form of number of racemes.
14. *Pleiochasium* It is a compound dichasium in which each dichasium has three lateral branches. It is also called *polychasium* or *multiparous cyme.*
15. *Raceme* An indeterminate, unbranched inflorescence with a single axis and the flowers arranged along the main axis on pedicels.
16. *Scorpioid Cyme* A zig-zag cymose inflorescence that appears to coil like a scorpion's tail. In this type, the branches develop alternately on opposite sides of the rachis.
17. *Secund* Flowers arranged only on one side of the rachis.
18. *Solitary Axillary* Single-flowered; flower attached in the axis.
19. *Solitary Terminal* Single-flowered; flower attached at the apex, and not in the axis.
20. *Spadix* A fleshy or thick, spikelike inflorescence with very small flowers usually enclosed in a spathe.
21. *Spike* Elongate, unbranched, indeterminate inflorescence with sessile flowers.
22. *Spikelet* A small spike, the basic inflorescence unit of Cyperaceae and Gramineae.
23. *Thyrse* A compact and compound panicle having an indeterminate main axis and laterally determinate axes.
24. *Umbel* An inflorescence in which all the pedicels are of the same length and arise from the same point.
25. *Umbellet* A secondary umbel in a compound umbel.
26. *Verticillaster* Whorled dichasial cymes arranged at the nodes of an elongate axis, e.g. Labiatae.

20.7 FLOWER (PARTS AND TYPES[1])

1. *Accessory Whorls* Calyx and corolla.
2. *Achlamydous* Without calyx and corolla.
3. *Actinomorphic* Flowers with radial symmetry; or flowers which can be bisected into similar halves along two or more planes.
4. *Amorphic* Flowers without symmetry.
5. *Androecium* All stamens of a flower.
6. *Androgynophore* Stripe bearing both androphore and gynophore together.
7. *Androphore* Elongated internodal part between corolla and androecium.
8. *Anthophore* Elongated internodal part between calyx and corolla.
9. *Bisexual or Hermaphrodite* Flowers having both the sex organs.

[1]Terms arranged alphabetically,

10. *Bract* Modified leaf which develops in the axil of• the flower.
11. *Bracteate* With bract.
12. *Bracteolate* With bracteole.
13. *Bracteole* Small leaf borne on pedicel of flowers.
14. *Calyx* Whorl of sepals.
15. *Carpel* A unit of gynoecium; or floral organ that bears ovules.
16. *Complete* Flowers having all the four floral whorls, i.e. sepals, petals, androecium and gynoecium.
17. *Corolla* Whorl of petals.
18. *Dimerous, Trimerous, Tetramerous and Pentamerous* Flowers in which various floral parts are arranged in groups of two, three, four and five, are called dimerous, trimerous, tetramerous and pentamerous, respectively.
19. *Ebracteate* Without bracts.
20. *Epicalyx* Leaves resembling sepals below the true calyx.
21. *Epigynous* Flowers with inferior ovary; or those having the floral parts situated above the ovary.
22. *Essential Whorls* Androecium and gynoecium.
23. *Gynoecium or Pistil* Group of all carpels of a flower.
24. *Gynophore* Elongated axis between androecium and gynoecium.
25. *Haplomorphic* Petals or tepals coloured.
26. *Hypanthium* Fused basal portion of sepals, petals or stamens around the ovary.
27. *Hypogynous* Flowers with superior ovary; or those having the floral parts situated below the ovary.
28. *Monochlamydous* Flowers with only one whorl.
29. *Monoecious* Having separate male and female flowers on the same individual.
30. *Pedicel* The stalk of the flower.
31. *Pedicellate* Flower with pedicel.
32. *Perianth* When there is no differentiation of calyx and corolla.
33. *Perigynous* Flowers with half-inferior ovary; or those having floral parts situated around the ovary.
34. *Petal* Individual unit of corolla.
35. *Pleomorphic* Actinomorphic flowers with number of reduced parts.
36. *Polyphore* A receptacle having several distinct carpels.
37. *Receptacle, Thalamus or Torus* Tip of the axis bearing floral appendages.
38. *Sepal* Individual unit of calyx.
39. *Sessile* Without stalk.
40. *Stamen* Individual unit of androecium.

41. *Tepal* Individual unit of perianth.
42. *Unisexual* Flower with only one sex, i.e. either male or female.
43. *Zygomorphic* Flowers with bilateral symmetry; or flowers which can be bisected into similar halves in only one plane; or flowers with parts usually irregular and reduced in number.

20.8 CALYX[1]

1. *Anterior Lobes* Abaxial lobes; or lobes that remain away from axis.
2. *Aposepalous* With separate sepals.
3. *Asepalous* Without sepals.
4. *Bilabiate* Consisting of two lips.
5. *Caducous* Sepals which wither or fall soon.
6. *Campanulate* Bell-shaped.
7. *Chorisepalous* With separate sepals.
8. *Cupulate* Cup-like.
9. *Dorsal Side* Abaxial, or back, or lower side of the sepals.
10. *Fringed* Modified margin of sepals or petals.
11. *Gamosepalous* Fused sepals.
12. *Infundibuliform* Funnel shaped.
13. *Pappus* Reduced, scaly or hairy calyx of Asteraceae.
14. *Persistent* Sepals which persist even in the fruit.
15. *Petalloid* Coloured sepals, except green.
16. *Polysepalous* When sepals are free.
17. *Posterior Lobe* Adaxial lobe; or lobe that remains next to axis.
18. *Sepal* A unit of calyx.
19. *Spurred* When one or more sepals are produced into spur.
20. *Synsepalous* With fused sepals.
21. *Tubular* Like a tube; cylindrical.
22. *Urceolate* Urn-shaped.
23. *Ventral Side* Upper or top side of sepal or petal.

20.9 AESTIVATION[1] (Fig. 20.10)

1. *Aestivation* The way in which sepals and petals are arranged in the bud condition.
2. *Imbricate Aestivation* Out of the five sepals or petals, one external, one internal, and other three are partly external and partly internal.

[1] Terms arranged alphabetically.

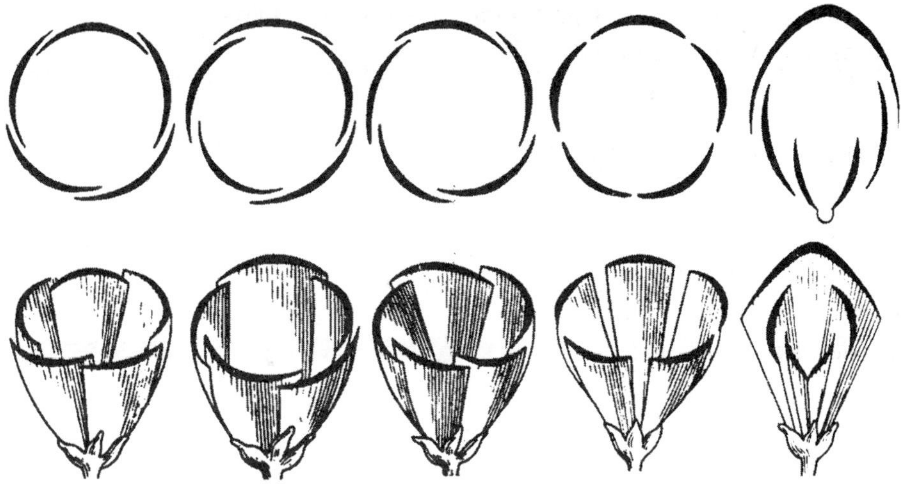

Fig. 20.10 Types of aestivation.

3. *Quincuncial Aestivation* Out of the five sepals or petals, two external, two internal, and remaining one is partly external and partly internal.
4. *Twisted Aestivation* When margins of each part are overlapped regularly, i.e. one edge of the sepal or petal are overlapped by the preceding part.
5. *Valvate Aestivation* When the sepals or petals meet edge to edge without overlapping each other.
6. *Vexillary Aestivation* Out of the five sepals or petals the posterior one is the largest and covers the two lateral sepals or petals, and the latter in turn overlap the two smallest and anterior sepals or petals, e.g. Papilionaceae.

20.10 COROLLA

20.10.1 General Terms[1]

1. *Apetalous* Without petals.
2. *Apopetalous* Having separate petals.
3. *Carina* Keel, or two fused petals of a papilionaceous flower.
4. *Choripetalous* Apopetalous.
5. *Gamopetalous* With fused petals.
6. *Lodicule* Scale-like perianth part of Gramineae.
7. *Petal* An individual unit of corolla.
8. *Polypetalous* With free petals.

[1]Arranged alphabetically.

9. *Standard or Vexillum* Wider petal in a papilionaceous flower.
10. *Sympetalous* With fused petals.
11. *Wing* Lateral petals, e.g. Fabaceae.

20.10.2 Types of Corolla[1] (Fig. 20.11)

1. *Bilabiate or Bilipped* Zygomorphic gamopetalous corolla with two unequal lips or divisions.
2. *Calcarate* Spurred corolla.
3. *Campanulate* Bell-shaped.
4. *Carinate* Keeled.

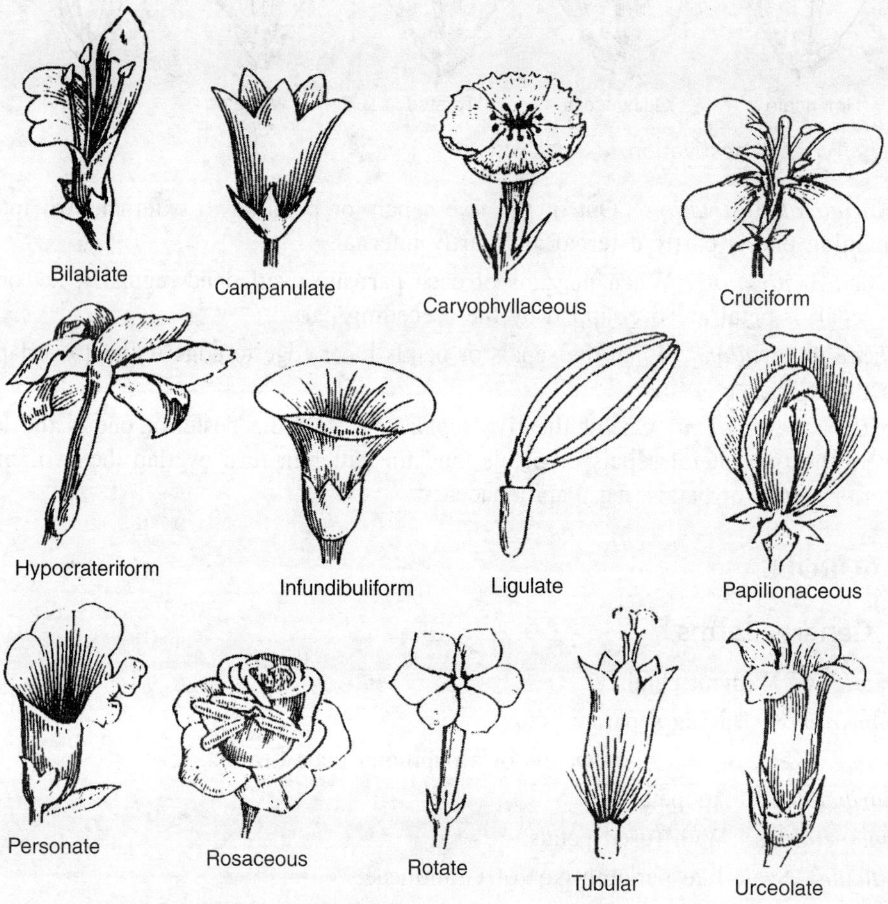

Fig. 20.11 Types of corolla.

[1] Terms arranged alphabetically.

5. *Caryophyllaceous* When all the five petals of the corolla are clawed and have their limbs spreading out.
6. *Cruciate or Cruciform* All the four petals are arranged in the form of a cross, e.g. Cruciferae.
7. *Hypocrateriform* Gamopetalous, salver-shaped corolla i.e. basal portion narrow and tubular having abruptly expanding flat apical portion.
8. *Infundibuliform* Funnel-shaped.
9. *Ligulate* Zygomorphic, gamopetalous, strap-shaped corolla with a short narrow tube and strap-like upper portion.
10. *Papilionaceous or Butterfly-like* When five petals of the corolla are arranged in a butterfly like shape. It consists of a posterior largest *vexillum,* two lateral *alae,* and two posterior fused petals called *keel,* e.g. Papilionaceae.
11. *Personate* It is also a bilabiate corolla but here the corolla mouth is closed because two lips are very close to one another.
12. *Rosaceous* When five or more petals are spreading like those of rose, e.g. Rosaceae.
13. *Rotate* Wheel-shaped gamopetalous corolla having narrow corolla tube, and the limbs of petals are at right angle to the lobe.
14. *Salverform* Trumpet-shaped.
15. *Tubular* Cylindrical; petals fused to form a tube.
16. *Urceolate* Urn- or pitcher-shaped.

20.11 PERIANTH[1]

1. *Gamotepalous* With fused tepals.
2. *Lodicule* Scalelike perianth part of Gramineae.
3. *Perianth* Collective term applied for calyx and corolla.
4. *Petalloid* Resembling with petals.
5. *Polytepalous* With free tepals.
6. *Sepalloid* Resembling with sepals.
7. *Tepal* Individual unit of perianth.

20.12 ANDROECIUM[1] (Fig. 20.12)

1. *Adnate* When filament runs throughout the entire length of the anther from the base to the top (Fig. 20.12).
2. *Allagostemonous* Condition in which stamens are attached to petals and torus alternately.
3. *Alternipetalous* Stamens present alternately with the petals.
4. *Anther* Pollen-bearing portion of stamen.

[1]Terms arranged alphabetically.

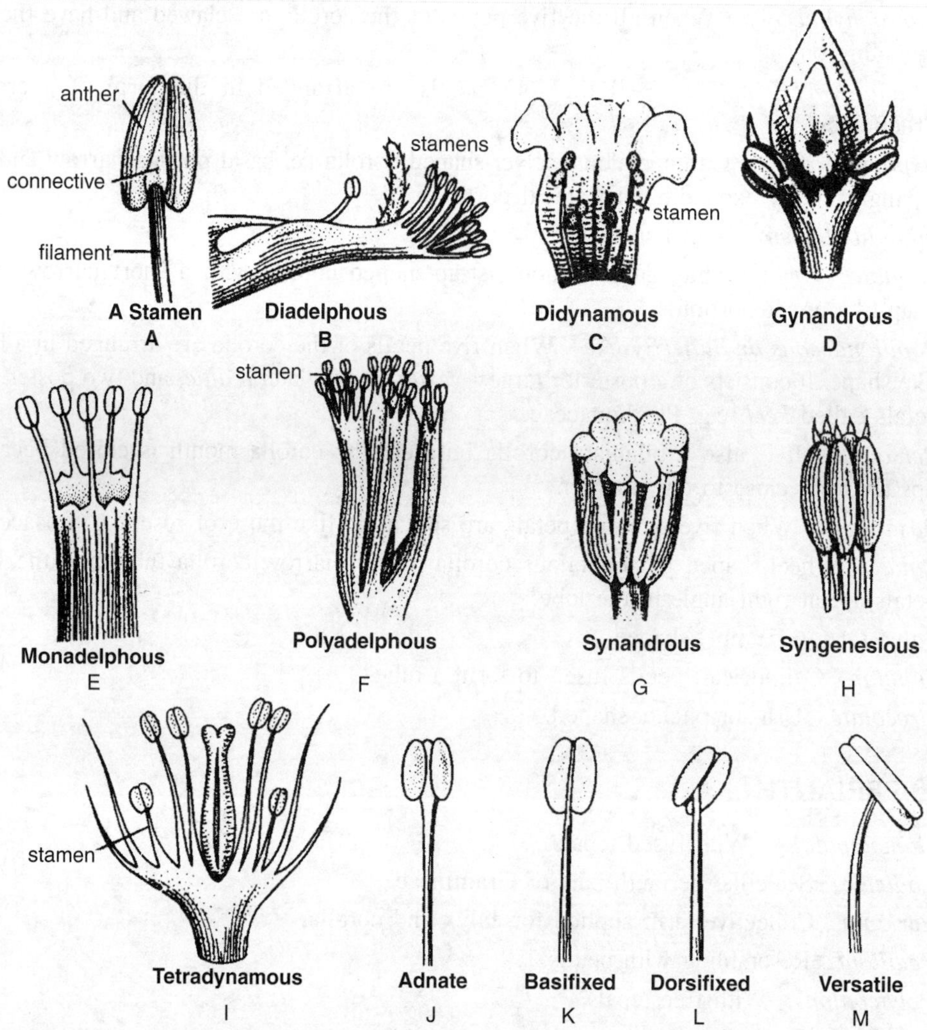

Fig. 20.12 A–I: Types of androecium; J–M: Fixation of filament.

5. *Antipetalous* Stamens opposite to petals.
6. *Antiphyllous* Stamens opposite to tepals.
7. *Antisepalous* Stamens opposite to sepals.
8. *Apostemonous* With separate stamens.
9. *Appendicular* Stamen with a modified or protruding connective.
10. *Basifixed* Filament attached with the base of the anther lobe.
11. *Connective* Filament extension of the anther between thecae.
12. *Diadelphous* With the filaments of stamens united or connate in two groups with their anthers being free.

13. *Diandrous* Flowers with two stamens.
14. *Didymous* With stamens in two equal pairs.
15. *Didynamous* When two out of the four stamens are larger and remaining two are smaller in size (Fig. 20.12).
16. *Diplostemonous* When stamens are arranged in two whorls, of which the outer whorl is opposite to sepals and inner whorl is opposite to petals.
17. *Dorsifixed* When the filament is attached on the dorsal side of the anther lobes (Fig. 20.12).
18. *Epipetalous* When the stamens remain attached to the petals.
19. *Epiphyllous* When the stamens remain attached with the tepals.
20. *Episepalous* When the stamens remain attached with the sepals.
21. *Extrorse* Stamen which dehisces longitudinally outward.
22. *Filament* Stalk of the stamen.
23. *Gynandrous* When stamens are fused with gynoecium.
24. *Introrse* Stamen which dehisces longitudinally inward.
25. *Laminar* When stamens are leaflike with no distinct anther and filament.
26. *Laterose* Stamen which dehisces longitudinally and laterally.
27. *Locule* Chamber of an anther.
28. *Monadelphous* When filaments of one group of stamens are fused but their anthers are free (Fig. 20.12).
29. *Monandrous* Flower with one stamen.
30. *Obdiplostemonous* When the stamens are arranged in two whorls, of which the outer whorl of stamens being opposite to petals.
31. *Petalloid* Stamens resembling with petals, showing no distinction between anther and filament.
32. *Petalostemonous* With filaments of stamens fused to petals with free anthers.
33. *Phaenantherous* With exerted stamens.
34. *Pollen grain* Young male gametophyte.
35. *Pollen sac* Male sporangium.
36. *Polyadephous* When filaments are united in several groups with their anthers remaining free (Fig. 20.12).
37. *Polyandrous* With several free stamens.
38. *Poricidal* Dehiscing through a pore at the theca apex.
39. *Stamen* Pollen-bearing floral organ of angiosperms.
40. *Staminode* Sterile stamen with reduced anthers.
41. *Synandrous* With anthers as well as their filaments being united in the form of one group.

42. *Syngenesious* With fused anthers and free filaments (Fig. 20.12).
43. *Tetrandrous* Flowers with four stamens.
44. *Tetradynamous* Two out of a total of six stamens are short while the remaining four are long.
45. *Theca* Half of the portion of anther containing two pollen sacs.
46. *Triandrous* Flowers with three stamens.
47. *Tridynamous* Stamens arranged in two equal groups of three.
48. *Valvular* When the stamen dehisces through a pore covered by a flap of tissue.
49. *Versatile* Dorsifixed but it appears as if the anther is swinging freely on the filament (Fig. 20.12).

20.13 GYNOECIUM[1] (Fig. 20.13)

20.13.1 Parts, Fusion Types, Carpel Parts and Types, Style Types, and Ovary Position

1. *Apocarpous* With free or separate carpels.
2. *Astylocarpellous* Without a stipe and a style.
3. *Astylocarpepodic* With a stipe and without a style.
4. *Astylous* Without any distinct style.
5. *Carpel* Organ of the flower that bears ovules.
6. *Epigynous* The condition in which the sepals, petals and androecium are attached to the floral tube above the ovary.
7. *Funiculus* The stalk with which the ovule is attached to the placentum.
8. *Geniculate Style* The style which bents abruptly.
9. *Gynobasic Style* The style which is attached at the base of the ovary in central depression.
10. *Heterostylous* With styles of different shapes or lengths.
11. *Homostylous* With styles of same shapes or lengths.
12. *Hypogynous* The condition in which sepals, petals and androecium are attached to the floral tube below the ovary.
13. *Inferior Ovary* The ovary in the epigynous condition.
14. *Locule* The cavity of the ovary.
15. *Monocarpellary to Polycarpellary* With one to several carpels; *bicarpellary* if two carpels are present, *tricarpellary* if three carpels are present, and *tetra*-and *pentacarpellary* if four and five carpels are present.
16. *Ovary* Ovule-bearing part of the gynoecium.

[1] Terms arranged alphabetically.

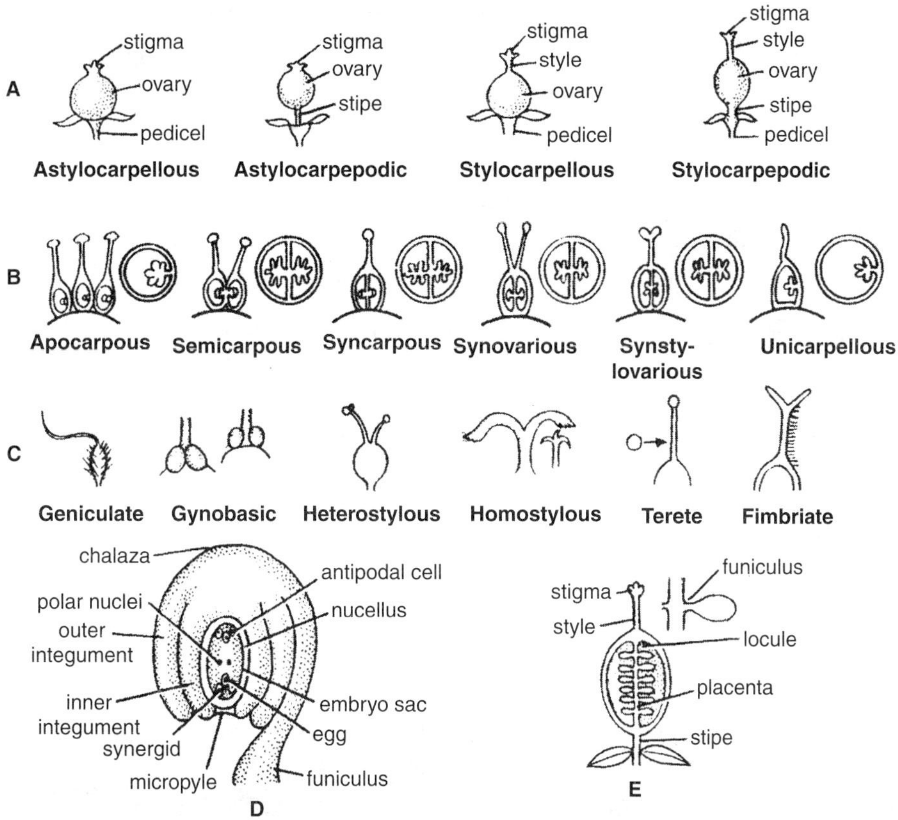

Fig. 20.13 A: Types of carpel; B: Types of gynoecium; C: Types of style; D: V.s. of an ovule and; E: V. s. of a carpel

17. *Ovule* Young seed made up of nucellus and integuments.
18. *Parts of a Carpel* Funiculus, locule, ovary, ovule, placenta, stigma, style.
19. *Parts of Gynoecium* Carpel, locule, ovary, ovule, placenta, stigma, style.
20. *Parts of Ovule* Chalaza, embryo sac, integuments, micropyle, nucellus, raphe.
21. *Perigynous* The condition in which the sepals, petals and androecium are attached to the floral tube around the ovary.
22. *Pistillode* A sterile carpel.
23. *Placenta* Ovule-bearing region of the wall of ovary.
24. *Raphe* Part of an ovule present in the form of a longitudinal ridge on its outer integument.
25. *Replum* False septum.
26. *Semicarpous* With fused ovaries of adjacent carpels and their free styles and stigmas.

27. *Semi-inferior* The position of ovary in the perigynous condition.
28. *Stigma* Uppermost, pollen-receptive part of the gynoecium.
29. *Stipe* Stalk of the gynoecium.
30. *Style* Long portion of gynoecium between ovary and stigma.
31. *Stylocarpellous* Carpel without a stipe and with a style.
32. *Stylocarpepodic* With a stipe and a style.
33. *Stylodious* With single free carpel.
34. *Superior* The position of ovary in the hypogynous condition.
35. *Syncarpous* With fused carpels.
36. *Synovarious* The condition when the ovaries of adjacent carpels are fused while their styles and stigmas separate.
37. *Synstylovarious* The condition when the ovaries and styles of adjacent carpels are fused while their stigmas separate.
38. *Terete* Cylindrical and elongate.
39. *Unicarpellous* With single carpel.
40. *Unilocular to Multilocular* Ovary with one chamber or locule is called *unilocular*; with two chambers is called *bilocular*; with three and four chambers are called *trilocular* and *tetralocular*; and with many locules is called *multilocular* (Fig. 20.14).

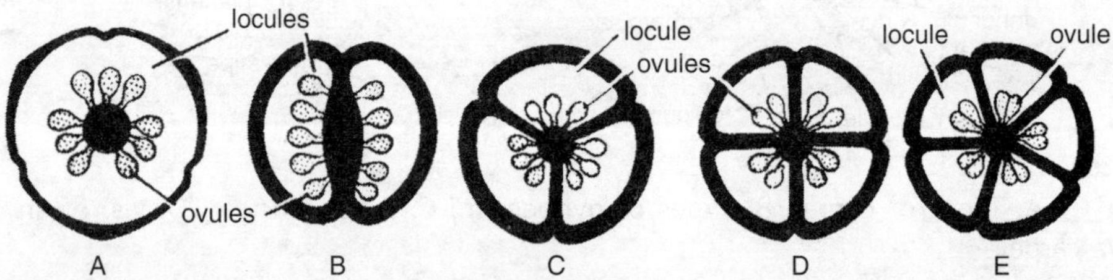

Fig. 20.14 Chambers of the ovary—A: Unilocular; B: Bilocular; C: Trilocular; D: Tetralocular, and; E: Pentalocular.

20.13.2 Types of Stigma (Fig. 20.15)

1. *Capitate* Like a cap or head.
2. *Clavate* Club-shaped.
3. *Crested* Possessing a terminal tuft or ridge.
4. *Decurrent* Long and extending downward.
5. *Diffuse* Spread over a wide surface.
6. *Discoid* Like a disc.
7. *Fimbriate* Fringed.
8. *Lineate* In the form of small lines.
9. *Lobed* Divided into some lobes.

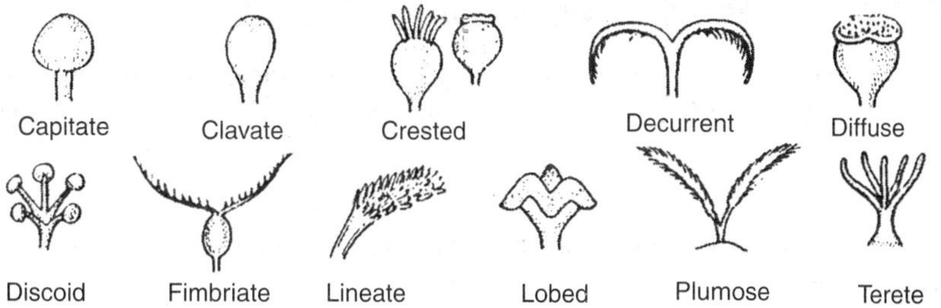

Fig. 20.15 Types of stigma.

10. *Plumose* Like a feather.
11. *Terete* Elongate and cylindrical.

20.14 PLACENTATION[1] (Fig. 20.16)

1. *Axile* The placentae develop along the central axis in a compound ovary with septa. It occurs in a bilocular to multilocular ovary.
2. *Basal* The placenta develops at the base of the ovary, i.e. directly on the thalamus. It occurs in a unilocular ovary.
3. *Free-central* The placenta develops along the central axis in a compound ovary without septa. It occurs in unilocular ovary.
4. *Laminar, or Laminate, or Superficial* The placenta develops over the inner surface of the ovary wall. It occurs in a multicarpellary ovary.
5. *Marginal* The placenta develops along the margin of the simple ovary. It occurs in mono-carpellary and unilocular ovary.
6. *Parietal* The placentae develop on the wall or intruding portions of a compound ovary. It occurs in bicarpellary to multicarpellary but unilocular ovary.
7. *Pendulous or Apical* The placenta develops at the top of the ovary and the ovules remain suspended in a pendulous manner.

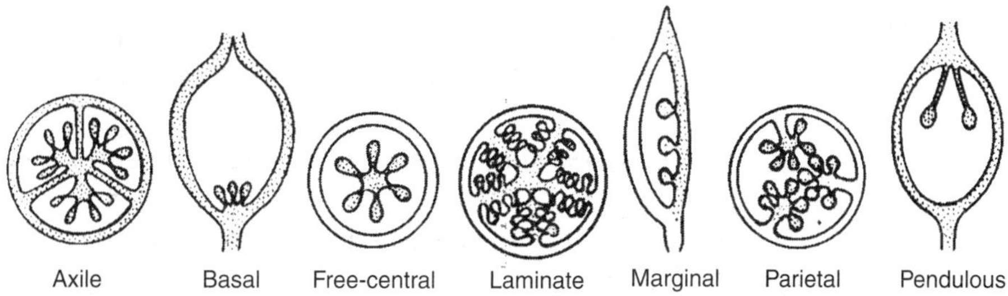

Fig. 20.16 Types of placentation.

[1] Terms arranged alphabetically.

20.15 FRUIT[1] (Figs. 20.17, 20.18, 20.19)

1. *Achene* Developing from a monocarpellary superior ovary, the achene is a one-seeded, dry, indehiscent, simple fruit having its seed attached to fruit wall only at one point, e.g. *Mirabilis, Clematis.*
2. *Amphisarca* A fleshy, simple, berry-like fruit with a woody rind, e.g. *Lagenaria.*
3. *Balausta* A simple, dry, many-seeded and many-loculed, indehiscent fruit with a tough pericarp, e.g. *Punica.*
4. *Berry* Fleshy, simple fruit with succulent pericarp, e.g. tomato, banana, *Vitis.*
5. *Bibacca* A multiple type of fruit formed as a fused double berry, e.g. *Lonicera.*
6. *Cacervulus* A dry, schizocarpic fruit which develops from bi- to multicarpellary, syncarpous, superior ovary with many locules, e.g. Labiatae.
7. *Calybium* A hard, unichambered, dry, indehiscent fruit derived from an inferior ovary, e.g. *Quercus.*
8. *Capsule* A dry, dehiscent fruit derived from a bi- to multicarpellary superior or inferior ovary, e.g. *Papaver.*
9. *Caryopsis* A dry, indehiscent fruit derived from monocarpellary ovary in which pericarp is inseparably fused with testa, e.g. Gramineae.
10. *Cremocarp* A dry, schizocarpic fruit derived from bicarpellary, syncarpous, inferior ovary with one ovule in each locule. It dehisces from two indehiscent, single-seeded mericarps attached with carpophore, e.g., Umbelliferae.
11. *Cypsela* A dry, indehiscent achene type of fruit derived from unilocular, inferior ovary, e.g. *Helianthus.*
12. *Drupe* A fleshy fruit with a stony endocarp and edible mesocarp, e.g. mango, *Prunus.*
13. *Druplet* A small drupe, e.g. *Rubus.*
14. *Etaerio of Achenes* Also called *achenecetum,* it is an aggregation of achenes, e.g. *Ranunculus.*
15. *Etaerio of Berries* Also called *baccacetum,* it is an aggregation of berries, e.g. *Artabotrys.*
16. *Etaerio of Drupes* Also called *drupecetum,* it is an aggregation of drupes, e.g. *Rubus.*
17. *Etaerio of Follicles* Also called *follicetum*, it is an aggregation of follicles, e.g. *Delphinium.*
18. *Etaerio of Samaras* Also called *samaracetum*, it is an aggregation of samaras, e.g. *Liriodendron.*
19. *Follicle* A dry, dehiscent fruit derived from a monocarpellary, superior ovary, and dehisces only along one suture, e.g. *Delphinium.*
20. *Hesperidium* A fleshy, thick-skinned berry with hard and leathery pericarp. It is derived from a polycarpellary, multilocular, superior ovary. Bulk of this fruit is derived from glandular hairs, e.g. *Citrus.*

[1] Terms arranged alphabetically.

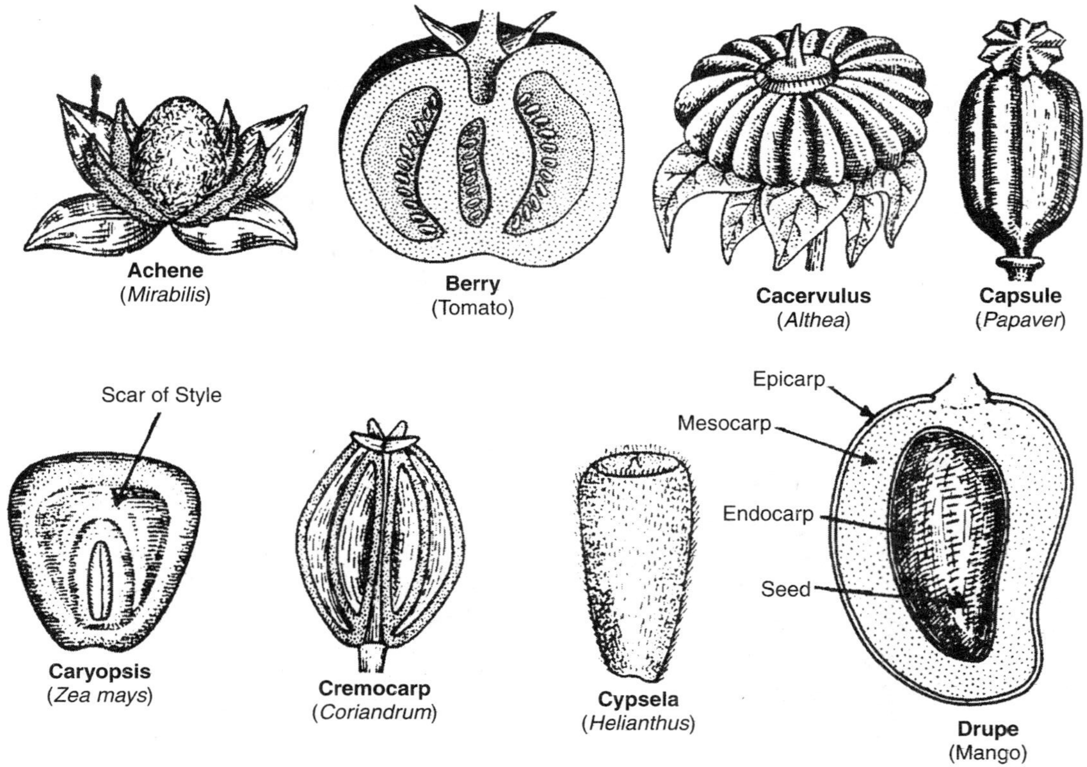

Fig. 20.17 Types of fruit (arranged alphabetically).

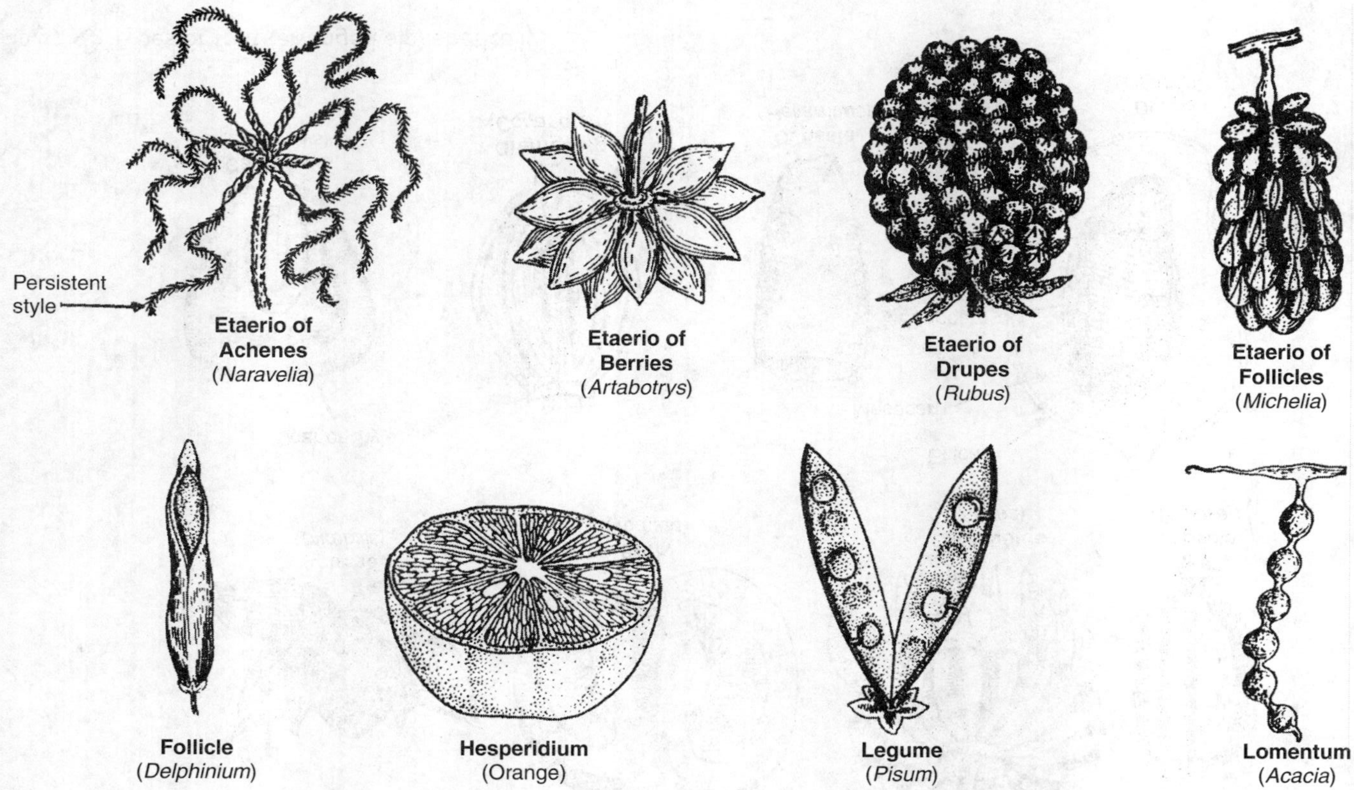

Fig. 20.18 Types of fruit (arranged alphabetically).

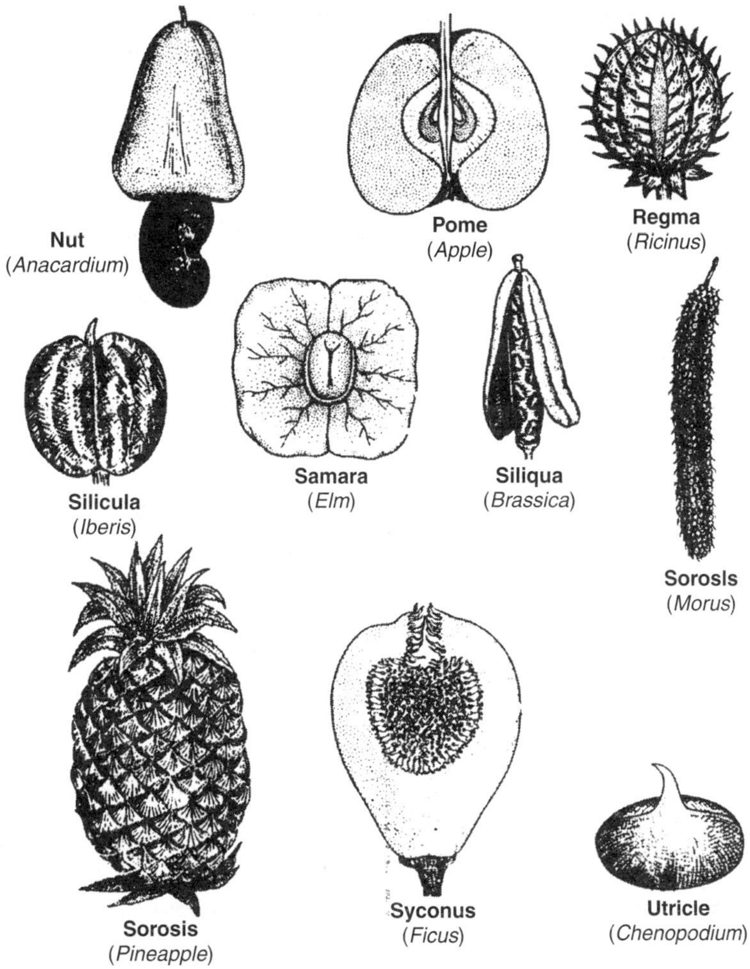

Fig. 20.19 Types of fruit (arranged alphabetically).

21. *Legume* A dry, dehiscent fruit derived from monocarpellary, superior ovary with marginal placentation. It dehisces along two sutures, e.g. *Pisum*.
22. *Lomentum* A dry, dehiscent legume that separates transversely between seed sections, e.g. *Acacia*.
23. *Nut* A dry, indehiscent, one-seeded fruit with a hard pericarp. It is derived from unilocular ovary, e.g. *Anacardium*.
24. *Nutlet* A small nut.
25. *Pepo* A berry with a leathery nonseptate rind, developing from tricarpellary, syncarpous, inferior ovary with parietal placentation, e.g. Cucurbitaceae.
26. *Pome* A fleshy fruit surrounded by the fleshy thalamus and developing from a two- or more-celled, syncarpous, inferior ovary, e.g. apple.

27. *Pyrene* Fleshy fruit with each seed covered by a bony endocarp, e.g. *Ilex*.
28. *Regma* Dry, schizocarpic fruit derived from tricarpellary, syncarpous, superior, trilocular ovary, and bears many spinous tubercles, e.g. *Ricinus*.
29. *Samara* Dry, indehiscent, simple fruit derived from bicarpellary, syncarpous ovary. Its pericarp is winged, e.g. *Elm*.
30. *Silicula* Dry, dehiscent fruit derived from two or more carpels. It dehisces along two sutures and leaves a persistent partition after dehiscence. It is as broad as, or even broader, than long, e.g. *Iberis*.
31. *Siliqua* Dry, dehiscent fruit derived from bicarpellary, syncarpous ovary with parietal placentation and a false septum. It is longer than broad, e.g. *Brassica*.
32. *Sorosis* A multiple fruit developing from the spadix or spike in which the flowers usually fuse by their succulent sepals, and axis bearing them becomes woody or fleshy forming a compact mass, e.g. pineapple.
33. *Syconus* A multiple fruit derived from hypanthodium type of inflorescence. Here the achenes develop on the inside of a hollowed-out fleshy receptacle, e.g. *Ficus*.
34. *Utricle* Dry, indehiscent, small, bladder-like, one-seeded fruit, e.g. *Chenopodium*.

Test Your Understanding

1. Differentiate between the following:
   (a) Shrub and tree
   (b) Adventitious root and tap root
   (c) Fusiform root and napiform root
   (d) Cladode and corm
   (e) Phyllode and phylloclade
   (f) Tendril and thorn
   (g) Bract and bracteole
   (h) Calyx and epicalyx
2. Make diagrams of the following leaf apices and explain each of them only in one sentence:
   (a) Acute
   (b) Acuminate
   (c) Mucronate
   (d) Obtuse
3. Explain the following types of leaf margins using only one sentence for each of them:
   (a) Dentate
   (b) Entire
   (c) Revolute
   (d) Serrate.
4. How will you differentiate between alternate, opposite and whorled type of phyllotaxies of leaf?
5. With the help of diagrams, differentiate between parallel and reticulate venations. What is the difference between unicostate and multicostate venations?
6. Explain briefly following types of inflorescence:
   (a) Corymb
   (b) Capitulum
   (c) Spadix
   (d) Verticillaster.

7. What do you mean by aestivation? Explain various types of aestivations with the help of suitable diagrams.
8. With reference to androecium, explain the following terms:
   (i) Basifixed (ii) Didynamous
   (iii) Monadelphous (iv) Obdiplostemonous
   (v) Syngenesious (vi) Staminode
9. Explain major types of placentations using suitable diagrams.
10. Give one example each of the plants bearing the following types of fruits:
    (a) Berry (b) Caryopsis
    (c) Hesperidium (d) Sorosis

Suggested Reading

Jones, S.B. Jr. and A.E. Luchsinger, 1987, *Plant Systematics* (2nd ed.), McGraw-Hill, New York.

Lawrence, G.H.M., 1951, *Taxonomy of Vascular Plants*, Macmillan Company, New York.

Sharma, O.P., 1993, *A Manual of Practical Botany Vol. II.* (5th ed.), Pragati Prakashan, Meerut.

Sugden, A., 1984, *Longman Illustrated Dictionary of Botany*, Longman, U.K.

SELECTED FAMILIES OF DICOTYLEDONS

Various workers have divided Dicotyledons into varying number of families. Bentham and Hooker (1862–1883) discussed 163 families under Dicotyledons in their *Genera Plantarum* while Armen Takhtajan (1964) divided Magnoliatae (or Dicotyledones) into 361 families in his book entitled *Flowering Plants: Origin and Dispersal*. Arthur Cronquist (1981) in his treatment *An Integrated System of Classification of Flowering Plants*, included 318 families under class Magnoliopsida (=Dicots) whereas Robert F. Thorne (1983) treated Dicotyledoneae (=Annonidae) as a subclass of class Angiospermae (=Annonopsida) and discussed 297 families under this subclass. Thorne's system of classification is published in *Nordic Journal of Botany* (1983).

The present chapter deals with the study of the taxonomic characters of 73 selected families of Dicotyledons. Those that have been selected were chosen because of their common availability, unusual characters, great economic importance, large number of recorded species, and/or showing typical representation of any particular order or group.

Compilation of characters of each family has been done from the available literature, including Lawrence (1951), Hutchinson (1973), Willis (1973), Heywood (1978), Radford (1986), and Jones and Luchsinger (1987). The *arrangement of families in this chapter follows that of Bentham and Hooker (1862–1883) because this old but natural system of classification still suits best to the students as well as teachers, specially in the laboratory.* However, the position of each family in some widely accepted systems of classification (Engler and Prantl, 1887–1915; Bessey, 1915; Takhtajan, 1969; Hutchinson, 1973; Cronquist, 1981; and Thorne, 1983) is also discussed separately under systematics and phylogeny. Economic importance of each family is also discussed in some detail.

21.1 DICOTYLEDONS

The dicotyledons include all those angiosperms in which the embryo possesses two cotyledons, leaves with reticulate venation and vascular bundles are open and arranged in one or more rings. These plants have secondary thickenings in the stems. Due to the presence of cambium, these plants may be either woody or herbaceous. Dicotyledons usually have pentamerous flowers. They possess a persistent primary root that develops into a tap root.

Bentham and Hooker (1862–1883) divided the class Dicotyledons into three subclasses, viz. Polypetalae, Gamopetalae and Monochlamydeae.

21.2 POLYPETALAE

The members of subclass Polypetalae contain flowers with free petals, and their perianth is usually in two whorls i.e. calyx and corolla. Polypetalae has been divided into 3 series, viz. Thalamiflorae, Disciflorae and Calyciflorae.

Series Thalamiflorae is characterised by (i) usually distinct sepals free from ovary, (ii) presence of many stamens, (iii) hypogynous flowers, (iv) superior ovary, and (v) absence of disc. Thalamiflorae includes 6 cohorts (=orders) and 34 orders (=families). Six cohorts are Ranales, Parietales, Polygalineae, Caryophyllineae, Guttiferales and Malvales.

Series Disciflorae is characterised by (i) distinct or united sepals, free or adnate to ovary, (ii) presence of disc, (iii) stamens hypogynous, usually definite, (iv) superior ovary. Disciflorae includes 4 cohorts (=orders) and 23 orders (=families). The cohorts are Geraniales, Olacales, Celastrales and Sapindales.

Series Calyciflorae is characterised by (i) usually inferior ovary, (ii) united or rarely free sepals, and (iii) flowers perigynous or epigynous. It includes 5 cohorts (=orders) and 27 orders (=families). The cohorts are Rosales, Myrtales, Passiflorales, Ficoidales and Umbellales.

Few general characters of different orders alongwith some detailed description of some representative families are discussed in the forthcoming text.

21.3 RANALES

21.3.1 General Characteristics

1. Flowers usually with numerous and indefinite number of stamens, and only rarely the stamens number is definite.
2. The gynoecium apocarpous and multicarpellate, i.e. composed of many unilocular unicarpellate pistils.
3. Embryo minute and seeds endospermic.
4. Floral parts typically spirally arranged, numerous and distinct.
5. Perianth often not clearly differentiated into calyx and corolla.

Bentham and Hooker included 8 orders (=families) under Ranales. These are Ranunculaceae, Mangnoliaceae, Annonaceae, Berberidaceae, Nymphaeaceae, Menispermaceae, Dilleniaceae and Calycanthaceae.

Engler and Diels, however, divided the order Ranales into 4 suborders containing 19 families, as under:

Suborder Nymphaeineae:	Nymphaeaceae, Ceratophyllaceae
Suborder Trochodendrineae:	Trochodendraceae, Cercidiphyllaceae
Suborder Ranunculineae:	Ranunculaceae, Lardizabalaceae, Berberidaceae, Menispermaceae.

Suborder Magnoliineae: Magnoliaceae, Himantandraceae, Calycanthaceae, Lactoridaceae, Annonaceae, Eupomatiaceae, Myristicaceae, Gomortegaceae, Monimiaceae, Lauraceae, Hernandiaceae.

Only *Magnoliaceae*, *Annonaceae*, *Ranunculaceae* and *Nymphaeaceae* are discussed in this text.

21.4 MAGNOLIACEAE (MAGNOLIA FAMILY)

21.4.1 Systematic Position

Polypetalae, Thalamiflorae, Ranales.

21.4.2 Field Recognition

Trees or shrubs; leaves simple, alternate; stipules leave a circular stipular scar; flowers large, actinomorphic; perianth parts 6–18; stamens many; apocarpous.

21.4.3 Selected Indian Genera

Magnolia, *Michelia*, *Liriodendron*, *Aclimandra*, *Talauma*, *Manglietia* and *Schizandra*.

21.4.4 Size, Distribution, and General Information

Named after a French botanist Pierre Magnol, Magnoliaceae is a family of about 12 genera and 230 species, mainly distributed in the warm tamperate regions of the Northern Hemisphere. The species are common in eastern Asia and eastern North America, and found also in Brazil, West Indies, and Malaysia. 36 species of this family have so far been reported from different parts of India.

21.4.5 Description of the Family

General Habit Deciduous or evergreen, often aromatic shrubs or trees; *Liriodendron tulipifera* reaches 20 to 65 m in height; *Michelia* and *Magnolia* are medium-sized trees or shrubs; *Schizandra* is a climber.

Root Tap root, branched.

Stem Aerial, woody, erect or climbing, branched; parenchyma contains oil glands.

Leaves Alternate, large, simple, entire, petiolate, pinnately-veined; stipulate, stipules large, protect young buds and leave a circular scar on falling; stipules absent in Wintereae; oil glands present.

Inflorescence Flowers usually solitary, either axillary (*Michelia*), or terminal (*Magnolia*); flowers crowded near the tips of branches in *Illicium*.

Flowers Large, showy, bracteate (*Magnolia*, *Michelia*) or ebracteate, pedicellate, bisexual but rarely unisexual in *Kmeria*; actinomorphic; hypogynous; floral parts arranged spirally on the elongated floral axis; some *Drimys* species have unisexual flowers while others have bisexual flowers.

Perianth Except in *Illicium* and some species of *Magnolia*, *Michelia and Drimys*, the perianth is undifferentiated into calyx and corolla. Perianth spiral to cyclic, in 3 or more series of 6 to 18 parts; polyphyllous. In *Drimys* sepals fused while petals free (Fig. 21.1). Perianth lobes are arranged in whorls of threes, valvate or imbricate in each whorl; 9–12 perianth lobes in *Magnolia* while 6–15 in *Michelia*. In *Euptelea*, perianth is absent.

Androecium Stamens numerous, free, spirally arranged on the basal part of floral axis; anthers long, dithecous, basifixed or adnate; introrse but extrorse in *Liriodendron*; dehiscence longitudinal; connective prolonged.

Gynoecium Carpels numerous, apocarpous (free), arranged spirally on the floral axis above the stamens; superior, unilocular, one or more ovules on a ventral suture; placentation parietal; style short, one; stigma one. In *Zygogynum* and *Pachylarnax*, the carpels are rarely united.

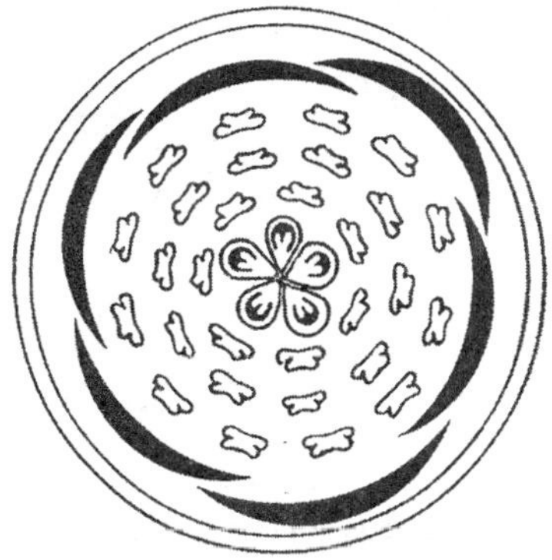

Fig. 21.1 Floral diagram of *Drimys*.

Fruits and Seeds Fruit of aggregate type. It may be a follicle (*Magnolia*), samara (*Liriodendron*), or rarely berry (*Schizandra*). *Seed* often suspended within the follicle by a thread-like funiculus, has a minute embryo in a large endosperm.

Pollination and Dispersal Pollination is entomophilous; dispersal takes place by those animals (e.g. birds, monkeys, etc.) which carry off the fruits.

General Floral Formula $\oplus$, ⚥ (rarely unisexual), $P_9 - \infty$, A ∞, G $\underline{\infty}$.

21.4.6 Economic Importance

- The Magnoliaceae are important as common ornamentals and as a source of wood.
- *Magnolia* species are commonly cultivated in gardens. *M. grandiflora* (Dulee Champa, or 'Bari Champa'), *M. fuscata* (Chidi Champa), *M. campbelli* ('Lal Champa'), *M. pumila* ('Jauhre Champa'), *M. pterocarpa* and *M. stellata* are known for their sweet, scented, beautiful flowers. Wood of *Magnolia acuminata* is used in making toys, tea boxes and quality furniture.
- *Michelia champaca* ('Champa') is a common garden ornamental plant, and is also the source of a volatile oil which is widely used in perfumery and cosmetics. Its flowers and fruits are used to cure kidney troubles and gonorrhoea. *M. fuscata* flowers yield an essential oil which is used as hair oil. *M. alba*, *M. excelsa* and *M. nilogrica* are other ornamental species with sweet-scented flowers.
- *Liriodendron tulipifera* provides a tall graceful posture and bears beautiful scented flowers. It yields good timber for furniture.
- *Talauma phellocarpa* trees yield good quality timber used in making tea boxes and furniture.
- *Manglietia hookeri* provides good furniture wood.

- Wood of *Pachylarnax* and *Aclimandra cathcartii* is also useful in making indoor furniture.
- *Illicium verum* yields an oil useful in colic disorders. *I. griffithii* fruits are used as condiment for flavoring curries.
- *Drimys winteri* bark is used as an astringent and stimulant.
- *Schizandra grandiflora* fruits are edible.

21.4.7 Systematics and Phylogeny (Affinities)

Family Magnoliaceae is divided into three tribes (Wintereae, Magnolieae and Schizandreae) by Bentham and Hooker (1867), three subfamilies (Magnolioideae, Illicioideae and Schizandroideae) by Rendel (1938), and only into two tribes (Magnolieae and Liriodendreae) by Jones and Luchsinger (1987). Several taxonomists treated families, such as Eupteleaceae, Illiciaceae, Schizandraceae, Trochodendraceae and Winteraceae as belonging to family Magnoliaceae. But Hutchinson (1959) has treated them as independent families, except Eupteleaceae, which was merged by him under Schizandraceae. Takhtajan (1969) also treated them as independent families.

Takhtajan (1969), Hutchinson (1973), Cronquist (1981) and several other taxonomists opine that Magnoliaceae is the most primitive family of angiosperms.

21.4.8 Description of Some Important Plants

1. Michelia champaca L. (Vern. "Champa"; Fig. 21.2)

Habit: A small evergreen tree. *Root:* Branched, tap root. *Stem:* Erect, aerial, branched, woody. *Leaf:* Simple, alternate; stipulate, stipule convolute; petiolate, ovate, entire, acute, unicostate reticulate. *Inflorescence:* Solitary axillary. *Flower:* Bracteate, pedicellate, complete, hermaphrodite, actinomorphic, hypogynous, fragrant, pale yellow. *Perianth:* Usually 9 perianth leaves arranged in three whorls of three each; either outer 3 leaves are sepalloid and inner 6 lobes are petalloid, or all the 9 perianth lobes are petalloid. *Androecium:* Stamens numerous, free, spirally arranged on a conically elongate receptacle, filaments short, basifixed, dithecous, extrorse. *Gynoecium:* Multicarpellary, apocarpous, carpels arranged spirally over the conical receptacle, superior, unilocular, many ovules, marginal placentation; style curved, stigma beaked. *Fruit:* Etaerio of follicles.

Floral Formulae: Br, ⊕, ⚥, K_3, C_{3+3}, A $\propto$, G $\underline{\propto}$; or Br, ⊕, ⚥, P_{3+3+3}, A $\propto$, G $\underline{\propto}$.

2. Magnolia grandiflora L. (Vern. "Dulee Champa")

Habit: A medium-sized, evergreen tree. *Root:* Branched, tap root. *Stem:* Aerial, erect, branched, solid, woody. *Leaf:* Alternate, simple; stipulate, stipules large and cover the bud; lanceolate, entire, acute, unicostate reticulate. *Inflorescence:* Solitary terminal. *Flower:* Ebracteate, complete, hermaphrodite, actinomorphic, hypogynous, fragrant, white. *Perianth:* 9 lobes, in three whorls of three each, valvate in each whorl, white. *Androecium:* Numerous stamens, polyandrous, arranged spirally on receptacle, dithecous, filament small, anther lobes cone-shaped, connective indistinct; introrse, dehiscence longitudinal. *Gynoecium:* Same as in *Michelia*. *Fruit:* Same as in *Michelia*.

Floral Formula: Ebr, ⊕, ⚥, P_{3+3+3}, A $\propto$, G $\underline{\propto}$.

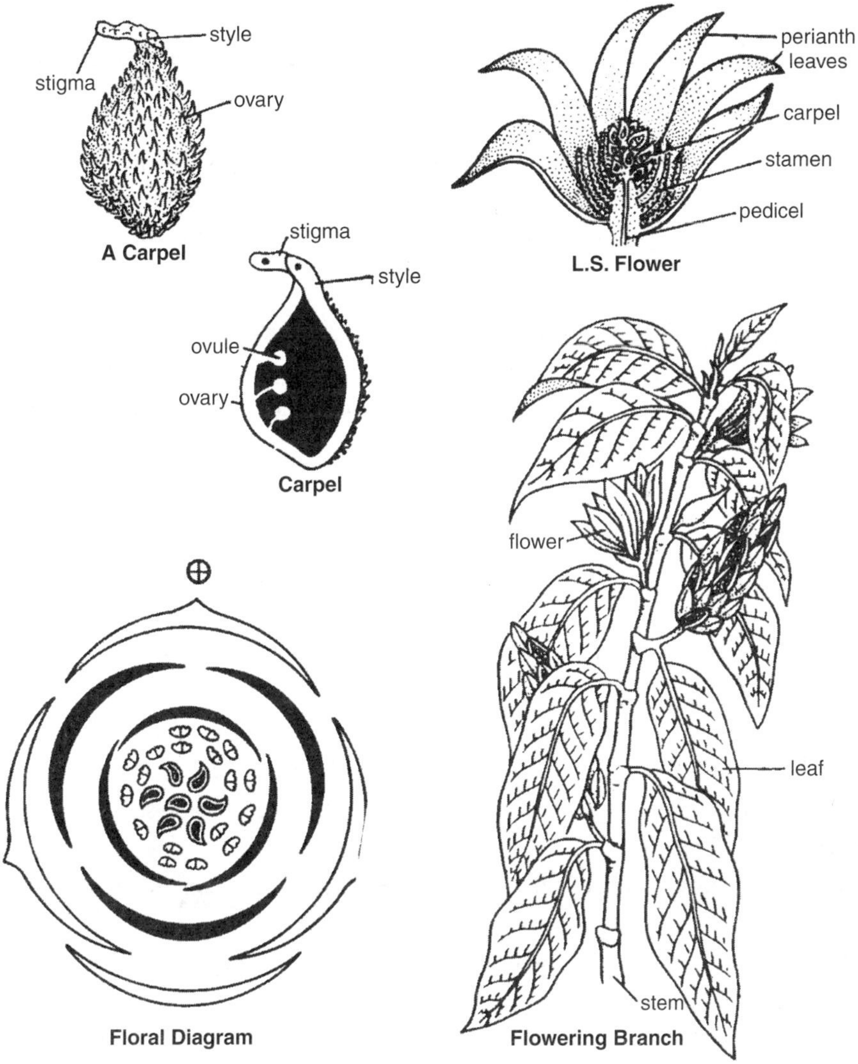

Fig. 21.2 *Michelia champaca* L.

21.5 ANNONACEAE (CUSTARD APPLE FAMILY)

21.5.1 Systematic Position

Polypetalae, Thalamiflorae, Ranales.

21.5.2 Field Recognition

Woody shrubs, trees or climbers; wood aromatic; leaves two-ranked, without stipules; perianth 3 + 3 + 3; stamens numerous, rarely few; gynoecium polycarpellary, apocarpous, superior ovary.

21.5.3 Selected Indian Genera

Annona squamosa (Vern. "Sharifa"), *Cananga odorata* (Ylang-ylang), *Artabotrys odoratissimus.*

21.5.4 Size, Distribution, and General Information

This largest family of order Magnoliales (130 genera and 2300 species, Jones and Luchsinger, 1987) is widely distributed in the Old World Tropics. Over 25 genera and 200 species have so far been reported from India, mostly from its peninsular region. *Annona* is represented by over 120 species, and *Xylopia* by about 150 species.

21.5.5 Description of the Family

General Habit Usually aromatic trees or shrubs, rarely liana-like, or woody climbers, e.g. species of *Oxymitra. Artabotrys* is a hook climber. Bark, leaves and floral parts contain oil ducts.

Root Extensively branched tap roots.

Stem Woody, branched, erect, rarely climbing, aerial, hard.

Leaves Simple, alternate, two-ranked, exstipulate, petiolate; apex acute, obtuse or even mucronate (*Annona mucronata*); ovate, obovate or lanceolate; coriaceous, glabrous; unicostate reticulate; gland-dotted, and so aromatic.

Inflorescence Solitary axillary or terminal, or leaf opposed; sometimes cauliflorous, i.e. flowers occur on stem (e.g. *Polyalthia fragrans*).

Flower Bracteate or ebracteate, ebracteolate, pedicellate, complete, hermaphrodite, rarely unisexual (*Stelechocarpus*); actinomorphic, rarely zygomorphic (*Monodora*); hypogynous.

Calyx Sepals 3, or sometimes 6 arranged in two whorls of three each; polysepalous, or rarely basally connate; valvate, rarely imbricate.

Corolla Generally 6 petals, arranged in two whorls of three each, sometimes only 3 petals; polypetalous; valvate or slightly imbricate.

Sometimes, sepals and petals are alike, and thus represent the *perianth.*

Androecium Stamens numerous, polyandrous; spirally arranged on large convex receptacle; filaments short and thick; connective continues beyond the anther and produces a dilated head; adnate; extrorse.

Gynoecium Carpels numerous, arranged spirally on the raised receptacle (*Polyalthia, Artabotrys*); only 1–3 carpels in *Cyathocalyx*; apocarpous; superior, unilocular, one to many anatropous ovules in the locule; parietal or sometimes basal or marginal placentation; style short or absent; stigma sessile, simple or trilobed (*Tetrameranthus*).

Fruits and Seeds Fruit an etaerio of berries; in *Annona*, the berries coalesce with fleshy receptacle; seeds large, endospermous.

Pollination and Dispersal Pollination generally by flies; dispersal by birds, bats or mammals.

Floral Formulae Br or Ebr, ⊕, ⚥, K_3 or $_{3+3}$, C_{3+3} or $_3$, A $\propto$, G $\underline{\propto}$

Or, Br or Ebr, ⊕, ⚥, P_{3+3+3}, A $\propto$, G $\underline{\propto}$.

21.5.6 Economic Importance

- *Annona squamosa* (Vern. "Sharifa" or "Sitaphal") is cultivated in Assam, West Bengal, Maharashtra, etc. for its edible fruits. Its leaves and seeds have insecticidal properties. *A. reticulata* (Vern. "Ramphal"), *A. muricata* (Vern "Maniphal"), *A. cherimola* (Vern. Hanumanphal), *A. diversifolia* are all grown for their edible fruits.
- *Artabotrys odoratissimus* ("climbing ylang-ylang") yields an ethereal oil and also cultivated for its scented flowers.
- *Polyalthia longifolia* and *P. pendula* (Vern. Ashok) are grown for their beautiful green drooping branches on road sides and in beautiful residential buildings. *P. longifolia* wood is used for making pencils, boxes, match box industry, etc.
- *Asimina triloba* is cultivated for its edible fruits in America.
- *Xylopia aethiopica* is the source of Guinea pepper. *X. parvifolia* wood is used in making plywood.
- *Cananga odorata* is cultivated as an ornamental tree because of its sweet-smelling flowers. An oil, obtained from its flowers, is used in perfume industry.
- *Unona pannosa* yields a strong fibre used for paper making and cordage.
- *Monodora myristica* is cultivated for its beautiful flowers. Its seeds contain oil used as a nutmeg substitute.
- *Uraria narum* roots yield an oil used as a medicine.
- Bark of *Goniothalamus* yields strong fibre.

21.5.7 Systematics and Phylogeny (Affinities)

The resemblances and differences between Magnoliaceae and Annonaceae indicate a close relationship between the two families. The two families resemble with each other in possessing (i) large solitary flowers having perianth leaves in more than one whorl, (ii) spirally arranged stamens, (iii) polycarpellary, apocarpous condition, (iv) large aggregate fruits, (v) seeds with large endosperm and small embryo, (vi) anatropous ovules, (vii) oil-containing cells; (viii) woody stems. etc. On the other hand, Annonaceae differs from Magnoliaceae in possessing (i) exstipulate leaves, (ii) extrorse stamens, (iii) ruminant endosperm, and (iv) sarcotesta in the seeds.

The above-mentioned resemblances and differences between Annonaceae and Magnoliaceae indicate that both of them are closely related, and have perhaps been derived from the same ancestral stock. Therefore, Hutchinson's (1973) placing them in two separate orders (Magnoliaceae in Magnoliales and Annonaceae in Annonales) is not advisable. Cronquist (1981) placed both of them under Magnoliales, and Thorne (1983) under suborder Annonineae of order Annonales, and this appears justified.

Because of some characters, Annonaceae are more advanced than Magnoliaceae.

Engler and Prantl (1887–1899) divided family Annonaceae into five tribes, viz. Uvarieae, Miluseae, Hexalobieae, Xylopeae, and Monodoreae.

21.5.8 Description of Some Important Plants

1. Artabotrys odoratissimus syn. A. uncinatus (Vern. Climbing Ylang-Ylang or Kantali Champa; Fig. 21.3)

Habit: Woody climber, climbing with the help of hooks developed on flower pedicels. *Root:* Branched, tap roots. *Stem:* Aerial, woody, branched, climbing, aromatic. *Leaf:* Simple, alternate, exstipulate, petiolate, elliptical, entire, acute apex, unicostate reticulate; oil ducts are present. *Inflorescence:* Solitary axillary, or rarely in groups. *Flower:* Bracteate, pedicellate, pedicel bears a hook; complete, hermaphrodite, actinomorphic, trimerous, hypogynous, spirocyclic. *Perianth:* Perianth lobes 9, arranged in 3 whorls of 3 each; sometimes, outer whorl of 3 lobes is green sepalloid, free, valvate, and inner 6 lobes of two whorls are petalloid, valvate, or slightly imbricate; perianth lobes are sweet-scented and gland-dotted. *Androecium:* Stamens numerous, polyandrous, spirally arranged on the receptacle; filaments short, connective prolonged beyond the anther; extrorse. *Gynoecium:* Multicarpellary, apocarpous, carpels spirally arranged on receptacle; ovary superior, unilocular, one to many anatropous ovules in each locule; marginal placentation. *Fruit:* Etaerio of berries. *Seed:* Large, endospermic.

Floral Formulae: Br, ⊕, ⚥, P_{3+3+3}, A ∞, G $\underline{\infty}$

Or, Br, ⊕, ⚥, K_3, C_{3+3}, A ∞, G $\underline{\infty}$.

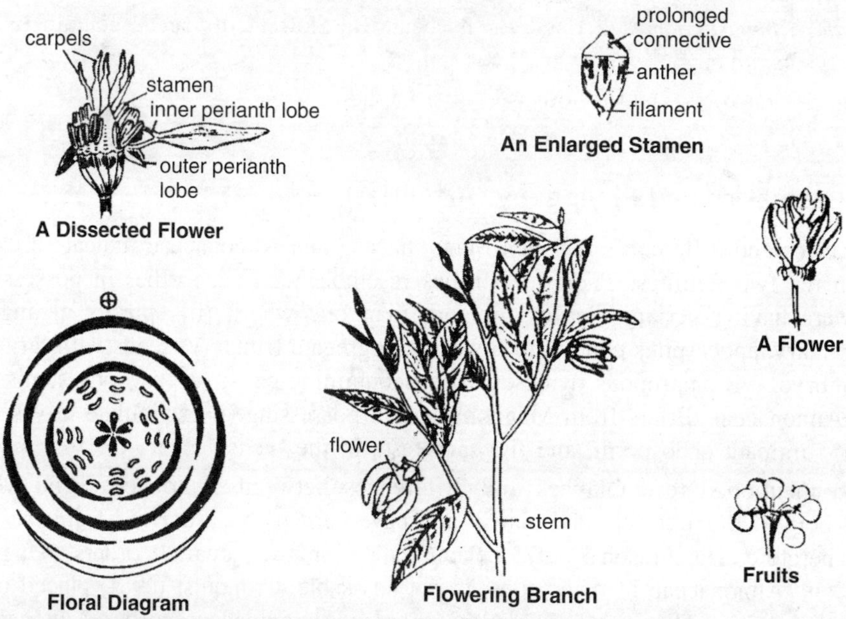

Fig. 21.3 *Artabotrys odoratissimus* R. Br.

2. Annona squamosa (Vern. "Sharifa")

Habit: A tall shrub or small tree. *Stem:* Erect, woody, branched. *Leaf:* Simple, alternate, exstipulate, oblanceolate, entire, obtuse. *Inflorescence:* Axillary, 2 or more flowers developing in the axis of leaf. *Flower:* Bracteate, pedicellate, hermaphrodite, actinomorphic, hypogynous, spirocyclic. *Calyx:* Sepals 3, poly- or gamosepalous, valvate. *Corolla:* Petals 3, or 6, arranged in two whorls of 3 each; polypetalous, valvate. *Androecium:* Anthers appendaged; other details similar to *Artabotrys odoratissimus*. *Gynoecium:* Multicarpellary, apocarpous; carpels arranged spirally on receptacle; superior, unilocular, only one ovule in the locule, basal placentation. *Fruit:* Etaerio of berries, or of achenes. *Seed:* Endospermic.

Floral Formula: Br, ⊕, ⚥, K_3 or (3), C_3 or 3 + 3, A $\propto$, G $\underline{\propto}$.

21.6 RANUNCULACEAE (BUTTERCUP FAMILY)

21.6.1 Systematic Position

Polypetalae, Thalamiflorae, Ranales.

21.6.2 Field Recognition

Herbs; leaves often divided or palmately compound with sheathing leaf bases; stamens numerous and spirally arranged; carpels numerous, distinct, spirally arranged; unilocular.

21.6.3 Selected Indian Genera

Ranunculus, *Delphinium*, *Aconitum*, *Actaea*, *Aquilegia*, *Anemone*, *Caltha*, *Clematis*, *Helleborus*, *Nigella*, *Paeonia*, *Thalictrum*.

21.6.4 Size, Distribution, and General Information

This family is represented by about 50 genera and 1900 species, chiefly distributed in the temperate regions of the Northern Hemisphere. Eastern Asia and eastern North America are the regions of their most common occurrence. The common genera, of which over 150 or more species have so far been reported, include *Ranunculus* (400), *Aconitum* (300), *Delphinium* (250), *Clematis* (250), *Anemone* (150) and *Thalictrum* (150). Some species occur in Arctic and Alpine zones. However, they are rare in tropics. 20 genera and over 165 species of this family have so far been reported from India. Plants of this family flower mainly during winters in our country.

21.6.5 Description of the Family

General Habit Generally annual or perennial herbs, rarely shrubs (*Xanthorhiza*) or vines (*Clematis*, *Naravelia*); perennials perennate by means of rhizome (*Cimicifuga*, *Helleborus*) or tuberous roots (*Aconitum*); some species are aquatic herbs (*Ranunculus aquatilis*).

Root Tap root as well as adventitious roots; tuberous roots of *Aconitum*, *Paeonia* and *Ranunculus* store food, swell and form tuber-like structures.

Stem Aerial as well as underground; herbaceous (*Ranunculus*); woody (*Paeonia*); climbing (*Clematis*); branched.

Leaves Usually basal and cauline; often palmately divided with sheathing leaf bases; usually exstipulate, but stipulate in *Trollius*, *Caltha*, *Thalictrum* and *Ranunculus*; mostly alternate, rarely opposite (*Clematis*); simple (*Ranunculus*), pinnately compound (*Clematis*), decompound (*Thalictrum*), or finely dissected into fine segments (*Delphinium ajacis*); aquatic species of *Ranunculus* show heterophylly.

Inflorescence Variously formed, from solitary (*Nigella*) to paniculate (*Thalictrum*), or in racemes (*Delphinium*) or cymes; a single terminal flower is produced in *Anemone* and *Eranthis*, but more often a cymose branching is seen.

Flower Bracteate (*Clematis*) or ebracteate (*Anemone*), bracteolate (*Delphinium*) or ebracteolate (*Ranunculus*), usually bisexual but unisexual in some species of *Thalictrum*; usually actinomorphic but zygomorphic in *Delphinium* and *Aconitum*; hypogynous; spurred in *Delphinium* and *Aquilegia*.

Calyx 3 to many sepals, usually 5 (*Ranunculus*), polysepalous, valvate or imbricate; sepals are petaloid in *Aconitum* and *Delphinium* and their posterior sepal is spurred. In *Nigella* and *Anemone* there is sometimes an involucre of green leaves below the flower, usually alternate with the calyx.

Corolla Usually 5 petals, polypetalous; nectariferous glands often present; sometimes petals are many or absent (*Caltha*); posterior petals are spurred.

In genera such as *Paeonia* (Fig. 21.4), *Clematis* (Fig. 21.5), *Actaea* and *Helleborus*, *perianth* is present i.e. there is no distinction between sepals and petals; innermost perianth leaves often bear nectaries (honey-leaves) or they are reduced to small nectariferous scales.

Androecium Stamens 5 to many, free, spirally arranged; anthers 2-celled, adnate, extrorse and longitudinally dehiscent; sometimes staminodes are present (*Aquilegia*); some outer stamens may be converted into honey-leaves (*Helleborus*); from 80 to 150 stamens in *Caltha*; stamens bright coloured in *Thalictrum*.

Gynoecium Usually numerous, free carpels with either one basal ovule or several anatropous ovules showing marginal placentation; ovary superior, unilocular; style 1; stigma 1; in *Actaea* there is only 1 carpel; in *Nigella* carpels are united, ovary pentalocular and placentation axile; in *Helleborus* carpels are only basally connate.

Fruits and Seeds Fruit may be an etaerio of follicles (*Aconitum*), etaerio of achenes (*Ranunculus*), a berry (*Actaea*), or capsule (*Nigella*); seeds with minute embryo and oily endosperm; *elaisome* (an appendage which contains oily substance attractive to ants) is present on the seeds of some species of *Helleborus*.

On the basis of the type of fruit and number of carpels, family Ranunculaceae is divided into two subfamilies:

1. *Helleboroideae:* Carpels with more than 1 ovule; fruit a berry or follicle.
2. *Ranunculoideae:* Carpels with 1 ovule; fruit etaerio of achenes, rarely a berry.

Pollination and Dispersal The flowers are usually protandrous in Ranunculaceae and therefore well-suited for insect pollination. Insects are also attracted because of the presence of nectar, petaloid calyx, conspicuous corolla and stamens. However, some species of *Thalictrum* are wind-pollinated, e.g. *T. minus*. Dispersal takes place by animals or birds (*Ranunculus*) or by wind (*Anemone*, *Clematis*).

General Floral Formula Br or Ebr, ⊕ or ⦶, ⚥, K_5 or $K_{3-\infty}$, C_5 or P_{4-6}, $A_{5-\infty}$, $G_{\underline{\infty}}$ or $\underline{1-\infty}$.

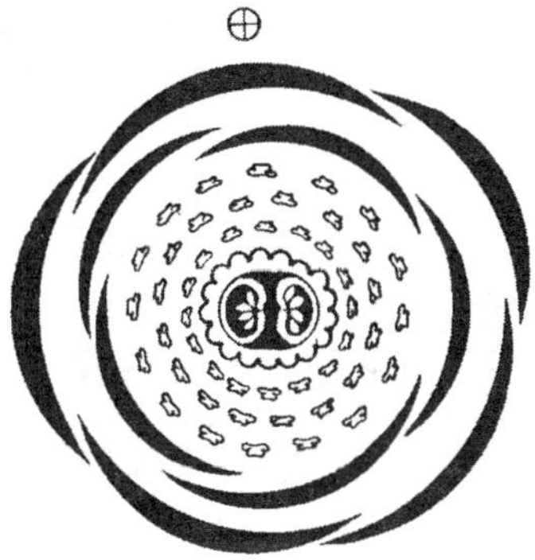

Fig. 21.4 Floral diagram of *Paeonia*.

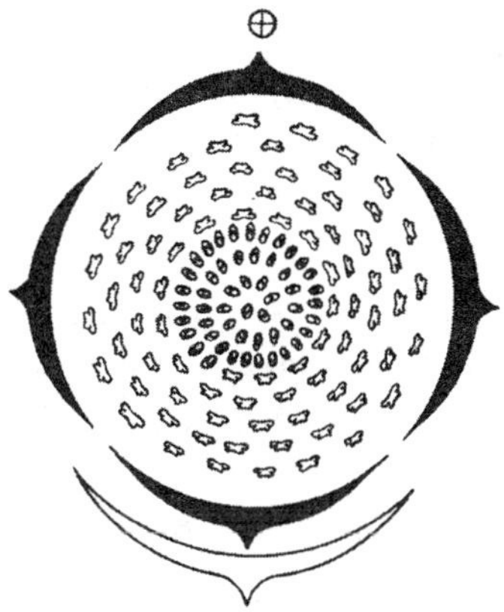

Fig. 21.5 Floral diagram of *Clematis*.

21.6.6 Economic Importance

Ranunculaceae are important as ornamental, and as drug plants. They are mostly poisonous owing to the presence of alkaloids.

(a) *Ornamentals* Many Ranunculaceae have showy flowers and are grown as *ornamentals* in our gardens. Some of such ornamentals along with their common name in parenthesis are *Aconitum napellus* (Aconite), *Anemone parviflora* (Wind flower), *Aquilegia coerulea* (Columbine), *Caltha palustris* (Marsh marigold), *Clematis virginiana* (Traveller's joy), *Clematis vitalba* (Virgins-bower), *Delphinium ajacis* (Larkspur), *Helleborus* (Christmas rose), *Nigella damascena* (Love-in-a-mist), *Paeonia emodi* (Peony), *Ranunculus* (Butter-cup, e.g. *R. asiaticus*, *R. muricatus*, *R. repens* etc.), and *Thalictrum alpinum* (Meadow-rue). Species of *Anemone*, *Clematis*, *Delphinium*, *Ranunculus* and *Thalictrum* are available in a wide range of colours and forms.

(b) *Medicinal Uses* Large number of plants of this family are of medicinal value, of which some are undermentioned:

1. *Aconitum napellus* roots are the source of aconite, used as a nerve sedative, and for rheumatism. Its dried roots are used to cure mild heart palpitation and gastritis. Deadly poisonous alkaloid, obtained from *A. ferox*, may cause death but is useful in snake bites.
2. *Actaea spicata* roots are used as a nerve sedative.
3. *Adonis aestivatis* and *A. scorbiculata* roots are used as heart stimulant.
4. *Anemone pulsatilla* is the source of the homoeopathic drug '*pulsatilla*' used to cure nervous disorders in women during menstrual periods.
5. *Clematis triloba* leaves are used to cure snake bites.
6. *Coptis teeta* roots are used in brain and liver disorders as well as in toothache and sore eyes.

7. *Delphinium ajacis* and *D. elatum* seeds are insecticidal while roots of *D. denudatum* are effective in toothache. Seeds, leaves and roots of *D. zatil* are used to cure jaundice and spleen. *D. vestitum* seeds are useful in cardiac disorders.
8. *Helleborus niger* flowers are given in the form of syrup to mad persons. This species is also useful in chronic skin infections.
9. *Nigella sativa* (vern. Kalonji or Kala Zeera) seeds are used as common condiment as well as in the treatment of cough, asthma and fever.
10. *Paeonia officinalis* tubers are used to cure colic and uterus disorders.
11. *Ranunculus muricatus* and *R. arvensis* juice is used in irregular fever. *R. aquatilis* and *R. arvensis* leaves and flowers are used to cure asthma and rheumatism. *R. ficaria* is used to cure piles while *R. falcatus* is poisonous and causes skin blisters.
12. *Thalictrum foliolosum* roots are used as diuretic, purgative and tonic.

21.6.7 Systematics and Phylogeny (Affinities)

Majority of the phylogenetists believe that Ranunculaceae is one of the most primitive families of dicotyledons because of the primitive features, such as (i) presence of scattered vascular bundles in the stem in some of its taxa, such as *Actaea* and *Thalictrum*, (ii) spiral arrangement of foliage and floral leaves, (iii) bisexual, hypogynous, actinomorphic flowers with apocarpous gynoecium, (iv) seeds with small embryo and oily endosperm, and (*v*) anatropous ovules. On the other hand, this family also shows several advanced characters such as (i) presence of unisexual and zygomorphic flowers in some genera, (ii) fusion of carpels (*Nigella*), and (iii) achene type of fruits (*Ranunculus*). Several phylogenetists believe that monocots have originated from dicots, and the point of their origin being Ranunculaceae or some allied groups of Ranales.

Ranunculaceae is divided into five tribes (Anemoneae, Clematideae, Helleboreae, Paeonieae and Ranunculeae) by Bentham and Hooker (1862–1883). However, Engler and Prantl (1887–1899) recognised only three tribes, viz. Anemoneae, Helleboreae and Paeonieae. Hutchinson (1973) separated tribe Helleboreae from Ranunculaceae and raised it to the rank of an independent family Helleboraceae. Hutchinson (1973), Takhtajan (1980) and Cronquist (1981) have also raised tribe Paeonieae to the rank of a family Paeoniaceae with the genus *Paeonia* as its sole representative.

Jones and Luchsinger (1987) have stated that in Ranunculaceae, the perianth varies greatly and is of importance in delimiting genera in this family.

Hickey and King (1988) divided family Ranunculaceae into 2 sub-families, i.e. Helleboroideae and Ranunculoideae. Helleboroideae is again divided into 2 tribes, i.e. Helleboreae and Delphinieae while Ranunculoideae into 3 tribes, i.e. Ranunculeae, Anemoneae and Clematideae.

21.6.8 Description of Some Common Plants

1. Ranunculus sceleratus L. (Vern. Celery-leaved Crow-foot; Fig. 21.6)

Habit: Annual herb. *Stem:* Branched, erect, glabrous and solid. *Leaf:* Simple, alternate, petiolate, exstipulate, leaf base sheathing, divided further into obovate or cuneate segments, multicostate reticulate.

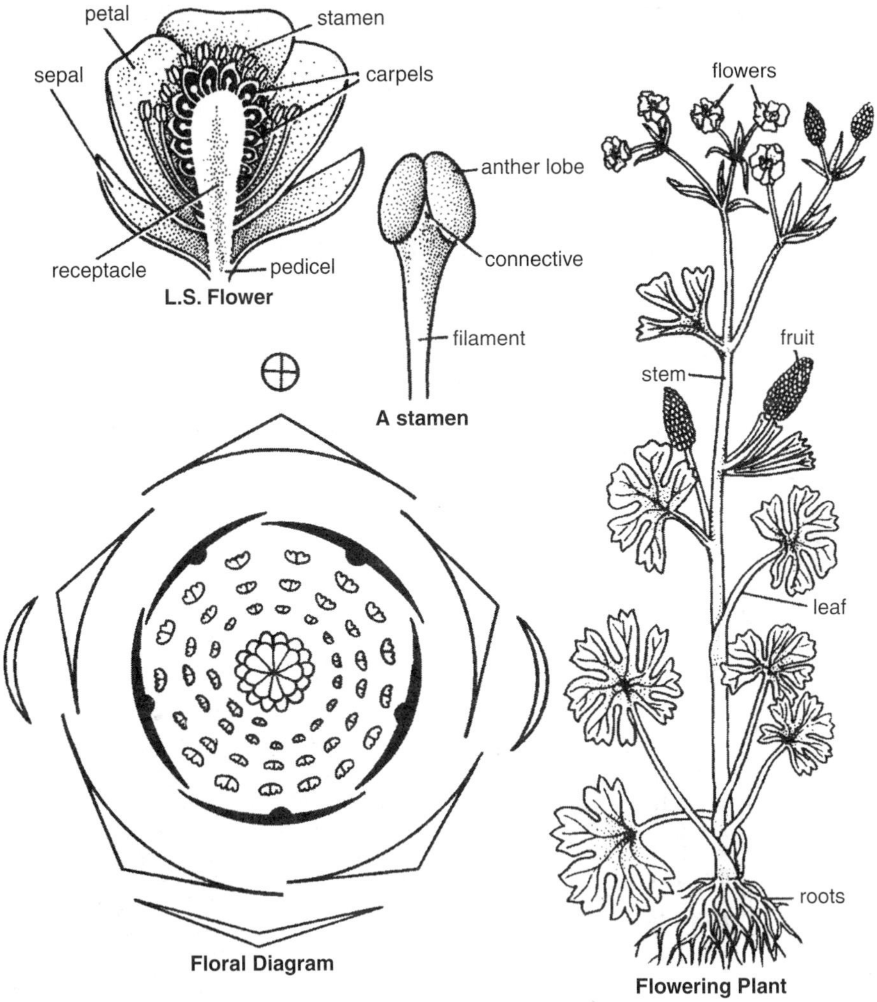

Fig. 21.6 *Ranunculus sceleratus* L.

Inflorescence: Cymose (dichasial cyme). *Flower:* Bracteate, bracteolate, pedicellate, complete, hermaphrodite, actinomorphic, pentamerous, hypogynous, spirocyclic. *Calyx:* 5 sepals, polysepalous, imbricate or quincuncial, petaloid, reflexed from the base. *Corolla:* 5 petals, polypetalous, imbricate, bright yellow, a nectary is present at the base of each petal. *Androecium:* Numerous stamens, polyandrous, spirally arranged; anthers dithecous, basifixed, extrorse. *Gynoecium:* Multicarpellary, apocarpous; each carpel superior, unilocular; one ovule in each locule; basal placentation; carpels arranged spirally over an oblong receptacle; style reduced; stigma fimbriate. *Fruit:* Etaerio of achenes. *Seed:* Endospermic.

Floral Formula: Br, Brl, ⊕, ⚥, K_5, C_5, A ∞, G $\underline{\infty}$.

2. Delphinium ajacis Auct. (D. ambiguum L., Consolida ambigua (L.) Ball. and Heyw; Vern. Larkspur; Fig. 21.7)

Habit: Cultivated, annual herb. *Stem:* Erect, branched, solid, green, hairy. *Leaf:* Simple or decompound, alternate, exstipulate, deeply dissected; each lobe is linear to elliptical, smooth, with acute or

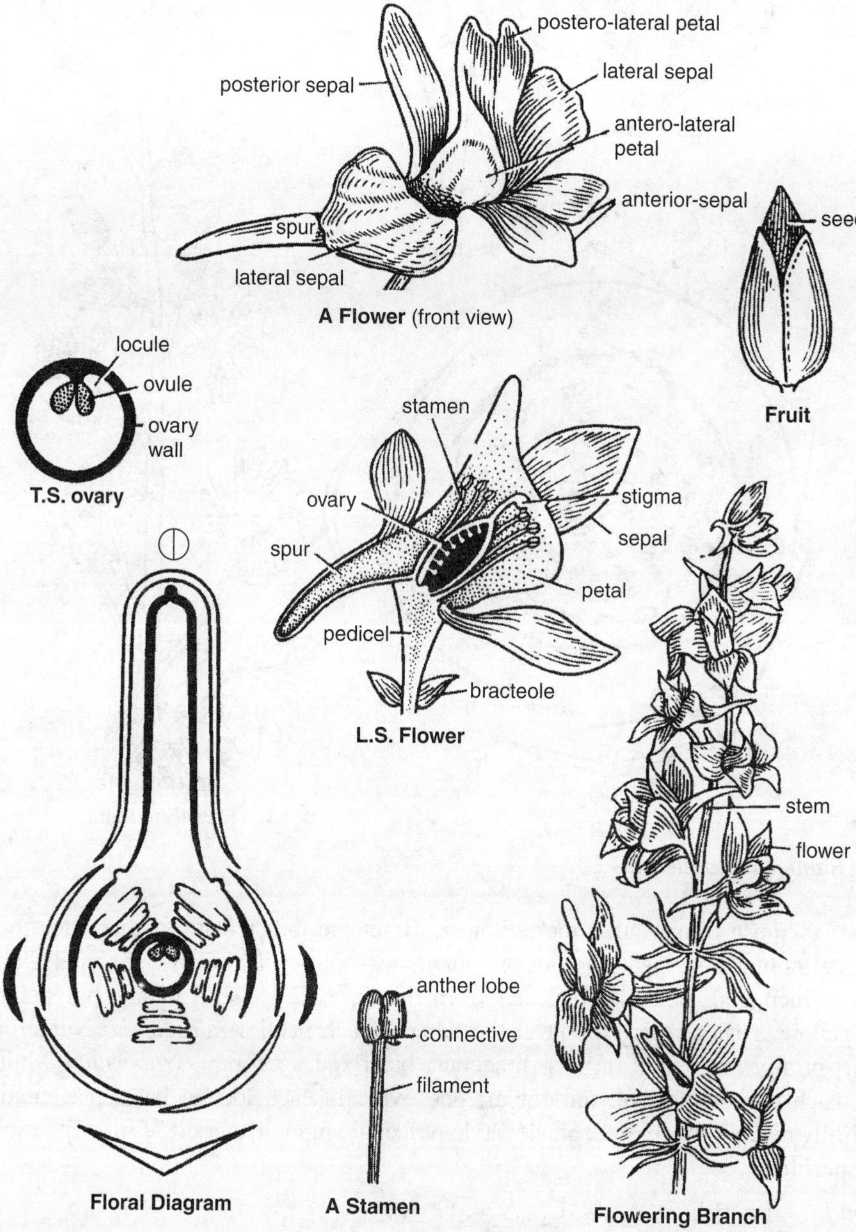

Fig. 21.7 *Delphinium ajacis* L.

acuminate apex. *Inflorescence:* Axillary raceme of 4 to 16 flowers. *Flower:* Bracteate, bracteolate, pedicellate, complete, hermaphrodite, zygomorphic, spurred posteriorly, pentamerous, hypogynous; bright blue, pink or white. *Calyx:* 5 sepals, polysepalous, petaloid; posterior odd sepal prolonged into a long spur; quincuncial or imbricate; blue or violet. *Corolla:* 4 petals, gamopetalous, valvate; two posterio-lateral petals fuse to form the spur, and this spur is situated inside the spur formed by the posterior sepal; sometimes only 2 petals. *Androecium:* Stamens numerous, usually 15, in 5 groups of 3 each, polyandrous; anthers adnate or basifixed, dithecous, extrorse. *Gynoecium:* Monocarpellary, superior, unilocular, with many ovules, marginal placentation; style reduced; stigma simple. *Fruit:* Follicle.

Floral Formula: Br, Brl, ⊕, ⚥, K_5, C(4) or (2), A_{15} or $\propto$, $G\underline{1}$.

21.7 NYMPHAEACEAE (WATER-LILY FAMILY)

21.7.1 Systematic Position

Polypetalae, Thalamiflorae, Ranales.

21.7.2 Field Recognition

Aquatic perennial herbs; leaves with long petiole, peltate or pseudopeltate, with latex; petals, stamens, and sometimes sepals numerous; marginal or parietal placentation.

21.7.3 Selected Indian Genera

Nymphaea, *Nelumbo*, *Euryale* and *Nuphar*.

21.7.4 Size, Distribution and General Information

This family of subclass Polypetalae is represented by only 8 genera and over 90 species. Members are cosmopolitan in distribution, except in very cold regions, and are favourite plants for garden pools. Flowers of water lily (*Nymphaea*) are variously coloured and open just above the water surface in ponds while that of crow lily (*Nuphar*) are golden yellow and extend up and out of the water in the pond. *Nelumbo* and *Nymphaea lotus* are the "*sacred lotus*" plants of the family. Indian lotus, found commonly in shallow lakes and ponds, is *Nelumbo nucifera*. *Nymphaea nouchali* is the Indian red water-lily. *Victoria* is a native of South America.

21.7.5 Description of the Family

General Habit Aquatic, freshwater, perennial herbs with large rhizomes; sometimes annual (*Euryale*); submerged or free-floating in ponds, shallow lakes, etc.

Root Adventitious, branched clusters of fibrous roots develop from the root stocks or rhizomes.

Stem Perennial, branched, rhizomatous, creeping in *Nymphaea* but somewhat erect in *Victoria* and *Brasenia*; full of aerenchyma, laticiferous vessels and intercellular spaces.

Leaf Simple, alternate, peltate or cordate, long petioled; submerged, floating or emersed (*Nelumbo*); large, reaching up to 2 metres in *Victoria*; smooth and waxy but lower portion prickly (*Euryale*); in *Nymphaea*, the mature leaves may be orbicular, sagittate or oval; *Cabomba* shows heterophilly.

Fig. 21.8 Floral diagram of *Nymphaea nouchali.*

Inflorescence Solitary and axillary.

Flower Solitary, bisexual, actinomorphic; large, showy, long-peduncled; usually hypogynous but epigynous in *Victoria* and perigynous in *Nymphaea*; often calyx and corolla are poorly differentiated and are represented by *perianth* of 6 to numerous free tepals.

Calyx 3 to many sepals, polysepalous; in *Nuphar* sepals are yellow and larger than petals; sepals 3 in *Brasenia*, 4 in *Nelumbo* and *Nymphaea* (Fig. 21.8) and 5 in *Barclaya*.

Corolla 3 to many petals, polypetalous; petals frequently grading into the stamens and therefore inner petals often form petaloid staminodia; white or variously coloured and showy; a nectary is present on abaxial side in *Nuphar*; imbricate.

Androecium 3 to many, distinct stamens; filaments usually extend as sterile appendage beyond the anther sacs; anthers dithecous, basifixed or adnate, introrse; dehiscence longitudinal; in *Victoria* the inner stamens are sterile; in *Nelumbo* and *Nymphaea* the stamens are spirally arranged.

Gynoecium Carpels 2 or 3 (*Cabomba*) to numerous (*Nelumbo, Nymphaea*); each carpel with 1 locule, numerous ovules, marginal or parietal placentation; ovary superior but inferior in *Victoria* and *Euryale*; stigmas mostly 1 and discoid or as many as 35 and radiate; numerous carpels of *Nymphaea* and *Nelumbo* remain embedded separately on upper flat surface of swollen receptacle.

Fruit and Seed Fruit a spongy berry (*Nymphaea*) or follicle (*Cabomba*); seeds 1 to many per locule; with straight embryo and starchy endosperm; seeds sometimes arillate (*Nymphaea*).

Pollination and Dispersal Large, too bright, showy and often scented flowers of this family show entomophily, i.e. pollinated by insects such as beetles, small flies, etc. Some *Nymphaea* species are self-pollinated while *Nuphar, Victoria* and remaining species of *Nymphaea* are cross-pollinated. Dispersal in this family mainly takes place by water. However, some species of *Cabomba* and *Nuphar* are dispersed by ducks.

General Floral Formula $\oplus$, ⚥, $K_{3-\infty}$, $C_{3-\infty}$, $A_{3-\infty}$, $\underline{G_{3-\infty}}$ or $\underline{(3-\infty)}$.

21.7.6 Economic Importance

- *Ornamentals:* Because of their beautiful, showy or sweet-scented flowers, several Nymphaeaceae members are of great ornamental value for aquatic gardening. Chief among them are sacred '*lotus of India*' or '*Kamal*' (*Nelumbo nucifera*), Indian 'red water-lily' (*Nymphaea nouchali*), Indian 'blue water-lily' (*Nymphaea stellata*), 'Pandharen-Kamal' (*Nymphaea alba*) and 'Royal water-lily' (*Victoria regia*). *Nelumbo lutea* has yellow flowers while *Brasenia schreberi* has purple flowers and *Nymphaea rubra* has red flowers. *Nymphaea coerulea* with blue flowers, *N. capensis* with sky-blue flowers, *N. odorata* with pink flowers and *N. mexicana* with bright yellow flowers are other species of ornamental value.

- *Edible Products:* (i) Seeds of '*gorgan nut*' or '*Makhana*' (*Euryale ferox*) are roasted and eaten; doctors recommend them for invalids. (ii) The vegetable '*Kamal-Kakri*' is actually the rhizome of *Nymphaea nouchali.* (iii) Fruiting torus, flowers, rhizomes, young leaves and petioles of *Nelumbo nucifera* are used as vegetable. (iv) Rhizomes of *Nymphaea stellata* are also edible. (v) Seeds of *Victoria regia* are also roasted and eaten.
- *Medicinal Value:* (i) Flowers of *Nelumbo nucifera* are recommended as a cardiac tonic and also used in curing diseases of liver, and skin and its dried and powdered rhizome to cure piles. (ii) Rhizomes of *Nymphaea nouchali* are used to cure dysentery and diarrhoea. (iii) Flowers of *Nelumbo nucifera* yield a valuable perfume. (iv) *Nymphaea stellata* is used to relieve cardiac pains.

21.7.7 Systematics and Phylogeny (Affinities)

On the basis of the floral construction, family Nymphaeaceae is divided into three subfamilies, i.e. Cabomboideae, Nelumboideae and Nymphaeoideae. Apocarpous condition is found in Cabomboideae and Nelumboideae while syncarpous condition is observed in Nymphaeoideae. Cabomboideae have cyclic trimerous flowers whereas Nelumboideae have spiral pleiomerous flowers. All these three subfamilies have been treated as independent families by Bessey (1915) under the names Cabombaceae and Nelumbaceae belonging to order Ranales, and Nymphaeaceae belonging to order Rhoeadales. Several taxonomists treated them on the line suggested by Bessey.

Cronquist (1981) included Nymphaeaceae under order Nymphaeales along with four more families (Nelumbonaceae, Barclayaceae, Cabombaceae and Ceratophyllaceae). However, Thorne (1983) treated Nelumbonaceae under order Nelumbonales while Nymphaeaceae and Cabombaceae under order Nymphaeales.

Bentham and Hooker treated Nymphaeaceae under order Ranales, and subfamily Cabomboideae forms a link between Nymphaeaceae and Ranunculaceae. Presence of trimerous flowers and scattered vascular bundles in Cabomboideae also suggest its affinity with monocotyledons. Nymphaeaceae show affinity with Berberidaceae in possessing scattered vascular bundles and arillate seeds, and also with Papaveraceae in having latex, large and discoid stigma and type of placentation:

Nymphaeales have been treated as probable ancestors of monocotyledons by Cronquist (1967) because of their aquatic occurrence, lack of vessels, absence of cambium, presence of uniaperturate pollen and laminar placentation.

21.7.8 Description of a Common Plant

1. Nelumbo nucifera Gaertn. ('Sacred lotus' or 'Kamal'; Fig. 21.9)

Habit: Aquatic, perennial herb. *Stem:* Branched, rhizomatous. *Leaves:* Simple, large, forming a triad, of which one is present on the lower side of rhizome and scaly, another is present on the upper side of rhizome, and third leaf is large and reaches up to the water surface; latter is long-petioled, stipulate, peltate; upper surface waxy; multicostate reticulate. *Inflorescence:* Solitary axillary. *Flower:* Ebracteate, pedicellate, pedicel long; large, complete, hermaphrodite, actinomorphic, hypogynous, showy. *Calyx:* 4 sepals, free, petaloid or greenish pink, imbricate. *Corolla:* Numerous petals, polypetalous, spirally arranged, pink. *Androecium:* Numerous stamens, polyandrous, spirally

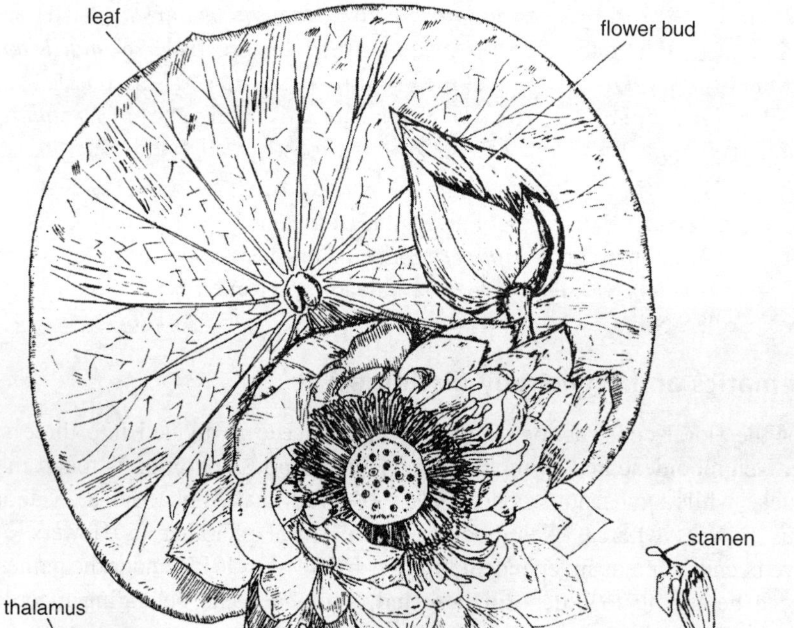

Fig. 21.9 *Nelumbo nucifera* Gaertn.

arranged; filaments long, anthers basifixed, connective is prolonged as an appendage. *Gynoecium:* Polycarpellary, apocarpous, superior; carpels embedded in the upper part of an obconically flat but spongy receptacle; unilocular, one ovule in each locule; stigma flat.

Floral Formula: Ebr, ⊕, ⚥, K_4, C ∞, A ∞, G $\underline{\infty}$.

21.8 PARIETALES

21.8.1 General Characteristics

1. Pentamerous flowers with their calyx showing imbricate aestivation.
2. Stamens as many as petals or more
3. Ovary tricarpellary, syncarpous, unilocular
4. Numerous ovules showing parietal placentation
5. Seeds endospermic

Bentham and Hooker divided cohort Parietales into 9 orders (=families), namely Cruciferae, Papaveraceae, Violaceae, Bixineae, Cannellaceae, Sarraceniaceae, Capparidaceae, Resedaceac and Cistineae. Engler and Diels, however, divided order Parietales into 10 suborders composed of 31 families. Lawrence (1951), while following Engler and Diels, however, commented that "available evidence indicates that the order is not a phylogenetic taxon, and the realignment of the families into several orders is to be expected. In some instances this may result in transfer of families to existing orders and in others in the establishment of new orders".

Only Papaperaceae, Fumariaceae, Capparidaceae, Cruciferae (Brassicaceae) and Violaceae are treated in this text.

21.9 PAPAVERACEAE (POPPY FAMILY)

21.9.1 Systematic Position

Polypetalae, Thalamiflorae, Parietales.

21.9.2 Field Recognition

Herbs; leaves alternate, often lobed or divided, exstipulate; flowers bisexual; poisonous latex usually present; petals rolled or crumpled in the bud; stamens numerous; ovary unilocular; fruit capsule.

21.9.3 Selected Indian Genera

Argemone, *Eschscholzia*, *Papaver.*

21.9.4 Size, Distribution and General Information

Poppy family is represented by about 26 genera and 280 species (Hickey and King, 1988) distributed chiefly in Northern Hemisphere. Western North America and eastern Asia are the main centres of its distribution. Representatives of this family also occur in Europe, South Africa, Australia and even in arctic regions.

Only 5 genera and about 20 species have so far been reported from India. *Meconopsis* with about 26 species grows wild in some parts of Himalayas. *Argemone mexicana* occurs wild in waste fields while *Papaver somniferum* (opium poppy) is cultivated widely in India. The drug "opium" is obtained by cutting notches in the half-ripened capsules, from which the latex exudes.

21.9.5 Description of the Family

General Habit Mostly annual or perennial herbs with milky or coloured latex; rarely woody shrubs (*Dendromecon*) or tree-like, reaching up to 10 metre high (*Bocconia arborea*).

Roots Extremely branched, usually surface feeder.

Stem Herbaceous, rarely woody; branched, erect, cylindrical; glabrous or sometimes hairy.

Leaves Alternate (rarely the upper ones whorled), basal or cauline; usually much divided or lobed and pinnately cut; exstipulate; with milky or coloured sap or latex; the latex is white in *Papaver*, yellow in *Argemone*, orange in *Chelidonium*, watery in *Eschscholzia*, and bright orange-red in *Sanguinaria*.

Inflorescence Solitary showy flowers in most of the species; racemes or cymes in *Meconopsis*; racemose panicles in *Bocconia*; umbellate clusters in *Chelidonium*.

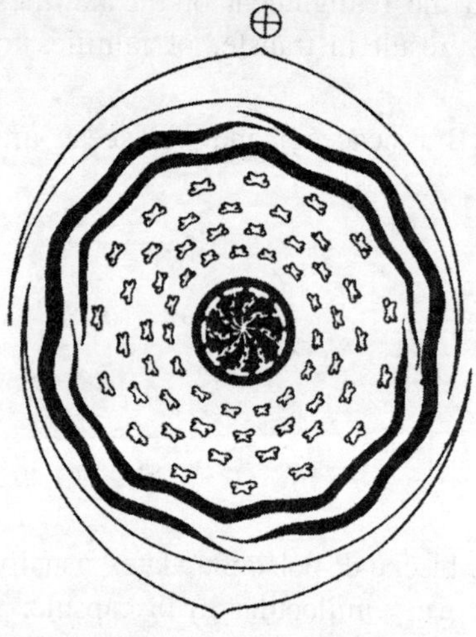

Fig. 21.10 Floral diagram of *Papaver rhoeas*.

Flower Ebracteate, pedicellate, bisexual, actinomorphic, complete, hermaphrodite, bi- or trimerous; large and showy; hypogynous but perigynous in *Eschscholzia*.

Calyx Sepals 2 to 3, polysepalous but connate in *Eschscholzia*; caducous i.e. fall off very soon; imbricate.

Corolla Petals 4 to 6 or more, polypetalous, often crumpled (full of folds) in the bud (Fig. 21.10); in one or two or rarely more whorls; imbricate; bright coloured; petals absent in *Bocconia* and *Macleaya*.

Androecium Stamens numerous, free, arranged in closed spiral; filaments short; anthers elongate, basifixed, extrorse; dehiscence longitudinal.

Gynoecium A compound pistil of 2 or more fused carpels, ovary superior, with one locule containing numerous ovules, parietal placentation; ovary half-inferior in *Eschscholzia*; ovary rarely becomes multilocular by fusion of intruding placentae; style reduced; stigmas as many as carpels; ovules anatropous or slightly campylotropous; rarely only one basal ovule in *Bocconia*.

Fruit and Seed Fruit a capsule, opening by valves (*Argemone*) or pores (*Papaver*); seeds numerous, with copious and oily endosperm and small embryo.

Pollination and Dispersal Pollination is entomophilous. Flowers are mostly large and conspicuous, but usually do not contain any nectar, and are visited mainly by pollen-seeking insects. In *Papaver*, the flowers are homogamous and both self- and cross-pollination take place. Dispersal of seeds is affected usually by wind or by animals, birds or human beings.

General Floral Formula Ebr, $\oplus$, ⚥, K_{2-3} or (2), C_{2+2} or 3 + 3, A ∞, $G\underline{(2-\infty)}$.

21.9.6 Economic Importance

- Opium poppy (*Papaver somniferum*), cultivated in the warmer countries for the universally known drug *opium*, is the most important plant of this family. Opium is obtained from the milky latex of unripe capsules of this plant. Morphine content ranges from 5–6% in unripe fruits, while only 0.8–1% in ripe fruits. Morphine, codeine, narcotine and papaverine are some of the alkaloids present in this milky latex. Opium is known for its sedative properties and its high doses first produce sleep and then depression, paralysis, unconsciousness and ultimately death.
- The seeds of *Papaver somniferum* contain no opium, and are used in baking and sprinkled on bread.
- An edible oil, obtained from the seeds of *P. somniferum*, is used in the preparation of animal and human foods, and also in the manufacture of soaps, paints and varnishes.
- Wrapping paper is prepared from the pulp obtained from opium poppy.
- Family is known for its popular garden plants, such as *Eschscholzia californica* (Californian Poppy), *Argemone mexicana* (Prickly poppy), *Macleaya cordata* (Plume poppy), *Meconopsis* (Blue poppy), *Dendromecon rigida* (Bush poppy), *Sanguinaria canadensis* (Blood root) and *Chelidonium* (Celandine).
- True poppies are species of *Papaver*, such as *P. nudicaule* (Iceland poppy, orange red flowers), *P. orientale* (Oriental poppy, bright red flowers), *P. rhoeas* (Corn poppy, red or deep purple flowers), etc.
- A non-edible oil, obtained from the seeds of *Argemone mexicana*, is used as an illuminant and in skin infections. It is also mixed to adulterate mustard oil.
- An emetic drug, 'Sanguinarine', is obtained from the rhizome and roots of *Sanguinaria canadensis*. It is used in dyspepsia i.e. indigestion.

21.9.7 Systematics and Phylogeny (Affinities)

Bentham and Hooker included Papaveraceae under order Parietales whereas Engler and Prantl (1887–1899) treated it under order Rhoeadales. Takhtajan (1980) also treated it under Rhoeadales whereas Cronquist (1981) included it under order Papaverales along with one more family Fumariaceae. Thorne (1983) treated it under suborder Papaverineae of order Berberidales.

Family Papaveraceae shows most close affinities with Ranunculaceae, mainly in having hemicyclic flowers, numerous stamens with extrorse anthers, numerous carpels, and the seeds with copious endosperm and very small embryo. Papaveraceae also shows affinities with Brassicaceae (Cruciferae) in possessing syncarpous, unilocular ovary and parietal placentation. Flowers of *Platystemon* establish a link between Papaveraceae and Nymphaeaceae.

21.9.8 Description of Some Common Plants

1. Argemone mexicana L. (Prickly Poppy or Peeli Kataili or Jangli Posth; Fig. 21.11)

Habit: A spiny or prickly annual herb. *Stem:* Herbaceous, erect, branched, cylindrical, solid, green, spiny, with yellow latex. *Leaf:* Simple, alternate, exstipulate, sessile, semi-amplexicaul; margin

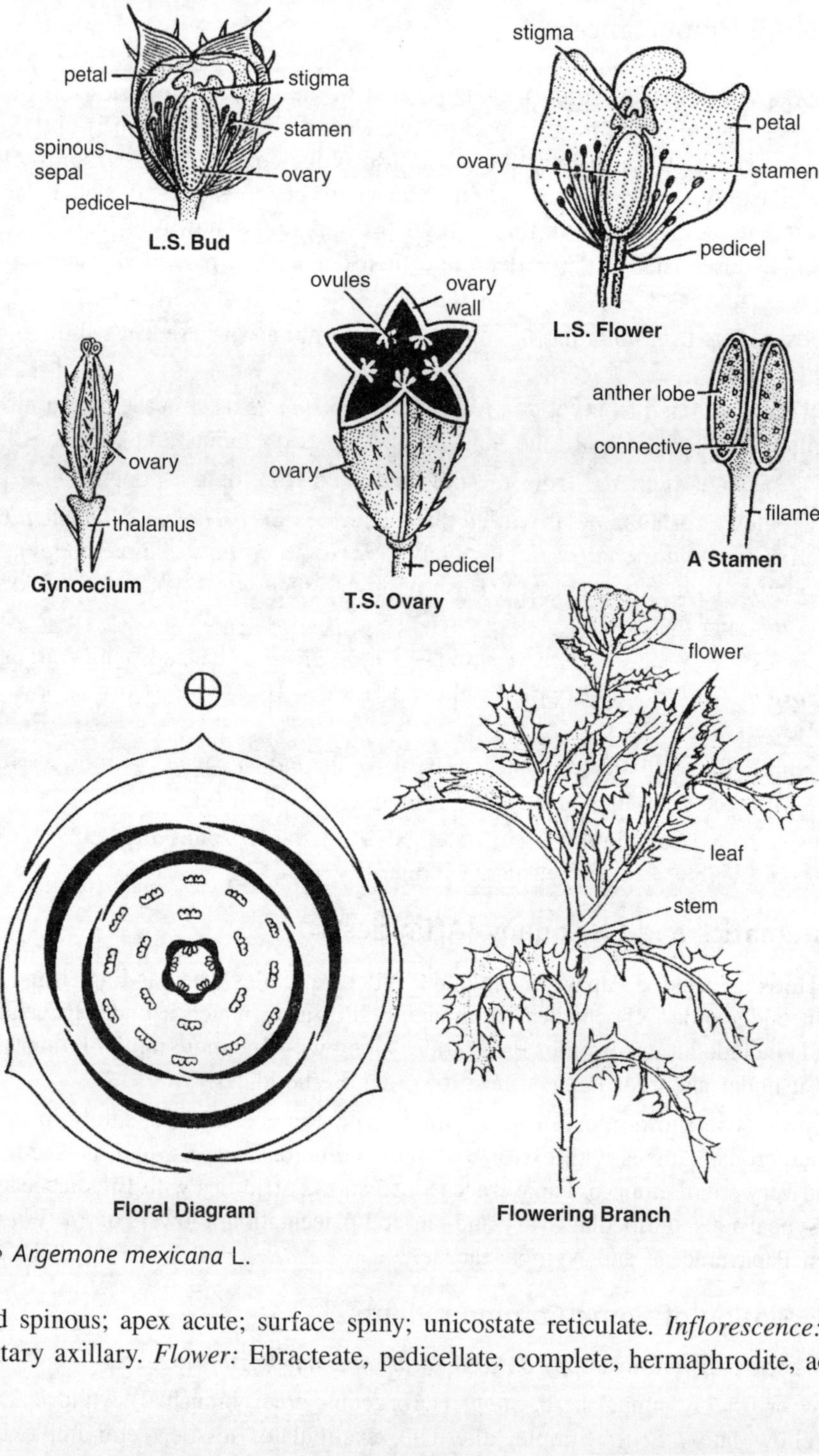

Fig. 21.11 *Argemone mexicana* L.

dissected and spinous; apex acute; surface spiny; unicostate reticulate. *Inflorescence:* Solitary terminal or solitary axillary. *Flower:* Ebracteate, pedicellate, complete, hermaphrodite, actinomorphic, trimerous, hypogynous, yellow. *Calyx:* Sepals 3, polysepalous, imbricate or twisted; spiny; caducous, contains a horn-like spine at the apex. *Corolla:* Petals 6, polypetalous, arranged in two whorls of three each, caducous, crumpled in the bud, yellow. *Androecium:* Stamens numerous, polyandrous, filaments long, dithecous, basifixed, extrorse. *Gynoecium:* 4 to 6 carpels, syncarpous, superior, unilocular, many ovules, parietal placentation; style reduced; stigma 4–6 lobed and hood like; ovary covered with prickles. *Fruit:* Capsule. *Seed:* Black, small, endospermic.

Floral Formula: Ebr, ⊕, ⚥, K_3, C_{3+3}, A ∞, G $(\underline{4-6})$.

2. Eschscholzia californica Chamb. (Californian Poppy; Fig. 21.12)

Habit: Herb. *Root:* Branched, tap root. *Stem:* Erect, herbaceous, branched, green, solid, glabrous. *Leaf:* Cauline and ramal, simple, alternate, exstipulate, very much dissected or decompound, leaf base sheathing; acute, glabrous; unicostate reticulate. *Inflorescence:* Solitary terminal or solitary axillary. *Flower:* Ebracteate, pedicellate, complete, hermaphrodite, actinomorphic, bimerous, perigynous,

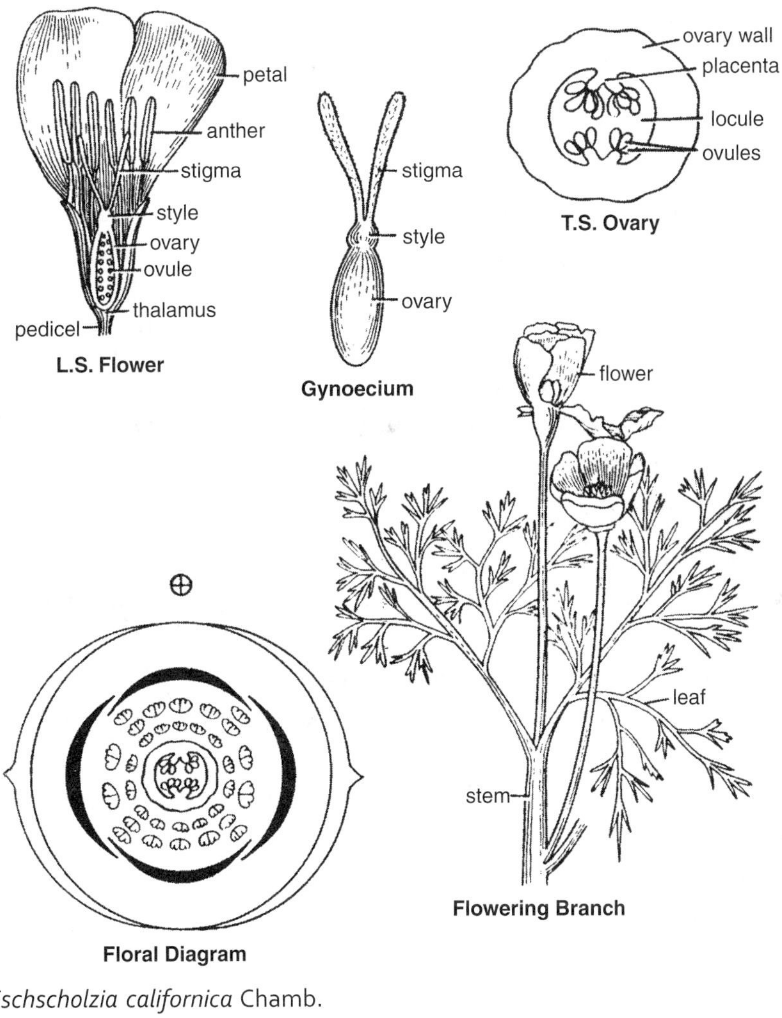

Fig. 21.12 *Eschscholzia californica* Chamb.

cyclic, yellow. *Calyx:* Sepals 2, raised on cup-like projection of floral axis; fused to form a pointed hood, specially in bud condition; caducous. *Corolla:* Petals 4, polypetalous, arranged in 2 whorls of 2 each, imbricate; bright yellow. *Androecium:* Stamens numerous, polyandrous, arranged at the brim of the cup; filaments short, dithecous, basifixed, extrorse. *Gynoecium:* Bicarpellary, syncarpous, half-inferior, unilocular, ovules numerous; parietal placentation; style short; stigma 2, each is bifid; out of 2 portions of stigma one is longer than the other. *Fruit:* Capsule.

Floral Formula: Ebr, ⊕, ⚥, $K_{(2)}$, C_{2+2}, A ∞, G(2)–.

21.10 FUMARIACEAE (FUMARIA FAMILY OR FUMITORY FAMILY)

21.10.1 Systematic Position

Treated as a tribe of family Papaveraceae by Bentham and Hooker.

21.10.2 Field Recognition

Herbs with watery juice but latex absent; leaves usually variously divided; flowers zygomorphic; petals spurred; stamens 6, in 2 groups; gynoecium bicarpellary.

21.10.3 Selected Indian Genera

Fumaria, Corydalis, Dicentra, Hypecoum.

21.10.4 Size, Distribution and General Information

Fumariaceae, often treated as a sub-family (Fumarioideae) of Papaveraceae, is represented by about 19 genera and over 400 species. It is distributed in temperate North America, Asia, Africa and Europe. 4 genera and over 40 species of this family have been reported from India. The three best known genera with their reported number of species are *Corydalis* (300), *Fumaria* (50) and *Dicentra* (22).

21.10.5 Description of the Family

General Habit Annual (*Fumaria indica*) or perennial herbs, with watery juice; sometimes climbers (*Corydalis, Fumaria officinalis*); instead of latex vessels of Papaveraceae, oil-containing sacs are present in Fumariaceae.

Root Branched tap roots, sometimes tuberous (*Corydalis*).

Stem Herbaceous; erect or climbing by leaf or petiole tendrils; sometimes tuberous (*Corydalis*) or bulbous (*Dicentra*); watery juice present.

Leaves Alternate, rarely opposite; simple with much divided blades (*Fumaria indica*); petiolate; sometimes with tendril-like leaf axes; unicostate reticulate.

Inflorescence Usually racemose racemes; rarely subumbellate (*Corydalis*)

Flower Bracteate, shortly pedicellate, complete, hermaphrodite; laterally or transversely zygomorphic; dimerous; hypogynous; cyclic.

Calyx Sepals 2, small, free; medianly placed; caducous or deciduous.

Corolla Petals 4, polypetalous, in 2 whorls of 2 each; one (*Fumaria*, *Corydalis*) or both (*Dicentra*) outer petals more or less spurred at base (Fig. 21.13A, B); two inner petals are smaller and often fused around the anthers; in *Hypecoum* the outer petals are trilobed at the apex while the inner petals are tripartite (Fig. 21.14).

Androecium Androecium consists of 2 groups or bundles, located opposite to the inner petals; each group divided into 3 parts at the apex, i.e. central part bearing a complete or dithecous anther and the 2 lateral parts, each bearing monothecous or half an anther (Fig. 21.15); stamens are 4 and antipetalous in *Hypecoum*; polyandrous; extrorse; a basal nectary is attached with either one (*Fumaria*, *Corydalis*) or both (*Dicentra*) the groups of stamens.

Some views regarding the *morphological nature of androecium* in Fumariaceae are undermentioned:

1. *Asa Gray* believed that each lateral group or bundle represents a single stamen which has become tripartite.
2. *De Candolle* opined that each lateral group of stamens consists of one lateral stamen and 2 half-stamens derived from 2 posterior and 2 anterior stamens of a 4-stamened androecium, as of *Hypecoum*.
3. *Arber* considered them as 6 stamens, of which 2 are normal with dithecous anthers and 4 are reduced with monothecous anthers.
4. *Eichler* has been of the view that Fumariaceae members have only 2 dithecous stamens, and the 4 monothecous parts attached with them are their stipular appendages.
5. *Norris* believed that each lateral group of stamens is a double structure, made up of 2 stamens with monothecous anthers. It is so because each group or bundle receives 2 bundles.

Gynoecium Bicarpellary, syncarpous, superior, unilocular, one to many ovules in locule, parietal placentation; style filiform; stigma capitate or slightly bilobed.

Fruits and Seeds Fruit a slender capsule or one-seeded nutlet; seeds black and shiny, often with an aril or caruncle, each contains a minute embryo and fleshy endosperm.

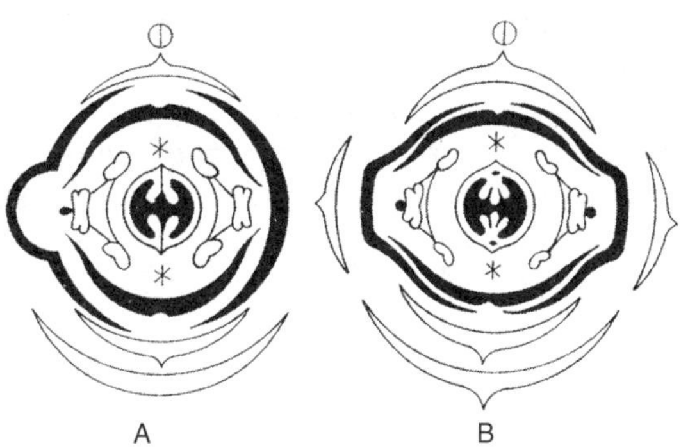

Fig. 21.13 Floral diagrams—A: *Corydalis*; B: *Dicentra*.

Fig. 21.14 Floral diagram of *Hypecoum*.

Pollination and Dispersal Nectary, attached with one (*Fumaria*, *Corydalis*) or both (*Dicentra*) the lateral groups of stamens, secretes honey which attracts the nectar-seeking insects helpful in pollination. Dispersal of seeds takes place by ants, birds or human being.

General Floral Formula Br, ⦶, ⚥, K_2, C_{2+2} or 2 + (2), $A_{(3)+(3)}$, $G_{\underline{(2)}}$.

21.10.6 Economic Importance

- Alkaloid protopine, found in *Hypecoum procumbens* and *H. leptocarpum*, is used in high blood pressure.
- Some species of *Corydalis* and *Dicentra* are grown as ornamental plants.
- *Corydalis racemosa* is used in eye diseases while *C. govaniana* roots are used against syphilis.
- Dried plants and seeds of *Fumaria indica* are used to cure fever and as blood purifier.

21.10.7 Systematics and Phylogeny (Affinities)

Botanists, such as Bentham and Hooker (1862–1883), Rendle (1925), Engler and Prantl (1931), Melchior (1964), etc. treated this family as a tribe or subfamily of family Papaveraceae. However, Hutchinson (1959) provided it the status of an independent family (Fumariaceae) under order Rhoeadales. Hutchinson's treatment is followed by most of the recent botanists, including Cronquist (1981) who discussed Fumariaceae and Papaveraceae as two independent families of order Papaverales. *Hypecoum* (Fig. 21.14) forms a connecting link between Fumariaceae and Papaveraceae. Moreover, Fumariaceae differ from Papaveraceae in showing absence of latex, and presence of variously divided leaves, zygomorphic flowers, spurred petals and stamens in two groups.

21.10.8 Description of Common Plant

1. Fumaria indica L. (= F. parviflora; Pit-papra, Fig. 21.15)

Habit: Annual herb. *Root:* Well-branched tap root. *Stem:* Herbaceous, branched, green, without latex. *Leaf:* Alternate, exstipulate, simple or decompound, petiolate, finely-dissected into several lobes or segments; each lobe entire, acute, unicostate reticulate. *Inflorescence:* Racemose raceme. *Flower:* Bracteate, pedicellate, complete, zygomorphic, dimerous, hypogynous, white or purple-pink. *Calyx:* Sepals 2, free, membranous, anterio-posterior, caducous, valvate. *Corolla:* Petals 4, arranged in 2 whorls of 2 each; one of the outer lateral petals is spurred; inner whorl petals are smaller and placed anterio-posterior. *Androecium:* Stamens in 2 groups; each group contains 1/2 + 1 + 1/2 stamens; filament of the stamen opposite to the spurred petal has a yellow-green nectary; basifixed, extrorse. *Gynoecium:* Bicarpellary, syncarpous, superior, unilocular, many ovules, parietal placentation, stigma lobed. *Fruit:* Nut.

Floral Formula: Br, ⦶, ⚥, K_2, C_{2+2}, $A\left(\frac{1}{2} + 1 + \frac{1}{2}\right) + \left(\frac{1}{2} + 1 + \frac{1}{2}\right)$, $G_{\underline{(2)}}$.

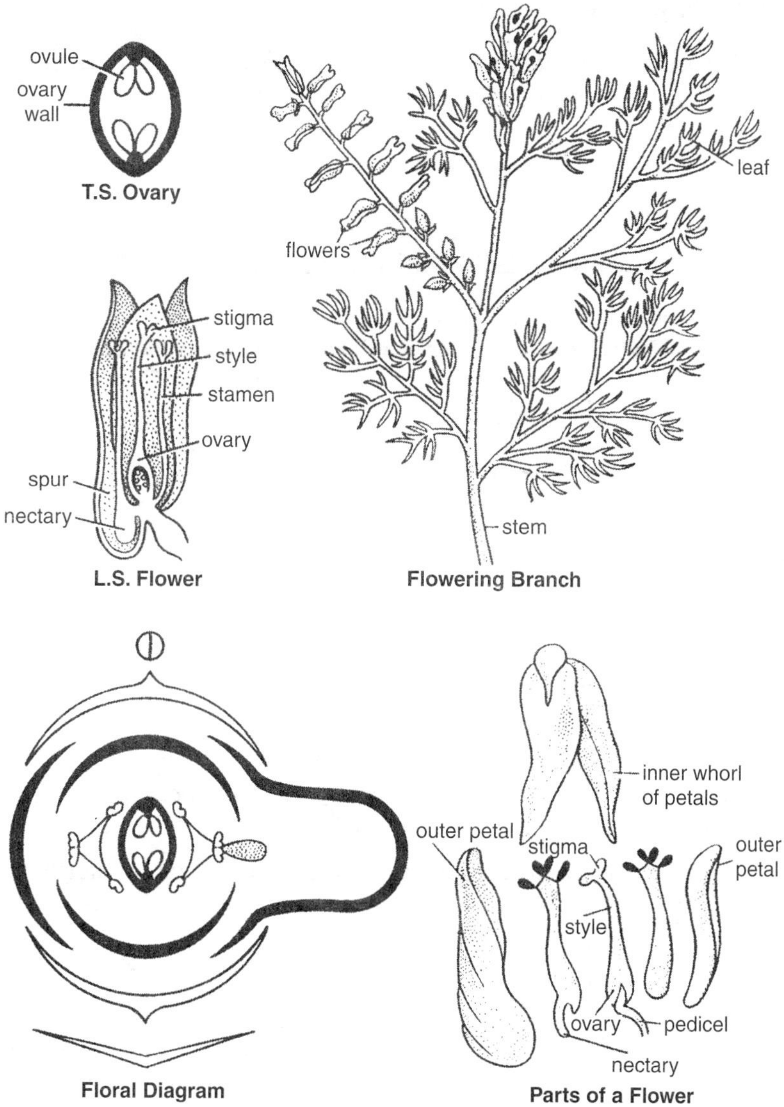

Fig. 21.15 ◈ *Fumaria indica* L.

21.11 ❁ CAPPARIDACEAE (THE CAPERS FAMILY)

21.11.1 Systematic Position

Polypetalae, Thalamiflorae, Parietales.

21.11.2 Field Recognition

Herbs, shrubs or climbers; leaves alternate, simple or compound; flowers usually actinomorphic; presence of androphore and gynophore; fruit capsule or berry.

21.11.3 Selected Indian Genera

Cleome, *Capparis*, *Crataeva*.

21.11.4 Size, Distribution and General Information

42 genera and over 900 species of this family are distributed mostly in the tropics and warm temperate regions of the world. Only 7 genera and over 65 species have so far been reported from India, mostly from western and southern parts. Quite a large number of plants occur in xeric conditions. Over 35 species of *Capparis* have been reported from India. *Roydsia* is very common in Assam.

21.11.5 Description of the Family

General Habit Herbs (*Cleome viscosa*, *C. gynandra*), shrubs (*Capparis*), lianas (*Maerua arenaria*), small tree (*Niebuhria linearis*) or tall tree (*Crataeva religiosa*). Mostly grow wild and are xerophytic.

Root Well-branched, deep feeder, tap root.

Stem Aerial, erect, herbaceous or woody (*Capparis*), rarely climbing (*Maerua arenaria*), hairy or covered with thorns (*Capparis aphylla*).

Leaf Alternate, simple (*Capparis zeylanica*, *Maerua*) or palmately compound (*Cleome*, *Crataeva*); usually stipulate; stipules may be foliaceous (*Cleome*) spinose (*Capparis*) or setaceous (*Niebuhria*); in *Capparis decidua* leaves are reduced, scaly and caducous.

Inflorescence Generally solitary or in racemes (*Cleome*, *Polanisia*), rarely shortly peduncled simple umbel (*Capparis sepiaria*).

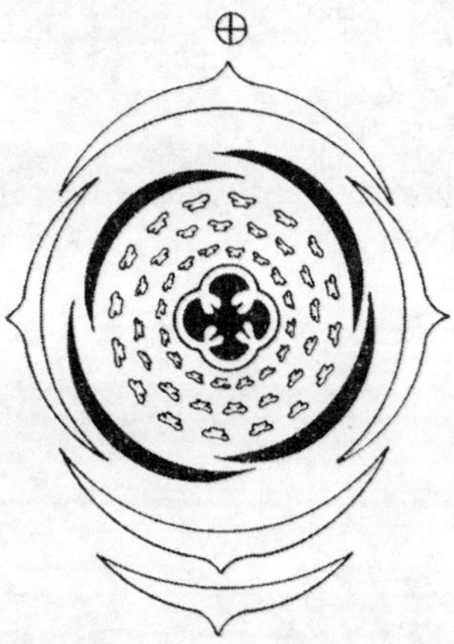

Fig. 21.16 Floral diagram of *Capparis*.

Flower Bracteate, ebracteolate, complete, hermaphrodite, actinomorphic, tetramerous, hypogynous; in some (*Pteropetalum*, *Emblingia*) flowers are medianly zygomorphic; incomplete due to the absense of petals in *Roydsia* and *Niebuhria*; a disc is usually present; thalamus of several genera is elongated into androphore and gynophore, collectively called *androgynophore*.

Calyx 4, free or rarely basally fused sepals, arranged in 2 whorls of 2 each; in *Capparis* posterior sepal forms a hood-like structure; valvate or imbricate; sepals are 5 and basally fused in *Emblingia*; in *Roydsia* 6 sepals are arranged in 2 whorls of 3 each.

Corolla 4, free, diagonally placed petals; valvate (*Cleome*) or imbricate (*Capparis*, Fig. 21.16); petals are 2 in *Cadaba* and *Emblingia* and none in *Roydsia*; in *Pteropetalum*, the 2 posterior petals are larger than the 2 anterior ones.

Androecium Stamens 4 (*Cleome tetrandra*) to numerous (*Crataeva, Capparis*), polyandrous; anthers dithecous, introrse; in some genera (*Cleome gynandra*) the receptacle of the flower between corolla and stamens gets elongated in the form of *androphore*; stamens are 6 in *Cleome spinosa* but never tetradynamous.

Gynoecium Usually bicarpellary, rarely tetra- to multicarpellary; syncarpous; in some genera the receptacle of the flower gets elongated between androecium and gynoecium in the form of *gynophore*; superior; unilocular, rarely falsely bilocular; a few to many ovules, parietal placentation; style short; stigma capitate or 2-lobed.

Fruit and Seed Fruit berry (*Capparis, Crataeva*), capsule (*Cleome, Polanisia*), drupe (*Roydsia*) or nut (*Emblingia*). Seeds non-endospermic, kidney-shaped, with folded embryo as in Brassicaceae.

Pollination Entomophilous.

General Floral Formula Br, Ebrl, ⊕, ⚥, K_{2+2}, C_4, $A_{4-\infty}$, $G\underline{(2-4)}$.

21.11.6 Economic Importance

Members of Capparidaceae are of little economic importance.

- *Ornamentals:* *Capparis sepiaria, C. spinosa, Cleome chelidonii, C. spinosa* (Capers), and *Roydsia suaveolens* are grown as ornamental plants.
- *Medicinal:* (i) *Capparis zeylanica* leaves are used against boils and stomach ache; (ii) *Cleome gynandra* (syn. *Gynandropsis pentaphylla*) leaves and seeds are used for expelling roundworms, and juice from its roasted leaves is used to cure earache; (iii) *Crataeva religiosa* bark is used for increasing appetite and reduction of bile secretion; (iv) *Maerua arenaria* roots are used as a tonic and stimulant.
- *Edibles:* (i) *Capparis spinosa* fruits are picked and form the *'Capers of Commerce'*; (ii) Unripe fruits of *Capparis aphylla* (syn. *C. decidua*) and *C. zeylanica* are also pickled and eaten in India; (iii) Seeds of *Cleome viscosa* ('Hurhur') are used in curries; (iv) Unripe fruits of *Maerua arenaria* are boiled and eaten.

21.11.7 Systematics and Phylogeny (Affinities)

Because of the presence of parietal placentation and commissural stigmatic lobes, Capparidaceae are related closely to the Parietalean families, such as Brassicaceae, Moringaceae, Papaveraceae, Resedaceae, etc.

Bentham and Hooker treated Capparidaceae under order Parietales while Engler and Prantl (1931) placed them under order Rhoeadales. Capparidaceae have been treated under a new name *Capparaceae* belonging to order Capparales by majority of the recent taxonomists, including Takhtajan (1969), Cronquist (1981) and Thorne (1983). Capparidaceae are usually divided into two tribes i.e. Cleomeae, including herbaceous plants and, Cappareae including woody plants.

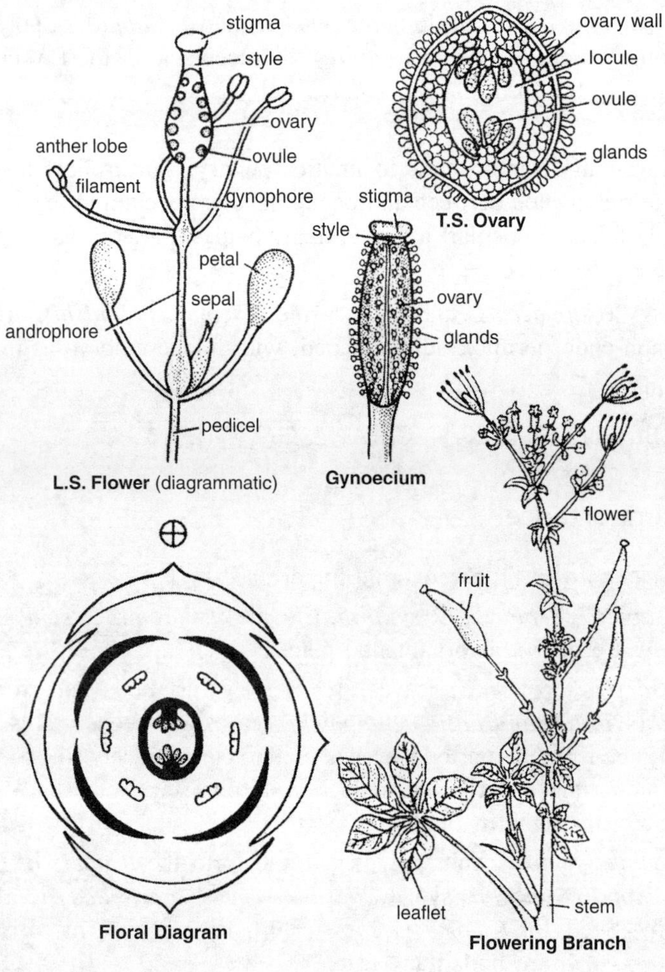

Fig. 21.17 *Cleome gynandra* D.C.

21.11.8 Description of a Common Plant

1. Cleome gynandra D.C. (syn. Gynandropsis pentaphylla L.; Fig. 21.17)

Habit: Annual, erect herb. *Stem:* Herbaceous, erect, branched hairy. *Leaf:* Alternate, exstipulate petiolate, palmately compound; leaflets sessile, ovate to obovate, acute, hairy, unicostate reticulate. *Inflorescence:* Corymbose raceme. *Flower:* Ebracteate, pedicellate, complete, hermaphrodite, actinomorphic, hypogynous, tetramerous; androphore and gynophore present. *Calyx:* 4 sepals, free, arranged in 2 whorls of 2 each, imbricate. *Corolla:* 4 petals, free, valvate, clawed, white. *Androecium:* 6 stamens, polyandrous, dithecous, dorsifixed, introrse; floral axis is elongated between corolla and stamens in the form of androphore; filaments long. *Gynoecium:* Bicarpellary, syncarpous, superior, unilocular, many ovules, parietal placentation; floral axis is elongated between androecium and ovary in the form of gynophore; style reduced; stigma capitate. *Fruit:* Capsule.

Floral Formula: Ebr, ⊕, ⚥, K_{2+2}, C_4, A_6, $G(\underline{2})$.

21.12 BRASSICACEAE OR CRUCIFERAE (MUSTARD FAMILY)

21.12.1 Systematic Position

Polypetalae, Thalamiflorae, Parietales.

21.12.2 Field Recognition

Herbs with a smelling watery juice; flowers of 4 sepals, 4 petals and 6 stamens; corolla cruciform; stamens tetradynamous; ovary with a false septum; fruit siliqua or silicula.

21.12.3 Selected Indian Genera

Alyssum, *Brassica*, *Capsella*, *Coronopus*, *Eruca*, *Iberis*, *Raphanus*, *Sisymbrium*.

21.12.4 Size, Distribution and General Information

This cosmopolitan family of about 375 genera and over 3200 species, occur chiefly in north temperate regions, specially in the Mediterranean region. About 50 genera and over 140 species have so far been reported from India. Some of the larger genera, along with their reported number of species from the world (Hickey and King, 1988) in parenthesis, include *Draba* (300), *Cardamine* (160), *Lepidium* (150), *Alyssum* (150), *Arabis* (120), *Erysimum* (100), *Sisymbrium* (90), *Heliophila* (75), *Rorippa* (70), *Brassica* (50), *Iberis* (30) and *Raphanus* (8).

21.12.5 Description of the Family

General Habit Annual, biennial, or perennial herbs with a pungent watery sap; herbage commonly covered with branched or stellate unicellular hairs; rarely subshrubs.

Root Tap root which in some genera gets swollen due to stored food and becomes fusiform (radish) or napiform (turnip).

Leaf Mostly simple, alternate, often dissected, exstipulate; rarely opposite or subopposite; pubescence of simple to stellate hairs; radical leaves form a rosette whereas cauline leaves are alternate or opposite.

Inflorescence Typically a raceme or corymb (*Iberis*) and nearly always without bracts or bracteoles.

Flower Ebracteate, ebracteolate, complete, bisexual, actinomorphic, rarely zygomorphic (*Iberis*, Fig. 21.18A); tetramerous; hypogynous.

Calyx 4 sepals, polysepalous, in 2 whorls of 2 each, outer whorl is anterio-posterior.

Corolla 4 petals, polypetalous, cruciform (arranged in a cross), alternate with sepals, often clawed; 2 anterior petals in *Iberis* are larger (Fig. 21.18A); petals reduced or scale-like in *Coronopus* (Fig. 21.18B); petals completely absent in *Lepidium*.

Androecium 6 stamens, arranged in 2 whorls; an outer whorl of 2 short stamens, and an inner whorl of 4 long stamens (i.e. tetradynamous); anthers dithecous (rarely monothecous); introrse,

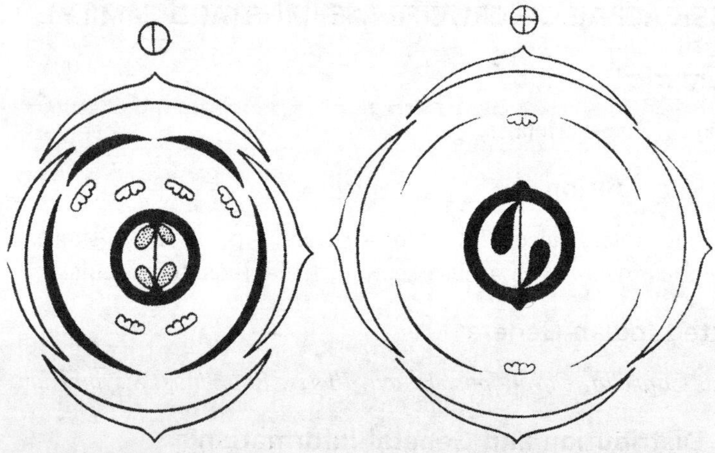

Fig. 21.18 Floral diagrams—A: *Iberis amara* and, B: *Coronopus didymus* syn. *Senebiera didyma*.

longitudinally dehiscent; stamens are only 4 in some species of *Nasturtium* while only 2 in *Coronopus didymus* (Fig. 21.18B) and *Lepidium* and up to 16 in *Megacarpaea polyandra*.

Gynoecium Bicarpellary, syncarpous, superior; unilocular but becoming bilocular due to the development of a false septum or *replum*; replum unites the two parietal placentae; numerous anatropous or campylotropous ovules; parietal placentation; styles 1 or obsolete; stigma capitate to bilobed; carpels 3 in *Lepidium sativum* and 4 in *Tetrapoma*.

Fruit and Seed Fruit siliqua (*Arabis*, *Brassica*, *Cheiranthus cheiri*) or silicula (*Capsella bursa-pastoris*, *Lunaria*), or occasionally a nut (*Crambe*) or rarely achene-like (*Isatis tinctoria*). Seeds are small, non-endospermic, with curved embryo and often mucilaginous testa. Fruits and seeds are characters of systematic importance in this family.

Pollination and Dispersal Pollination is entomophilous, and dispersal of seeds may be by birds, wind, water or cattles.

General Floral Formula Ebr, Ebrl, ⊕, ⚥, K_{2+2}, C_4, A_{2+4}, $G\underline{(2)}$

21.12.6 Economic Importance

Brassicaceae are of considerable economic importance for providing us a variety of vegetables, oils, food crops, ornamentals, and weeds.

1. *Vegetables:* Several common vegetables of daily use belong to this family. Some of them are Mooli (*Raphanus sativus*, radish), Shaljam (*Brassica rapa*, turnip), Phool Gobhi (*B. oleracea* var. *botrytis*, cauliflower), Band Gobhi (*B. oleracea* var. *capitata*, cabbage), Ganth

Gobhi (*B. oleracea* var. *gongylodes*, Knol-knol), Button Gobhi (*B. oleracea* var. *gemmifera*, Brussel's sprouts), Kadamsag (*B. oleracea* var. *acephala*), Sarson (*B. campestris* var. *sarson,* mustard) and Sengri (*B. sativus* var. *caudatus*).

2. *Oils:* Fatty oil, used for cooking, pickles, burning, massaging, etc. is obtained from the seeds of several species of this family. Mustard oil is obtained from several species of *Brassica*, such as *Brassica campestris* var. *sarson* (mustard or yellow sarson) and *B. campestris* var. *toria* (Toria or Indian rape). Oil is also obtained from the seeds of *B. juncea* var. *cuneifolia* (Rai), *B. nigra* (Kali Rai or black mustard), *B. alba* (white or Safed sarson) and *Eruca sativa* (Taramira). The oil cake left after the extraction of oil, is a good cattle feed.
3. *Ornamentals:* Among the well-known ornamentals are candytuft (*Iberis*), honest (*Lunaria*), sweet alyssum (*Lobularia*), wall flower (*Cheiranthus*), stocks (*Mathiola*), etc.
4. *Medicinal:* Majority of Crucifers produce an abundance of vitamin C. Characteristic pungent odor of its members is because of sulphur compounds. Cauliflower and a few other Crucifers have recently attracted botanists as possible anticancer food items.
5. *Weeds:* Troublesome weeds of cultivated fields and lawns include *Brassica arvensis*, *Capsella bursa-pastoris* (shepherd's purse), *Barbarea*, *Lepidium virginicum* (pepper grass), *Coronopus didymus* and *Sisymbrium*.

21.12.7 Systematics and Phylogeny (Affinities)

Cruciferae has been placed in order Parietales by Bentham and Hooker while Engler and Prantl placed it under order Rhoeadales. Hutchinson (1959) placed it under order Cruciales while Cronquist (1981) and Thorne (1983) placed it under order Capparales. Some taxonomists believe that Cruciferae has been derived from Papaveraceae whereas others believe it to be derived from Capparidaceae.

21.12.8 Description of a Common Plant

Brassica campestris var. sarson (Sarson; Fig. 21.19)

Stem: Herbaceous, erect, branched, smooth. *Leaf:* Alternate, exstipulate, simple, sessile or subsessile, lower leaves lyrate, upper ones ovate to lanceolate; serrate, acute, unicostate reticulate. *Inflorescence:* Racemose raceme. *Flower:* Ebracteate, ebracteolate, pedicellate, complete, actinomorphic, hermaphrodite, tetramerous, hypogynous, yellow. *Calyx:* 4 sepals, in 2 whorls of 2 each; free, slightly petaloid. *Corolla:* 4 petals, polypetalous, clawed, cruciform, valvate or imbricate. *Androecium:* 6 stamens, polyandrous, tetradynamous; in 2 whorls of outer 2 short stamens and inner 4 long stamens; dithecous, basifixed, introrse. *Gynoecium:* Bicarpellary, syncarpous, superior; unilocular but development of replum or false septum makes it bilocular; many ovules in each locule; parietal placentation; style reduced; stigma bilobed.

Floral Formula: Ebr, ⊕, ⚥, K_{2+2}, C_4, A_{2+4}, $G\underline{(2)}$.

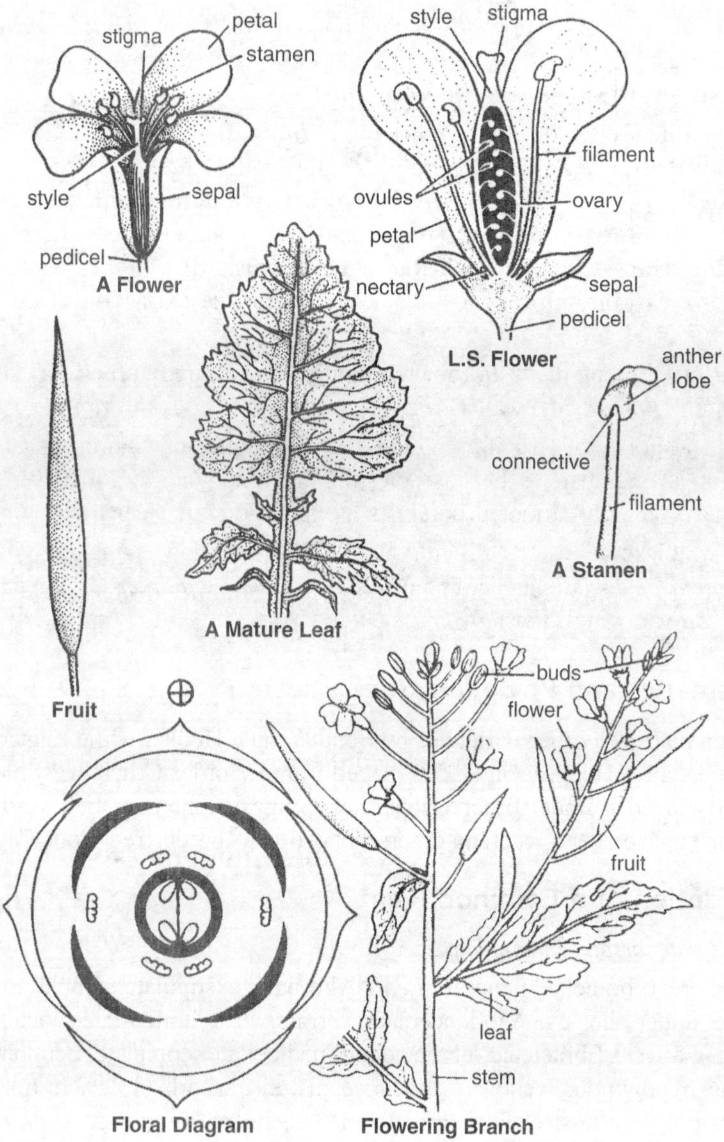

Fig. 21.19 *Brassica campestris* L. var. *sarson* Prain.

21.13 VIOLACEAE (VIOLET FAMILY)

21.13.1 Systematic Position

Polypetalae, Thalamiflorae, Parietales.

21.13.2 Field Recognition

Herbs or shrubs; pentamerous, zygomorphic flowers; spurred corolla; 1 of the 5 stamens is frequently spurred at the base; parietal placentation; fruit a capsule.

21.13.3 Indian Genera

Viola, *Hybanthus* and *Vahila*.

21.13.4 Size, Distribution and General Information

A family of about 22 genera and 900 species, Violaceae are distributed in both tropical and temperate regions. Only 3 genera (*Hybanthus*, *Vahila* and *Viola*) have been reported from India. Famous *pansies*, *violas* and *violets* belong to this family. Some of the larger genera with their approximate number of species in parenthesis include *Viola* (500), *Rinorea* (340) and *Hybanthus* (150). Majority of Indian Violaceae are found in Himalayas and hills of southern and western India.

21.13.5 Description of the Family

General Habit Usually perennial or annual herbs (*Viola*, *Hybanthus*), or shrubs (*Alsodeia*); rarely trees (*Rinorea*) or woody climbers (*Agatea*).

Root Branched, tap roots.

Stem Usually short, solid, erect, usually unbranched.

Leaves Alternate or rarely opposite (*Hybanthus*), simple; stipulate, stipules often large and leafy (some *Viola* species) or small.

Inflorescence Flowers solitary or arranged in racemose racemes (*Rinorea*) or panicles.

Flower Bracteate, invariably bracteolate, bisexual, zygomorphic (*Viola odorata*, Fig. 21.20) to actinomorphic (*Rinorea*); sometimes cleistogamous and hidden underground; pentamerous, hypogynous.

Calyx 5 sepals, persistent, polysepalous or slightly connate; uniformly sized; imbricate or quincuncial.

Corolla 5 petals, mostly unequal, polypetalous; anterior petal larger and often spurred to hold the nectar; imbricate; equal-sized petals in *Rinorea* and other arborescent genera.

Androecium 5 stamens, usually alternate with petals, free or connate, often forming a cylinder round the ovary; abaxial stamens often spurred at the base; filaments very short, connective usually prolonged at the apex, dithecous; introrse; 2 anterior stamens are spurred in *Viola* (Fig. 21.20).

Gynoecium Tricarpellary, syncarpous, superior, unilocular, ovules numerous, parietal placentation; style simple; stigma capitate, truncate or lobed.

Fruits and Seeds Fruit a loculicidal capsule or rarely a berry. Seeds endospermic; smooth, winged or with tomentum; with straight embryo; *elaiosome* is present in some species of *Viola*.

Pollination and Dispersal Pollination entomophilous in most of the Violaceae. However, cleistogamous flowers (*Viola bicolor*) are self-pollinated. Dispersal of seeds is usually by animals, birds, or by water.

General Floral Formula Br, Brl, ⦶ or ⊕, ⚥, K_5, C_5, A_5, $G\underline{(3)}$.

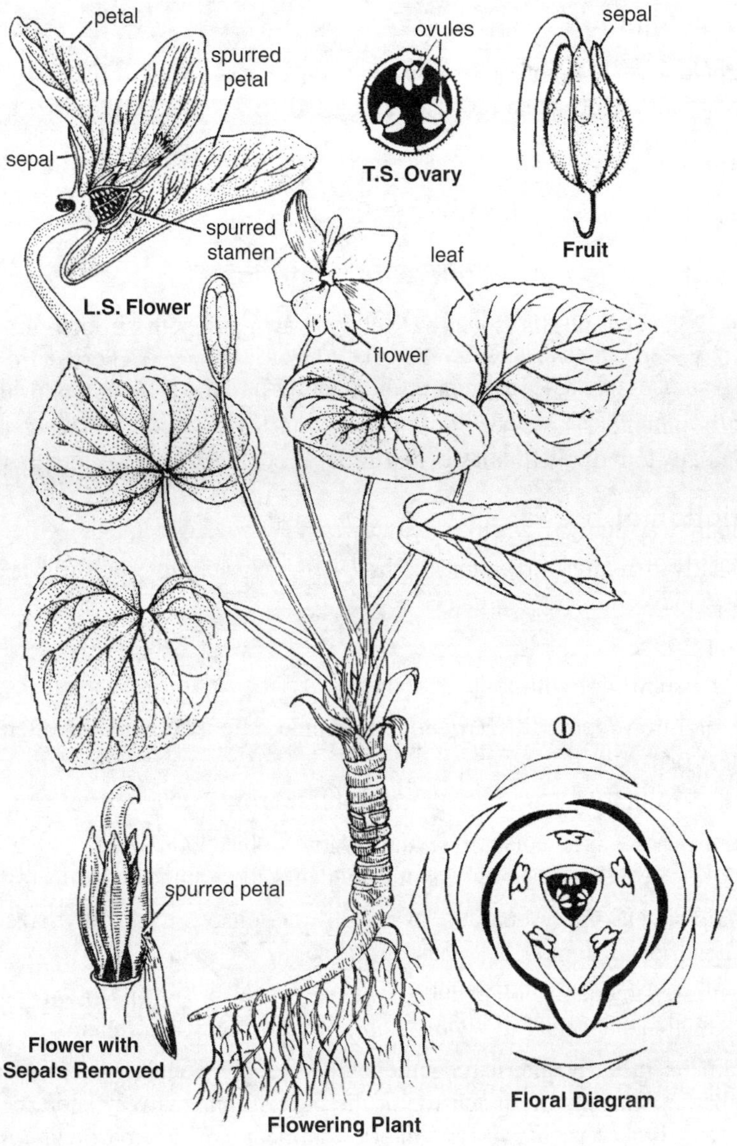

Fig. 21.20 *Viola odorata* L.

21.13.6 Economic Importance

- *Ornamentals:* Except for garden favourites, such as pansies, violas, and violets, the family is of little economic importance. Over 100 species of *Viola* are being grown as ornamental plants. Common cultivated violet is *Viola odorata* and common pansy is *V. tricolor.*

- *Medicinal:* Dried flowers of *Viola odorata* (Vern. Banafsha) are used against cough, influenza and lung troubles. Roots *of Hybanthus enneaspermus* are used against urinary tract infections and gonorrhoea, and also to cure bowel troubles of children.
- *Oil and Perfumery:* Oil obtained from the flowers of *Viola odorata* are used in perfumes and flavourings. Its leaves yield an essential oil used in scenting soaps.

21.13.7 Systematics and Phylogeny (Affinities)

Violaceae has been placed under order Parietales by Bentham and Hooker. Hutchinson (1959) placed it under Violales and opined that Violaceae are derived from Ranalian stock through Rhoeadales. Cronquist (1981) treated it as belonging to order Violales along with 23 more families while Thorne (1983) discussed Violaceae under suborder Violineae of order Violales belonging to superorder Violiflorae. Violaceae is divided into 2 subfamilies i.e. Violoideae and Leonioideae (Hickey and King, 1988).

21.14 CARYOPHYLLINEAE

21.14.1 General Characteristics

Caryophyllineae, as circumscribed by Bentham and Hooker, show following characteristics:

1. Flowers actinomorphic and bear calyx with 2–5 or rarely 6 sepals and usually as many petals.
2. Flowers show polysepalous and polypetalous conditions.
3. Stamens typically 5 or 10, or as many as petals, or twice as many as petals, arranged in 1 or 2 whorls, obdiplostemonous.
4. Ovary unilocular; numerous ovules, free-central placentation.
5. Seeds usually with curved embryo.

Cohort Caryophyllineae includes 4 orders (=families), namely Frankeniaceae, Caryophyllaceae, Portulacaceae and Tamariscineae. Only Caryophyllaceae and Portulacaceae are treated in the present text. Lawrence (1951), however, treated both Caryophyllaceae and Portulacaceae under order Centrospermae along with 8 more families, *viz.* Chenopodiaceae, Amaranthaceae, Nyctaginaceae, Phytoloccaceae, Gyrostemonaceae, Achatocarpaceae, Aizoaceae and Basellaceae.

21.15 CARYOPHYLLACEAE (PINK FAMILY)

21.15.1 Systematic Position

Polypetalae, Thalamiflorae, Caryophyllineae.

21.15.2 Field Recognition

Herbs with often swollen nodes; leaves opposite decussate; inflorescence dichasial cyme or flowers solitary; flowers obdiplostemonous; placentation free-central; fruit capsule.

21.15.3 Selected Indian Genera

Antigonon, Dianthus, Drymaria, Gypsophila, Saponaria, Silene, Spergula, Stellaria.

21.15.4 Size, Distribution and General Information

Represented by about 75 genera and over 2000 species (Jones and Luchsinger, 1987), Caryophyllaceae are cosmopolitan in distribution, and mainly found in the north temperate or warm temperate regions of the world. Over 20 genera and about 100 species have so far been reported from India. Some of the larger genera, along with their common names and number of species in parenthesis, include *Silene* (catchfly, 500), *Dianthus* (carnation, 300), *Arenaria* (sandwort, 250), *Gypsophila* (baby's breath, 125), *Stellaria* (chickweed, 100), *Cerastium* (mouse-ear chickweed, 100), *Drymaria* (50), *Spergularia* (40), *Saponaria* (30), etc.

21.15.5 Description of the Family

General Habit Mostly annual or perennial herbs, rarely suffrutescent shrubs; arctic forms are reduced and possess caespitose habit.

Stem Herbaceous; rarely with a basal woody part; nodes swollen; branching usually dichotomous but sometimes monopodial.

Leaf Simple, usually opposite decussate, rarely alternate (*Dysphania*), connected at the base by a transverse line; linear to lanceolate, entire; usually exstipulate, and if stipulate then the stipules are scarious (*Paronychia*); leaves highly reduced in arctic forms (*Lychnis*).

Inflorescence Flowers either solitary terminal (*Arenaria*) or arranged in dichasial cymes (cincinus-type).

Flower Bracteate, bracteolate (*Spergula*, Fig. 21.21A), complete, hermaphrodite but rarely unisexual (*Lychnis*); actinomorphic; hypogynous but perigynous in *Arenaria*; pentamerous but tetramerous in *Sagina*; many bracts appear like epicalyx in *Dianthus*.

Calyx 5 sepals, polysepalous (Fig. 21.21A) or united into a tube at the base (Fig. 21.21B); rarely 4 sepals (*Sagina*); quincuncial or imbricate and persistent.

Corolla 5 petals, free; sometimes 4, rarely minute or none (*Sagina*); often notched (*Cerastium*) at the tip; differentiated into a claw and limb; caryophyllaceous; petals deeply bifid in *Stellaria media* (Fig. 21.23); a ligule-like outgrowth is present on the petals in some genera of subfamily Silenoideae.

Androecium Stamens typically 5 or 10, or as many as petals, or twice as many as petals; arranged in 1 or 2 whorls; obdiplostemonous i.e. stamens of outer whorl lie opposite the petals while that of inner whorl alternate with the petals; polyandrous or basally connate; staminodia sometimes present (*Stellaria media, Arenaria*); stamens of outer whorl are absent in *Arenaria*; in *Stellaria media* some or all the stamens of outer whorl and even 2–3 of inner whorl are missing or present in the form of staminodes; only 8 stamens in 2 whorls in *Sagina*; 5 stamens of inner whorl are absent in *Paronychia*; anthers 2-celled, longitudinally dehiscent; introrse.

Floral developmental studies, however, indicate that instead of real obdiplostemony, Caryophyllaceae show *false obdiplostemony* created by the mechanical pushing-out of the inner whorl of stamens.

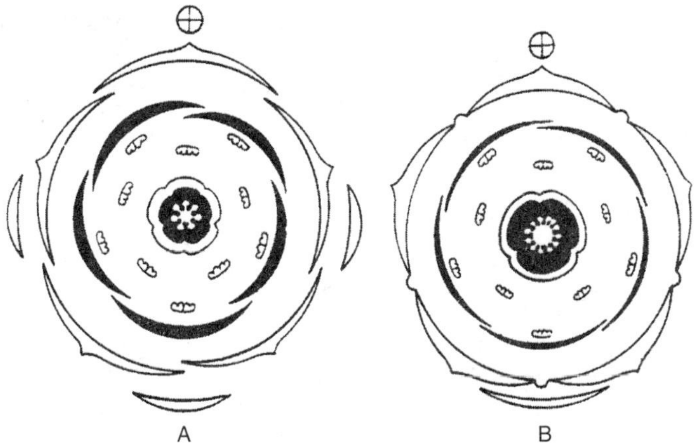

Fig. 21.21 Caryophyllaceae: Floral diagrams—A: *Spergula arvensis* and B: *Silene conoidea*.

Gynoecium Carpels 2 (*Dianthus*, *Saponaria*), 3 (*Silene*) or 5 (*Lychnis*); syncarpous; superior but slightly inferior in *Arenaria*; unilocular; numerous ovules; free-central placentation; single basal ovule in *Paronychia*; styles and stigmas as many as the carpels.

Fruits and Seeds Fruit is usually a dry capsule opening by valves or teeth; rarely in achene or utricle; very rarely a berry (*Cucubalus*). Seeds with usually curved embryo around the perisperm.

Pollination and Dispersal Pollination is entomophilous. Seeds can not escape from fruit until shaken by wind or animals.

General Floral Formula Br, Brl, ⊕, ⚥, K_5, C_5, A_{5-10}, $G\underline{(2-5)}$.

21.15.6 Economic Importance

- *Ornamental plants* Amongst several ornamentals of the family, some important ones include *Dianthus caryophyllus* (carnation), *D. berbatus* (sweet William), *D. plumarius* (pink), *D. sinensis* (rainbow pink), *Gypsophila paniculata* (baby's breath), *Silene pendula* (catchfly), *Lychnis alba* (white campion), *Saponaria vaccaria* (cowcockle), *Arenaria grandiflora*, *Cerastium arvense* and species of *Petrocoptis*.
- *Medicinal value* (i) *Dianthus anatolicus* is used to cure intermittent fever; (ii) *D. sinensis* is used as a diuretic and also against gonorrhoea; (iii) *Lychnis coronaria* is used against diseases of liver and lungs; (iv) *Silene apetala* is useful in eye troubles; (v) *Spergula arvensis* seeds are used in lung tuberculosis; (vi) *Saponaria vaccaria* sap is used against urinary bladder diseases; (vii) *Stellaria media* leaves are used against inflammations of digestive and respiratory tracts; (viii) The extract of the entire plant of *Stellaria semivestita* has anticancerous properties.
- Fodder for cattles is obtained from *Spergula arvensis* and several other species of this family.

- Roots of *Saponaria vaccaria* (syn. *Vaccaria pyramidata*), *Gypsophila*, etc. contain saponin and used for washing wood and silk.
- Perfume is obtained from the flowers of *Dianthus caryophyllus.*
- *Drymaria cordata* is used to check soil erosion.
- Troublesome weeds of the family include *Stellaria media*, *Cerastium*, *Silene conoidea*, *Saponaria vaccaria* and *Polycarpaea*.

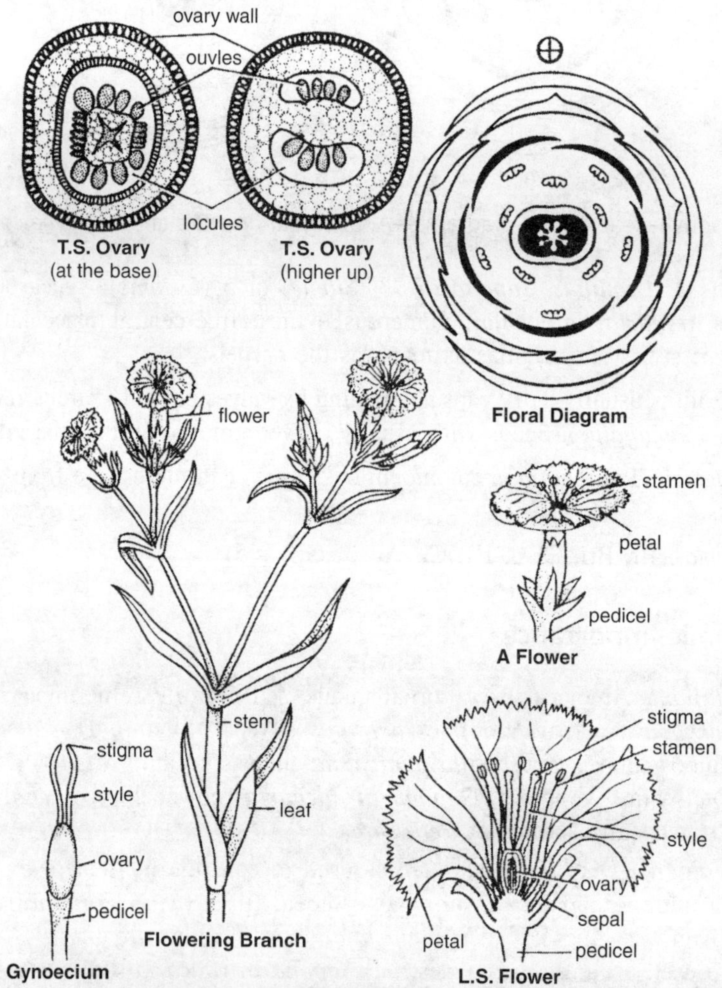

Fig.21.22 *Dianthus caryophyllus* L.

21.15.7 Systematics and Phylogeny (Affinities)

This family belonging to order Caryophyllales (Bentham and Hooker) is usually classified into two subfamilies *viz.* Alsinoideae (with free sepals) and Silenoideae (with fused sepals). Tutin *et al.*

(1964) divided it in their *Flora Europaea* into three subfamilies, i.e. Alsinoideae, Paronychioideae and Silenoideae. Cronquist (1981) placed it under order Caryophyllales along with 11 more families whereas Thorne (1983) discussed Caryophyllaceae under order Chenopodiales of Centrospermae.

Usually, Caryophyllaceae is considered to have been derived from Ranales. However, Eichler considered it to have originated from Phytolaccaceae. Dickson (1936) believed that Caryophyllaceae probably originated from Geraniaceae.

21.15.8 Description of Some Common Plants

1. Dianthus caryophyllus (Carnation, Fig. 21.22)

Habit: Annual, ornamental herb. *Stem:* Aerial, erect, solid, herbaceous. *Leaf:* Simple, exstipulate, opposite decussate, sessile, united at the base, lanceolate, entire, acute, unicostate reticulate. *Inflorescence:* Solitary terminal or dichasial cyme. *Flower:* Multibracteate, bracts appear like epicalyx; pedicellate, complete, hermaphrodite, actinomorphic, hypogynous, pentamerous. *Calyx:* 5 sepals, gamosepalous, tubular, quincuncial or valvate. *Corolla:* 5 petals, polypetalous, caryophyllaceous, twisted. *Androecium:* 10 stamens, arranged in 2 whorls of each, polyandrous, obdiplostemonous; anthers dithecous, dorsifixed, introrse. *Gynoecium:* Bicarpellary, syncarpous, superior; ovary unilocular at base but bilocular at top; ovules many, free-central placentation; styles 2, long; stigma simple. *Fruit:* Capsule.

Floral Formula: Br, ⊕, ⚥, $K_{(5)}$, C_5, A_{5+5}, $G\underline{(2)}$.

2. Stellaria media (Chickweed, Fig. 21.23)

Habit: Annual, herbaceous common weed. *Leaf:* Simple, opposite decussate, exstipulate, ovate or oblong; entire, acute, unicostate reticulate. *Inflorescence:* Dichasial cyme. *Flower:* Bracteate, ebracteolate, rarely 2 bracteoles are present; pedicellate, complete, actinomorphic, hermaphrodite, pentamerous, hypogynous. *Calyx:* 5 sepals, free, quincuncial, green. *Corolla:* 5 petals, free, petals deeply bilobed appearing as if they are 10; imbricate or valvate. *Androecium:* Stamens 10 or reduced to 5 or 8, leaving the rest abortive in the form of staminodes; present in 2 whorls; obdiplostemonous; anthers dithecous, basifixed, introrse. *Gynoecium:* Tricarpellary, syncarpous, superior, unilocular, numerous ovules, free-central placentation; styles 3 but reduced; stigmas 3. *Fruit:* Capsule.

Floral Formula: Br, ⊕, ⚥, K_5, C_5, $A_{10 \text{ or } 5-8}$, $G\underline{(3)}$.

21.16 PORTULACACEAE (PURSLANE FAMILY)

21.16.1 Systematic Position

Polypetalae, Thalamiflorae, Caryophyllineae.

21.16.2 Field Recognition

Annual or perennial herbs with usually fleshy leaves; scarious stipules; sepals 2, persistent; petals 4–6; ovary unilocular; basal placentation.

21.16.3 Selected Indian Genera

Portulaca, Talinum.

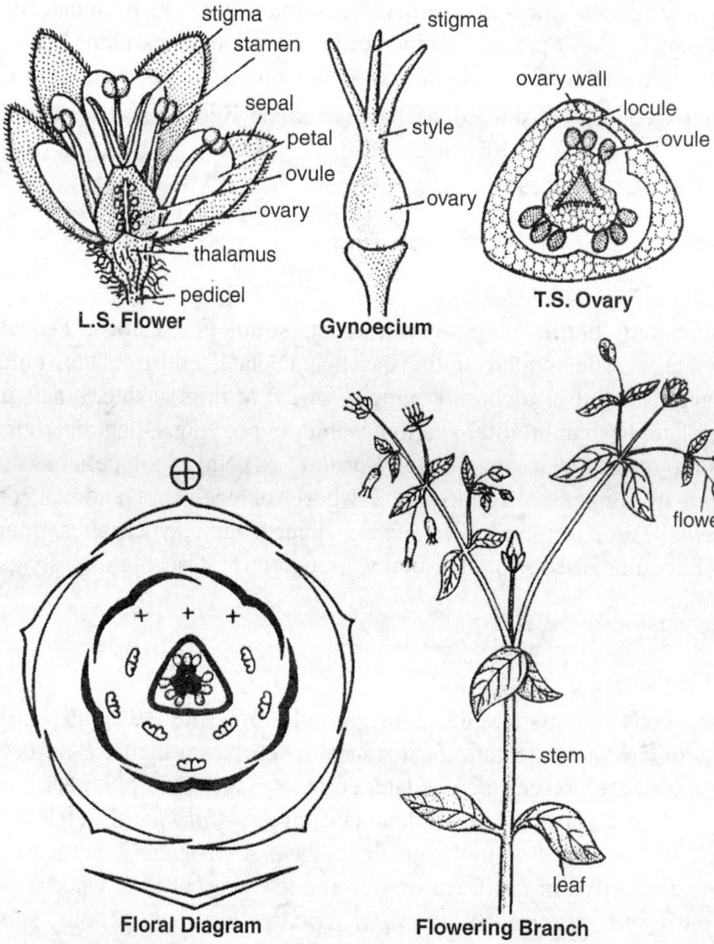

Fig. 21.23 *Stellaria media* L.

21.16.4 Size, Distribution and General Information

Portulacaceae is represented by about 20 genera and 500 species (Jones and Luchsinger, 1987). Though cosmopolitan, the members are mainly distributed in western North America and southern South America. Only 7 species of Portulacaceae have so far been reported from India. Some larger genera of the family along with their number of species in bracket include *Portulaca* (purslane, 200), *Calandrinia* (red maids, 150), *Anacampseros* (70), *Talinum* (50), *Claytonia* (spring beauty, 35), *Lewisia* (20) and *Montia* (15).

21.16.5 Description of the Family

General Habit Mostly annual, sometimes perennial herbs or suffrutescent shrubs.

Root Branched, tap root; fleshy in *Lewisia*.

Leaves Alternate or opposite; often succulent, simple; stipulate, stipules sometimes in the form of hairs or scales; exstipulate in *Claytonia*.

Inflorescence Usually cymose, or racemose, or solitary flowers.

Flower Bracteate or ebracteate, bisexual, actinomorphic, showy, complete, hypogynous or perigynous.

Calyx Sepals usually 2, often persistent, free or united at the base. Some regard sepals as bracteoles, and according to them the flowers have petaloid perianth.

Corolla 4 to 6 petals, free or united at the base; caducous; usually imbricate, but valvate in *Claytonia*; showy.

Androecium 4 to many, free stamens; usually 5 stamens, opposite to petals; anthers dithecous, introrse, dehiscing longitudinally; stamens 2 in *Montia* and 8–10 in *Portulaca oleracea*.

Gynoecium Usually tricarpellary, syncarpous, superior but partly inferior in *Portulaca*; unilocular; 2 to many campylotropous ovules on a central basal placenta; styles 2–5; stigmas 2–5 or as many as carpels.

Fruit and Seed Fruit usually a dehiscent capsule; rarely a nut and indehiscent. Seeds with embryo curved around the perisperm.

Pollination and Dispersal Pollination entomophilous and seeds are dispersed by animals or rains.

General Floral Formula Br or Ebr, ⊕, ⚥, K_2 or (2), C_{4-6}, $A_{4-\infty}$, $G\underline{(3)}$.

21.16.6 Economic Importance

Portulacaceae is of little economic importance except of some well-known garden ornamentals, such as *Portulaca grandiflora* for its pink, red, yellow or white flowers, *P. afra* for its rose-coloured flowers, and *P. oleracea* for its showy flowers. Other ornamental plants of the family include several species of *Calandrinia*, *Talinum* and *Lewisia*. *Portulaca oleracea* is sometimes used as a pot herb and in salads and also in the ailments of kidney and urinary bladder. *P. quadrifida* proves effective in cough and asthma.

21.16.7 Systematics and Phylogeny (Affinities)

Portulacaceae belongs to order Caryophyllineae (Bentham and Hooker). Cronquist (1981) discussed it under Caryophyllales. However, Thorne (1983) placed it under suborder Portulacineae of order Chenopodiales of Centrospermae.

Portulacaceae is related closely to the Cactaceae and Aizoaceae of order Caryophyllales (Hickey and King, 1988).

Recent anatomical findings of the family suggest that its 2 sepals are actually the bracts and its petals are thus the sepals.

21.16.8 Description of a Common Plant

1. Portulaca oleracea L. (Common Purslane; Fig. 21.24)

Habit: Herb with prostrate or ascending branches. *Stem:* Aerial, branched, succulent. *Leaves:* Simple, opposite or alternate, stipulate, sessile or subsessile, apex obtuse or truncate, entire. *Inflorescence:* Flowers solitary or in terminal clusters. *Flower:* Ebracteate, complete, hermaphrodite, actinomorphic, perigynous. *Calyx:* 2 sepals, free, anterio-posterior. *Corolla:* 5 petals, free, quincuncial. *Androecium:* 8–12 stamens, usually 10, polyandrous, in one whorl, adnate, dithecous, introrse. *Gynoecium:* Tri- to pentacarpellary, syncarpous, half-inferior, unilocular, many ovules, free-central placentation; styles 3–5. *Fruit:* Capsule.

Floral Formula: Ebr, ⊕, ⚥, K_2, C_5, $A_{8-12 \text{ or } 10}$, $G_{(3-5)}$.

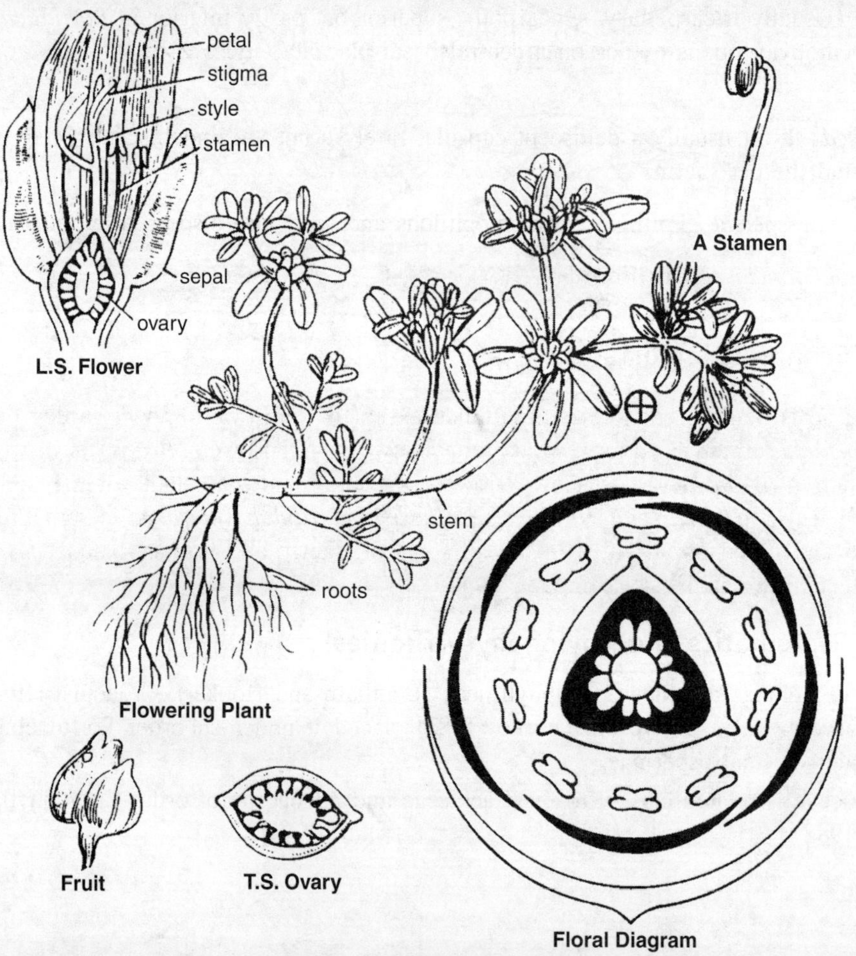

Fig. 21.24 *Portulaca oleracea* L.

21.17 MALVALES

21.17.1 General Characteristics

1. Flowers bisexual and actinomorphic, and only rarely zygomorphic.
2. Flowers usually pentamerous with their calyx valvate.
3. Petals usually as many as sepals; epicalyx usually present.
4. Stamens numerous and usually monadelphous.
5. Ovary tri- to multicarpellary.
6. Placentation usually axile.

Cohort Malvales of Bentham and Hooker includes 3 orders = (families) namely Malvaceae, Sterculiaceae and Tiliaceae. Engler and Diels, however, treated the order Malvales to be composed of 4 suborders and 8 families as under:

1. *Elaeocarpineae:* Elaeocarpaceae
2. *Chlaeneae:* Chlaenaceae
3. *Malvineae:* Tiliaceae, Malvaceae, Bombacaceae, Sterculiaceae
4. *Scytopetalineae:* Scytopetalaceae

Hutchinson, however, included only Malvaceae under order Malvales, and treated Sterculiaceae, Bombacaceae and Tiliaceae in a new order Tiliales.

Tiliaceae, Malvaceae, Sterculiacene and Bombacaceae have been treated in this text.

21.18 MALVACEAE (MALLOW FAMILY)

21.18.1 Systematic Position

Polypetalae, Thalamiflorae, Malvales.

21.18.2 Field Recognition

Shrubs or herbs, frequently with stellate trichomes; sepals persistent; monadelphous stamens; anthers monothecous; carpels 5 or more.

21.18.3 Selected Indian Genera

Gossypium, *Hibiscus*, *Althaea*, *Malvastrum*, *Sida*, *Malva*, *Abutilon*.

21.18.4 Size, Distribution and General Information

A family of about 85 genera and 1,000 to 1,500 species, Malvaceae are distributed widely in tropical and temperate regions. 22 genera and about 125 species of Malvaceae have so far been reported from India. Some of the larger genera, along with their common names and/or number of reported species in parenthesis, include *Hibiscus* (rose mallow, 300), *Sida* (200), *Pavonia* (200), *Abutilon* (Indian mallow, 100), *Alcea* (60), *Malva* (mallow, 40), *Lavatera* (25), *Gossypium* (cotton, 20), and *Althaea* (12).

The family is recognized by *Hibiscus rosa-sinensis* (rose of China) because of its beautiful large flowers and hundreds of its known cultivated varieties. Cotton (*Gossypium*), the most important plant of this family from the commercial viewpoint, has been cultivated in India since last 5,000 years.

21.18.5 Description of the Family

General Habit Mostly annual (*Althaea*, *Malva*) or perennial herbs, sometimes growing very tall (*Lavatera*); shrubs (*Gossypium*, *Hibiscus rosa-sinensis*); or trees (*Hibiscus elatus*, *Thespesia populnea*, *Kydia*); often with mucilaginous sap.

Root Profusely branched, tap roots.

Stem Erect, branched, sturdy; young portions often covered with stellate trichomes; decumbent in *Malva rotundifolia.*

Leaf Simple, often palmately lobed and palmately veined; alternate; stipulate, stipules often deciduous; petiolate; ovate, cordate or reniform; margin toothed or entire; venation reticulate and multicostate divergent.

Inflorescence Usually cymose or flowers solitary in the leaf axil (*Hibiscus rosa-sinensis*); raceme in *Althaea rosea.*

Flower Bracteate; bracteolate, bracteoles often form calyx-like *epicalyx*; complete, bisexual, actinomorphic, hypogynous, pentamerous; flowers rarely unisexual and the species are dioecious (*Napaea*); variously coloured.

Epicalyx Outside the calyx, several bracteoles often form a calyx-like whorl called *epicalyx* (Fig. 21.27). Epicalyx consists of 3 bracteoles (*Malvastrum tricuspidatum*, *Malva*), or 5–9 bracteoles (*Althaea rosea*), or several bracteoles (*Hibiscus radiatus*); bracteoles are free (*Malvastrum*) or fused (*Althaea*); epicalyx or involucre of bracteoles is absent in *Abutilon* (Fig, 21.26) and *Sida* (Fig, 21.25).

Fig. 21.25 Floral diagram of *Sida cordifolia.*

Calyx 3 to 5 but usually 5 sepals, free or basally connate, valvate; tubular in *Hibiscus.*

Corolla 5 petals, free but basally adnate to the staminal column; usually twisted but sometimes imbricate; showy and variously coloured.

Androecium Numerous stamens, monadelphous; anthers 1-celled or monothecous; reniform; extrorse; dehiscence longitudinal; filaments of anthers fuse to form a staminal column around the ovary.

Gynoecium A compound pistil of 1 to many carpels (often pentacarpellary, syncarpous); ovary superior; 2 to 5 or more locules, rarely 1 locule; 1 to numerous ovules in each locule; axile placentation; style 1 and branched above, or as many as the carpels; stigmas as many or twice as many as carpels, discoid or capitate.

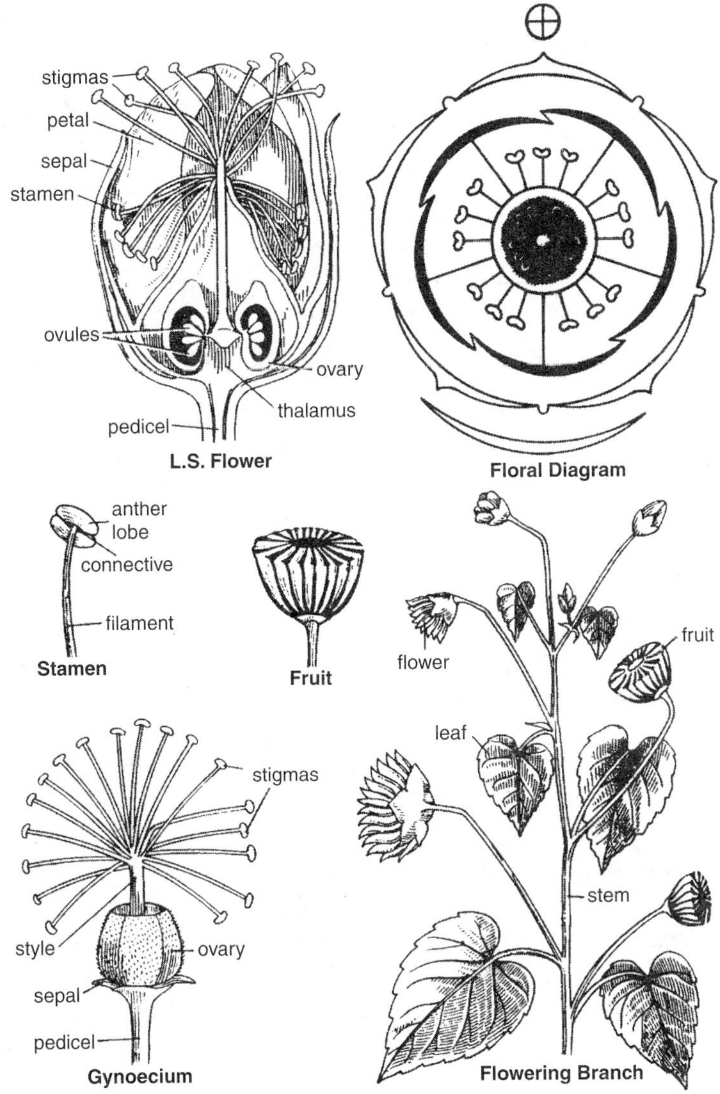

Fig. 21.26 *Abutilon indicum Sweet.*

Fruit and Seed Fruit usually a capsule (*Gossypium*) or schizocarp (*Malva*), or rarely a berry (*Malvaviscus*) or samara. Seeds with curved embryo and scanty or no endosperm.

Pollination and Dispersal Majority of Malvaceae are insect pollinated mainly because of their large-sized flowers, brightly coloured corolla and extrafloral nectaries (in some). Dispersal of seeds takes place mainly by wind (*Gossypium*), or sometimes by water (*Malva*), or by animals because of the presence of hooked spines (*Urena lobata*).

General Floral Formula Br, $\oplus$, ⚥, $Epik_{3-9 \text{ or } (3-9)}$, $K_{(5)}$, $\widehat{C_{5 \text{ or } (5)}, A}(\infty)$, $G(\underline{2-5})$ or $(\underline{1-\infty})$.

21.18.6 Economic Importance

- *Cotton:* Cotton, manufactured from the dense mass of long hairs attached to the seeds of several species of *Gossypium*, is the most important product of this family. Cotton comes from *Gossypium barbadense, G. hirsutum*, G. *arboreum* and *G. herbaceum.*
- *Other Fibres:* *Hemps*, which are the bast fibres (i.e. obtained from the phloem fibres of the stem), are obtained from *Hibiscus elatus* (Cuba bast), *H. sabdariffa* (Rozella hemp), *H. cannabinus* (Deccan hemp), *Sida rhombifolia* (Queensland hemp), etc. Bags, ropes, cords, etc. are made from the fibres obtained from *Abutilon asiaticum, A. theophrasti, Hibiscus falcatus, H. tiliaceous, Abelmoschus moschatus, Urena lobata*, etc.
- *Oils:* (i) Cotton seed oil, which is edible and also used for the manufacture of soaps, lubricants, etc., is obtained from the seeds of *Gossypium*. The oil cake is an excellent cattle feed. (ii) An essential oil, used in perfumery, is obtained from the seeds of *Abelmoschus moschatus*. (iii) A fatty oil, obtained from the seeds of 'Patsan' (*Hibiscus cannabinus*), is used in the manufacture of paints, varnishes and linoleum, and its refined oil is edible.
- *Ornamentals:* Malvaceae are known universally for their ornamental plants, chief among which include *Hibiscus rosa-sinensis* (China rose or shoe flower), *H. mutabilis* (cotton rose), *H. Schizopetalus, H. syriacus* (Rose of Sharon), *H. sabdariffa, H. collinus, Althaea rosea* (Hollyhock), *Malva sylvestris* (Mallow), *Malvaviscus arboreus, Thespesia populnea*, etc.
- *Vegetables:* (i) Lady's finger or okra, a famous vegetable, is the fruit of *Abelmoschus esculentus* (vern. Bhindi); (ii) a delicious 'chutney' is prepared from the sepals of *Hibiscus cannabinus* and *H. sabdariffa*; (iii) flower buds of *Hibiscus syriacus* are acidic and used as vegetable.
- *Timber:* Wood of *Thespesia populnea* is used in making boats while that of *Hibiscus elatus* is utilized in furniture making.
- *Medicinal Value:* (i) Roots of *Abutilon indicum* are used against fever while that of *Abelmoschus moschatus* in stomach ache, and that of *Althaea officinalis* and *Hibiscus rosa-sinensis* against cough; (ii) roots of *Althaea rosea* are used in the treatment of dysentry; (iii) bark of *Gossypium* is used for stopping haemorrage in ladies, specially after child birth; (iv) roots of *Malva verticillata* are used against whooping cough; (v) leaves and seeds of *Malva sylvestris* are used in fever; (vi) roots of *Malachra capitata* are used in rheumatism; (vii) roots and bark of *Urena repanda* are used to cure hydrophobia, a dreaded disease caused by bite of mad dog.
- *Some Other Uses:* (i) 'Hina', a well-known Indian perfume, is obtained from the roots of *Pavonia odorata*; (ii) toys, pencils and matchsticks are manufactured from the wood of *Thespesia populnea*; (iii) a blue dye is obtained from the leaves of *Althaea rosea*; (iv) a well-known drink is prepared from the calyx of *Hibiscus sabdariffa*.

21.18.7 Systematics and Phylogeny (Affinities)

Malvaceae, placed under order Malvales by Bentham and Hooker, has been retained under the same order (Malvales) by almost all recent taxonomists including Cronquist (1981) and Thorne (1983). Schumann (1964) divided it into 4 tribes viz. Malopeae, Malveae, Ureneae and Hibisceae.

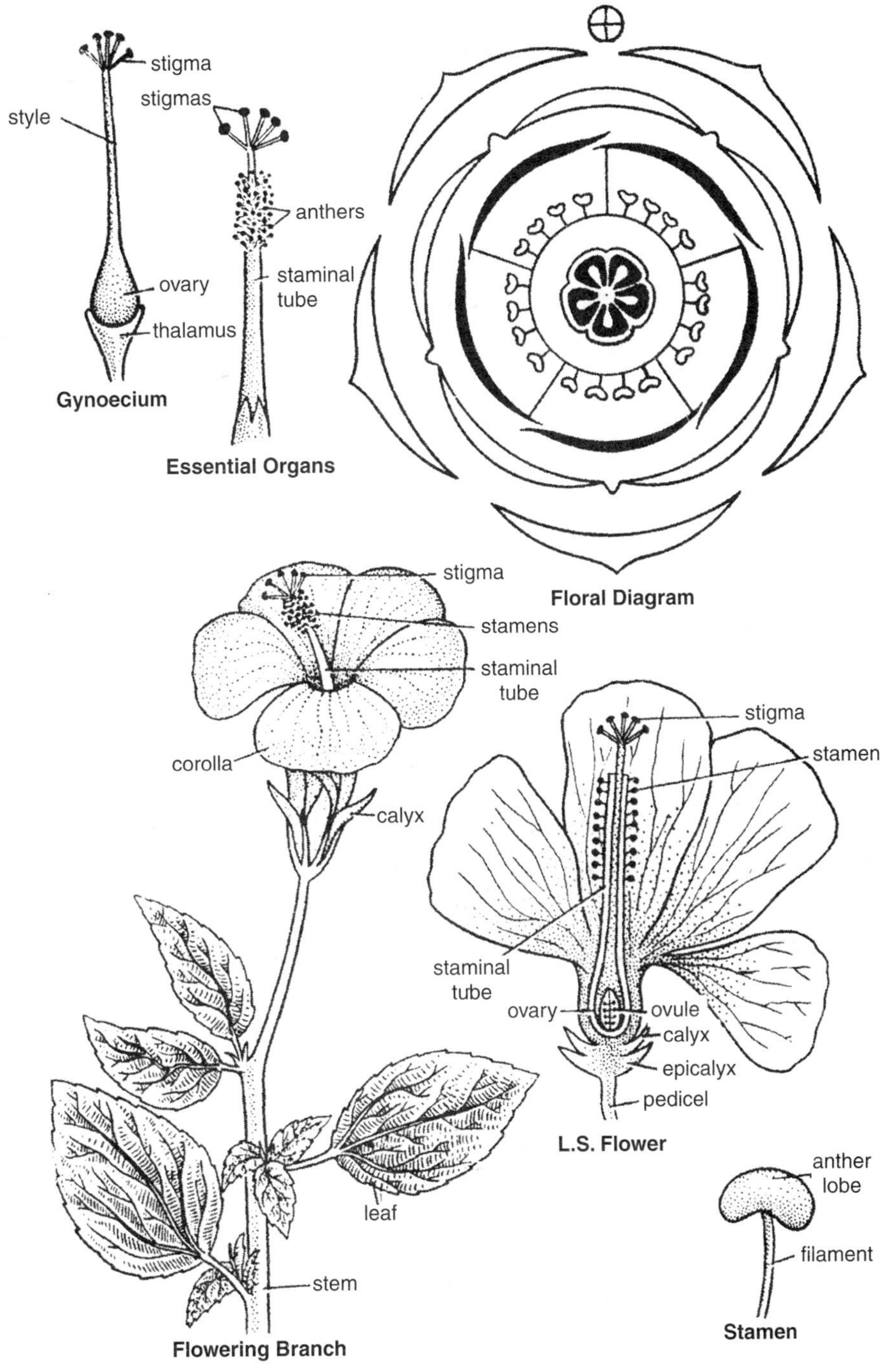

Fig. 21.27 ◆ *Hibiscus rosa-sinensis* L.

Malvaceae are related to Bombacaceae but members of the former bear monadelphous androecium while that of the latter bear polyadelphous condition.

In several respects Malvaceae also show affinities with Sterculiaceae and Tiliaceae. However, the former bears monothecous anthers whereas the members of Sterculiaceae and Tiliaceae possess dithecous anthers.

21.18.8 Description of Some Common Plants

1. Abutilon indicum Sweet. (Vern. Kanghi, Fig. 21.26)

Habit: Annual herb. *Stem:* Erect, herbaceous, branched, solid, hairy. *Leaf:* Simple, alternate, petiolate, stipulate, serrate, acute, multicostate reticulate. *Inflorescence:* Solitary axillary. *Flower:* Bracteate, ebracteolate, pedicellate, complete, hermaphrodite, actinomorphic, pentamerous, hypogynous. *Calyx:* 5 sepals, gamosepalous, valvate. *Corolla:* 5 petals, free but slightly united at the base and adnate to staminal tube, twisted. *Androecium.* Numerous monadelphous stamens forming staminal tube around style, monothecous, extrorse. *Gynoecium:* Polycarpellary, syncarpous, superior, multilocular, one ovule in each locule, axile placentation; style long; stigmas as many as carpels. *Fruit:* Capsule.

Floral Formula: Br, Ebrl, ⊕, ⚥, K(5), $\widehat{C_{5 \text{ or } (5)}, A}(\propto)$, $G(\underline{\propto})$.

2. Hibiscus rosa-sinensis L. (Chinese-rose or Gurhal, Fig. 21.27)

Habit: Typical ornamental shrub of the family. *Leaf:* Simple, alternate, petiolate, stipulate, ovate, serrate, acute. *Inflorescence:* Solitary axillary. *Flower:* Ebracteate, complete, hermaphrodite, actinomorphic, large, red, showy. *Epicalyx:* 5–8 bracteoles around the calyx form epicalyx. *Calyx:* 5 sepals, fused, valvate. *Corolla:* 5 petals, free but adnate to the staminal tube, twisted, red, petals with sinuous margins. *Androecium:* Same as that of *Abutilon*. *Gynoecium:* 5 fused carpels, superior ovary, pentalocular, one to many ovules in each locule, axile placentation; styles 5, long, united below and passing through the staminal tube; stigmas 5 and discoid. *Fruit:* Capsule.

Floral Formula: Ebr, ⊕, ⚥, $Epik_{5-8}$, $K_{(5)}$, $\widehat{C_5, A}(\underline{\propto})$, $G(\underline{5})$.

21.19 STERCULIACEAE (STERCULIA FAMILY)

21.19.1 Systematic Position

Polypetalae, Thalamiflorae, Malvales.

21.19.2 Field Recognition

Generally shrubs or trees with stellate hairs; leaves alternate, stipulate; sepals and petals 5, sometimes petals modified or absent; stamens generally in 2 whorls, of which outer whorl is reduced to staminodes; gynoecium generally pentacarpellary, syncarpous; some bear androgynophore.

21.19.3 Selected Indian Genera

Abroma, Helicteres, Sterculia, Theobroma, Pterospermum, Melochia.

21.19.4 Size and Distribution

Represented by about 60 genera and 700 species, Sterculiaceae are distributed exclusively in tropical and subtropical parts of the world. About 18 genera and 90 species of this family have been reported from India.

21.19.5 Description of the Family

General Habit Usually shrubs or trees with soft wood; rarely herbs with mucilage (*Waltheria*, *Melochia*); sometimes lianous (*Buettneria*).

Stem Well-branched, woody, with its younger parts covered with stellate hairs.

Leaf Simple, alternate or subopposite; stipulate, stipules caducous; pulvinous petiole; entire or palmately lobed (Fig. 21.28) or compound; covered with stellate hairs.

Inflorescence Axillary or terminal compound cymose (Fig. 21.28); sometimes cauliflorous (*Theobroma*) i.e. flowers borne on stem.

Flower Bracteate, bracteolate (*Melochia*), bisexual, sometimes unisexual (*Cola*); actinomorphic, hypogynous, pentamerous; androgynophore present in some (*Mansonia*).

Calyx 3–5 sepals, free or basally connate, hairy, petaloid (*Sterculia*), valvate.

Corolla 5 petals, but occasionally absent in *Sterculia* (Fig. 21.28), polypetalous; hooded in *Theobroma*; twisted or imbricate.

Androecium Usually 5 to 10 stamens, free or monadelphous; in two whorls, of which outer whorl of stamens is antisepalous and reduced to staminodes or scales or even absent, and filaments of inner whorl of stamens connate (monadelphous); anthers 2-celled; dehiscence longitudinal; stamens are 20 in *Pentapetes* and *Pterospermum*, of which 5 antipetalous stamens are reduced to staminodes.

Gynoecium Bi- to pentacarpellary, syncarpous, superior ovary raised over an androgynophore in *Cola*, *Mansonia*; bi- to pentalocular; 2 to many ovules in each locule; axile placentation; monocarpellary and unilocular condition is seen in *Waltheria*; style simple or divided into lobes, each lobe terminates into a stigma.

Fruit and Seed Fruit is a capsule (*Pterospermum*), follicle (*Helicteres*), or drupe (*Theobroma*). Seeds are arillate, often winged, with straight or curved embryo and abundant endosperm.

Pollination and Dispersal Pollination generally entomophillous, and seeds are dispersed generally by wind and also sometimes by animals.

General Floral Formula Br, Brl, ⊕, ⚥, $K_{3-5 \text{ or } (3-5)}$, C_5, $A_{5+(5)}$, $G_{\underline{(2-5)}}$.

21.19.6 Economic Importance

- Chocolate, cocoa and cocoa-butter are made from the seeds of *Theobroma cacao*, which is rich in alkaloids thein and theobromine.

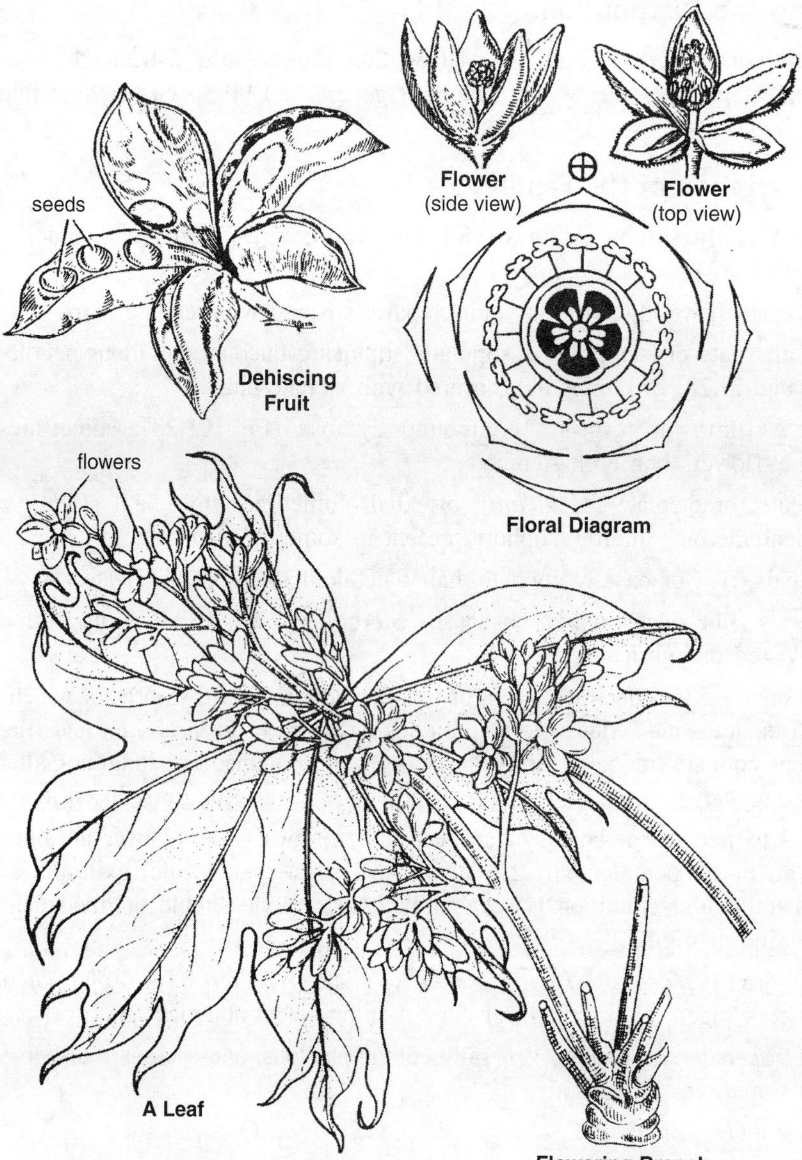

Fig. 21.28 *Sterculia villosa* Roxb.

- Soft drinks are prepared from the cola nuts (*Cola acuminata*), which contain thein and caffeine.
- Fruits of *Guazuma* and seeds of *Sterculia* are edible whereas valuable fibres are obtained from the bark of *Sterculia urens*, *S. villosa*, *Helicteres isora* and *Guazuma ulmifolia*.
- Wood of *Heritiera minor* is utilised for domestic purposes.

- Flowers of *Pterospermum acerifolium* are used as disinfectant and insect-repellant.
- Ornamental plants of Sterculiaceae, grown usually in the gardens, include species of *Abroma, Brachychiton, Dombeya, Firmiana, Guazuma, Pentapetes, Pterospermum* and *Reevesia*.
- Useful timber, utilized for toys, match-sticks, tea boxes, etc. is obtained from species of *Eriolaena candllei, Heritiera littoralis, Petrospermum diversifolium, Sterculia foetida* and *S. villosa*.
- Gum of commercial importance is obtained from several species of *Sterculia*.
- Diabetes is cured by the juice of the roots of *Helicteres isora*, whereas roasted seeds of *Guazuma ulmifolia* are used in stomach troubles. Fruits and seeds of *Sterculia balanghas* have laxative properties.

21.19.7 Systematics and Phylogeny (Affinities)

Placed under order Malvales by most of the taxonomists (Bentham and Hooker, Engler and Prantl, Takhtajan, Cronquist, Thorne, etc.), Sterculiaceae are treated under Tiliales by Hutchinson. This family differs from Malvaceae and Tiliaceae in the absence of epicalyx and presence of definite number of stamens. Because of the presence of characters, such as stellate hairs, tendency of unisexuality, and arillate seeds, Sterculiaceae show affinities with Euphorbiaceae. However, Sterculiaceae are generally considered to have been derived from the same stock as the Tiliaceae.

21.20 TILIACEAE (LIME FAMILY OR BASSWOOD FAMILY)

21.20.1 Systematic Position

Polypetalae, Thalamiflorae, Malvales.

21.20.2 Field Recognition

Woody trees or shrubs with strong phloem fibres, rarely herbs; leaves alternate, simple, stipulate; flowers bisexual, actinomorphic, usually pentamerous, arranged in cymes or panicles; stamens 5 to many, often polyadelphous; axile placentation; fruit drupe or capsule.

21.20.3 Selected Indian Genera

Tilia, Corchorus, Triumfetta, Grewia.

21.20.4 Size, Distribution and General Information

Tiliaceae includes about 50 genera and 550 species distributed in tropical and temperate regions, mainly in South East Asia and Brazil. *Tilia* is mainly distributed in temperate regions. About 15 genera and over 100 species of Tiliaceae are found in India. Some of the larger genera of the family, along with the number of their approximately reported species in parenthesis, are *Grewia* (150), *Triumfetta* (150), *Corchorus* (100), *Tilia* (50) and *Sparmannia*. *Grewia subinaequalis* (Phalsa), *Corchorus capsularis* (*Jute*) and *Elaeocarpus ganitrus* (Rudraksh) are some of the well-known species of Tiliaceae.

21.20.5 Description of the Family

General Habit Usually trees (*Tilia*) or shrubs (*Grewia*), rarely herbs (species of *Triumfetta*, *Corchorus*) or climbers (*Plagiopteron*).

Stem Woody, branched, contains strong phloem fibres; often with stellate pubescence.

Leaf Usually alternate, simple, stipulate; rarely opposite (*Plagiopteron*); petiolate; margin dentate or serrate; covered with stellate or branched hairs; leaf blade often oblique with larger side towards the branch.

Inflorescence Cymose; rarely axillary racemes (*Elaeocarpus*); greatly condensed dichasial cymes in *Corchorus*.

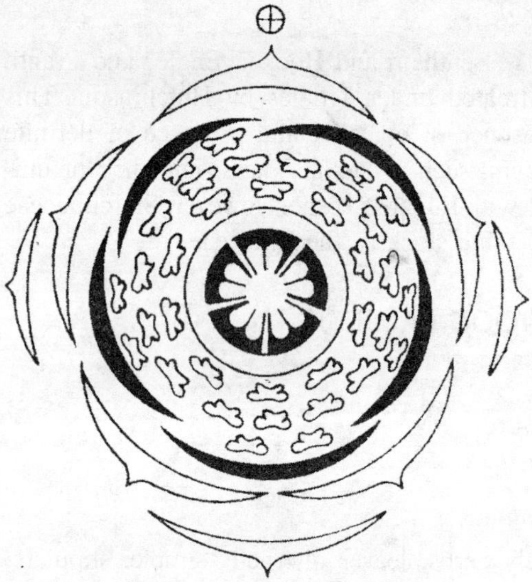

Fig. 21.29 Floral diagram of *Tilia*.

Flower Bracteate; bracteolate (Fig. 21.29), bracteoles often appear like epicalyx; pedicellate, complete, hermaphrodite but rarely unisexual; actinomorphic, hypogynous; usually pentamerous; sometimes with androgynophore (*Duboscia*, *Grewia*).

Calyx Usually 5 sepals, rarely 3 or 4; polysepalous but gamosepalous in *Chartacalyx*; generally hairy, valvate; nectar-secreting glands at the base of sepals in *Tilia*.

Corolla Usually 5 petals, rarely less or even absent (*Chartacalyx*); polypetalous, imbricate; usually yellow coloured but rarely sepaloid (*Elaeocarpus*); often glandular at base.

Androecium Usually 10 or more stamens, free or often united by the basal part of their filaments into 5 or 10 bunches (polyadelphous); some of the stamens may modify into petaloid staminodes (*Pentace*); anthers 2-celled, dehiscing by longitudinal slits or apical pores; in *Grewia* androgynophore is present.

Gynoecium Bi- to multicarpellary, syncarpous, superior; 2 to many loculed, one to many ovules in each locule, axile placentation; style one dividing into as many stigmas as number of carpels; in *Elaeocarpus*, septa in the ovary get dissolved to make it unilocular.

Fruit and Seed Fruit variable, fleshy or dry, and dehiscent or indehiscent; fruit usually a capsule (*Corchorus*, Fig. 21.30) or schizocarp. Seeds with straight embryo; endosperm copious or scanty.

Pollination and Dispersal Pollination entomophilous, usually by flies or bees; entomophily is favoured because of protandry and presence of nectar or scent secreted in the spoon-like glands at the base of the sepals in some (*Tilia*).

General Floral Formula Br, Brl, ⊕, ⚥, K_5, C_5, $A_{10} - \infty$, $G(\underline{2 - \infty})$.

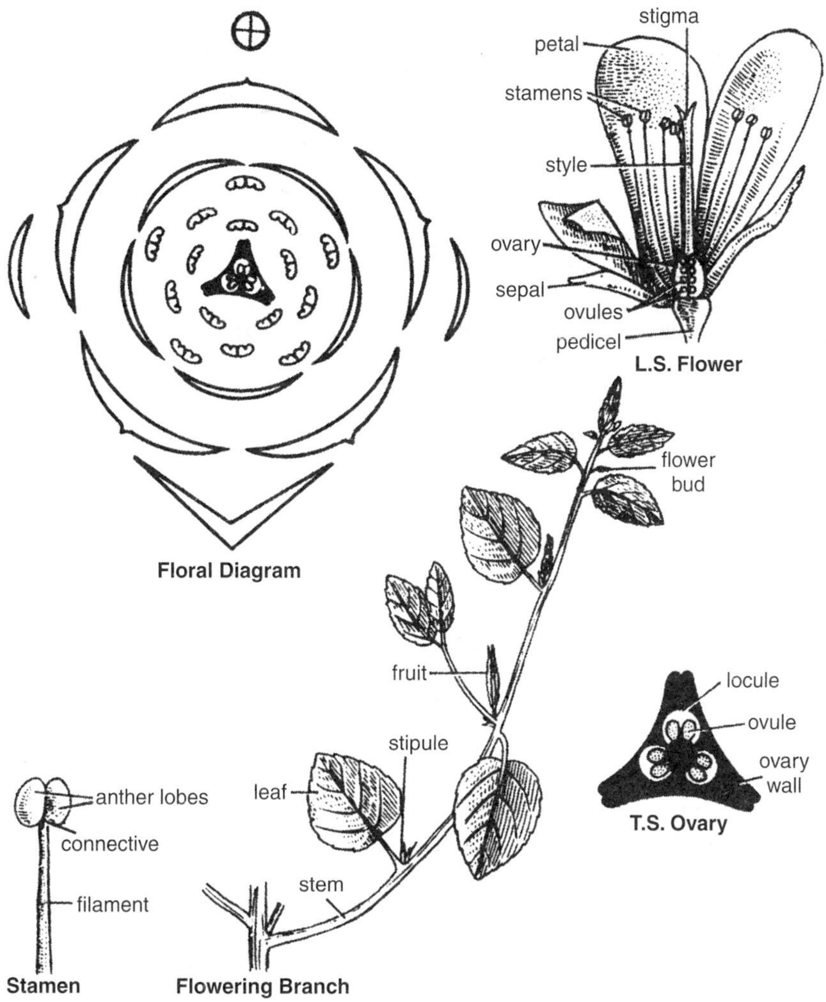

Fig. 21.30 *Corchorus aestuans* L.

21.20.6 Economic Importance

- *Jute* Two species of *Corchorus* (*C. capsularis* and *C. olitorius*) are grown widely in warm areas as the source of the jute fibre. Jute is obtained from the bast or phloem fibres. It is cultivated extensively in Bangladesh, and West Bengal in India.
- *Timber* Several *Tilia* species are grown for their timber in temperate regions. Species of *Berrya*, *Grewia*, *Elaeocarpus* and *Erinocarpus* are also grown for similar purposes.
- *Ornamentals* Several Tiliaceae are planted for ornament in parks, large gardens and on roadsides because of their fragrant flowers. Common among them are *Sparmannia africana* (House Lime), *Tilia americana* (Basswood), *T. europaea* (European Linden),

T. tomentosa (White Linden), *Grewia elastica*, *G. serrulata*, *Elaeocarpus ganitrus* and *Muntingia calabra*.

- *Edibles* Fruits of *Grewia asiatica* syn. *G. subinaequalis* (Phalsa), *G. tenax*, *G. tilifolia* and *G. villosa* are edible.
- *Medicinal Value* (i) Dried leaves and decoction of roots of *Corchorus capsularis* and *C. olitorius* are used to cure dysentery and diarrhoea; (ii) root bark of *Grewia asiatica* proves effective in rheumatism while its fruits help in digestion; (iii) *G. microcos* plants are used in eczema; (iv) roots of *G. sclerophylla* are prescribed in cough; (*v*) wood of *G. tiliaefolia* is emetic and taken as an antidote to opium poisoning; (vi) roots of *G. villosa* are used in diarrhoea; (vii) leaves, flowers and fruits of *Triumfetta bartramia* are used against gonorrhoea.
- *Fodder* Leaves of *Grewia glabra*, *G. oppositifolia*, *G. sapida* and G. *tenax* are used as fodder for cattles.

21.20.7 Systematics and Phylogeny (Affinities)

Tiliaceae belongs to order Malvales according to most of the botanists (Bentham and Hooker, 1862–1883; Rendle, 1925; Cronquist, 1981; and Thorne, 1983). However, Hutchinson (1959) placed this family under order Tiliales. Tiliaceae is closely allied to Malvaceae. However, members of Tiliaceae possess dithecous anthers and their stamens are nearly distinct whereas those of Malvaceae have monothecous anthers and their androecium is monadelphous. Tiliaceae is similar to Sterculiaceae in habit, leaves and stipules but differs in possessing indefinite androecium. According to some botanists, Tiliaceae is allied to some genera of Euphorbiaceae which have free stamens attached upon a gynophore.

21.20.8 Description of a Common Plant

1. Corchorus aestuans Linn. syn. C. acutangulus Lamk. (Fig. 21.30)

Stem: Branched, aerial, herbaceous, solid, hairy, dark red. *Leaf*: Simple, alternate, petiolate, stipulate, ovate, serrate, acute, unicostate reticulate. *Inflorescence:* Axillary cyme. *Flower:* Bracteate, bracteolate, pedicellate, complete, hermaphrodite, actinomorphic, hypogynous, pentamerous. *Calyx:* 5 sepals, free, valvate. *Corolla:* 5 petals, polypetalous, valvate, yellow. *Androecium:* 10–20 stamens in 2 to 4 whorls of 5 stamens in each whorl, polyandrous, dithecous, basifixed or dorsifixed, introrse. *Gynoecium:* Tricarpellary, syncarpous, superior ovary, trilocular, 2 ovules in each locule, axile placentation; style single; stigma trifid, each stigmatic lobe is again bifurcated. *Fruit:* Capsule.

Floral Formula: Br, Brl, ⊕, ⚥, K_5, C_5, A_{10-20}, $G(\underline{3})$.

21.21 BOMBACACEAE (BOMBAX FAMILY)

21.21.1 Systematic Position

Treated as a subfamily of Malvaceae by Bentham and Hooker.

21.21.2 Field Recognition

Exclusively arborescent or large-sized trees; simple or palmately compound leaves; stipules caducous; large-sized flowers; anthers dithecous or monothecous; pollen grains always smooth; fruit capsule.

21.21.3 Selected Indian Genera

Adansonia, *Bombax*, *Ceiba*, *Salmalia*, *Cullenia*.

21.21.4 Size, Distribution and General Information

A small family of about 22 genera and 180 species, Bombacaceae are mainly distributed in tropical regions. *Bombax malabaricum* syn. *Salmalia malbarica* (Simbal) is very common in India whereas *Durio zibethinus* is indigenous to Malayan archipelago.

21.21.5 Description of the Family

General Habit Tall trees with large, thick and spreading branches (*Adansonia*, *Ceiba pentandra*).

Stem Aerial, erect, branched, tall, thick; a few have extremely light wood (*Ochroma lagopus*).

Leaf Alternate, simple (*Cullenia*) or palmately compound; stipulate, stipules deciduous; petiolate.

Inflorescence Solitary axillary or in axillary clusters.

Flower Bracteate, rarely ebracteate, pedicellate, complete, bisexual, actinomorphic, large, showy, hypogynous.

Calyx 5 sepals, gamosepalous, valvate (Fig. 21.31); leathery in texture.

Corolla 5 petals, polypetalous, imbricate or twisted; petals rarely absent (*Cullenia excelsa*).

Androecium Stamens 5 (*Ceiba pentandra*) to numerous (*Salmalia malabarica*), free or monadelphous (*Adansonia*, *Ceiba*) or polyadelphous (*Bombax malbaricum*; Fig. 21.31); staminodes often present; anthers 1-, 2- or more-celled containing smooth pollen grains; stamens very long in *Pachira insignis*; dehiscence longitudinal; extrorse.

Gynoecium 2–5 carpels, syncarpous, ovary superior, 2- to 5-locular, 2 or more ovules in each locule, axile placentation; style long and simple; stigmas 1 to 5.

Fruit and Seed Fruit capsule. Seeds endospermic (*Salmalia*) or non-endospermic (*Cullenia*); bear silky hairs in *Ceiba pentandra*.

General Floral Formula Br, ⊕, ⚥, $K_{(5)}$, C_5, $A_{5-\infty}$ or $(5-\infty)$, $G(\underline{2-5})$.

21.21.6 Economic Importance

1. Fruits of *Adansonia digitata* provide us a good beverage.
2. Wooly outgrowth present on the pericarp of the fruits of *Bombax ceiba* is used for stuffing pillows while the floss obtained from the seed hairs of *Ceiba pentandra* is the source of 'Kapok'.

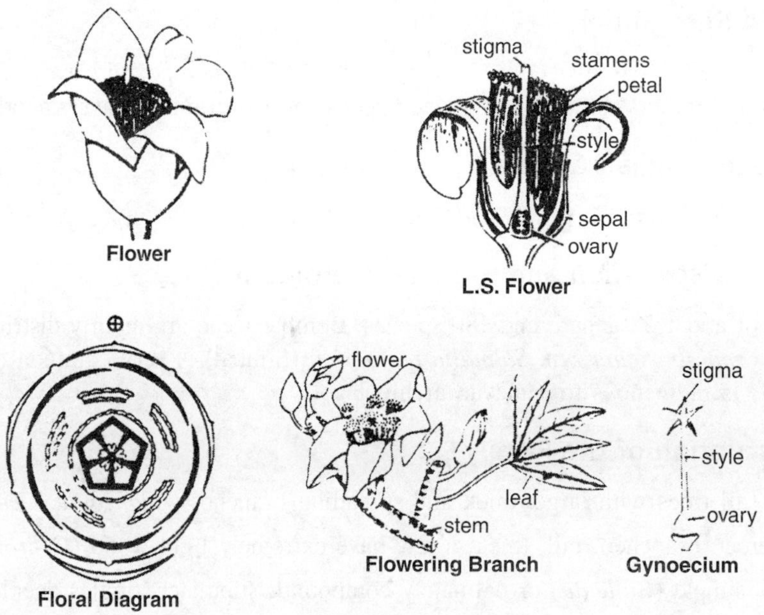

Fig. 21.31 *Bombax malbaricum* D.C.

3. The flower buds and fruits of *Bombax ceiba* and *Durio zibethinus* are edible. Root decoction of *D. zibethinus* is used for fever.
4. World's lightest commercial wood with a specific gravity of only 0.12 is obtained from *Ochroma lagopus* and *O. pyramidala*.
5. Wood obtained from *Cullenia excelsa* is used for making pencils, plywood, packing cases, etc.

21.21.7 Systematics and Phylogeny (Affinities)

In old classical literature of taxonomy, Bombacaceae has been considered as a taxon under family Malvaceae. Bentham and Hooker also treated it as a subfamily of Malvaceae. However, in majority of the recent taxonomic works including that of Cronquist (1981) and Thorne (1983), Bombacaceae has been treated as an independent family of order Malvales similar to that of Malvaceae. Moreover, Bombacaceae differs from that of Malvaceae in (i) being exclusively arborescent, (ii) often possessing a prickly trunk, (iii) bearing the dithecous anthers in some and monothecous in others, and (iv) always having smooth pollen grains.

In several anatomical and floral characters Bombacaceae shows close affinities with Malvaceae. However, several of its genera show close relationship with Dilleniaceae on the basis of their stamen morphology. Takhtajan (1969) and Cronquist (1981) show close relationship of Bombacaceae with Malvaceae, Sterculiaceae and Tiliaceae.

21.22 GERANIALES

21.22.1 General Characteristics

1. Stamens usually twice as many as sepals.
2. Disc present; usually disc is annular, adnate to stamens or reduced to glands.
3. Ovary multicarpellary and syncarpous.
4. Style often persistant in fruit.
5. Seeds usually non-endospermic or endosperm scanty.

According to Bentham and Hooker, Cohort Geraniales includes 11 orders (=families), of which three major ones are Geraniaceae, Rutaceae and Meliaceae.

Engler and Diels divided order Geraniales into 6 suborders (Geraniineae, Malpighiineae, Polygalineae, Dichapetalineae, Triococceae and Callitrichineae) and 21 families. Lawrence opined that the "evidence from all fields of Botany indicates that the Geraniales (*sensu* Engler) are not a natural taxon".

Geraniaceae, Oxalidaceae, Rutaceae and Meliaceae are treated in this text.

21.23 GERANIACEAE (GERANIUM FAMILY)

21.23.1 Systematic Position

Polypetalae, Disciflorae, Geraniales.

21.23.2 Field Recognition

Herbs, sometimes suffrutescent; leaves palmately lobed or dissected; 5-merous flowers; stamens with filaments united at base; beaked or lobed fruits with elastic dehiscent schizocarp.

21.23.3 Selected Indian Genera

Geranium and *Pelargonium*.

21.23.4 Size, Distribution and General Information

Geraniaceae is a family of about 11 genera (Jones and Luchsinger, 1987) and 780 species (Hickey and King, 1988), which are cosmopolitan in distribution and found widely in temperate and tropical regions. Only 3 genera and about 25 species have been reported from India, mainly from Himalayas. Some of the larger genera along with their approximately reported number of species in parenthesis include *Geranium* (400), *Pelargonium* (250), *Erodium* (90) and *Monsonia* (25). Well-known 'geranium-oil' is distilled from *Pelargonium odoratissimum*.

21.23.5 Description of the Family

General Habit Mostly herbs, sometimes undershrubs; suffrutescent or arborescent; stems often fleshy (*Pelargonium*); hairy.

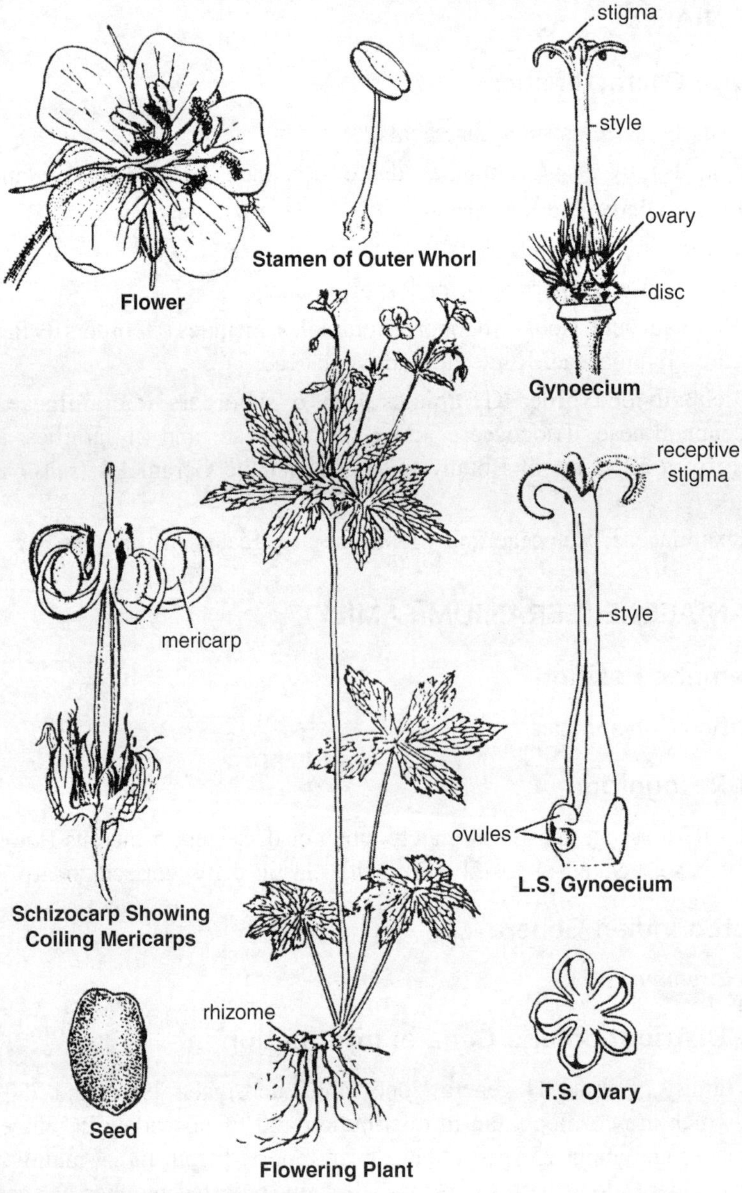

Fig. 21.32 *Geranium maculatum* L.

Leaf Simple and lobed or dissected, or compound; opposite or alternate; stipulate; venation mostly palmate.

Inflorescence Usually cymose, sometimes umbellate (*Pelargonium*), or flowers solitary.

Flower Bracteate, bracteolate, complete, hermaphrodite, actinomorphic but zygomorphic in *Pelargonium*, hypogynous, pentamerous; often very attractive.

Calyx Usually 5 sepals, free or rarely united, imbricate; sepals 4 in *Vivania* and 8 in *Dirachma*; persistent; in zygomorphic flowers the dorsal sepal is sometimes spurred and the spur is adnate to the pedicel.

Corolla 5 petals, free, alternate with sepals, imbricate or twisted; rarely petals are 2, 4, 8 or 0; often alternating with nectariferous glands.

Androecium 5 to 15 stamens; filaments often united at the base; in 1 to 3 whorls; obdiplostemonous; staminodes often present (*Erodium*); anthers usually versatile, 2-celled, longitudinally dehiscent; a nectary is present at the base of each stamen of inner whorl in *Geranium*.

Gynoecium Pentacarpellary, syncarpous, superior ovary, pentalocular, one to two ovules in each locule, axile placentation; rarely 3 to 8 carpels; styles 3 to 5, slender and beak like; stigmas 3 to 5.

Fruit and Seed Fruit capsular, dehiscing into 1 to 2-seeded 3 to 5 mericarps (Fig. 21.32), the styles adhering to the ovarian beak. Seeds usually with curved embryos; endosperm scanty or absent.

Pollination and Dispersal Pollination mainly by bees which visit the flowers in search of nectar. In *Geranium*, the dispersal of seeds takes place by the awn, which suddenly curls upwards, throwing the seeds at some distance from the plant (Fig. 21.33).

General Floral Formula Br, Brl, ⊕, ⚥, K_5, C_5, A(5–15), $G(\underline{5})$.

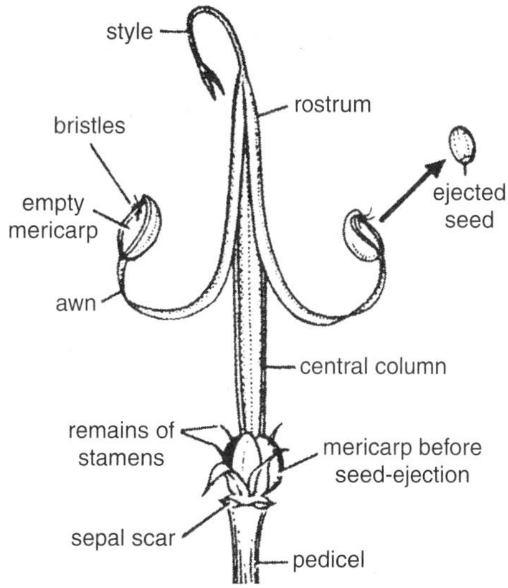

Fig. 21.33 Mechanism of seed dispersal in *Geranium pratense*.

21.23.6 Economic Importance

- Ornamental value of Geraniaceae is because of the beautiful and attractive flowers of several of its members, including *Pelargonium hortorum*, *P. zonale*, *Geranium wallichianum*, *G. pratense*, *Erodium cicutarium*, etc.
- Well-known 'geranium oil', used in perfumery, is obtained from *Pelargonium odoratissimum*.
- Roots and leaves of *Geranium robertianum* are used in the treatment of diarrhoea whereas the rootstocks of *G. wallichianum* are used in eye troubles. Roots and leaves of *G. nepalense* are used in the treatment of kidney troubles.
- Several species of *Erodium* and *Geranium* are weedy.

21.23.7 Systematics and Phylogeny

Geraniaceae belongs to order Geraniales, series Disciflorae and subclass Polypetalae of Dicotyledons (Bentham and Hooker). Majority of taxonomists believe it to belong to order Geraniales. Thorne (1983) placed Geraniaceae under suborder Geranineae of order Geraniales belonging to superorder Geraniiflorae. Usually Geraniales are considered to have derived from Ranales. However, Takhtajan (1969) stated that Geraniales are "clearly connected with Rutales, especially with the Rutaceae".

21.24 OXALIDACEAE (WOOD-SORREL FAMILY)

21.24.1 Systematic Position

Treated as a tribe of Geraniaceae by Bentham and Hooker.

21.24.2 Field Recognition

Usually palmately compound leaves with a sour taste due to the presence of oxalic acid; pentamerous flowers with monadelphous stamens; styles 5; fruit loculicidal capsule.

21.24.3 Indian Genera

Oxalis and *Biophytum.*

21.24.4 Size, Distribution and General Information

The family includes only 8 genera and about 900 species, distributed mainly in tropical and subtropical regions. Only a few species extend into temperate regions. Only 2 genera and about a dozen species have so far been reported from India. Most common representative in India is *Oxalis corniculata* (Indian Sorrel). Chief genera with their approximately reported species are *Oxalis* (800), *Biophytum* (70), *Averrhoa* (2) and *Eichleria* (2). *Biophytum sensitivum* is sensitive to touch like that of *Mimosa pudica* of Mimosoideae.

21.24.5 Description of the Family

General Habit Mostly perennial herbs, rarely shrubs or trees (*Averrhoa*); often suffrutescent or scapose and produce fleshy rhizomes or bulbous tubers; juice of the plants is sour because of the presence of oxalic acid.

Leaf Alternate, pinnately or palmately compound; exstipulate; often long-petioled; leaflets obcordate and characteristically folding at night; sometimes sensitive to touch (*Biophytum sensitivum*); leaves sometimes replaced by phyllodes (*Oxalis bupleurifolia*).

Inflorescence Solitary or subumbellate, rarely racemose or cymose.

Flower Bracteate, bracteolate, complete, hermaphrodite, actinomorphic, pentamerous, hypogynous; often on long peduncles.

Calyx 5 sepals, polysepalous (Fig. 21.34) or slightly united; imbricate or quincuncial, persistent.

Corolla 5 petals, free or slightly united at the base, short-clawed; imbricate or twisted.

Androecium Stamens 10, fused at the base (monadelphous), obdiplostemonous, introrse; rarely 5 stamens are without anthers.

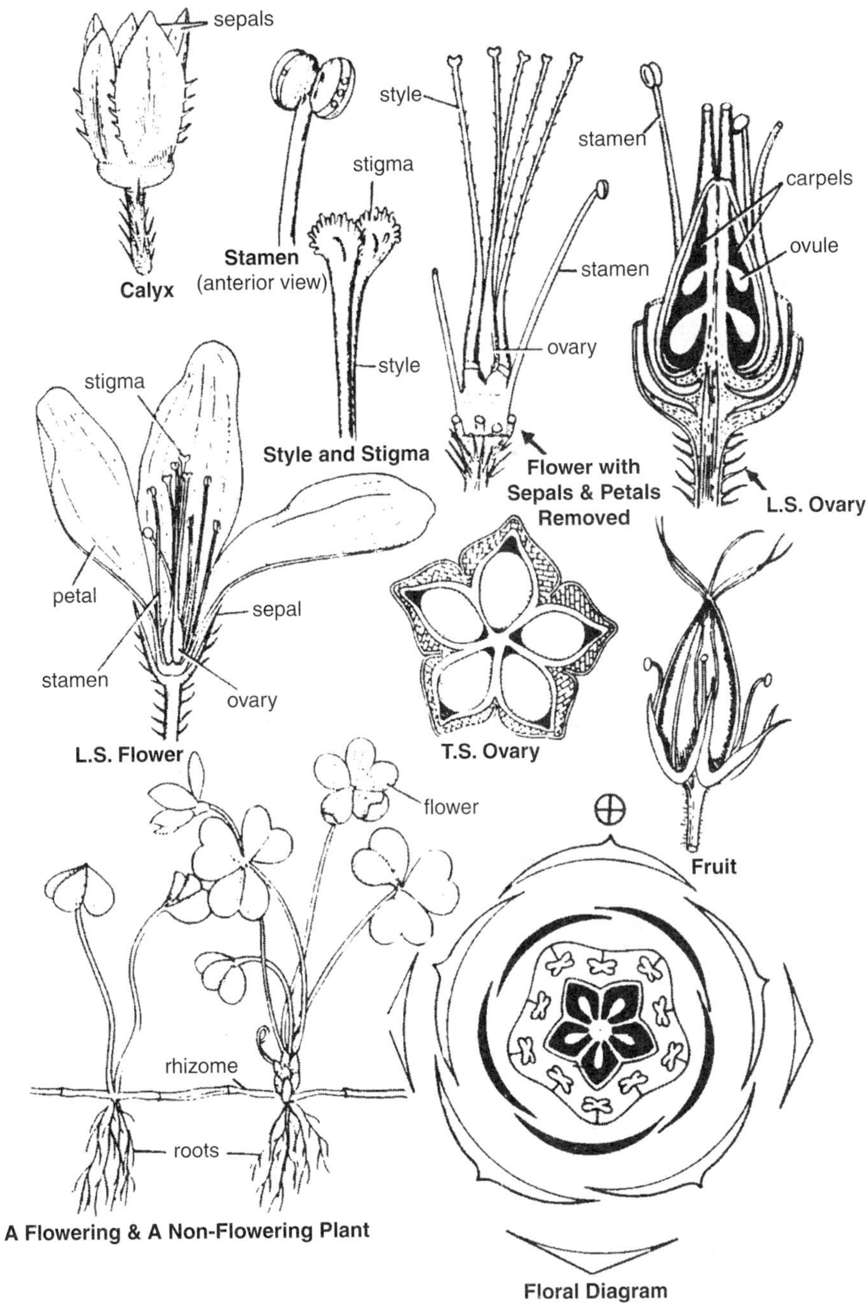

Fig. 21.34 *Oxalis acetosella* L.

Gynoecium Pentacarpellary, syncarpous, ovary superior, pentalocular, one or more anatropous ovules in each locule, axile placentation; styles 5, free and persistent.

Fruit and Seed Fruit a loculicidal capsule, rarely a berry (*Averrhoa*). Seeds with straight embryo and fleshy endosperm.

Pollination and Dispersal Pollination is usually by insects. Seeds are often discharged by the elastic separation of fleshy aril of the seed coat from an inner harder layer.

General Floral Formula Br, Brl, ⊕, ⚥, K_5, C_5, A(10), $G(\underline{5})$.

21.24.6 Economic Importance

The family is of little importance. The leaves of *Oxalis* (sour grass) are chewed by the children for their pleasant sour taste due to oxalic acid. Tuberous roots of *O. tuberosa* are used as food in India and South America. Leaves of *O. acetosella* (common wood sorrel) are used as salads and stems of *O. pescaprae* as a vegetable. Ornamental plants of the family include *Oxalis articulata* (bright pink or white flowers), *O. deppei, O. hedysaroides,* and both the species of genus *Averrhoa* (evergreen trees). *Biophytum sensitivum* is sensitive to touch and grown in gardens as a curiosity, as its leaves are sensitive like those of *Mimosa pudica* of Mimosoideae.

21.24.7 Systematics and Phylogeny (Affinities)

Bentham and Hooker treated Oxalidaceae only as a tribe Oxalideae under family Geraniaceae, belonging to order Geraniales, series Disciflorae of subclass Polypetalae. However, all recent taxonomists, including Cronquist (1981) and Thorne (1983) etc. treat it as an independent family belonging to order Geraniales. Hutchinson placed the arborescent genus *Averrhoa* under a separate family Averrhoaceae. Most of the taxonomists believe that Oxalidaceae have close affinities with Geraniaceae.

21.25 RUTACEAE (CITRUS FAMILY OR RUE FAMILY)

21.25.1 Systematic Position

Polypetalae, Disciflorae, Geraniales.

21.25.2 Field Recognition

Distinguished by the presence of translucent, very clear dots in the leaves, Rutaceae are usually shrubs or trees with aromatic oil glands; inflorescence cymose; outer stamens usually opposite the petals; ovary deeply lobed and elevated on a disc.

21.25.3 Selected Indian Genera

Citrus, Aegle, Murraya, Ruta, Zanthoxylum, Limonia.

21.25.4 Size, Distribution and General Information

A family of about 150 genera and over 1500 species, Rutaceae are distributed in both tropical and temperate regions, especially in Australia and South Africa. About 25 genera and over 80 species of this family have so far been reported from India. Known throughout the world for its several juicy citrus fruits, such as our present-day Oranges, Lemons, Grape-fruits, etc., Rutaceous members have high percentage of vitamin C as well as several alkaloids. Some of its larger genera along with their approximate number of reported species in bracket include *Fagara* (250), *Zanthoxylum* (200), *Ruta* (60), *Glycosmis* (60), *Eriostemon* (32), *Atalantia* (18), *Citrus* (12), *Murraya* (12) and *Aegle* (3).

21.25.5 Description of the Family

General Habit Shrubs or trees, and rarely herbs (*Boenninghausenia albiflora*, *Ruta graveolens*); often xerophytic and aromatic or strong-smelling.

Root Branched tap roots.

Stem Erect, woody, branched; often with spines or thorns (*Aegle*, *Citrus*); usually aromatic.

Leaf Exstipulate; alternate (*Citrus*, *Aegle*) or opposite (*Evodea*); simple or compound; compound leaves may be pinnately or palmately compound; sometimes reduced to spines (*Citrus*, *Aegle*); usually gland-dotted (Fig. 21.35).

Inflorescence Generally cymose; rarely raceme (*Feronia*); sometimes axillary (*Citrus*).

Flower Bracteate, bracteolate (*Ruta*) or ebracteolate, bisexual or rarely unisexual (*Toddalia*, *Zanthoxylum*), actinomorphic or rarely zygomorphic (*Dictamnus*, *Cusparia*), pentamerous or rarely tetramerous, hypogynous; cup-like nectariferous disc is usually present between ovary and stamens; rarely trimerous (*Triphasia trifoliata*).

Calyx 4 to 5 sepals, free (*Ruta*) or united (*Murraya*, Fig. 21.35); valvate (*Murraya*) or imbricate or quincuncial (*Citrus medica*); cup-like in *Correa speciosa*; rarely sepals absent in some species of *Zanthoxylum*.

Corolla 4 to 5 petals, sometimes none; usually polypetalous; imbricate or rarely valvate; bell-shaped (*Correa*); white, yellow or red; gamopetalous in *Correa* and *Almeidea*.

Androecium Usually 8 or 10 stamens; obdiplostemonous; anthers dithecous; introrse; rarely numerous stamens (*Citrus*, *Aegle*); filaments usually free, but in *Citrus* filaments of numerous stamens are united in several groups showing polyadelphous condition (Fig. 21.36); stamens of outer whorl change into staminodes in *Tetractomia*.

Gynoecium Usually 4 to 5 carpels, rarely less or indefinite; carpels usually united or sometimes free at base and united above by style; tetra- to pentalocular, 1 to 2 ovules in each locule, axile placentation; ovary superior; rarely ovary unilocular with parietal placentation (*Feronia*).

Fruit and Seed Fruit hesperidium (*Citrus*), capsule (*Flindersia*), samara (*Ptelea*), drupe (*Skimmia*), or rarely schizocarp, or follicle. Seeds without endosperm, with curved or straight embryo. Polyembryony, with as many as 13 embryos (*Citrus*), is common.

Pollination and Dispersal Pollination entomophilous. Flies, bees, etc. are attracted because of coloured corolla and nectar secreted by the disc. Dispersal of seeds takes place mainly by animals and also by man.

General Floral Formula Br, ⊕, ⚥, K_{4-5} or (4–5), C_{4-5}, A_{8-10}, $G(\underline{4-5})$.

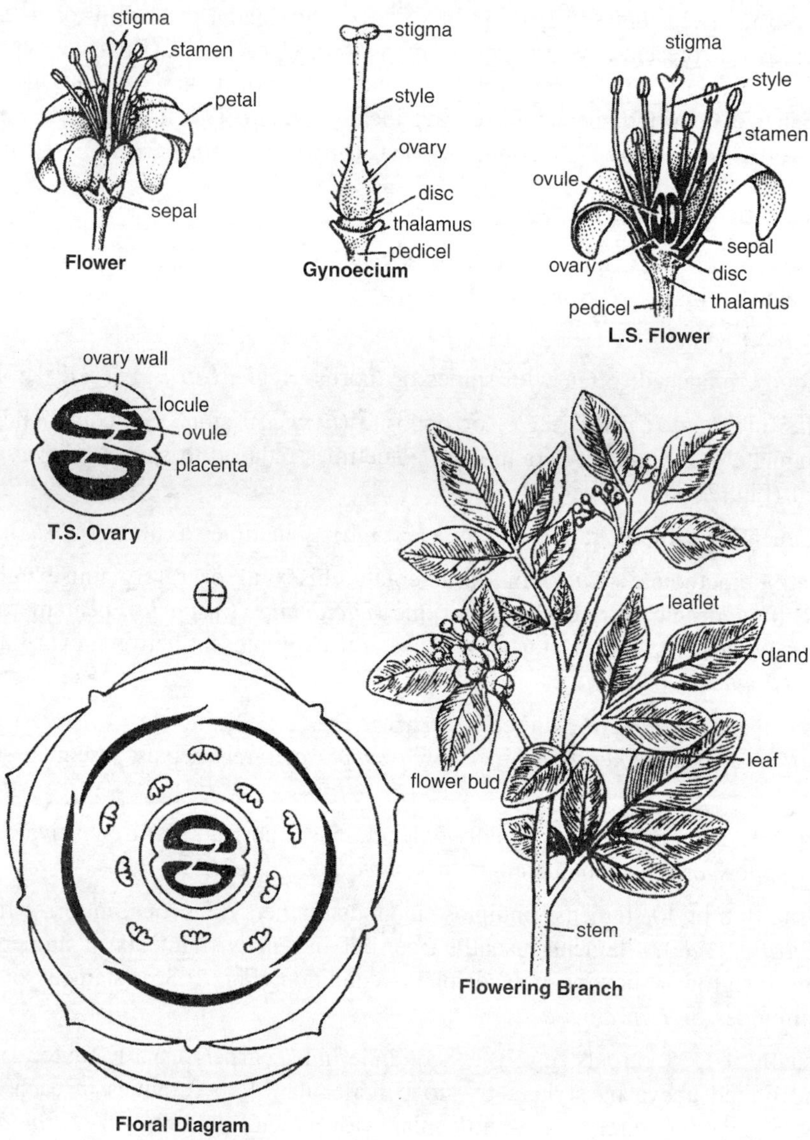

Fig. 21.35 *Murraya paniculata* (L.) Jacq.

21.25.6 Economic Importance

- *Fruits:* Rutaceae provides some of the world's most important, tasty and juicy citrus fruits, which are the richest source of vitamin C. Some of them are *Citrus aurantifolia* (Lime, Kaghzi Nimbu), *C. aurantium* (Sour orange, Khatta) *C. limettioides* (Sweet lime, Meetha Nimbu), *C. limon* (Lemon, Galgal), *C. maxima* (Shaddock, Chakotra), *C. medica* (Citron,

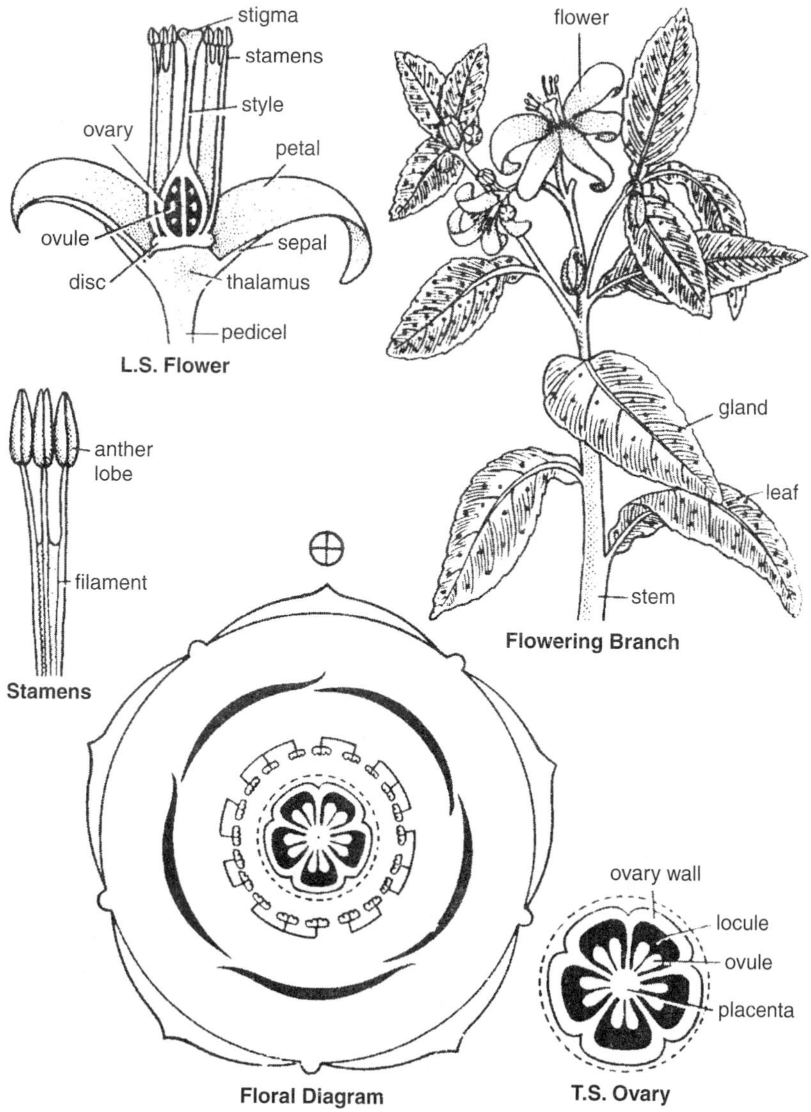

Fig. 21.36 *Citrus medica* L.

Bara Nimbu), *C. paradisi* (grapefruit), *C. reticulata* (Orange, Shantra) and *C. sinensis* (sweet orange, Malta or Mussambi). Some other famous fruits of the family include *Aegle marmelos* (Bael, Bilva), *Evodia fraxinifolia* (Kanukpa), *Feronia limonia* (Wood-apple, Kaith), *Fortunella japonica* (Kumquat) and *Glycosmis pentaphylla* (Ban Nimbu).

- *Condiments:* Bark of *Zanthoxylum alatum* (Tejpat) is a famous condiment and is also used to clean teeth and as a carminative and stomachic. Leaves of *Z. limonella* are used as condiment and its fruits are digestive and appetizing. Fruits of *Z. nitidum* are used as condiment and its roots are used in toothache. The leaves of *Murraya koenigii* are mixed with a little turmeric to make curry powder in India. Leaves of *Evodia lunuranthenoa* are also used as a condiment.

- *Medicinal Value:* (i) A poultice of bark of *Acranychia laurifolia* is applied on ulcers and sores; (ii) fruit pulp of *Aegle marmelos* is a mild laxative whereas its roasted fruits are used against diarrhoea and dysentry, and its root and stem bark in intermittent fever; (iii) oil from the fruits of *Atalantia monophylla* is useful in paralysis and rheumatism; (iv) leaves of *Barosma betulina* are used in curing kidney diseases; (v) bark of *Cusparia febrifuga* is used in malaria; (vi) bark of *Dictamnus albus* is useful in nervous diseases and intermittent fevers; (vii) ripe fruits of *Feronia limonia* are taken as a cardiac tonic; (viii) roots of *Glycosmis pentaphylla* are used in fever; (ix) bark and roots of *Murraya koenigii* are used in intestinal disorders; (x) seeds of *Peganum harmala* are given in asthma, neuralgia and rheumatism; (xi) leaves of *Pilocarpus pinnatifolius* are used in kidney troubles; (xii) dried leaves of *Ruta graveolens* are used as nerve stimulant; (xiii) root bark of *Toddalia asiatica* is a potent antimalarial drug; (xiv) seeds of *Zanthoxylum alatum* are used in cholera; (xv) roots of *Z. nitidum* are used in toothache.
- *Ornamental Plants:* Amongst the ornamental shrub and tree genera of Rutaceae include *Phellodendron*, *Ptelea*, *Choisya*, *Skimmia* and *Zanthoxylum*. Perennial herbs of ornamental value include *Dictamnus albus* and *Ruta graveolens*. The fragrant-flowered *Murraya paniculata* is commonly grown for decorative purposes. Some other ornamentals include *Fortunella japonica*, *Glycosmis pentaphylla*, *Luvunga scandens* and *Poncirus trifoliata*.
- *Timber:* The wood obtained from *Choisya*, *Chloroxylon swietennia*, *Feronia limonia*, *Glycosmis pentaphylla*, *Murraya paniculata* and *Zanthoxylum alatum* is used for veneering, making walking sticks, agricultural implements, etc.
- *Oils:* Essential oils, obtained from the leaves and fruit rind of various species of *Citrus*, *Feronia*, *Murraya*, etc., are used in medicine and perfumery. Seeds of *Citrus bergamia* are the source of famous 'bergamot oil'.

21.25.7 Systematics and Phylogeny (Affinities)

Rutaceae belongs to order Geraniales, series Disciflorae and subclass Polypetalae according to Bentham and Hooker, who divided it into four tribes *viz.* Aurantieae, Ruteae, Toddalieae and Zanthoxyleae. Takhtajan (1969) and Thorne (1983) placed it under order Rutales whereas Cronquist (1981) placed it under order Sapindales. Recently, Hickey and King (1988) divided Rutaceae into 5 subfamilies *viz.* Rutoideae, Dictyolomatoideae, Spathelioideae, Toddalioideae and Aurantioideae, and stated that "Rutaceae is most closely related to Zygophyllaceae and Cneoraceae on the one hand and to Meliaceae, Burseraceae and Simaroubaceae on the other, all included in the order Sapindales".

21.25.8 Description of Some Common Plants

1. Murraya paniculata (L.) Jacq. (syn. M. exotica; Fig. 21.35)

Stem: Herbaceous, lower portions woody, branched, erect, solid. *Leaf*: Alternate, petiolate, exstipulate, compound, unipinnate and imparipinnate, 3–7 leaflets; leaflets ovate to elliptical or lanceolate, entire, obtuse; unicostate reticulate; gland-dotted. *Inflorescence:* Dichasial cyme. *Flower:* Bracteate, pedicellate, complete, actinomorphic, hermaphrodite, pentamerous, hypogynous, white; a nectar-secreting disc is present below the ovary. *Calyx:* 5 sepals, gamosepalous, valvate. *Corolla:* 5 petals, free, imbricate. *Androecium:* 10 stamens, arranged in 2 whorls of 5 each, free, dithecous, basifixed, introrse. *Gynoecium:* Bicarpellary, rarely 3 carpels, syncarpous; superior, bilocular, 1 to 2 ovules in each locule, axile placentation; a hypogynous nectariferous disc is present; stigma bifid. *Fruit:* Berry.

Floral Formula: Br, ⊕, ⚥, $K_{(5)}$, C_5, A_{5+5}, $G(\underline{2})$.

2. *Citrus medica (Citron or Bara Nimbu; Fig. 21.36)*

Stem: Same as in *Murraya*. *Leaf:* Exstipulate, leaf base swollen, simple, alternate, entire, ovate, obtuse or acute. *Inflorescence: Solitary* axillary or axillary umbellate cyme. *Flower:* Almost same as in *Murraya*. *Calyx:* 5 sepals, free or united, valvate. *Corolla:* Same as in *Murraya*. *Androecium:* Numerous stamens, polyadelphous, attached round a disc, anthers dithecous, dorsifixed or basifixed, introrse. *Gynoecium:* 5 to many carpels, syncarpous, superior, multilocular, one or more ovules in each locule, axile placentation; a nectariferous disc is present. *Fruit:* Hesperidium.

Floral Formula: Br or Ebr, ⊕, ⚥, $K_{(5)}$, C_5, $A(\infty)$ polyadelphous, $G(\underline{5-\infty})$.

21.26 MELIACEAE (MAHOGANY FAMILY)

21.26.1 Systematic Position

Polypetalae, Disciflorae, Geraniales.

21.26.2 Field Recognition

Evergreen or deciduous trees or shrubs; leaves exstipulate, alternate, pinnately compound; inflorescence axillary, panicles or cymose; flowers bisexual, actinomorphic, hypogynous; sepals and petals 4–5; stamens 8–10, monadelphous; a disc is usually present between gynoecium and androecium.

21.26.3 Selected Indian Genera

Azadirachta, *Melia*, *Swietenia*, *Toona*, *Walsura*.

21.26.4 Size, Distribution and General Information

A family of about 50 genera and about 1400 species, Meliaceae is exclusively a tropical region family of the world. About 20 genera and over 75 species of Meliaceae are reported from India. Common Indian species include *Azadirachta indica* (Neem), *Melia azedarach* (Bakain) and *Toona ciliata* (Toon) whereas *Swietenia* is a timber-yielding species of tropical America. *Xylocarpus* is a mangrove genus of Meliaceae.

21.26.5 Description of the Family

General Habit Trees or shrubs, evergreen (*Appanomixis*) or deciduous (*Melia*, *Toona*), contain hard scented wood.

Stem Woody, solid, erect, branched, hard.

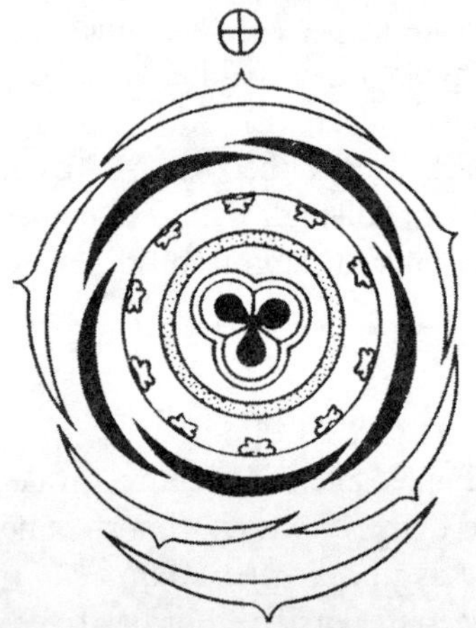

Fig. 21.37 Floral diagram of *Azadirachta indica* Juss.

Leaf Alternate; pinnately compound, twice or thrice pinnate in *Melia*, decompound, or rarely simple; exstipulate, do not possess pellucid dots; leaflets usually with entire or serrate margin and oblique at the base.

Inflorescence Axillary panicles or cymose.

Flower Bracteate (Fig. 21.37), bracteolate or ebracteolate; bisexual, but rarely unisexual and polygamodioecious (*Appanomixis*); actinomorphic, hypogynous; usually with an annular or a tubular disc between ovary and stamens.

Calyx 4 to 5 sepals, usually basally connate or free; small; imbricate but rarely valvate or contorted.

Corolla Usually 4 to 5 petals, rarely few or more than 5, free or united; imbricate but rarely valvate; usually adnate to the staminal tube.

Androecium Usually 8 to 10 stamens, rarely less than 8 or more than 10; monadelphous, rarely free (*Toona*, *Cedrela*); filaments unite to form a columnar tube; anthers dithecous, sessile, dehiscing longitudinally; a disc is usually present between androecium and gynoecium.

Gynoecium Bi- to pentacarpellary, syncarpous, superior; 2 to 5 locules, 2 or rarely more (up to 12 in *Swietenia*) ovules in each locule; axile placentation; style short or ill-developed; stigma capitate or discoid or lobed.

Fruit and Seed Fruit a berry, capsule, or rarely a drupe. Seeds often winged (*Toona*, *Khaya*), with or without endosperm; rarely arillate (*Appanomixis*); wings absent in *Azadirachta*, *Carapa* and *Melia*.

Pollination and Dispersal Pollination generally entomophilous; insects are attracted because of the nectar secreted by the intrastaminal disc. Seeds are dispersed mainly by birds or other animals. However, winged seeds are dispersed by wind.

General Floral Formula Br, ⊕, ⚥, $K_{(4-5)}$, C_{4-5}, $A_{(8-10)}$, $G(\underline{2-5})$.

21.26.6 Economic Importance

- *Timber:* Excellent timber trees of Meliaceae include *Amoora wallichii*, *Azadirachta indica* (Neem tree or Margosa tree), *Carapa granatum*, *Cedrela odorata*, *Chikrassia tabularis*, *Chloroxylon swietenia*, *Dysoxylum malabaricum* (white cedar), *Khaya senegalensis* (African Mahogany), *Melia azedarach* (Bakain), *Sandoricum koetjape*, *Soymida febrifuga*, *Swietenia mahogani* (True Mahogany), *Toona ciliata* syn. *Cedrela toona* (Toon wood), etc.
- *Medicinal Value:* (i) Seeds of Neem or Margosa tree (*Azadirachta indica*) are the source of margosa oil, used in skin diseases; (ii) bark of *A. indica* is used in malarial fever while its tender twigs are used as 'datoon' and are very effective in pyorrhea; (iii) leaf juice of *Melia azedarach* is anthelmintic and its seeds are used in rheumatism; (iv) fruits of *Aglaia*

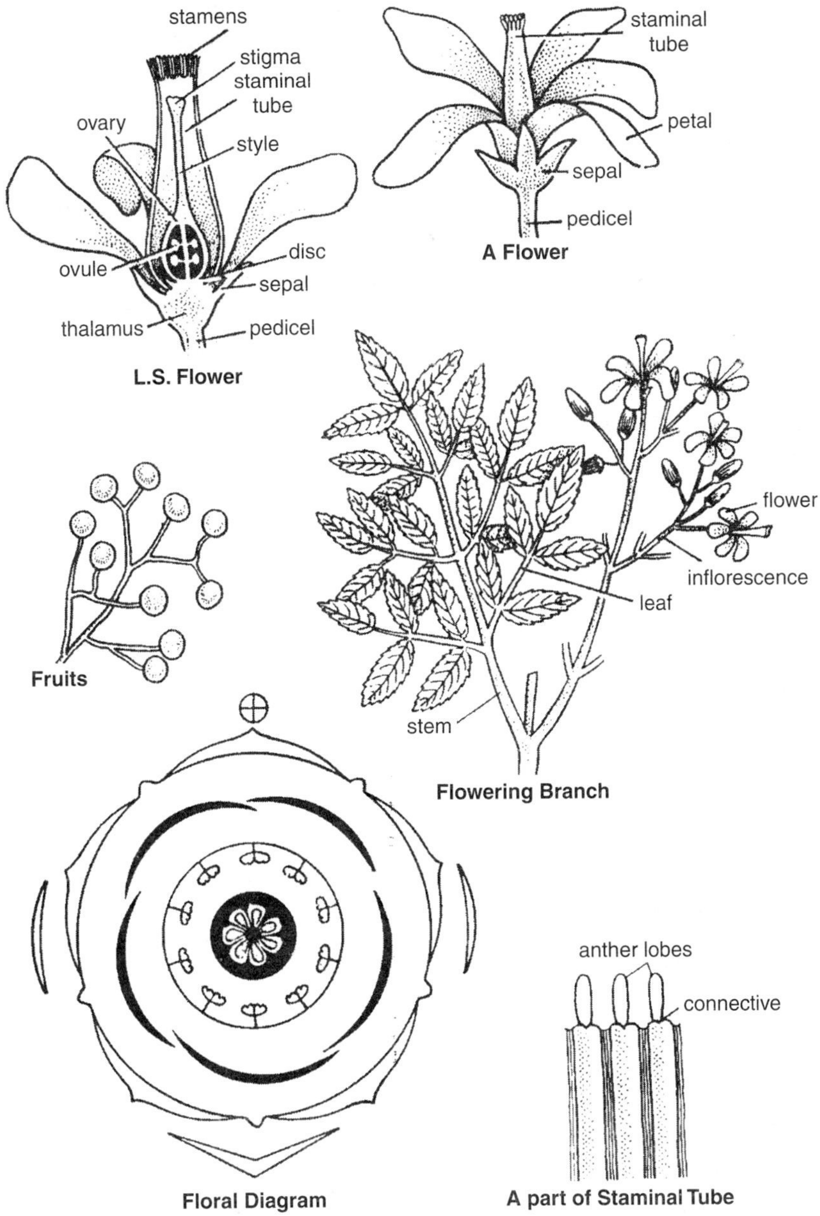

Fig. 21.38 ◈ *Melia azedarach L.*

odorata are used in leprosy and inflammation; (v) bark of *Carapa granatum* are used in dysentery and stomach troubles; (vi) wood oil of *Dysoxylum malabaricum* is used in eye and ear diseases; (vii) bark of *Soymida febrifuga* is a bitter tonic and used in intermittent fever; (viii) bark of *Walsura piscidia* is stimulant, expectorant and used in skin diseases.

- *Minor Uses:* (i) Ornamental plants of the family include *Aglaia odorata, Melia composita, Walsura trijuga,* etc.; (ii) Cedar oil used in microscopy is obtained from *Cedrela odorata*; (iii) flowers of *Chukrasia* yield a red-coloured dye.

21.26.7 Systematics and Phylogeny (Affinities)

Bentham and Hooker placed Meliaceae under order Geraniales of series Disciflorae. Meliaceae is related to Rutaceae on one hand and to Burseraceae and Simarubaceae on the other. Bessey and Engler also included Meliaceae under Geraniales and believed that it originated from Malvales. However, Hallier was of the opinion that Meliaceae was derived from Rutaceae. Meliaceae has been placed under order Meliales by Hutchinson while under Rutales by Takhtajan. Cronquist (1981) placed Meliaceae under Sapindales while Thorne (1983) placed it under Rutales.

21.26.8 Description of a Common Plant

Melia azedarach (syn. Bakain, Fig. 21.38)

Habit: A big tree. *Stem:* Woody, erect, branched, solid. *Leaf*: Alternate, exstipulate, compound, bipinnate and imparipinnate; leaf base pulvinus; pinnae ovate to lanceolate; serrate, unicostate reticulate. *Inflorescence:* Axillary panicle cyme. *Flower:* Bracteate, bracteolate, pedicellate, complete, actinomorphic, hermaphrodite, pentamerous, hypogynous; contains a nectariferous disc below ovary. *Calyx*: 5 sepals, fused, valvate. *Corolla:* 5 petals, free, imbricate. *Androecium:* 10 stamens, monadelphous; filaments form a staminal tube having a ten-toothed apex; dithecous, basifixed, introrse. *Gynoecium:* 5–8 or many carpels, syncarpous, superior, 5 to 8 or many-loculed, 1 or 2 ovules in each locule, axile placentation; stigma lobed or capitate; a nectariferous disc is present below the ovary. *Fruit:* Drupe.

Floral Formula: Br, Brl, ⊕, ⚥, $K_{(5)}$, C_5, $A_{(10)}$, $G(\underline{5-\infty})$.

21.27 CELASTRALES

21.27.1 General Characteristics

1. Leaves simple accept Ampelideae; often armed with spines or climbing by hooks or tendrils.
2. Flowers actinomorphic and bisexual.
3. Stamens as many or lesser than that of petals, and only rarely twice as many as petals.
4. Stamens usually opposite the petals.
5. Disc present.

Bentham and Hooker included 4 orders (=families) under Celastrales. These are Ampelideae (=Vitaceae), Celastrineae, Rhamneae and Stackhousieae.

Engler and Diels included Vitaceae and Rhamnaceae under Rhamnales and families Celastraceae and Stackhousiaceae under suborder Celastrineae of order Sapindales.

Only Rhamnaceae and Vitaceae are treated in this text.

21.28 RHAMNACEAE (BUCKTHORN FAMILY)

21.28.1 Systematic Position

Polypetalae, Disciflorae, Celastrales.

21.28.2 Field Recognition

Shrubs or trees, often armed with spines or climbing by hooks or tendrils; leaves stipulate; flowers perigynous; stamens opposite the petals; fruit drupe or capsule; seeds hard.

21.28.3 Selected Indian Genera

Zizyphus, Helinus, Berchemia, Ventilago.

21.28.4 Size and Distribution

Rhamnaceae comprises 58 genera and about 900 species. The plants of this family are cosmopolitan in distribution. Some of the larger genera of the family along with their approximate number of reported species are *Rhamnus* (150), *Phylica* (150), *Zizyphus* (100), *Ceanothus* (55), *Frangula* (50), and *Colletia* (17).

21.28.5 Description of the Family

General Habit Mostly trees or shrubs, often climbing by their hooks (*Ventilago*) or tendrils (*Gouania, Helinus*) or twining stems (*Ventilago, Berchemia*); some thorny (*Colletia, Condalia*); rarely herbs.

Leaf Simple, stipulate, alternate or rarely opposite; stipules modify into spines in *Zizyphus*; usually pinninerved but palminerved in *Zizyphus*.

Inflorescence Axillary corymbs or cymose; solitary axillary in *Microrhamnus*.

Flower Bracteate, bisexual or rarely unisexual, actinomorphic, sometimes apetalous (*Colletia*), pentamerous or tetramerous; perigynous or epigynous; usually small; hypogynous in *Berchemia scandens*.

Calyx 4 or 5 sepals, free or basally connate, valvate.

Corolla 4 or 5 petals, free, usually concave and often clawed at the base, small, valvate; sometimes absent (*Colletia*); petals often inserted on the calyx tube to form a hypanthium.

Androecium Stamens as many as petals, opposite the petals (Fig. 21.39) and enclosed by them, inserted on the hypanthium or develop on the margin of intrastaminal disc; anthers dithecous; dehiscence longitudinal.

Fig. 21.39 Floral diagram of *Zizyphus jujuba*.

Gynoecium Carpels 2–4, syncarpous or free, bi- to tetralocular, ovary superior or half-inferior, one or 2 ovules in each locule, basal placentation; style simple or divided; stigma simple or bilobed.

Fruit and Seed Capsule (*Phylica, Pomaderis*), drupe (*Rhamnus, Zizyphus*) or samaroid (*Ventilago*), or nut. Seeds hard, large, with straight embryo and little or no endosperm.

Pollination and Dispersal Pollination mostly entomophilous, and majority of the dry fruits are dispersed by wind.

General Floral Formula Br, ⊕, ⚥, K_{4-5}, C_{4-5}, A_{4-5}, $G(\underline{2-4})$.

21.28.6 Economic Importance

- *Ceanothus* (Californian Lilacs) is the most commonly grown genus for decorative purposes because of its beautiful blue, pink or white flowers. Some other ornamental plants include species of *Berchemia*, *Colletia*, *Discaria*, *Paliurus*, etc.
- *Zizyphus mauritiana* (Baer of India), *Z. nummularia* (Jhar Ber), *Z. jujuba*, *Z. vulgaris*, *Sageretia oppositifolia*, *Hovenia dulcis* (Japanese Raisin-tree) are well known for their edible fruits.
- Green and yellow-coloured dyes of commercial importance are obtained from various species of *Rhamnus.* A red dye is obtained from root bark of *Ventilago maderaspatana.*
- Cascara sagarda bark, a purgative, is obtained from *Rhamnus purshiana.*
- Gun powder is prepared from the wood of *Zizyphus mauritiana.*
- *Ceanothus americanus* (New Jersey Tea) leaves are used as a substitute of tea.
- Valuable timber is obtained from *Frangula alnus*, *Maeosopis emini* and *Zizyphus chloroxylon.* Wood of *Gouania domingensis* is used for making chew-sticks.

21.28.7 Systematics and Phylogeny (Affinities)

Bentham and Hooker placed Rhamnaceae under order Celastrales of series Disciflorae. But majority of the other taxonomists place it under order Rhamnales (Takhtajan, 1969; Cronquist, 1981; Thorne, 1983). Rhamnaceae shows affinity with Vitaceae from the standpoint of floral construction. However, it differs from Vitaceae by its small petals, receptacle, endocarp and always simple leaves. Some believe Rhamnaceae to be close to Celastraceae but the main distinction between the two is the presence of antipetalous stamens in Rhamnaceae.

21.29 VITACEAE (GRAPE FAMILY)

21.29.1 Systematic Position

Polypetalae, Disciflorae, Celastrales.

21.29.2 Field Recognition

Climbing habit; terminal buds of plants develop into tendrils; inflorescence opposite the leaves; stamens usually opposite petals; ovary bilocular; axile placentation; fruit berry.

21.29.3 Selected Indian Genera

Ampelocissus, *Cayratia*, *Cissus*, *Leea* and *Vitis*.

21.29.4 Size, Distribution and General Information

A small family of only 12 genera and 700 species, Vitaceae are mostly distributed in tropical and subtropical regions of the world. 8 genera and about 70 species of Vitaceae have been reported from India, mostly growing in plains. Most important and universally known plant of the family is *Vitis vinifera* (Grape vine). Some larger genera of Vitaceae along with their approximately reported species are *Cissus* (350), *Ampelocissus* (95), *Tetrastigma* (90), *Vitis* (60 to 70) and *Parthenocissus* (15).

21.29.5 Description of the Family

General Habit Most genera are vines or climbing shrubs; rarely erect shrubs, or small tree (*Carameriana*); species of *Leea* are erect herbs, shrubs or even trees.

Stem Hard, woody, with nodes often swollen or jointed; tendrils are leaf-opposed and considered as the metamorphosed apical part of the sympodial stem by Eichler. (*Tendril* is a negatively geotropic organ. The tips or ends of tendrils expand to form sticky and mucilaginous, ball-like structures, which help them to adhere firmly to the support.)

Leaf Alternate, but lower leaves are sometimes opposite; simple or compound (pinnately or palmately); stipules petiolar or absent; with pellucid punctate dots.

Inflorescence Flowers borne opposite the leaves in racemose, paniculate or cymose manner.

Flower Bracteate, usually bracteolate, actinomorphic, bisexual or unisexual; 4 to 5-merous, hypogynous; minute; peduncle is flat and ribbon-like in *Pterisanthes*.

Calyx 4 to 5 sepals, rarely 3 to 7, fused; very small, cupular, valvate.

Corolla 4 to 5 petals, free or fused; caducous and fall easily when the flowers open; valvate; in *Vitis* petals are free near base and united by the apices.

Androecium 4 to 5 stamens, opposite the petals (Fig. 21.40); arise from the base of an intrastaminal disc which is well-developed, annular or lobed; anthers free or connate, dithecous, introrse; dehiscence longitudinal; in *Leea*, the filaments unite and form a 5-lobed disc.

Gynoecium 2 to 8 carpels, syncarpous, superior, 2 to 8-loculed, 1 or 2 ovules in each locule, axile placentation; style one and short; stigma discoid or capitate.

Fruit and Seed Fruit berry, often with watery juice. Seeds with copious endosperm and straight embryo.

Pollination and Dispersal Pollination is entomophilous and dispersal takes place usually by birds, animals or even man.

General Floral Formula Br, Brl, ⊕, ⚥, $K_{(4-5)}$, C_{4-5}, A_{4-5}, $G(\underline{2-8})$.

21.29.6 Economic Importance

Vitaceae are important for the table grapes, wine grapes, raisins, grape juice, and for several ornamental and medicinal purposes.

- *Fruits:* Principal economic plant is *Vitis vinifera* (Grape vine, Fig. 21.40), whose several varieties are grown for their edible fruit, the source of raisins, wines, sultanas, and currants. Other species of *Vitis* known for their less palatable edible fruits include *V. aestivalis* and *V. labrusca*. Fruits of *Leea crispa* and *L. macrophylla* are also edible.

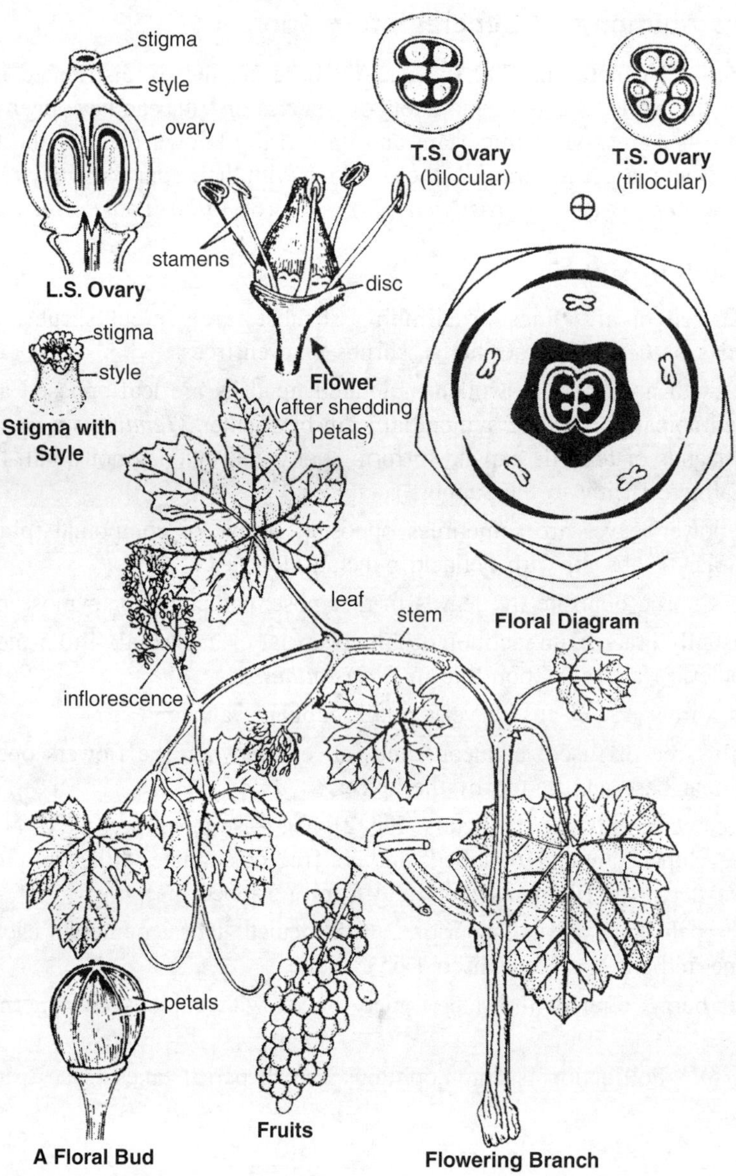

Fig. 21.40 *Vitis vinifera* L.

- *Ornamental Vines:* *Parthenocissus tricuspidata* (Japanese creeper), *P. quinquefolia* (Virginia creeper) and several species of *Ampelopsis*, *Cissus*, *Rhoicissus*, etc. are cultivated as ornamental vines on walls.
- *Medicinal Value:* (i) Leaves and roots of *Cayratia carnosa* are used in high fever; (ii) tubers of *Cissus adnata* are used as blood purifier; (iii) the essential oil obtained from the plant

of *Leea aequata* are used against tuberculosis; (iv) roots of *L. indica* are used in diarrhoea; (v) tubers of *L. macrophylla* are used in ringworm disease.

- *Minor Uses:* (i) Indian lac insect is grown on *Leea crispa* and *L. edgeworthi*; (ii) leaves of several species of *Parthenocissus* are used as fodder; (iii) a strong cordage fibre is obtained from the stems of *Cissus adnata* and *C. quadrangularis.*

21.29.7 Systematics and Phylogeny (Affinities)

Bentham and Hooker included Vitaceae (= Ampelideae) under order Celastrales of series Disciflorae whereas majority of the other authors (Hutchinson, Takhtajan, Cronquist, Thorne, etc.) place it under order Rhamnales. Vitaceae is closely allied to Rhamnaceae due to their similar floral structure. However, it differs from Rhamnaceae by its climbing habit, berried fruit, small embryo and albuminous seeds.

21.30 SAPINDALES

21.30.1 General Characteristics

1. Usually woody shrubs or trees.
2. Leaves usually compound and exstipulate.
3. Flowers small, often irregular, pentamerous and hypogynous.
4. Stamens definite, usually inserted on a prominent receptacular or extrastaminal disc.
5. Ovary tricarpellary, syncarpous, superior, usually trilocular.
6. Ovules pendulous with the dorsal raphe and micropyle upward or erect with ventral raphe and micropyle downward.

Sapindales includes Sapindaceae, Anacardiaceae and Sabiaceae according to Bentham and Hooker. Engler and Diels treated Sapindales as composed of 11 suborders and 23 families, of which some major ones are Buxaceae, Anacardiaceae, Celastraceae, Salvadoraceae, Sapindaceae and Sabiaceae.

Only Sapindaceae and Anacardiaceae are treated in this text.

21.31 SAPINDACEAE (SOAPBERRY FAMILY)

21.31.1 Systematic Position

Polypetalae, Disciflorae, Sapindales.

21.31.2 Field Recognition

Usually trees or shrubs with pinnately compound leaves; flowers small, pentamerous and hypogynous; sepals imbricate; petals glandular; an extra-staminal disc is usually present; ovary tricarpellary, syncarpous.

21.31.3 Selected Indian Genera

Litchi, *Sapindus*, *Acer*, *Aesculus*, *Dodonaea* and *Schleichera*.

21.31.4 Size, Distribution and General Information

A family of about 155 genera and over 2220 species, Sapindaceae are distributed mainly in tropical and subtropical regions of the world. About 25 genera and 75 species of this family have been reported from India, mainly from tropical Eastern Himalayas and Western Peninsular India. *Sapindus saponaria* (soapberry tree) provides the common name 'Soapberry family' to this family. *Litchi chinensis* (Litchi), *Acer saccharinum* (Sugar Maple), *Paullinia* (Guarana) and *Cardiospermum* (Balloon vine) are other famous genera of the family.

21.31.5 Description of the Family

General Habit Trees or shrubs, rarely herbs, sometimes tendril-bearing vines (*Cardiospermum*) or lianas (*Serjania*, *Paullinia*); tendrils being the flowerless branches of inflorescence.

Leaves Usually alternate, pinnately compound and exstipulate; very rarely opposite (*Acer*, Fig. 21.41A); simple (*Cardiospermum*); small and deciduous stipules present in climbing species; often contain resin or latex cells.

Inflorescence Flowers minute and usually arranged in racemose to paniculate unilateral cymes or a thyrsus; plants commonly polygamodioecious (i.e. bear apparently bisexual but functionally unisexual flowers); metamorphosed sterile branches of inflorescence become coiled like watch-spring and appear like tendrils.

Flower Minute, bracteate or ebracteate, bracteolate (*Acer*), bisexual or unisexual, actinomorphic (*Litchi*, *Aphania*, *Sapindus*) or obliquely zygomorphic (*Cardiospermum*, *Erioglossum*), pentamerous or rarely tetramerous, hypogynous.

Calyx 5 sepals, poly- or gamosepalous, imbricate or rarely valvate; sometimes only 4 sepals due to the fusion of 3rd and 5th sepals.

Corolla 5 petals, sometimes 3, or even absent (*Schleichera*), polypetalous, valvate (*Acer*) or imbricate (*Sapindus*); petals equal or unequal-sized and often bear scale-like or hairy appendage at the base.

Androecium Stamens 8 (*Acer*) or 10 (*Sapindus*, Fig. 21.41E), free and inserted inside a prominent receptacular or extrastaminal disc; only 5 stamens in *Turpinia*, and numerous stamens in *Deinbollia*; filaments often hairy; zygomorphic flowers often have eccentric androecium; anthers 2-celled, basifixed or versatile and introrse. In between petals and stamens is often present an annular disc.

Gynoecium Tricarpellary, syncarpous; ovary superior; trilocular with one or two or rarely more ascending ovules in each locule, axile placentation; style simple or divided; rarely the placentation is parietal.

Fruit and Seed Fruits may be dry or fleshy and highly variable from capsule, berry, nut, drupe, samara, or schizocarp in different genera. Seeds are non-endospermic with a curved embryo; often arillate.

Pollination and Dispersal Pollination is by insects such as flies, bees, etc. Dispersal takes place by wind, birds or other animals.

General Floral Formula Br or Ebr, ⊕, ⚥, K_5, C_5 or zero, A8 – 10, $G(\underline{3})$.

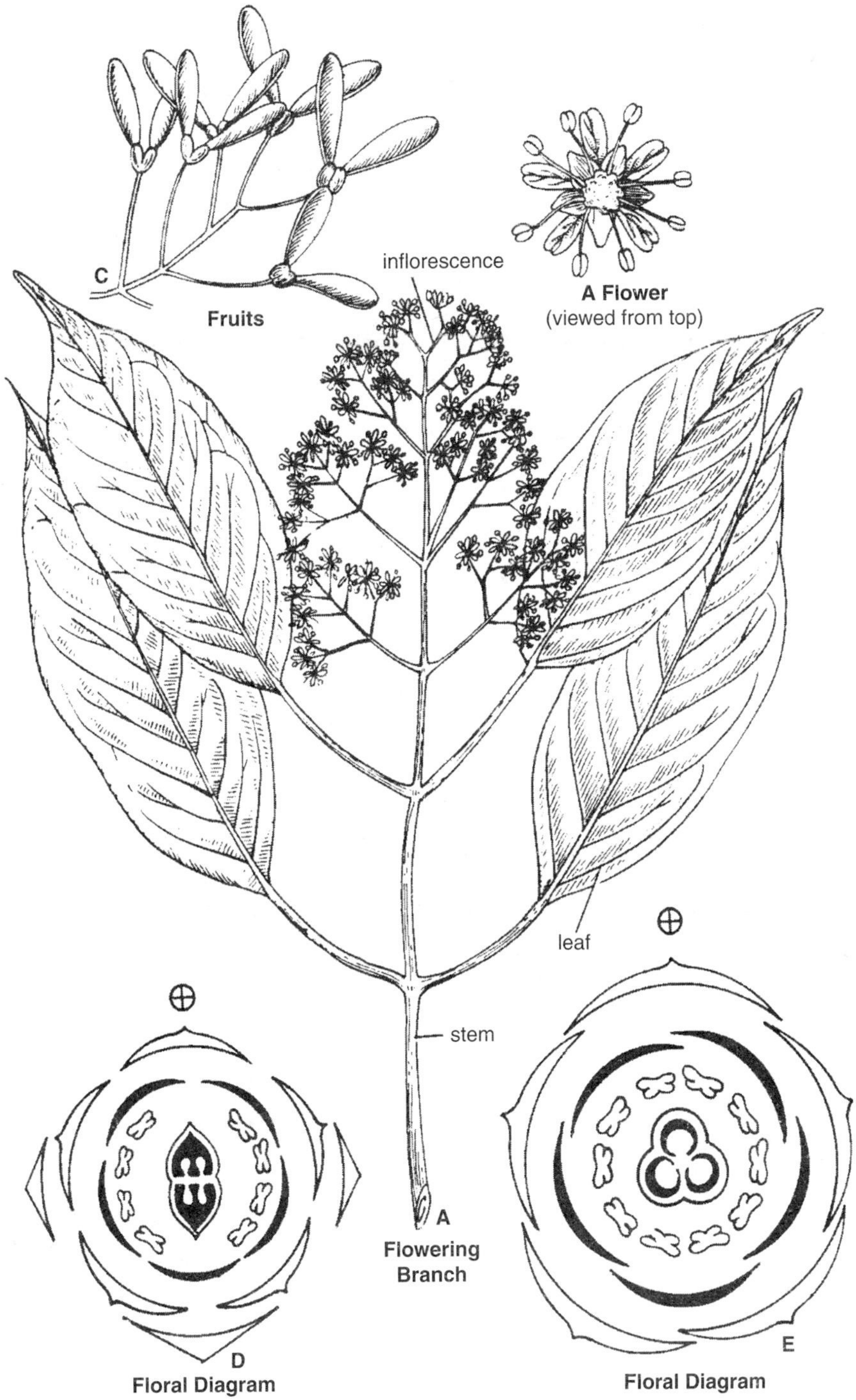

Fig. 21.41 A–D: *Acer oblongum* Wall, ex D, C.; E: Floral diagram of *Sapindus*.

21.31.6 Economic Importance

The economic importance of Sapindaceae lies in the fact that it provides several edible fruits, e.g. Litchi; saponin, guarana, akee, plus some ornamental plants.

- Delicious edible fruits are obtained from *Litchi chinensis* (Litchi), *Blighia sapida* (Akee), *Euphoria longana* (Anshphal), *Melicocca bijuga* and *Nephelium lappaceum* (Rambutan). The edible part of Litchi is sweet pulpy aril which surrounds the seed.
- *Acer hippocastanum* (Horse chestnut tree) fruits are given to horses as well as used for manufacture of starch; *A. indica is* a timber-yielding plant and its fruits and seeds are used in rheumatic complaints; *A. saccharinum* (Sugar maple) is the source of maple sugar and timber.
- *Dodonaea viscosa* (Villayati Mehndi), an evergreen shrub, is a popular ornamental hedge plant.
- *Nephelium lappaceum* (Rambutan of Malaya) fruits are edible and the fat obtained from its seeds is used in making candles and soap.
- *Paullinia cupana* (Guarana) seeds are used like cacao in Brazil to make a drink having high caffeine content.
- *Sapindus mukorossi* (Ritha or Soapnut tree) and *S. saponaria* (Soapberry tree) fruits contain saponin. On being moistened, saponin forms a lather and used as a soap. Ritha is used for washing hairs and clothes. Saponin is used in the preparation of soaps, tooth pastes, shampoos as well as in some insecticides.
- *Schleichera oleosa* yields edible fruit and this tree is used for rearing lac insects. It also provides a good timber and the oil obtained from its seeds is used for cooking.

21.31.7 Systematics and Phylogeny (Affinities)

Sapindaceae was placed under order Sapindales (after Celastrales and before Rosales) of series Disciflorae by Bentham and Hooker. Order Sapindales was placed between Rutales and Rhamnales by Rendle (1925), Geraniales and Rhamnales by Lawrence (1951), and between Polygalales and Geraniales by Cronquist (1981). Thorne (1983) discussed Sapindaceae under order Rutales, suborder Sapindineae.

Sapindaceae is related to Anacardiaceae in habit and general floral structure. However, the flowers are mostly irregular in Sapindaceae and regular in Anacardiaceae. In the characters, such as the presence of one ovule in each locule, tricarpellary ovary, often unisexual flowers and arillate seeds, family Sapindaceae resembles Euphorbiaceae.

Sapindales are generally thought to be evolved parallel to Rutales and Meliales.

Bentham and Hooker divided Sapindaceae into five subfamilies (Sapindeae, Acerineae, Dodoneae, Meliantheae and Staphyleae). However, the recent trend is to divide it into only two subfamilies *viz.* Sapindoideae and Dodonaeideae.

21.32 ANACARDIACEAE (CASHEW FAMILY)

21.32.1 Systematic Position

Polypetalae, Disciflorae, Sapindales.

21.32.2 Field Recognition

Trees or shrubs; pentamerous flowers; presence of intrastaminal disc; presence of resin ducts; unilocular ovary; fruit usually drupe.

21.32.3 Selected Indian Genera

Mangifera, *Rhus*, *Anacardium*, *Spondias*, *Buchanania* and *Lannea*.

21.32.4 Size, Distribution and General Information

A family of about 80 genera and over 600 species (Jones and Luchsinger, 1987), Anacardiaceae are mostly tropical in distribution but found also in temperate regions of Mediterranean, America and eastern Asia. 23 genera and over 115 species of Anacardiaceae have been reported from India. Some larger genera of Anacardiaceae, along with their common names and approximately reported species in parenthesis, include *Rhus* (Sumac or poison ivy, 250), *Semecarpus* (Dhobis nut, 50), *Mangifera* (Mango, 40), *Schinus* (Pepper tree, 30), *Anacardium* (Cashew nut, 15), *Pistacia* (Pistachio nut, 10), *Spondias* (Hog plum, 10) and *Cotinus* (Smoke tree, 3). *Mangifera indica* (mango) is the most familiar plant of the family.

21.32.5 Description of the Family

General Habit Usually trees and shrubs, rarely climbers or woody vines; containing resin ducts with plenty of gum and acrid juice.

Leaves Usually alternate (opposite in *Dobinea*, *Bouea*), simple or pinnately compound; exstipulate or stipules obscure.

Inflorescence Axillary or terminal panicles.

Flower Bracteate, bisexual or sometimes unisexual, pentamerous, actinomorphic; hypogynous, rarely perigynous (*Holigarna*) or epigynous; small and numerous.

Calyx Usually 5 sepals, sometimes 3 to 7; free or basally connate or adnate to gynoecium, or variously divided; imbricate or quincuncial.

Corolla Usually 5 petals, sometimes 3 to 7; or even absent (*Pistacia*); free or rarely fused; imbricate or rarely valvate.

Disc A cupular, extrastaminal or intrastaminal, nectariferous disc is usually present; rarely the disc is absent (*Pistacia*).

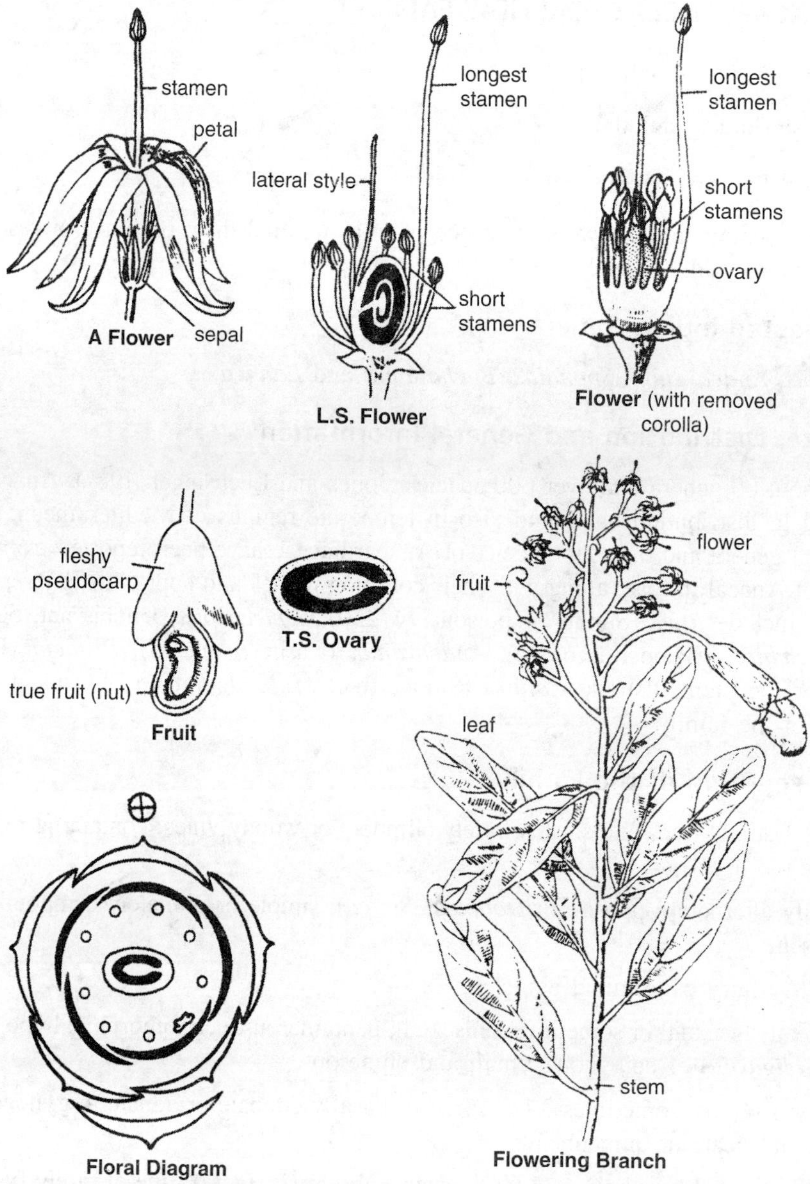

Fig. 21.42 *Anacardium occidentale* L.

Androecium 10 stamens (or stamens double the number of petals), in 2 whorls; usually fewer stamens (1 stamen and 6–9 staminodes or ill-developed stamens in *Anacardium*, Fig. 21.42; 1 stamen and 4 staminodes in *Mangifera*, Fig. 21.43; 5 stamens in one whorl in *Rhus*, Fig. 21.44); androecium

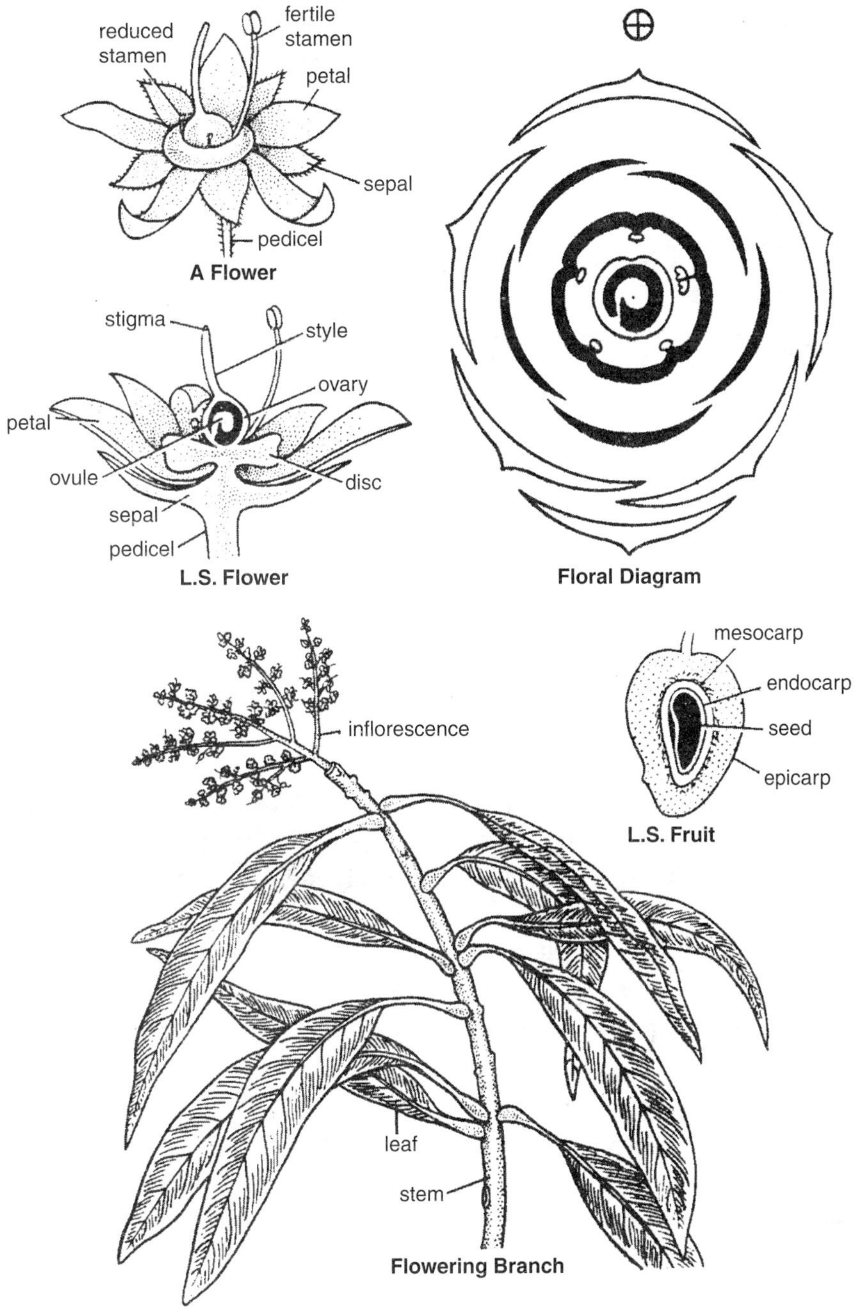

Fig. 21.43 ◈ *Mangifera indica* L.

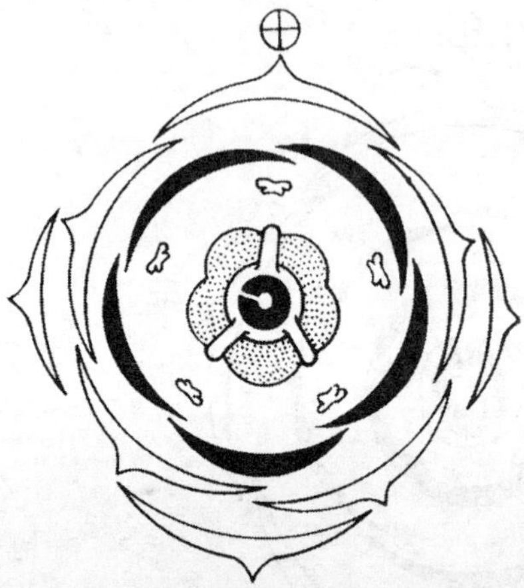
Fig. 21.44 Floral diagram of *Rhus parviflora*.

usually arise from beneath a disc; filaments free or rarely basally connate (*Anacardium*); anthers basifixed or dorsifixed, introrse.

Gynoecium A compound pistil of 1 to 5 fused carpels to form a unilocular or multilocular superior ovary, having a solitary ovule in only one of the ovary chamber; sometimes 3 carpels; styles 1–5; each with a capitate stigma; only one carpel in *Anacardium* and *Mangifera*; in *Drimycarpus* the ovary is inferior; placentation basically axile.

Fruit and Seed Fruit usually a drupe (*Mangifera*) with resinuous mesocarp; in *Anacardium* the fruit is a nut and remains situated on a pyriform fleshy mass developed by the elongation of disc and top of pedicel. Seed contains curved embryo with very little or no endosperm.

General Floral Formula Br, ⊕, ⚥, K_5, C_5, A_{10}, $G(\underline{1-5})$.

21.32.6 Economic Importance

Anacardiaceae is important for the mango, cashew, pistachio, resins, oils, tannic acids, chironji, lacquers, several ornamentals, as well as for causing dermatitis in humans.

- *Mangifera indica* (mango) is universally known for its delightful and delicious fruit as well as for providing us pickles and timber.
- *Anacardium occidentale* (Cashew-nut) is famous for its edible nuts (seeds). The fruit and seed coat of the cashew contain a poison. However, its seeds are edible and highly priced.
- *Buchanania lanzan* (Chironji) seeds are commonly used in sweet-meats in India. Its bark is used for tanning.
- *Harpephyllum caffrum* (kaffir plum) is famous for its edible fleshy fruits.
- *Holigarna arnottiana* provides useful timber for match industry and packing cases.
- *Lannea coromandelica* bark is used for tanning and provide a gum used in calico printing.
- *Parishia insignis* provides useful timber for railway sleepers.
- *Pistacia vera* (pistachio-nuts) is the source of famous edible nuts used in sweet-meats, ice-creams, etc. *P. lentiscus* (mastic tree) yields a mastic resin used in chewing gums, alcoholic beverages, etc.
- *Rhus vernicifera* (Lacquer tree) is grown for its milky poisonous latex. Contact with this plant may cause dermatitis in humans. Tannic acids are obtained from *R. coriaria*. *R. parviflora* (Tatri) fruits are used as tamarinds.
- *Schinopsis lorentzii* is used in leather-tanning industry.
- *Semecarpus anacardium* (Dhobis-nut) fruits provide a black ink used for dyeing textiles.

- *Spondias axillaris* and *S. pinnata* (Hog plum) fruits are edible and made into pickles.
- *Toxicodendron vernicifera* provides us resins and oils.
- Some ornamental plants of Anacardiaceae include *Cotinus coggygria*, *Rhus typhina* and *Spondias pinnata*.

21.32.7 Systematics and Phylogeny (Affinities)

Bentham and Hooker, Engler and Prantl, Cronquist and several other taxonomists placed Anacardiaceae under Sapindales while Takhtajan (1969) and Thorne (1983) placed it under order Rutales.

Majority of the phylogenists trace the affinity of Anacardiaceae with Sapindaceae. The family is, however, related closely to Aceraceae, Hippocastanaceae, Sapindaceae, Meliaccac, Rutaceae and Burseraceae. The recent trend is to divide Anacardiaceae into 5 tribes *viz.* Anacardieae, Spondiadeae, Rhoeae, Semecarpeae and Dobineeae (Hickey and King, 1988).

21.33 ROSALES

21.33.1 General Characteristics

1. Flowers bisexual, generally cyclic and typically pentamerous.
2. Flowers hypogynous to epigynous with common perigyny.
3. Stamens commonly arranged in many whorls.
4. Gynoecium apocarpous to syncarpous but the styles general distinct.
5. Ovary generally inferior, less-commonly half-inferior or superior.
6. Thalamus discoid or concave or flask-shaped.

Bentham and Hooker included 9 orders (=families) under Rosales, of which the major ones are Leguminosae, Rosaceae and Saxifragaceae. Order Rosales was considered by Engler and Diels to be composed of 17 families, chief amongst which are Crassulaceae, Saxifragaceae, Hamamelidaceae, Rosaceae and Leguminosae.

Only Leguminosae, Rosaceae and Saxifragaceae are treated in this text.

21.34 LEGUMINOSAE (LEGUME FAMILY)

21.34.1 Systematic Position

Polypetalae, Calyciflorae, Rosales.

Traditional approach is to treat all legumes as one large, somewhat heterogenous family, the Leguminosae under order Rosales. Bentham and Hooker followed this approach and divided family Leguminosae into three subfamilies (Papilionaceae, Caesalpinieae and Mimoseae). However, majority of the recent taxonomists (Hutchinson, 1973; Takhtajan, 1980; Cronquist, 1981, etc.) treat all these three as three separate families (Mimosaceae, Caesalpiniaceae and Fabaceae or Papilionaceae) under order Fabales (Takhtajan, 1980; Cronquist, 1981) or Leguminales (Hutchinson, 1969). The traditional approach of Bentham and Hooker is followed in this book, and the three have been treated as subfamilies under the names Mimosoideae, Caesalpinioideae and Papilionoideae. However, some *generalised characters* of Leguminosae are given in the following paragraph:

Leguminosae, represented by about 600 genera and over 12000 species, is the third largest family of the flowering plants. Plants are herbs, shrubs, trees, or vines. *Leaves* simple or pinnately compound, often trifoliate, rarely palmately compound, alternate or opposite, usually stipulate; at the base of petiole or petiolule is often present a pulvinus; rarely leaves are reduced to scales. *Inflorescence* racemose raceme, panicle, spike, head, or flowers solitary. *Flower* bisexual, actinomorphic (Mimosoideae) or zygomorphic (Caesalpinioideae and Papilionoideae, Fig. 21.45), hypogynous or rarely perigynous. *Calyx* of usually 5 connate sepals, imbricate or rarely valvate. *Corolla* of 5 petals, rarely less, polypetalous or 2 anterior petals fuse along their lower margins to form keel, equal or unequal, imbricate or rarely valvate. *Androecium* of 10 stamens (numerous in Mimosoideae), rarely less than 10, polyandrous, or monadelphous, or diadelphous, usually longitudinally dehiscent. *Gynoecium* monocarpellary, ovary superior or rarely semi-inferior, unilocular, marginal placentation, style and stigma one and simple. *Fruit* usually a legume, or lomentum. *Seeds* usually with a shining leathery testa; endosperm very scanty or even absent.

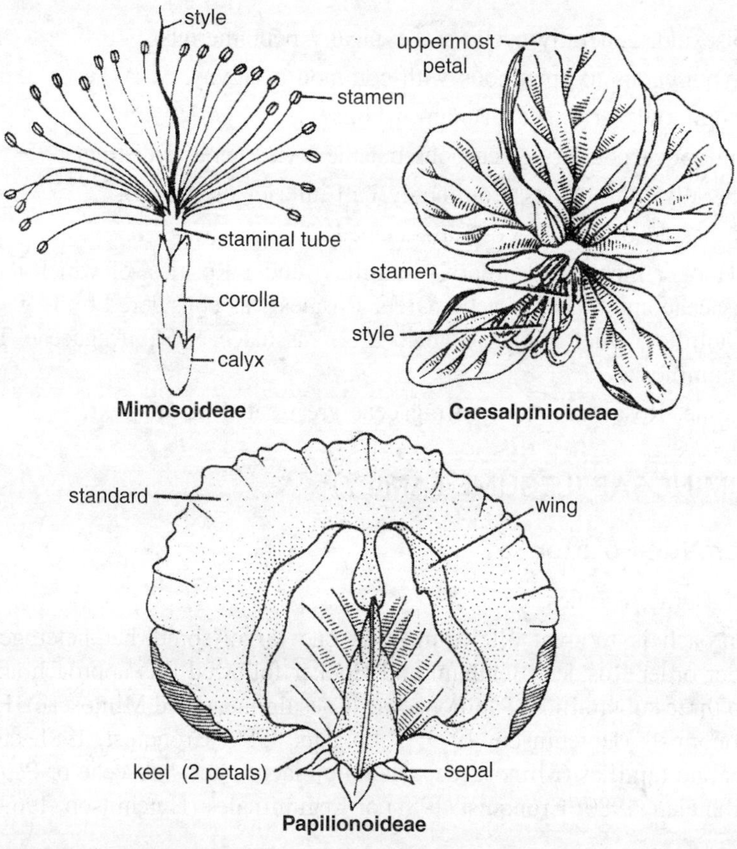

Fig. 21.45 Flowers of Leguminosae.

21.34.2 Key to Subfamilies

Flowers actinomorphic; perianth aestivation valvate or rarely imbricate; stamens 10 or usually more. .... *Mimosoideae*

Flowers zygomorphic; perianth aestivation usually imbricate; stamens between 5–10.

Corolla aestivation ascending-imbricate; posterior petal innermost; petals 5 and polypetalous; stamens polyandrous or monadelphous. .... *Caesalpinioideae*

Corolla aestivation descending imbricate; posterior petal outermost; petals 5 but anterior petals basally fused; stamens monadelphous or diadelphous. .... *Papilionoideae.*

21.35 SUBFAMILY MIMOSOIDEAE OR MIMOSEAE (MIMOSA FAMILY)

21.35.1 Systematic Position

Polypetalae, Calyciflorae, Rosales, Leguminosae.

21.35.2 Field Recognition

Trees or shrubs, rarely herbs; leaves bipinnately compound; stipules usually represented by a pair of thorns; flowers pentamerous, actinomorphic; stamens 10 to many; filaments coloured and often much longer than corolla.

21.35.3 Selected Indian Genera

Acacia, *Albizzia*, *Leucaena*, *Mimosa.*

21.35.4 Size, Distribution and General Information

Represented by about 56 genera and 2800 species, Mimosoideae are predominantly tropical and subtropical in distribution and almost absent in the colder regions of the world. About 15 genera and 75 species of this subfamily have so far been reported from India. Some of the larger genera along with their number of approximately reported species include *Acacia* (750), *Mimosa* (450), *Albizzia* (100), *Desmanthus* (40), *Prosopis* (40), *Schrankia* (30), and *Neptunia* (11). *Mimosa pudica* (sensitive plant) is the best known plant of this subfamily.

21.35.5 Description of the Family

General Habit Mostly medium-sized trees or shrubs (*Albizzia*, *Acacia*, *Dicrostachys*, *Parkia*, *Xylia*, etc.), few are herbs (*Neptunia oleracea*) or woody and tendril-bearing climbers (*Entada*).

Leaves Mostly bipinnate; some species of *Acacia* appear to possess simple leaves, which are actually phyllodes or leaf-like petioles having no blade; stipulate, stipules often modify into thorns (*Acacia*); sensitive or possess sleep-movement character in *Mimosa pudica* (Fig. 21.46) and *Neptunia oleracea*.

Inflorescence Flowers often grouped in tight clusters, or racemose head or capitulum, sometimes spicate or racemose (*Prosopis*).

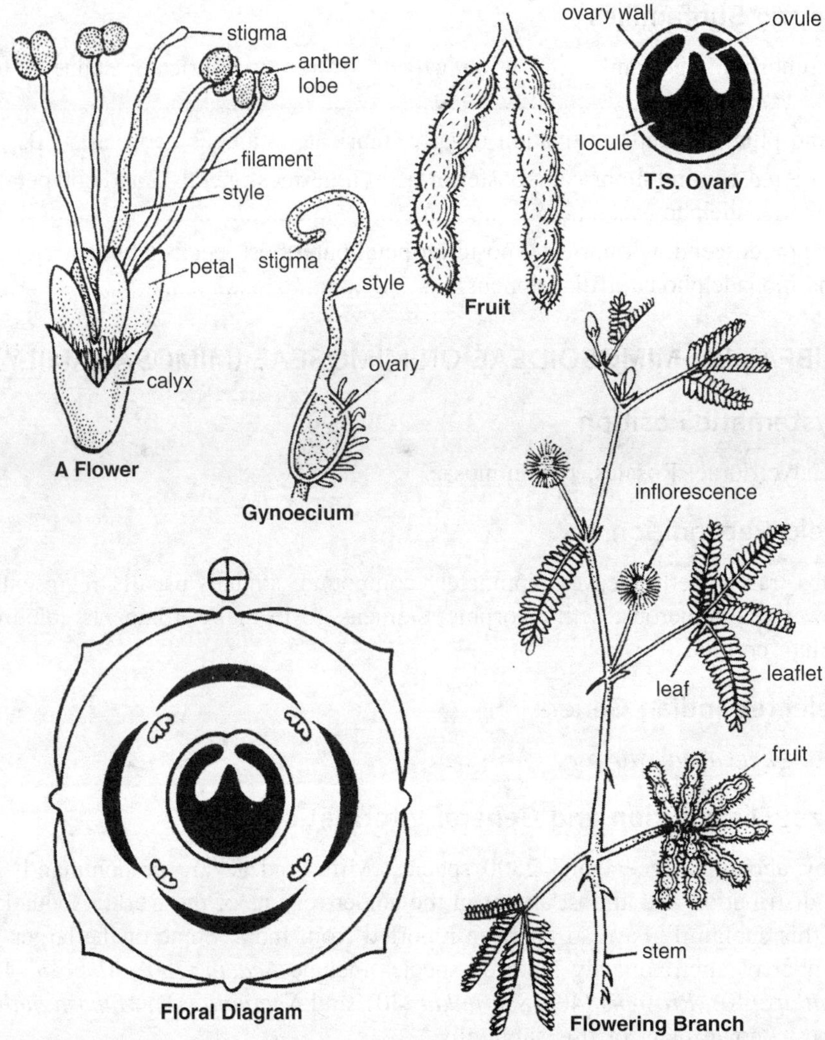

Fig. 21.46 *Mimosa pudica* L.

Flower Bracteate, actinomorphic, bisexual, pentamerous, sometimes tetra-, tri- or hexamerous; hypogynous, rarely perigynous.

Calyx 5 sepals, fused into a 5-lobed tube; valvate or rarely imbricate (*Parkia*); sometimes 4 diagonally placed sepals (*Mimosa*).

Corolla 5 petals, fused (*Acacia*, *Albizzia*) or free (*Parkia*, *Mimosa*, *Dicrostachys*); valvate; 4 petals in *Mimosa* (Fig. 21.46).

Androecium 4 to numerous stamens, show much variation in number and cohesion; extend beyond the petals and provide a fluffy soft appearance to the floral cluster; stamens are 4 or as many as petals (*Mimosa*, Fig. 21.46), or double the number of petals and diplostemonous (*Prosopis*), or

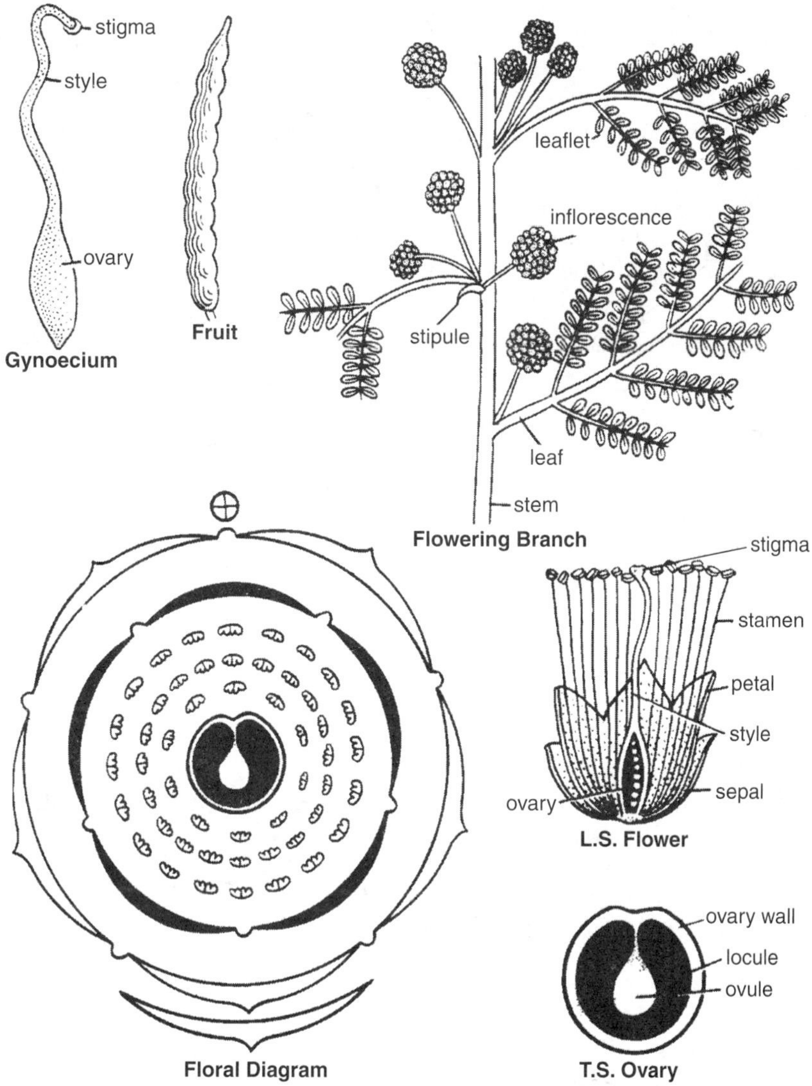

Fig. 21.47 *Acacia nilotica* L.

numerous and free (*Acacia*, Fig. 21.47), or numerous and monadelphous (*Albizzia*); dithecous, basifixed, introrse.

Gynoecium Monocarpellary; ovary superior, unilocular; ovules numerous, marginal placentation; style and stigma simple.

Fruit A legume, sometimes indehiscent, lomentum in *Acacia*.

Seed Non-endospermic or with scanty endosperm; embryo straight.

General Floral Formula Br, ⊕, ⚥, $K_{(5)}$, $C_{5 \text{ or } (5)}$, $A_{4 \text{ or } 10 \text{ or } \infty}$, $G\underline{1}$.

21.35.6 Economic Importance

- *Ornamental Plants:* *Acacia arabica* syn. *A. nilotica* (Babool or Kikar), *A. concinna* (Rassaul), *A. farnesiana* (Vilayati kikar), *A. melanoxylon* (Australian Acacia), *Albizzia lebbek* (Siris), *Leucaena leucocephala* (Subabul), *Mimosa pudica* (sensitive plant) and *Pithecellobium dulce* are the commonly grown ornamental plants.
- *Catechu of Commerce or Katha:* Katha, used for colouring khaki clothes and taken with betel leaves, is obtained from the heartwood of *Acacia catechu.*
- *Gum Arabic:* True gum arabic, used in medicine, confectionery and textile industry, is obtained from *Acacia senegal* and *A. stenocarpa.*
- *Tannins:* Bark of several species of *Acacia* (*A. catechu, A. decurrens, A. leucophloea, A. pennata*, etc.) is rich in tannins and is used in tanning of animal skins.
- *Timber:* Valuable timber for furniture, cabinet work, agricultural implements and other similar purposes is obtained from various species of *Acacia* (*A. catechu, A. ferriginea, A. leucophloea, A. nilotica*), *Albizzia* (*A. amara, A. chinensis, A. lebbek, A. procera*), *Adenanthera pavoniana* and *Xylia xylocarpa.*
- *Edible Products:* Beans of *Entada phaseoloides*, leaves of *Neptunia oleracea* and seeds of *Pithecellobium dulce* are edible. Fruits and seeds of *Prosopis cineraria* and *Samanea saman* are used as fodder.
- *Miscellaneous Uses:* (i) *Acacia concinna* (Shikakai) pods are used for washing hairs; (ii) *A. decurrens* and *Xylia xylocarpa* wood pulp is used for making wrapping paper; (iii) *A. farnesiana* flowers provide us a delicate perfume; (iv) *A. nilotica* plants are used as host for lac insects; (v) *A. sinuata* pods are used for manufacturing of shampoos; (vi) *Adenanthera pavoniana* seeds are used as weights by jewellers; (vii) *Entada phaseoloides* seeds are used for washing hairs.

21.35.7 Description of a Common Representative

1. Acacia nilotica (L.) Del. syn. A. arabica Willd (Kikar or Babool; Fig. 21.47)

Habit: Medium-sized tree. *Leaf:* Alternate; stipulate, stipules modify into long thorns; bipinnate and paripinnate. *Inflorescence:* Axillary cymose head. *Flower:* Bracteate, complete, actinomorphic, hermaphrodite, hypogynous, pentamerous, numerous, small. *Calyx:* 4 or 5 sepals, fused basally, valvate. *Corolla:* 4 or 5 petals, free or connate at the base, valvate. *Androecium:* Stamens numerous, polyandrous, filaments long; anthers dithecous, basifixed, introrse. *Gynoecium:* Monocarpellary, ovary superior; unilocular; many ovules, marginal placentation. *Fruit:* Lomentum.

Floral Formula: Br, $\oplus$, ⚥, $K_{(4-5)}$, $C_{4-5 \text{ or } (4-5)}$, A_{∞}, $G\underline{1}$.

21.36 SUBFAMILY CAESALPINIOIDEAE OR CAESALPINIEAE (CASSIA FAMILY)

21.36.1 Systematic Position

Polypetalae, Calyciflorae, Rosales, Leguminosae.

21.36.2 Field Recognition

Trees, shrubs, or herbs; flowers zygomorphic; corolla aestivation ascending-imbricate; posterior petal innermost; petals 5 and polypetalous; stamens usually 10, rarely numerous.

21.36.3 Selected Indian Genera

Bauhinia, Caesalpinia, Cassia, Delonix, Parkinsonia, Saraca, Tamarindus.

21.36.4 Size, Distribution and General Information

Represented by about 180 genera and 3000 species, members of subfamily Caesalpinioideae are distributed mostly in tropical and subtropical countries. About 23 genera and 85 species of this subfamily have been reported from India. Some of the largely represented genera along with their number of approximately reported species include *Cassia* (500–600), *Bauhinia* (300), *Senna* (250), *Chamaecrista* (250), *Caesalpinia* (100), *Gleditsia* (11) and *Cercis* (7) as stated by Jones and Luchsinger (1987) and Hickey and King (1988).

21.36.5 Description of the Family

General Habit Mostly trees (*Tamarindus*) and shrubs (*Parkinsonia*); rarely herbs (*Cassia tora*), or prickly climbers (*Caesalpinia sepiaria*); *Cassia* comprises of all the three habits i.e. herbs (*C. pumila*), shrubs (*C. timoriensis*) and tall trees (*C. javanica*).

Leaves Usually alternate, pinnate or bipinnately compound; rarely simple (*Bauhinia*); stipules sometimes modify into spines (*Parkinsonia*).

Inflorescence Usually raceme, or spikes, rarely corymb; or flowers solitary.

Flower Bisexual, zygomorphic, pentamerous, hypogynous or perigynous; rarely unisexual (*Gleditsia*).

Calyx Usually 5 sepals, free or basally connate, imbricate or valvate; odd sepal anterior; sometimes 2 posterior sepals fuse (*Tamarindus indica*, Fig. 21.48A); sepals petaloid in *Saraca indica*.

Corolla 5 petals, polypetalous, ascending-imbricate aestivation; showing posterior petal innermost; usually the petals are unequal-sized; petals greatly reduced in *Dialium*; petals absent in *Ceratonia*.

Androecium Usually 10 free stamens, arranged in 2 alternate whorls of 5 each (diplostemonous); rarely numerous and united below; often a few stamens reduced to staminodes (e.g., number of perfect stamens is only 3 in *Tamarindus*; 3 to 5 in *Bauhinia*, Fig. 21.48B; and 5 to 7 in *Cassia*, Fig. 21.49; while their all remaining stamens are reduced to staminodes); filaments often unequal; anthers dithecous, introrse and dehisce by apical pores (*Cassia*) or longitudinal slits (*Delonix*, *Parkinsonia*).

Gynoecium Monocarpellary, superior; unilocular ovary, with two alternating rows of anatropous or campylotropous ovules on marginal placentation; style single, ending into a simple stigma.

Fruit and Seeds Fruit legume, sometimes indehiscent; often winged. Seeds endospermic or non-endospermic.

General Floral Formula ⊕, ⚥, $K_{5 \text{ or } (5)}$, C_5, $A_{10 \text{ or } (\infty)}$, $G\underline{1}$.

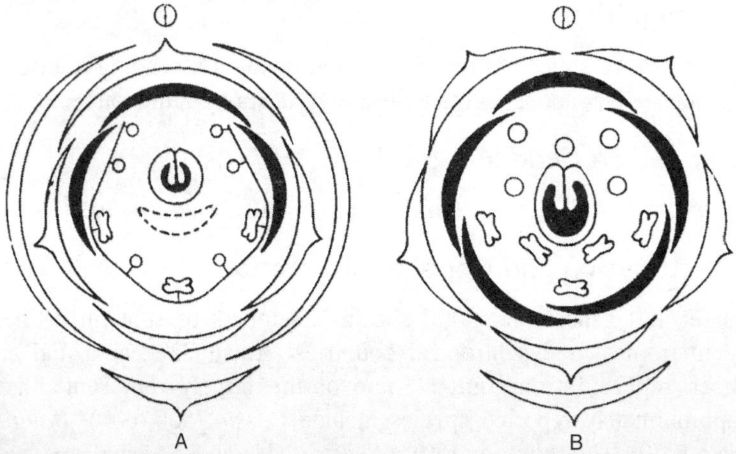

Fig. 21.48 A: Floral diagram of *Tamarindus indica*; B: Floral diagram of *Bauhinia variegata*.

21.36.6 Economic Importance

Subfamily Caesalpinioideae is important as a source of several ornamentals, drugs, dyes and timber.

- *Ornamental Plants:* *Amherstia nobilis* (Noble Amherstia), *Bauhinia acuminata* (Safed Kachnar), *B. corymbosa* (white-flowers), *B. galpini* (scarlet flowers), *B. monandra* (hedge plant), *B. purpurea* (purple-flowered Kachnar or Orchid tree), *B. racemosa* (pink-flowered Kachnar), *B. tomentosa* (yellow-flowered), *B. variegata* (variegated-flowered Kachnar), *Caesalpinia pulcherrima* (orange or yellow-flowered Poinciana), *Cassia alata*, *C. fistula*, *C. javanica*, *C. marginata*, *C. roxburghii*, *C. siamea*, *Cercis siliquastrum* (Judas tree), *Delonix alata* (white Gulmohar), *D. regia* (Gulmohar or Royal Poinciana), *Parkinsonia aculeata* (Vilayati Kikar) and *Saraca indica* (Ashok tree).
- *Medicinal Value:* (1) *Caesalpinia bunduc* seeds are used in diarrhoea and rheumatism, (ii) *Cassia acutifolia* leaves are used as purgative, (iii) *C. alata* leaves are used in skin infections, (iv) *C. fistula* fruit pulp is taken as a purgative. (v) *C. occidentalis* seeds and leaves are used in skin infections; (vi) *C. sophera* leaf decoction is beneficial in acute bronchitis.
- *Haematoxylin:* The best known nuclear stain, haematoxylin, is obtained from the heartwood of *Haematoxylon campechianum* (logwood).
- *Timber:* Valuable timber, useful in making agricultural implements, is obtained from *Cassia fistula*, *Copaifera pubiflora*, *Hardwickia binata* (Anjan), *Hymenaea courbaril* and *Tamarindus indica* (Tamarind).
- *Tannin:* Tannin of commercial importance is obtained from the bark and seeds of *Bauhinia malabarica*, *B. purpurea*, *B. racemosa*, *B. vahlii*, *B. variegata*, *Caesalpinia coriaria*, *C. digyna* and *C. sepiaria*.
- *Edible Products:* (i) Floral buds of *Bauhinia purpurea* (Kachnar) are cooked as vegetable, (ii) leaves and tender seeds of *B. malabarica* are also eaten, (iii) roasted seeds of *Cassia*

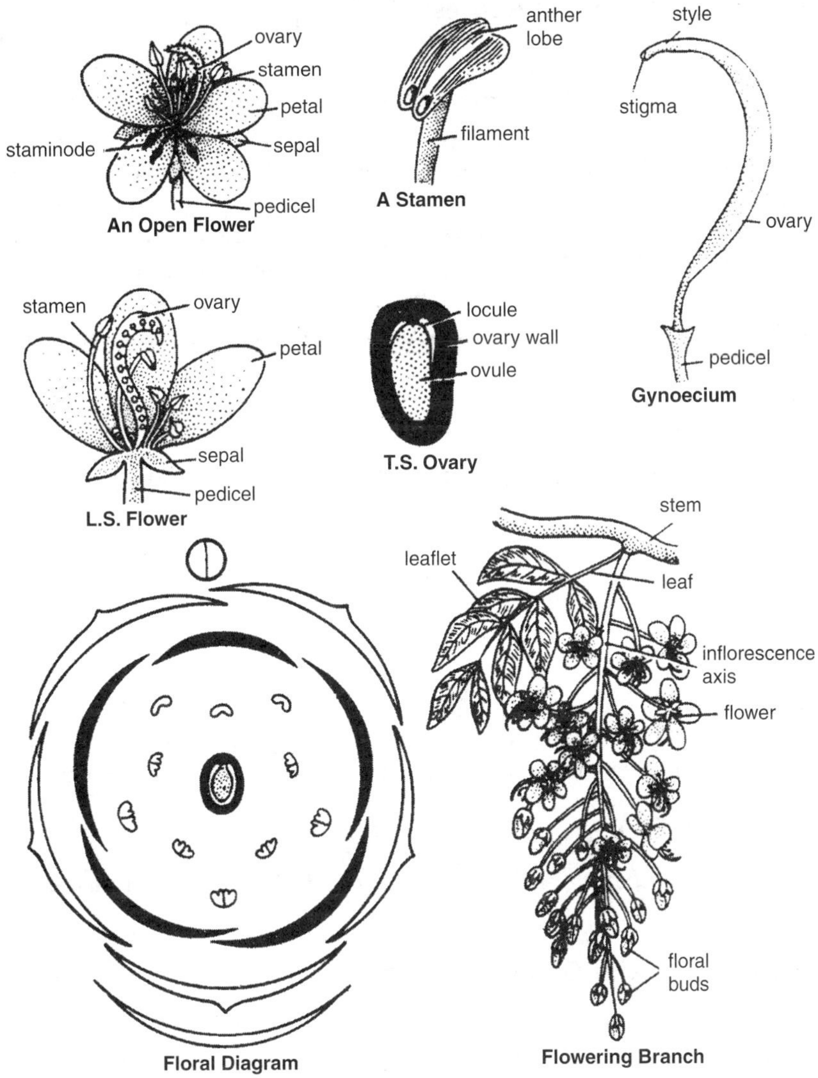

Fig. 21.49 *Cassia fistula* L.

occidentalis are used in place of coffee while that of *Gymnocladus canadensis* are used as Kentucky coffee, (iv) acidic fruit pulp of *Tamarindus indica* is used for sauces, curries, chutneys, etc.

- Gum is obtained from *Bauhinia racemosa*, *B. variegata* and *Copaifera pubiflora*.
- *Fibre* for making ropes is obtained from the bark of *Bauhinia racemosa* and *B. tomentosa*.
- Red dye for making red ink is obtained from the heartwood of *Caesalpinia sepiaria*.

21.36.7 Description of a Common Representative

Cassia fistula Linn. (Amaltas; Fig. 21.49)

Habit: An ornamental cultivated tree. *Leaf:* Stipulate, petiolate, unipinnate and paripinnately compound; leaflets 6–9, opposite, ovate, entire, acute, unicostate reticulate. *Inflorescence:* Racemose raceme. *Flower:* Bracteate, pedicellate, zygomorphic, hermaphrodite, pentamerous, hypogynous. *Calyx:* 5 sepals, free, petalloid (yellowish-green), quincuncial or imbricate, odd sepal anterior. *Corolla:* 5 petals, polypetalous, ascending imbricate, yellow, posterior petal small. *Androecium*: 10 stamens, polyandrous; 3 anterior stamens large, 4 stamens of inner whorl medium-sized, and 3 posterior stamens reduced into staminodes; anthers dithecous, dorsifixed, introrse. *Gynoecium:* Monocarpellary, superior, unilocular, marginal placentation; ovary sickle shaped; stigma reduced and curved. *Fruit:* Legume.

Floral Formula: Br, ⊕, ⚥, K_5, C_5, $A_{3+4+3\ \text{staminodes}}$, $G\underline{1}$.

21.37 SUBFAMILY PAPILIONOIDEAE (PEA FAMILY OR BEAN FAMILY)

21.37.1 Systematic Position

Polypetalae, Calyciflorae, Rosales, Leguminosae.

21.37.2 Field Recognition

Herbs, shrubs, or trees; leaves simple or imparipinnately compound or palmate; flowers zygomorphic; corolla papilionaceous (i.e. consisting of a standard, 2 wings and a keel); corolla aestivation descending-imbricate; posterior petal outermost; stamens monadelphous or diadelphous.

21.37.3 Selected Indian Genera

Arachis, Cicer, Crotalaria, Dalbergia, Dolichos, Glycine, Indigofera, Lathyrus, Pisum, Vicia, Vigna.

21.37.4 Size, Distribution and General Information

The subfamily Papilionoideae contains about 500 genera and over 10,000 species. It is distributed all over the globe excepting arctic regions, and is largely represented in warm temperate regions of both the Northern and Southern Hemispheres. Over 100 genera and 800 of its species have so far been reported from India. *Astragalus* (Milk Vetch), the largest genus of this subfamily, contains about 2,000 species. Some of the other larger genera, of which 100 or more species have been reported, include *Indigofera* (700, Indigo), *Crotalaria* (600, Rattlebox), *Desmodium* (450, Beggar-lice), *Dalbergia* (300, Shisham), *Trifolium* (300, Clover), *Phaseolus* (240, Cultivated beans), *Lupinus* (200, Lupine), *Vicia* (150, Broad bean), *Lathyrus* (130, Sweet pea), *Psoralea* (130, Snakeroot), *Medicago* (110, Alfalfa), *Erythrina* (100, Coral bean), and *Vigna* (100, Southern pea).

Takhtajan (1969) treated subfamily Papilionoideae as an independent family of order Fabales but instead of Papilionaceae he calls the family as *Fabaceae*. Cronquist (1981) and Jones and Luchsinger (1987) also used the name *Fabaceae* for Papilionaceae.

21.37.5 Description of the Family

General Habit Mostly herbs; however, shrubs, trees, and climbers are also common; *Clitoria*, *Dolichos*, *Phaseolus*, etc. are twiners; *Crotalaria juncea* is an example of shrub; *Millettia* is a woody climber; *Sesbania* is a small tree whereas species of *Dalbergia*, *Erythrina*, *Pterocarpus*, etc. are large trees.

Roots Lateral branches of roots of most Papilionoideae contain nodules or tubercles, which contain nitrogen-fixing bacteria (*Rhizobium*) having the ability to fix the atmospheric nitrogen.

Leaves Usually alternate, stipulate, compound, with pulvinous leaf base; stipules are adnate (*Medicago*, *Trifolium*), or free lateral (*Cajanus*, *Sesbania*), or foliaceous (*Pisum*); compound leaves may be unifoliate (*Desmodium gangeticum*), or bifoliate (*Zornia*), or trifoliate (*Medicago*, *Trifolium*, *Trigonella*); paripinnate (*Sesbania*), or imparipinnate (*Robinia*); in *Indigofera cordifolia* and *I. linifolia*, the leaves are simple; terminal leaflet modifies into a tendril in *Lathyrus*, *Pisum* and *Vicia*.

Inflorescence Usually racemose raceme (*Melilotus alba*), or head or capitulum (*Trifolium*), or spike (*Uraria*), rarely an umbel, or panicles (*Dalbergia*); sometimes the flowers are solitary (some species of *Lathyrus*).

Flowers Bracteate, sometimes bracteolate (*Sesbania*, *Astragalus*), complete, hermaphrodite, zygomorphic, pentamerous; hypogynous or sometimes perigynous.

Calyx 5 sepals, more or less united in a tube and persistent; valvate or imbricate; invariably the odd sepal is anterior in position.

Corolla 5 petals, very unequal and *papilionaceous* (i.e. 1-posterior petal is largest and called *standard* or *vexillum*, 2 lateral petals are lanceolate and slightly curved and called *wings*, and 2 anterior petals are asymmetrical and more or less united to form a boat-shaped structure and called *keel*); aestivation is vexillary or descending-imbricate i.e. standard covers the wings, and the wings cover the keel; all petals have a claw at the base.

Androecium 10 stamens, diadelphous, usually 9+1 (Fig. 21.50) or sometimes 5+5; in case of 9+1 arrangement, 9 stamens are united and 1 is free, and the free stamen is always posterior in position; often the posterior stamen is sterile or even absent (*Arachis hypogaea*, *Dalbergia sissoo*) and the remaining 9 stamens show monadelphous condition; anthers are dithecous, introrse and dehisce by longitudinal slits; rarely all the 10 stamens are free (*Sophora*).

Gynoecium Almost same as in Mimosoideae and/or Caesalpinioideae.

Fruit and Seed Fruit usually a legume, sometimes indehiscent; rarely a lomentum (*Desmodium*); matures inside the ground in *Arachis hypogaea*. Seeds usually with food reserves in cotyledons.

General Floral Formula Br, ⊕, ⚥, $K_{(5)}$, $C_{1+2+(2)}$, $A_{(9)+1}$, $G\underline{1}$.

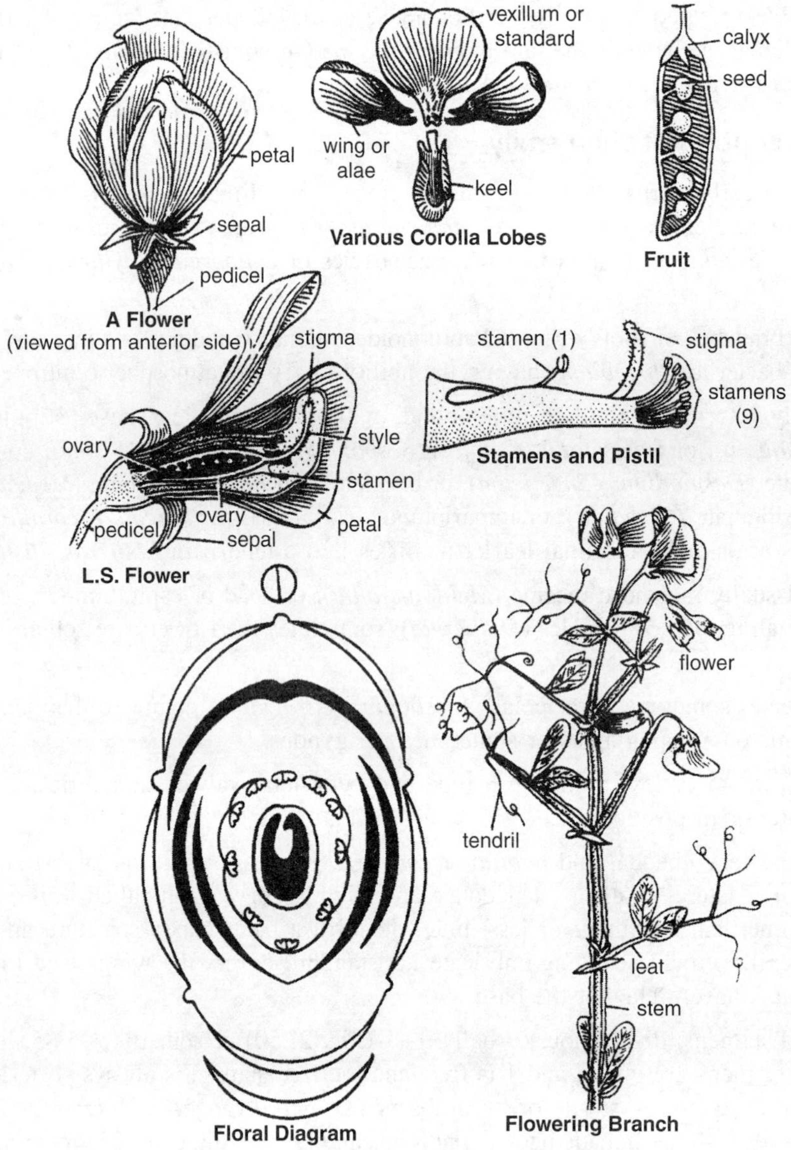

Fig. 21.50 *Lathyrus odoratus* L.

21.37.6 Economic Importance

Subfamily Papilionoideae is of great economic importance as a source of high-protein food (pulses), ornamentals, oil, forage, timber, dye, etc. Some of the universally known economically important plants of this subfamily include *Glycine max* (soybean, source of oil and high-protein meal), *Pisum sativum* (garden pea), *Cicer arietinum* (gram or Chana), *Medicago sativa* (alfalfa, world's best forage crop), *Arachis hypogaea* (groundnut, source of edible seeds, oil, peanut butter),

Indigofera tinctoria (Indigo plant, source of indigo dye), and *Lathyrus odoratus* (sweet pea, well-known ornamental).

- *Pulses and Vegetables*[1]*—Source of Proteins and Starch of our Food:* These include (i) Arhar or pigeon pea (*Cajanus cajan*), (ii) Bankla or field bean (*Vicia faba*), (iii) Chana or gram (*Cicer arietinum*), (iv) Gwar or cluster bean (*Cyamopsis tetragonoloba*), (v) Lobia or lima bean (*Phaseolus lunatus*), (vi) Masoor or lentil (*Lens culinaris* syn. *L. esculenta*), (vii) Matar or pea (*Pisum sativum*), (viii) Moth or dew gram (*Vigna aconitifolia* syn. *Phaseolus aconitifolius*), (ix) Mung or green gram (*Vigna radiata* syn. *Phaseolus radiatus*), (x) Rajmah or cowpea (*Vigna unguiculata*), (xi) Sem or bean (*Lablab purpureus* syn. *Dolichos lablab*), (xii) Soybean or Soya (*Glycine max* syn. *Glycine soja*), (xiii) Urd or black gram (*Vigna mungo* syn. *Phaseolus mungo*), and (xiv) Vilayati Sem or kidney bean (*Phaseolus vulgaris*).
- *Ornamental Plants*[2]*—Source of Beautification: Clianthus dampieri* (Glory pea), *Clitoria ternatea* (Butterfly pea, blue-flowered climber), *Erythrina indica* (Indian coral tree, orange red-flowered), *Lathyrus odoratus* (sweet pea, variously-coloured flowers), *Lupinus hirsutus* (Lupine, variegated-flowered annual), *Robinia pseudoacacia* (Black locust, white-flowered tree), *Sesbania grandiflora* (Sesban), *S. sesban* (Jait), *Sophora japonica* (Japanese Pagoda tree, multicoloured flowers), and *Wisteria sinensis* (Chinese Wisteria, variously-coloured flowers).
- *Oils:* *Arachis hypogaea* (groundnut or peanut or Moongphali; source of edible peanut oil; seeds are eaten after roasting; oil is used for preparing soaps, cosmetics; oil cake is used as a fodder), *Pongamia pinnata* (Pongam oil; oil obtained from its seeds is used for illumination, soap preparation and medicinal purposes), *Psophocarpus tetragonolobus* (Goa bean; seeds yield the oil used for soap making and cooking), and *Spatholobus roxburghii* (oil from its seeds is used for anointing and cooking).
- *Fodder:* Several plants of this family are of high forage value and grown as fodder for cattles. These include *Medicago sativa* (Alfalfa or Rizka), *M. abscura*, *M. hispida*, *Trifolium repens* (Clovers), *T. pratense*, *Dolichos biflorus*, *Phaseolus aconitifolius*, *Pisum sativum*, *Vicia benghalensis* and *V. sativa*.
- *Timber:* Valuable timber used for furniture, cabinet works and other building materials is obtained from *Baphia nitida*, *Dalbergia sissoo* (Shisham), *D. latifolium* (Kala Shisham), *D. melanoxylon* (African black wood), *Pterocarpus indicus* (Malay Padauk), *P. dalbergioides* (Andaman redwood), *P. marsupium* (Indian Kino tree), and *P. santalinus* (Red sandalwood, or Lal Chandan).
- *Medicinal Value:* Several plants[3] of this subfamily are used as the source of drugs. Some of them, along with their parts utilized and the name of the disease, to be cured, in parenthesis, include (i) *Abrus precatorius* (*Ratti;* leaves and roots in cough and cold); (ii) *Crotalaria albida* (roots used as purgative); (iii) *Glycyrrhiza glabra* (*Mulhatti* or *liquorice*; roots in cough and sore throat); (iv) *Krameria triandra* (roots in chronic diarrhoea); (v) *Moghania grahamiana*

[1] Arranged alphabetically in order of their Hindi local names.
[2] Arranged alphabetically in order of their botanical names.
[3] Botanical names of plants arranged alphabetically.

(roots are anthelmintic); (vi) *M. strobilifera* (roots induce sleep); (vii) *Physostigma venenosum* (seeds in eye troubles); (viii) *Psoralea corylifolia* (seeds used in leucoderma and leprosy), (ix) *Teramnus labialis* (entire plant *used* in tuberculosis); and (x) *Uraria lagopodioides* (leaves and roots used in intermittent fever).

- *Insecticides:* Roots of *Derris elliptica*, *D. ferruginea* and *D. trifoliata*, roots and seeds of *Milletia extensa*, entire plant of *Spatholobus roxburghii*, and roots of *Tephrosia vogelii* possess insecticidal properties.
- *Sola Hats and Toys:* Stuffing material of sunhats is the soft wood of *Aeschynomene aspera* and *A. indica*. Pith of these plants is used for making toys.
- *Dyes*[1]*:* (i) Orange-red dye is obtained from the flowers of *Butea monosperma* (Dhak); (ii) blue dye is obtained from the flowers and seeds of *Clitoria ternatea*; (iii) black dye is produced from the flowers of *Crotalaria striata*; (iv) red dye is obtained from the flowers of *Erythrina variegata*; (v) *indigo*, the famous blue dye, used for dyeing cotton clothes and in making paints and printing ink, is obtained from *Indigofera tinctoria* and some other species; (vi) yellow dye is obtained from the pods of *Psoralea pricata*; and (vii) red dye is produced from the wood of *Pterocarpus santalinus*.
- *Green Manuring:* Because of the presence of nitrogen-fixing bacteria in the lateral roots, several members of this subfamily are grown for enriching the soil with nitrogen and then ploughed as "green manure" in the field. Some of such members include *Crotalaria juncea*, *Medicago sativa*, *Tephrosia vogelli*, *Trifolium pratense* and *Trigonella foenum-graecum*.
- *Fibre:* Some of the fibre-yielding plants[2] of this subfamily, used for making cords, ropes, bags, etc. include *Butea monosperma* (Dhak), *Crotalaria juncea* (Sunnhemp or Sani), *Erythrina variegata*, *Millettia auriculata*, *Sesbania bispinosa* (Dhaincha), *S. cannabina* and *S. sesban*.
- *Gums and Resins:* Gums and resins obtained from some of the plants of this subfamily along with their utility in parenthesis include *Astragalus gummifer* (in confectionery, cosmetics and textile industry), *A. strobiliferus* (in glazing and calico-printing), *Cyamopsis tetragonoloba* (in food and textile industry), and *Myroxylum balsamum* (in medicine and perfumery).
- *Honey Production:* *Lathyrus odoratus*, *Medicago sativa*, *Pisum sativum* and *Trifolium pratense* are some of the plants of this subfamily which are used for artificial culture of bees for honey production.

21.37.7 Description of Some Common Plants

1. Lathyrus odoratus Linn. (Sweet pea or Phool Matar, Fig. 21.50)

Habit: Annual, cultivated, garden ornamental, climbing herb. *Leaf:* Alternate, petiolate; stipulate, stipule leafy; pinnately compound, imparipinnate; upper leaflets modify into tendrils. *Inflorescence:* Racemose or solitary axillary flowers. *Flower:* Bracteate, complete, hermaphrodite, zygomorphic, pentamerous, perigynous. *Calyx:* 5 sepals, fused, odd sepal anterior. *Corolla:* 5 petals, papilionaceous

[1] Botanical names of plants arranged alphabetically.

[2] Botanical names arranged alphabetically.

corolla (consisting of 1 standard or vexillum, 2 wings or alae and 2 fused petals forming keel or carina); descending imbricate. *Androecium:* 10 stamens, diadelphous with 9 stamens united and 10th posterior stamen free, dithecous, basifixed, introrse. *Gynoecium:* Monocarpellary, semi-inferior, unilocular, marginal placentation. *Fruit:* Legume.

Floral Formula: Br, ⊕, ⚥, $K_{(5)}$, $C_{1+2+(2)}$, $A_{(9)+1}$, G1–.

2. *Crotalaria medicaginea Lamk. (Fig. 21.51)*

Habit: Annual herb. *Leaf:* Alternate, petiolate; stipulate, stipules free-lateral; trifoliate; leaf base pulvinus; obovate, apex emarginate. *Inflorescence:* Racemose raceme. *Flower:* Same as in *Lathyrus. Calyx:* Same as in *Lathyrus. Corolla:* Same as in *Lathyrus. Androecium:* 10 stamens, monadelphous, dithecous, dorsifixed, introrse. *Gynoecium:* Almost same as in *Lathyrus. Fruit:* Legume.

Floral Formula: Br, ⊕, ⚥, $K_{(5)}$, $C_{1+2+(2)}$, $A_{(10)}$, G1–.

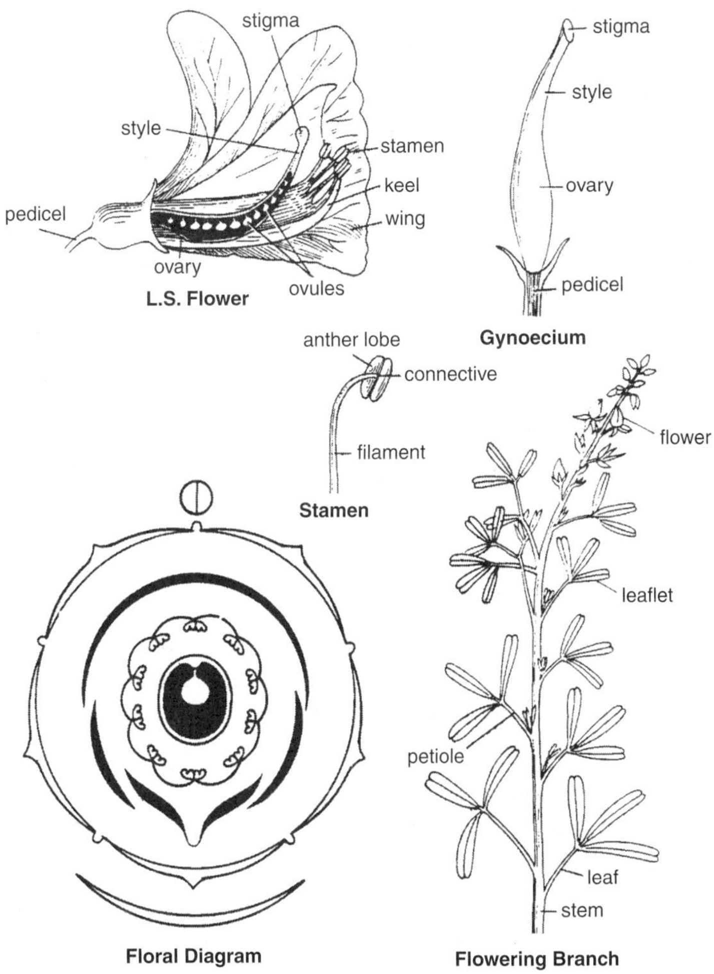

Fig. 21.51 *Crotalaria medicaginea* Lamk.

21.38 ROSACEAE (ROSE FAMILY)

21.38.1 Systematic Position

Polypetalae, Calyciflorae, Rosales.

21.38.2 Field Recognition

Perennial herbs, shrubs, or trees; leaves stipulate; flowers pentamerous, actinomorphic; stamens numerous and usually in multiple of 5; ovary superior to inferior; fruit various types; seeds non-endospermic.

21.38.3 Selected Indian Genera

Rosa, Pyrus, Prunus, Eriobotrya, Potentilla, Rubus, Spiraea.

21.38.4 Size, Distribution and General Information

A family of approximately 100 genera and 3000 species which are cosmopolitan in distribution but particularly abundant in Eastern Asia, North America and Europe. About 25 genera and 230 species of Rosaceae have so far been reported from India. Some of the largely represented genera along with their common names and approximately reported species include *Potentilla* (Cinquefoil, 500), *Prunus* (peach, plum, almond, cherry, apricot, 430), *Rosa* (Rose, 250), *Rubus* (Raspberry, blackberry, 250), *Alchemilla* (250), *Crataegus* (Hawthorn, 200), *Sorbus* (Chokeberry, 100), *Acaena* (100), and *Spiraea* (100).

Named after the universally famous flowering plant, rose, the Rosaceae show great diversity in types of fruit (Fig. 21.52).

21.38.5 Description of the Family

General Habit Mostly trees (*Prunus, Pyrus, Eriobotrya*) and shrubs (*Spiraea*), and only some are herbs (*Fragaria indica, Potentilla supina*); some are prickly climbers (*Rosa moschata, Rubus*) or prickly shrubs (*Rosa damascena*).

Stem Herbaceous or woody, usually with thorns or prickles.

Leaves Mostly alternate, simple (*Prunus, Nuttalia*) or compound; usually stipulate, stipules paired and sometimes adnate to the petiole (*Rosa, Rubus*), sometimes exstipulate (*Spiraea*); compound leaves may be palmately compound (*Fragaria*) or pinnately compound (*Rosa*); extrafloral nectaries present on the back of leaves in *Prunus laurocerasus*.

Inflorescence Highly variable from solitary flowers to racemose and cymose clusters; solitary (*Rosa*), racemose raceme (*Agrimonia, Prunus mystifolia*), corymb (*Spiraea*), head or spike (*Poterium*), cymes (*Spiraea, Fragaria*), or thyrsoid panicles (*Eriobotrya*).

Flower Mostly bracteate, bracteolate; usually actinomorphic but zygomorphic in *Hirtella* and *Parastemon*, usually bisexual but unisexual in *Poterium sanguisorba* and *Pygeum*, usually complete

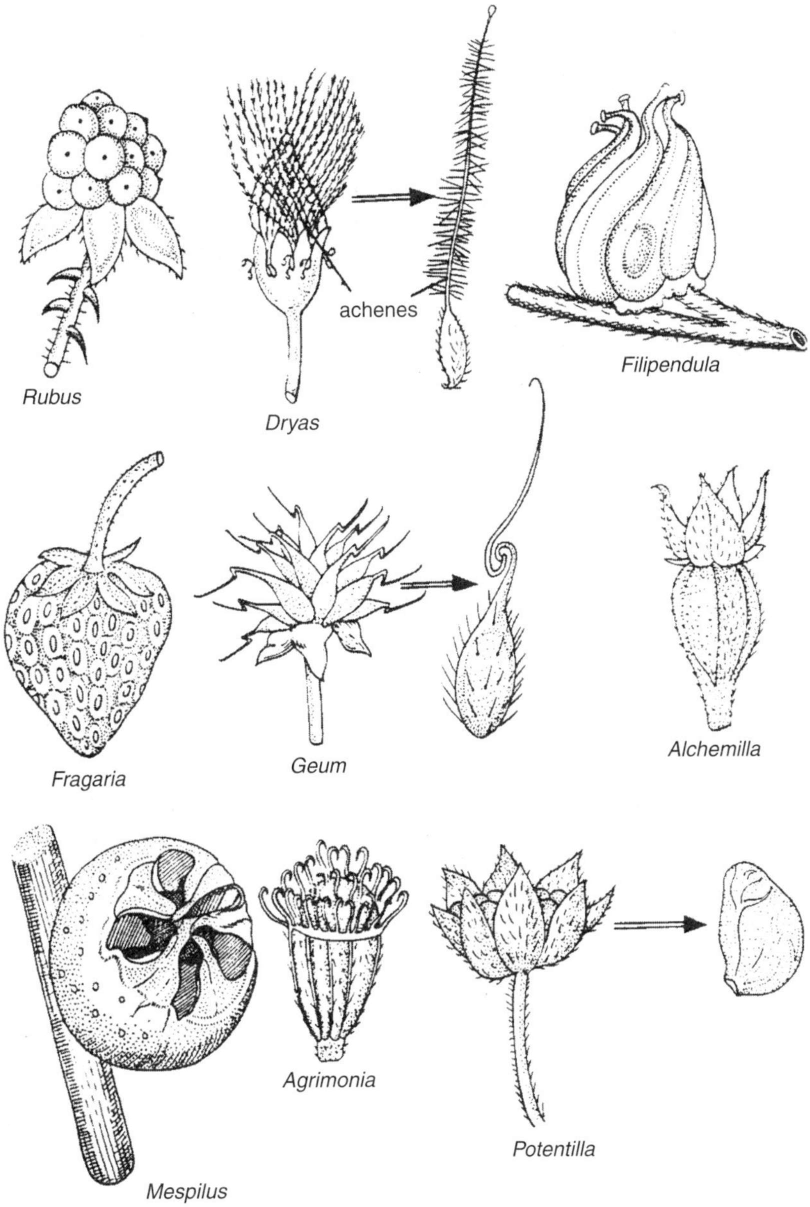

Fig. 21.52 Types of fruit in Rosaceae.

but apetalous in *Alchemilla*, pentamerous; usually perigynous (*Rosa*) or epigynous (*Pyrus*) and only seldom hypogynous (*Prunus*); epicalyx usually present; usually the floral receptacle along with the basal adnate portions of perianth is variously modified into hypanthium; the hypanthium is often lined with a nectariferous disc.

Epicalyx Often present in several genera (*Fragaria*, *Geum*, *Potentilla supina*) and alternate with calyx lobes.

Calyx 5 sepals, free or basally connate; valvate or imbricate; sepals often foliaceous (*Rosa*).

Corolla 5 petals, usually attached to the rim of hypanthium; petals are numerous, large and variously coloured in cultivated species of *Rosa*; petals absent in *Alchemilla* and *Poterium*.

Androecium Usually numerous, distinct stamens in whorls of 5, sometimes only 5 or 10 stamens; anthers basifixed or dorsifixed, dithecous, introrse and dehisce by longitudinal slits; usually a honey-secreting disc is present in between stamens and carpels. Undermentioned are some of the commonly seen arrangements of stamens in some Rosaceous genera, though there may also be some deviations from these arrangements:

1. Only 1 stamen in *Alchemilla arvensis* (Fig. 21.53A).
2. Only 4 stamens alternating with sepals in *Alchemilla vulgaris* (Fig. 21.53B).
3. Only 10 stamens with their filaments joined together (monadelphous) in *Acioa guianensis*.
4. Only 15 stamens arranged in an outer whorl of 5 pairs of antisepalous stamens and an inner whorl of 5 antipetalous stamens in *Nuttalia*.
5. Only 20 stamens arranged in an outer whorl of 5 antisepalous pairs followed by a middle whorl of 5 antipetalous and an inner whorl of 5 antisepalous stamens in *Pyrus* (Fig. 21. 53C).
6. Only 25 stamens arranged in an outer whorl of 5 antipetalous pairs, a middle whorl of 5 antisepalous pairs and an inner whorl of 5 antipetalous stamens in *Potentilla* (Fig. 21.53D).
7. Only 30 stamens arranged in an outer whorl of 5 antisepalous pairs followed by a middle whorl of 10 antipetalous and an inner whorl of 10 antisepalous stamens in *Prunus*.
8. Only 30 stamens arranged in an outer whorl of 5 antipetalous pairs followed by a 2nd and 3rd whorl of 10 stamens each in antisepalous and antipetalous positions in *Spiraea*.
9. Numerous stamens arranged in an outer whorl of 5 antipetalous pairs followed by numerous whorls of 10 stamens each in alternating positions in *Rosa* (Fig. 21.54).

Gynoecium Highly variable from 1 to numerous pistils; apocarpous to syncarpous; present at the base of the floral cup, or on the sides of floral cup, or enclosed by the floral cup; pistils are either free from the floral cup or adnate to the floral cup; ovary superior to inferior with intermediate stages; styles and stigmas as many as carpel number.

Monocarpellary (*Alchemilla*, *Chrysobalanus*, *Prunus*), or pentacarpellary and apocarpous (*Spiraea*), or multicarpellary and apocarpous (*Fragaria*, *Rosa*, *Rubus*), or 5 carpels fused only in the basal region (*Nuttalia*), or carpels united completely with each other as well as with the floral cup (*Pyrus*); if the receptacle is convex or dome-shaped, the ovary is superior (e.g. *Fragaria*, *Potentilla*, *Rubus*); if receptacle is cup-shaped and the carpels are developing on its inner surface, the ovary is half-inferior (e.g. *Prunus*, *Rosa*, *Spiraea*); and if the carpels are fused with one another and also with the floral cup (e.g. *Pyrus*), the ovary is inferior; placentation is marginal or axile.

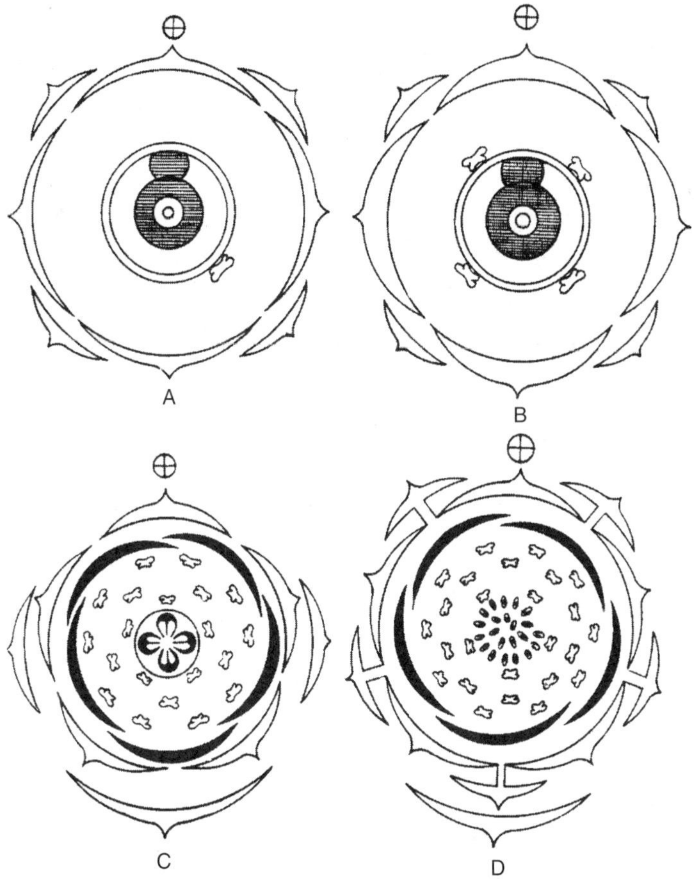

Fig. 21.53 A–D Floral diagrams of *Alchemilla arvensis* (A); *A. vulgaris* (B); *Pyrus* (C) and *Potentilla* (D).

Fruit Various types (Fig. 21.52), varying from drupe (*Prunus*), pome (*Pyrus*, *Malus*, *Crataegus*, *Sorbus*), follicles (*Spiraea*, *Sorbaria*, *Physocarpus*), etaerio of achenes (*Potentilla*, *Fragaria*, *Rosa*), or etaerio of drupes (*Rubus*).

Seed Endosperm usually absent.

General Floral Formula Br, ⊕, ⚥, $K_{5 \text{ or } (5)}$, C_5, A_{∞}, $G\underline{1-\infty}$ –, or $G\overline{1-\infty}$, or $G(\infty)$–.

21.38.6 Economic Importance

- *Edible Fruits:* Variety of delicious fruits belong to Rosaceae, and this family should better be called as the "family of edible fruits". Some of them[1] include almond or Badam (*Pyrus*

[1]Arranged alphabetically according to their English names along with the Hindi names of some of them.

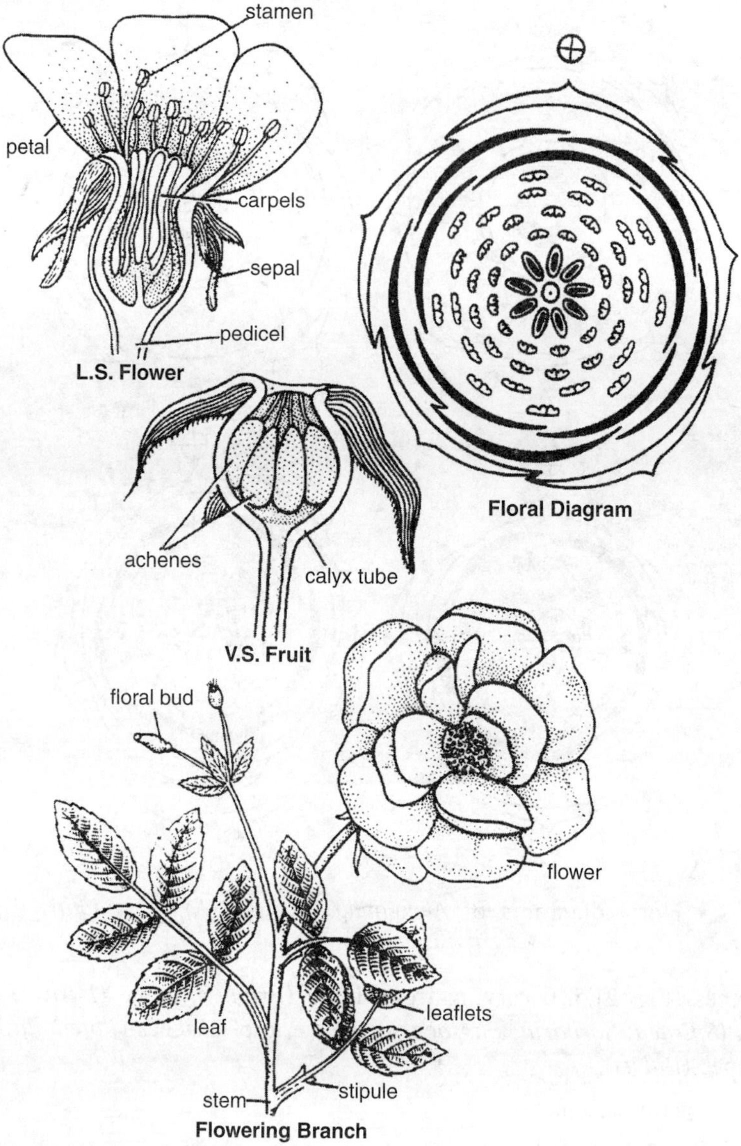

Fig. 21.54 *Rosa indica* L.

amygdalus), alpine strawberry (*Fragaria vesca*), apple or Seb (*Pyrus malus*), apricot or Khubani or Zardalu (*Prunus armeniaca*), blackberry or Vilaiti Alucha (*Rubus fruticosus*), blackcherry (*Rubus molucannus*), Ceylon raspberry (*Rubus laciocarpus*), cherry laurel (*Prunus laurocerasus*), Chinese pear (*Pyrus pyrifolia*), dewberry (*Rubus caesius*), Japanese plum (*Prunus saticina*), Loquat (*Eriobotrya japonica*), peach or Aadu (*Prunus persica*), pear or Nakh (*Pyrus communis*), plum or Aloocha or Alubukhara (*Prunus domestica* sub. sp. *instita*), quince (*Cydonia oblonga*), raspberry (*Rubus idaeus*), rowan (*Sorbus aucuparia*), sour cherry or Alu-balu (*Prunus cerasus*), strawberry (*Fragaria chiloensis* and *F. vesca*), sweet cherry (*Prunus avium*), wild Himalayan cherry or Padam (*Prunus cerasoides*) and yellow Himalayan raspberry (*Rubus ellipticus*).

- *Ornamental Plants:* Rose (*Rosa*), the best ornamental plant of the world and represented by its innumerable varieties, belongs to this family. Some of the commonly grown species of *Rosa* include *R. alba*, *R. banksiae*, *R. centifolia*, *R. chinensis*, *R. damascena*, *R. indica* and *R. moschata*. Some other garden ornamentals of Rosaceae are *Acioa guianensis*, *Crataegus mollis*, *Exochorda racemosa* and several species of *Physocarpus*, *Potentilla*, *Sorbus* and *Spiraea*.
- *Medicines and Perfume:* (i) 'Gulkand', prepared from rose (*Rosa*) petals is a good tonic, laxative as well as useful in tonsilitis; (ii) root-stocks of *Geum urbanum* are used in diarrhoea and chronic dysentery; (iii) female flowers of *Hagenia abyssinica* are the source of an antihelmintic drug, cusso; (iv) decoction of leaves of *Potentilla anserina* is used against arthritis and kidney stones; (v) dried bark of *Pyrus pyrifolia* is useful in cough and cold; (vi) valuable perfume is obtained from the petals of *Rosa damascena* and other rose species.
- *Bark and Wood:* Bark of soaptree (*Quillaja saponaria*) is powdered and used in shampoos and hair tonics, while walking sticks and tool handles are prepared from the wood of *Docynia indica*, *Prunus cerasoides*, *Pyrus pashia* and *P. communis*.

21.38.7 Systematics and Phylogeny (Affinities)

Rosaceae, included under order Rosales of series Calyciflorae of subclass Polypetalae by Bentham and Hooker, are perhaps allied to Calycanthaceae in the order Laurales and to Myrtaceae (Hickey and King, 1988). Some believe Rosaceae to be allied to Saxifragaceae. Hutchinson believed that Rosaceae is derived from the same stock as Dilleniaceae. Rosaceae is also related to Leguminosae, the flowers of Chrysobalanoideae and Prunoideae having monocarpellary pistil and in Chrysobalanoideae the flowers being zygomorphic.

Fockey in *Pflanzenfamilien* divided Rosaceae into 6 subfamilies (Spiraeoideae, Pomoideae, Rosoideae, Neuradoideae, Prunoideae and Chrysobalanoideae) whereas Hutchinson divided Rosaceae into 20 tribes. However, majority of the recent workers (Jones and Luchsinger, 1987; Hickey and King, 1988, etc.) divide Rosaceae into 4 well-marked subfamilies as under:

1. *Spiraeoideae* (fruit a follicle or capsule; carpels usually 2–5, free or connate at base, not sunk in hypanthodium), e.g. *Spiraea*, *Sorbaria*, *Physocarpus*.
2. *Maloideae* (fruit a fleshy pome; ovary inferior, carpels 2–5, united to inner wall of hypanthium), e.g. *Pyrus*, *Malus*, *Crataegus*, *Sorbus*.
3. *Rosoideae* (fruit often dry; gynoecium of usually 10 or more pistils), e.g. *Rosa*, *Potentilla*, *Fragaria*, *Rubus*, *Alchemilla*.
4. *Prunoideae* (fruit drupe; carpel usually 1, free of hypanthium), e.g. *Prunus*.

21.39 SAXIFRAGACEAE (SAXIFRAGE FAMILY)

21.39.1 Systematic Position

Polypetalae, Calyciflorae, Rosales.

21.39.2 Field Recognition

Mostly perennial herbs; leaves alternate, usually deciduous and exstipulate; flowers pentamerous; ovary superior to inferior, often half-inferior; fruit usually a capsule or berry; seeds with abundant endosperm.

21.39.3 Selected Indian Genera

Ribes, *Hydrangea*, *Saxifraga*, *Deutzia*.

21.39.4 Size, Distribution and General Information

A family of about 80 genera and 1250 species (Hickey and King, 1988), Saxifragaceae are almost cosmopolitan in their distribution. They are, however, found mainly in north-temperate and arctic regions. Some of the larger genera are *Saxifraga* (370; Fig. 21.55), *Ribes* (150), *Hydrangea* (80), *Philadelphus* (75), *Escallobonia* (60), *Chrysoplenium* (55), *Parnassia* (50) and *Deutzia* (50).

21.39.5 Description of the Family

General Habit Perennial herbs and shrubs, and a few small trees; only rarely annual herbs; in *Saxifraga urbium* the stem is short, procumbent and rooting at the nodes.

Leaves Simple, alternate and exstipulate; compound (*Astilbe*), deeply lobed (*Francoa*); opposite (*Hydrangea*); usually deciduous; fleshy, spathulate to obovate to form basal rosettes in *Saxifraga urbium*.

Inflorescence Usually racemose or cymose; rarely solitary; panicle with loosely arranged flowers in *Saxifraga urbium*.

Flower Usually bisexual and actinomorphic; sometimes unisexual (*Ribes*) and zygomorphic (some species of *Saxifraga*); receptacle flat or hollowed to various depths so that perianth lobes and stamens may be perigynous or epigynous.

Calyx Usually 5 sepals, sometimes 4; poly- or gamosepalous; sometimes deeply coloured and petaloid.

Corolla Petals usually of same number as that of sepals, polypetalous; developing on a receptacle or on an hypanthium; alternisepalous; sometimes absent.

Androecium Stamens usually of same number as that of petals, or twice as many to the petals; alternipetalous or obdiplostemonous; free; anthers usually dithecous, dehiscing longitudinally; nectar glands or staminodes often present (*Parnassia*).

Gynoecium Bicarpellary, syncarpous; uni- to pentalocular, ovary superior to inferior, often half-inferior; axile placentation, but sometimes parietal; numerous anatropous ovules; styles and stigmas as many as the carpels.

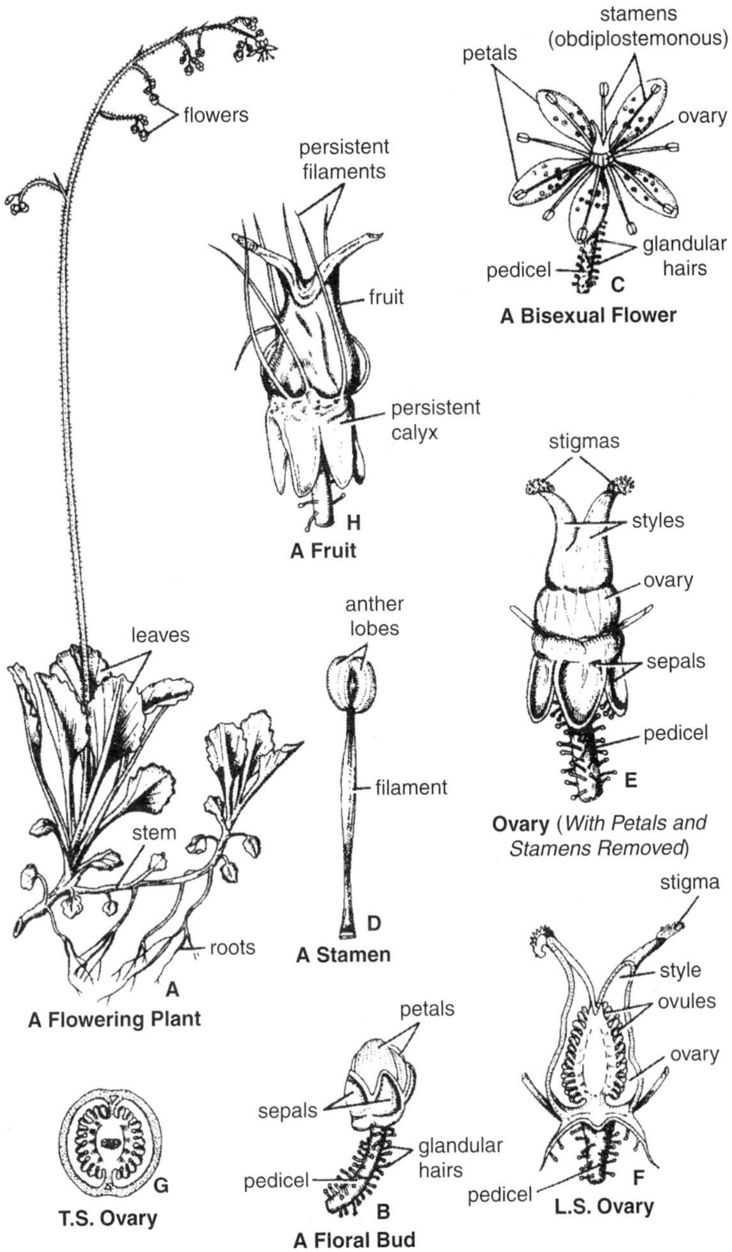

Fig. 21.55 *Saxifraga urbium*

Fruit A capsule or berry.

Seed With abundant endosperm surrounding a small embryo.

General Floral Formula Br, ⊕, ⚥, K_5, C_5, A_5 or 5 + 5, G $(\underline{2})$.

21.39.6 Economic Importance

Except for rock gardens (*Saxifraga*) or perennial ornamentals (*Astilbe*), Saxifragaceae are not of much importance. Many species of *Astilbe, Bergenia, Heuchera* and *Saxifraga* are grown widely as garden ornamentals. Ornamental flowering shrubs of Saxifragaceae belong to several species of *Deutzia, Escallonia, Hydrangea, Philadelphus* and *Ribes*. Bush edible fruits of gooseberries (*Ribes*) and white currant (*R. sativum*), black currant (*R. nigrum*) and red currants (*R. rubrum*) are also of some importance.

21.39.7 Systematics and Phylogeny (Affinities)

Saxifragaceae shows close affinities with Rosaceae, and it is difficult to separate the two. Saxifragaceae members, however, have (i) more abundant endosperm, (ii) exstipulate leaves instead of stipulate leaves of Rosaceae, and (iii) lesser numbers of pistils and stamens than Rosaceae. Saxifragaceae also shows close relationship with Crassulaceae. However, Crassulaceae differs from it (i) in the floral parts of a regular numerical plant, and (ii) pistils subtended by scale-like glands.

Hickey and King (1988) placed all herbaceous members of Saxifragaceae in four subfamilies, viz. (i) Astilboideae (e.g. *Astilbe*), (ii) Saxifragoideae (e.g. *Saxifraga*), (iii) Francooideae (e.g. *Francoa*) and, (iv) Parnassioideae (e.g. *Parnassia*) and all woody members in three subfamilies, viz. (i) Ribesoideae (e.g. *Ribes*), (ii) Hydrangeoideae (e.g. *Hydrangea*), and (iii) Escallonioideae (e.g. *Escallonia*), Earlier, Engler (1930) treated Saxifragaceae as composed of 15 subfamilies, of which most of them were treated as independent families by several workers, including Hutchinson.

21.40 MYRTALES

21.40.1 General Characteristics

1. Leaves simple, exstipulate, generally glandular-punctate; usually opposite and only sometimes alternate.
2. Flowers actinomorphic, usually bisexual.
3. Ovary 2 to 5 or more carpelled, syncarpous, usually inferior.
4. Usually axile or sometimes parietal placentation.
5. Different members show transition from perigyny to epigyny.
6. Flowers show distinct development of an hypanthium.
7. Internally, the stem generally contains internal phloem or intraxylary phloem.

Bentham and Hooker included Combretaceae, Myrtaceae, Lythraceae and 3 more families under Myrtales. Engler and Diels subdivided order Myrtiflorae (= Myrtales) into 4 suborders and 23 families, of which some major ones are Lythraceae, Punicaceae, Combretaceae, Myrtaceae and Onagraceae.

Only Combretaceae, Myrtaceae and Lythraceae are discussed in this text.

21.41 COMBRETACEAE (COMBRETUM FAMILY)

21.41.1 Systematic Position

Polypetalae, Calyciflorae, Myrtales.

21.41.2 Field Recognition

Trees or shrubs, often lianous; leaves simple, exstipulate; tubular receptacle along with adnate tubular base of calyx form hypanthium; ovules 4–6 in a single locule, all ovules suspended from the locule apex by slender funiculi.

21.41.3 Selected Indian Genera

Combretum, *Lumnitzera*, *Quisqualis*, *Terminalia*.

21.41.4 Size, Distribution and General Information

Combretaceae contains about 20 genera and 600 species distributed mainly in the tropical and subtropical countries of the world. 8 genera and about 45 species of this family have been reported from India, chiefly from Assam and West Bengal. *Combretum* and *Terminalia* are two largest genera of the family. *Lumnitzera littoria* is found as a mangrove plant in Sundarban and other tidal forests of India.

21.41.5 Description of the Family

General Habit A family of exclusively arborescent taxa, consisting of tall trees (*Terminalia*), or woody twiners of lianas, such as *Combretum* and *Quisqualis*.

Stem Solid with large mucilaginous sacs and abundant tannin.

Leaves Simple, alternate or sometimes opposite (*Quisqualis*); exstipulate; margin entire.

Inflorescence Racemose, or paniculate, or simple or branched spikes (*Quisqualis*, Fig. 21.56); rarely racemose heads (*Anogeissus*).

Flower Bracteate, bisexual (rarely unisexual); actinomorphic but occasionally zygomorphic; pentamerous; epigynous; tubular floral receptacle along with the adnate tubular base of calyx form a hypanthium; rarely flowers are tri- to octomerous.

Calyx Usually 5 sepals, fuse to form a tube; sometimes 4 or 8 sepals; valvate, rarely imbricate.

Corolla Usually 5 or as many petals as sepals, free; located alternately with sepals; valvate, imbricate, or twisted (*Quisqualis*); petals absent in *Terminalia* (Fig. 21.57).

Androecium Usually 10 or twice as many as the petals, arranged in 2 alternate whorls; stamens of the outer whorl are antisepalous; sometimes only 4 or 5 stamens; filaments inflexed in bud; anthers dithecous, versatile, dehiscing longitudinally.

Gynoecium Carpels 4–5 or less, syncarpous; ovary inferior, unilocular; angular or ribbed, with as many angles or ribs as calyx lobes; ovules 2–6, anatropous, pendulous from the top of the locule; style slender and solitary; stigma pointed or capitate; ovary is surrounded by nectariferous disc.

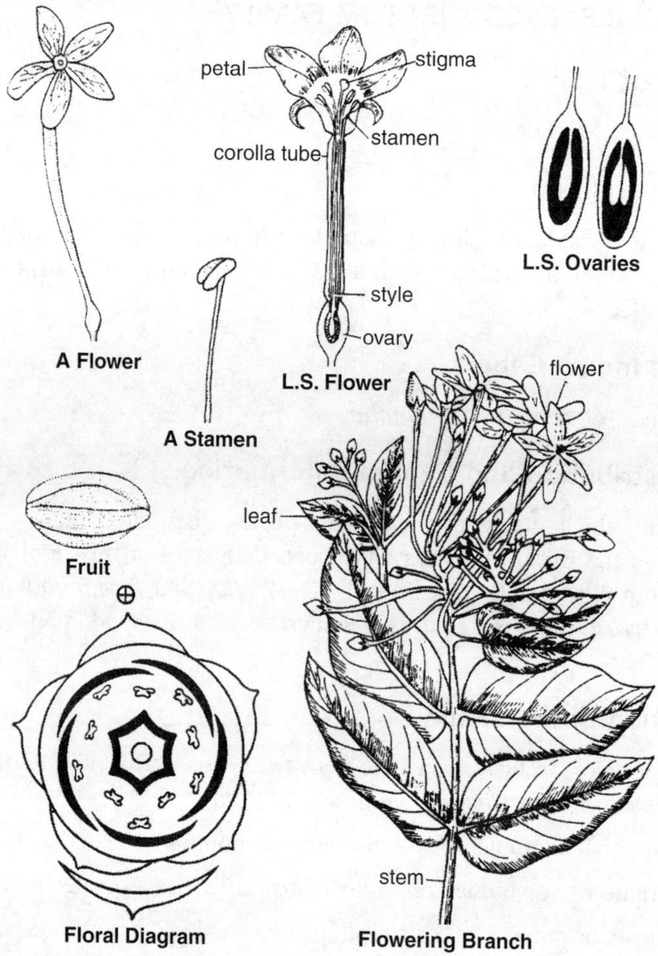

Fig. 21.56 *Quisqualis indica* L.

Fruit and Seed Fruit leathery, one-seeded drupe; often winged. Seed non-endospermic; embryo with folded or convolute cotyledons.

General Floral Formula Br, ⊕, ⚥, $K_{(5)}$, C_5, A_{5+5}, $G(\overline{4-5})$.

21.41.6 Economic Importance

Family is of little importance for domestic purposes.

- *Ornamental Value:* Favourite ornamental plants of this family include *Bucida buceras* (black-olive tree), few species of *Combretum* which are scandant shrubs, *Quisqualis indica* (Rangoon creeper), and *Terminalia arjuna* (Arjun).

- *Timber:* Valuable timber used in construction work is obtained from *Anogeissus acuminata*, *A. latifolia*, and several species of *Terminalia*, such as *T. alata*, *T. bialata*, *T. catappa*, *T. ivorensis* (indigo timber), *T. myriocarpa*, *T. paniculata*, *T. procera* and *T. tomentosa*.
- *Medicinal Value:* (i) Bark of *Terminalia arjuna* is highly valued as a cardiac tonic; (ii) *Terminalia bellirica* ("Bahera" or Belleric myroblans) and *T. chebula* ("Hararh" or black myroblans) are the two constituents of "Triphala", the famous Ayurvedic Indian preparation along with *Phyllanthus emblica* ('Aamla') of family Euphorbiaceae as its 3rd constituents; (iii) Leaves of *Calycopteris floribunda* are used in dysentery and malarial fever; (iv) Leaves of *Combretum acuminatum* are anthelmintic while that of *C. roxburghii* are used in malarial fever.

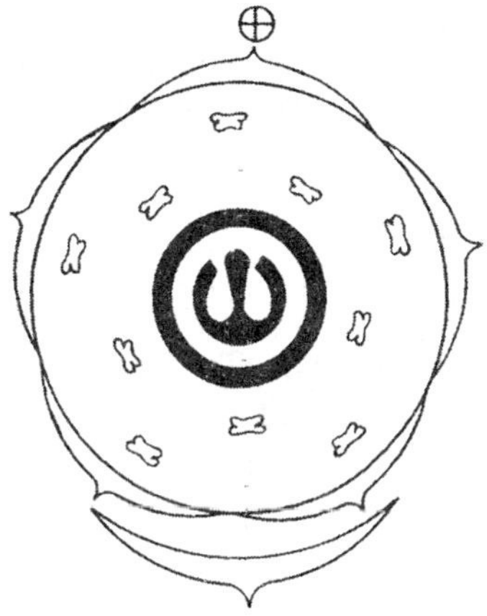

Fig. 21.57 Floral diagram of *Terminalia bellirica*.

- *Tanning and Dyeing:* The bark and fruits of *Terminalia alata*, *T. arjuna*, *T. catappa*, *T. chebula*, *T. ivorensis* and *T. citrina* are used for tanning and dyeing purposes.
- *Edible Nut:* Nuts of *Terminalia catappa* (Indian almond) are edible.

21.41.7 Systematics and Phylogeny (Affinities)

Bentham and Hooker placed Combretaceae under order Myrtales, series Calyciflorae and subclass Polypetalae. Most of the other taxonomists (Takhtajan, 1969; Cronquist, 1981; Thorne, 1983) also placed Combretaceae under order Myrtales. From the affinities point of view, Combretaceae is closely related to Myrtaceae and Rhizophoraceae, and to some other families of Myrtales to some extent.

21.42 MYRTACEAE (MYRTLE FAMILY)

21.42.1 Systematic Position

Polypetalae, Calyciflorae, Myrtales.

21.42.2 Field Recognition

Large trees or shrubs; leaves glandular-punctate; stamens numerous with their anther connectives gland-tipped; ovary inferior; axile or parietal placentation.

21.42.3 Selected Indian Genera

Callistemon, *Eucalyptus*, *Myrtus*, *Psidium*, *Syzygium*.

21.42.4 Size, Distribution and General Information

A family of about 140 genera and 3,400 species, distributed mainly in subtropical and tropical regions of the world. Family is particularly abundant in Australia where some *Eucalyptus* trees are world's tallest angiosperms. Myrtaceae is represented in India by about 15 genera and over 170 species. Beautiful pinkish flowers of *Myrtus communis* are considered highly sacred among Greeks, Egyptians, Jews, and Persians, and used in religious rites and ceremonies. Some of the larger genera of Myrtaceae along with their number of reported species include *Eugenia* (1000), *Eucalyptus* (500), *Syzygium* (500), *Myrcia* (500), *Psidium* (140), *Myrtus* (100) and *Melaleuca* (100).

21.42.5 Description of the Family

General Habit Trees and shrubs, varying from small creepers to giant *Eucalyptus* reaching to a towering height of 300–400 feet; lysigenous cavities containing ethereal oils are present in young stem, leaves, floral parts and fruits.

Leaves Usually opposite, but rarely alternate; simple; exstipulate; usually entire, gland-dotted; evergreen; and coriaceous.

Inflorescence Usually cymose, sometimes racemose, rarely of solitary flowers; multichasial cyme in *Eucalyptus*; spikes in *Callistemon*; solitary axillary in *Psidium guajava* and *Myrtus communis*; racemose raceme in *Barringtonia*.

Flower Bracteate; often with 2 bracteoles; bisexual, actinomorphic, penta-or tetramerous; epigynous (*Eucalyptus, Psidium*) or perigynous (*Tristania*); in several genera, the floral receptacle unites with the gynoecium at the base and often prolonged into a hypanthium.

Calyx 4 or 5 sepals, free or united; valvate or quincuncial; sepals in *Eucalyptus* are thrown off unopened as a lid; sepals contain oil glands.

Corolla 4 to 5 petals, free (*Callistemon citrinus*) or united (*Eucalyptus*); imbricate; often nearly circular; in *Eucalyptus citridora* corolla fuses with calyx to form a cup called operculum (Fig. 21.58); few *Eugenia* species lack petals.

Androecium Stamens numerous, free (*Psidium*), or sometimes basally connate (*Callistemon*); usually bent inwards in bud; connectives often gland-tipped; anthers usually 2-celled, versatile, introrse and dehiscing longitudinally or sometimes apically; sometimes stamens are grouped into as many clusters as petals and are obdiplostemonous (*Syzygium*); stamens variously coloured.

Gynoecium 2 to 5 or more carpels, syncarpous; ovary inferior or semi-inferior; with one to many (often 2–5) locules; 2 to many anatropous or campylotropous ovules in each locule; usually axile or rarely parietal placentation; style one, long and simple; stigma capitate.

A nectar-secreting, circular disc is present on the inside of the upper part of the floral cup.

Fruit and Seed Fruit variable; fleshy berry (*Psidium, Eugenia*), capsule (*Callistemon, Eucalyptus*), nut (*Darwinia*), or drupe (*Pimenta*). Seeds with little or no endosperm.

Pollination and Dispersal Pollination entomophilous. Insects are attracted because of nectar and variously coloured stamens. Seeds are dispersed by wind, birds or other animals.

General Floral Formula Br, Brl, ⊕, ⚥, $K_{4-5 \text{ or } (4-5)}$, $C_{4-5 \text{ or } (4-5)}$, $A\infty$, $G_{\overline{(2-5)}}$.

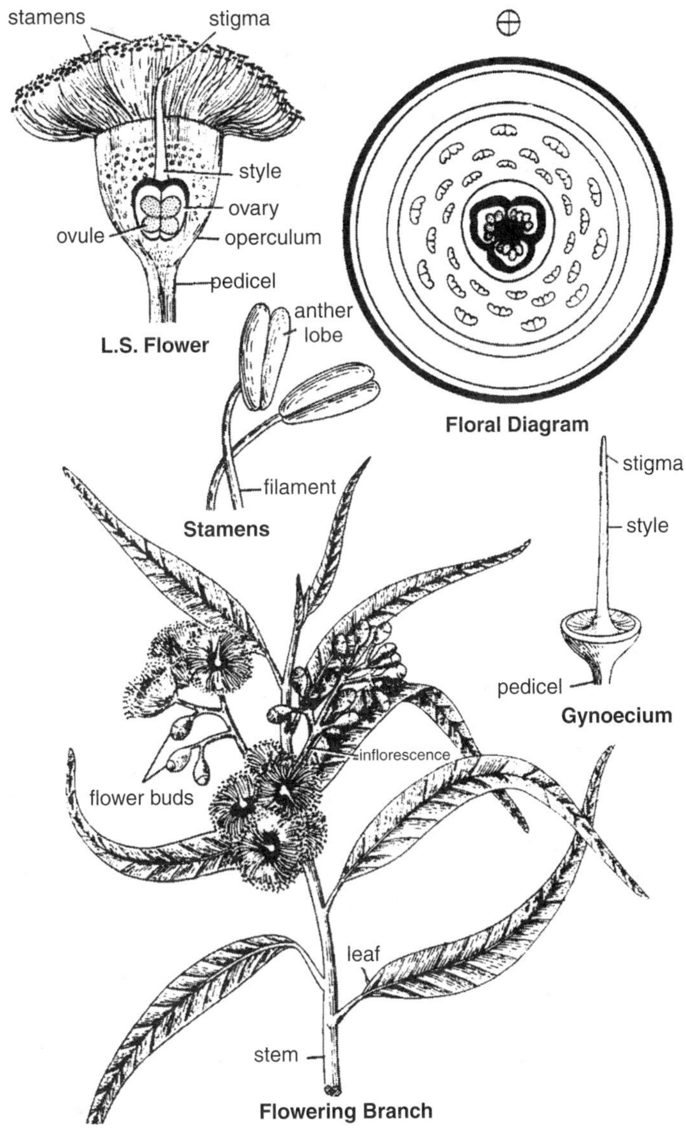

Fig. 21.58 *Eucalyptus citridora* Hook.

21.42.6 Economic Importance

- *Edible Fruits: Psidium guajava* (Amrood or Guava), *P. cattleianum* (Strawberry guava), *Eugenia uniflora* (Surinam-cherry), *Feijoa sellowiana* (Feijoa), *Myriciaria cauliflora* (Jaboticabo fruit) and *Syzygium cumini* syn. *Eugenia jambolana* (Jamun or Jambolan) are some of the common delicious fruits. 'Jamun' fruits are useful for diabetic patients.

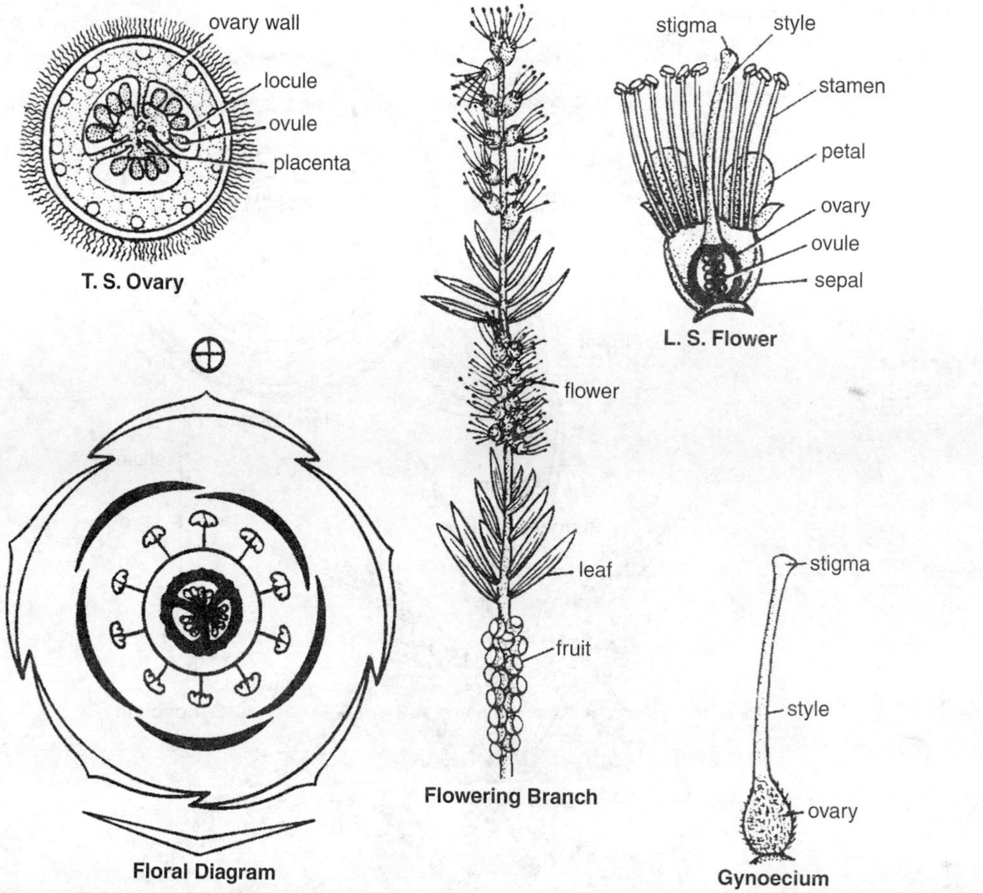

Fig. 21.59 *Callistemon lanceolatus* D.C.

- *Oil Plants:* (i) Eucalyptus oil, obtained from *Eucalyptus globulus*, *E. citridora* and several other species is used widely in medicine, perfumery and other industries, (ii) clove oil is obtained from the dried flower buds of *Syzygium aromaticum* (Laung or clove tree); (iii) cajuput oil is obtained from the leaves of *Melaleuca leucadendron*, (iv) oil of bay rum is obtained from leaves and flower buds of *Pimenta racemosa*.
- *Spices:* Cloves (Laung) are the dried flower buds of *Syzygium aromaticum* while allspice are the unripe berries of *Pimenta dioica*.
- *Ornamental Plants:* Hundreds of species of *Eucalyptus* are grown as ornamental plants. Other garden ornamentals of Myrtaceae include *Callistemon lanceolatum* (bottle brush or Laila Majnu plant), *C. linearis*, *C. viminalis*, *Myrtus communis* (Myrtle), *Leptospermum laevigatum* (Australian tea tree), *Melaleuca leucadendron* (cajuput tree), and *Tristania conferta* (Brisbane box).
- *Timber:* Valuable timber is obtained from several species of *Eucalyptus*, *Eugenia*, *Barringtonia* and *Coreya*.

- *Paper:* *Eucalyptus* wood is used for paper manufacturing.
- *Gum:* Valuable gum is obtained from *Eucalyptus globulus* (blue gum), *E. maculata* (citron gum) and *E. rostrata* (red gum).

21.42.7 Systematics and Phylogeny (Affinities)

Almost all taxonomists place Myrtaceae under order Myrtales. Myrtaceae is allied to Rhizophoraceae, Lythraceae, Combretaceae and Onagraceae, and to some extent to Melastomaceae and Sonneratiaceae.

Myrtaceae is divided into 2 subfamilies i.e. (i) *Myrtoideae* (fruit berry, rarely a drupe; leaves always opposite), and (ii) *Leptospermoideae* (fruit dry, leaves opposite or alternate).

21.42.8 Description of a Common Plant

Callistemon citrinus Skeels syn. C. lanceolatus D.C. (Bottle brush or Laila Majnu, Fig. 21.59).

Habit: Large tree. *Leaf:* Simple, alternate, exstipulate, subsessile, lanceolate, gland-dotted. *Inflorescence:* Long pendant spike. *Flower:* Bracteate, subsessile, complete, hermaphrodite, actinomorphic, pentamerous, epigynous, crimson-yellow to red. *Calyx:* 5 sepals, fused, imbricate. *Corolla:* 5 petals, polypetalous, imbricate or quincuncial. *Androecium:* Stamens numerous, filaments long, bright red and united at the base to form a staminal sheath; dithecous, dorsifixed, introrse. *Gynoecium:* Tricarpellary, syncarpous, inferior, trilocular, many ovules in each locule, axile placentation. *Fruit:* Capsule.

Floral Formula: Br, ⊕, ⚥, $K_{(5)}$, C_5, A_∞ (basally fused), $G_{(\bar{3})}$.

21.43 LYTHRACEAE (LOOSESTRIFE FAMILY)

21.43.1 Systematic Position

Polypetalae, Calyciflorae, Myrtales.

21.43.2 Field Recognition

Herbs to trees; leaves simple, opposite; hypanthium present; corolla crumpled; ovary superior; seeds non-endospermic.

21.43.3 Selected Indian Genera

Lawsonia, *Punica*,[1] *Lagerstroemia* and *Woodfordia*.

21.43.4 Size, Distribution and General Information

A family of about 25 genera and 550 species, Lythraceae are cosmopolitan in distribution. About 12 genera and over 50 species have been reported from India. Some of the larger genera along with their approximately reported species are *Cuphea* (250), *Lagerstroemia* (43) and *Lythrum* (35).

[1] Several taxonomists include *Punica* under a separate unigeneric family Punicaceae.

21.43.5 Description of the Family

General Habit Herbs, shrubs (*Lawsonia inermis*, *Woodfordia fruticosa*), or trees (*Sonneratia*); rarely spinescent (*Lawsonia*).

Leaves Usually opposite, or whorled; simple; exstipulate or with minute stipules; margin entire.

Inflorescence Racemose raceme, panicle (*Lagerstroemia*), or cyme (*Punica*).

Flower Bisexual, actinomorphic or zygomorphic (*Cuphea*); usually tetra- to hexamerous; hypanthium present; sepals appearing like lobes of hypanthium.

Epicalyx Hypanthium is sometimes subtended by the connate pairs of bracts representing epicalyx (*Lythrum*).

Calyx 4 to 8 sepals; sometimes forming a hollow tube; valvate.

Corolla Petals of same number as sepals (i.e. 4 to 8), polypetalous, crumpled in bud (Fig. 21.60A); petals insert on the margin of the receptacular cup and alternate with sepals; sometimes petals absent (*Rotala*, *Peplis*).

Androecium Stamens typically twice as many as sepals (Fig. 21.60B), and only rarely fewer or more; inserted often very low down on the calyx tube or hypanthium; stamens generally arranged in 2 whorls; filaments unequal; anthers dithecous, introrse and dehisce longitudinally.

Gynoecium 2–6 carpels, syncarpous, 2–6 locular, numerous ovules in each locule, axile placentation; ovary superior or half-inferior; style simple, stigma capitate; some genera show heterostyly (*Lythrum*).

Fruit and Seeds Fruit usually a dry capsule. Seeds non-endospermic.

General Floral Formula ⊕, ⚥, $K_{4-8 \text{ or } (4-8)}$, C_{4-8}, A_{8-16}, $G_{\underline{(2-6)} \text{ or } (2-6)-}$.

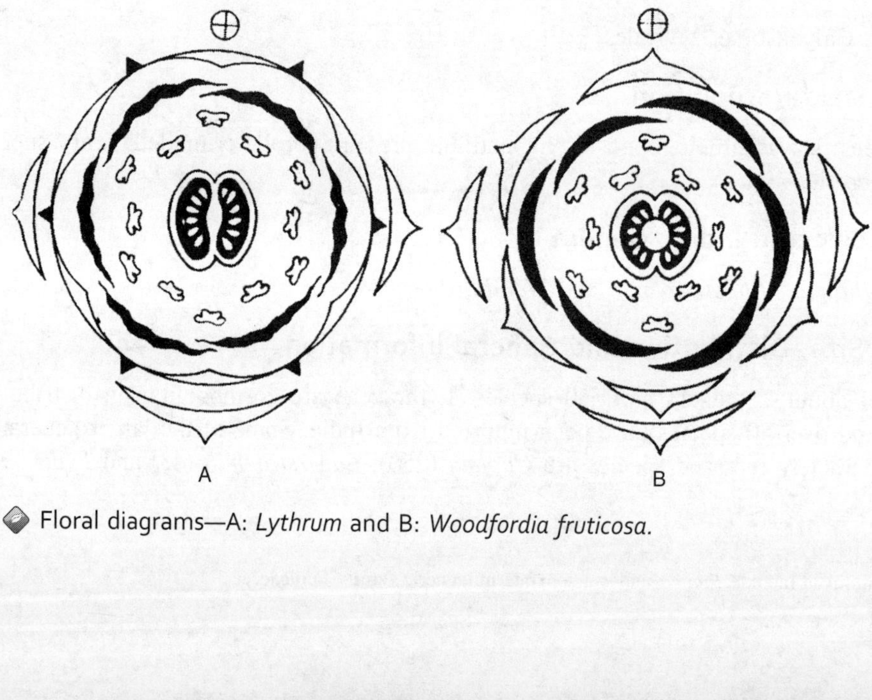

Fig. 21.60 Floral diagrams—A: *Lythrum* and B: *Woodfordia fruticosa*.

21.43.6 Economic Importance

- *Lawsonia inermis* (Mehndi or Henna plant) leaves are the source of dye, henna, used for staining fingers, nails, hairs, and also wool, silk and leather. Leaves and flowers of *Woodfordia fruticosa* are also used for tanning and dyeing silk.
- *Punica granatum* (Pomegranate), a small tree, is cultivated for its edible juicy fruits.
- Timber for making boats and other similar articles is obtained from *Lagerstroemia speciosa*, *L. indica*, *L. lanceolata*, etc.
- Ornamental plants of Lythraceae include *Cuphea lanceolata* (red-flowered), *C. ignea* (cigar-flower), *Punica granatum*, *Lawsonia inermis* (hedge-plant), *Lagerstroemia indica* (pinkish flowers), *L. speciosa* (purple flowers) and several species of *Lythrum*.

21.43.7 Systematics and Phylogeny (Affinities)

Lythraceae is closely allied to Myrtaceae, Onagraceae and Melastomaceae. Bentham and Hooker treated genus *Punica* under Lythraceae. However, in several recent taxonomic treatments *Punica* is placed in an independent unigeneric family Punicaceae.

21.44 PASSIFLORALES

21.44.1 General Characteristics

1. Flowers regular; usually unisexual and only sometimes bisexual.
2. Flowers pentamerous, usually epigynous and only rarely perigynous.
3. Ovary is usually tricarpellary, syncarpous, inferior, unilocular or trilocular.
4. Placentation parietal or axile.
5. Fruit pepo or capsule and seeds non-endospermic.

Bentham and Hooker included Cucurbitaceae, Begoniaceae, Passifloraceae and four more families under Passiflorales. Engler and Diels did not use the name Passiflorales, and treated only Cucurbitaceae under order Cucurbitales, and Begoniaceae and Passifloraceae under order Parietales along with 29 more families.

Only Cucurbitaceae, Begoniaceae and Passifloraceae have been treated in this text.

21.45 CUCURBITACEAE (GOURD FAMILY)

21.45.1 Systematic Position

Polypetalae, Calyciflorae, Passiflorales.

21.45.2 Field Recognition

Climbing or prostrate, tendril-bearing herbs; flowers usually unisexual and pentamerous; ovary inferior; fruit pepo or berry.

21.45.3 Selected Indian Genera

Citrullus, Coccinia, Cucurbita, Lagenaria, Luffa, Momordica, Trichosanthes.

21.45.4 Size, Distribution and General Information

A family of about 110 genera and between 650 to 850 species, Cucurbitaceae are distributed mainly in tropical and subtropical regions of the world. Its 37 genera and about 100 species have been reported from India. Some of its largely represented genera along with their number of approximately reported species include *Momordica* (45, bitter gourd), *Cucumis* (25, Cucumber, muskmelon), *Cucurbita* (15, pumpkin, gourd, squash), *Lagenaria* (6, bottle gourd), *Luffa* (6, vegetable sponge or Loofah), and *Echinocystis* (15, wild cucumber).

21.45.5 Description of the Family

General Habit Annual or perennial prostrate herbs or vines usually with spirally coiled tendrils; very rarely shruby (*Acanthosicyos*) or arborescent (*Dendrosicyos*).

Stem Herbaceous, usually pentangular with ridges and furrows; contains two rings of bicollateral vascular bundles.

Leaves Alternate, exstipulate, simple but often palmately or pinnately lobed; long-petioled; palmately veined; tendrils present (*tendrils*, interpreted variously as stipules, leaves, shoots, flower stalks, stems or even as roots, are perhaps stipular in origin).

Inflorescence Axillary and basically cymose or the flowers solitary; plants are monoecious or dioecious.

Flower Usually unisexual and only rarely bisexual (*Schizopepon*); actinomorphic; pentamerous; epigynous or rarely perigynous (*Actinostemma*); female flowers contain staminodes while male flowers contain pistillode.

(a) Male Flower

Calyx 5 sepals, tubular, gamosepalous, rarely distinct; fused to ovary wall in female flower; valvate or imbricate.

Corolla 5 petals, usually united to form a campanulate or salver-shaped corolla; rarely petals are free (*Fevillea*); usually petals alternate with the sepals; imbricate or valvate; yellow or white.

Androecium 1–5 stamens (typically 5, more or less united into a column, but very variable in structure or highly modified into a number of types of cohesion and twisting of anthers and filaments); usually 5 stamens appear as 3 stamens, with 2 groups of 2 fused together forming 2 compound stamens, plus 1 simple stamen. Some of the androecium types found in Cucurbitaceae are:

(i) In *Fevillea*, there are 5 free stamens alternating with polypetalous coralla.

(ii) In *Thladiantha*, filaments of 4 stamens coherent below in 2 pairs, the 5th stamen remaining free.

(iii) In *Sicydium*, the condition is same as in *Thladiantha* but filaments of the pairs become united at the base.

(iv) In *Bryonia*, *Citrullus and Momordica*, the coherent filaments and anthers of all the 5 stamens unite completely so that the androecium appears to be made up of 3 stamens; of these, 2 have 4 cells due to complete union and the 3rd one has only 2 cells.

(v) In *Cucurbita*, the anthers are all united and spirally twisted, and filaments free below and united above or completely united. Such folded anthers are called *conduplicate.*

(vi) In *Cyclanthera* the stamens completely unite with the lobes of the anthers forming two horizontal rings surrounding a peltate mass of connectives.

Gynoecium Represented by rudimentary pistillodes.

(b) Female Flower

Calyx and Corolla Same as in male flower.

Androecium Represented only by staminodes, if any.

Gynoecium Usually tricarpellary, syncarpous, ovary inferior. According to Engler, Eichler, Wettstein and Willis, the ovary is trilocular, the carpels intrude and meet at the centre and the placentation is axile. However, Rendle opined that the ovary is unilocular and the placentation is parietal. In some cases, carpels range from 1 to 10; ovules numerous and anatropous; style simple or 3-parted; stigmas as many as carpels.

Fruit and Seed Fruit a pepo or capsule. Seeds often flat, numerous, with straight embryo and no endosperm; seeds are winged in *Zanonia*.

General Floral Formulae

(a) *Male Flower:* $\oplus$, ♂, $K_{(5)}$, $C_{(5)}$, $A_{5 \text{ or } (2)+\underline{(2)}+1}$, G_0.

(b) *Female Flower:* $\oplus$, ♀, $K_{(5)}$, $C_{(5)}$, A_0, $G(\bar{3})$.

21.45.6 Economic Importance

- *Edible Fruits and Vegetables:* Cucurbitaceae is a family of several edible fruits which are either cooked or eaten raw in our daily food. Some of them nclude (i) *Benincasa hispida* (petha-kaddoo or white gourd), (ii) *Citrullus vulgaris* (Tarbooj or watermelon), (iii) *C. vulgaris* var. *fistulosus* (Tinda or round gourd), (iv) *Coccinia cordifolia* (Kanduri or Kovai fruit), (v) *Cucumis anguria* (Gherkin), (vi) *C. melo* (Kharbooja or melon), (vii) *C. melo* var. *momordica* (Phoont or Kachra), (viii) *C. melo* var. *utilissimus* (Kakdi). (ix) *C. sativus* (Kheera or Cucumber), (x) *Cucurbita maxima* (Red pumpkin), (xi) *C. moschata* (Sitaphal or pumpkin), (xii) *C. pepo* (Field pumpkin), (xiii) *Lagenaria siceraria* syn. *L. vulgaris* (Lauki or bottle gourd), (xiv) *Luffa acutangula* (Kali Tori), (xv) *L. cylindrica* syn. *L. aegyptiaca* (Ghia Tori), (xvi) *Momordica charantia* (Karela or bitter gourd), (xvii) *M. cochinchinensis* (Kakrol or Bhat Karela), (xviii) *Trichosanthes anguina* (Chichenda or snake gourd), and (xix) *T. dioica* (Palwal or Parwal).
- *Ornamental Plants:* Some species of *Benincasa*, *Coccinia*, *Ecballium*, *Luffa*, *Sechium* and *Trichosanthes* are of ornamental value.
- Some *Luffa* species provide sponges for oil filters and bathing.

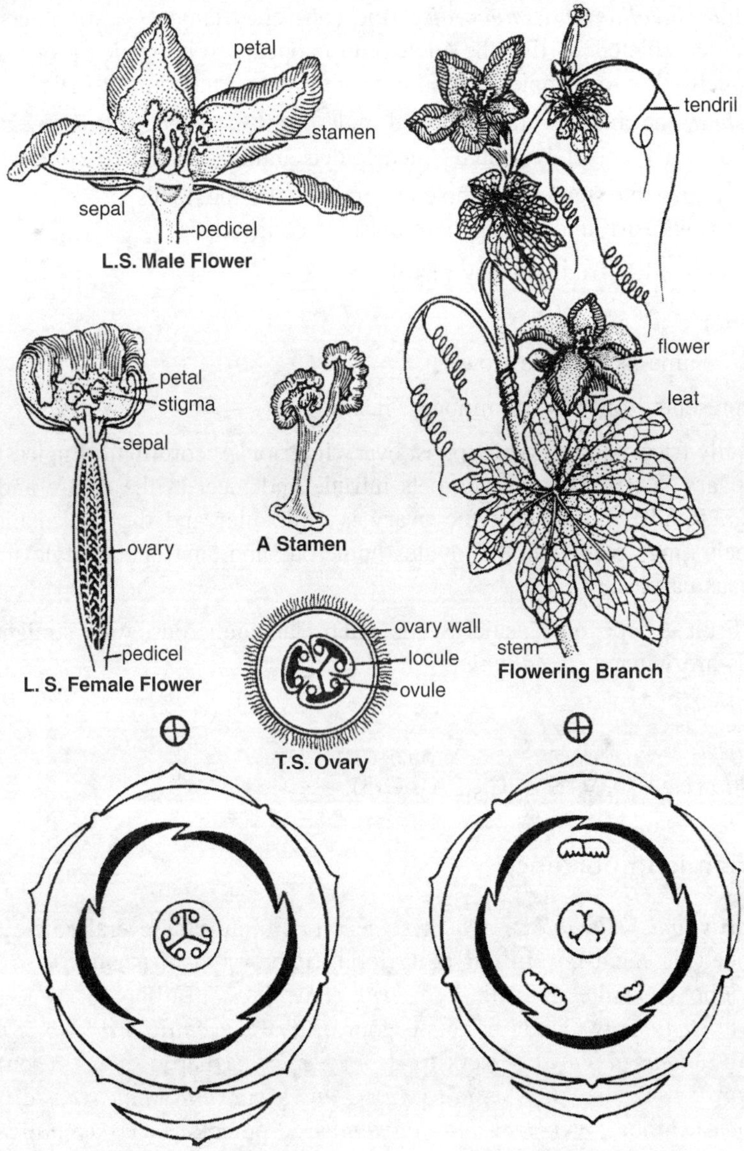

Fig. 21.61 *Luffa cylindrica* (L). Roem. syn. *L aegyptiaca* Mill.

- Musical instruments and some decorative articles are prepared from fruits of *Lagenaria* and *Benincasa*.
- Seeds of *Citrullus vulgaris* and *Cucumis melo* are highly nutritious and used in preparing sweets.

21.45.7 Systematics and Phylogeny (Affinities)

Systematic position of Cucurbitaceae is still highly controversial and its affinities are not very clear among taxonomists. Its characters such as herbaceous nature, presence of bicollateral vascular bundles, unisexual epigynous flowers, curved anther lobes, etc. make it a truly advanced family. Engler, therefore, placed it in a monotypic order Cucurbitales and thought it to be closely allied to Campanulaceae. Bentham and Hooker placed it under order Passiflorales. Rendle thought it to belong to order Peponiferae while Takhtajan (1969) placed it under Cucurbitales in between Passiflorales and Begoniales, and mentioned that Cucurbitaceae is "very closely related to the family Passifloraceae". Cronquist (1981) placed it under order Violales along with 23 more families while Thorne (1983) placed it under suborder Cucurbitineae of order Violales along with 16 more families. Cronquist considers Violales to be a natural taxon derived from Theales.

21.45.8 Description of a Common Plant

Luffa cylindrica (L.) M. Roem. syn. L. aegyptiaca Mill. (Ghia Tori; Fig. 21.61)

Habit: A tendril-bearing herb. *Stem:* Pentangular. *Leaf:* Alternate, simple, exstipulate, palmately lobed. *Inflorescence:* Male flowers arranged in raceme clusters; female flowers solitary. *Flower:* Bracteate, incomplete, actinomorphic, unisexual, pentamerous. *MALE FLOWER—Calyx:* 5 sepals, fused, valvate. *Corolla:* 5 petals, gamopetalous or distinct, imbricate, yellow. *Androecium:* 5 stamens appearing as if 3, of which 4 stamens are fused in 2 groups of 2 each, and 5th stamen is free. *Gynoecium:* Absent or represented by rudimentary pistil. *FEMALE FLOWER—Calyx and Corolla:* Same as in male flower. *Androecium:* Absent. *Gynoecium:* Tricarpellary, syncarpous, inferior, unilocular or divided incompletely into 3 locules, many ovules, parietal placentation; styles 3; stigma forked. *Fruit:* Pepo. *Floral Formulae:* Same as mentioned in general description of family.

21.46 BEGONIACEAE (BEGONIA FAMILY)

21.46.1 Systematic Position

Polypetalae, Calyciflorae, Passiflorales.

21.46.2 Field Recognition

Unisexual flowers; the staminate flowers are zygomorphic with numerous stamens arranged in many whorls; the pistillate flowers contain inferior, angled or winged ovary with lobed or forked placentae, generally twisted stigmas; seeds generally non-endospermic and contain oily embryo.

21.46.3 Common Indian Genus

Begonia.

21.46.4 Size, Distribution and General Information

Only 5 genera and about 920 species of this family have been reported so far. *Begonia*, with about 900 species, is largest genus distributed widely in tropics and subtropics. *Hillebrandia* found in Hawaii and *Symbegonia* found in New Guinea are both monotypic genera. 3 species each of *Begoniella* and *Semibegoniella* have also been reported.

21.46.5 Description of the Family

General Habit Mostly perennial, monoecious herbs or low shrubs; erect, creeping or climbing with succulent stems and thick rhizomes or tubers.

Stem Succulent, thick, rhizomatous or tuberous; somewhat jointed; in *Begonia semperflorens* the stem is glabrous, fleshy, reddish-green (Fig. 20.62) attaining a height of 15 to 45 cm.

Leaves Simple, basal or alternate in 2 ranks; usually asymmetrical; stipulate, stipules large, membranous and deciduous; petiolate, palmately-nerved; in *B. semperflorens* the leaves are roundish-ovate, somewhat oblique at the base and have serrulate or ciliate margins.

Inflorescence Axillary dichasial cyme with unisexual flowers.

Flowers Unisexual; often zygomorphic, sometimes actinomorphic.

(a) Male Flower

Perinath 2, valvate, petalloid, free sepals, and 2 usually smaller, valvate petals.

Androecium Stamens 2 to numerous, in many whorls, free or basally connate; anthers dithecous, basifixed.

Gynoecium Absent or rudimentary.

(b) Female Flower

Perinath 2 to many, imbricate, petalloid tepals not differentiated into colyx and corolla.

Androecium Absent.

Gynoecium Tricarpellary, syncarpous; inferior ovary; trilocular, with numerous anatropous ovules on axile placentation; styles 2–5 free or basally connate; in *Hillebrandia* the ovary is half-inferior; stigmas strongly papillose and often twisted.

Fruit Usually a loculicidal, winged capsule, rarely berry.

Seeds Numerous, minute, Non-endospermic with straight oily embryo.

General Floral Formulae

Male flower Φ, ♂, P_{2+2}, A2 – $\propto$, G_0.

Female flower $\oplus$, ♀, $P_{2-\propto}$, A0, $G(\bar{3})$.

21.46.6 Economic Importance

Except that of great horticultural or ornamental value of genus *Begonia*, the Begoniaceae contain no plant of much economic importance. Large number of species of *Begonia* are grown as decorative house plants. They are commonly grown for summer bedding in lawns of gardens.

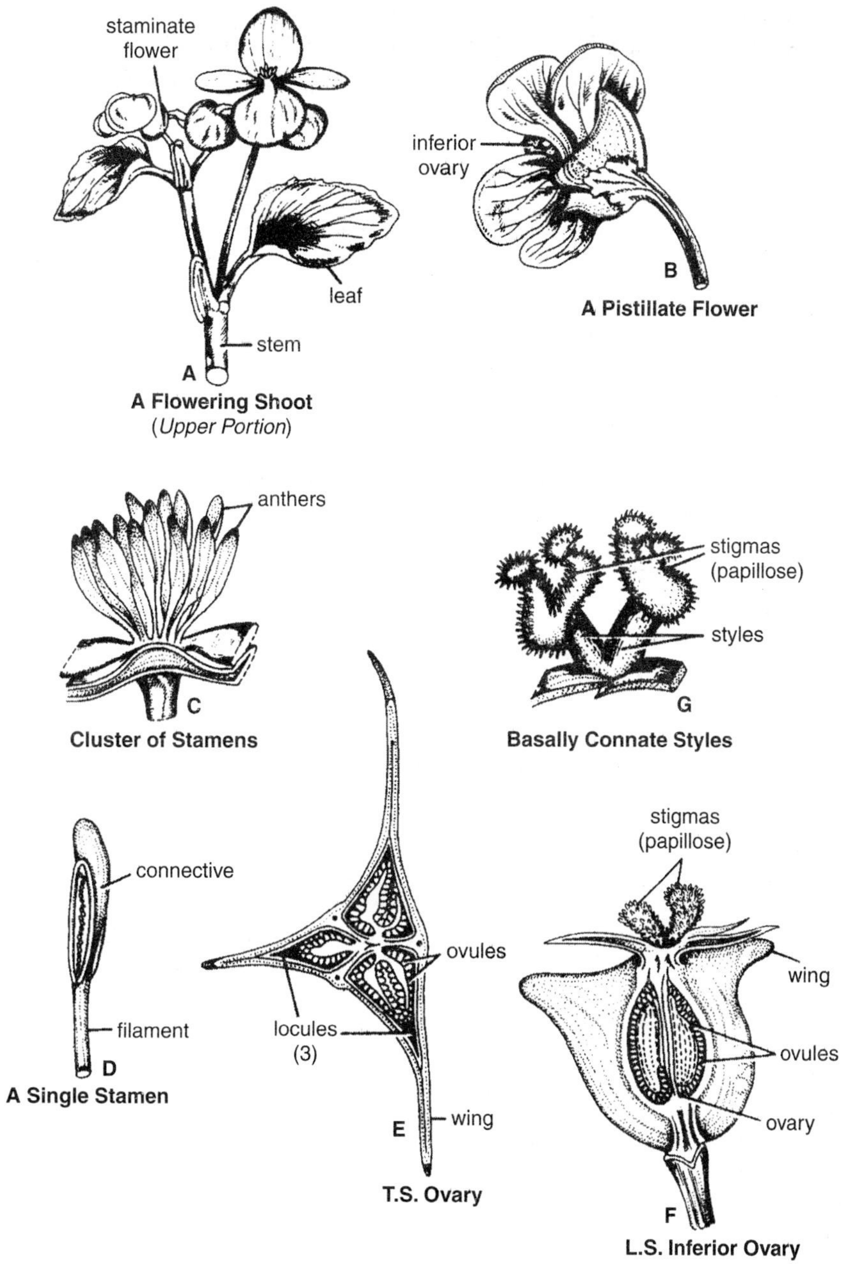

Fig. 21.62 *Begonia semperflorens* Link and Otto.

21.46.7 Systematics and Phylogeny

Engler and Diels treated Begoniaceae under suborder Begoniineae of order Parietales, and Lawrence (1951) also followed it. Bessey placed it under order Loasales along with families of two other suborders (Datiscineae and Ancistrocladineae) of Parietales. Randle transferred Begoniaceae under order Cucurbitales and treated it with the Cucurbitaceae. Bentham and Hooker treated Begoniaceae under order Passiflorales of series Calyciflorae of Polypetalae and discussed it along with Cucurbitaceae and five more families. Hickey and King (1988) described Begoniaceae under order Violales of Dicotyledons.

21.47 PASSIFLORACEAE (PASSION-FLOWER FAMILY)

21.47.1 Systematic Position

Polypetalae, Calyciflorae, Passiflorales.

21.47.2 Field Recognition

Passifloraceae members can be recognised by the combination of the usually climbing habit, the uniflowered peduncles often in pairs, the variously modified corona, androgynophores, and the seeds with often mucilaginously pulpy aril.

21.47.3 Selected Indian Genera

Passiflora, *Adenia*.

21.47.4 Size, Distribution and General Information

A medium-sized family of 16 genera and over 650 species distributed mainly is tropical and warm temperate regions. Of the three large genera, *Passiflora* (500) is distributed chiefly in USA, Asia and Australia, while *Adenia* (92) is found mainly in tropical and South Africa, Indo-Malaysia and South-West Arabia, and *Typhostemma* (35) in tropical and South Africa

21.47.5 Description of the Family

General Habit Chiefly shrubs, or herbs; often climbers or lianous with axillary tendrils.

Stem Herbaceous, weak.

Leaves Alternate; simple or compound; stipulate; tendrillar, tendrils opposite the leaves; in Blue Passion flower (*Passiflora caerulea*, Fig. 20.63) the leaves are palmate, 5- or 7-lobed with axillary tendrils.

Inflorescence Modified dichasial cyme (*Passiflora caerulea*, Fig. 21.63), in which the central flower is represented by a tendril, one of the lateral outgrowths develops into a flower, and the opposite lateral outgrowth remains undeveloped.

Flowers Bracteate; bisexual or rarely unisexual; actinomorphic; axillary and usually in pairs.

Receptacle of various shapes, usually hollowed and generally with a central androgynophore; usually terminated by outgrowths of petalloid or staminodial appearance forming a fleshy, cup-shaped *corona*. Puri (1948) interpreted that corona is made up of (i) the hypanthodium, and (ii) all associated structures situated between the perianth lobes and the stamens.

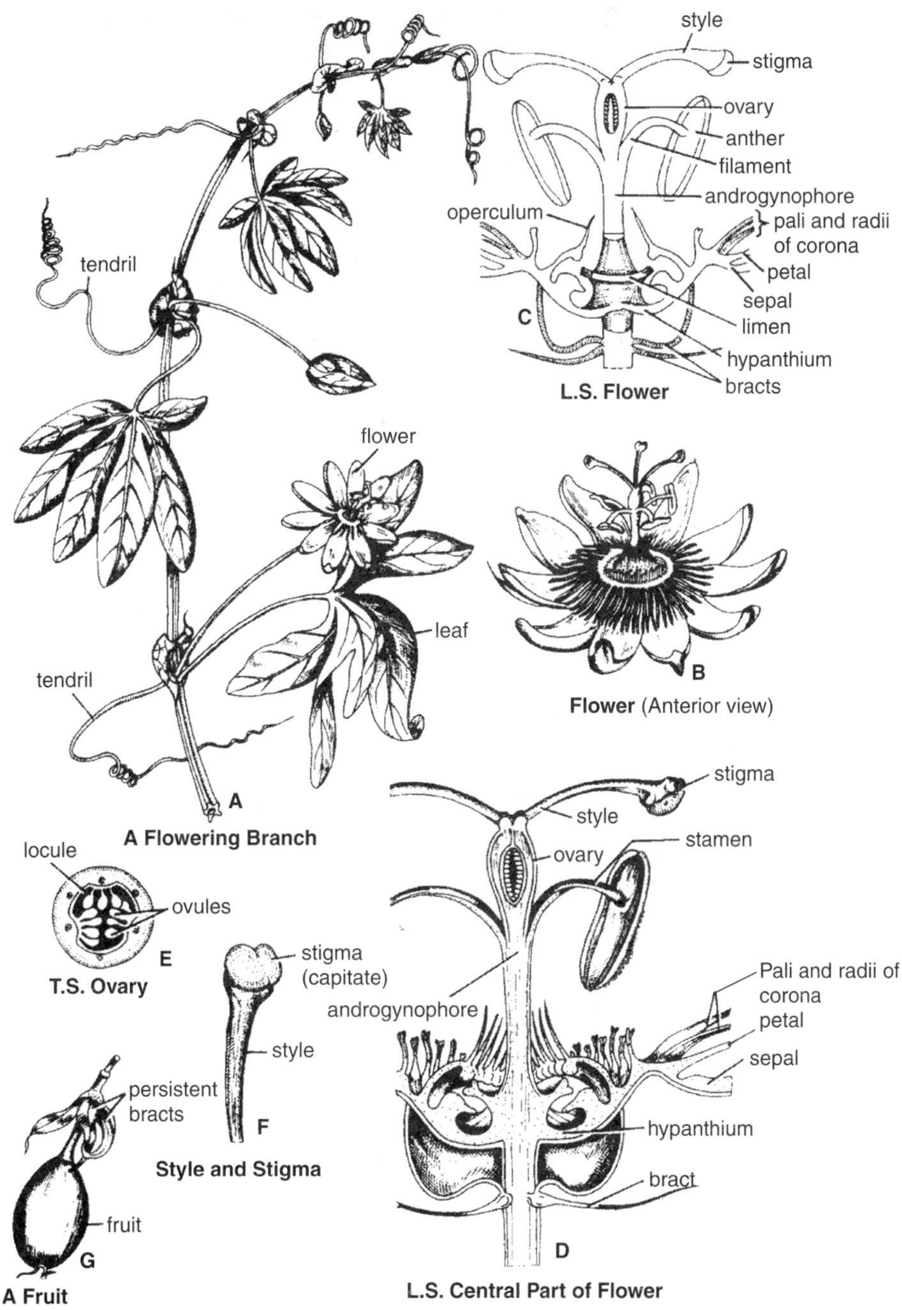

Fig. 21.63 *Passiflora caerulea* L. (Blue Passion-flower).

Calyx 3–5 sepals, free or basally connate, often petalloid; imbricate; persistent.

Corolla 3–5 petals, free or basally connate, sometimes petals absent; imbricate.

Androecium 3–5 or more stamens, sometimes upto 10, usually opposite the petals, developing from the base of corona or from receptacle rim or hypogynous from the apex of gynophore, thus forming an androgynophore; anthers free or basally connate, dithecous, dehiscing longitudinally; some genera show the presence of staminodes.

Gynoecium 3–5 carpels, syncarpous; ovary superior, unilocular, containing many anatropous ovules, parietal placentation; pistil often raised on gynophore or more commonly on androgynophore; styles as many as carpels, free or all connate; stigmas 3–5, often capitate or discoid.

Fruit A capsule or a berry.

Seeds With straight embryo, and fleshy aril and endosperm.

General Floral Formula Br, $\oplus$, ⚥, K3–5 or (3 – 5), C3–5 or (3 – 5), A3 – 5 or (3 – 10), G$(\underline{3-5})$.

21.47.6 Economic Importance

- Ornamental plants of domestic importance include several species of *Passiflora* (e.g. *P. edulis, P. ligularis, P. laurifolia, P. caerulea, P. quadrangularis* etc. Most of these species also provide edible fruits. Several of these species are cultivated for their attractive flowers and also grown in glasshouses.
- *Passiflora edulis* is widely known in tropical regions for its edible fruits (Passion fruit or Purple Granadilla).
- Some species of Passifloraceae produce cyanide and are thus *deadly poisonous.* So, care must be taken while eating the fruits of unidentified species.

21.47.7 Systematics and Phylogeny (Affinities)

Bentham and Hooker treated Passifloraceae under order Passiflorales of series Calyciflorae of Polypetalae along with families like Cucurbitaceae and Begoniaceae. Hutchinson, however, treated both Cucurbitaceae and Begoniaceae under order Cucurbitales while Passifloraceae under order Passiflorales quite close to Cucurbitales.

21.48 FICOIDALES

21.48.1 General Characteristics

1. Flowers usually solitary, showy, each with numerous undifferentiated floral parts.
2. Flowers mostly actinomorphic, bisexual and epigynous.
3. Numerous stamens arranged spirally or in clusters.
4. Ovary syncarpous, usually inferior, unilocular, parietal placentation.

5. Mostly prickly or spiny, fleshy succulent plants.
6. Seeds with scanty or no endosperm, often arillate with curved or straight embryo.

Ficoidales includes only Cactaceae and Ficoideae according to Bentham and Hooker. Engler placed the single family Cactaceae in the Opuntiales, an order he considered to have been derived from the Parietales. Ficoideae, also sometimes named as Aizoaceae, has been treated by Lawrence (1951) to belong to order Centrospermae along with 9 more families, chief among which are Chenopodiaceae, Amaranthaceae, Nyctaginaceae, Portulacaceae and Caryophyllaceae.

Only Cactaceae is discussed here. Families like Chenopodiaceae, Amaranthaceae and Caryophyllaceae have been discussed elsewhere in this text.

21.49 CACTACEAE (CACTUS FAMILY)

21.49.1 Systematic Position

Polypetalae, Calyciflorae, Ficoidales.

21.49.2 Field Recognition

Mostly prickly or spiny, fleshy succulent plants; flowers usually solitary, showy, each with numerous floral parts; ovary unilocular, inferior; parietal placentation.

21.49.3 Selected Indian Genera

Opuntia, *Mammillaria*, *Cereus*, *Echinocactus*.

21.49.4 Size, Distribution and General Information

Cactaceae includes approximately 85 genera and 2,000 species. But different authorities mention the number of genera differently between 30 to 200. Members are distributed chiefly in semi-desert or drier regions of tropical and subtropical America, and also in Africa, Mauritius, Seychelles, India, Sri Lanka, and Australia. Mainly the Cactaceae are the succulents of New World. Main species native of Old World is *Rhipsalis*. Some of the larger genera along with their number of reported species are *Mammillaria* (200–300), *Opuntia* (250), *Echinocereus* (75), *Rhipsalis* (60), *Cereus* (50), *Cephalocereus* (48), *Echinopsis* (35), *Epiphyllum* (21) and *Pereskia* (20).

21.49.5 Description of the Family

General Habit Fleshy, succulent or xerophytic; perennial herbs, shrubs, or trees of diverse forms; with spines or bristles or both.

Roots Root system generally shallow, with elongated, slender but fleshy roots.

Stem Fleshy, of various shapes; rarely bearing normal leaves, usually provided with spines or barbed bristles (*glochidia*) which are generally considered to be modified leaves; spines and glochidia develop from small cushion-like structures called *areoles*; in several genera, the stem assumes leaf-like flattened form (e.g. *Epiphyllum*, *Rhipsalis*).

Leaves Simple, highly reduced or scale-like, or even absent in many genera, or flat and fleshy; spines or tuft of bristles develop from areoles.

Inflorescence Flowers mostly solitary, showy and borne upon or near the areoles; only rarely corymbose or paniculate (*Pereskia*).

Flower Exceptionally beautiful, large, showy but only very rare (*Mammillaria*, *Opuntia*, *Cereus*, *Echinocereus*, *Schlumbergera*); sessile; mostly bisexual; actinomorphic or nearly zygomorphic; epigynous.

Perianth Perianth parts numerous (Fig. 21.64), showing a gradual transition from sepals to petals; spirally arranged; sepals often petalloid; petals epigynous and present in several series; often fuse to form a perianth tube or hypanthium.

Androecium Stamens numerous, in several series or groups, epipetalous or inserted at the base of petals; anthers 2-celled, basifixed, introrse; longitudinally dehiscent.

Gynoecium Two to many carpels, syncarpous; ovary inferior; unilocular, numerous anatropous ovules; parietal placentation, very rarely basal placentation; style 1 and simple; stigmas often as many as the carpels.

Fruit and Seeds Fruit usually a many-seeded berry, often spiny, scaly, bristly or glochidiate. Seeds with scanty or no endosperm; often arillate, with curved or straight embryo.

General Floral Formula $\oplus$, ⚥, $\overbrace{P_{\infty \text{ or } (\infty)}, A_{\infty}}$, $G_{\overline{(2-\infty)}}$.

21.49.6 Economic Importance

- Cactaceae are mainly of ornamental value. Some commonly grown ornamental cacti include species of *Cereus*, *Cephalocereus*, *Echinocactus*, *Echinocereus*, *Mammillaria*, *Opuntia*, *Rebutia* and *Stenocereus*.

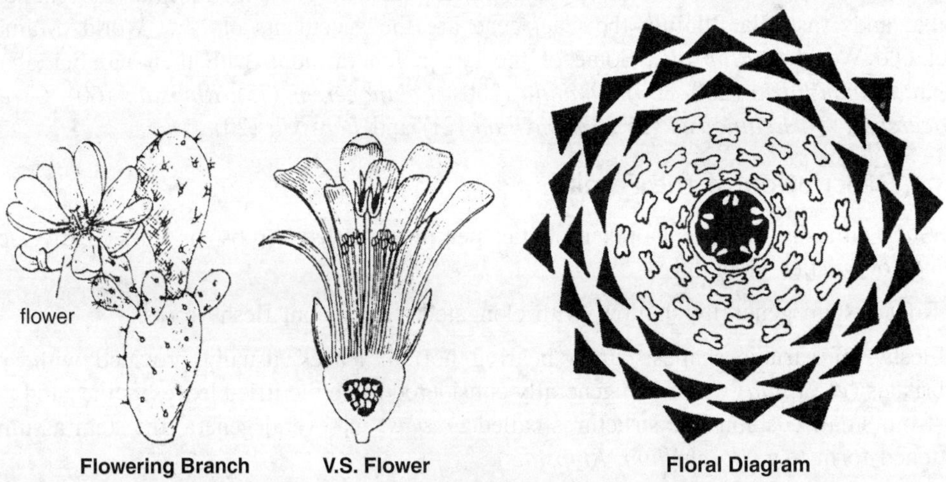

Fig. 21.64 *Opuntia dillenii* Haw.

- Fruits of *Nopalea cochenillifera*, *Lophophora williamsii* and several *Opuntia* species are edible.
- Stems of several cacti are used as emergency fodder.
- A red dye, cochineal, used in cosmetics, is obtained from several species of *Opuntia*.
- Some species of *Cereus* and *Opuntia* have medicinal properties.

21.49.7 Systematics and Phylogeny (Affinities)

Systematic position and relationships of family Cactaceae are highly controversial. Bentham and Hooker treated Cactaceae under order Ficoidales and placed in between Passiflorales and Umbellales. Hutchinson (1973) included it under order Cactales (placed in between Cucurbitales and Tiliales), Cronquist (1981) included it under order Caryophyllales (placed in between Casuarinales and Polygonales), and Thorne (1983) discussed it as the last family of order Chenopodiales.

Engler placed Cactaceae under order Opuntiales and opined this order to have been derived from Parietales. Hutchinson considered that Cactaceae is allied to Cucurbitaceae while Maheshwari and others suggested on the basis of the evidence of embryology and floral morphology that Cactaceae show close affinities to the families included under Centrospermae or Caryophyllales. On the basis of the recent discovery of the presence of the betalain pigments rather than anthocyanins, taxonomists such as Cronquist (1981) and Jones and Luchsinger (1987) also support the placement of Cactaceae under order Caryophyllales.

Britton and Rose (1919–1923), in their 4 volume works entitled *"Cactaceae"*, divided the family into following three tribes:

1. *Pereskieae* (leaves broad, flat; glochids absent; flowers stalked), e.g. *Pereskia*.
2. *Opuntieae* (leaves, if present, more or less terete, small; glochids present; flowers sessile), e.g. *Opuntia*.
3. *Cacteae* or *Cereeae* (leaves absent or rudimentary; glochids absent; flowers sessile), e.g. *Mammillaria*, *Cereus*, *Echinopsis*, *Epiphyllum*, *Rhipsalis*, *Schlumbergera*.

21.50 UMBELLALES

21.50.1 General Characteristics

1. Plants show simplification of their floral parts.
2. Flowers arranged in simple or compound umbel.
3. Flowers are usually bisexual, actinomorphic, pentamerous and epigynous.
4. Carpels reduced to 2, syncarpous, ovary inferior.
5. Each locule contains a single anatropous and pendulous ovule.
6. Style usually possesses a swollen base called stylopodium.
7. Fruit usually a dry schizocarp, called cremocarp, splitting into two mericarps.

Bentham and Hooker placed 3 families (Umbelliferae, Araliaceae and Cornaceae) under Umbellales. Engler and Diels also divided order Umbelliflorae into same 3 families.

Only Umbelliferae and Araliaceae are treated in the present text.

21.51 UMBELLIFERAE OR APIACEAE (CARROT FAMILY)

21.51.1 Systematic Position

Polypetalae, Calyciflorae, Umbellales.

21.51.2 Field Recognition

Aromatic herbs; leaves compound and leaf base sheathing or pericladial; inflorescence simple or compound umbel; inferior ovary; stylopodium present; fruit schizocarp.

21.51.3 Selected Indian Genera

Centella, Coriandrum, Daucus, Ferula, Foeniculum.

21.51.4 Size, Distribution and General Information

A family of about 275 genera and 3,000 species, Umbelliferae are cosmopolitan in distribution and chiefly found in North Temperate Zone of the World. Over 50 of its genera and about 200 species have been reported from India. Some of its larger genera with over 100 or more reported species include *Eryngium* (230), *Pimpinella* (150), *Bupleurum* (150), *Ferula* (133), *Peucedanum* (120), and *Hydrocotyle* (100).

The great Greek philosopher Socrates was put to death by using a weed of this family, *Conium maculatum* (poison hemlock).

21.51.5 Description of the Family

General Habit Mostly biennial or perennial herbs with stout stems and hollow internodes; some are prostrate herbs (*Centella*, *Hydrocotyle*), few species of *Bupleurum* and *Trachymene* are somewhat shruby; emit aromatic smell due to the presence of essential oil or oleoresin in all organs.

Leaves Alternate, generally pinnately compound or decompound, much divided, exstipulate, with sheathing leaf bases; rarely leaves are entire and simple (*Bupleurum*, *Centella*, *Hydrocotyle*) or palmately compound (*Sanicula*, *Astrantia*); bi- or tripinnately divided (*Coriandrum*, *Foeniculum*).

Inflorescence Usually a compound umbel subtended by an involucre of bracts; or simple umbel (*Astrantia*, *Hydrocotyle*); or cymose head (*Eryngium*); compound umbels of majority of genera are formed from partial umbels (each one often subtended by an involucre of bracteoles called involucel).

Flower Usually bisexual, actinomorphic, pentamerous, epigynous; rarely unisexual because the last umbellets to develop are sometimes staminate.

Calyx 5 sepals, free or united; very small and narrow, usually adnate to the ovary; valvate or rarely imbricate (*Astrantia*).

Corolla 5 petals, polypetalous; valvate (*Centella*, Fig. 21.65) or rarely imbricate; petals alternate with sepals; usually white or yellow; some deciduous, rarely absent; often bilobed; petal lobes of peripheral flowers of *Coriandrum* are unequal.

Androecium 5 free stamens, alternate with the petals, inserted on an epigynous disc; anthers dithecous, basi- or dorsifixed, introrse, dehiscence longitudinal.

Gynoecium Bicarpellary, syncarpous, carpels medianly-placed; ovary inferior, bilocular; each locule contains a single anatropous and pendulous ovule; axile placentation; styles 2, often with a thickened or swollen base (*stylopodium*) appearing like an epigynous disc; stigmas 2.

Fruit and Seeds Fruit a dry schizocarp (called *cremocarp*) splitting into 2 mericarps, often ribbed. Two mericarps separate from below upward from a stalk called carpophore. Each mericarp contains a single seed. On the outer surface of each mericarp are usually present 5 primary ridges, of which 2 lateral ridges are at the edges where the splitting takes place. In the furrows between the ridges are often present oil canals called *vittae*. Fruit structure is an important diagnostic character in the taxonomy of this family. *Seeds* with a minute embryo and copious endosperm.

Fig. 21.65 Floral diagram of *Centella asiatica*.

General Floral Formula $\oplus$, ⚥, $K_{(5) \text{ or } 5}$, C_5, A_5, $G_{(\bar{2})}$.

21.51.6 Economic Importance

- *Vegetables:* Certain plants of Apiaceae, which provide us vegetables, which are either eaten raw or cooked, include *Apium graveolens* (Celery), *Daucus carota* (Gajar or Carrot), *Pastinaca sativa* (Parsnip) and *Petroselinum crispum* (Parseley).
- *Condiments, Spices and Flavouring Agents:* Mature fruits and seeds of several plants are used as spices, condiments and for flavouring biscuits, breads, cheese, etc. Common species are *Anethum graveolens* (Sowa or Dill), *Apium graveolens* (Celery seeds or Ajmud), *Carum carvi* (Caraway or Shiajeera), *Coriandrum sativum* (Dhania or Coriander), *Cuminum cyminum* (Zeera or Cumin), *Foeniculum vulgare* (Saunf or Fennel), *Trachyspermum ammi* syn. *Carum capticum* (Ajwain), *T. roxburghianum* (Ajmud). Almost all these plants are carminative, stimulant and used in stomach disorders as well as in many dishes in India, China and several other countries.
- *"Heeng":* 'Heeng' of commerce or Asafoetida is the dried latex obtained after making incisions in the roots and rootstocks of *Ferula assafoetida* and *F. narthex*. Heeng is an oleoresin and used as a condiment as well as in medicines. It stimulates respiratory as well as nervous and intestinal systems.
- *Poison:* *Conium maculatum* (poison hemlock) is a deadly poison and was used to kill the great Greek philosopher, Socrates. *Cicuta virosa* (water hemlock) is also poisonous.
- *Medicinal Value:* *Centella asiatica* (Brahmi Booti) is the king of medicinal plants. It is used as a brain tonic and also in madness and leprosy. *Ferula sumbule* is used in curing hysteria.

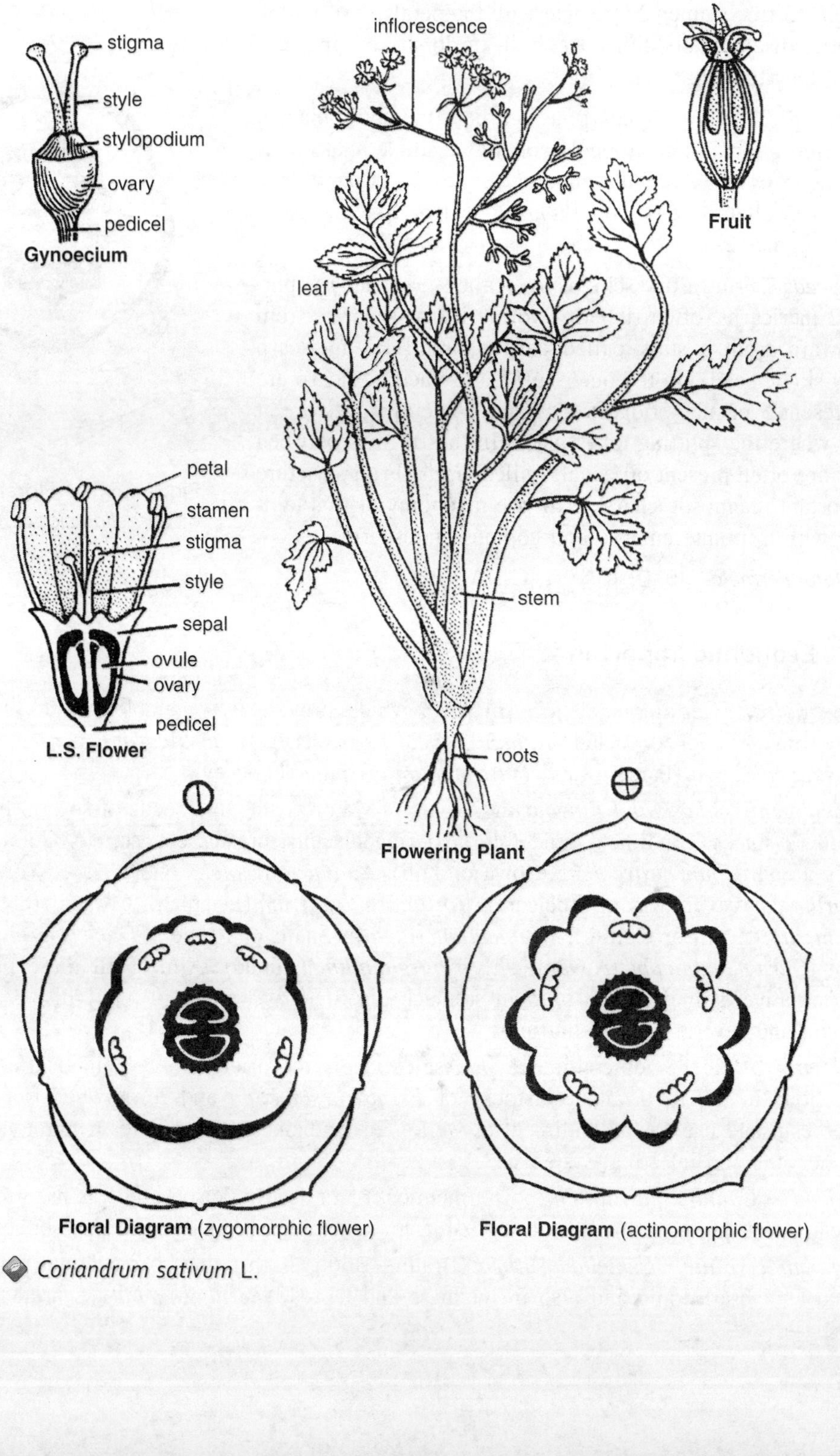

Fig. 21.66 *Coriandrum sativum* L.

- *Ornamental Plants:* Some of the ornamentals of Apiaceae include *Angelica archangelica* (Angelica), *Aegopodium podograria, Heracleum maxima* and *Trachymene caerulea.*

21.51.7 Systematics and Phylogeny (Affinities)

Bentham and Hooker placed Umbelliferae under order Umbellales. Pollen characters of Umbelliferae and Araliaceae show close affinities between these two families. However, Hutchinson opined that Araliaceae is a primitive family and should be placed under Araliales of Lignosae while Umbelliferae is a much advanced family of Herbaceae. In reduction of calyx and bicarpellary, syncarpous ovary with 2 separate styles, Umbelliferae resembles Asteraceae (Compositae). Hutchinson opined that Umbelliferae is derived from Saxifragales. However, Cronquist believed that Umbelliferae originated from Sapindales.

Engler divided Umbelliferae into 3 subfamilies and 12 tribes as under:

I. *Hydrocotyloideae:*	1. Hydrocotyleae, 2. Mulineae.
II. *Saniculoideae:*	3. Saniculeae, 4. Legoiceae.
III. *Apioideae:*	5. Echinophoreae, 6. Scandiceae, 7. Coriandreae, 8. Smyrnieae, 9. Ammieae, 10. Peucedaneae, 11. Laserpitieae, 12. Dauceae.

21.51.8 Description of a Common Plant

Coriandrum sativum L. (Dhania or Coriander, Fig. 21.66)

Habit: Annual aromatic herb. *Leaf:* Alternate, exstipulate, decompound; upper leaves more finely dissected; leaf base sheathing; unicostate reticulate. *Inflorescence:* Compound umbel. *Flower:* Bracteate or ebracteate, complete, hermaphrodite, central flowers actinomorphic while peripheral flowers are zygomorphic, pentamerous, epigynous. *Calyx:* 5 sepals, gamosepalous, free at the tips, valvate; anterior sepals may be larger in outer peripheral flowers. *Corolla:* 5 petals, free, bilobed, valvate; in central actinomorphic flowers the petal lobes are equal-sized; in peripheral zygomorphic flowers one anterior petal has 2 large equal-sized lobes, two lateral petals have 2 unequal-sized lobes, and two posterior petals have 2 equal-sized small lobes. *Androecium:* 5 stamens, alternipetalous, free, dithecous, dorsifixed, introrse. *Gynoecium:* Bicarpellary, syncarpous, inferior, bilocular, one ovule in each locule, axile placentation; styles 2, arise on a nectar-secreting disc called stylopodium; stigmas 2, capitate. *Fruit:* Cremocarp, splitting into 2 mericarps.

Floral Formulae:

(a) *Peripheral flowers:* ⦶, ⚥, $K_{5 \text{ or } (5)}$, C_5, A_5, $G_{(\bar{2})}$.

(b) *Central flowers:* $\oplus$, ⚥, $K_{5 \text{ or } (5)}$, C_5, A_5, $G_{(\bar{2})}$.

21.52 ARALIACEAE (ARALIA OR GINSENG FAMILY)

21.52.1 Systematic Position

Polypetalae, Calyciflorae, Umbellales.

21.52.2 Field Recognition

Leaves alternate, often large and compound; usually umbellate inflorescence; actinomorphic, pentamerous flowers; inferior ovary with one ovule in each locule; fruit berry or drupe.

21.52.3 Selected Indian Genera

Aralia, *Helwingia*, *Panax*.

21.52.4 Size, Distribution and General Information

A family of about 70 genera and over 800 species, Araliaceae are distributed chiefly in tropical region, mainly in tropical America and Indomalaysia. In India, *Helwingia himalaica* occurs commonly in regions of Darjeeling while *Panax fruticosum* and *P. pseudo-ginseng* in Meghalaya and Khasi Hills. Some other major genera along with their approximately reported species are *Schefflera* (200), *Aralia* (35), *Hedera* (15, Fig. 21.67), *Mackinlaya* (12) and *Panax* (8). Members of Araliaceae show close resemblances with that of Apiaceae.

21.52.5 Description of the Family

General Habit Herbs, shrubs, sometimes vines or trees; some appear like palms; a few are root climbers (e.g. *Hedera*).

Stem Solid, often prickly; In *Hedera helix*, the woody stem is evergreen, climber or ground-carpeting and clothed densely by adhesive roots (Fig. 20.67).

Leaves Large-sized, compound, alternate, rarely opposite or whorled; glabrous; dark green; palmately tri- or pentalobed (*Hedera helix*); stipulate, stipules usually modified as a membranous border of petiolar base or linguliform.

Inflorescence Flowers small, forming umbels or heads, often grouped into compound inflorescences; umbels sometimes racemose, corymbose paniculate or umbellate.

Flowers Small, bracteate, bisexual or unisexual (either on separate plants or occasionally male and female flowers on the same plant); actinomorphic; usually epigynous and pentamerous; whitish or greenish.

Calyx Usually 5, very small sepals, fused or adnate to ovary representing only small teeth-like structures; cupuliform.

Corolla Usually 5 but sometimes 3–10 petals; free, caducous; often valvate.

Androecium Stamens 5, sometimes 3, generally equal to the number of petals and alternipetalous; attached to the disc; anthers dithecous, dorsifixed, dehiscing longitudinally; anthers versatile in *Hedera helix* (Fig. 21.67).

A nectariferous epigynous disc, covering top of the ovary and usually fused with the stylar bases, is present.

Gynoecium Pentacarpellary, syncarpous, usually inferior ovary; rarely 1 to many fused carpels with half-inferior or superior ovary; pentalocular, each locule with one pendulous and anatropous ovule; axile placentation; styles as many as carpels, free or united; sometimes even absent, making the stigmas then sessile.

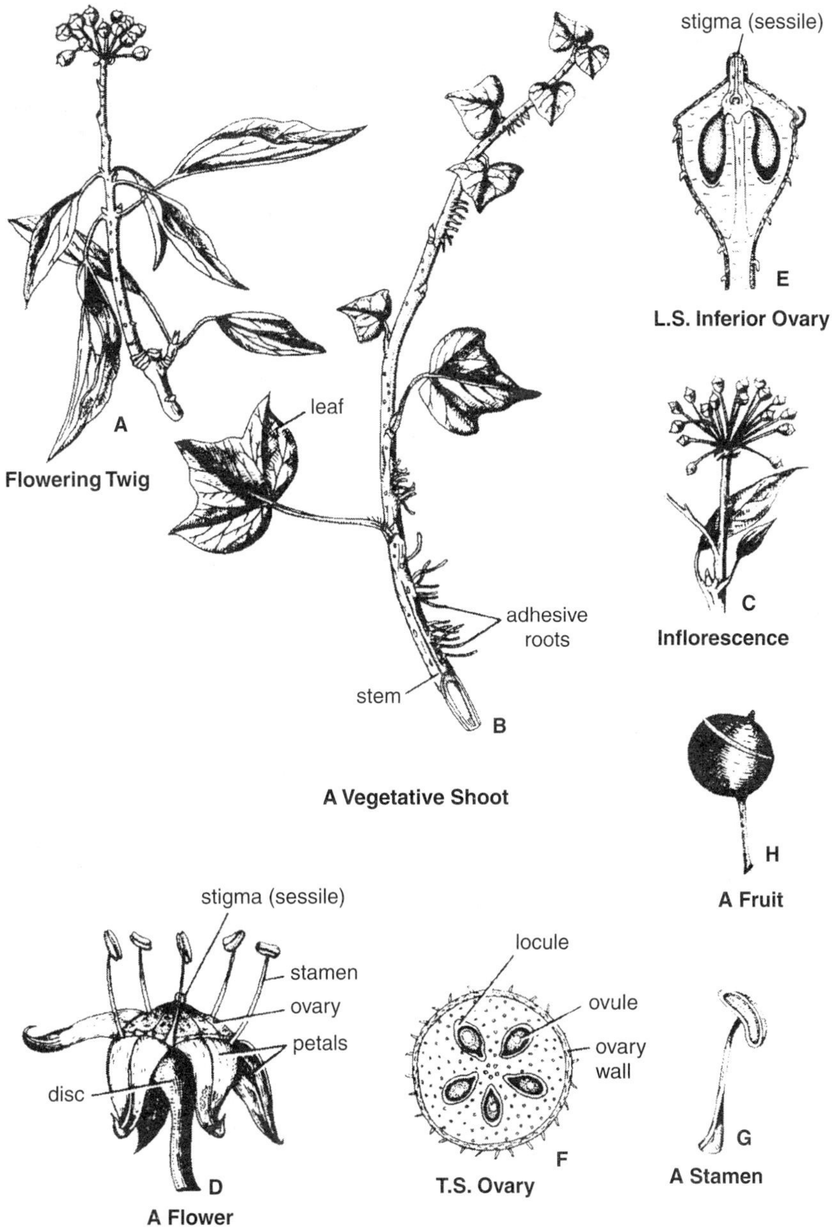

Fig. 21.67 *Hedera helix.*

Fruit Berry, rarely a drupe; with as many seeds as carpels.

Seed Seeds with small embryo and copious endosperm.

General Floral Formula Br, ⊕, ⚥, K(5), C5, A5, $G(\bar{5})$.

21.52.6 Economic Importance

Araliaceae is not of much economic importance, except that of following major aspects:

- It provides some evergreen climbing vines (e.g. *Hedera helix*) and beautiful ornamental shrubs and small trees belonging to several species of the genera like *Aralia, Acanthopanax, Schefflera, Fatsia, Kalopanax* and *Polyscias*. Canary Island ivy (*Hedera canariensis*) is also a well known plant of ornamental value.
- Chinese rice paper is obtained from the pith of rice-paper plant (*Tetrapanax papyriferus*).
- Ginseng roots, used in several medicinal preparations, are obtained from *Panax ginseng* and *P. quinquefolius*. Ginseng is widely used in China as an stimulant and aphrodisiac.
- Species of *Dizygotheca* and *Schefflera* are cultivated indoors as pot plants. Several cultivars have been developed from *Hedera helix* showing wide range of variegation and leaf shape.

21.52.7 Systematics and Phylogeny (Affinities)

Bentham and Hooker included three families under Umbellales, namely Cornaceae, Araliaceae and Umbelliferae (Apiaceae). Hutchinson, however, discussed Cornaceae under order Cunoniales and Araliaceae under order Araliales of division Lignosae, besides discussing Umbelliferae under order Umbellales of division Herbaceae. According to him, Araliales have been derived from Cornaceae.

Hickey and King (1988) divided family Araliaceae into three tribes, viz. (1) Scheffllereae (corolla having valvate aestivation and petals with broad base), e.g. *Hedera*, *Schefflera*, (2) Aralieae (corolla having imbricate aestivation and petals with broad base, e.g. *Aralia*, and (3) Mackinlayeae (corolla having valvate aestivation and petals are shortly clawed).

Hallier recognised only 2 families under order Umbelliflorae, namely Cornaceae and Umbelliferae and included all Araliaceae members under Umbelliferae.

21.53 GAMOPETALAE

The members of subclass Gamopetalae of Dicotyledons contain their flowers with partially or completely fused petals of their corolla. It is divided into 3 series, viz. Inferae, Heteromerae and Bicarpellatae.

Series Inferae is characterised by the presence of inferior ovary and contains 3 cohorts, viz. Rubiales, Asterales and Campanales.

Series Heteromerae is characterised by the (i) presence of superior ovary; (ii) androecium of one or two series, and (iii) presence of more than 2 carpels. It contains 3 cohorts, viz. Ericales, Primulales and Ebenales.

Series Bicarpellatae contains (i) superior ovary, (ii) androecium of only one series, and (iii) only 2 carpels. It contains 4 cohorts, viz. Gentianales, Polemoniales, Personales, and Lamiales.

21.54 RUBIALES

21.54.1 General Characteristics

1. Leaves are usually simple and opposite decussate.
2. Flowers are arranged in cymose inflorescences, generally dichasial cymes.
3. Flowers gamopetalous, usually actinomorphic but sometimes also zygomorphic.
4. Stamens epipetalous and inserted in the corolla tube.
5. Ovary bicarpellary, syncarpous, with 1 to many ovules in each locule; inferior ovary.

Bentham and Hooker included only two families in Rubiales. These are Rubiaceae and Caprifoliaceae. Engler and Diels, however, also included 3 more families along with Rubiaceae and Caprifoliaceae. These are Adoxaceae, Dipsacaceae and Valerianaceae.

Only Rubiaceae and Caprifoliaceae are treated in this text.

21.55 RUBIACEAE (COFFEE FAMILY OR MADDER FAMILY)

21.55.1 Systematic Position

Gamopetalae, Inferae, Rubiales.

21.55.2 Field Recognition

Herbs (mostly in north temperate regions), shrubs or trees; leaves stipulate, opposite or whorled; flowers tetra- or pentamerous; stamens alternipetalous and as many as corolla lobes; usually bicarpellary; ovary inferior.

21.55.3 Selected Indian Genera

Cinchona, *Coffea*, *Gardenia*, *Ixora*, *Mussaenda*, *Oldenlandia*, *Rubia*.

21.55.4 Size, Distribution and General Information

A family of about 500 genera and 6500 species, Rubiaceae are chiefly distributed in tropics, but some are distributed in temperate or even arctic (*Galium*) regions. About 75 genera and 275 species have been reported from India, chiefly in tropical and subtropical eastern Himalayas. Some of the largely represented genera with their approximately reported species (Hickey and King, 1988) include *Psychotria* (700), *Galium* (400), *Ixora* (400), *Gardenia* (250), *Mussaenda* (200), *Asperula* (200), *Cephaelis* (180), *Rondeletia* (120), *Coprosoma* (90), *Guettarda* (80), *Rubia* (60), *Coffea* (40) and *Cinchona* (40). Two most important plants of the family are *Coffea arabica* (coffee-yielding plant) and species of *Cinchona* (quinine-yielding plant).

21.55.5 Description of the Family

General Habit Mostly trees (*Cinchona officinalis*) or shrubs (*Mussaenda luteola*, *Hamelia patens*); however, those in north temperate regions are mostly herbs (e.g. species of *Galium*, *Oldenlandia*); sometimes hook climber (*Uncaria*) or vines.

Leaves Simple, entire; opposite decussate or whorled; stipulate, stipules usually interpetiolar and rarely intrapetiolar, sometimes foliaceous (*Galium*) and become as large as the leaf blades so that the leaves appear whorled; or sometimes reduced to glandular setae (*Pentas*).

Inflorescence Basically a dichasial cyme, dichasia sometimes grouped into globose heads (*Adina*, *Nauclea*); rarely flowers solitary (*Gardenia*, *Randia*).

Flower Bracteate, sometimes bracteolate; bisexual; usually actinomorphic, rarely slightly zygomorphic; penta- or tetramerous; usually epigynous, only rarely perigynous (*Synaptantha*) or very rarely hypogynous (*Gaertnera*).

Calyx 4 or 5 sepals, polysepalous, valvate; rarely greatly reduced sepals (*Morinda*); in *Mussaenda* one of the sepals of some flowers becomes enlarged and bright coloured (Fig. 21.68).

Corolla 4 or 5 petals, gamopetalous; usually salverform, rotate, or infundibular; valvate (*Ixora*, *Mussaenda*), twisted (*Gardenia*) or imbricate (*Rondeletia*).

Androecium 4 or 5 stamens, epipetalous, inserted in the corolla tube or at its mouth; alternate with the corolla lobes; anthers 2-celled, basifixed, introrse, dehiscing longitudinally.

Gynoecium Usually bicarpellary, syncarpous, inferior ovary; bilocular, with 1 to many anatropous ovules in each locule; axile placentation; style simple or bifid; stigma capitate; epigynous disc often present; 5 carpels in *Gardenia* and *Hamiltonia*; ovary half-inferior in *Synaptantha,* or even rarely superior in *Gaertnera*.

Fruit and Seeds Fruit a capsule (*Cinchona*, *Oldenlandia*), berry (*Mussaenda*), schizocarp (*Galium*), or even drupe. Seeds with a small straight or curved embryo in endosperm; sometimes winged.

General Floral Formula Br, ⊕, ⚥, $K_{4 \text{ or } 5}$, $\overbrace{C_{(4) \text{or } (5)}, A_{4 \text{ or } 5}}$, $G_{(\bar{2})}$.

21.55.6 Economic Importance

- *Ornamental Plants:* Some of the common ornamental plants grown in gardens and roadsides include *Anthocephalus cadamba* (Kadam), *Cephalanthus occidentalis, Galium verum* (Ladies Bedstraw), *Gardenia jasminoides* (Jasmine), *G. lucida, Hamelia patens, Hamiltonia suaveolens* (Padera), *Ixora arborea, I. coccinia, Morinda tinctoria, Mussaenda frondosa, M. luteola, Pentas lanceolata* and *Rondeletia*.
- *Coffee:* Coffee, the most popular nonalcoholic, caffeine-containing beverage of the world, comes mainly from roasted and powdered seeds of *Coffea arabica*. Other coffee-producing species are *C. liberica* and *C. robusta*.
- *Quinine:* This famous drug, used throughout the world in the treatment of malaria, comes from the bark of several species of *Cinchona*, such as *C. calisaya*, *C. ledgeriana* and *C. officinalis*.
- *Other Medicinal Plants:* (i) *Cephaelis ipecacuanha* (Ipecac) roots are used against amoebic dysentery and pyorrhoea, (ii) *Gardenia gummifera* provides gum used as a carminative stimulant, (iii) *Randia tinctoria* fruit pulp is emetic and anthelmintic, (iv) *Paederia foetida* is used in indigestion and stomach troubles.

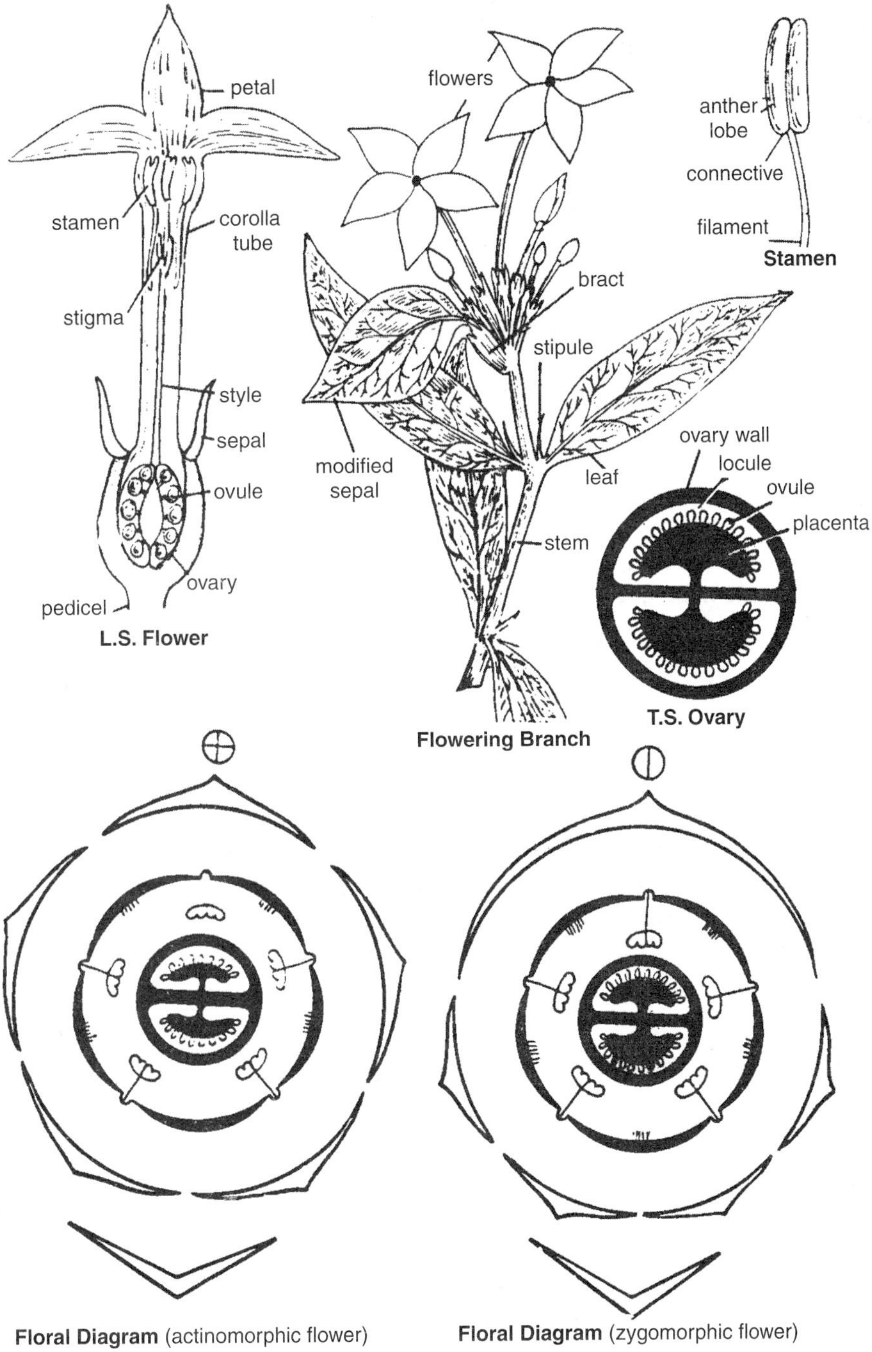

Fig. 21.68 *Mussaenda luteola* Dilile.

- *Dyes:* Roots of *Rubia tinctoria* (Madder) contain dye alizarin and purpurin, while that of several species of *Morinda* contain red (*M. tinctoria*, *M. bracteata*) and yellow (*M. citridora*) dye. Red dye is also obtained from the root bark of *Oldenlandia umbellata*.
- *Timber:* Useful timber is obtained from several species of this family, such as *Adina cordifolia* (Haldu), *Anthocephalus cadamba*, *Ixora ferrea*, *Mitragyna parviflora* and *Randia spinosa*.

21.55.7 Systematics and Phylogeny (Affinities)

Bentham and Hooker treated Rubiaceae under order Rubiales while Takhtajan (1969) and Thorne (1983) discussed it under order Gentianales.

Affinities of Rubiaceae are controversial. Similar type of alkaloids bring Rubiaceae close to Loganiaceae. But in Loganiaceae the ovary is superior while it is inferior in Rubiaceae. In possessing opposite leaves, cymose inflorescence and inferior ovary Rubiaceae comes closer to Caprifoliaceae. However, leaves are exstipulate in Caprifoliaceae. In possessing epigynous flowers, cymose inflorescence, epigynous disc and bicarpellate ovary, Rubiaceae comes closer to Umbelliferae and Cornaceae.

21.55.8 Description of a Common Plant

Mussaenda luteola Dilile (vern. Bedina, Fig. 21.68)

Habit: An ornamental, perennial shrub. *Leaf:* Opposite decussate, simple; stipulate, stipules interpetiolar; subsessile, acute; unicostate reticulate. *Inflorescence:* Dichasial cyme. *Flower:* Bracteate, pedicellate, complete, hermaphrodite; younger flowers actinomorphic, older flowers zygomorphic; pentamerous, epigynous. *Calyx:* 5 sepals, free but sometimes fused, valvate; in zygomorphic flowers one of the sepals is modified into a large, yellow, leaf-like structure. *Corolla:* 5 petals, fused, valvate, hypocrateriform, throat and mouth of corolla hairy. *Androecium:* 5 stamens, polyandrous, epipetalous, alternipetalous; dithecous, basi- or dorsifixed, introrse. *Gynoecium:* Bicarpellary, syncarpous, inferior, bilocular, with many ovules on T-shaped placentae; axile placentation; style long; stigma bilobed. *Fruit:* Berry.

Floral Formula: Br, ⊕ or ⊕, ⚥, K_5, $\overbrace{C_{(5)}, A_5}$, $G_{(\overline{2})}$.

21.56 ❀ CAPRIFOLIACEAE (HONEYSUCKLE FAMILY)

21.56.1 Systematic Position

Gamopetalae, Inferae, Rubiales.

21.56.2 Field Recognition

Shurbs or lianas, rarely herbs; opposite; exstipulate leaves; flower tetra- or pentamerous, epigynous; calyx fused to the ovary; inferior ovary; multicarpellate gynoecium.

21.56.3 Selected Indian Genera

Lonicera, Viburnum, Wiegela, Abelia.

21.56.4 Size, Distribution and General Information

Caprifoliaceae is a family of about 15 genera and more than 530 species, distributed mostly in north-temperate regions and tropical mountains of the world. Some of the larger genera of the family are *Lonicera* (200, honeysuckle) distributed mainly in the North America and Eurasia, *Viburnum* (200) distributed in temperate regions and subtropics, *Sambucus* (40) showing a cosmopolitan distribution, *Abelia* (30) found chiefly in Himalayas to east Asia and Mexico, and *Symphoricarpos* (18) distributed in North America and China.

Sweet nectar, obtained from the flowers of *Lonicera* (honeysuckle; Fig. 21.69), is enjoyed by millions of children throughout the world.

21.56.5 Description of the Family

General Habit Shrubs, sometimes lianas (*Lonicera* sp.); only rarely herbs (*Triosteum*); *Lonicera periclymenum* (Fig. 21.69), a woody climber, reaches upto 6 m.

Leaves Simple but sometimes pinnately compound (*Sambucus*); opposite; usually exstipulate but sometimes stipulate in *Sambucus*, stipules reduced to nectariferous glands in *Viburnum*; margin entire but occasionally lobed in some members; ovate to elliptical in *Lonicera periclymenum* (Fig. 21.69).

Inflorescence Cymose or its modifications.

Flowers Usually bisexual; actinomorphic or sometimes zygomorphic (*Lonicera periclymenum*, Fig. 21.69); in some members, the 2-flowered cymes are bracteolate; in some species of *Viburnum*, the flowers are sterile or neutral.

Calyx Sepals 4 or 5, gamosepalous; usually fused to the ovary; sometimes polysepalous.

Corolla Petals 5, sometimes 4, gamopetalous, imbricate; sometimes bilabiate or variable in form; epigynous.

Androecium Stamen 5, epipetalous, attached on corolla tubes; alternipetalous; sometimes stamens 4 due to suppression of posterior stamen as in *Linnaea*; anthers usually introrse as well as extrorse as in *Sambucus*.

Gynoecium Compound pistil of usually 3–5 carpels; syncarpous; ovary inferior; locules as many as carpels, each containing 1 or more pendulous ovules, axile placentation; style terminal, often slender; stigmas capitate and as many as carpels, distinct or united.

Fruit A berry or drupe; sometimes an achene or capsule.

Seeds With abundant fleshy endosperm, usually each with a small and straight embryo.

General Floral Formula $\mathrm{Brl}, \oplus, ⚥, K_{(5)}, \overparen{C_{(5)}, A_5}, G_{\overline{(3-5)}}$.

21.56.6 Economic Importance

Caprifoliaceae is important for a number of ornamental shrubs, vines and edible fruits.

- Sweet nectar obtained from the flowers of honeysuckle (*Lonicera*) has been enjoyed by the children of countless generations.

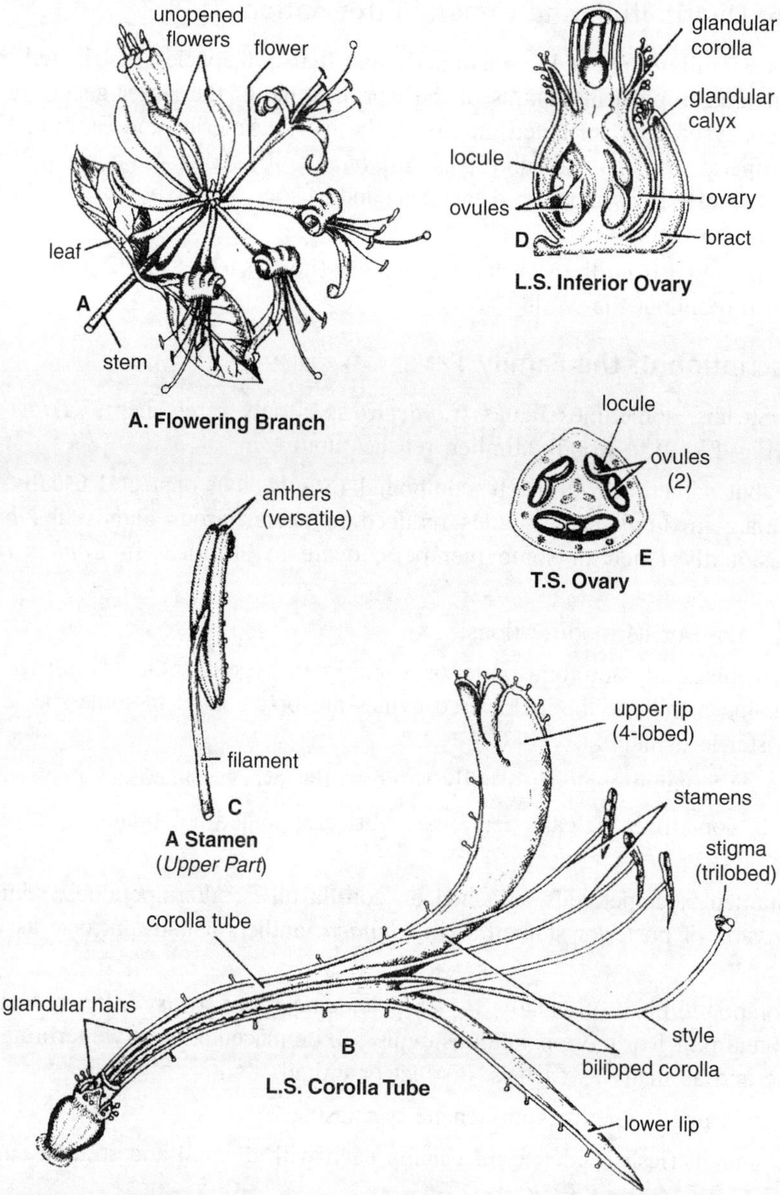

Fig. 21.69 *Lonicera periclymenum* L.

- Fruits of elderberry (*Sambucus nigra*) are used to make elderberry wine.
- Over 200 species of several genera (*Viburnum*, *Lonicera*, *Abelia*, *Leycesteria*, *Kolkwitzia*, *Symphoricarpos*, *Diervilla* and *Linnaea* are used as showy ornamental shrubs.

- Fruits of *Viburnum trilobum* (highbush cranberry) are edible and also used in preparing jelly.
- *Lonicera japonica* is a noxicus troublesome weed and should be taken with all care.

21.56.7 Systematics and Phylogeny (Affinities)

Treated under order Rubiales of series Inferae of Gamopetalae by Bentham and Hooker, honeysuckle family (Caprifoliaceae) has been included under order Dipsacales of subclass Asteridae by Cronquist (1981) along with three other families, viz. Adoxaceae, Valerianaceae and Dipsacaceae. Most phylogenists have treated the monotypic genus *Adoxa* in a separate family Adoxaceae. The genus *Sambucus*, sometimes placed in a separate family Sambucaceae, should be treated only under Caprifoliaceae because of its several similarities with the family, such as (i) pinnate leaves, and (ii) extrorse anthers.

21.57 ASTERALES

21.57.1 General Characteristics

1. Herbs, shrubs or vines; only rarely trees.
2. Inflorescence an involucrate head or capitulum.
3. Stamens epipetalous, syngenesious or synandrous or free.
4. Bicarpellary, syncarpous; ovary inferior, unilocular, one ovule in the locule.
5. Placentation basal.

Bentham and Hooker included Compositae and 3 more families under Asterales. Engler and Diels, however, included Compositae under order Campanulatae along with 5 more families (Campanulaceae, Brunoniaceae, Calyceraceae, Goodeniaceae and Stylidaceae).

Only Compositae (=Asteraceae) has been discussed in this text. Campanulaceae has been treated under Campanales and has been discussed with this order elsewhere in this text.

21.58 COMPOSITAE OR ASTERACEAE (SUNFLOWER OR ASTER FAMILY)

21.58.1 Systematic Position

Gamopetalae, Inferae, Asterales.

21.58.2 Field Recognition

Herbs, vines, or shrubs, rarely trees; inflorescence an involucrate head or capitulum; pappus often present; stamens syngenesious; carpels 2, united; ovary inferior; fruit cypsela.

21.58.3 Selected Indian Genera

Ageratum, Aster, Chrysanthemum, Dahlia, Eclipta, Helianthus, Launaea, Sonchus, Vernonia.

21.58.4 Size, Distribution and General Information

Compositae is one of the largest families of flowering plants, represented by about 1100 genera and over 20,000 species (Jones and Luchsinger, 1987), which are worldwide in distribution. Radford (1986) has mentioned the number of genera to be "One thousand to 2000". About 140 of its genera and over 700 species have been reported from India. Some of the largely represented genera with their approximately known number of species (Jones and Luchsinger, 1987; Hickey and King, 1988) in brackets include *Senecio* (2000+), *Eupatorium* (1200), *Vernonia* (1000), *Hieracium* (1000), *Centaurea* (600), *Aster* (500), *Helichrysum* (500), *Cousinia* (400), *Artemisia* (400), *Baccharis* (400), *Mikania* (250), *Bidens* (230), *Crepis* (200), *Inula* (200), *Achillea* (200), *Gnaphalium* (200), *Chrysanthemum* (200), *Anthemis* (200), *Erigeron* (200), *Cirsium* (150), *Ligularia* (150), *Helianthus* (110), *Lactuca* (100), *Hypochoeris* (100), and *Carduus* (100).

21.58.5 Description of the Family

General Habit Usually annual or perennial herbs, but some are shrubs and only a few are trees (*Leucomeris, Vernonia arborea*); *Senecio*, the largest genus of the family includes herbs (*S. vulgaris*), shrubs (*S. magnificus*), trees (*S. cruentus*) and even climbers (*S. scandens*); majority of the temperate zone Compositae are herbaceous; sap watery or milky; usually with rhizomes, stolons, tubers, or fleshy roots; plants often with spiny or various types of vestiture or surface coverings, such as tomentose, lanate, pannose, strigose, etc.; few are xerophytes, e.g. *Hoplophyllum*.

Leaves Usually alternate, sometimes opposite (*Dahlia, Helianthus*), rarely whorled (*Eupatorium*); mostly simple, sometimes needle-like or reduced to scales; margins often pinnately or palmately lobed or divided; only rarely truly compound; exstipulate; frequently in basal rosettes; leaves usually with oil passages; some contain latex (*Lactuca, Crepis*).

Inflorescence The primary inflorescence is a *head* or *capitulum* with many (rarely 1) flowers (called *florets*) borne on a conical, flat, concave, or convex *receptacle* (Fig. 21.70); each *floret* is often subtended by a receptacular bract called *pale* or *chaff*; the receptacle is often subtended by an involucre of bracts called *phyllaries*; bracts or phyllaries are of various shapes, sizes, and textures, and present in one or more series. Usually the heads or capitula are arranged in various ways, such as racemes, panicles, corymbs, or compound heads; in *Echinops*, the heads are compound, but each smaller head contains only one flower. The capitulum may be *heterogamous* or *homogamous*.

Flower Flowers are actinomorphic or zygomorphic; bisexual, or unisexual, or even neutral (both stamens and carpels are aborted); pentamerous; epigynous.

In *heterogamous capitulum inflorescence* (e.g. *Helianthus*, Fig. 21.71), the central florets are *disc florets* which are bisexual and actinomorphic; and peripheral florets are *ray florets* which are ligulate, generally female or neutral and zygomorphic.

In *homogamous capitulum inflorescence*, all florets are similar, usually bisexual, actinomorphic or zygomorphic, rarely all unisexual.

Undermentioned are the *five basic plans* of florets found within the heads or capitula of Compositae:

1. Tubular or disc florets, corolla regular, florets perfect.
2. Tubular or disc florets, corolla regular, florets pistillate.

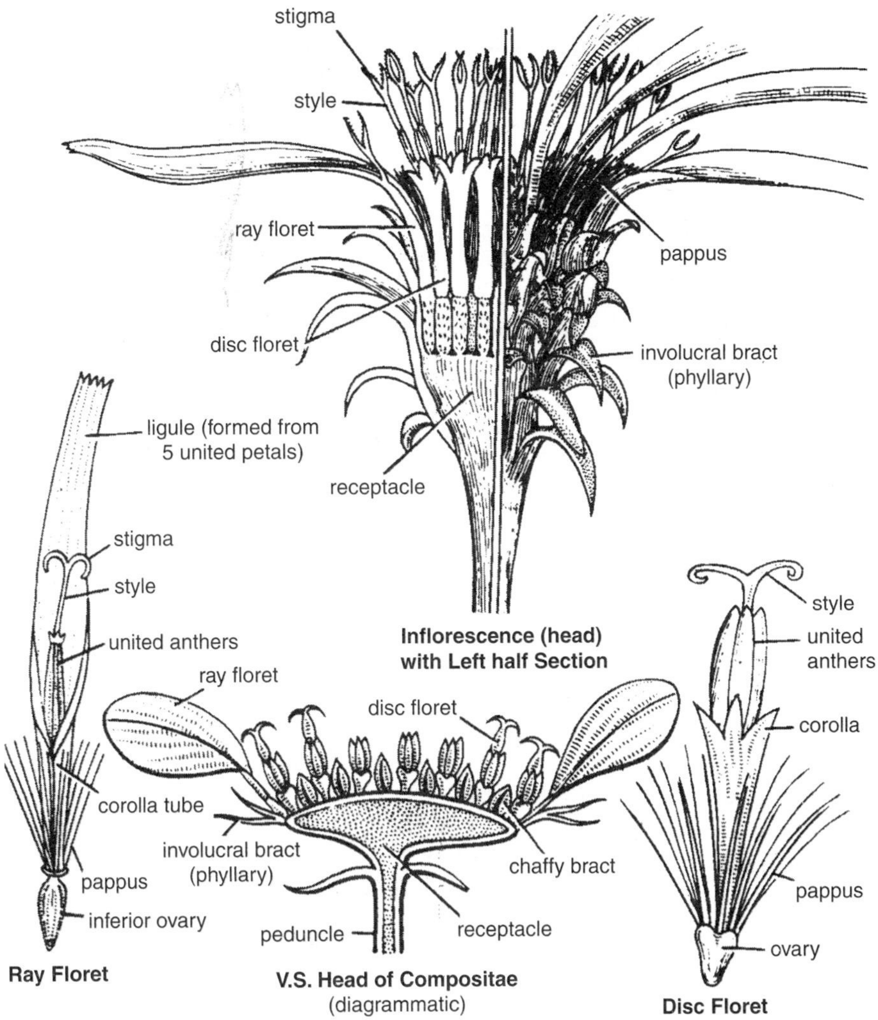

Fig. 21.70 Inflorescence and florets of Compositae.

3. Ligulate or ray florets, zygomorphic, pistillate.
4. Ligulate or ray florets, zygomorphic, sterile.
5. Ligulate or ray florets, zygomorphic, perfect.

Calyx Represented by *pappus* of bristles, awns, or scales; or absent; epigynous.

Corolla 5 petals, gamopetalous, valvate; represented by 3 basic types: (i) 5-lobed and tubular, (ii) ligulate with 3 to 5 teeth, (iii) bilabiate with 3 lobes in the upper lip and 2 lobes in the lower lip. In *Mutisia*, all florets are bilabiate while in *Lactuca* and *Cichorium* all florets are ligulate.

Androecium 5 stamens, epipetalous, alternating with corolla lobes; anthers nearly always united into a tube around the style and filaments free i.e. syngenesious; dithecous, connective often

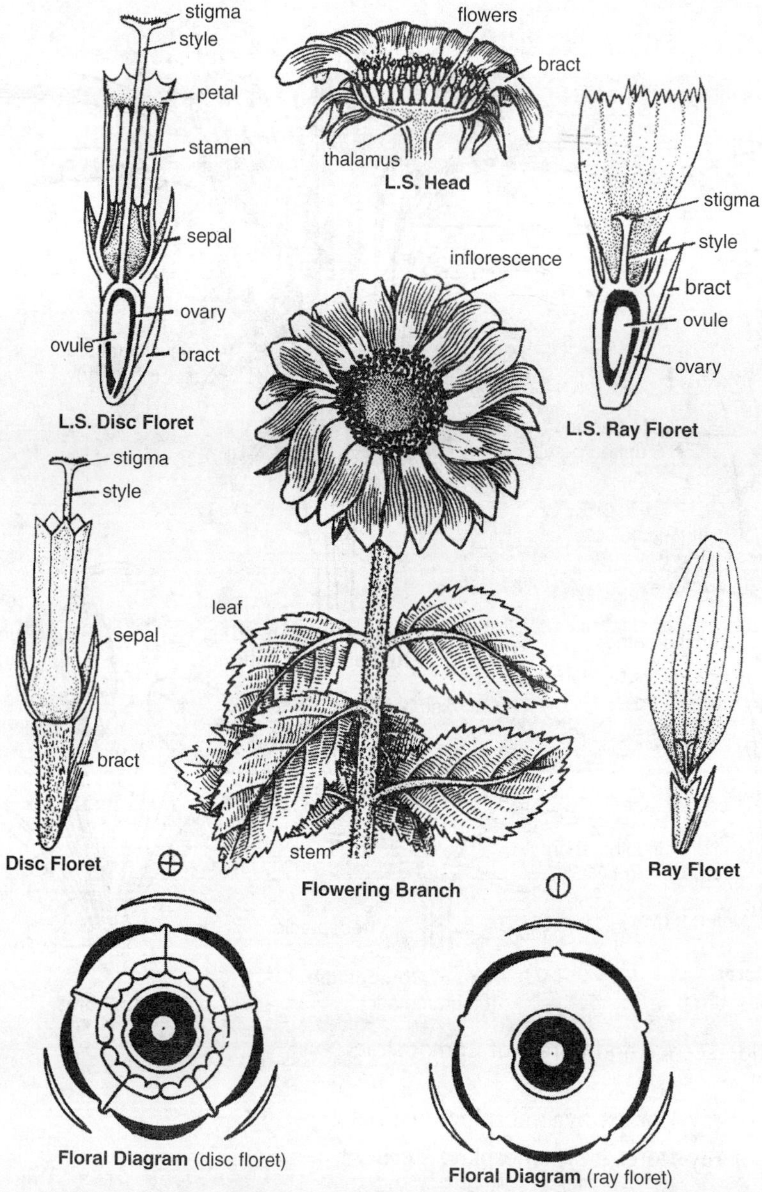

Fig. 21.71 *Helianthus annuus* L.

prolonged, introrse, longitudinally dehiscent; in *Silybium* the stamens unite by anthers as well as by filaments.

Gynoecium Bicarpellary, syncarpous; ovary inferior; unilocular; one ovule, anatropous; basal placentation; style 1; usually 2 branched stigma of diverse forms; an epigynous nectar-secreting disc is also present at the base of the corolla tube surrounding the style.

Fruit and Seeds Fruit a cypsela, often crowned by the persistent pappus, sometimes enclosed by persistent bracts. Seed is 1 per fruit; with large and straight embryo and no endosperm.

Pollination The pollination mechanism is entomophilous and noteworthy in Compositae. A single insect may pollinate numerous flowers of a capitulum in single visit. Insects are attracted by the conspicuous ligulate ray florets of the capitulum. This also benefits the often inconspicuous disc florets. Ray florets also protect the nectar from rain. Cross pollination is promoted also by the protandrous condition of the androecium. In case of failure of cross-pollination, self-pollination may also take place.

General Floral Formula of a Bisexual Flower Φ or ⊕, ⚥, K_{pappus}, $\overbrace{C_{(5)}, A_5}$, $G_{(\overline{2})}$.

21.58.6 Economic Importance[1]

- *Ornamental Plants:* Well-known ornamental species of the family include *Aster amellus, A. grandiflorus, Calendula officinalis, Centaurea moschata, Chrysanthemum carinatum, C. coronarium, C. indicum, Coreopsis grandiflora, Cosmos bipinnatus, Dahlia excelsa, D. pinnata, Dimorphotheca sinuata, Gaillardia pulchella, Gynura aurantiaca, Helianthus annuus* (Sunflower or Surajmukhi), *Helichrysum petiolatum, Tagetus patula* (French Marigold or Genda), *Zinnia elegans* and *Z. linearis.*
- *Disease-Causing Plants:* (i) *Ambrosia artemissifolia* (Rugweed)—pollen cause hay fever, (ii) *Eupatorium urticaefolium* causes milk sickness in animals, (iii) *Pyrethrum hysterophorus* (carrot grass)—pollen cause skin allergy.
- *Medicinal plants:* (i) *Anthemis nobilis* dried capitula are used against dyspesia, (ii) *Artemisia cina* flower heads provide the drug santonin, an antidote against intestinal worms, (iii) *Blumea balsamifera* leaves are used against excitement and insomnia while *B. lacera* leaves are given to cure cholera, (iv) *Grindelia camporum* heads are used in bronchitis and whooping cough, (v) *Inula helium* leaves are effective against tuberculosis, (vi) *Lactuca virosa* leaves have sedative properties, (vii) *Sphaeranthus indicus* capitula are used for curing stomach ache and piles, (viii) *Spilanthes paniculata* capitula are chewed to relieve toothache, (ix) *Tanacetum vulgare* leaves are effective in chronic ulcers and rheumatism, (x) *Taraxacum officinale* roots and rhizome provide the drug "taraxacum", used as a mild laxative, and (xi) *Tussilago farfara* leaves are used in asthama, cough and colds.
- *Insecticides:* Powdered dry capitula of *Centrantherum anthelmenticum, Chrysanthemum coccineum* and *Pyrethrum cinerariifolium* are used in preparing insecticides.
- *Edible Products:* (i) Tuberous roots of *Cynara scolymus, Dahlia tuberosa, Helianthus tuberosus* (Jerusalem artichoke) and *Tragopogon porrifolius* (vegetable oyster) are edible and used as food by man, (ii) *Lactuca sativa* (Salad or Garden lettuce) leaves are eaten raw as salad, (iii) roasted seeds of *Taraxacum officinale* are used in place of coffee.
- *Oils:* (i) Oil obtained from seeds of *Carthamus tinctorius* (safflower or Kusum) is used for manufacture of soaps, paints, varnishes etc., (ii) hair-oil, 'Bhrangraj', is obtained from

[1] Plants mentioned alphabetically in the subheadings.

the leaves of *Eclipta alba*, (iii) fatty oil, obtained from the seeds of *Helianthus annuus*, is edible, (iv) a strong, aromatic essential oil, used in perfumery, is obtained from the seeds of *Tagetus patula* (French Marigold), (v) oil from the seeds of *Xanthium strumarium* is used as an illuminant.

- *Dye:* A red dye safflower, obtained from *Carthamus tinctorius*, is used for colouring candles, butter, and several liquors.

21.58.7 Systematics and Phylogeny (Affinities)

Compositae is usually divided into 13 tribes. These tribes, along with one common example of each of them, are (1) Heliantheae (*Helianthus*), (2) Astereae (*Aster*), (3) Anthemideae (*Chrysanthemum*), (4) Arctotideae (*Arctotis*), (5) Inuleae (*Inula*), (6) Senecioneae (*Senecio*), (7) Calenduleae (*Calendula*), (8) Eupatorieae (*Eupatorium*), (9) Vernonieae (*Vernonia*), (10) Cynareae (*Carthamus*), (11) Mutisieae (*Mutisia*), (12) Liabeae (*Liabum*), and (13) Lactuceae (*Lactuca*).

In possessing pentamerous flowers, syngenesious anthers and inferior ovary, Compositae is related to Campanulaceae and Goodeniaceae. In all these three families the ovary is usually bicarpellary. In possessing inferior bilocular ovary, Compositae also comes near to Rubiaceae and Stylidiaceae. The characters, such as dense inflorescence and inferior ovary, also trace the affinity of Compositae with Dipsacaceae and Valerianaceae. It is presumed that Compositae originated from Rubiaceae.

21.58.8 Whether Compositae Occupy Highest Position Among Angiosperms?

Undermentioned are some *advanced characters*, because of which some workers regard Compositae to be the most highly evolved taxon among angiosperms:

1. Cosmopolitan distribution from arctic to tropical zone.
2. Herbaceous habit.
3. Capitulum inflorescence.
4. Presence of large number of flowers in a small space which makes it possible to pollinate several flowers easily by a single insect.
5. Presence of both actinomorphic and zygomorphic flowers.
6. Presence of both unisexual as well as bisexual flowers.
7. Presence of calyx in the form of pappus.
8. Gamopetalous corolla.
9. Syngenesious anthers.
10. Protandry of androecium.
11. Reduction of carpels to 2 and number of ovules to only one.
12. Presence of inferior ovary and basal placentation.
13. Cypsela fruit.
14. Elaborate mechanism of dispersal.

21.58.9 Description of a Common Plant

1. Ageratum conyzoides L. (Goat weed, Fig. 21.72)

Habit: Annual herb. *Leaf:* Simple, opposite, lower leaves alternate; exstipulate, ovate, serrate, acute, hairy; unicostate reticulate. *Inflorescence:* Compound capitulum; different capitula arranged in corymbose manner; inflorescence homogamous with all flowers tubular; outside each capitulum is present involucre of bracts. *Flower:* Complete, actinomorphic, hermaphrodite, pentamerous, epigynous. *Calyx:* 5 sepals, polysepalous; valvate; reduced to long scaly structures. *Corolla:* 5 petals, gamopetalous; valvate; tubular. *Androecium:* 5 stamens, epipetalous, syngenesious; filaments short;

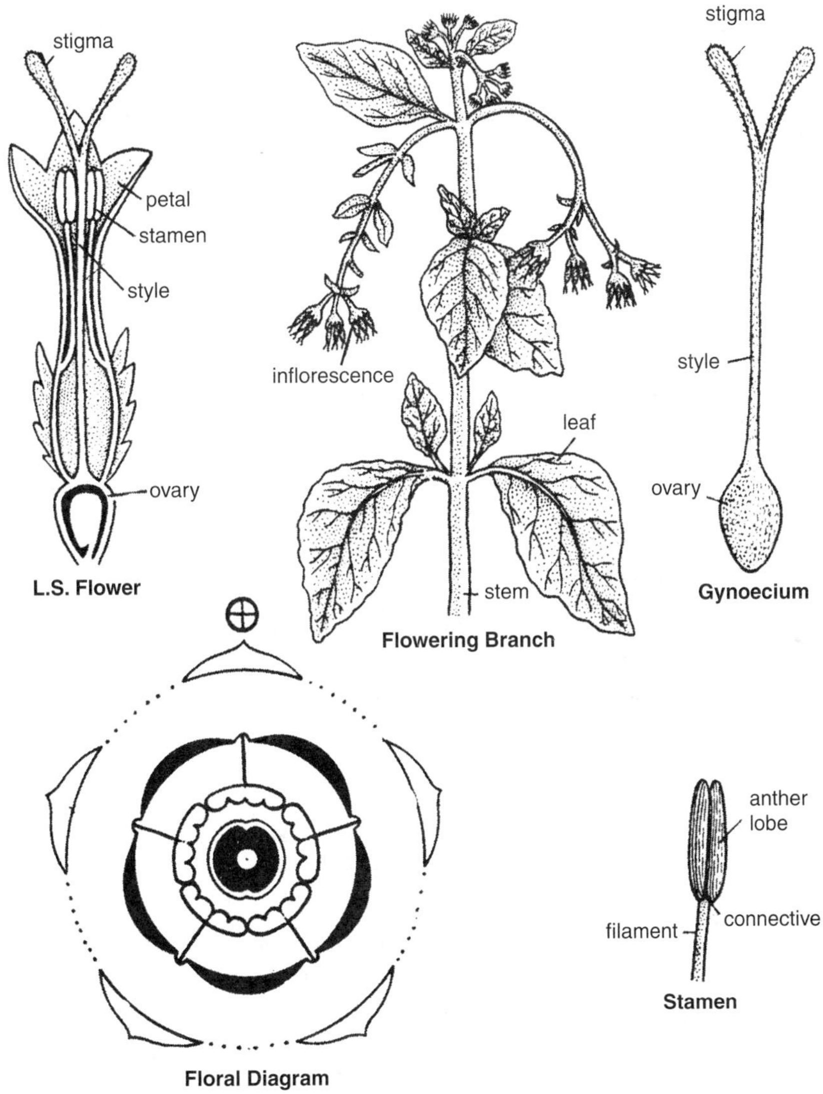

Fig. 21.72 *Ageratum conyzoides* L.

anthers dithecous, basifixed, introrse. *Gynoecium:* Bicarpellary, syncarpous, inferior ovary, unilocular, one ovule in the locule, basal placentation; style long; stigma bifid. *Fruit:* Cypsela.

Floral Formula: $\oplus$, ⚥, $K_{5\ (pappus)}$, $\widehat{C_{(5)}, A_5}$, $G_{(\bar{2})}$.

21.59 CAMPANALES

21.59.1 General Characteristics

1. Flowers typically pentamerous with 5 sepals, 5 petals and 5 stamens.
2. Corolla gamopetalous.
3. Stamens usually in a single whorl, anthers are dithecous and often coherent to connate.
4. Gynoecium is bi- to pentacarpellary, syncarpous.
5. Ovary inferior; often unilocular but sometimes bi- to multilocular; when unilocular, the locule contains a single ovule.

Cohort Campanales includes Campanulaceae and 3 more orders (=families) according to Bentham and Hooker. Engler and Diels, however, placed 6 families in the order Campanulatae, viz. Campanulaceae, Brunoniaceae, Calyceraceae, Goodeniaceae, Stylidaceae and Compositae.

Only Campanulaceae of Campanales has been treated here. Compositae of order Campanulatae (according to Engler and Diels) has been discussed under order Asterales elsewhere in this text.

21.60 CAMPANULACEAE (BELLFLOWER FAMILY)

21.60.1 Systematic Position

Gamopetalae, Inferae, Campanales.

21.60.2 Field Recognition

Largely herbaceous, rarely shrubs or trees; leaves simple, alternate and exstipulate; flowers showy, bisexual, epigynous, pentamerous; stamens frequently united, epipetalous and show connation of their anthers or filaments; ovary inferior, with axile placentation.

21.60.3 Selected Indian Genera

Campanula, *Lobelia*, *Centropogon*, *Cyphia*.

21.60.4 Size, Distribution and General Information

A family of about 70 genera and over 2000 species, Campanulaceae are widely distributed in temperate and sub-tropical regions of the world, specially in tropical mountains. The chief genera of the family are *Campanula* (300), *Lobelia* (235), *Centropogon* (230), *Siphocampylus* (215), *Wahlenbergia* (150), *Cyphia* (50), *Cyanea* (50), *Phyteuma* (40) and *Lightfootia* (40).

Several taxonomists treat Campanulaceae as a family distinct from Lobeliaceae. However, majority of phylogenists, except Hutchinson, treat Campanulaceae comprising of 3 subfamilies (namely, Campanuloideae, Cyphioideae and Lobelioideae, as has also been treated here in this text.

21.60.5 Description of the Family

General Habit Annual or mostly perennial herbs or subshrubs, rarely trees (e.g. *Clermontia*); contain milky or watery sap.

Leaves Simple, alternate or rarely opposite; exstipulate; contain latex; in *Siphocampylus* the leaves are whorled; with long petiole in *Campanula rotundifolia* (Fig. 21.73).

Inflorescence Usually racemose, sometimes cymose as in *Canarina*; if cymose, it may be a determinate monochasial or dichasial cyme; sometimes the flowers are arranged in involucral heads or they are solitary axillary (*Campanula rotundifolia*, Fig. 21.73).

Flower Bracteate, often bracteolate, complete, hermaphrodite, actinomorphic or zygomorphic, pentamerous, epigynous or rarely hypogynous (e.g. *Cyananthus*); blue, regular and bell-shaped in *Campanula rotundifolia* (Fig. 21.73).

Calyx Sepals 5, rarely 3–10, fused with the ovary; valvate or imbricate.

Corolla Petals 5, gamopetalous, valvate; campanulate or tubular or bilabiate; in bilabiate flowers, the corolla is zygomorphic and split down on one side; petal lobes are 6 in *Canarina* while only 3–4 in *Edrianthus*; instead of gamopetalous, the corolla is polypetalous in *Phyteuma* and even absent in some species of *Specularia*; often blue or violet.

Androecium Stamens 5 or as many as corolla lobes, alternipetalous, distinct or variously connate; bases of filaments often expand and form a dome-shaped chamber over epigynous disc; epipetalous or distinct; anthers dithecous, distinct or connate, introrse.

Gynoecium Bi- to pentacarpellary, syncarpous; inferior; only rarely half-inferior in some species of *Lobelia* and *Edrianthus* or even superior as in *Cyananthus;* usually two to many locules, with many ovules in each locule on axile placentation; style slender and simple; stigmas usually 2–5 or as many as carpels.

Fruits A capsule showing different ways of dehiscence in different genera; sometimes a berry (e.g. *Centropogon*).

Seeds With fleshy abundant endosperm and small and straight embryo.

General Floral Formula Br, Brl, $\oplus$ or Φ, ⚥, $K_{(5)}$, $\overset{\frown}{C_{(5)}, A_5}$ or $A_{(5)}$, $G_{\overline{(2-5)}}$.

21.60.6 Economic Importance

- Several members of Campanulaceae, with their showy flowers, are grown as garden ornamentals. Chief among them are species, hybrids and cultivars of *Campanula* (bell flower), *Codonopsis*, *Edrianthus*, *Lobelia*, *Phyteuma*, *Platycodon*, *Specularia* and *Sympyandra*.
- The drug "Lobelia" is obtained from the dried leaves and tops of Indian tobacco (*Lobelia inflata*), a small annual with blue flowers. The plant, as such, is poisonous and cultivated in Kerala, Nilgiris and Darjeeling. The drug is used as an expectorant, antispasmodic, emetic and also in chronic bronchitis and asthma.

21.60.7 Systematics and Phylogeny (Affinities)

Most taxonomists as well as phylogenists divide family Campanulaceae into three sub-families as under:

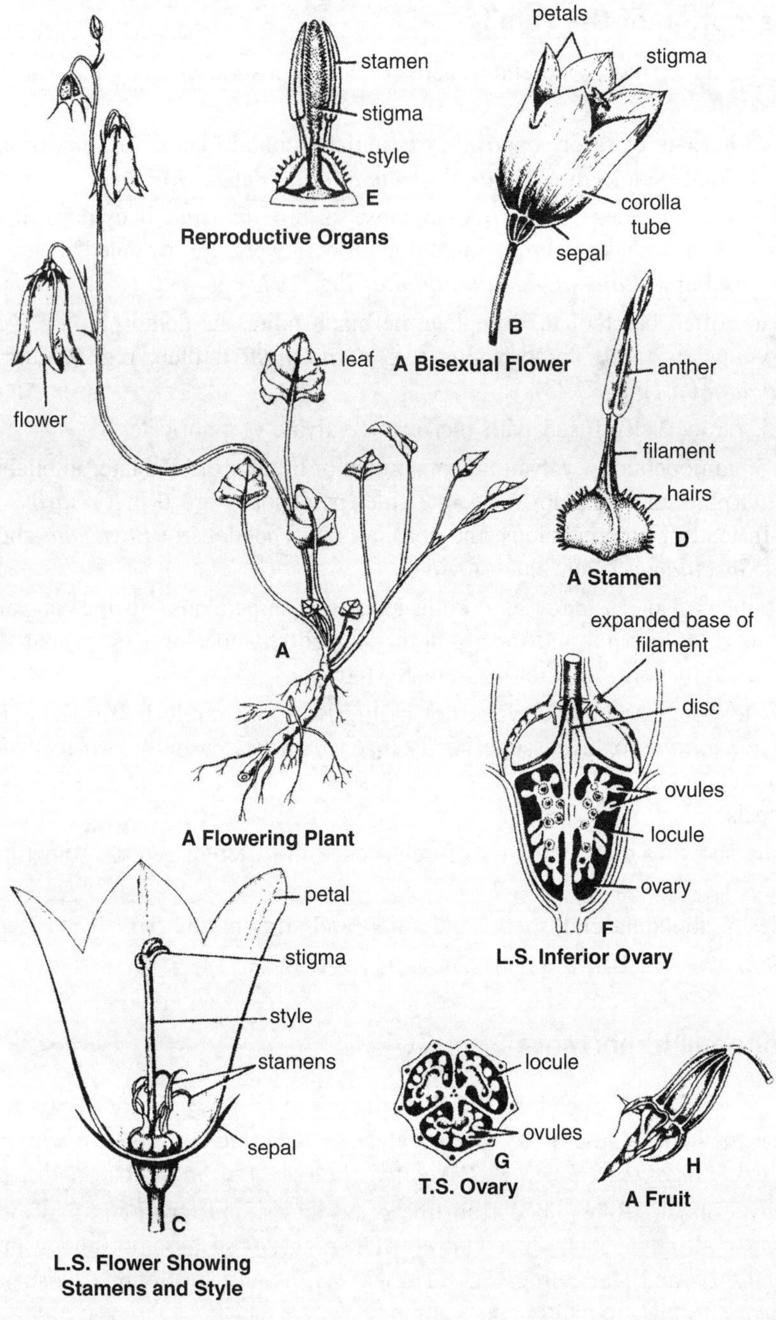

Fig. 21.73 *Campanula rotundifolia* L.

(1) Campanuloideae, having actinomorphic and only rarely zygomorphic flowers with usually free anthers, e.g. *Campanula, Phyteuma, Jasione.*

(2) Cyphioideae, having zygomorphic flowers with their stamens having their filaments united and anthers free, e.g. *Cyphia, Nemacladus.*

(3) Lobelioideae, having zygomorphic flowers and their stamens having their anthers united, e.g. *Lobelia, Centropogon.*

Some taxonomists, however, believe that members of Lobelioideae should be treated under independent family Lobeliaceae due to several striking differences between *Lobelia* and *Campanula.*

21.61 ERICALES

21.61.1 General Characteristics

1. Flowers generally actinomorphic, hypogynous and pentamerous.
2. Petals generally basally connate or sometimes distinct and free.
3. Stamens obdiplostemonous and usually inserted at the edge of a hypogynous nectariferous disc.
4. Gynoecium usually pentacarpellary, syncarpous with superior ovary.
5. Numerous anatropous ovules in each locule; axile placentation.
6. Seeds very minute with copious endosperm.

Cohort Ericales, according to Bentham and Hooker, contains Ericaceae and 5 more families. Engler and Diels, however, divided order Ericales into 2 suborders and 4 families i.e. suborder Ericineae with 3 families (Ericaceae, Clethraceae, and Pyrolaceae) and suborder Epacridineae with only 1 family Epacridaceae.

Only Ericaceae is discussed in this text.

21.62 ERICACEAE (HEATH FAMILY)

21.62.1 Systematic Position

Dicotyledons, Gamopetalae, Heteromerae, Ericales.

21.62.2 Field Recognition

Predominantly shrubs; leaves alternate; flowers urceolate or campanulate; the stamens usually distinct, mostly twice as many as of corolla lobes, developing from a nectariferous disc and only rarely adnate to petals; anthers usually opening by apical pores; ovary typically tetralocular to multilocular.

21.62.3 Selected Indian Genera

Rhododendron hookerii, Gaultheria fragrantissima, Lyonia ovalifolia.

21.62.4 Size, Distribution and General Information

Represented by about 125 genera and 3500 species (Jones and Luchsinger, 1987), members of Ericaceae are cosmopolitan in their distribution. They are, however, almost absent from desert areas and from Australasia where it is replaced by members of the allied family Epacridaceae (Hickey and King, 1988). Ericaceae are mainly confined in the tropical regions, especially in acidic soils on peat, swamps and woodlands. They are mainly associated with endotrophic mycorrhiza.

Major genera of the family with their approximately known number of species in parenthesis are *Rhododendron* (850), *Erica* (600, heath), *Vaccinium* (450, blueberry or cranberry), *Arctostaphylos* (70, manzanita) and *Gaylussacia* (50, *huckleberry*). Several handsome cultivars have been produced by horticulturists by crossing various species of *Rhododendron*.

21.62.5 Description of the Family

General Habit Woody shrubs, rarely small trees (*Arbutus*), or trailing or scrambling vines attaining a length upto 20 metres.

Leaves Simple, usually alternate and evergreen or coriaceous; sometimes opposite or whorled; exstipulate; entire, elliplical to oblong (*Rhododendron ponticum*, Fig. 21.74); sometimes linear and in whorls of 4 (*Erica herbacea*).

Inflorescence Flowers solitary in axils or in axillary or terminal clusters; racemose racemes or panicles; flowers form terminal umbel-like racemes in *Rhododendron ponticum* (Fig. 21.74) or leafy and one-sided raceme in *Erica herbacea*.

Flowers Bracteate, bracteolate, usually bisexual (*Epigaea*) and actinomorphic but slightly zygomorphic as in *Rhododendron ponticum*; often showing wide range of colours.

Calyx Usually 4 to 7 sepals, generally gamosepalous but sometimes polysepalous; usually persistent.

Corolla 4 to 7 petals, gamopetalous but sometimes distinct petals; urceolate, campanulate or salver-shaped; convolute or imbricate aestivation.

Androecium Stamens as many or generally twice as many as petals developing from the base of a nectar-secreting disc; disc is hypogynous in some members (e.g. *Rhododendron*) and epigynous in *Gaultheria*; stamens free or fused at the base; obdiplostemonous; anthers dithecous, frequently appendaged; anthers usually opening by apical pores; introrse.

Gynoecium Usually 4 to 5 carpels, sometimes upto 10 carpels, syncarpous; usually superior ovary, with 4 to 5 locules, locules generally opposite the corolla lobes; numerous anatropous ovules; axile placentation; style simple, usually 1, conical to filiform; stigma usually capitate; sometimes ovary inferior (e.g. *Vaccinium*).

Fruit A capsule, berry or drupe.

Seed Usually small with cylindrical embryo and copious or fleshy endosperm.

General Floral Formula Br, Brl, ⊕, ⚥, $K_{(4-7)}$, $C_{(4-7)}$, $A_{4-7 \text{ or } 8-14}$, $G_{\underline{(4-5)}}$.

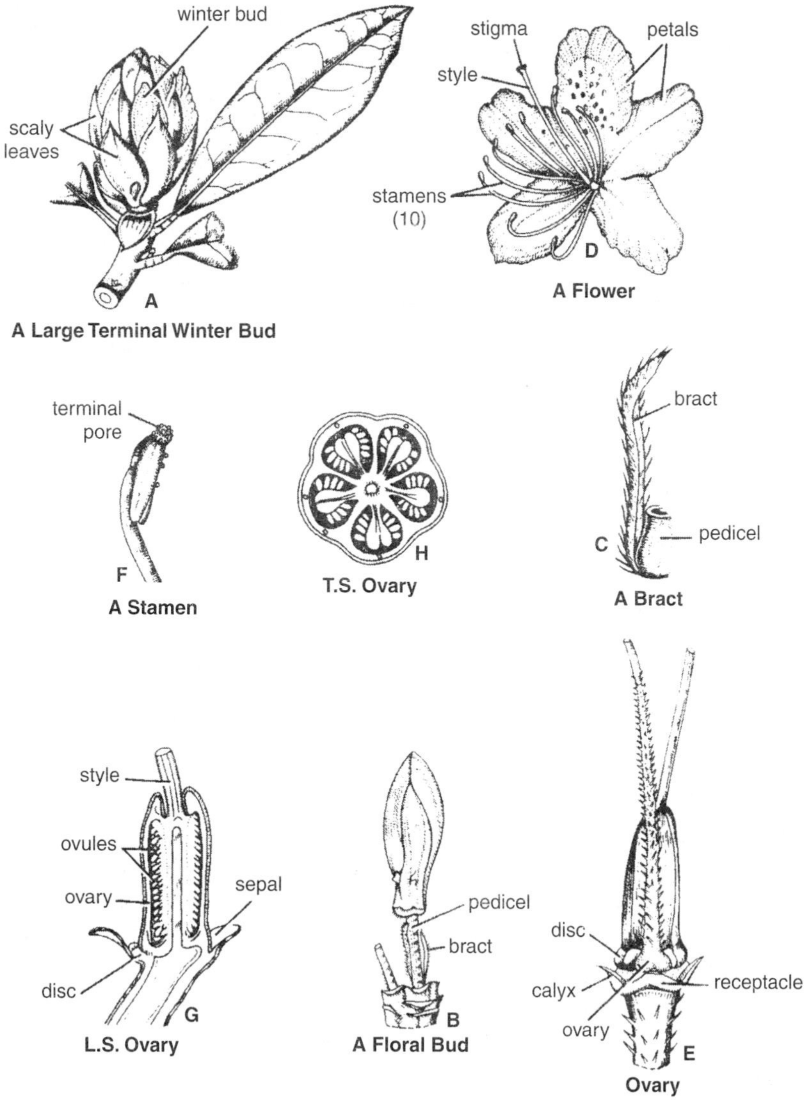

Fig. 21.74 *Rhododendron ponticum.*

21.62.6 Economic Importance

From the point of view of economic importance, heath-family is important for its ornamentals and for blueberries and cranberries.

- Ornamental shrubs are obtained from several species of *Erica*, *Rhododendron*, *Kalmia*, *Leucothoe* and *Pieris*. Several beautiful hybrids and cultivars of horticultural value are obtained from *Erica* (Heath), *Rhododendron*, *Arbutus*, *Pernettya* and *Gaultheria*. They are commonly grown in gardens and glasshouses.

- The fruits of several species of *Vaccinium* are edible being variously known as cranberries (*V. macrocarpon*), blueberries, bilberries and cowberries.
- Huckleberry, an edible fruit, comes from the genus *Gaylussacia.*
- Some of the suffrutescent perennials or subshrubs are Mayflower (*Epigaea*), bearberry (*Arctostaphylos uva-ursi*) and wintergreen (*Gaultheria procumbens*).
- Foliage of *Gaultheria shallon* is sold in the market as "lemon leaf".
- The so-called "briar-pipes" are obtained from the burls of *Erica arborea.*

21.62.7 Systematics and Phylogeny (Affinities)

Some taxonomists divide family Ericaceae, as circumscribed above, into two independent families, viz. Ericaceae and Vacciniaceae, with the former having superior ovary and capsular fruit and the latter possessing partially or completely inferior ovary and berry-type of fruit. Engler and Diels divided Ericaceae into four subfamilies, namely Rhododendroideae, Arbutoideae, Vaccinioideae and Ericoideae. Hickey and King (1988) divided Ericaceae into four subfamilies as under:

1. *Rhododendroideae* Possessing winter buds with scales; petals caducous; stamens without any appendages, fruit septicidal capsule, and winged seeds, e.g. *Rhododendron.*

2. *Ericoideae* Possessing no scales or winter buds; persistent corolla lobes; leaves needle-like; stamens with appendages; fruit loculicidal capsule, or nut, and; seeds with no wings, e.g. *Erica.*

3. *Vaccinioideae* Possessing winter buds with scales; leaves broad, flat and not needle-like; petals caducous; stamens with appendages; inferior ovary; fruit a capsule, berry or drupe, and; seeds with no wings; e.g. *Vaccinium.*

4. *Epigaeoideae* Dioecious plants; leaves cordate; stamens with no appendages; expanded stigma with 5 lobes; e.g. *Epigaea.*

Bentham and Hooker treated Ericaceae under order Ericales of series Heteromerae of Gamopetalae while Engler and Prantl discussed Ericales under subclass Metachlamydeae of Dicotyledoneae. Hutchinson discussed order Ericales under Division Lignosae of subphylum Dicotyledones while Takhtajan (1980) treated order Ericales under superorder Ericanae of subclass Dilleniidae of class Magnoliopsida. Arthur Cronquist (1981) discussed Ericales under subclass Dilleniidae of Magnoliopsida. Thorne (1983) included Ericaceae under order Ericales of superorder Theiflorae of subclass Annonidae (Dicotyledoneae).

21.63 PRIMULALES

21.63.1 General Characteristics

1. Flowers actinomorphic, bisexual, hypogynous, gamopetalous and pentamerous.
2. Stamens usually opposite the petals and epipetalous.
3. Unilocular, superior ovary with generally free-central or basal placentation.

4. The ovules are bi-integumented.
5. Seeds with straight embryo and copious endosperm.

Bentham and Hooker included Primulaceae, Plumbaginaceae and Myrsinaceae under cohort Primulales. Engler and Diels, however, divided order Primulales into 3 families, viz. Theophrastaceae, Myrsinaceae and Primulaceae. They discussed Plumbaginaceae under order Plumbaginales as also did by recent workers like Takhtajan (1980), Cronquist (1981) and Dahlgren (1983).

Only Primulaceae and Plumbaginaceae have been treated in this text.

21.64 PRIMULACEAE (PRIMULA OR PRIMROSE FAMILY)

21.64.1 Systematic Position

Gamopetalae, Heteromerae, Primulales.

21.64.2 Field Recognition

Perennial herbs, only rarely shrubs; flowers pentamerous; petals fused; stamens opposite the petals; free-central placentation.

21.64.3 Selected Indian Genera

Primula, *Anagallis*, *Androsace*, *Cyclamen* and *Lysimachia*.

21.64.4 Size, Distribution and General Information

Primrose family is represented by about 30 genera and 1000 species which are cosmopolitan but chiefly distributed in northern hemisphere. Some of the larger genera with their approximately reported species include *Primula* (500), *Lysimachia* (200), *Dodecatheon* (50), *Dionysia* (41), *Anagallis* (30), *Cyclamen* (15), *Salomus* and *Soldanella* (11).

21.64.5 Description of the Family

General Habit Mostly perennial herbs, only rarely annual herbs or shrubs; plants generally perennate by means of rhizome (*Primula*) or tubers.

Leaves Mostly simple, exstipulate; sometimes variously lobed or dissected; show all the three types of phyllotaxy i.e. alternate, opposite, or whorled; basal or cauline; often gland-dotted or farinose.

Inflorescence Flowers often borne on scapes in terminal or axillary position; solitary axillary (*Anagallis*), or umbellate (*Primula*), or terminal racemes (*Lysimachia vulgaris*).

Flower Bracteate, ebracteolate, complete; actinomorphic, rarely zygomorphic (*Coris*); bisexual; usually pentamerous, rarely trimerous (*Pelletiera*) or 5–9-merous (*Trientalis*); hypogynous; often heterostyled.

Calyx Usually 5 sepals, gamosepalous (Fig. 21.75), rarely 3 to 9 sepals; valvate; persistent; foliaceous.

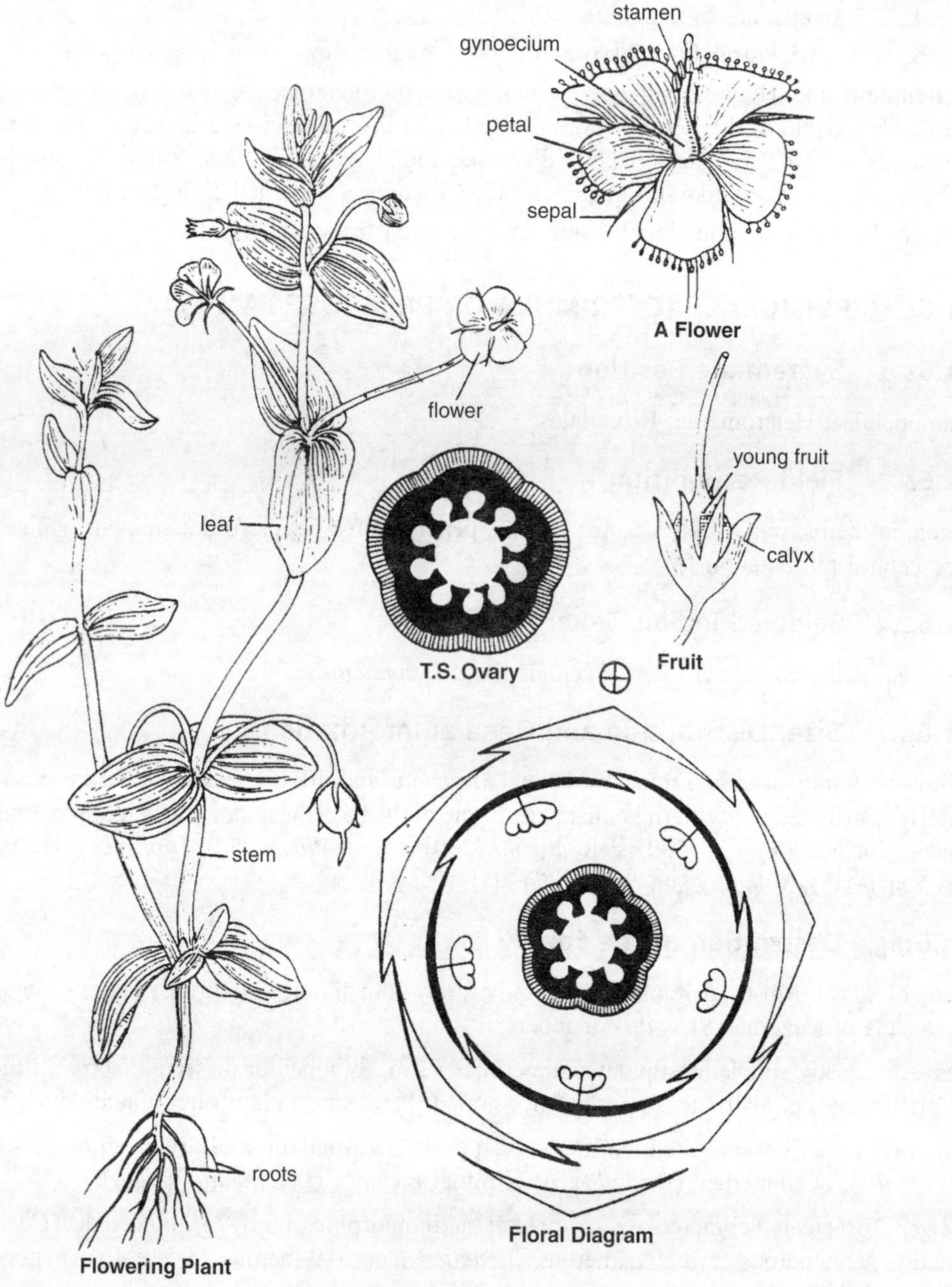

Fig. 21.75 *Anagallis arvensis* L.

Corolla Usually 5 petals, gamopetalous; imbricate or quincuncial; rarely 3 to 9 petals; usually rotate (*Anagallis*) to salverform (*Primula*); polypetalous in *Pelletiera*, and petals absent in *Glaux*; petal lobes are reflexed sharply downward in *Cyclamen* (Fig. 21.76) and *Dodecatheon*.

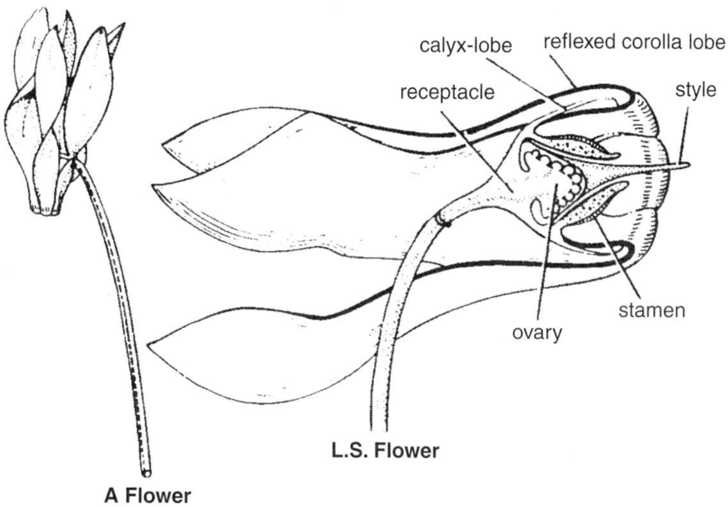

Fig. 21.76 Entire and L.S. flower of *Cyclamen*.

Androecium Stamens 5 or as many as corolla lobes, epipetalous; opposite the corolla lobes; sometimes 5 staminodes are also present opposite the sepals (*Samolus*, *Soldanella*); anthers dithecous, basifixed, introrse, dehiscing longitudinally.

Gynoecium Pentacarpellary, syncarpous; ovary superior, rarely half-inferior (*Samolus*); unilocular, numerous ovules; free-central placentation; style and stigma simple; heterostyly common.

Fruit and Seeds Fruit usually a 5-valved capsule or pyxis, variously dehiscent, generally dehiscing by teeth at the tip. Seeds with small and straight embryo, in hard or fleshy endosperm.

Pollination and Dispersal Majority of the *Primula* species are cross-pollinated. Dispersal of seeds in *Primula* is by wind but in a few species (*Primula vulgaris*) seeds are dispersed by ants.

General Floral Formula Br, ⊕. ⚥, $K_{(5)}$, $\overparen{C_{(5)}, A_5}$, $G_{(\underline{5})}$.

21.64.6 Economic Importance

The family is of no specific economic importance except that of some well-known decorative plants it contains. Some species of *Primula* may cause contact dermatitis. *Anagallis arvensis* (Fig. 21.75) is used for curing snakebite, leprosy, gout, and even hydrophobia. *Cyclamen purpurascence* contains glucoside cyclamin which is poisonous. Some of the commonly cultivated ornamental plants of Primulaceae are the species of *Primula* (primrose), *Androsace* (rock jasmine), *Cyclamen* (Fig. 21.76), *Dodecatheon* (shooting star), *Douglasia*, *Lysimachia* (loosestrile) and *Omphalogramma*.

21.64.7 Systematics and Phylogeny (Affinities)

The family is divided into 4 tribes:

1. *Primuleae* (Ovary superior, corolla imbricate)
2. *Cyclamineae* (Ovary superior, corolla convolute, plants tuberous)
3. *Lysimachieae* (Ovary superior, corolla convolute, plants not tuberous)
4. *Samoleae* (Ovary half-inferior).

Because of free-central placentation, Primulaceae is related to Caryophyllaceae and also to Theaceae to some extent. Primulaceae is also allied to Myrsinaceae and Theophrastaceae, and all these 3 families are included in one order Primulales by many taxonomists. Bessey, Hallier and Hutchinson opined that Primulaceae and Primulales originated from Caryophyllales. However, Cronquist considered that Primulaceae and Primulales originated from Theales.

21.65 PLUMBAGINACEAE (LEADWORT FAMILY)

21.65.1 Systematic Position

Gamopetalae, Heteromerae, Primulales.

21.65.2 Field Recognition

Perennial herbs or shrubs, sometimes scandent or lianous; pentamerous flowers are polypetalous or gamopetalous; 5-styled pistil; unilocular and uniovulate ovary; fruit dry, 1-seeded, surrounded by calyx.

21.65.3 Selected Indian Genera

Plumbago, Armeria, Ceratostigma.

21.65.4 Size, Distribution and General Information

A family of about 19 genera and 775 species (Hickey and King, 1988), mostly of semiarid regions, found very commonly in Mediterranean and central Asiatic regions. It shows a cosmopolitan distribution, especially along sea-shores and on salt steppes. Plumbaginaceae shows some relationship with Primulaceae but differs from the latter in possessing solitary ovule and free styles. Most phylogenists, however, believed that it has close affinities with caryophyllaceous taxa.

21.65.5 Description of the Family

General Habit Perennial herbs or shrubs; sometimes lianous; sometimes tufted and cushion-like (*Armeria maritima*).

Stem Well-branched, stout, woody.

Leaf Simple, exstipulate, entire; either forming a basal rosette of linear, punctate, glabrous, single-veined leaves (Fig. 21.77), or arranged alternately on well-branched stems; water glands or chalk-glands are present on the leaf surfaces.

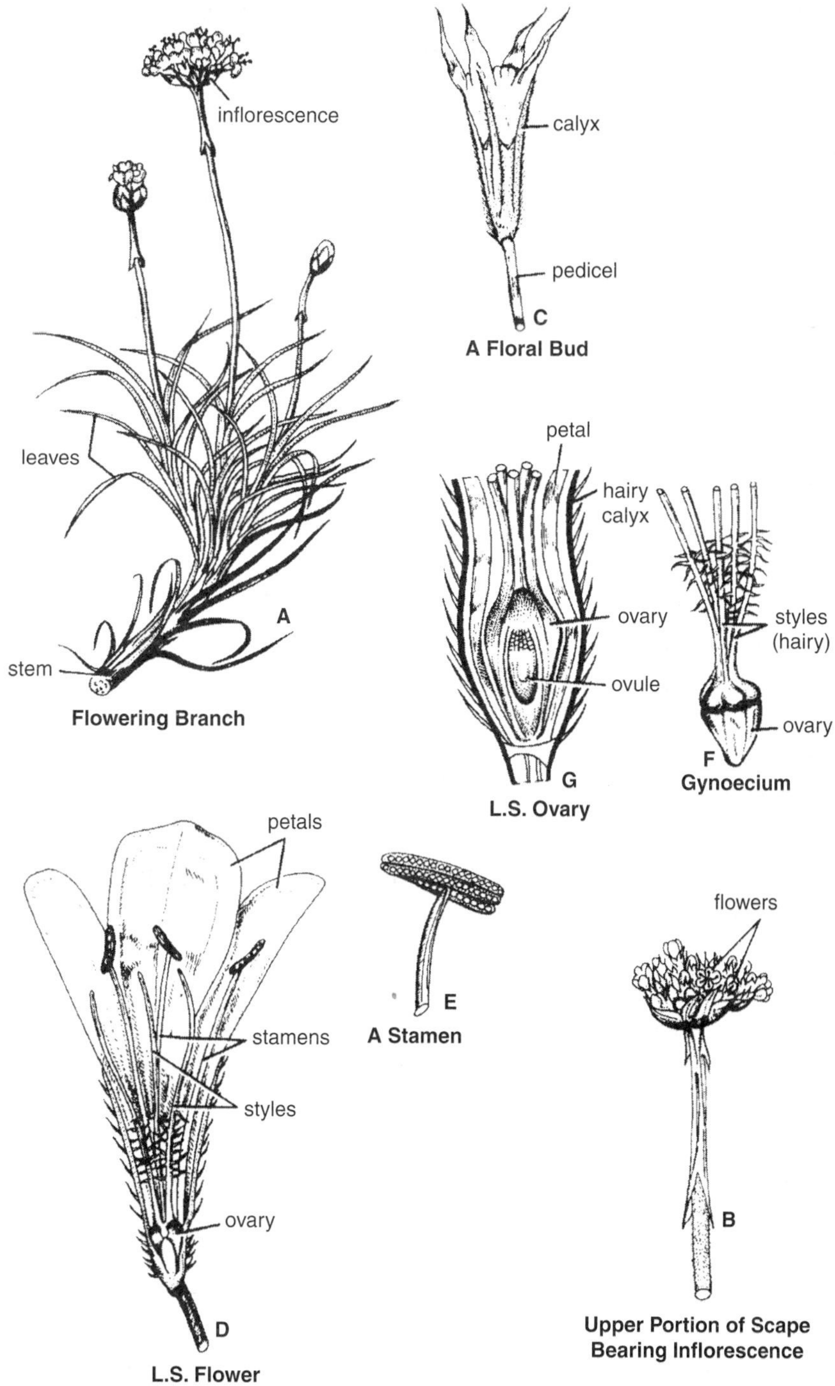

Fig. 21.77 ◈ *Armeria maritima* (Mill.) Willd.

Inflorescence Racemose (*Limonium*), cymose or capitular; in *Armeria maritima* (Fig. 21.77) the inflorescence is an erect scape supporting a terminal capitulum of fragrant, rose-pink or white flowers.

Flower Bracteate; bracts scarious, sometimes forming an involucre; bisexual, actinomorphic; pentamerous.

Calyx 5 sepals, united, persistent; sepals sometimes with smaller secondary lobes; often 5-10 ribbed or angled or winged.

Corolla 5 petals, gamopetalous; contorted or imbricate; 5 petal lobes sometimes quite deep, appearing to be polypetalous.

Androecium 5 stamens, epipetalous, opposite to corolla lobes; anthers 2-celled, dehiscing longitudinally, introrse.

Gynoecium Pistil 1, ovary superior, unilocular, 5-carpelled, syncarpous, usually 5-lobed or 5-ribbed; ovule single, pendulous, anatropous; basal placentation; styles 5, opposite to sepals, basally united, often hairy or glandular; stigmas 5, filiform.

Fruit Dry, 1-seeded utricle, often enclosed within persistent calyx.

Seed Embryo straight, with floury or crystalline-granular endosperm.

General Floral Formula Br, $\oplus$, ⚥, $K_{(5)}$, $\overparen{C_{(5) \text{ or } 5,} A_5}$, $G_{\underline{(5)}}$.

21.65.6 Economic Importance

Except of the ornamental value of some plants, the family Plumbaginaceae is of little importance. Ornamental plants include some species of *Armeria* (Thrift or Sea pink), *Limonium* (Satice), *Plumbago, Ceratostigma* and *Acantholimon*. Dried floral branches of *Limonium* are used in floral decoration while *Plumbago indica* and *P. auriculata* are cultivated as semi-climbing subshrubs in greenhouses. Some species of *Plumbago* and *Limonium* are used as medicine. *Plumbago europaea* and *P. scandens* are used to treat dental ailments. Extracts from the leaves and roots of *P. zeylanica* are used to treat some skin diseases.

21.65.7 Systematics and Phylogeny (Affinities)

Plumbaginaceae shows some relationship with Primulaceae due to a common floral plan with antipetalous stamens, unilocular ovary and bitegmic ovule but is distinguished from Primulaceae by its solitary ovule and free styles. Hutchinson considered it as one of the two families of Primulales. Lawrence believed that there is evidence that the Plumbaginales (having a single family Plumbaginaceae) "may have evolved from stocks ancestral to the Primulales, and most phylogenists have been of the opinion that they have close affinities with the caryophyllaceous taxa". However, both orders (Primulales and Plumbaginales) have been "derived from Centrospermae or their ancestors".

21.66 EBENALES

21.66.1 General Characteristics

1. Flowers gamopetalous.
2. Stamens usually in 2–3 whorls, epipetalous.

3. Ovary superior.
4. Carpels more than two, usually 4 or 5, syncarpous, superior.
5. Ovary usually with as much locules as number of carpels, each locule is generally uniovulate.
6. Placentation axile.

Bentham and Hooker included 3 families in Ebenales, namely Sapotaceae, Ebenaceae and Styraceae. Engler and Diels, however, treated the order Ebenales as composed of 2 suborders and 7 families; suborder Sapotineae (Sapotaceae, Hoplestigmataceae) and suborder Diospyrineae (Ebenaceae, Diclidantheraceae, Symplocaceae, Styracaceae, Lissocarpaceae).

Only Sapotaceae is discussed here in some details.

21.67 SAPOTACEAE (SAPOTA FAMILY)

21.67.1 Systematic Position

Gamopetalae, Heteromerae, Ebenales.

21.67.2 Field Recognition

Plants with milky latex; leaves exstipulate; stamens in 2–3 whorls; ovary superior; axile placentation; fruit berry.

21.67.3 Selected Indian Genera

Achras, *Madhuca*, *Manilkara*, *Mimusops*.

21.67.4 Size, Distribution and General Information

Family includes about 40 genera and over 800 species, with their main distribution in tropical regions of the world. Only 12 genera and about 50 species have been reported from India, mainly from eastern and southern regions. Because of Sapodilla (*Achras sapota*), the family is also called "Sapodilla family".

21.67.5 Description of the Family

General Habit Mostly trees or shrubs, with milky sap present in the laticiferous ducts or sacs in their vegetative parts.

Leaves Simple, alternate, petiolate; exstipulate but rarely stipulate (*Madhuca*); lanceolate, obcordate, or elliptical; entire; tomentose (*Madhuca*); unicostate reticulate.

Inflorescence Usually cymose clusters; sometimes solitary axillary (*Chrysophyllum*) or solitary terminal; rarely cauliflorous (*Dichopsis*).

Flower Bracteate (*Madhuca*) or ebracteate (*Achras*), pedicellate, bisexual, actinomorphic, hypogynous.

Calyx 4 to 8 sepals, usually gamosepalous, sometimes free; imbricate or valvate; 6 sepals of *Palaquium, Mimusops and Achras* are arranged in 2 series of 3 each; in *Madhuca indica* 4 sepals are arranged in 2 whorls of 2 each.

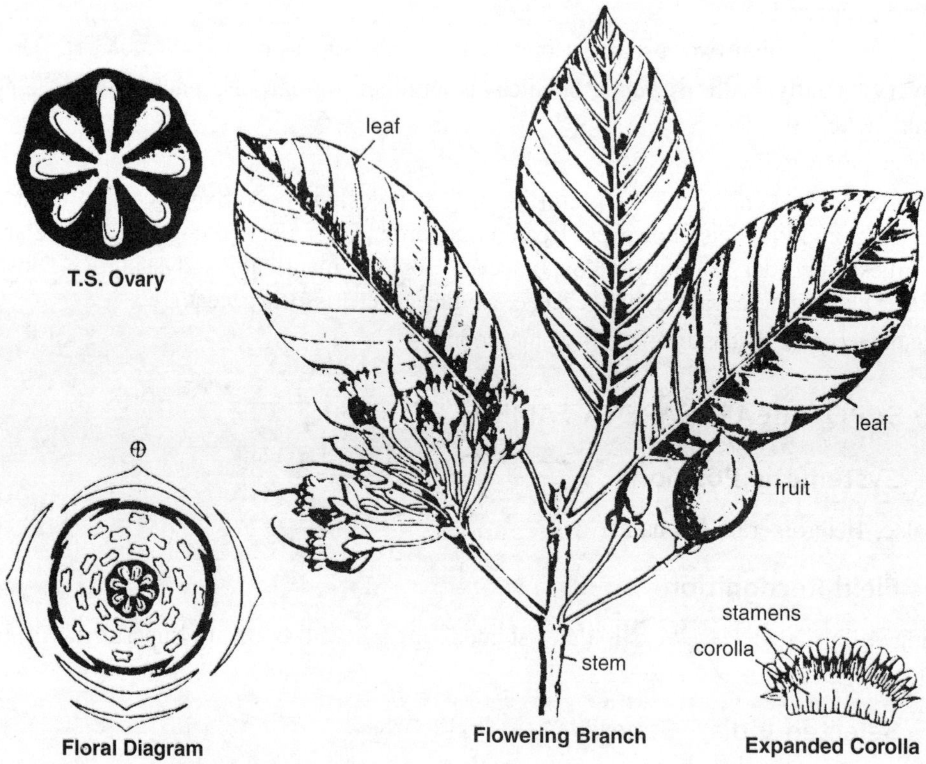

Fig. 21.78 *Madhuca indica* Gmel.

Corolla Petals usually as many as sepals, gamopetalous, imbricate; sometimes petals contain dorsal or lateral appendages; sometimes petals are twice as many as sepals (*Madhuca indica*, Fig. 21.78); only 6 petals in *Achras sapota*.

Androecium Stamens many, usually in 2 or 3 whorls of 4–5 each; often the stamens of innermost whorl are fertile while that of all others are reduced to staminodes (*Mimusops*); epipetalous, polyandrous; filaments short; anthers dithecous, basifixed or dorsifixed, extrorse, with their connective often prolonged beyond anther lobes, dehiscing longitudinally.

Gynoecium Carpels varying from 1–14 but usually 4 or 5, syncarpous, superior; usually with as many locules as the number of carpels, each locule uniovulate, ovule anatropous, axile placentation; style 1, often with apically lobed stigma.

Fruit and Seeds Fruit often berry; seeds usually endospermic, with straight embryo.

General Floral Formula Br, ⊕, ⚥, $K_{(4-8)}$, $\widehat{C_{(4-8)\text{ or }}A_{\infty}}$, $G_{\underline{(1-14)}}$.

21.67.6 Economic Importance

- *Chewing gum:* Chicle for chewing gum is obtained from the coagulated resinuous latex obtained from the bark of *Achras sapota* (Sapodilla plum)

- *Gutta percha:* Gutta percha, used for insulation and other similar purposes, is obtained from several genera, such as *Palaquium* (*P. gutta*), *Payena*, *Mimusops*, etc.
- *Edible fruits:* Notable edible fruits of Sapotaceae include *Achras sapota* (Sapodilla plum or Chiku), *Manilkara hexandra* (*Khirni*), *M. kauki* (Kauki), *Chrysophyllum cainito* (star apple), *Pouteria campechina* (eggfruit), *Calocarpum sapota* (Sapote plum), etc. Dried corolla of *Madhuca indica* is eaten by man.
- *Timber:* Valuable timber is obtained from *Manilkara hexandra*, *Sideroxylon*, *Madhuca indica* (Mahua), *Xantolis tomentosa*, etc.
- *Medicinal value:* (i) Chicle gum, obtained from *Achras sapota,* is used in dental surgery, (ii) Seeds and dried fruits of *Mimusops elengi* (Maulsari) are highly effective in curing piles, (iii) Oil from seeds of *Madhuca indica* is effective in skin diseases and rheumatism.
- *Oil:* Oil, obtained from seeds of *Madhuca indica* and *Diploknema butgracea,* is used in the manufacture of soaps and candles.

21.67.7 Systematics and Phylogeny (Affinities)

Sapotaceae is usually divided into 2 tribes, viz. *Palaquieae* (petal lobes without appendage) and *Mimusopeae* (petal lobes with appendage).

Sapotaceae is allied closely to Ebenaceae, and both these families belong to order Ebenales. However, Hallier placed Sapotaceae under a monotypic order Sapotales, while Ebenaceae under Santalales. Majority of the phylogenists consider that both Sapotales and Santalales originated from the same ancestral stock, i.e. Linaceae.

21.68 GENTIANALES

21.68.1 General Characteristics

1. Leaves opposite, simple or pinnately compound and exstipulate.
2. The flowers are bisexual, actinomorphic and hypogynous.
3. Corolla gamopetalous and corolla lobes usually convolute.
4. Androecium of one series; stamens adnate to or near corolla base, epipetalous and alternipetalous.
5. Gynoecium bicarpellary and ovary superior.

Bentham and Hooker included 6 families under Gentianales viz. Oleaceae, Apocynaceae, Asclepiadaceae, Salvadoraceae, Loganiaceae and Gentianaceae. Engler and Diels used the name "Contortae" for Gentianales and divided it into 2 suborders containing 6 families: (i) suborder Oleineae (Oleaceae) and, (ii) suborder Gentianineae (Desfontaineaceae, Loganiaceae, Gentianaceae, Apocynaceae, Asclepiadaceae).

Oleaceae, Asclepiadaceae, Apocynaceae and Loganiaceae have been discussed in this text.

21.69 OLEACEAE (OLIVE FAMILY)

21.69.1 Systematic Position

Gamopetalae, Bicarpellatae, Gentianales.

21.69.2 Field Recognition

Trees or shrubs; leaves opposite; flowers bi- to tetramerous; gamosepalous and gamopetalous; stamens 2; ovary bilocular.

21.69.3 Selected Indian Genera

Nyctanthes[1], *Jasminum, Fraxinus, Myxopyrum, Osmanthus* and *Olea*.

21.69.4 Size, Distribution and General Information

A family of approximately 29 genera and 600 species, Oleaceae are cosmopolitan but occur in great diversity in temperate and tropical Asia. Some of the largely represented genera of olive family and their approximately reported species are *Jasminum* (300), *Chionanthus* (125), *Linociera* (80), *Fraxinus* (70), *Ligustrum* (40), *Syringa* (30) and *Olea* (20).

21.69.5 Description of the Family

General Habit Shrubs (*Syringa*) or trees (*Olea, Nyctanthes*), occasionally vines (*Myxopyrum smilacifolium*).

Leaves Simple (*Nyctanthes*) or pinnately compound (*Fraxinus*); usually opposite, rarely alternate (some species of *Jasminum*); often entire, exstipulate; unicostate reticulate venation.

Inflorescence Axillary or terminal racemose, paniculate or thyrsiform.

Flower Bracteate (*Olea*) or ebracteate (*Jasminum*), pedicellate, usually bisexual but unisexual in some species of *Olea* and *Fraxinus* (Fig. 21.79G); actinomorphic, bi- to hexamerous but usually tetramerous, hypogynous.

Calyx 4 sepals, gamosepalous, valvate; rarely sepals absent (*Fraxinus excelsior*, Fig. 21.71F); sometimes 4–15 toothed; persistent.

Corolla Usually 4 united petals, valvate or imbricate; sometimes 4–9 lobed (some species of *Jasminum*); rarely petals absent (*Fraxinus micrantha*); sometimes deeply lobed or divided and seemingly polypetalous.

Androecium Usually 2 stamens, epipetalous, alternate with the carpels; anthers dithecous, the cells usually back to back; often apiculate because of the extension of the connective; dehiscing longitudinally; stamens are 4 in *Hesperelaea* and *Tessarandra*.

Gynoecium Usually bicarpellary, syncarpous; ovary superior, bilocular; with usually 2 anatropous ovules in each locule; axile placentation; style simple, stigma capitate or bilobed.

Fruit and Seeds Fruit drupe (*Olea*), berry (*Ligustrum*), samara (*Fraxinus*, Fig. 21.79H), or capsule (*Forsythia*) with 1–4 seeds. Seeds with straight embryo and little or no endosperm.

General Floral Formula Br or Ebr, $\oplus$, ⚥, $K_{(4)}$, $\overset{\frown}{C_{(4)}, A_2}$, $G_{(\underline{2})}$.

[1] Some workers treat *Nyctanthes* as a member of sub-family Nyctanthoideae of family Verbenaceae.

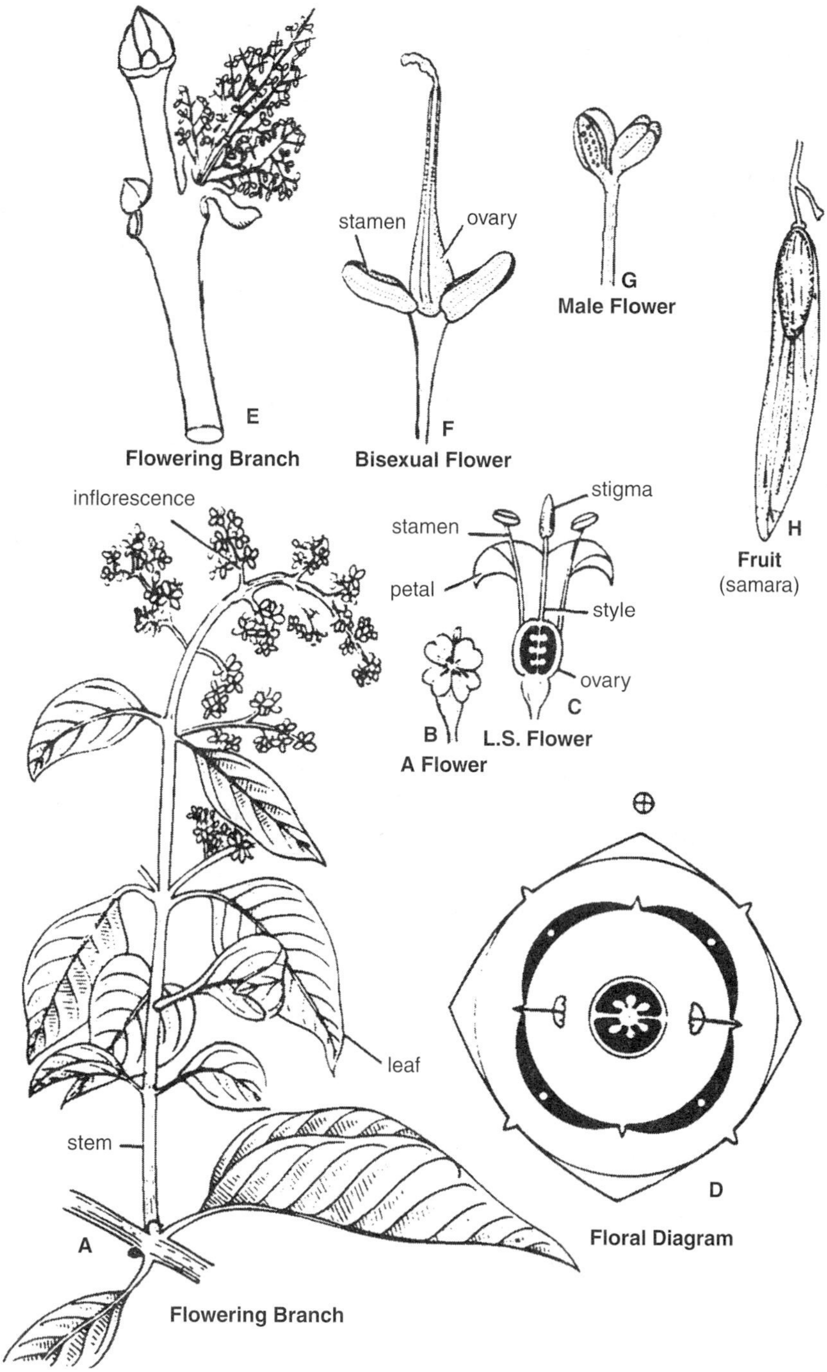

Fig. 21.79 A–D: *Nyctanthes arbortristis*, and E–H: *Fraxinus excelsior*.

21.69.6 Economic Importance

Olive family is of considerable economic importance as a source of olives, olive oil, ash lumber, dyes and several ornamental plants.

- Notable ornamentals are various species of *Jasminum* (Jasmine), *Syringa* (lilac), *Forsythia* (golden bells), *Ligustrum* (privet), *Osmanthus* (fragrant olive), *Phillyrea, Chiananthus* (fringe tree) and *Nyctanthes.*
- Olive (*Olea europaea*) has been a source of food since ancient times, and the oil extracted from its fruit is of high value.
- Flowers of several species of *Jasminum* are the source of an oil used in perfumery.
- Valuable timber is obtained from various species of *Fraxinus* (ash lumber), *Notalaea* and *Olea.*
- An orange dye, used for colouring silk and cotton, is obtained from the flowers of various species of *Nyctanthes.* Leaves of *N. arbortristis* (Har Singar, Fig. 21.79A–D) are useful in fever and rheumatism.

21.69.7 Systematics and Phylogeny (Affinities)

Family Oleaceae is divided into 2 subfamilies, i.e. *Jasminoideae* (ovules generally erect) and *Oleoideae* (ovules pendulous). To some extent, Oleaceae is related to Loganiaceae. In certain characters it is also allied to Apocynaceae. According to Hutchinson, Oleaceae shows affinities to Melastomaceae and Rubiaceae. However, Cronquist placed Oleaceae under Scrophulariales and believed it to be allied to Buddleiaceae. According to Takhtajan, Oleaceae belongs to a separate order Oleales and is closely related to some families of order Celastrales. Takhtajan regards that both Oleales and Celastrales have a common origin from Saxifragales.

21.70 ASCLEPIADACEAE (MILKWEED FAMILY)

21.70.1 Systematic Position

Gamopetalae, Bicarpellatae, Gentianales.

21.70.2 Field Recognition

Herbs or vines with milky sap; leaves exstipulate and opposite or whorled; flowers pentamerous with bicarpellate gynoecium; corona, pollinia, translators and corpuscula are present; fruit follicle; silky hairs cover the seeds.

21.70.3 Selected Indian Genera

Asclepias, Calotropis, Cryptostegia, Leptadenia, Tylophora.

21.70.4 Size, Distribution and General Information

A family of about 250 genera and 2000 species (Jones and Luchsinger, 1987), distributed mainly in tropical and subtropical regions and a few also in temperate regions. About 50 genera and over 250 species of Asclepiadaceae have been reported from India. Some of the larger genera along with their number of approximately reported species are *Hoya* (200, wax plant), *Ceropegia* (160), *Asclepias* (150, milkweed), *Oxypetalum* (150), *Matelea* (130, milkweed vine), *Dischidia* (80), *Stapelia* (75, carrion flower) and *Periploca* (10).

21.70.5 Description of the Family

General Habit Erect or twining perennial herbs (*Asclepias*), shrubs (*Calotropis procera*), vines (*Leptadenia*), or rarely small trees (*Calotropis gigantea*); usually with milky sap; often fleshy or cactus-like (*Stapelia*), sometimes epiphytic (*Dischidia*). Rootstock fleshy, tuberous, woody, or sometimes absent.

Leaves Simple; usually opposite decussate (*Calotropis procera*, Fig. 21.82), sometimes alternate or in whorls; reduced or even absent in succulent species; usually exstipulate, or stipules minute if present; margins generally entire; often have a waxy coating.

Inflorescence Mostly cymose (dichasial to multichasial cyme), but also racemose or umbellate (*Stapelia*).

Flower Bracteate, often bracteolate, complete; actinomorphic, rarely zygomorphic (*Ceropegia*); hermaphrodite, pentamerous, hypogynous, cyclic.

Calyx 5 free or basally united sepals forming a short tube; valvate, imbricate, or quincuncial.

Corolla 5 petals, gamopetalous; contorted, imbricate or valvate; corolla tube often short and with lobes; corolla tube is zygomorphic in *Ceropegia*, funnel form in *Cryptostegia* and campanulate in *Gymnema*; often the corolla tube is with a *corona*, made up of a ring of scales, hairs or processes.

Androecium 5 stamens, epipetalous, rarely distinct, pollen in waxy pollinia.

Typically, the filaments of stamens are connate into a short tube and their anthers are usually adherent or adnate to the stigmatic area of the gynoecium to form a complex structure called *gynostegium* or *gynandrium* (Fig. 21.80A, B). Anthers are 2-celled and their pollens often united within each anther sac into a structure called *pollinium*. Two pollinia, one each from the adjacent anthers, are united together by two arms (*retinaculae* and *translators*) attached to a central gland called *corpusculum* (Fig. 21.80C).

A nectariferous *corona*, consisting of 5 hoods, may be present. These hoods are sometimes mistaken for petals. A beak or hood may be associated with each hood of corona (Fig. 21.80B).

Gynoecium Bicarpellary; the carpels are free below but united by their apices into the single 5-lobed stigma; ovary superior; ovary of each carpel is unilocular; numerous ovules in each locule, marginal placentation (Fig. 21.81A); styles 2; stigma 1 and 5-lobed.

Fruit and Seeds Fruit is a pair of follicles (Fig. 21.81B), of which often one aborts. Seeds numerous, with tufts of silky hairs called coma (Fig. 21.81C); embryo straight, large, nearly as long as the seed, with small and thin endosperm.

Pollination It is entomophilous, and butterflies are the chief pollinating agents which visit the brightly coloured flowers in search of nectar. When a butterfly alights on a flower in search of nectar, its legs tend to slip down the slits in the column and hook onto the band connecting a pair of pollinia. As the butterfly flies away, this connecting band twists sharply. This brings the pollinia

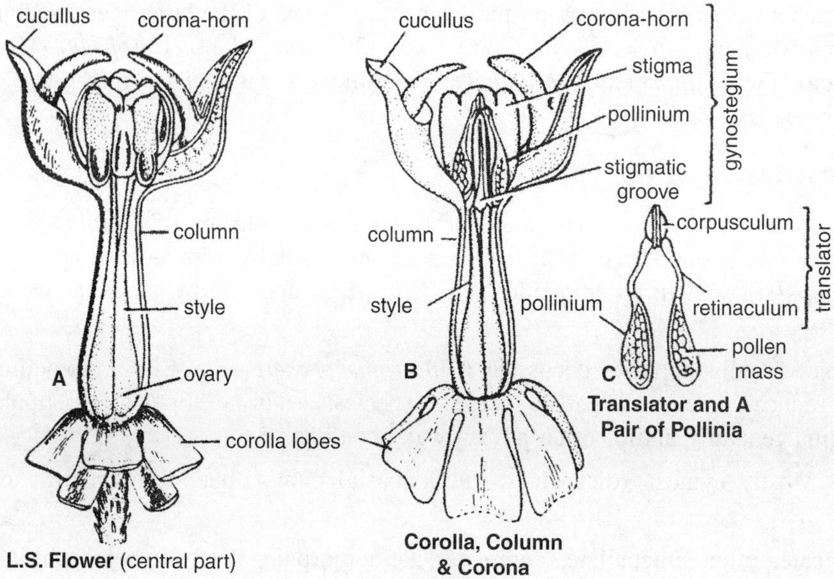

Fig. 21.80 A–C Floral details of *Asclepias curassavica*.

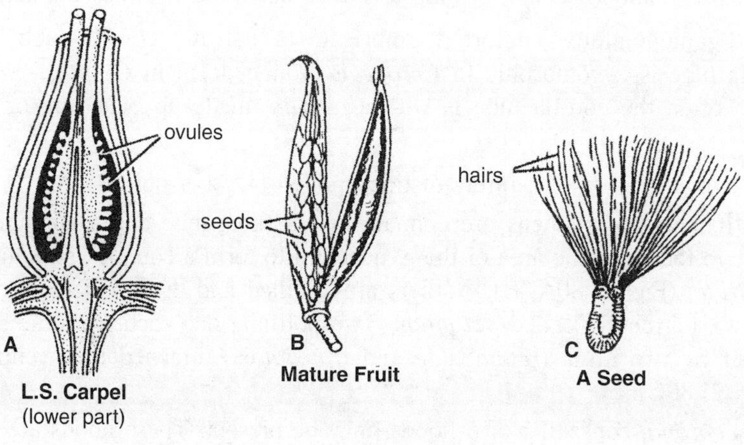

Fig. 21.81 A–C Carpel, fruit and seeds of *Asclepias curassavica*.

close together. When the butterfly arrives at another flower, the pollinia are in such a position that they lodge in a slit in the column of the flower and break off, and thus distributing the pollen grains over the stigma.

General Floral Formula Br, ⊕, ⚥, $K_{(5)\ \text{or}\ 5}$, $\overparen{C_{(5)}, A_{(5)}}$, $\underline{G}_{(2)}$.

21.70.6 Economic Importance

- Ornamental plants of this family grown in glasshouses include several species of *Asclepias*, *Caralluma*, *Ceropegia*, *Cryptostegia*, *Hoya*, *Huernia*, *Oxypetalum*, *Periploca*, *Stapelia* and *Stephanotis*.

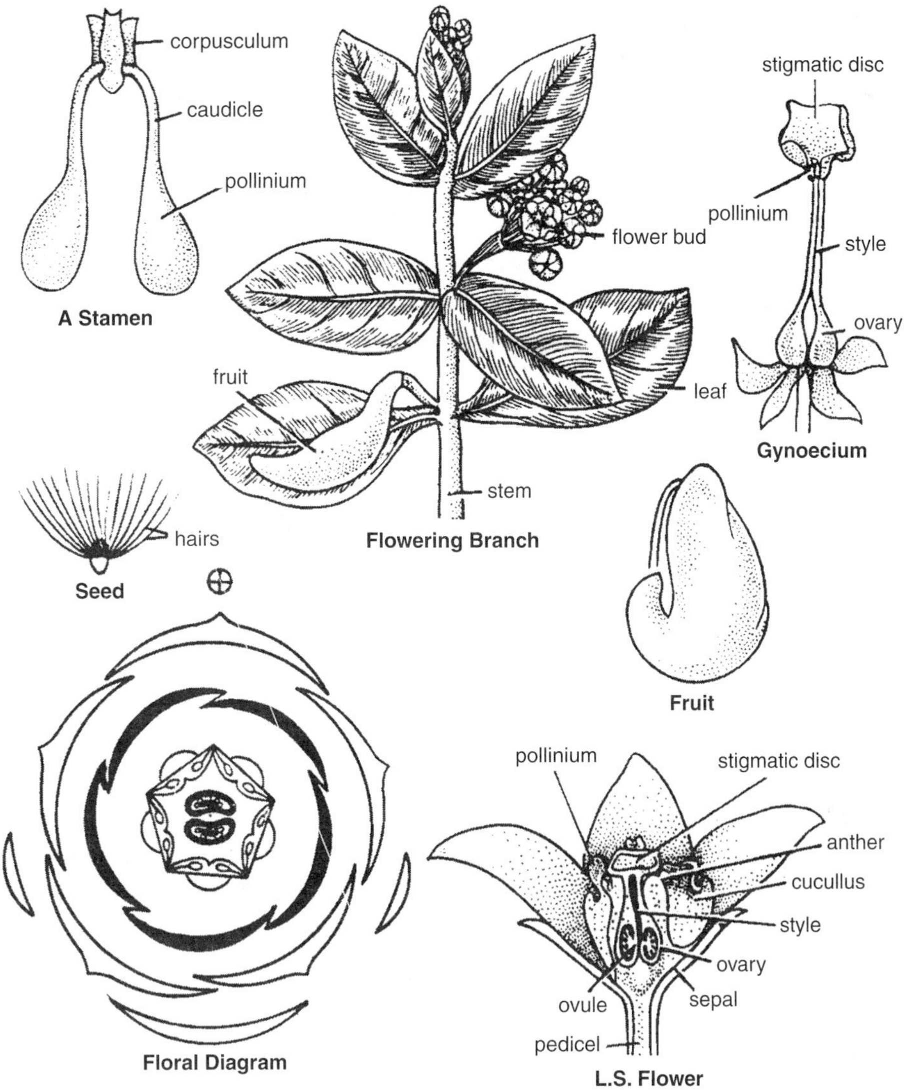

Fig. 21.82 *Calotropis procera* R. Br.

- Fibre, used for cordage, twines, fishing nets, etc., is obtained from *Asclepias curassavica* (blood-flower), *Calotropis gigantea*, *Cosmostigma racemosa*, *Leptadenia pyrotechnica*, *Marsdenia roylei*, *M. tinctoria* and *Tylophora tenuis*.
- Indigo-like dye is obtained from *Marsdenia tinctoria*.
- *Cryptostegia grandiflora* has been cultivated commercially as a source of rubber.
- Latex obtained from species of *Matelea* is used as an arrow poison.
- Certain plants of Asclepiadaceae are of medicinal importance. These include, (i) *Calotropis procera* (roots are used against cough), (ii) *Gymnema sylvestre* (used as a laxative

and diuretic), (iii) *Hemidesmus indicus* (roots are used in urinary and skin diseases), (iv) *Holostemma rheedianum* (roots are used in cough and cold), (v) *Pergularia daemia* (used as an emetic and expectorant), (vi) *Sarcostemma acidum* (infusion of roots is used as an antidote for snakebite), and (vii) *Tylophora indica* (roots are used in curing bronchitis, whooping cough and asthama).

21.70.7 Systematics and Phylogeny (Affinities)

On the basis of pollen characters, Asclepiadaceae is divided into two subfamilies, viz. (i) *Periplocoideae* (pollen grains in tetrads; filaments free), and (ii) *Cynanchoideae* (pollen grains united in pollinia; filaments united in a tube).

Asclepiadaceae is most closely allied to Apocynaceae, and in majority of the systems of classification both these families are placed side by side. However, the characters such as presence of translators and presence of gynostegium are absent in Apocynaceae.

21.70.8 Description of a Common Plant

Calotropis procera R. Br. (Madar or Aak, Fig. 21.82)

Habit: A herbaceous or semishrubby weed with its upper portions covered with wooly hairs; contains milky latex. *Leaf:* Simple, opposite decussate, exstipulate, sessile or subsessile, ovate to oblong, entire, acute, unicostate reticulate. *Inflorescence:* Dichasial cyme. *Flower:* Bracteate, bracteolate, pedicellate, hermaphrodite, actinomorphic, pentamerous, hypogynous, cyclic. *Calyx:* 5 sepals, distinct, quincuncial. *Corolla:* 5 petals, fused, twisted, coloured. *Androecium:* 5 stamens, united with stigma to form gynostegium; each stamen is represented by two pollinia with their caudicles or retinaculae; pollinia of the adjacent anthers are joined by their retinaculae to corpusculum in a groove, to form a unit called translator; at the back of each stamen is present a coronary outgrowth. *Gynoecium:* Bicarpellary, superior; ovaries free but the upper portions of styles and stigma fused; each ovary is unilocular, many ovules in the locule, marginal placentation; styles 2; stigmatic head pentagonal. *Fruit:* Follicle. *Floral Formula:* Br, Brl, ⊕, ⚥, K_5, $C_{(5)}$, $\overbrace{A_{(5)}, G_2}$.

21.71 APOCYNACEAE (DOGBANE FAMILY)

21.71.1 Systematic Position

Gamopetalae, Bicarpellatae, Gentianales.

21.71.2 Field Recognition

Plants with milky sap; leaves opposite or whorled, exstipulate; flowers pentamerous; stamens usually epipetalous; corona, pollinia and corpusculum of Asclepiadaceae absent; fruit follicle and seeds with hairs.

21.71.3 Selected Indian Genera

Apocynum, Catharanthus, Nerium, Plumeria, Tabernaemontana, Thevetia, Rauvolfia, Vinca.

21.71.4 Size, Distribution and General Information

A family of about 200 genera and 2000 species (Jones and Luchsinger, 1987) of mostly tropical and subtropical distribution, but a few species distributed also in temperate regions. About 30 genera and over 60 species of Apocynaceae have been reported from India. Some of the larger genera along with their approximately reported species in bracket are *Rauvolfia* (100), *Tabernaemontana* (100), *Parsonsia* (100), *Aspidosperma* (80), *Prestonia* (65), *Strophanthus* (60), *Apocynum* (7), *Plumeria* (6), *Vinca* (5) and *Catharanthus* (5).

21.71.5 Description of the Family

General Habit Trees (*Alstonia*, *Holarrhena*), shrubs (*Thevetia*, *Nerium*), herbs (*Catharanthus*), woody twiners (*Allamanda*), or even succulent (*Adenium*); usually twining shrubs with milky sap in laticiferous vessels.

Leaves Simple, exstipulate; opposite decussate or sometimes alternate (*Thevetia*) or in whorls of 3; entire; rarely stipulate (*Tabernaemontana*).

Inflorescence A panicle, cyme, or raceme, or flowers solitary (*Catharanthus*); usually dichasial cyme.

Flower Bracteate, bracteolate, complete, hermaphrodite, actinomorphic, pentamerous; hypogynous but rarely perigynous or epigynous (*Plumeria*).

Calyx 5 sepals, usually gamosepalous, sometimes free (*Thevetia*, *Catharanthus*) or deeply lobed; valvate (*Catharanthus*), quincuncial (*Thevetia*); odd sepal posterior; rarely only 4 sepals; sometimes glandular appendages (*Squamellae*) present on the adaxial side.

Corolla 5 petals, united into a tube, usually salver-shaped or funnel-shaped; twisted or rarely valvate; rarely only 4 petals; often hairy within or contain some corona-like outgrowths at the mouth of corolla tube.

Androecium 5 stamens, epipetalous; alternate with the corolla lobes; filament short; attached at the mouth or deeper in the corolla tube; anthers are dithecous, basifixed, often sagittate; introrse; dehiscence longitudinal; anthers sometimes basally awned; sometimes bear hairy appendage (*Nerium*) over the lobes (Fig. 21.83).

Gynoecium Usually bicarpellary, carpels apically united; superior or rarely half-inferior (*Plumeria*); 1- or 2-locular, with 2 to many anatropous and pendulous ovules in each locule; marginal placentation; style one and simple; stigma is characteristically thickened, massive and bilobed; receptive surface of stigma is situated below; rarely the carpels are 3–5.

A nectariferous disc is often present around or at the base of the gynoecium in several genera, e.g. *Thevetia*, *Catharanthus*, *Allamanda* and *Rauvolfia*.

Fruit and Seeds Fruit usually consisting of 2 follicles, or a berry (*Landolphia*), drupe (*Cerbera*), capsule (*Allamanda*), or 2 indehiscent mericarps. Seeds often flat with straight embryo and a crown of hairs.

Pollination and Dispersal Pollination is entomophilous. Insects are attracted mainly because of often large and conspicuous petals and the presence of nectar. Because the receptive stigmatic surface is located on the underside of the stigma, the cross-pollination is the rule of the family. Self-pollination is absent.

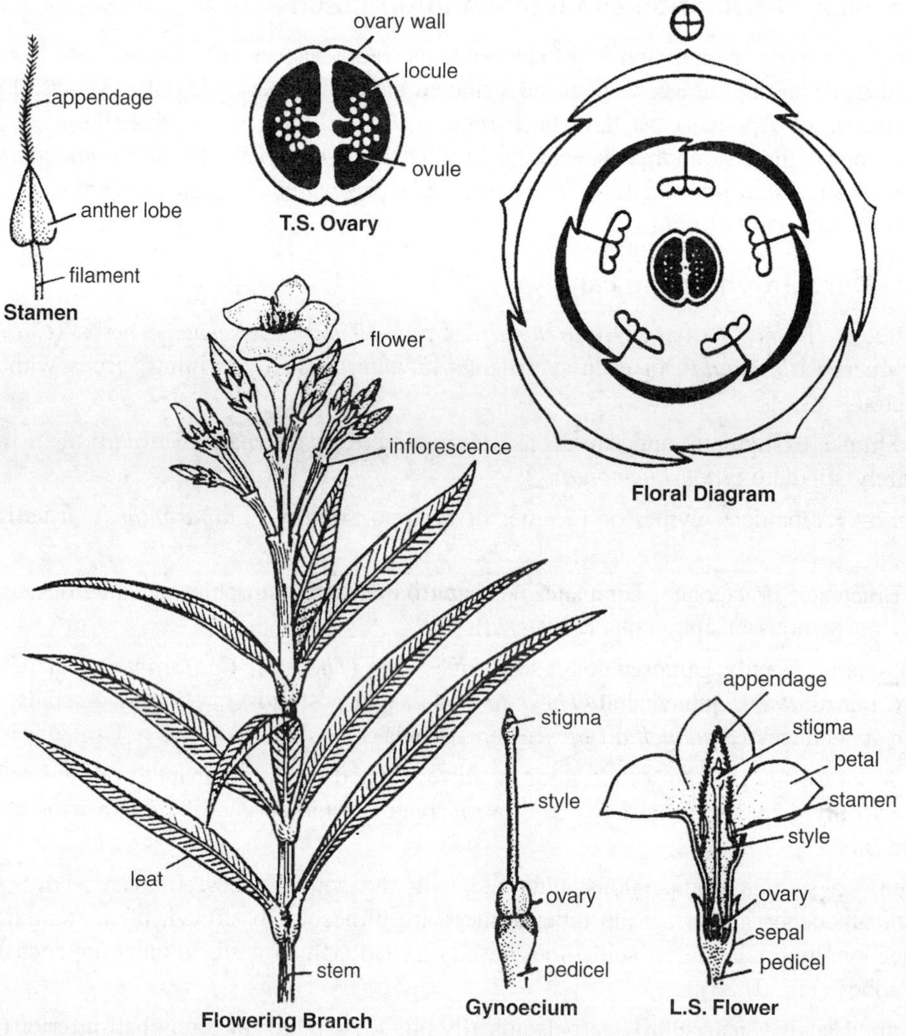

Fig. 21.83 *Nerium indicum* Mill.

Dispersal is affected mainly by wind because of winged seeds and presence of crown of hairs.

General Floral Formula $Br, Brl, \oplus, ⚥, K_{(5)}, \overparen{C_{(5)}, A_5}, G_{(2)}$.

21.71.6 Economic Importance

The family is of importance for its several ornamentals and for drugs.

- Ornamental plants[1] include *Allamanda neriifolia* (golden yellow flowers), *Alstonia scholaris* (pink flowers), *Beaumontia grandiflora* (large white flowers), *Carrisa carandas* (white-pink

[1] Plants arranged alphabetically.

flowers), *Catharanthus roseus* syn. *Vinca rosea* (Sadabahar, rosy-purple flowers), *Mandevilla taxa* (pink flowers), *Nerium indicum* syn. *N. odorum* (Kaner, white-pinky flowers), *Plumeria* (white, red, pink or purple flowers), *Tabernaemontana divaricata* (Chandni, white flowers), *Thevetia peruviana* (Peeli Kaner, yellow flowers), and *Vallaris solanacea* (Dudhi Bel).

- Plants of medicinal value include (i) *Rauvolfia serpentina* (Sarpgandha) used in the treatment of hypertension and mental disorders; drug is obtained from the dried roots; the alkaloid reserpine of the plant can lower the blood pressure and tranquilize mental patients suffering from schizophrenia, (ii) *Thevetia peruviana* (seeds are used in rheumatism), (iii) *Vallaris solanacea* (latex is useful in toothache and inflated gums), (iv) *Holarrhena antidysentrica* (root and bark is used in curing dysentery), (v) *Alstonia scholaris* (bark is used in malaria and dysentry), (vi) *Cerbera manghas* (latex is used as an emetic and purgative), (vii) *Strophanthus hispidus* (seeds yield the drug strophanthin), and (viii) *Wrightia tomentosa* (bark and roots are used as antidote to snakebite).
- Rubber is obtained from the coagulated latex of several species of *Carpodinus*, *Funtumia*, *Hancornia*, *Landolphia* and *Rhynchodia*.
- Poisonous plants of the family belong to *Acokanthera*, *Nerium*, *Strophanthus* and *Thevetia*. If one leaf of *Nerium oleander* or *Thevetia peruviana* is ingested, it may be fatal to an adult human. Poisoned arrows are made from the bark extract of *Acokanthera abyssinica* and seeds of several species of *Strophanthus*.
- Fruits of *Carissa carandas* (Karonda) are edible and used in pickles.
- An indigo-like dye is obtained from the seeds of *Wrightia tinctoria*.

21.71.7 Systematics and Phylogeny (Affinities)

The family is closely allied to Asclepiadaceae and both are included in the same order almost in all systems of classification. However, Apocynaceae lack the pollinia, corona and translator-corpusculum of Asclepiadaceae.

21.71.8 Description of a Common Plant

Catharanthus roseus (L.) Don. syn. Vinca rosea Reichb. (Periwrinkle or Sadabahar; Fig. 21.84)

Habit: Erect, everblooming ornamental plant with milky latex. *Leaf:* Simple, opposite decussate, exstipulate, subsessile or petiolate, elliptic-obovate, entire, mucronate, unicostate reticulate. *Inflorescence:* Dichasial cyme or solitary axillary. *Flower:* Ebracteate, pedicellate, complete, hermaphrodite, actinomorphic, pentamerous, hypogynous, rosy-purple or pink. *Calyx:* 5 sepals, polysepalous, valvate. *Corolla:* 5 petals, gamopetalous, twisted, hypocrateriform. *Androecium:* 5 stamens, epipetalous, inserted at the mouth of corolla tube, filaments short; anthers sagittate, dithecous, basifixed or dorsifixed, introrse. *Gynoecium:* Bicarpellary, syncarpous; ovaries free and superior; many ovules, marginal placentation; two hypogynous nectaries are present, one on anterior and another on posterior side of the ovary. *Fruit:* Follicle.

Floral Formula: Ebr, ⊕, ⚥, K_5, $\overbrace{C_{(5)}, A_5}$, $G_{(\underline{2})}$.

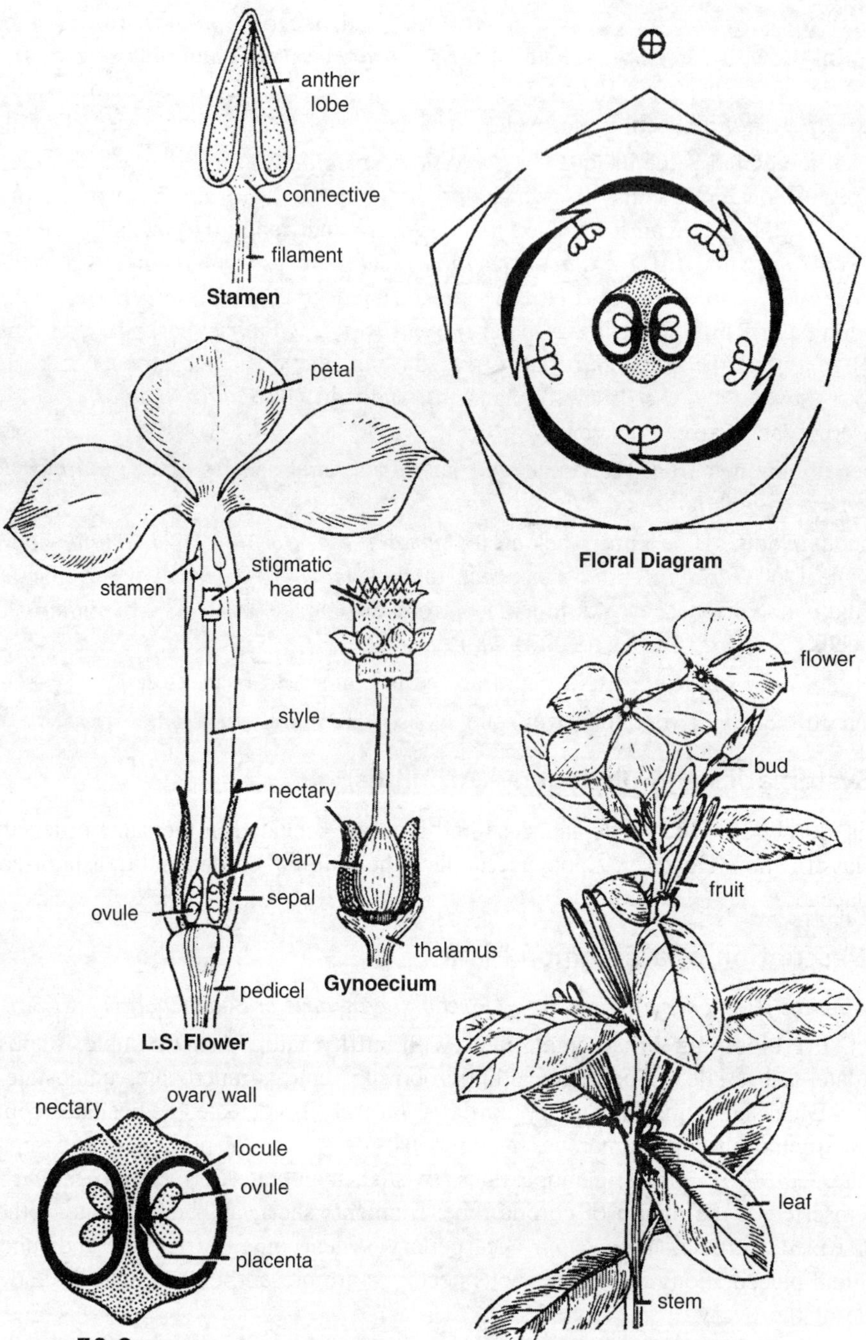

Fig. 21.84 *Catharanthus roseus* (L.) Don. syn. *Vinca rosea* Reichb.

21.72 LOGANIACEAE (LOGANIA FAMILY)

21.72.1 Systematic Position

Gamopetalae, Bicarpellatae, Gentianales

21.72.2 Field Recognition

Leaves opposite, stipulate; ovary superior, bilocular; axile placentation; most members show the presence of internal phloem.

21.72.3 Selected Indian Genera

Logania, Strychnos, Buddleia, Fagraea.

21.72.4 Size, Distribution and General Information

A family of over 32 genera and about 800 species, Loganiaceae are distributed mainly in tropical regions. A few members are also found in warm temperate regions. Some of the larger genera along with their approximately reported species are *Strychnos* (200), *Buddleia* (150), *Geniostoma* (60), *Nuxia* (40), *Spigelia* (35) and *Logania* (25).

21.72.5 Description of the Family

General Habit Herbs, shrubs or trees; several members are climbers.

Leaves Usually simple, opposite; stipulate, the stipules often much reduced; in a few species of *Buddleia*, the leaves are alternate or even whorled; lanceolate and finely toothed in *Buddleia davidii* (Fig. 21.85).

Inflorescence Usually cymose.

Flower Bracteate, bracteolate, bisexual, actinomorphic; mostly tetra- to pentamerous.

Calyx Sepals 4 or 5, gamosepalous, imbricate.

Corolla Petals 4 or 5, gamopetalous; rarely 10-lobed corolla; showing various types of aestivation; a crown of hairs often present at the mouth of corolla tube; corolla lobes have irregular margins as in *Buddleia davidii* (Fig. 21.85).

Androecium Stamens 4 or 5, generally as many as corolla lobes, epipetalous, alternipetalous; rarely stamens twice as many as petals, sometimes only 1 stamen (*Usteria*); anther lobes dithecous; dehiscing longitudinally.

Gynoecium Usually bicarpellary, syncarpous; superior or rarely half-inferior (e.g. *Mitreola*) ovary, bilocular; usually numerous anatropous or amphitropous ovules in each locule on axile placentation; the ovary is unilocular in *Strychnos* or incompletely bilocular in *Fagraea*; style 1, rarely 2 (e.g. *Cynoctonum*); stigmas 1–2.

Fruit Usually a septicidal capsule, rarely drupe or berry.

Seeds Endospermic with small and straight embryo; sometimes winged .

General Floral Formula Br, Brl, ⊕, ⚥, $K_{(4-5)}$, $\overset{\frown}{C_{(4-5)\text{ or }}A_{4-5}}$, $G_{(\underline{2})}$.

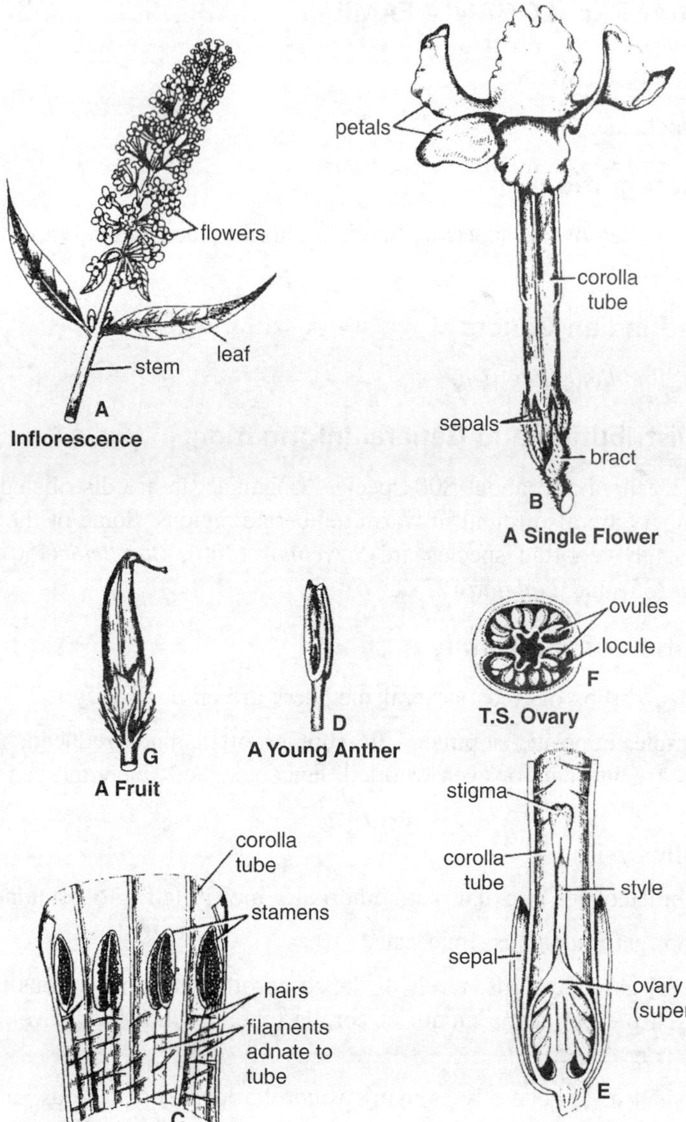

Fig. 21.85 *Buddleia davidii* Franch.

21.72.6 Economic Importance

- *Strychnos nux-vomica*, the source of alkaloids like strychnine and brucine, is well-known as a drug plant. The famous drug nux-vomica, obtained from its seeds, is used as a tonic, stimulant and in the treatment of nervous disorders and paralysis, but used always in small doses because it is a virulent poison.

- The drug, curare, used in shock therapy, as a muscle relaxant and also in surgical operations and tetanus, is obtained from the bark of *Strychnos toxifera*.
- Fruits of *Strychnos spinosa* are edible.
- Several species of *Buddleia* (butterfly bush), *Gelsemium* (yellow jessamine), *Spigelia* (Indian pink) and *Logania* are grown for decorative purposes.

21.72.7 Systematics and Phylogeny (Affinities)

Bentham and Hooker included 6 families (Oleaceae, Salvadoraceae, Asclepiadaceae, Loganiaceae, Apocynaceae and Gentianaceae) under order Gentianales. Hutchinson, however, discussed Loganiaceae and Oleaceae under order Loganiales of division Lignosae. Most taxonomists included Loganiaceae in the Contortae, except Hallier, who placed it in Tubiflorae. Bessey suggested that Gentianales arose from Geraniales, which separated into two phyletic lines, one giving rise to Loganiales and the other to the Polemoniales.

21.73 POLEMONIALES

21.73.1 General Characteristics

1. It includes mostly herbs and climbers; only some are shrubs and trees.
2. Leaves are usually simple and alternate.
3. Flowers actinomorphic, only rarely zygomorphic; bisexual, hypogynous, pentamerous.
4. Corolla gamopetalous, imbricate.
5. Stamens are epipetalous and alternipetalous; usually adnate to the mostly elongated corolla tube.
6. Gynoecium with 2–5 carpels; syncarpous, superior.

Bentham and Hooker included five families in Polemoniales. These are Boraginaceae, Polemoniaceae, Hydrophyllaceae, Convolvulaceae and Solanaceae. Engler and Diels, however, included all these five families under order Tubiflorae, to which they divided into 8 suborders and 23 families. They included families Convolvulaceae and Polemoniaceae under suborder Convolvulineae, families Hydrophyllaceae and Boraginaceae under suborder Boragineae, and to family Solanaceae under suborder Solanineae. Several families (e.g. Verbenaceae, Labiatae, Scrophulariaceae, Bignoniaceae, Pedaliaceae and Acanthaceae), discussed elsewhere in this text, have also been included by Engler and Diels under order Tubiflorae.

Only Boraginaceae, Convolvulaceae, Solanaceae and Polemoniaceae have been discussed in this text.

21.74 BORAGINACEAE (BORAGE FAMILY)

21.74.1 Systematic Position

Gamopetalae, Bicarpellatae, Polemoniales.

21.74.2 Field Recognition

Bristly or hispid herbs; leaves alternate; inflorescence helicoid cyme; flowers actinomorphic, pentamerous; ovary 4-lobed; style gynobasic; fruit of 4-nutlels.

21.74.3 Selected Indian Genera

Heliotropium, *Cynoglossum*, *Trichodesma*, *Ehretia* and *Cordia*.

21.74.4 Size, Distribution and General Information

A family of approximately 100 genera and over 2,000 species, distributed in both tropical and temperate regions but mainly in the Mediterranean region. About 37 genera and over 150 species of Boraginaceae have been reported from India, mainly from the temperate regions and alpine Himalayas. Some of the larger genera along with their number of approximately reported species are *Heliotropium* (250), *Tournefortia* (150), *Cryptantha* (100), *Plagiobothrys* (100), *Eritrichium* (65), *Lithospermum* (60), *Lappula* (55), *Cynoglossum* (50–60), *Mertensia* (50), *Myosotis* (50) and *Amsinckia* (50).

Myosotis, commonly named as "forget-me-not", perhaps derived its name from the "last words of a gallant gentleman who drowned while crossing a stream to collect the bouquet of these sky-blue flowers for his girl friend" (Jones and Luchsinger, 1987).

21.74.5 Description of the Family

General Habit Usually bristly or hispid, hairy, herbaceous (*Heliotropium*) plants with fleshy roots or rhizomes; sometimes shrubs (some species of *Tournefortia*), trees (*Ehretia*, *Cordia*) or lianas (some species of *Cordia*).

Leaves Usually simple, alternate, exstipulate; margin usually entire; often covered by rough hairs; rarely opposite (*Trichodesma*); usually narrow and sessile (*Heliotropium*).

Inflorescence Usually a scorpioid or helicoid cyme or a coiled "cincinnus' uncoiling as flowers open so that newly opened flowers face the same direction; sometimes panicle or corymb (*Ehretia*, *Cordia*).

Flower Bracteate, bisexual, actinomorphic (rarely zygomorphic, e.g. *Lycopsis, Echium*); pentamerous, hypogynous.

Calyx 5 sepals, distinct (*Lithospermum*) or basally connate to form a short or long tube; quincuncial (*Heliotropium*) or imbricate, or rarely valvate:

Corolla 5 petals, gamopetalous, often lobed; imbricate (*Heliotropium*) or twisted; tubular, infundibuliform, campanulate or rotate (*Trichodesma*); often with projecting scales (*Heliotropium*) in throat in the alternate position with the stamens (Fig. 21.86).

Androecium 5 stamens, epipetalous, alternate with petals; anthers dithecous; sagittate or conical; basifixed, introrse, longitudinally dehiscent; nectariferous disc sometimes present at the base of filaments.

Gynoecium Bicarpellary, syncarpous; ovary superior, bilocular, with 2 ovules in each locule; often becoming 4-locular and deeply 4-lobed by the growth of false septum, with each locule containing

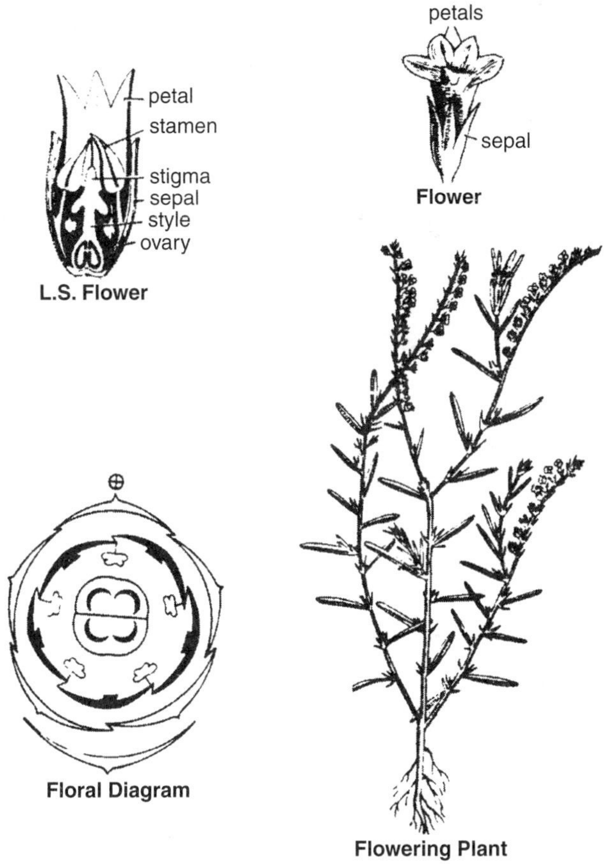

Fig. 21.86 *Heliotropium strigosum* Willd.

one ovule; ovules anatropous, erect, ascending or horizontal; placentation axile, sometimes appearing basal; style usually gynobasic; stigma typically one.

Fruit and Seeds Fruit usually of 4 achenes (nutlets), or a drupe, or a nut. Seeds with erect or curved embryo and usually non-endospermic.

Pollination It is entomophilous. Insects are attracted because of the presence of nectar, colour of the flowers and floral aggregation on the upper part of the inflorescence branches.

General Floral Formula Br, ⊕, ⚥, $K_{(5) \text{ or } 5}$, $\overparen{C_{(5)}, A_5}$, $G_{(\underline{2})}$.

21.74.6 Economic Importance

Except that of some ornamental and medicinal value, the Borage family is not of much economic importance.

- Ornamental plants[1] of the family include species of *Anchusa* (Alkanet), *Borago* (Borage), *Cordia* (Geiger tree), *Cynoglossum* (Hound's-tongue), *Echium* (Viper's Bugloss), *Heliotropium*

[1] Plants arranged alphabetically.

(Heliotrope), *Mertensia* (Blue bells), *Myosotis* (Forget-me-not), *Omphalodes* (Creeping forget-me-not), *Pulmonaria* (Lungwort) and *Symphytum* (Comfrey).

- Plants[1] of medicinal value include (i) *Cordia myxa* (bark and fruits are used in coughs), (ii) *Ehretia macrophylla* (leaves are used in cough and stomach trouble), (iii) *Heliotropium indicum* (leaf-decoction is used in fever, and roots in cough), (v) *Lithospermum officinale* (seeds are used in urinary bladder diseases), and (v) *Onosma bracteatum* (leaves and flowers are used as tonic and diuretic).
- Edible plants of the family are (i) *Cordia dichotoma* (Lassora), of which the unripe fruits are used as vegetable, and (ii) *Cordia gharaf* (Gondhi), of which fruits are edible.
- Ratanjot, a red dye, is obtained from the roots of *Onosma hispida.*
- Wood of various species of *Ehretia* is used in making match boxes, brush backs, tea boxes, and other similar articles.

21.74.7 Systematics and Phylogeny (Affinities)

Willis (1973) divided Boraginaceae into 2 sub-families viz. *Heliotropioideae* (style terminal; fruit drupe) and *Boraginoideae* (style gynobasic; fruit achenes). Boraginoideae is further divided into 5 tribes, i.e. Cynoglosseae, Eritricheae, Boragineae, Lithospermeae and Echieae.

Takhtajan (1969) placed Boraginaceae under Polemoniales, while Cronquist (1981) placed it under order Lamiales, and Thorne (1983) under Solanales. Hutchinson placed Boraginaceae under a monotypic order Boraginales. The family is allied closely to Polemoniaceae, Hydrophyllaceae, Solanaceae and Convolvulaceae. According to Hallier, Boraginaceae is derived directly from Annonaceae.

21.75 CONVOLVULACEAE (MORNING GLORY FAMILY)

21.75.1 Systematic Position

Gamopetalae, Bicarpellatae, Polemoniales.

21.75.2 Field Recognition

Usually climbing or twining vines with milky sap; leaves alternate, exstipulate; flowers axillary and pentamerous; corolla tubular, infundibuliform or salverform; stamens 5, epipetalous; gynoecium bicarpellary.

21.75.3 Selected Indian Genera

Argyreia, Convolvulus, Cuscuta, Evolvulus, Ipomoea.

21.75.4 Size, Distribution and General Information

A family of about 55 genera and 1650 species, distributed in both tropical and temperate regions of the world. About 20 genera and over 150 species have been reported from India. Some of the largely represented genera with their number of approximately reported species are *Ipomoea* (500, Morning

[1] Plants arranged alphabetically.

glory), *Convolvulus* (250, blindweed, or Wild Morning glory), *Cuscuta* (170, Dodder), *Jacquemontia* (120), *Evolvulus* (100) and *Calystegia* (25).

21.75.5 Description of the Family

General Habit Annual or perennial herbs, shrubs, or rarely trees (some species of *Erycibe*); often climbing (*Ipomoea*) with somewhat milky sap; some with tuberous roots (*Ipomoea batata*) or stems, others with rhizomes; *Cuscuta* is the stem parasite, twining round the host and sending haustorial roots; some are xerophytic and thorny; others are aquatic (*Ipomoea aquatica*).

Leaves Simple, alternate, exstipulate, petiolate; entire or variously lobed; cordate (*Ipomoea purpurea*), hastate (*I. aquatica*), deeply emarginate (*I. biloba*), palmately dissected (*I. palmata*) or pinnately dissected (Q*uamoclit pennata*); greatly reduced to small scales (*Cuscuta*).

Inflorescence Typically an axillary dichasial cyme, or flowers solitary axillary; sometimes paniculate (*Porana*).

Flower Bracteate, bracts often large and showy, sometimes forming involucre; bracteolate, complete, hermaphrodite, actinomorphic, pentamerous, hypogynous; rarely unisexual and such species are then dioecious.

Calyx 5 sepals, usually free; persistent; quincuncial (*Convolvulus*, *Ipomoea*), or fused and valvate (*Cuscuta*); 4 sepals in *Hildebrandtia.*

Corolla 5 united petals; often infundibuliform (*Ipomoea*), or salverform, or campanulate (*Convolvulus*); induplicately valvate or twisted in bud; 5 coronary outgrowths are present at the base of the corolla and alternate with the stamens in *Cuscuta*; 4 petals in *Hildebrandtia.*

Androecium 5 stamens, epipetalous, alternipetalous, inserted towards the base of corolla tube; anthers dithecous, dorsifixed or basifixed, introrse, longitudinally dehiscent; sagittate in *Ipomoea*; 4 stamens in *Hildebrandtia*; filaments are usually of different length, i.e. heterodynamous.

Gynoecium Usually bicarpellary (rarely 3 to 5 carpels), syncarpous; superior; usually 2 locules (rarely 4 locules formed by false septation), 1 or 2 ovules in each locule, axile placentation; style usually 1; stigmas 1 or 2; unilocular ovary with parietal placentation is present in *Erycibe:* often a cup-shaped or ring-like nectariferous disc is present below the ovary.

Fruit and Seed Fruit usually a capsule (*Ipomoea*), rarely a nut or berry. Seeds with large and straight or curved embryo, and scanty and cartilaginous endosperm; sometimes hairy.

General Floral Formula Br, Brl, ⊕, ⚥, K_5, $\overbrace{C_{(5)}, A_5}$, $G_{(\underline{2})}$.

21.75.6 Economic Importance

The family is important as a source of food, for drugs, for several ornamentals, etc.

- Plants used for edible purposes include (i) *Ipomoea batata* (sweet potato or Shakarkandi), of which fleshy roots are eaten because of their rich content of sugar and starch, and are also used for production of industrial alcohol, pectin, starch and sugar-syrup; (ii) *Ipomoea aquatica* (Nari-ka-Sag), of which leaves and young shoots are used as vegetable, (iii) *Calonyction*

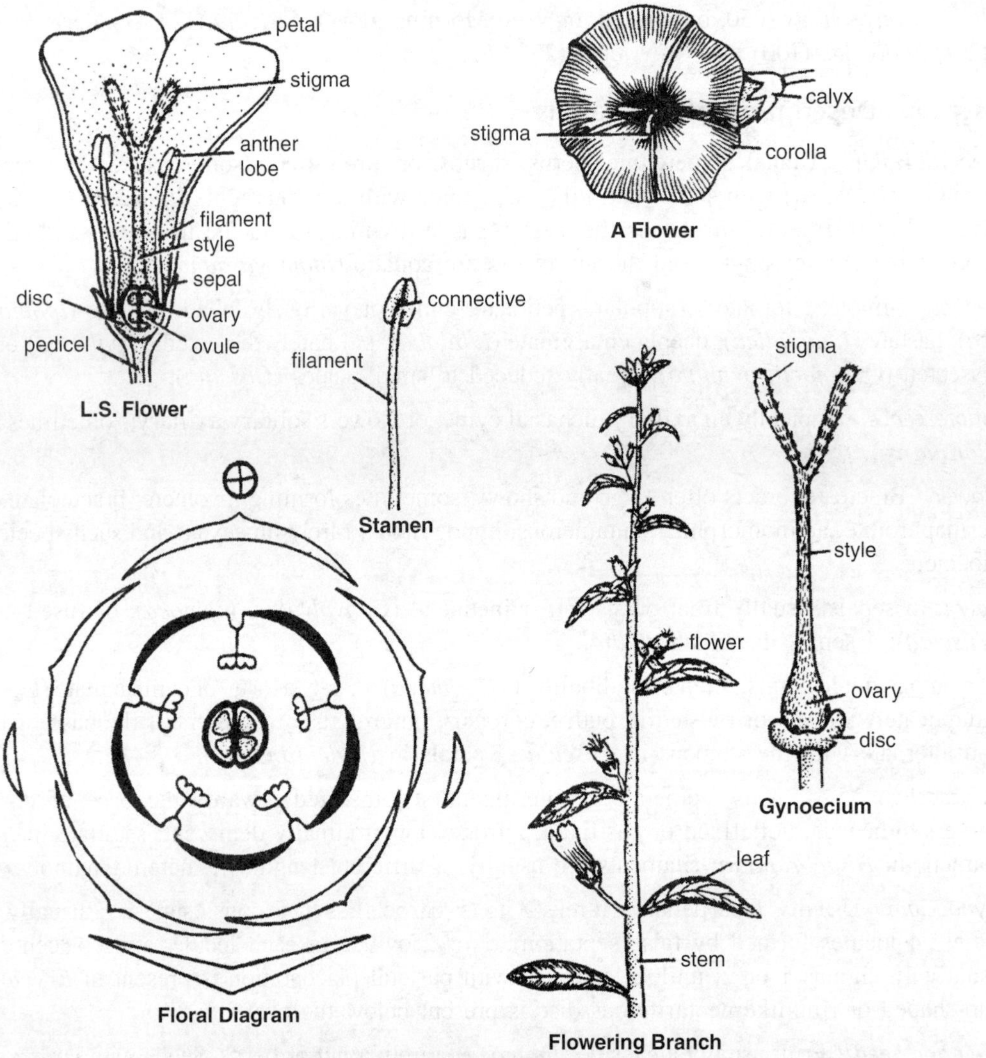

Fig. 21.87 *Convolvulus pluricaulis* Choisy.

muricatum, of which floral pedicels are eaten, (iv) *Calystegia sepium*, of which roots are cooked and eaten, (v) *Rivea hypocrateriformis*, of which young shoots and leaves are used as vegetable.

- Plants of medicinal value include (i) *Evolvulus alsinoides*, used as a bitter tonic, (ii) *Exogonium purga*, yields the drug 'jalap', used as a purgative, (iii) *Ipomoea violacea*, of which seeds contain *d*-lysergic acid amide and are hallucinogenic.
- Plants of ornamental value include (i) *Argyreia speciosa* (elephant creeper), *Calonyction aculeatum* (moon flower), *Ipomoea carica* (railway creeper), *I. coccinea* (star *Ipomoea*), *I. fistulosa*, *I. lobata*, *I. purpurea* (morning glory), *I. quamoclit* (Cyperus vine), *I. tuberosa* (wood rose), *I. violacea* (heavenly blue), and *Porana paniculata* (Christmas vine).

- Plants used as sand binder are *Ipomoea biloba, I. fistulosa* and *I. pescaprae. I. fistulosa* is also used as a hedge plant.
- Plants which become troublesome weeds are *Ipomoea arvensis* and *Cuscuta reflexa.*

21.75.7 Systematics and Phylogeny (Affinities)

The family is usually divided into 2 subfamilies, viz. *Convolvuloideae* (leafy autophytes; interstaminal scales absent) and *Cuscutoideae* (leafless total parasites; interstaminal scales present). However, Bentham and Hooker divided Convolvulaceae into five tribes i.e. Convolvuleae, Dichondreae, Nolaneae, Cresseae and Cuscuteae. Several taxonomists place *Cuscuta* under an independent unigeneric family Cuscutaceae, and some also treat *Dichondra* and related genera in an independent family Dichondraceae.

Convolvulaceae is related closely to Solanaceae and also to Polemoniaceae, Boraginaceae and to some extent to Hydrophyllaceae and Nolanaceae. According to Takhtajan, it is more allied to Polemoniaceae, Boraginaceae and Hydrophyllaceae, and thus belongs to Polemoniales. Cronquist (1981) and Thorne (1983) treated it under Solanales.

21.75.8 Description of Common Plants

1. Convolvulus pluricaulis Choisy (Fig. 21.87)

Habit: A perennial, prostrate herb. *Leaf:* Simple, alternate, exstipulate. *Inflorescence:* Solitary axillary or dichasial cyme. *Flower:* Bracteate, bracteolate, actinomorphic, hermaphrodite, pentamerous, hypogynous, light purple. *Calyx:* 5 sepals, free, quincuncial. *Corolla:* 5 petals, united, infundibuliform, valvate. *Androecium:* 5 stamens, epipetalous, alternipetalous, filaments of different sizes; dorsifixed. *Gynoecium:* 2 carpels, syncarpous, superior, bilocular, carpels medianly placed, 2 or more ovules in each locule, axile placentation; stigma bifid and hairy; a nectariferous disc is present below the ovary. *Fruit:* Capsule.

Floral Formula: Br, Brl, ⊕, ⚥, K_5, $\overparen{C_{(5)}, A_5}$, $G_{(\underline{2})}$.

2. Cuscuta reflexa Roxb. (Amar Bel or Akash-Bel; Fig. 21.88)

Habit: A twining total parasite. *Stem:* Weak, twiner, pale-green, develop haustoria at the point of contact with the host. *Leaf:* Absent. *Inflorescence:* Flowers solitary or in racemose clusters. *Flower:* Bracteate, ebracteolate, hermaphrodite, actinomorphic, pentamerous, small, pale-green. *Calyx:* 5 sepals, fused, valvate. *Corolla:* 5 petals, united, campanulate, valvate, with 5 coronary outgrowths at the base of corolla. *Androecium:* Same as in *Convolvulus*. *Gynoecium:* Style very much reduced, disc red coloured; other details same as in *Convolvulus*. *Floral Formula:* Br, Ebrl, ⊕, ⚥, $K_{(5)}$, $\overparen{C_{(5)}, A_5}$, $G_{(\underline{2})}$.

21.76 ❁ SOLANACEAE (POTATO OR NIGHTSHADE FAMILY)

21.76.1 Systematic Position

Gamopetalae, Bicarpellatae, Polemoniales.

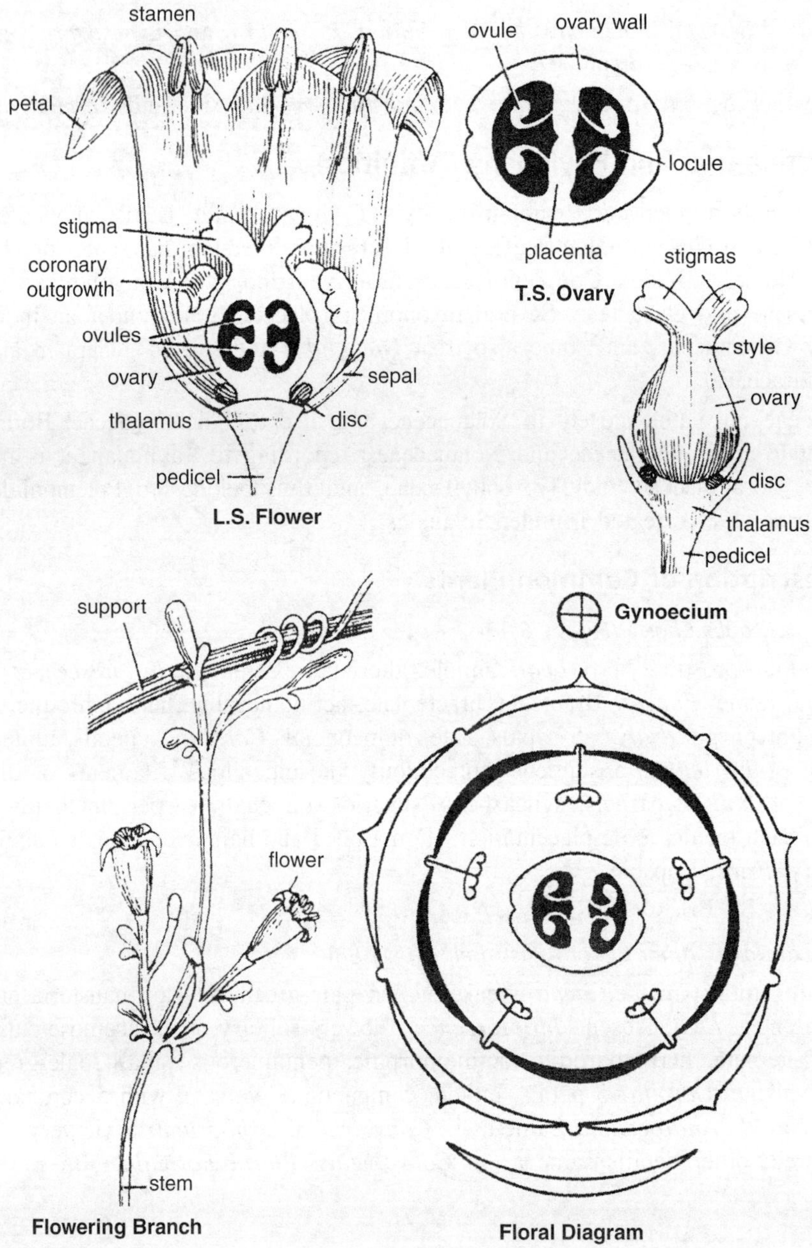

Fig. 21.88 *Cuscuta reflexa* Roxb.

21.76.2 Field Recognition

Herbs or shrubs, rarely trees; leaves alternate, exstipulate; flowers bisexual, actinomorphic, pentamerous; ovules bilocular; carpels obliquely placed; placentation axile; placenta obliquely placed; fruit berry or capsule.

21.76.3 Selected Indian Genera

Atropa, Capsicum, Cestrum, Datura, Lycopersicon, Nicotiana, Petunia, Physalis, Solanum, Withania.

21.76.4 Size, Distribution and General Information

A family of about 90 genera and 2800 species distributed in both tropical and temperate regions. Central and South America are the chief centres of distribution where over 40 genera are found. About 15 genera and over 90 species have been reported from India. Some of the larger genera along with their approximately worldover reported species (Hickey and King, 1988) are *Solanum* (1700), *Cestrum* (150), *Physalis* (100), *Lycium* (80–90), *Nicotiana* (66), *Capsicum* (50), *Petunia* (40), *Hyoscyamus* (20) and *Datura* (10). *Solanum tuberosum* (potato) and *Nicotiana tabacum* (tobacco) are the two most utilized plants of Solanaceae.

21.76.5 Description of the Family

General Habit Mostly annual (*Physalis minima*), biennial, or perennial herbs (*Withania somnifera*), sometimes shrubs (*Brunfelsia*), or trees (*Solanum giganteum, S. verbascifolium*); rarely vines (*Lycium sinensis*) or lianas.

Stem Herbaceous or woody; erect or twining, or creeping; sometimes modified into tubers (*Solanum tuberosum*); often with bicollateral vascular bundles.

Leaves Usually simple, alternate, exstipulate; entire; sometimes variously dissected or lobed; opposite or sub-opposite in upper part of the plant; pinnately compound in tomato and potato; unicostate reticulate.

Inflorescence Generally cymose, or sometimes solitary flowers (*Datura stramonium*); sometimes extra-axillary scorpioid cyme called *rhipidium* as in *Solanum nigrum*; or umbellate cyme (*Withania somnifera*).

Flower Bracteate (*Petunia*) or ebracteate (*Withania*), pedicellate, bisexual; actinomorphic or weakly zygomorphic due to oblique position of the ovary; pentamerous, hypogynous.

Calyx 5 sepals, gamosepalous, valvate, persistent; rarely the sepals are 4 or 6, and often enlarging to envelope the fruit (*Physalis, Withania*).

Corolla 5 petals, gamopetalous, rotate to tubular (*Solanum, Cestrum*) or bell-shaped (*Atropa*) or infundibuliform (*Petunia*); usually alternate with sepals; rarely bilipped and zygomorphic (*Schizanthus*); usually valvate, sometimes convolute (*Datura*).

Androecium 5 stamens, epipetalous, alternipetalous; filaments usually of unequal length; stamens are only 2 in *Schizanthus*, and 4 and didynamous in *Salpiglosis*; anthers dithecous; usually connivent i.e. forming a cone by coming close together but not actually fusing (e.g. *Solanum*); anthers opening either lengthwise or by terminal pores. .

Gynoecium Bicarpellary, syncarpous, carpels obliquely placed in the flower; ovary superior, bilocular, numerous anatropous or slightly amphitropous ovules in each locule; axile placentation; often the placentae are swollen; sometimes carpels divided by a false septum; style simple, with entire or

2-lobed stigma; sometimes the carpels are more than 2 (*Capsicum*) and also the number of locules is 3 to 5 (*Nicandra*); a hypogynous disc usually present.

Fruit and Seeds Fruit usually a berry (*Physalis*, *Withania*) or septicidal capsule (*Datura*). Seeds with copious endosperm and straight or curved embryo.

Pollination and Dispersal Pollination is entomophilous. Bright and showy petals, presence of nectar secreted by nectariferous disc and strongly scented flowers of several members of this family are the chief attractions for insects. Flowers are protogynous and both self-pollination and cross-pollination occur. Only self-pollination occurs in cleistogamous flowers of *Salpiglossis*. Dispersal of seeds takes place mainly by birds and animals and sometimes by water (*Atropa*).

General Floral Formula Br or Ebr, ⊕, ⚥, $K_{(5)}$, $\overparen{C_{(5)}, A_5}$, $G_{(\underline{2})}$.

21.76.6 Economic Importance

- *Ornamental Plants:* Common plants[1] of ornamental value of Solanaceae are *Brunfelsia calycina* (fragrant flowers), *Cestrum diurnum* (Day Jasmine), *C. nocturnum* (Night Jasmine), *Hyoscyamus niger* (Henbane), *Nicotiana alata* (white-flowered), *Petunia hybrida* (pink-flowered), *P. violacea* (white-flowered), *Salpiglossis sinuata*, *Schizanthus pinnatus*, *S. retusus*, *Solanum dulcamara* (violet-coloured), *S. jasminoides* (white and blue-coloured), *S. seaforthianum* (bluish-purple coloured).
- *Tobacco:* *Nicotiana tabacum* (tobacco) leaves are dried, made into tobacco, and used universally in cigarette, bidi, cigar, pipes, hukkah as well as for chewing and snuffing. It is also used in medicine as a sedative and anti-spasmodic. *N. rustica* is also used for similar purposes.
- *Food Plants:* (i) Tubers of *Solanum tuberosum* (potato or Aalu) are used throughout the world as a common vegetable and also used for the production of starch, dextrin, several alcohols and some other industrial products, (ii) fruits of *Lycopersicon lycopersicum* syn. *Solanum lycopersicum* are the tomato (Tamatar), used as a delicious vegetable and eaten raw, (iii) fruits of *Solanum melongena* (Brinjal, eggplant or Baingan) are eaten as vegetable, (iv) chillies, *'Mirch'* or red pepper are the fruits of *Capsicum annuum* and *C. frutescens*, (v) *Physalis peruviana* (Raspberry or Cape Gooseberry) yield the delicious edible fruits.
- *Medicinal Plants*[2]*:* (i) *Atropa belladona* (Belladona or deadly nightshade) roots yield a powerful alkaloid 'atropine' used in belladona plasters, tinctures etc., for relieving pain, and also for dilating pupils of eyes for eye-testing, (ii) *Datura stramonium* (Datura) leaves and flowers are the source of the drug "stramonium", used in asthama, and its seeds are deadly poisonous, (iii) *Hyoscyamus niger* (Henbane) leaves yield the alkaloid hyoscyamine used in treating asthama and whooping cough, (iv) *Mandragora autumnalis* roots are used as sedative and hypnotic, (v) *Nicotiana tabacum* leaves are used as sedative, antispasmodic and also in iradicating animal pests, such as lices, (vi) *Solanum surattense* roots are used in

[1] Plants arranged alphabetically.
[2] Plants arranged alphabetically.

asthma and leaves in rheumatism, (vii) *Withania coagulans* fruits are used in curing asthma and liver troubles, (viii) *Withania somnifera* (Asgandh) roots are used in curing cough and rheumatism.

21.76.7 Systematics and Phylogeny (Affinities)

Variously placed under Polemoniales (Bentham and Hooker), Solanales (Hutchinson, Cronquist, Thorne) and Scrophulariales (Takhtajan), the family Solanaceae has been divided by Bentham and Hooker into 5 tribes (Solaneae, Atropeae, Hyoscyameae, Cestrineae and Salpiglossideae). Wettstein divided Solanaceae into 5 different tribes namely Nicandreae, Solaneae, Datureae, Cestreae and Salpiglossideae.

Solanaceae is allied closely to Convolvulaceae in both having pentamerous flowers, persistent calyx and bicarpellary ovary. However, Solanaceae possess obliquely placed ovary, and are predominantly erect whereas these characters are absent in Convolvulaceae. In certain characters Solanaceae is closely allied to Scrophulariaceae, Boraginaceae and Polemoniaceae. Hutchinson has raised the tribe Salpiglossideae to the rank of an independent family Salpiglossidaceae.

21.76.8 Description of a Common Plant

Solanum nigrum L. (Black Nightshade or Makoi, Fig. 21.89)

Habit: A small annual herb. *Leaf:* Simple, alternate but opposite in the floral region; petiolate, exstipulate, ovate, entire or slightly lobed, acute, unicostate reticulate. *Inflorescence:* Extra-axillary scorpioid cyme called rhipidium. *Flower:* Ebracteate, pedicellate, hermaphrodite, actinomorphic, pentamerous, hypogynous, white. *Calyx:* 5 sepals, fused, valvate, persistent. *Corolla:* 5 petals, gamopetalous, valvate. *Androecium:* 5 stamens, free, epipetalous, filaments short; anthers conniving and forming an envelope around the style, dithecous, basifixed, with apical pores. *Gynoecium:* Bicarpellary, syncarpous, superior, bilocular, many ovules in each locule, axile placentation, septum oblique and placentae are highly swollen; style long and hairy at the base; stigma bifid. *Fruit:* Berry.

Floral Formula: Ebr, ⊕, ⚥, $K_{(5)}$, $\overbrace{C_{(5)}, A_5}$, $G_{(\underline{2})}$.

21.77 POLEMONIACEAE (PHLOX FAMILY)

21.77.1 Systematic Position

Gamopetalae, Bicarpellatae, Polemoniales.

21.77.2 Field Recognition

Milky latex absent; flowers pentamerous; calyx gamosepalous; stamens inserted at different levels on corolla tube; ovary typically tricarpellate; ovules and seeds numerous.

21.77.3 Selected Indian Genera

Phlox, Polemonium, Gilia.

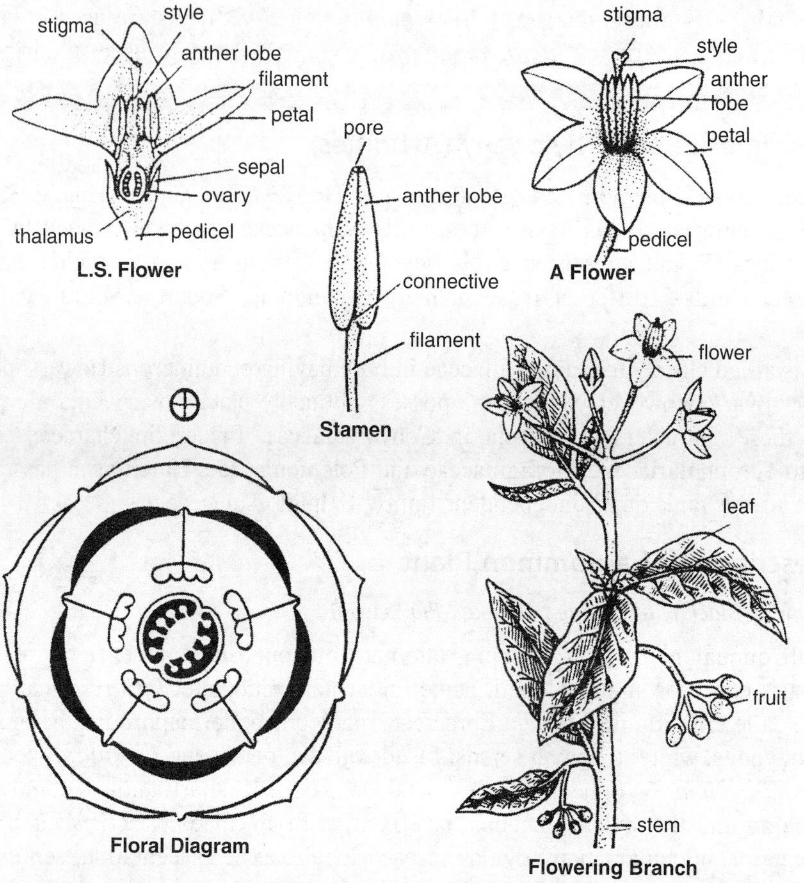

Fig. 21.89 *Solanum nigrum* L.

21.77.4 Size, Distribution and General Information

A family of about 18 genera and over 300 species, Polemoniaceae are distributed chiefly in North America, some in South America, Europe and North Asia. Some of the large genera along with the approximate number of their reported species are *Gilia* (120), *Phlox* (77), *Polemonium* (52), *Cobaea* (18) and *Collomia* (15). Colourful corollas of Polemoniaceae, specially of *Phlox*, have attracted the attention of many and are thus adopted into the gardens throughout the world.

21.77.5 Description of the Family

General Habit Annual or perennial herbs, rarely shrubs, small trees, or twining vines; milky latex or coloured sap absent.

Stem Erect, hollow, angled (*Polemonium caeruleum*, Fig. 21.90).

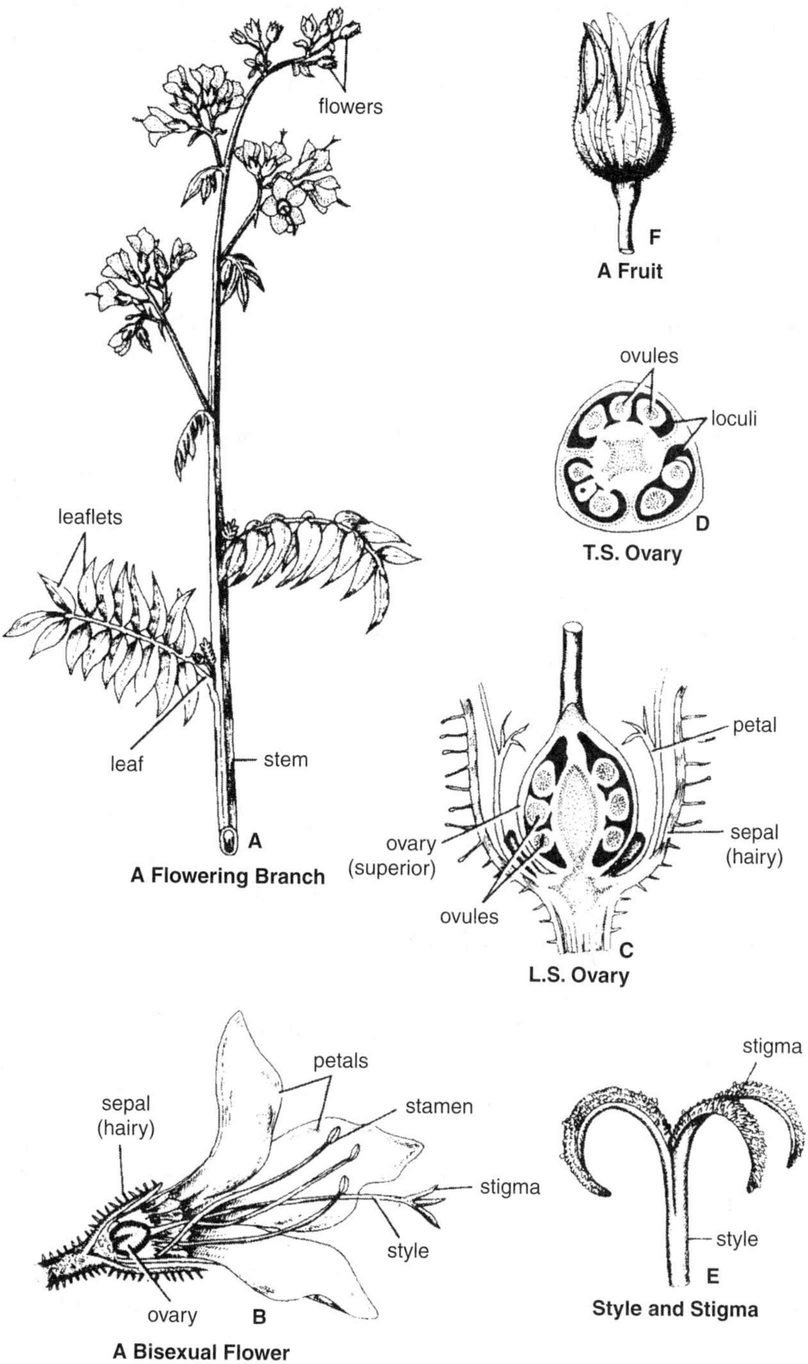

Fig. 21.90 *Polemonium caeruleum* L.

Leaves Alternate or opposite; simple or compound; exstipulate; stipules, if present are large and foliaceous (*Cobaea*); in *Polemonium caeruleum* the leaves are pinnate, lower petiolate while the upper ones are sessile; 6–12 pairs of lateral leaflets are lanceolate to oblong in shape with acuminate apex (Fig. 21.90).

Inflorescence Usually cymose, corymbose to capitate; rarely flowers solitary and axillary.

Flowers Bracteate, bracteolate, complete, hermaphrodite; actinomorphic, rarely slightly zygomorphic (*Bonplandia*); pedicellate, pentamerous and usually showy.

Calyx Sepals 5, gamosepalous; valvate or imbricate; persistent.

Corolla Petals 5, gamopetalous, campanulate (*Cobaea*), funnel-shaped or salver-shaped; usually convolute; corolla tube usually well-developed.

Androecium Stamens 5, epipetalous and alternipetalous; attached on the corolla tube at various heights; anthers dithecous; an intrastaminal disc usually present.

Gynoecium A compound pistil of usually 3 united carpels; trilocular, one to numerous ovules in each locule; axile placentation; ovary superior; style, filiform; stigmas 3 or rarely 2.

Fruit Usually a loculicidal capsule; sometimes septicidal (*Cobaea*); rarely indehiscent.

Seeds Usually with abundant and fleshy endosperm; in *Cobaea*, however, seeds have no endosperm; seeds sometimes covered by mucilaginous coat (*Collomia*).

General Floral Formula $\mathrm{Br}, \mathrm{Brl}, \oplus, ⚥, K_{(5)}, \overgroup{C_{(5)}, A_5}, G_{(\underline{3})}$.

21.77.6 Economic Importance

Polemoniaceae is not of much economic importance except that of a few ornamentals grown as annuals, biennials, perenniales or vines. Some of such plants of ornamental value belong to *Phlox*, *Polemonium*, *Cobaea*, *Gilia* and *Linanthus*. Beautiful flowers come from *Phlox drummondii* (annual), *P. paniculata* (perennial) and *Cobaea scandens* (climber), etc.

21.77.7 Systematics and Phylogeny (Affinities)

Bentham and Hooker placed Polemoniaceae and four more families (Boraginaceae, Hydrophyllaceae, Convolvulaceae and Solanaceae) under order Polemoniales of series Bicarpellatae of subclass Gamopetalae. Along with 18 more families, all these five families have been placed under order Tubiflorae by Engler and Prantl. Hutchinson included only 3 families (Polemoniaceae, Cuscutaceae and Hydrophyllaceae) under order Polemoniales of division Herbaceae.

Phylogeny of Polemoniaceae has been perplexing botanists since long. According to Hallier, Polemoniaceae has been "probably derived from Linaceae". Bessey considered it to be derived from Boraginaceous stock whereas Rendle believed Polemoniaceae to have been derived from sympetalous relatives of Rosales. According to Hutchinson, however, Polemoniales have been derived from Geraniales. Some morphologists, however, believe that Polemoniaceae are closely related to both Caryophyllaceae and Geraniaceae (Dawson, 1936).

21.78 PERSONALES

21.78.1 General Characteristics

1. Mostly herbs or undershrubs, rarely trees.
2. Leaves are exstipulate.
3. Flowers zygomorphic and hypogynous.
4. Number of stamens are generally less than corolla lobes, and epipetalous.
5. Stamens are usually didynamous.
6. Internal phloem is present in members of some families.
7. Fruit usually a capsule.

Bentham and Hooker included 8 families under Personales. These are Scrophulariaceae, Orobanchaceae, Lentibulariaceae, Bignoniaceae, Pedaliaceae, Columelliaceae, Gesneriaceae and Acanthaceae. Engler and Diels discussed all these families under order Tubiflorae.

Only Scrophulariaceae, Bignoniaceae, Pedaliaceae and Acanthaceae have been discussed in this text.

21.79 SCROPHULARIACEAE (SNAPDRAGON FAMILY)

21.79.1 Systematic Position

Gamopetalae, Bicarpellatae, Personales.

21.79.2 Field Recognition

Mostly herbs or shrubs, rarely trees; flowers pentamerous; zygomorphic, corolla bilipped; usually with 2 or 4 fertile stamens, epipetalous; ovary bilocular, superior.

21.79.3 Selected Indian Genera

Antirrhinum, Digitalis, Linaria, Lindenbergia, Mazus, Verbascum, Veronica.

21.79.4 Size, Distribution and General Information

Represented by about 220 genera and 3500 species (Hickey and King, 1988), Scrophulariaceae are cosmopolitan in their distribution. About 60 genera and 360 species have been reported from India, chiefly from the Himalayas. Some of the genera along with their approximately reported species from worldover are *Pedicularis* (500), *Calceolaria* (300–400), *Scrophularia* (300), *Verbascum* (300), *Veronica* (300), *Penstemon* (280), *Linaria* (150), *Castilleja* (150), *Selago* (150), *Sutera* (130), *Mimulus* (100), *Antirrhinum* (40) and *Digitalis* (20). The common names of the family (snapdragon, foxglove or figwort family) are because of its genera *Antirrhinum* (snapdragons), *Digitalis* (foxglove) and *Scrophularia* (figwort).

21.79.5 Description of the Family

General Habit Mostly herbs or undershrubs, and only a few are big shrubs (some species of *Veronica*) or trees (*Paulownia*); several are climbers (*Maurandia*), and some are saprophytic or parasitic (*Hyobanche*); *Lathrea* is a chlorophyll-less root parasite while *Euphrasia* and *Pedicularis* are semi-parasitic.

Leaf Simple or rarely pinnately compound; exstipulate; alternate (*Verbascum*), opposite (*Mimulus*) or whorled (*Russelia*); *Ambulia* shows heterophylly; both stem and leaves are generally hairy.

Inflorescence Variable from racemose (*Mazus*) or cymose (*Verbascum*), or flowers solitary axillary (*Linaria, Lindenbergia*); spike (*Digitalis*), panicle (*Scrophularia himalayensis*).

Flower Bracteate, bracteolate, bisexual; zygomorphic but sometimes nearly actinomorphic (*Verbascum*); upper leaves and bracts are sometimes brightly coloured (*Castilleja*); hypogynous; usually with an annular disc.

Calyx Usually 5 united sepals; imbricate, quincuncial or valvate; sepals are 4 in *Veronica* and *Scoparia*; sometimes 2 anterior sepals are united (*Calceolaria*).

Corolla Usually 5 united petals; often bilipped (2/3); sometimes only 4 petals are present, and the one placed anteriorly is smaller than the others (*Veronica*, Fig. 21.91A); frequently the anterior petals are spurred (*Linaria*) or with a gibbous (*Antirrhinum*, Fig. 21.91B) to saccate base; in *Verbascum* corolla is regular with all the 5 equal petals (Fig. 21.91C).

Androecium 4 stamens (*Mazus*), sometimes only 2 stamens (*Veronica*); rarely all the 5 stamens are fertile (*Verbascum*); didynamous, epipetalous; sometimes the posterior stamen is represented by a staminode (*Antirrhinum, Scrophularia, Penstemon*); fertile stamens are alternate with corolla lobes; anthers dithecous, occasionally connivent, dehiscence mostly longitudinal and rarely poricidal.

Gynoecium Bicarpellary, syncarpous, superior, bilocular, numerous anatropous or amphitropous ovules in each locule, axile placentation; carpels usually anterio-posteriorly placed; style simple; stigma bilobed; nectar-secreting disc present below this ovary.

Fruit and Seeds Fruit usually a capsule, rarely a berry (*Leucocarpus*) or drupe surrounded by persistent calyx. Seeds numerous, small, endospermic with straight or slightly curved embryo.

Pollination and Dispersal Flowers usually adapted to insect-pollination. Insects are attracted because of coloured corolla and nectar-secreting hypogynous disc. Flowers which have long-tubed corolla (e.g. *Scrophularia*) are pollinated by wasps while those having short-tubed corolla (e.g. *Veronica*, *Verbascum*) are pollinated by bees. Dispersal of seeds usually takes place by water, birds or animals.

General Floral Formula Br, Brl, ϕ, ⚥, $K_{(5)}$, $\overset{\frown}{C_{(5)}, A_{4 \text{ or } 2}}$, $G_{(\underline{2})}$.

21.79.6 Economic Importance

- Ornamental plants[1] of the family include *Angelonia angustifolia*, *Antirrhinum majus* (*snapdragon*), *Calceolaria scabiosaefolia* (slipperwort), *Castilleja* (Indian paintbrush), *Chelone*

[1] Plants arranged alphabetically.

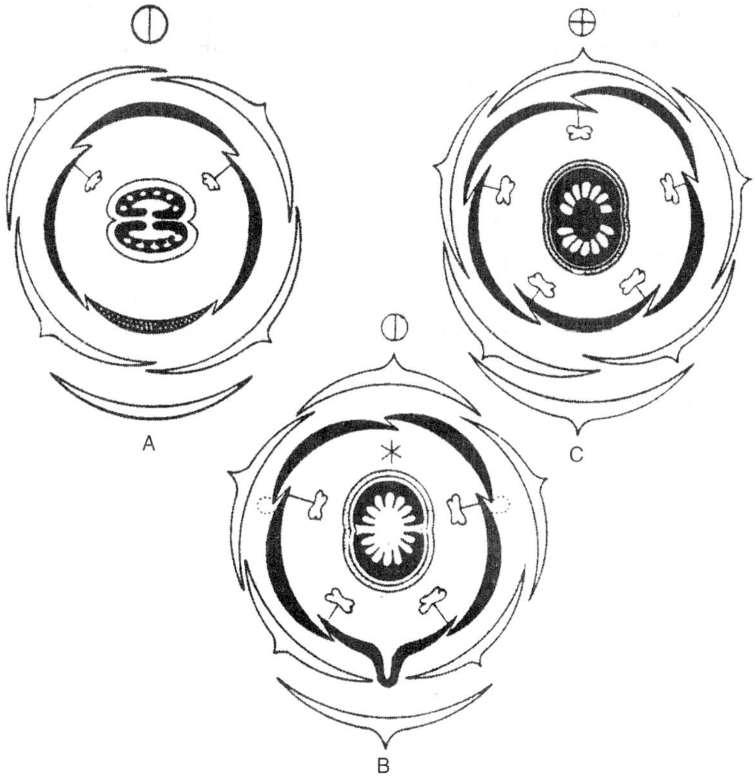

Fig. 21.91A–C Floral diagrams of *Veronica anagallis* (A), *Antirrhinum orontium* (B) and *Verbascum chinense* (C).

glabra (turtlehead), *Collinsia grandiflora*, *Digitalis purpurea* (foxglove), *Linaria vulgaris* (toadflax), *Mimulus luteus* (monkey flower), *Nemesia versicolor* (Nemesia), *Paulownia tomentosa*, *Pedicularis canadensis* (housewort), *Russelia juncea* (coral blow), *Torenia asiatica* and *Veronica bungifolia* (speedwell).

- Medicinal plants[1] of the family include (i) *Bacopa monnieri* (used in epilepsy and insanity), (ii) *Digitalis purpurea* (dried leaves are used in congested heart failure), (iii) *Herpestis cuneifolia* (used in rheumatism), (iv) *Picrorhiza kurroa* (used as a laxative and cathartic), (v) *Scoparia dulcis* (used in fever, cough and bronchitis), and (vi) *Verbascum sinensis* (used in dysentery).

21.79.7 Systematics and Phylogeny (Affinities)

Scrophulariaceae has been variously placed under Personales (Bentham and Hooker, and Hutchinson), Scrophulariales (Takhtajan, Cronquist) and Bignoniales (Thorne). Wettstein divided Scrophulariaceae into 3 subfamilies viz. Verbascoideae, Scrophularioideae and Rhinanthoideae.

[1] Plants arranged alphabetically.

Scrophulariaceae is closely allied to Solanaceae and Convolvulaceae having exstipulate leaves, persistent gamosepalous calyx and bicarpellary superior ovary. It is also closely related to Bignoniaceae having zygomorphic flowers, exstipulate leaves and hypogynous disc. In several characters Scrophulariaceae is closely allied also to Labiatae and Acanthaceae.

21.79.8 Description of a Common Plant

1. Mazus japonicus (Thunb.) Kuntze syn. M. rugosus Lour. (Fig. 21.92)

Habit: Small annual herb. *Leaf:* Radical, simple, exstipulate, sessile but lamina narrowed basally into the petiole, spathulate, dentate. *Inflorescence:* Racemose raceme raised on a scape. *Flower:* Bracteate, pedicellate, complete, hermaphrodite, zygomorphic, bilipped, pentamerous, hypogynous. *Calyx:* 5 sepals, poly- or gamosepalous, valvate or quincuncial. *Corolla:* 5 petals, united; bilabiate (2/3) personate, posterior lip bilobed while anterior lip trilobed. *Androecium:* 4 stamens, polyandrous, epipetalous,

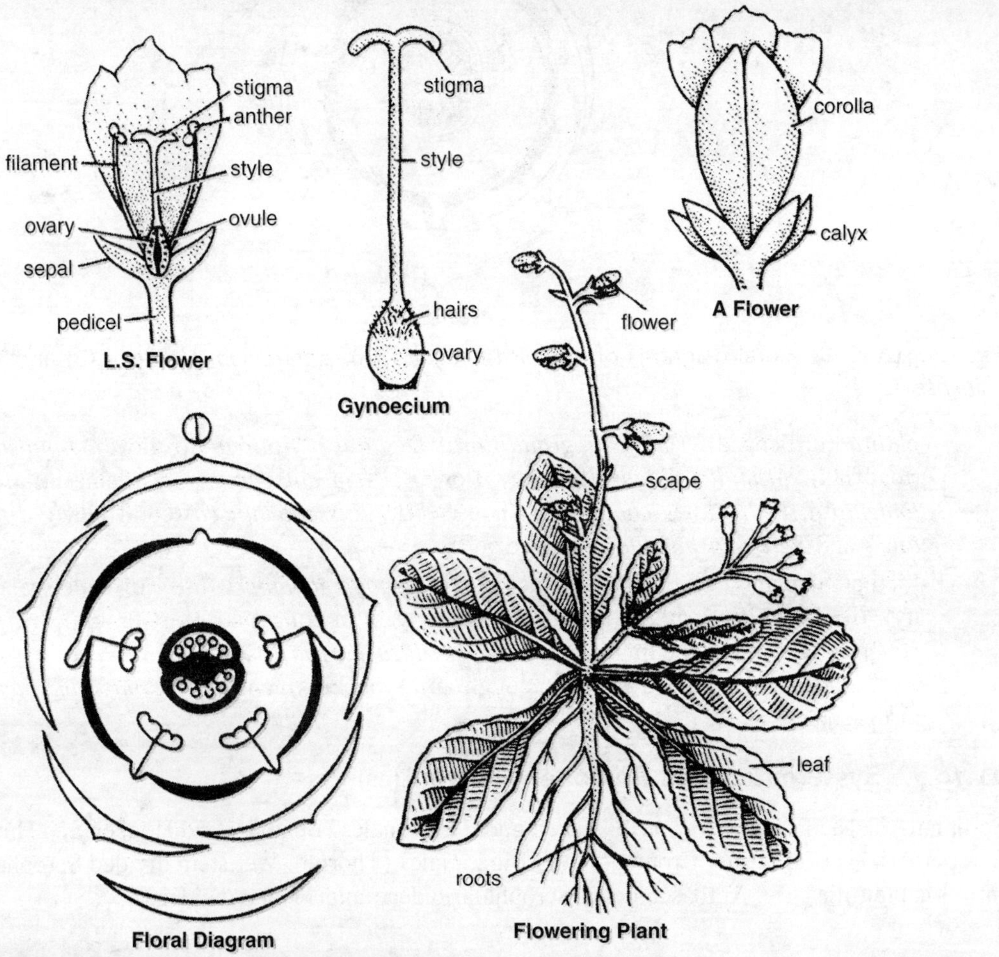

Fig. 21.92 *Mazus japonicus* Thunb.

didynamous, filaments curved apically; anthers dithecous, basifixed, introrse. *Gynoecium:* Same as in family description. *Fruit:* Capsule.

Floral Formula: Br, Ebrl, ⊕, ⚥, K_5, $\overset{\frown}{C_{(2/3)}\ A_{2+2}}$, $G_{(\underline{2})}$.

21.80 BIGNONIACEAE (BIGNONIA FAMILY)

21.80.1 Systematic Position

Gamopetalae, Bicarpellatae, Personales.

21.80.2 Field Recognition

Woody lianas, shrubs or trees; leaves usually opposite and compound; inflorescence cymose; flowers zygomorphic; fruit capsule.

21.80.3 Selected Indian Genera

Jacaranda, Kigelia, Millingtonia, Tecoma, Heterophragma, Oroxylum and *Dolichandrone.*

21.80.4 Size, Distribution and General Information

Represented by about 120 genera and over 650 species, Bignoniaceae are distributed mainly in tropical and subtropical, and a few in temperate regions. About 15 genera and 40 species are found in India. Some of the larger genera along with their approximate number of worldover reported species are *Tabebuia* (100), *Arrabidaea* (70), *Jacaranda* (50), *Anemopaegma* (30), *Tecoma* (16), *Catalpa* (11) and *Crescentia* (5).

21.80.5 Description of the Family

General Habit Mostly woody lianas or climbers, climbing with the help of tendrils; or trees or shrubs; only rarely herbs; some climb with the help of aerial roots (*Campis radicans*), others are twiners (*Pandorea, Tecomaria*) and most genera are tendril climbers.

Anatomically, wedges of phloem develop in stem in between secondary xylem.

Leaf Usually opposite, pinnately compound; exstipulate; terminal leaflet usually modifying into tendril; tendrils may be simple or branched, hooked, or provided with some adhesive discs; sometimes the leaves are simple and rarely whorled (*Diplanthera*).

Inflorescence Usually cymose (dichasial cyme) with bracts and bracteoles.

Flower Bracteate, bracteolate, complete, hermaphrodite, zygomorphic, hypogynous, often showy.

Calyx Usually 5 united sepals, 5-toothed, often campanulate, valvate.

Corolla Usually 5 united petals, campanulate or funnel-shaped, imbricate; sometimes bilipped (2/3) with the upper lip of 2 petals and lower lip of 3 petals.

Androecium Typically 4, epipetalous stamens, didynamous, alternipetalous, with a posterior staminode; sometimes 2 (*Catalpa*, Fig. 21.93) and rarely 5 (*Oroxylum*) stamens; usually the anther cells one above the other; dithecous, introrse, longitudinal dehiscence.

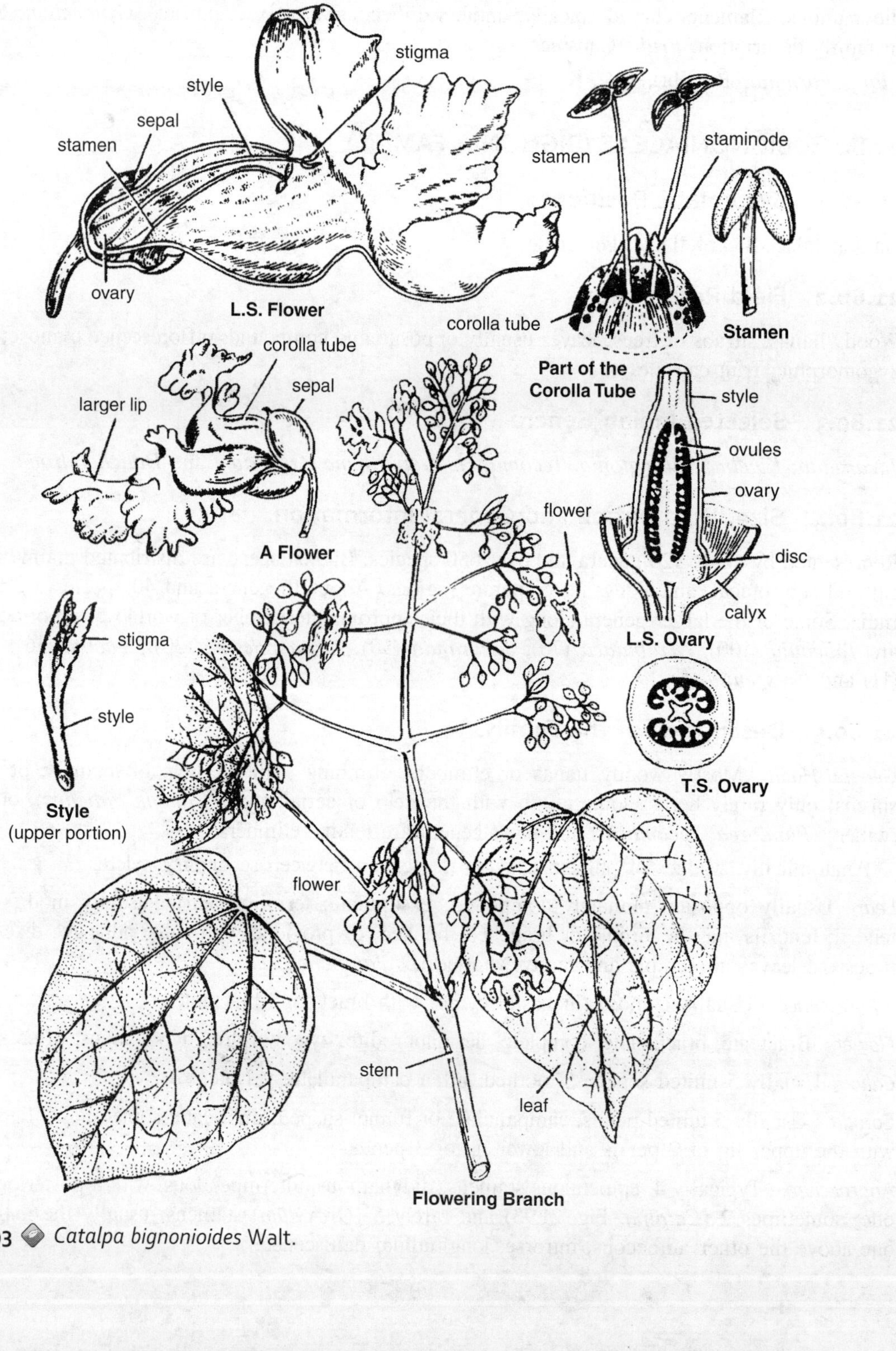

Fig. 21.93 *Catalpa bignonioides* Walt.

Gynoecium Bicarpellary, syncarpous, superior, bilocular, numerous anatropous ovules in each locule, axile placentation; sometimes unilocular (*Kigelia*) with parietal placentation; style simple, terminal; stigma bilipped; a hypogynous nectar-secreting disc is present.

Fruit and Seeds Fruit usually a capsule, sometimes fleshy and indehiscent. Seeds often winged, non-endospermic, with straight embryo.

General Floral Formula Br, Brl, ⊕, ⚥, $K_{(5)}$, $\widehat{C_{(5)}, A_{4+1\ \text{staminode}}}$, $G_{(\underline{2})}$.

21.80.6 Economic Importance

- Ornamental plants of the family include handsome trees (*Catalpa bignonioides, Haplophragma adenophyllum, Jacaranda mimosaefolia, Kigelia pinnata, Millingtonia hortensis, Spathodea campanulata* and *Tabebuia spectabilis*), shrubs (*Bignonia unguis-cati*), beautiful tendril-climbers (*Adenocalymna alliaceum, Bignonia venusta* syn. *Pyrostegia venusta*), aerial rootlet climbers (*Campis grandiflora* and *C. radicans*) and herbaceous genus *Incarvillea.*
- Some of the timber-yielding plants of the family include species of *Catalpa* (Fig. 21.93), *Dolichandrone, Haplophragma, Millingtonia, Oroxylum, Radermachera, Spathodea, Stereospermum, Tabebuia* and *Tecoma.*
- An inferior quality of cork is prepared from the bark of *Millingtonia hortensis* (Akash Neem or Indian cork tree).
- *Crescentia cujete* (Calabash-tree) is grown for its gourd-like fruits, which, when hollowed out and dried, are used for holding liquids.

21.80.7 Systematics and Phylogeny (Affinities)

Variously placed under Personales (Bentham and Hooker), Bignoniales (Hutchinson, Thorne) and Scrophulariales (Takhtajan, Cronquist), the family Bignoniaceae has been divided into 4 tribes (Bignonieae, Tecomeae, Eccremocarpeae and Crescentieae) by Schumann in Engler and Prantl's *Pflanzenfamilien.* In possessing didynamous stamens, bicarpellary ovary, zygomorphic flowers and capsular fruits, Bignoniaceae is closely allied to Scrophulariaceae. In some respects it is also allied to Acanthaceae and Pedaliaceae.

21.81 PEDALIACEAE (BENNE FAMILY)

21.81.1 Systematic Position

Gamopetalae, Bicarpellatae, Personales.

21.81.2 Field Recognition

Herbs, rarely shrubs; leaves opposite or the upper ones alternate; flowers usually solitary axillary; flowers bisexual, zygomorphic; calyx and corolla 5 and united; stamens generally 4 and didynamous; a small staminode present; fruit capsule, often prickly.

21.81.3 Selected Indian Genera

Sesamum (Fig. 21.94), *Martynia*, *Pedalium*.

21.81.4 Size, Distribution and General Information

Represented by about 15 genera and about 65 species, the members of this family are distributed largely on sea shores and desert areas occurring in Indo-Malaysian regions, Madagascar, South Africa and tropical Australia. In India, the family is represented by three genera, occurring chiefly in south Indian states. *Sesamum indicum* (Sesame or Gingelly) is cultivated as an oil-seed crop throughout the warmer parts of India.

21.81.5 Description of the Family

General Habit Annual or perennial herbs or undershrubs, with mucilage containing glandular hairs; sometimes roots are tuberous (*Martynia*).

Leaf Opposite or spirally arranged, sometimes the upper leaves alternate; simple, exstipulate; entire, lobed or divided.

Inflorescence Usually solitary axillary flowers, sometimes in three-flowered axillary cymes; sometimes terminally racemose (*Martynia*).

Flower Bisexual, zygomorphic, hypogynous; characteristic glands are present at the base of floral stalk; these are actually metamorphosed flowers.

Calyx Usually 5 sepals, more or less united; rarely sepals distinct (*Martynia*).

Corolla Petals 5, united; tubular-ventricose; limb obscurely bilabiate, 5-lobed and the lobes imbricate.

Androecium Stamens 4, didynamous, epipetalous on corolla tube; small posterior staminode represents fifth stamen; two fertile stamens and remaining three staminodes are present in *Martynia*; anthers dithecous, introrse and dehisce longitudinally.

Gynoecium Bicarpellary, syncarpous, superior; ovary bilocular or becomes tetralocular due to the formation of false septa; one to many ovules in each locule; axile placentation; ovary sometimes unilocular with two parietal placentae in *Martynia*; style one, slender; stigma bilobed; at the base of ovary is present a nectar-secreting disc.

Fruit and Seed Fruit capsule or sometimes nut, commonly spiny; two long horns are present in *Martynia*. Seeds are non-endospermic or with a thin endosperm possessing a straight embryo.

General Floral Formula EBr, ⊕, ⚥, $K_{(5)}$, $\overset{\frown}{C_{(5)}, A}_{4+1\ \text{staminode}}$, $G_{(2)}$.

21.81.6 Economic Importance

- "Sesame oil", a well-known edible oil, is extracted from the seeds of *Sesamum indicum* (Sesame or Til; Fig. 21.94). It is widely used in confectionary, in the manufacture of several cosmetics, insecticides and soaps and also for making margarine. Hindus use the seeds in

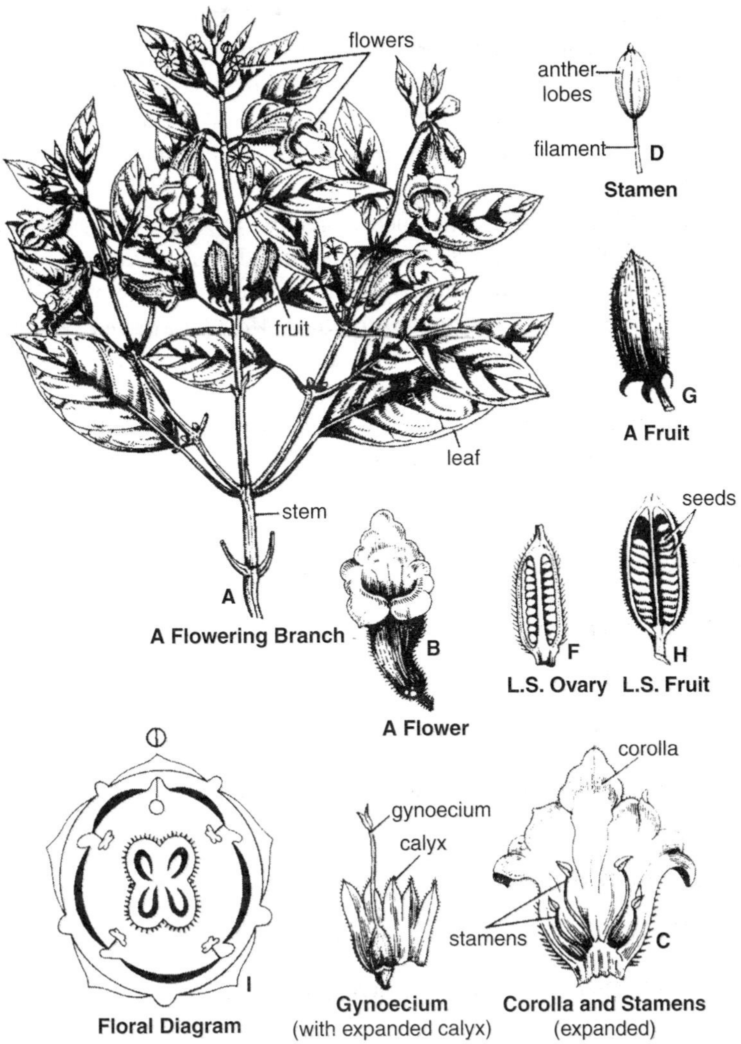

Fig. 21.94 *Sesamum indicum* L.

many of their religious ceremonies. Seeds are widely used as nourishing food and also as flavouring agents. Oil is also extracted from *Sesamum angustifolium* and used for similar purposes.

- Leaves of *Pedalium murex* and *Ceratotheca sesamoides* are used as vegetables.

21.81.7 Systematics and Phylogeny (Affinities)

Family Pedaliaceae has been placed differently under order Personales by Bentham and Hooker, Tubiflorae by Engler and Prantl, Scrophulariales by Takhtajan (1980) and Cronquist (1981) and under order Bignoniales by Hutchinson (1973) and Thorne (1983). Bentham and Hooker (1862) divided family Pedaliaceae into four tribes namely Martynieae, Pedalieae, Sesameae and Pretreae. Many botanists treat the tribe Martynieae as a separate family, Martyniaceae, due to the presence of characters such as unilocular ovary, parietal placentation and horned capsules. Members of Pedaliaceae resemble quite closely to Bignoniaceae. Some taxonomists also treat *Martynia* and other related genera in family Bignoniaceae.

As far as the chemical features are concerned, saponins are lacking in all members of family except *Sesamum indicum*. Tannins are absent in *S. indicum* and *Pedalium murex* while they are doubtfully present in *Sesamum laciniatum* and *Martynia annua*. Steroids are present in *S. indicum* and *S. laciniatum* while they are absent in *Pedalium murex* and *Martynia annua*. Leucoanthocyanins are present in *Sesamum laciniatum* and *Martynia annua* while they are absent in *Sesamum indicum* and *Pedalium murex*. All these facts indicate that *Martynia annua* resembles other taxa of Pedaliaceae, hence its inclusion in the family seems correct. Das *et al.* (1985), however, opined to treat *Martynia* in an independent family Martyniaceae on the basis of their studies of the distributional pattern of phenolic acids.

21.82 ACANTHACEAE (ACANTHUS FAMILY)

21.82.1 Systematic Position

Gamopetalae, Bicarpellatae, Personales.

21.82.2 Field Recognition

Herbs, shrubs or trees; leaves opposite decussate; flowers zygomorphic; corolla bilipped (2/3); seeds with jaculators.

21.82.3 Selected Indian Genera

Acanthus, Adhatoda, Barleria, Eranthemum, Justicia, Peristrophe, Ruellia, Thunbergia.

21.82.4 Size, Distribution and General Information

A family of over 250 genera and 2500 species, Acanthaceae are distributed mainly in tropics, but also in Mediterranean region, Australia and USA. Central America, Brazil, Africa and Indo-Malaysia are the 4 chief centres of distribution. About 70 genera and over 340 species have been reported from India. Some of the largely represented genera along with the approximate number of their worldover reported species *are Justicia* (300), *Strobilanthes* (250), *Barleria* (230), *Thunbergia* (200), *Aphelandra* (200), *Decliptera* (150), *Blepharis* (100), *Staurogyne* (80), *Mendoncia* (60), *Crossandra* (60), and *Acanthus* (50).

21.82.5 Description of the Family

General Habit Mostly perennial herbs or shrubs, rarely trees; some are twiners (*Thunbergia, Mendoncia*), or xerophytes (*Barleria, Blepharis*), or halophytes (*Acanthus ilicifolius*), or spiny shrubs (*Acanthus montanus*).

Leaf Simple, opposite decussate, exstipulate, entire, rarely pinnately lobed with small spines on the margins (*Acanthus mollis*); cystoliths, appearing as protuberances or streaks, are usually common on stems and leaves.

Inflorescence Usually a dichasial cyme, frequently condensed in the leaf axils; sometimes racemose spike (*Adhatoda vasica*), or flowers solitary axillary (*Thunbergia*).

Flower Usually bracteate, bracteolate, pedicellate or subsessile, complete, hermaphrodite, zygomorphic, hypogynous; bracteoles are often large, coloured and more or less enclosing the flower; in *Peristrophe* (Fig. 21.95) out of the 2 bracts the posterior one is larger than the anterior one, and 4 laterally placed bracteoles cover the calyx.

Calyx Usually 5 united (*Justicia*) or free (*Adhatoda*) sepals; valvate (*Justicia*, Fig. 21.96) or quincuncial (*Adhatoda*); sometimes sepals are 4 or rarely 3; rarely the calyx is much reduced and divided into teeth-like structures (*Thunbergia*).

Corolla Usually 5 lobed, gamopetalous, imbricate or twisted; bilipped (2/3) or bilabiate personate, of which the upper lip is bifid or bilobed and the lower lip is trilobed; sometimes the upper lip is not developed (*Acanthus*); corolla is tubular in *Thunbergia* and *Ruellia*.

Androecium Usually 4 didynamous stamens (*Ruellia*, *Thunbergia*), or only 2 stamens (*Adhatoda*, *Justicia*, *Peristrophe*), epipetalous, usually exserted; rarely 5 stamens (*Pentstemonacanthus*); frequently 1–3 staminodes present; anthers bilobed with often one smaller lobe than the other, connective often long (Fig. 21.95); introrse, longitudinal dehiscence.

Gynoecium Bicarpellary, syncarpous; carpels median; ovary superior, bilocular, one to many anatropous ovules in each locule, axile placentation; style narrow and long; stigmas 2, of which the anterior one is often larger; usually with a hypogynous, nectar-secreting disc.

Fruit and Seeds Fruit usually a capsule, loculicidal to the very base; rarely a drupe (*Mendoncia*). Seeds 1 to many, of which the funiculus develops into a hook-like retinaculum or *"jaculator"*; often non-endospermic.

Pollination Flowers are suited for insect pollination because of coloured bilabiate corolla and abundant nectar in hypogynous disc. Protandry favours cross-pollination.

General Floral Formula Br, Brl, Φ, ⚥, $K_{(5) \text{ or } 5}$, $\widehat{C_{(2/3)}, A_{2 \text{ or } 4}}$, $G_{(\underline{2})}$.

21.82.6 Economic Importance

- Ornamental plants include *Aphelandra aurantiaca*, *Barleria cristata*, *B. gibsoni*, *B. lupulina*, *B. montana*, *Beloperone guttata* (shrimp plant), *Crossandra infundibuliformis*, *Eranthemum bicolor*, *E. nervosum*, *Fittonia gigantea*, *Jacobinia carnea*, *J. tinctoria*, *Justicia gendarussa*, *Ruellia brittoniana*, *R. tuberosa*, *Strobilanthes dyerianus*, *S. isophyllus*, *Thunbergia coccinia*, *T. erecta*, *T. fragrans* and *T. grandiflora*.
- Plants of medicinal value include (*i*) *Adhatoda vasica* (Vasaka or Basak)–leaves are useful in cough, asthma and bronchitis, (ii) *Asteracantha longifolia*–seeds and roots are used against cough, (iii) *Barleria prionitis*–leaves and roots are used against bronchitis and cough, (iv) *Barleria cristata*–seeds are used as an antidote for snakebite, (v) *Hygrophila spinosa*-leaves and roots are used against jaundice and rheumatism, (vi) *Peristrophe*

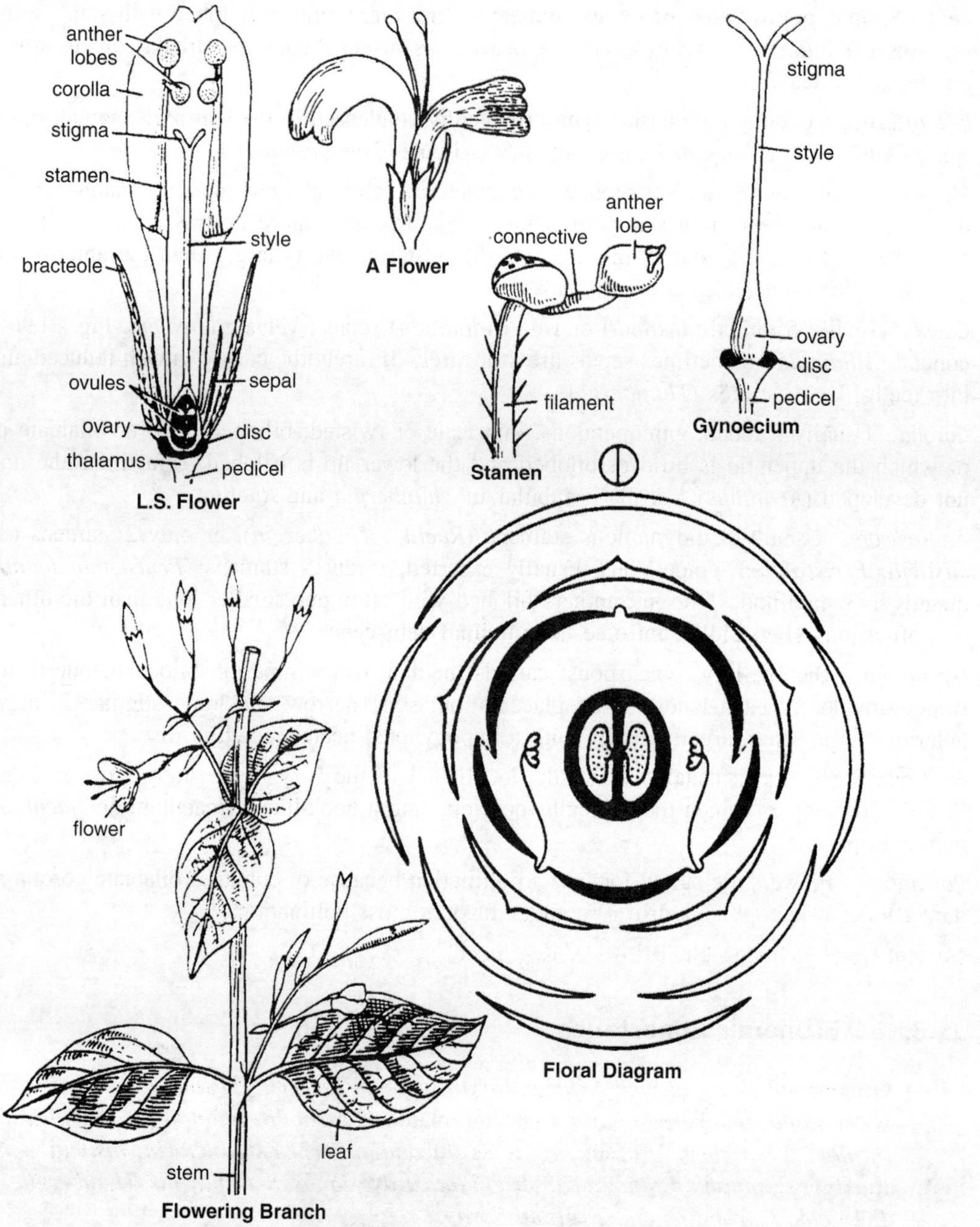

Fig. 21.95 *Peristrophe bicalyculata* Nees.

bicalyculata—leaves and roots are used against snakebite, (vii) *Ruellia prostrata*—leaf juice is used in earache.

- A blue dye is obtained from the leaves of *Strobilanthes cusia*.

21.82.7 Systematics and Phylogeny (Affinities)

Acanthaceae is varyingly placed under Personales (Bentham and Hooker, and Hutchinson), Tubiflorae (Engler and Prantl) and Scrophulariales (Takhtajan, Cronquist). It is usually divided into 4 subfamilies (Nelsonioideae, Mendoncioideae, Thunbergioideae and Acanthoideae).

Subfamily Nelsonioideae is very close to family Scrophulariaceae, and its genera are included in that family by several workers. Subfamilies Mendoncioideae and Thunbergioideae are intermediate between Bignoniaceae and Acanthaceae and are usually considered as independent families Acanthaceae is considered to have been derived from Scrophulariaceae.

21.82.8 Description of a Common Plant

Justicia gendarussa Burm. (Fig. 21.96)

Habit: A perennial herb. *Leaf:* Simple, stipulate, opposite decussate, subsessile, lanceolate, obtuse, unicostate reticulate. *Inflorescence:* Dichasial cyme arranged in racemose manner. *Flower:* Bracteate, bracteolate, pedicellate, complete, zygomorphic, pentamerous, hypogynous. *Calyx:* 5 sepals, united, valvate. *Corolla:* 5 petals, united; 2/3 bilabiate personate, of which posterior lip is bilobed while anterior lip is trilobed; valvate. *Androecium:* 2 stamens, polyandrous, epipetalous, anther lobes situated at unequal heights; dithecous, basifixed, introrse. *Gynoecium:* Same as in description of the family. *Fruit:* Capsule.

Floral Formula: Br, Brl, ⊕, ⚥, $K_{(5)}$, $\overbrace{C_{(2/3)}, A_2}$, $G_{(\underline{2})}$.

21.83 LAMIALES

21.83.1 General Characteristics

1. Mostly herbs, some are shrubs and only rarely trees.
2. Leaves are generally simple, opposite or rarely whorled, and exstipulate.
3. Flowers mostly zygomorphic and pentamerous; bilabiate in Labiatae.
4. Gynoecium bicarpellary, syncarpous, superior; generally deeply 4-loobed.
5. Placentation usually axile.
6. Style gynobasic.
7. Fruit schizocarpic.
8. Seeds with scanty or no endosperm and straight embryo.

Bentham and Hooker included 4 families (Labiatae, Verbenaceae, Selaginaceae and Myoporaceae) under Lamiales. Engler and Diels, however, included all these families, except Selaginaceae under order Tubiflorae. Actually, it was Bessey who separated Labiatae and Verbenaceae as of a distinct order, the Lamiales, on the basis of corolla zygomorphy and gynoecial characters.

Only Labiatae and Verbenaceae of Lamiales have been discussed in this text.

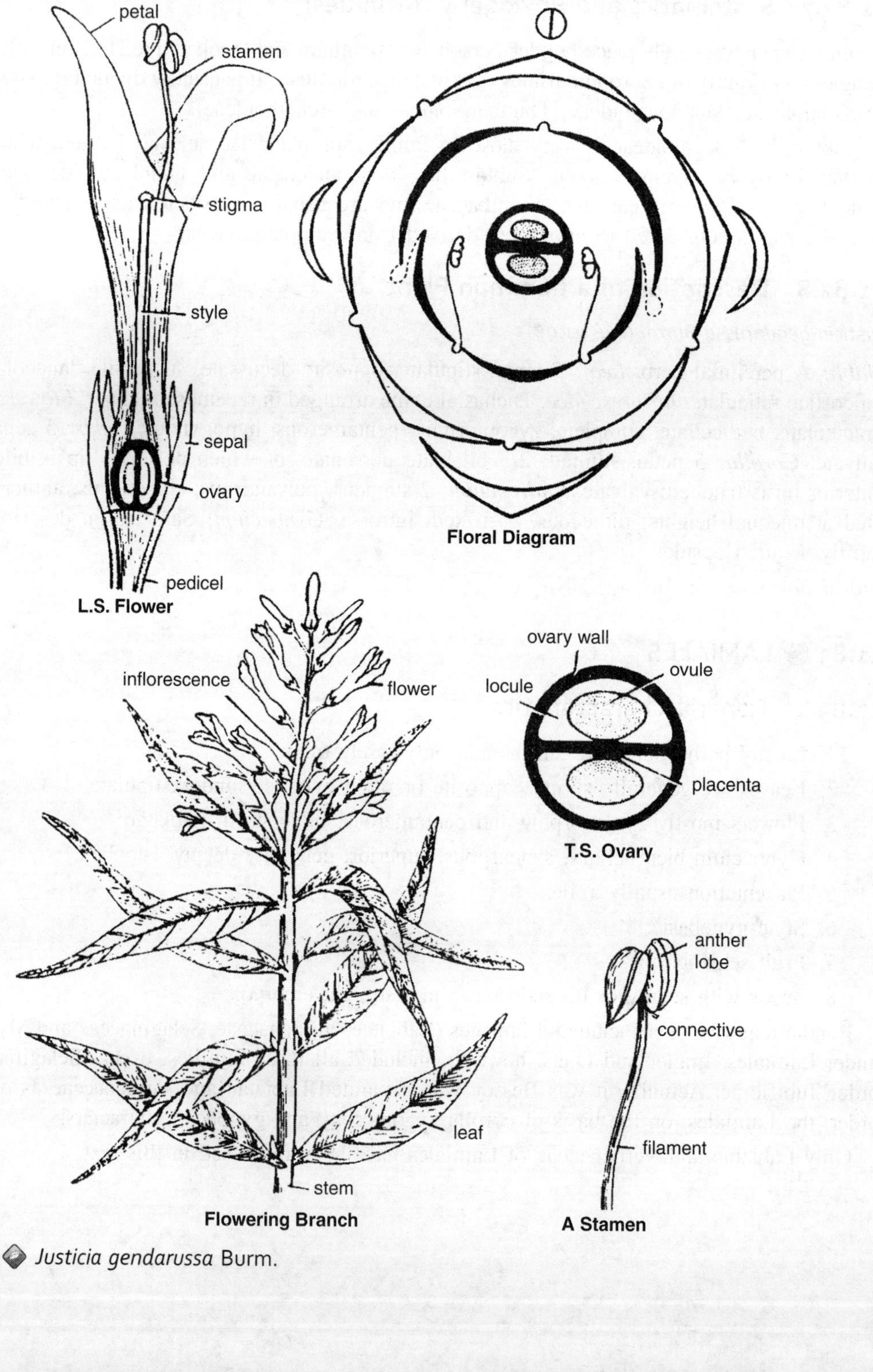

Fig. 21.96 *Justicia gendarussa* Burm.

21.84 LABIATAE OR LAMIACEAE (MINT FAMILY)

21.84.1 Systematic Position

Gamopetalae, Bicarpellatae, Lamiales.

21.84.2 Field Recognition

Aromatic herbs or shrubs with square stems; leaves opposite decussate; inflorescence verticillaster; flowers zygomorphic and pentamerous; ovary deeply 4-lobed; style gynobasic; fruit schizocarpic.

21.84.3 Selected Indian Genera

Coleus, Lamium, Lavandula, Leucas, Mentha, Ocimum, Salvia.

21.84.4 Size, Distribution and General Information

A family of about 200 genera and 3500 species, Labiatae are cosmopolitan in their distribution and their chief centre is the Mediterranean region. About 65 genera and over 400 species have been reported from India, mainly distributed in North Western India and South India. Some largely represented genera along with the approximate number of their worldover reported species and common names of some of them are *Salvia* (500–700, sage), *Hyptis* (400), *Thymus* (300–400, thyme), *Teucrium* (300, germander), *Stachys* (300), *Scutellaria* (200, skullcap), *Coleus* (200, Jacob's coat), *Ocimum* (150, basil), *Nepeta* (150, catnip), *Lavandula* (28, lavender), *Mentha* (25, mint) and *Rosmarinus* (3, rosemary).

21.84.5 Description of the Family

General Habit Usually perennial or annual terrestrial herbs with square stems, sometimes shrubs (*Orthosiphon*), rarely trees (*Hyptis*, *Leucosceptrium*), or vines (*Scutellaria*); plant parts usually covered with glandular hairs; usually with aromatic oils.

Stem Usually quadrangular and covered with glandular hairs.

Leaf Usually simple, opposite, exstipulate; often hairy and with epidermal glands secreting volatile oils; sometimes whorled (*Dysophyla*); rarely pinnately dissected or compound.

Inflorescence Often cymose, the cymes at the nodes condensed into a false whorl or *verticillaster*; rarely flowers are in simple racemes (*Scutellaria*), or solitary and axillary; very rarely in heads (*Hyptis*).

Flower Usually bracteate, bracteolate, complete, hermaphrodite, zygomorphic, hypogynous; rarely nearly actinomorphic (some species of *Mentha*).

Calyx 5 sepals, gamosepalous or united into a funnel-shaped or campanulate tube; tube often ribbed; persistent in fruit; valvate or imbricate; usually 2-lipped (1/4 in *Ocimum* and 3/2 in *Salvia*).

Corolla 5 united petals; 2-lipped (4/1 in *Ocimum* i.e. 4 in posterior lip and 1 in anterior lip, or 2/3 in *Salvia*, *Leucas*); corolla is usually differentiated into a tube and a limb; valvate (*Ocimum*), imbricate (*Salvia*) or contorted.

Androecium Usually 4 stamens (*Ocimum*), didynamous, epipetalous; sometimes only 2 stamens (*Salvia*), in each of which the fertile anther lobe is attached at the anterior end of very long connective while at the posterior end is present the sterile anther lobe (Fig. 21.97); filament is usually free and rarely connate (*Coleus*); dithecous, dorsifixed, introrse; dehiscence longitudinal; usually the missing stamens are present in the form of staminodes.

Gynoecium Bicarpellary, syncarpous; ovary superior, deeply 4-lobed; bilocular when young but becomes quadrilocular due to the formation of false septum; 2 ovules in each locule in bilocular condition, but in quadrilocular condition one anatropous ovule is present in each locule; axile placentation; style gynobasic; stigma bifid; a 4-lobed, hypogynous, nectar-secreting disc is present.

Fruit and Seeds Fruit schizocarp of 1 to 4 nutlets, each nutlet containing one seed; rarely a drupe. Seeds with scanty or no endosperm and straight embryo.

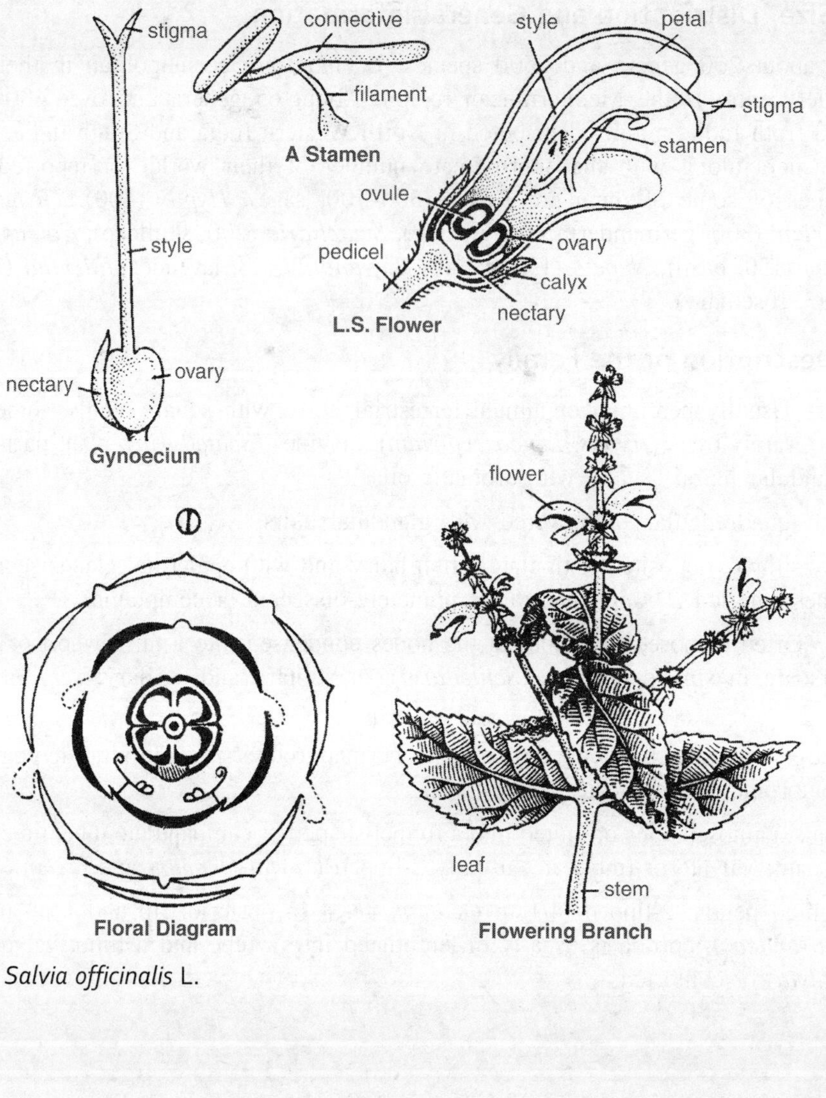

Fig. 21.97 *Salvia officinalis* L.

Pollination Pollination is entomophilous. Coloured bracts (*Salvia*), corolla and nectar in several other genera attract the insects. Usually the lower lip of the corolla forms the platform for the visiting insects. The long connective in *Salvia* also proves helpful in shedding pollens from the fertile anther lobe. Cross-pollination is effected because the flowers are usually protandrous. Long-tubed flowers of *Monarda* and several other genera are pollinated by butterflies and moths.

General Floral Formula Br, Brl, Φ. ⚥, $K_{(5) \text{ or } (1/4) \text{ or } (3/2)}$, $\overline{C_{(4/1) \text{ or } (2/3)},\ A_{2 \text{ or } 4 \text{ or } 2+2}}$, $G_{(\underline{2})}$.

21.84.6 Economic Importance

- *Garden Ornamentals:* *Coleus blumei*, *C. thyrsoideus*, *Lavandula officinalis*, *Ocimum basilicum* (sweet basil), *Rosmarinus officinalis* (rosemary), *Salvia coccinea*, *S. officinalis* (garden sage), *S. splendens* (scarlet sage), and *Thymus vulgaris* (thyme).
- *Volatile Oils:*
  - (i) Basil oil, obtained from *Ocimum basilicum* (sweet basil), *O. canum* (Ram Tulsi) and *O. sanctum* (holy basil or Tulsi), is used in variety of medicines and perfumery.
  - (ii) Lavender oil (*Lavandula latifolia*, *L. spica*, *L. vera*) is used in perfumery and cosmetics.
  - (iii) Mint oil (*Mentha arvensis*, *M. longifolia*, *M. spicata*) is used in nausia and vomiting.
  - (iv) Origanum oil (*Origanum vulgare*) is used in cosmetics and soap industry.
  - (v) Patchouli oil (*Pogostemon benghalense*, *P. cablin*) is used in perfumery.
  - (vi) Pennyroyal oil (*Mentha pulegium*) is used in preparing synthetic menthol.
  - (vii) Peppermint oil (*Mentha piperata*) is used in pharmacy and confectionery.
  - (viii) Perilla oil (*Perilla frutescens*) is used in paints, varnishes and printing ink.
  - (ix) Rosemary oil (*Rosmarinus officinalis*) is used in medicines and perfumery.
  - (x) Sage oil (*Salvia splendens*) is used in soap industry.
  - (xi) Sweet Majoran oil (*Majorana hortensis*) is used in perfumery and liquor industry.
- *Medicinal Plants:* (i) *Ocimum sanctum* (Tulsi) leaves are used for cough, cold and fever, (ii) *Mentha spicata* (Podina) leaves, used for preparing 'Chutney', help in indigestion and rheumatism, (iii) *Thymus vulgaris* leaves are effective against hookworms, and (iv) *Mentha longifolia* leaves are used in rheumatic pain and some heart diseases.
- Camphor is obtained from the oil of *Ocimum kilimandscharicum.*
- Root tubers of some species of *Coleus* and *Ocimum* are edible.

21.84.7 Systematics and Phylogeny (Affinities)

Labiatae has been placed under order Lamiales in majority of the systems of classification. Briquet in Engler and Prantl's *Pflanzenfamilien* divided Labiatae into 8 subfamilies (Ajugoideae, Prostantheroideae, Prasioideae, Scutellarioideae, Lavanduloideae, Lamioideae, Ocimoideae and Catopherioideae).

Labiatae is closely allied to Verbenaceae as well as to Boraginaceae. In some minor characters it also shows affinities with Scrophulariaceae. Hutchinson opined that order Lamiales, to which Labiatae belongs, is the most highly evolved order among Dicotyledoneae.

21.84.8 Description of a Common Plant

Ocimum basilicum (Sweet Basil or Ban Tulsi, Fig. 21.98)

Habit: Cultivated aromatic herb. *Stem:* Herbaceous, erect, hairy, quadrangular. *Leaf:* Simple, opposite decussate, petiolate, exstipulate, ovate, serrate, hairy, unicostate reticulate. *Inflorescence:* Verticillaster. *Flower:* Bracteate, pedicellate, complete, hermaphrodite, hypogynous, zygomorphic, pentamerous. *Calyx:* 5 sepals, fused, 1/4 bilabiate, valvate. *Corolla:* 5 united petals, 4/1 bilabiate, valvate. *Androecium:* 4 stamens, epipetalous, didynamous, posterior stamen absent; anthers dithecous, dorsifixed, introrse. *Gynoecium:* Same as in general description of the family. *Fruit:* Schizocarpic (carcerulus), consisting of 4 nutlets.

Floral Formula: Br, Φ, ⚥, $K_{(1/4)}$, $\overbrace{C_{(4/1)}\ A_{2+2}}$, $G_{(\underline{2})}$.

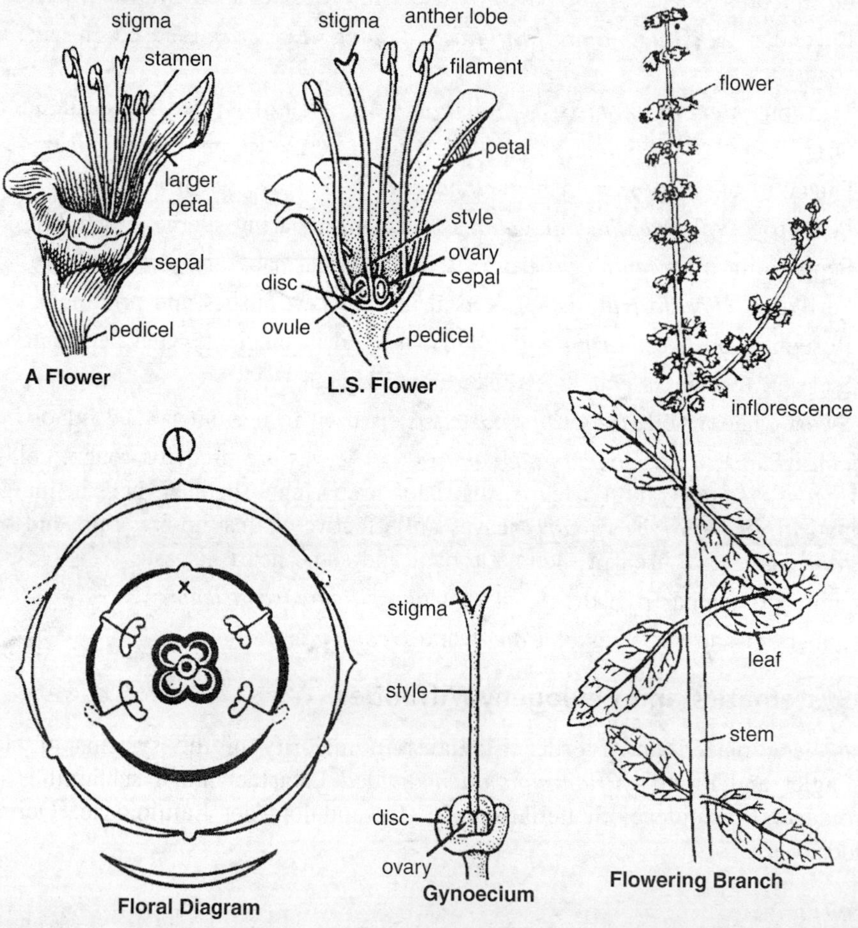

Fig. 21.98 *Ocimum basilicum* L.

21.85 VERBENACEAE (VERBENA FAMILY)

21.85.1 Systematic Position

Gamopetalae, Bicarpellatae, Lamiales.

21.85.2 Field Recognition

Herbs, shrubs, or small trees; stem usually quadrangular; leaves usually opposite; flowers zygomorphic, pentamerous; ovary not lobed; style terminal.

21.85.3 Selected Indian Genera

Clerodendrum, Duranta, Lantana, Tectona, Verbena, Vitex.

21.85.4 Size, Distribution and General Information

A family of about 99 genera and 3151 species (Radford, 1986), distributed chiefly in tropics and subtropics. About 22 genera and over 30 species have been reported from India. Some of the genera along with their number of approximately reported species and common names of some of them are *Clerodendrum* (400, glory-bower), *Verbena* (250, vervain), *Vitex* (250, chaste tree), *Lippia* (220, frog-fruit), *Lantana* (150), *Callicarpa* (140, beauty-berry), *Stachytarpheta* (100) and *Tectona* (3, teak).

21.85.5 Description of the Family

General Habit Herbs, shrubs (*Lantana*) or trees (*Tectona*); often lianas (*Clerodendrum, Vitex*), sometimes of a xerophytic nature (several species of *Verbena*); a few are mangrove shrubs (*Avicennia*).

Stem Herbaceous or woody, twining or sprawling in climbers; often quadrangular.

Leaf Usually simple, sometimes pinnately (*Peronema*) or palmately (*Vitex*) compound; usually opposite, rarely alternate or whorled; exstipulate; highly reduced in xerophytic species (*Verbena*).

Inflorescence Racemose raceme (*Duranta*), or umbel (*Lantana*) or a spike; often with an involucre of coloured bracts; or consisting of dichotomous cymes, or panicled cymes (*Tectona*).

Flower Bracteate, often bracteolate (*Duranta, Lantana*), complete, hermaphrodite, zygomorphic (rarely actinomorphic, e.g. *Physopsis*), pentamerous, hypogynous.

Calyx Usually 5 united sepals, persistent; usually lobed or toothed; valvate; rarely the sepals are 4, 6 or 8; sometimes coloured.

Corolla Usually 5 united petals, or as many petals as sepals; often bilipped with a narrow tube; rarely campanulate; typically lobed with the lobes generally unequal, usually imbricate.

Androecium Usually 4 stamens, didynamous, epipetalous, alternate with the corolla lobes; anthers dithecous, introrse; longitudinally dehiscent; posterior stamen is generally missing or present in the form of staminode; only 2 fertile stamens and 3 staminodes are present in *Oxera*; in *Tectona* all the 5 stamens are fertile.

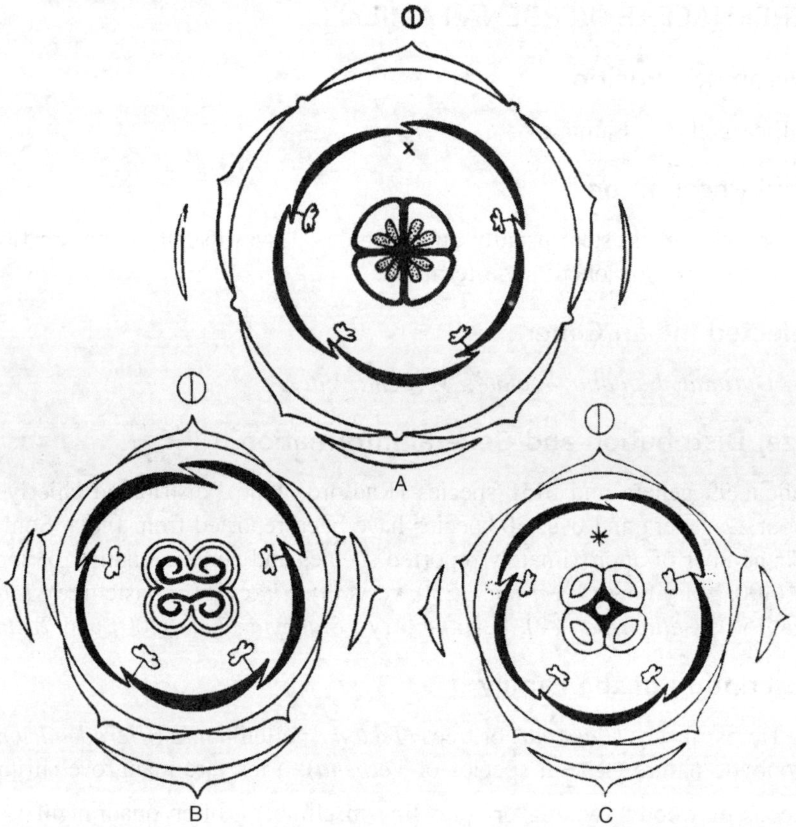

Fig. 21.99A–C Verbenaceae: Floral diagrams–A: *Duranta plumiri*; B: *Clerodendrum indicum,* and C: *Verbena.*

Gynoecium Usually bicarpellary, syncarpous; ovary superior; originally bilocular but divided into 4 locules by the formation of false septum in each locule; 2 ovules in each carpel i.e. 1 ovule in each locule after septation; sometimes 2 ovules in each locule (Fig. 21.99A); axile placentation; ovules anatropous to orthotropous; style terminal and only rarely sunk in between the lobes of the ovary; stigma lobes usually as many as carpels; rarely carpels are 4 (*Duranta*) or 5 (*Geunsia*); ovary is unilocular with free-central placentation in *Avicennia*; in *Clerodendrum* and *Verbena* the ovary is unilocular with parietal placentation when young but on maturity the placentae unite at the centre and the placentation becomes axile (Fig. 21.99B, C).

Fruit and Seeds Fruit is generally a drupe, and rarely a capsule, berry, or schizocarp (nutlets). Seeds non-endospermic with straight embryo.

Pollination and Dispersal Pollination is mainly by butterflies and bees, and seeds are dispersed mainly by birds or animals.

General Floral Formula Br, Brl, ϕ, ⚥, $K_{(5)}$, $\overparen{C_{(5)}, A_{2+2}}$, $G_{\underline{(2)}}$.

21.85.6 Economic Importance

- *Timber-yielding Plants:* Teak (*Tectona grandis*), one of the most famous timber-yielding plants of the world, belongs to Verbenaceae. It is used for high quality furnitures, heavy constructions, ship building, musical instruments and several other similar purposes. Some other timber-yielding plants of Verbenaceae are *Gmelina arborea*, *Premna benghalensis* and *Vitex altissima*.
- Common garden ornamentals include *Aloysia triphylla*, *Caryopteris incana*, *Clerodendrum bungei*, *C. fragrans*, *C. interne*, *C. siphonanthus*, *C. splendens*, *C. thomsoniae*, *C. trichotomum*, *Duranta repens* syn. *D. plumieri* (golden dewdrop), *Holmskioldia sanguinea*, *Lantana camara*, *Petrea volubilis* (Queen's wreath), and several species of *Verbena* (Vervain) such as *V. bipinnatifida* and *V. hybrida*.
- Verbena oil, used in perfumery, is obtained from the leaves of *Aloysia triphylla*.

21.85.7 Systematics and Phylogeny (Affinities)

Verbenaceae was placed under Lamiales by majority of the workers including Bentham and Hooker, Takhtajan, Cronquist, and Thorne. However, Hutchinson placed it under order Verbenales. Briquet divided Verbenaceae into 7 tribes (Stilbeae, Verbeneae, Chiloantheae, Viliceae, Caryopterieae, Symphoremeae and Avicenneae).

Verbenaceae is closely allied to Labiatae, and to some extent to Boraginaceae and Scrophulariaceae. However, Hutchinson opined that Verbenaceae and Labiatae are not at all related.

21.85.8 Description of a Common Plant

Lantana camara Linn. (Fig. 21.100)

Habit: A small shrub. *Leaf:* Simple, exstipulate, opposite decussate, petiolate, ovate, crenate or serrate, acute, unicostate reticulate. *Inflorescence:* Umbel or a compound spike with flat top. *Flower:* Bracteate, bracteolate, hermaphrodite, zygomorphic, pentamerous, hypogynous. *Calyx:* 5 sepals, united, valvate. *Corolla:* 5 petals, united, quincuncial; corolla bilipped (4/1) in which 1 anterior petal is large and pointed; white, yellow, red or pale-purple. *Androecium:* 4 stamens, didynamous, epipetalous; posterior stamen is generally missing; dithecous, basifixed, introrse. *Gynoecium:* Bicarpellary, syncarpous, superior, bilocular, 1 ovule in each locule, axile placentation; style simple; stigma knob-like. *Fruit:* Drupe.

Floral Formula: Br, Brl, ⊕, ⚥, $K_{(5)}$, $\overparen{C_{(4/1)}, A}_{2+2}$, $G_{(\underline{2})}$.

21.86 PLANTAGINACEAE (PLANTAGO FAMILY)

21.86.1 Systematic Position

Gamopetalae, Bicarpellatae, Ordines Anomali.

Or

Metachlamydeae (Sympetalae), Plantaginales (Engler and Prantl).

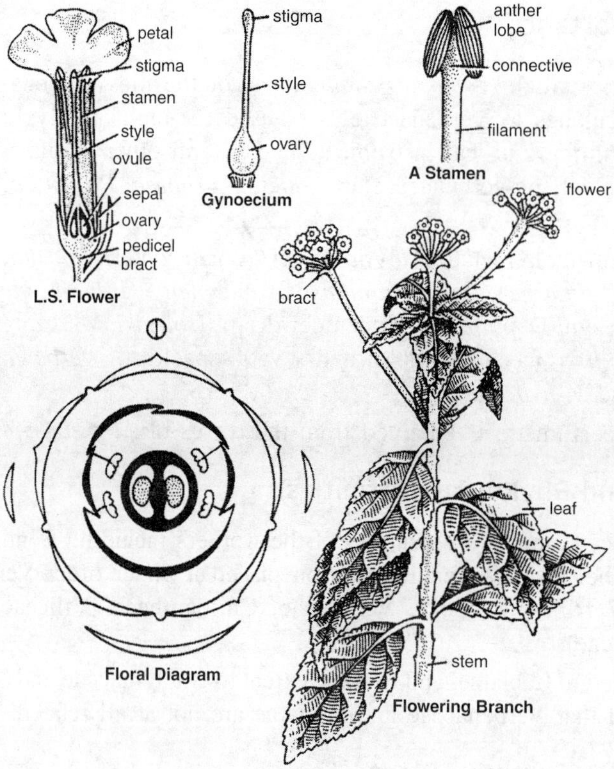

Fig. 21.100 *Lantana camara* Roxb.

Or

Dicotyledones, Herbaceae, Plantaginales (Hutchinson).

Or

Magnoliopsida, Asteridae, Plantaginales (Cronquist, 1981).

21.86.2 Field Recognition

Herbs; leaves in a basal rosette with apparent parallel venation and often sheathing at base; spicate or capitate inflorescence on stout or wiry scapes; flowers usually tetramerous; petals membranous; stamens often exerted.

21.86.3 Common Indian Genus

Plantago.

21.86.4 Size, Distribution and General Information

A family of only 3 genera (*Plantago*, *Littorella* and *Bougueria*) and over 250 species, Plantaginaceae are cosmopolitan in their distribution. *Plantago* (Plantain) with more than 200 species is the largest genus. *Bougueria* is a monotypic Andean genus.

21.86.5 Description of the Family

General Habit Annual or perennial herbs or rarely well-branched subshrubs.

Leaves Basal, simple, alternate or rarely opposite; often sheathing at the base, narrow, parallel-veined with no distinction into stalk and blade; exstipulate; in *Plantago lanceolata* the leaves form a basal rosette (Fig. 21.101), often lanceolate to ovate, entire, 3–5 nerved and sessile.

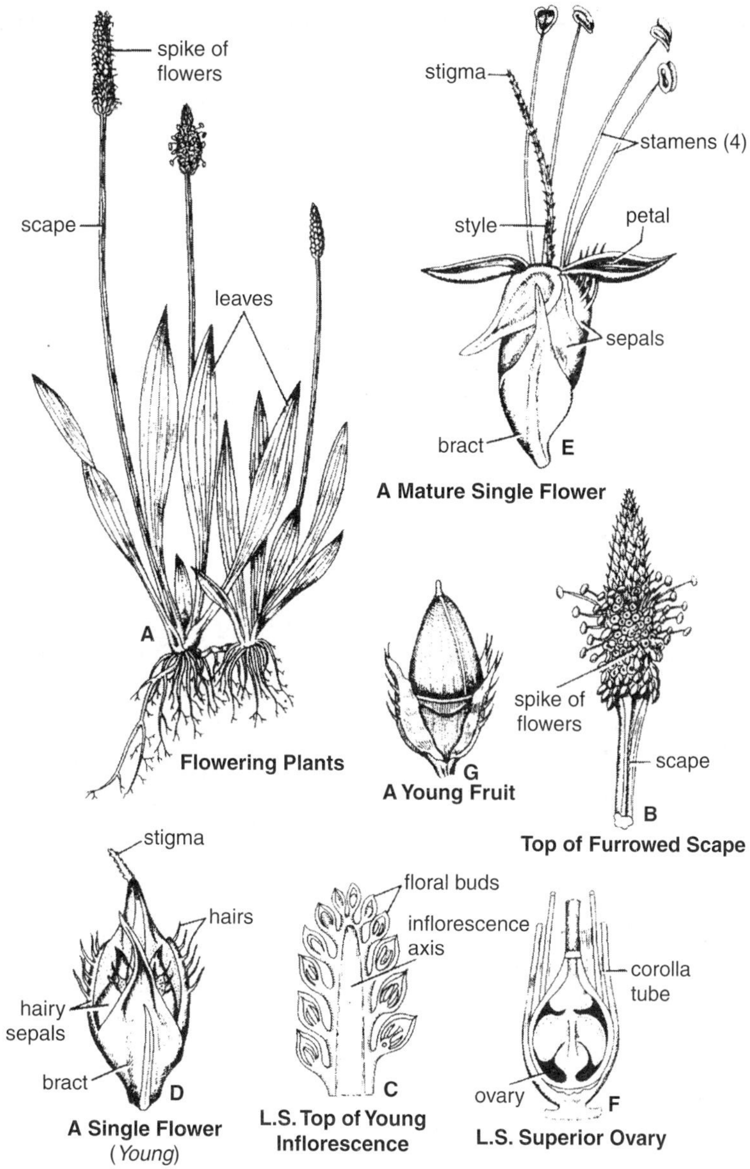

Fig. 21.101 *Plantago lanceolata* L.

Inflorescence Scapose, capitate or spicate present on stout or wiry scapes: scapes usually twice as long as leaves (*Plantago*, Fig. 21.101).

Flowers Usually small, inconspicuous, bracteate, bisexual, actinomorphic, hypogynous, tetramerous

Calyx Sepals 4, gamosepalous, imbricate; sometimes deeply-divided and membranous.

Corolla Petals 4, gamopetalous, imbricate and membranous.

Androecium Stamens 4, rarely 1–2, epipetalous, alternipetalous; anthers dithecous, versatile, large; dehiscing generally longitudinally.

Gynoecium Bicarpellary, syncarpous, ovary superior, 1 to 4 locules with 1 or more ovules in each locule on basal or axile placentation; usually only 2 locules; style simple, filiform and bifid.

Fruit A capsule or bony nutlet (e.g. *Littorella*) surrounded by persistent calyx.

Seeds With fleshy endosperm and small straight embryo.

General Floral Formula Br, Brl, ⊕, ⚥, $K_{(4)}$, $\widehat{C_{(4)}, A_4}$, $G_{\underline{(2)}}$.

21.86.6 Economic Importance

Members of Plantaginaceae are not of much economic importance except of the following:

- Several species of *Plantago* are troublesome garden weeds, specially in lawns.
- Seeds of *Plantago psyllium* are used as a laxative. The mucilaginous seed coats of *P. indica* and *P. ovata* are also used for this purpose.
- Leaves of some species of *Plantago* have occasionally been used as food.

21.86.7 Systematics and Phylogeny (Affinities)

Plantaginaceae is an anomalous taxon and hence of doubtful relationship. Due to this Bentham and Hooker treated this family under Ordines Anomali. Bessey included it in Primulales as allied to and originating from Plumbaginaceae. Hallier discussed it under Tubiflorae as related to Scrophulariaceae. Hickey and King (1988) also regarded Plantaginaceae as derived from a pentamerous type of flower as that of *Veronica* of Scrophulariaceae, and due to this the two families are considered as being allied. Both Lawrence and Rendle retained it under order Plantaginales as also by Hutchinson and Cronquist. Some taxonomists believe that Plantaginaceae represents "an offshoot from the Polemoniales or perhaps from the Primulales" (Porter 1959).

21.87 MONOCHLAMYDEAE

Instead of free or distinct petals of Polypetalae and completely or partially fused petals of Gamopetalae, members of Monochlamydeae have no petals. Monochlamydeae is directly divided into following 8 series with no cohorts by Bentham and Hooker:

Series 1. Curvembreae, characterised by coiled embryo and generally one ovule. It includes 7 orders (=families), viz. Chenopodiaceae[1], Polygonaceae[1], Nyctaginaceae, Amaranthaceae[1], Illeceraceae, Phytolacaceae and Batideae.

[1] Discussed in this text.

Series 2. Multiovulatae-aquaticae, characterised by their aquatic or immersed habit and presence of syncarpous pistil and many ovules. It includes Podostemaceae.

Series 3. Multiovulate-terrestris, characterised by their terrestrial habit and presence of syncarpous pistil and many ovules. It includes families such as Nepenthaceae, Aristolochiaceae[1], and Cytinaceae.

Series 4. Microembryeae, characterised by the presence of very minute embryo in copious endosperm, apocarpous or syncarpous condition, and usually single ovule. It includes Piperaceae[1], Chloranthaceae, Myristicaceae and Monimiaceae.

Series 5. Daphnales, characterised usually by monocarpellary gynoecium, ovary with one carpel and one ovule, bisexual flowers and mostly woody plant. It includes Proteaceae, Lauraceae, Thymeleaceae, Elaeagnaceae and Penaeaceae.

Series 6. Achlamydosporeae, characterised by usually inferior ovary, single locule with 1–3 ovules; seeds without a seed coat; often parasitic habit. It includes Loranthaceae[1], Santalaceae and Balanophoraceae.

Series 7. Unisexuales, characterised by unisexual flowers. It includes 9 families, viz. Euphorbiaceae[1], Urticaceae[1], Moraceae[1], Plantanaceae, Leitneriaceae, Juglandaceae, Balanopsidaceae, Myricaceae and Casuarinaceae[1].

Series 8. Ordines anomali, includes families of uncertain relationship. These bear generally unisexual flowers but their plants are not closely related to others. It includes families like Ceratophyllaceae, Salicaceae[1], Lacistemaceae and Empetraceae.

21.88 CHENOPODIACEAE (GOOSEFOOT FAMILY)

21.88.1 Systematic Position

Monochlamydeae or Apetalae, Curvembryeae.

21.88.2 Field Recognition

Halophytic, xerophytic or succulent, annual or perennial herbs or shrubs; leaves and stem mealy to touch; flowers monochlamydous, small and greenish; placentation basal; fruit urticle.

21.88.3 Selected Indian Genera

Basella, *Beta*, *Chenopodium*, *Kochia*, *Spinacia* and *Suaeda*.

21.88.4 Size, Distribution and General Information

A family of about 102 genera and over 1400 species, Chenopodiaceae are cosmopolitan in distribution, found chiefly in saline and xeric habitats. About 20 genera and over 55 species represent the family in India. Some of the genera along with their common names and approximate number of their worldover reported species are *Atriplex* (*200*, saltbush), *Salsola* (*150*, glassworts), *Chenopodium* (150, goosefoot), *Kochia* (*90*, summer-cypress), *Salicornia* (35, glassworts), *Beta* (6, sugarbeet or Chukander), and *Spinacia* (3, spinach or Palak).

[1] Discussed in this text.

21.88.5 Description of the Family

General Habit Predominantly halophytic (*Atriplex*, *Salsola kali*) or xerophytic; annual or perennial herbs or shrubs, only rarely small trees (*Haloxylon ammodendron*), or sometimes succulent (*Salicornia*); mostly adapted to grow in salty or alkaline soil; plant parts covered by hairs.

Roots Deep-rooted; become thick, fleshy and store large amount of sugar in some (e.g. *Beta vulgaris*).

Stem Jointed, or sometimes becomes fleshy and leafless in succulents (e.g. *Salicornia*); shows abnormal secondary growth by forming interxylary phloem.

Leaf Simple, alternate, exstipulate; rarely opposite (*Haloxylon*, *Salicornia*); fleshy in some (*Suaeda*) or reduced to scales; often covered with hairs which provide a characteristic mealy appearance to the plants.

Inflorescence Usually cymose clusters (*Chenopodium*) arranged in terminal or axillary dense spikes; often dense, dichasial or unilateral cymes.

Flower Bracteate, complete, bisexual, or sometimes unisexual (*Grayia*), and the species may be monoecious (*Sarcobatus*) or dioecious; actinomorphic, hypogynous but rarely perigynous (*Beta*); monochlamydous; very small; greenish; *Atriplex* has male, female and sometimes also bisexual flowers.

Perianth Usually 5 tepals, sometimes varying between 2 to 5; free or connate at the base; persistent in the fruit; quincuncial or imbricate; mostly fleshy, small, green and resemble sepals; tepals differ in shape and size (*Alexandra*, *Corispermum*); petals absent.

Androecium Usually 5 or as many stamens as tepals, polyandrous, opposite the tepals (Fig. 21.102); anthers bent inwards in bud; filaments usually free or subulate; anthers dithecous, introrse, longitudinally dehiscent; only 1 or rarely 2 stamens in *Salicornia*; anthers open by terminal pores in *Basella*.

Gynoecium Usually bi-or tricarpellary, syncarpous, superior or rarely half-inferior (*Beta*); unilocular, with one campylotropous ovule, basal placentation; styles 1 to 3; stigmas 2–3 or rarely more.

Fruit and Seeds Fruit usually a small, round utricle, nut or achene. Seed single in each fruit, lens- or kidney-shaped; with curved or spirally coiled embryo, usually surrounding the endosperm; sometimes non-endospermic

General Floral Formula Br, ⊕, ⚥, P_5, A_5, $G_{\underline{(2 \text{ or } 3)}}$.

21.88.6 Economic Importance

- *Sugar:* Tuberous napiform roots of *Beta vulgaris* (sugarbeet or Chukander) contain about 20% sugar, and are the chief source of sugar in several temperate countries.
- *Edible Plants:* (i) Green leaves of *Atriplex hortensis* (Orache), *Chenopodium album* (Bathua), *C. bonus-henricus* (Good King Henry), *Beta benghalensis* (Gobru Palak) and *Spinacia oleracea* (Spinach or Palak) are used as famous vegetables. (ii) Seeds of

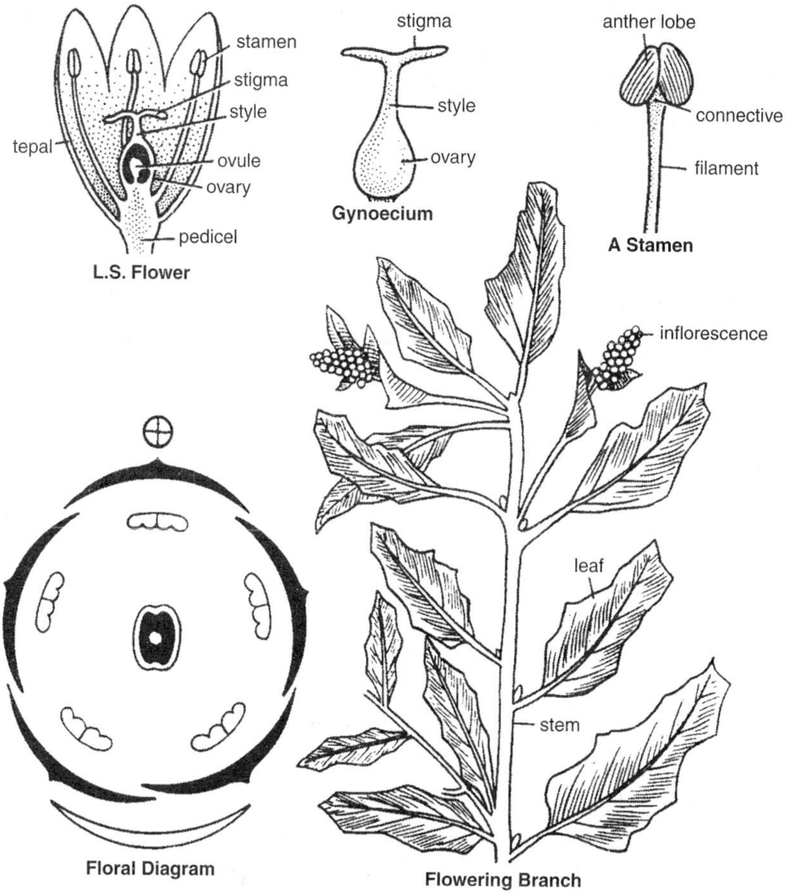

Fig. 21.102 Chenopodium album L.

Chenopodium album and *C. quina* are ground into flour and eaten. (iii) Roots of *Beta vulgaris* are eaten as 'salad'. (iv) Seeds of *Chenopodium quina* are boiled and eaten with rice.

- Fodder for cattles is obtained from *Atriplex hortensis*, *Chenopodium album*, *Kochia indica* and *Salsola foetida*.
- Ornamental plants of the family are *Kochia scoparia* (Burning bush), *Suaeda maritima*, *Chenopodium amaranticolor* and *Atriplex hortensis*.
- *Chenopodium ambrosioides* is used against hookworms.

21.88.7 Systematics and Phylogeny (Affinities)

Treated under series Curvembryeae by Bentham and Hooker, Chenopodiaceae was placed under Centrospermae by Engler and Prantl (1931), under Chenopodiales by Hutchinson (1959) and

Thorne (1983), and under Caryophyllales by Takhtajan (1969) and Cronquist (1981). Bentham and Hooker divided Chenopodiaceae into two subfamilies (Chenopodieae and Baselleae) while Ulbrich (1934) in Engler and Prantl's *Pflanzenfamilien* divided it into 8 sub-families (Polycnemoideae, Betoideae, Chenopodioideae, Corispermoideae, Salicornioideae, Sarcobatoideae, Suaedoideae and Salsoloideae).

Chenopodiaceae is closely allied to Amaranthaceae, from which it differs in not possessing membranous perianth and connate stamens.

21.89 AMARANTHACEAE (AMARANTHUS OR PIGWEED FAMILY)

21.89.1 Systematic Position

Monochlamydeae or Apetalae, Curvembryeae.

21.89.2 Field Recognition

Mostly herbs; dense or congested inflorescence; flowers small with dry scarious bracts; stamens connate, at least at the base; fruit utricle or capsule.

21.89.3 Selected Indian Genera

Achyranthes, *Alternanthera*, *Amaranthus*, *Celosia*, *Digera*, *Gomphrena* and *Pupalia*.

21.89.4 Size, Distribution and General Information

A family of about 65 genera and 900 species, Amaranthaceae are mostly distributed in tropical but also in temperate regions. About 18 genera and over 50 species have been reported from India. Some of the larger genera with the number of their approximately reported species and common names are *Alternanthera* (200, alligator weed), *Gomphrena* (100, globe amaranth), *Iresine* (80, gizzard plant), *Amaranthus* (60, pigweed), and *Celosia* (60, Celosia).

21.89.5 Description of the Family

General Habit Mostly annual or perennial weedy herbs, rarely woody shrubs or climbing shrub (*Deeringia amaranthoides*).

Stem Anomalous secondary growth occurs by the formation of collateral vascular bundles from cambium in the pericycle. In the scattered vascular bundles, the fascicular cambium is functionless.

Leaf Simple, alternate (*Amaranthus*) or opposite decussate (*Achyranthes*, Fig. 21.103), exstipulate, usually entire or nearly so, often thickly covered with hairs.

Inflorescence Dense or congested, axillary or terminal spikes (*Amaranthus*), heads, racemes; or flowers solitary; often spiny because of the firm spiny tips on the inflorescence bracts; usually the basic unit is either a 3-flowered dichasium unit (*Celosia*) or 7 to 11-flowered dichasium units (*Pupalia*).

Flower Bracteate, bracts scaly, membranous, dry or scarious; bracteolate, bracteoles 2, dry, membranous, large and lateral; usually bisexual, or when unisexual (*Amaranthus*, Fig. 21.104B, C) the species are mostly dioecious or polygamodioecious; actinomorphic, pentamerous, hypogynous, incomplete, apetalous, very small.

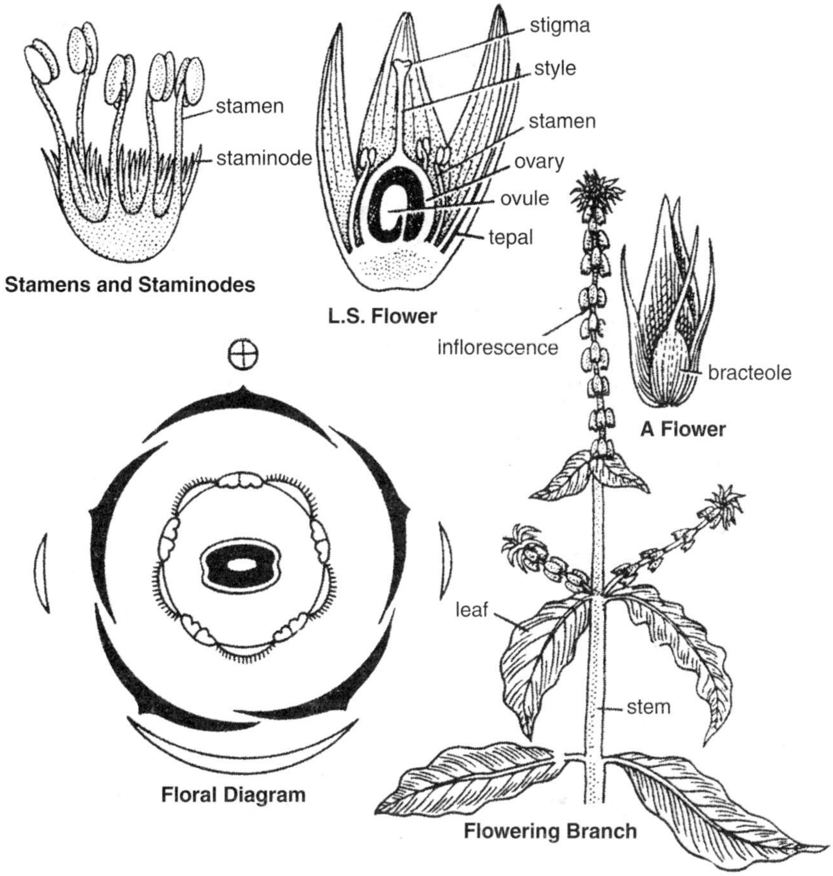

Fig. 21.103 *Achyranthes aspera* L.

Perianth Uniseriate, consisting of 5, free or basally connate tepals; green or sepalloid, dry or membranous; imbricate; only 3 tepals in *Amaranthus viridis*, but 5 free tepals in *A. spinosus* (Fig. 21.104).

Androecium Usually 5 stamens present opposite the tepals; often monadelphous (*Celosia*, *Achyranthes*); separate in *Amaranthus*, often with equal number of staminodes alternating with the anthers (*Celosia*, *Achyranthes*); staminodes are fimbriated in *Achyranthes*; filaments united at the base into a short tube, but in *Amaranthus* the filaments are free; anthers dithecous, dorsifixed (*Achyranthes*) or versatile (*Amaranthus*); introrse; longitudinally dehiscent; only 1 or 2 stamens in *Nothosaerva*.

Gynoecium Bi- to tricarpellary, syncarpous, superior, unilocular, with a single pendulous or campylotropous ovule in the locule, basal placentation; rarely several ovules (*Celosia*, Fig. 21.104A) are present; styles 1 to 3; stigma 1 to 3, capitate.

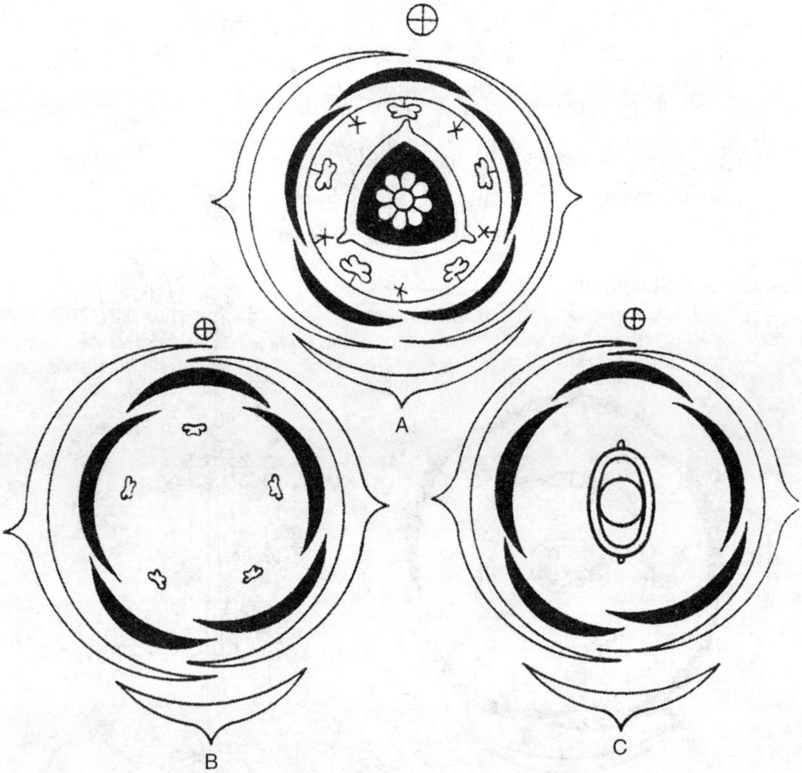

Fig. 21.104A–C Amaranthaceae: Floral diagrams—A: *Celosia*; B and C: *Amaranthus* spinosus, male and female flower, respectively.

Fruit and Seeds Fruit usually a utricle (*Achyranthes*) or nutlet (*Digera*), or a capsule (*Celosia*), or rarely a berry or drupe. Seeds with curved embryo and mealy endosperm.

Pollination and Dispersal Pollination is entomophilous because numerous small flowers are aggregated in dense or congested inflorescence. Dispersal takes place either by wind or by animals.

General Floral Formula Br, Brl, ⊕, ⚥, P_5, $A_{(5+5\ \text{Staminodes})}$, $G_{\underline{(2\ \text{or}\ 3)}}$.

21.89.6 Economic Importance

- Ornamental plants of the family known for their multicoloured foliage and beautiful inflorescence are *Amaranthus caudatus* (Foxtail), *A. tricolor*, *A. salicifolius*, *Celosia cristata* (cocks comb; white, purple or yellowish-red panicles), *Deeringia amaranthoides*, *Gomphrena globosa* (globe amaranth), *Iresine herbtsii* (purplish-red foliage) and *I. lindenii* (red foliage).
- Edible plants of the family, of which the leaves and young shoots are used as vegetable, are *Amaranthus blitum* (Amaranth or Chaulai), *A. caudatus*, *A. hybridus*, *A. spinosus* (Kantili

Chaulai), *A. tricolor* (*Bari* Chaulai), *A. viridis*, *Celosia argentea* and *Digera muricata* (Lehsua). Grains of *Amaranthus caudatus* are roasted and eaten as 'Ramdana'.

- Inflorescence paste of *Achyranthes aspera* (Latjeera or Chirchita) is used as an antidote against snake and scorpion bites.
- Widespread troublesome weeds of this family belong to several genera including *Acnidia*, *Amaranthus*, *Gomphrena* and *Iresine*.

21.89.7 Systematics and Phylogeny (Affinities)

Amaranthaceae was differently placed under series Curvembreae (Bentham and Hooker), Centrospermae (Engler and Prantl), Chenopodiales (Hutchinson, 1959; Thorne, 1983) and Caryophyllales (Takhtajan, 1969; Cronquist, 1981). Bentham and Hooker divided Amaranthaceae into three tribes viz. Celosieae, Amaranteae and Gomphreneae.

The family is closely related to Chenopodiaceae, from which it differs in having dry and membranous perianth, quite conspicuous and scarious bracts and bracteoles, and usually connate stamens. Hutchinson opined that Amaranthaceae evolved from Caryophyllaceous ancestors.

21.90 POLYGONACEAE (BUCKWHEAT OR SMARTWEED FAMILY)

21.90.1 Systematic Position

Monochlamydeae or Apetalae, Curvembryeae.

21.90.2 Field Recognition

Herbs with swollen nodes; ochreate stipule present; tepals petalloid and often biseriate; petals absent; fruit usually lens-shaped or triangular.

21.90.3 Selected Indian Genera

Antigonon, *Coccoloba*, *Fagopyrum*, *Polygonum*, *Rheum*, *Rumex*.

21.90.4 Size, Distribution and General Information

Represented by about 40 genera and 1000 species, Polygonaceae are distributed chiefly in north temperate regions, and a few also in tropical, arctic and Southern Hemisphere. About 10 genera and over 100 species have been reported from India. Some of the larger genera along with their number of approximately reported species are *Polygonum* (300, smartweed), *Eriogonum* (250, false buckwheat), *Rumex* (200, dock), *Coccoloba* (150, seagape), *Rheum* (50, rhubarb), *Chorizanthe* (50), *Fagopyrum* (15) and *Muehlenbeckia* (15).

Polygonum orientale, an ornamental plant, is commonly known as "kiss-me-over-the-garden-gate" (Jones and Luchsinger, 1987).

21.90.5 Description of the Family

General Habit Mostly herbs, some shrubs (*Polygonum hydropiper*) or climbers (*Antigonon leptopus*), and only rarely trees (*Triplaris*, *Coccoloba uvifera*).

Stem Often with swollen nodes; sometimes geniculate i.e. bent like a knee; phylloclades in some species (*Muehlenbeckia platyclada*).

Leaf Usually simple, alternate, stipulate; stipules ochreate i.e. stipules united into a sheath (ochrea) clasping the stem above the leaf base, a diagnostic feature of Polygonaceae; rarely the leaves are opposite or whorled; usually entire but lobed in *Rumex acetosella*; leaves radical in *R. hastatus*; rarely exstipulate (*Koenigia islandica*).

Inflorescence Primarily racemose, but partial inflorescences usually cymose; clusters of panicled raceme (*Rumex*); or racemose forming a terminal panicle (*Polygonum*); or cymose umbels (*Eriogonum*).

Flower Bracteate, pedicellate; usually bisexual but rarely unisexual; when unisexual, the species may be monoecious or dioecious; actinomorphic; small and crowded on the inflorescence axis; usually trimerous but rarely dimerous (*Oxyria digyna*), hypogynous.

Perianth 3 to 6, free or basally connate tepals, often in two whorls (*Rumex*); monochlamydous; often petalloid; in acyclic flowers (*Polygonum glabrum*) 5 tepals are present; imbricate; often persistent, enlarged and becoming membranous in fruit; petals absent.

Androecium 6 to 9 stamens, in two series (rarely less or more), free or united at the base; anthers 2-locular; dehiscence longitudinal; in several genera the outer stamens are introrse while the inner stamens are extrorse; only 5 to 8 stamens (5 introrse and 2 extrorse) in *Polygonum glabrum* (Fig. 21.105).

Gynoecium Usually tricarpellary, rarely bicarpellary (*Oxyria*) or tetracarpellary; syncarpous, superior, unilocular, one orthotropous ovule, basal placentation; style 1; stigmas 2 to 4, fringed in *Rumex*; usually with an annual, nectar-secreting, hypogynous disc is present.

Fruit and Seeds Fruit a lens-shaped, angular, usually triangular achene or nut (*Polygonum*, *Rumex*) with a persistent perianth forming a membranous wing. Seeds with a curved embryo and plenty of endosperm.

Pollination It is with the help of wind in *Rumex* and some species of *Polygonum*, or by insects in *Rheum* and several other species of *Polygonum*.

General Floral Formula Br, ⊕, ⚥, P_{3+3}, A_{6-9}, $G_{(3)}$.

21.90.6 Economic Importance

Except of some medicinal and ornamental value, the family is not of much importance. Some of its economic aspects are undermentioned:

- Common ornamental plants are *Antigonon leptopus* (climber with bright pink or white flowers), *Muehlenbeckia platyclada* syn. *Coccoloba platyclada*, *Polygonum alpinum*, *P. glabrum* (pink-flowers), *P. orientale*, *P. paniculatum* and *P. blebejum* (pink flowers).
- Some of the medicinal plants of family include (i) *Polygonum aviculare*, used against diabetes and rheumatism, (ii) *Rheum emodi* and *R. officinale* roots and rhizome yield the drug 'rhubarb' used as a laxative, purgative and stomach tonic, (iii) *Rumex acetosella* leaves are said

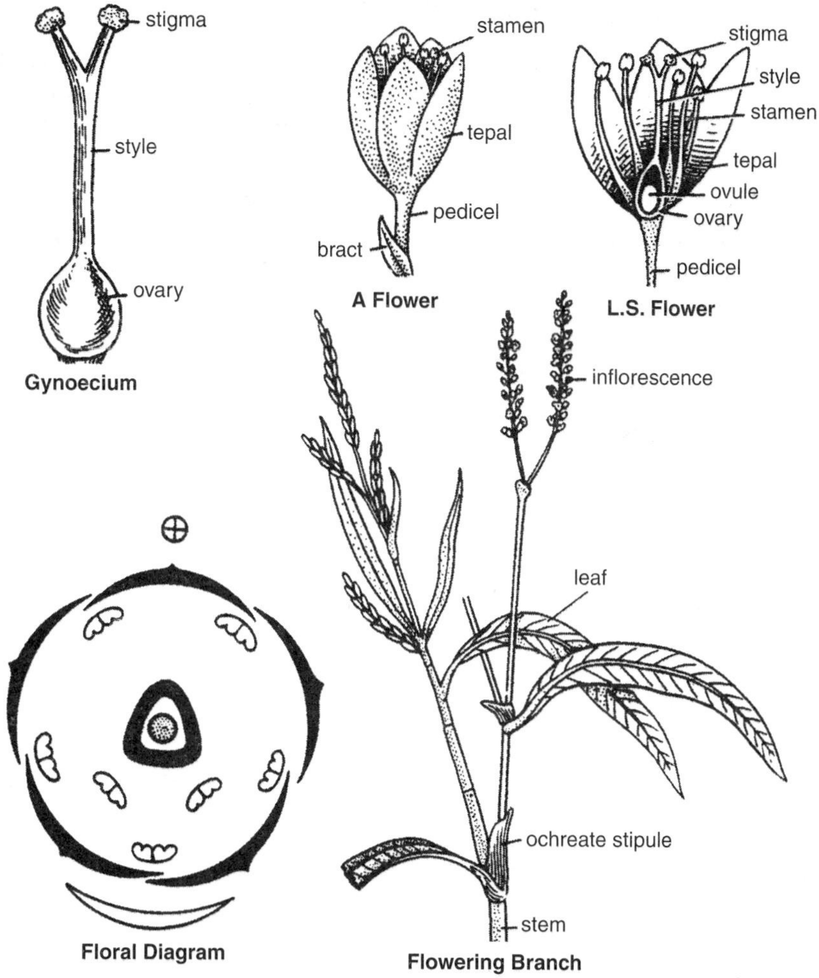

Fig. 21.105 *Polygonum glabrum* Willd.

to be effective in tumours and cancer, and (iv) *Rumex acetosa*, used in bronchial diseases and also as a blood purifier; (v) *Polygonum glabrum* seeds are used in leucorrhoea.

- Leaves of *Oxyria digyna*, *Rumex acetosella* and *R. acetosa* (sorrel) are eaten as vegetable or salad.
- *Fagopyrum esculentum* (buckwheat or 'Kuttu') seeds are edible because they are rich source of protein. Its flowers are very rich in honey. *F. tataricum* (duckwheat) is also used for the similar purposes.
- *Rumex hastatus* leaves are used as condiment.
- *Polygonum tinctorium* is the source of a blue dye.

21.90.7 Systematics and Phylogeny (Affinities)

In most of the systems of classification (Hutchinson, 1959; Takhtajan, 1969; Cronquist, 1981; Thorne, 1983), the family Polygonaceae is placed under order Polygonales. Bentham and Hooker placed it under series Curvembryeae of Monochlamydeae. The family is divided into 3 subfamilies, (i) Rumicoideae (flowers cyclic and endosperm not ruminate), (ii) Polygonoideae (flowers acyclic and endosperm not ruminate), and (iii) Coccoloboideae (flowers acyclic and endosperm ruminate).

Polygonaceae is allied to families such as Amaranthaceae, Chenopodiaceae and Caryophyllaceae. According to Hutchinson Polygonaceae and families of Caryophyllales were derived from a common ancestor which originated from Ranales.

21.91 ARISTOLOCHIACEAE (BIRTHWORT FAMILY)

21.91.1 Systematic Position

Monochlamydeae, Multiovulatae Terrestris.

21.91.2 Field Recognition

Plants terrestrial herbs or shrubs (mainly lianas); leaves alternate, often cordate; several stamens (6–36) showing adonation to style; 6-loculed or more or less inferior ovary; flowers lacking corolla; perianth usually an enlarged and a petalloid gamosepalous calyx often of trumpet-like or bell-like shapes; ovules many.

21.91.3 Selected Indian Genera

Aristolochia, *Asarum*.

21.91.4 Size, Distribution and General Information

A family of 8–10 genera and about 600 species distributed in tropical and temperate America, Asia, Europe and Africa; absent in Australia; *Aristolochia* (500) and *Asarum* (70) are the two larger genera found mostly in tropical and temperate regions.

21.91.5 Description of the Family

General Habit Herbs or shrubs, the latter mostly twining lianas.

Stem Softly woody with broad medullary rays.

Leaf Usually simple, alternate, petiolate, often ovate-cordate and palmately-veined; entire; exstipulate.

Inflorescence Flowers in axillary clusters (*Aristolochia*, Fig. 21.106), solitary or arranged in racemes or cymes.

Flower Bisexual, actinomorphic (*Asarum*) or zygomorphic (*Aristolochia*), epigynous.

Perianth Usually a petaloid gamosepalous calyx, variously 3-lobed or unilateral, often bizarrely coloured; corolla usually absent but sometimes an inner whorl of 3 minute teeth (or vestigial corolla?)

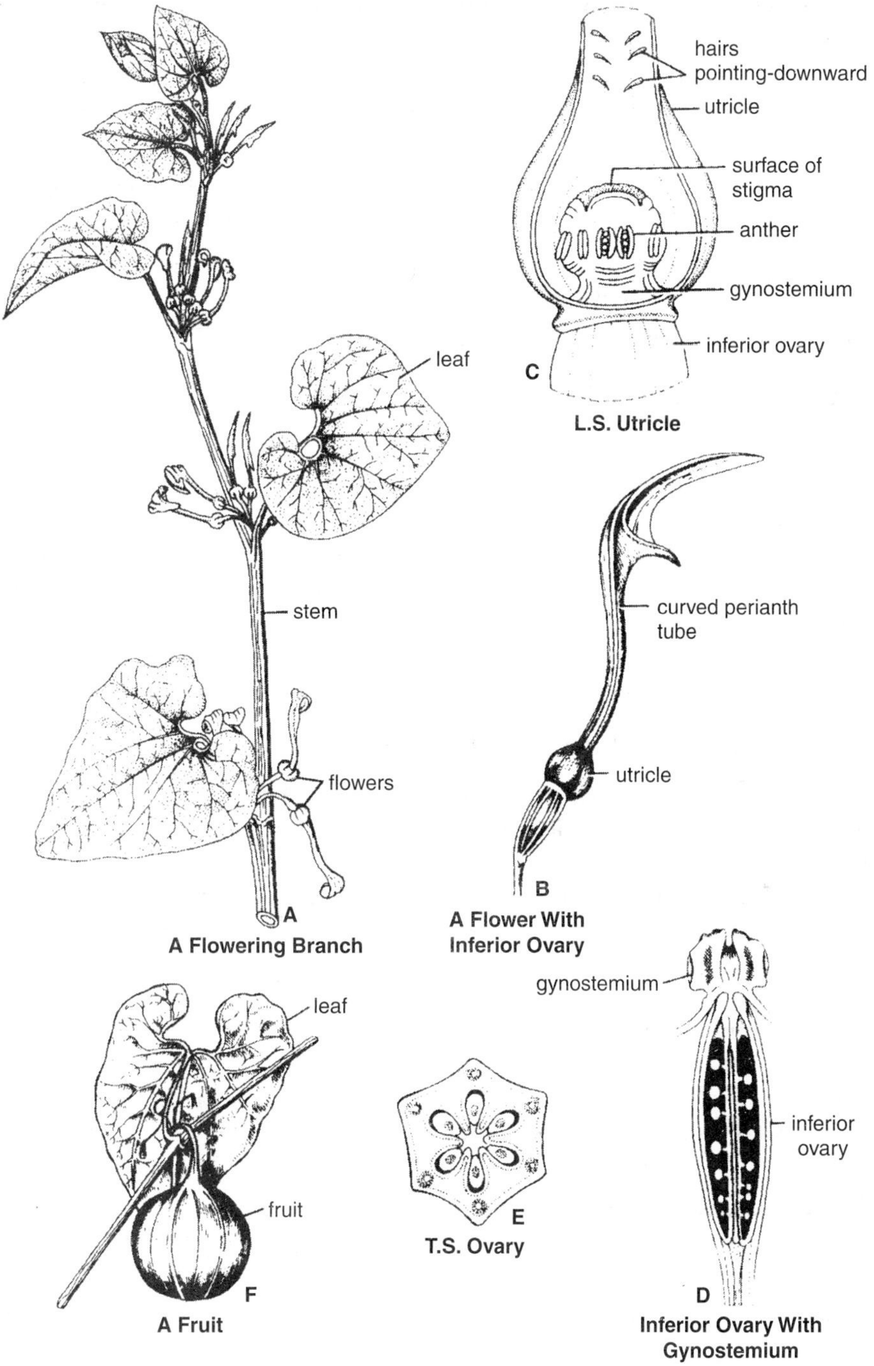

Fig. 21.106 *Aristolochia clematitis* L.

present; in *Aristolochia clematitis* (Fig. 21.106) three united perianth lobes form a yellow tube with a brownish limb.

Androecium Stamens 6–36, free, or united with the style to form staminal column (gynostemium); filaments short and thick; anthers free or adnate to style, dithecous, dehiscing longitudinally.

Gynoecium Compound pistil of 4–6 united carpels; with mostly 4–6 locules with numerous anatropous, horizontal or pendulous ovules in each locule; axile or parietal placentation; ovary inferior, rarely half-inferior; style 1, short and stout; with 4–6 stigmas.

Fruit and Seed Fruit a many-seeded or septicidal capsule, often dehiscing basally i.e. parachute-like. Seed with small embryo, rich in endosperm.

General Floral Formula EBr, ⊕ or %, ⚥, $P_{(3)}$, $\widehat{A_{6-36}}$, $G_{\overline{(4-6)}}$.

21.91.6 Economic Importance

Not being of much economic importance, many species of two genera (*Asarum* and *Aristolochia*) of the family have some horticultural value and cultivated as ornamentals. *Aristolochia macrophylla* is often grown as ornamental vine and called Dutchman's pipe because of its bent perianth tube and brownish flowers. *A serpentina* (Virginia snakeroot) has been used medicinally and serve as a cure for snake-bites. *Asarum* (wild ginger) also includes several species of horticultural interest. Roots of *Aristolochia indica* are used by snake-charmers for catching snakes. Some other species of ornamental value include *A. clematitis* (Fig. 21.106), *A. gigas*, *A. grandiflora*, *A. ornithocephala* and *Asarum europaecum*.

21.91.7 Systematics and Phylogeny (Affinities)

Aristolochiaceae has been included under series Multiovulatae Terrestris of Monochlamydeae by Bentham and Hooker. However, it has been treated under order Aristolochiales of subclass Archichlamydeae by Engler and Prantl and under the same order (Aristolochiales) of class Magnoliopsida by Takhtajan (1980) and Cronquist (1981). Hickey and King (1988) have mentioned that inspite of many attempts to find a suitable position in the system for the Aristolochiaceae, "it has proved a difficult family to place, and there is still a divergence of opinion on this matter. It is, however, probable that it is connected with the Annonaceae, a family in the Magnoliales, through the genera *Thottea* and *Apama*". Engler divided the Aristolochiaceae into three tribes, namely Sarumeae (flowers regular and solitary), Bragantieae (flowers regular but in cymes or racemes) and Aristolochieae (flowers irregular). Hutchinson (1948) placed Aristolochiaceae as a terminal derivative from the Ranales via herbaceous members of the Berberidaceae.

21.92 PIPERACEAE (PEPPER FAMILY)

21.92.1 Systematic Position

Monochlamydeae or Apetalae, Microembryeae

21.92.2 Field Recognition

Succulent herbs a shrubs, rarely trees; petals absent; stamens 1–10; ovary unilocular; embryo very minute.

21.92.3 Selected Indian Genera

Piper, Peperomia.

21.92.4 Size, Distribution and General Information

According to some authors, Piperaceae is composed of 12 genera and about 1400 species inhabiting mostly rainforests. Hickey and King (1988), however, mention that Piperaceae is a family of only 8 genera and 3100 species. *Piper*, distributed mainly in tropics, is represented by about 2000 species, while *Peperomia* is distributed both in tropics and subtropics with its 1000 species. Fresh leaves of *Piper betle* are chewed with betlenuts and used as a masticatory in most of the Asiatic countries; *Peperomia incana* (Fig. 21.107) is often grown as a pot plant in greenhouses, and it roots readily from stem cuttings.

21.92.5 Description of the Family

General Habit Mostly slender (*Peperomia pellucida*) or stout (*P. tetraphylla*) herbs, found in waste places; *P. incana* (Fig. 21.107) is a fleshy subshrub, reaching about 30 cm in height; often climbing; more rarely small trees.

Stem Herbaceous, sometimes succulent, often climbing; distinct vascular strand scattered and not arranged in concentric rings.

Leaf Alternate, rarely opposite or whorled; simple, petiolate, entire, sometimes fleshy; stipules, if present, are lateral adnate; in *Peperomia incana* (Fig. 21.107) leaves are stiff, broadly cordate, covered with hairs.

Inflorescence A dense spike or umbellate spikes; spikes of usually unisexual flowers in *Piper.*

Flower Minute, bracteate, usually bisexual, devoid of perianth; hypogynous, regular; in *Peperomia incana* (Fig. 21.107) the flowers are minute and deeply sunk into the fleshy axis of the inflorescence, posing difficulty even in seeing the floral parts.

Perianth Absent.

Androecium Stamens 1–10; anthers 2-celled (*Piper*), confluent into one in *Peperomia*; dehiscence longitudinal.

Gynoecium Carpels 1–5, united; ovary superior, unilocular; basal orthotropous ovule, one in each locule; stigmas 1–5.

Fruit A small drupe.

Seeds Very small, embryo minute; with abundant mealy perisperm.

Pollination Pollination process is little known in Piperaceae. According to Hickey and King (1988) Piperaceae are considered to be closely related to the insect-pollinated Magnoliaceae showing entomophily. The flowers are protogynous, which encourages cross-pollination. Due to glutinous nature of pollen grains and scented flowers, at least *Peperomia resediflora* is insect-pollinated.

General Floral Formula Br, ⊕, ⚥, P_0, A_{1-10}, $G_{\underline{(1-5)}}$.

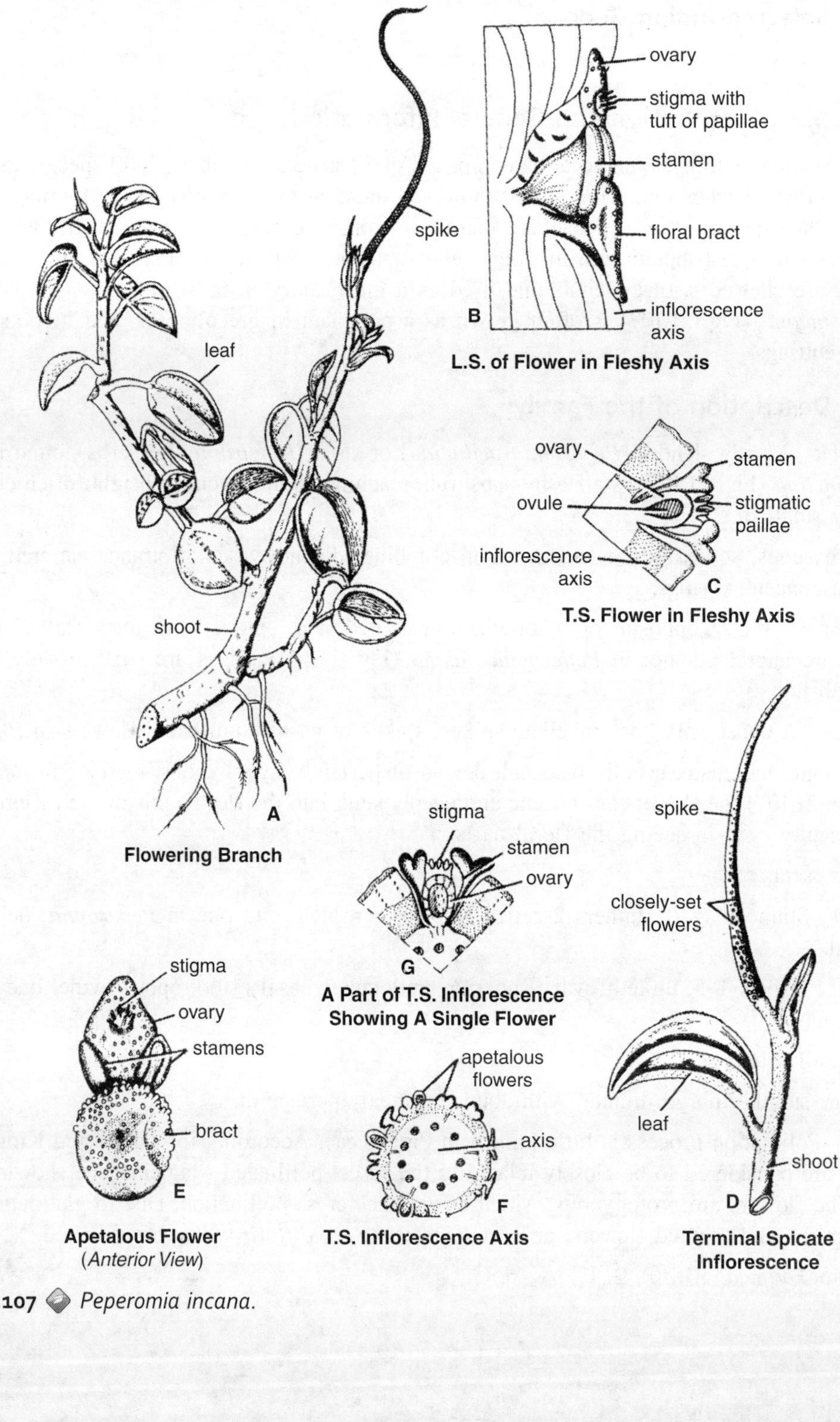

Fig. 21.107 *Peperomia incana.*

21.92.6 Economic Importance

- *Piper betle* (Betel or Paan) is used as a common masticatory in many Asiatic countries. Its fresh leaves are chewed with betlenuts.
- The whole drief fruits of *Piper nigrum* are known as "black pepper". If the outer layer is removed, the fruits are called "white pepper".
- Dried unripe fruits of *Piper cubeba* are used for flavouring as the source of "cubebs", while the roots of *P. methysticum* are used for making an intoxicating Polynesian beverage "Kava". *P. caninum* yields a type of volatile oil called "Kababchini". Wood of *P. chaba* forms a pungent condiment. "Long pepper" is obtained from *P. longum* and *P. retrofactum.* In Mexico, walking sticks are made from the stems of *P. tiliaefolium.*
- Many species of *Peperomia* and a few of *Piper* and *Macropiper* are grown for their ornamental value. *Peperomia* species are cultivated as house plants for their variegated foliage.

21.92.7 Systematics and Phylogeny

Bentham and Hooker included Piperaceae under series Microembreae of Monochlamydeae or Apetalae while Engler and Prantl discussed it under order Piperales of subclass Archichlamydeae of class Dicotyledoneae. Hutchinson (1973) treated Piperales under division Herbaceae of Dicotyledons. The minute-flowered, spicate inflorescence of several members of Piperaceae bring it close to Betulaceae and Corylaceae. Because of the entomophilous nature, however, Piperaceae are close to Magnoliaceae.

21.93 LORANTHACEAE (MISTLETOE FAMILY)

21.93.1 Systematic Position

Monochlamydeae or Apetalae, Achlamydosporae.

21.93.2 Field Recognition

Usually aerial parasitic habit; cup-shaped receptacle; inferior ovary; ovules not differentiated from the placenta.

21.93.3 Selected Indian Genera

Dendrophthoe, Loranthus, Viscum.

21.93.4 Size, Distribution and General Information

Family is represented by about 36 genera and 1300 species and distributed in both tropical and temperate regions. About 10 genera and over 52 species have been reported from India. Some of the largely represented genera along with the number of their approximately reported species are *Loranthus* (600), *Phoradendron* (190), *Struthanthus* (75), *Viscum* (70), *Dendrophthora* (55) and *Psittacanthus* (50).

21.93.5 Description of the Family

General Habit Mostly semiparasitic herbs or small shrubs attached to their hosts by sucking haustoria; rarely erect, terrestrial trees (e.g. *Nuytsia*).

Roots Modified adventitious roots present in the form of haustoria; *Nuytsia* is rooted in the earth; in *Viscum* the root often branches within the host tissue; haustoria penetrate the bark and sapwood of the host.

Stem Usually dichotomously branched, often with swollen nodes.

Leaf Usually opposite or whorled, rarely alternate; exstipulate; sometimes reduced to scales; usually evergreen, thick, leathery, entire and persistent.

Inflorescence Usually cymose, or a panicle, raceme, spike; or flowers solitary with cup-shaped receptacle.

Flower Usually bracteate, bracteolate; bisexual (*Loranthus*) or unisexual (*Viscum*, Fig. 21.108); usually actinomorphic, rarely slightly zygomorphic (*Dendrophthoe*); epigynous; flowers usually in groups of 3 (or 2, by abortion of the central flower).

Perianth In 1 or 2 whorls of 4 to 6, free or united, large tepals developing from the margin of a cup-shaped receptacle; tepals, if united, form a tube; sepaloid (*Viscum*) or petaloid (*Loranthus*); *calyculus*, an outgrowth present in the form of a small rim or fringe below the perianth, develops in several genera (*Loranthus*, *Nuytsia*, *Struthanthus* and *Psittacanthus*).

Androecium Stamens equal in number to perianth lobes, adnate to perianth lobes or at their base; generally opposite the tepals; anthers 2-celled or transversely multiloculate; dehiscing either transversely or longitudinally or even by terminal pores.

Gynoecium 3–5 carpels, syncarpous, inferior, unilocular, the ovules not differentiated from the placenta; carpels usually sunk in or fused with the receptacle; basal placentation; style simple or absent; stigma 1 and often sessile.

Fruit and Seeds: Fruit usually 1 to 3-seeded berry or drupe. Seeds usually surrounded by viscin, a sticky substance; endospermic with straight embryo.

General Floral Formula: Br, Brl, ⊕, ⚥ or unisexual, $P_{4-6 \text{ or } (4-6)}$, A_{4-6}, $G_{\overline{(3-5)}}$.

21.93.6 Economic Importance

Plants of Loranthaceae are of no specific economic importance. Only *Viscum album* (Mistletoe, Fig. 21.108) and some species of *Phoradendron* (e.g. *P. flavescens*) are of ornamental value and used for decorative purposes, specially at Christmas time.

21.93.7 Systematics and Phylogeny (Affinities)

Most of the taxonomists place Loranthaceae under order Santalales (Lawrence, 1951; Takhtajan, 1969; Cronquist, 1981; Thorne, 1983). However, Bentham and Hooker discussed Loranthaceae under series Achlamydosporae of Metachlamydeae. In Engler's *Syllabus der Pflanzenfamilien*, Loranthaceae is divided into 2 subfamilies viz. Loranthoideae (calyculus or 2 bracteoles present) and Viscoideae (calyculus absent). Loranthaceae shows close affinities with Santalaceae.

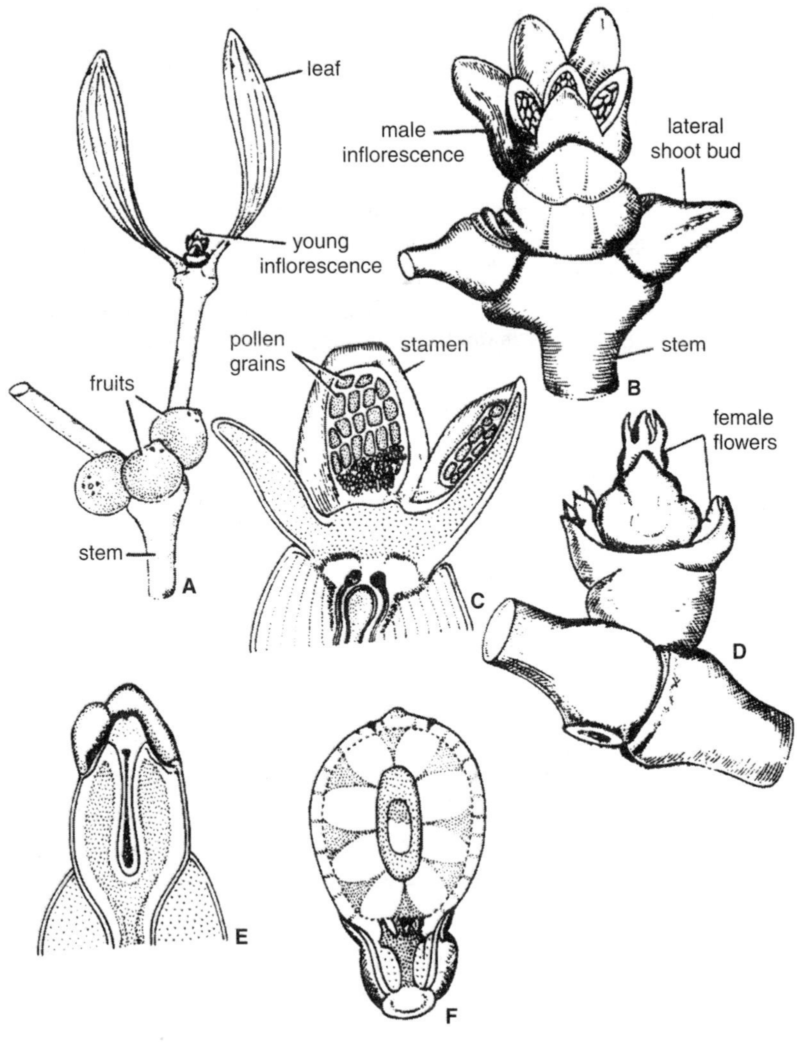

Fig. 21.108 *Viscum album* L (A: Female plant; B: Male inflorescence; C: L.S. male flower; D: Female inflorescence; E: L.S. female flower, and F: L.S. fruit).

21.94 EUPHORBIACEAE (SPURGE FAMILY)

21.94.1 Systematic Position

Monochlamydeae or Apetalae, Unisexuales.

21.94.2 Field Recognition

Herbs, shrubs, or small trees; milky latex present; flowers apetalous, unisexual; female flowers usually tricarpellary and ovary trilocular; fruit capsule or regma.

21.94.3 Selected Indian Genera

Croton, *Euphorbia*, *Hevea*, *Jatropha*, *Manihot*, *Phyllanthus* and *Ricinus*.

21.94.4 Size, Distribution and General Information

A family of about 300 genera and 7500 species (Jones and Luchsinger, 1987), Euphorbiaceae are cosmopolitan in distribution. They are abundant in both tropical and temperate regions, and not well-represented in arctic regions. Over 60 genera and about 350 species have been reported from India. Some of the common genera along with the approximate number of their worldover reported species as well as the common names of some of them are *Euphorbia* (2000, spurge or poinsettia), *Croton* (750), *Phyllanthus* (600), *Acalypha* (450, chenille plant), *Jatropha* (175), *Manihot* (170), *Sapium* (120), *Tragia* (100, nose-burn), *Clutia* (70), *Manadenium* (47), *Ricinocarpos* (16), *Hevea* (12, Para-rubber) and *Ricinus* (1, castor bean).

21.94.5 Description of the Family

General Habit Mostly shrubs (*Ricinus communis*, *Jatropha gossypifolia*) or trees (*Emblica officinalis*), or sometimes herbs (*Phyllanthus nirurii*), or climbers (*Tragia involucrata*); some are xerophytic (many species of *Euphorbia*) or cactus-like or phylloclades; some are marshy (*Coperonia*); usually the plants contain milky sap or latex.

Stem Herbaceous or woody; becomes cactus-like in several species of *Euphorbia* (e.g. *E. neriifolia*, *E. royleana*, *E. trigona*); in *E. xylophylloides* stem becomes flattened and leaf-like (phylloclade).

Leaf Mostly simple (*Euphorbia pulcherrima*), alternate, stipulate; often reduced or deciduous as in several xerophytic species of *Euphorbia*; sometimes palmately lobed (*Ricinus*, *Jatropha*), or deeply so in *Manihot*; rarely opposite (*Euphorbia hirta*, *Choriophyllum*) or whorled (*Mischodon*); stipules sometimes present in the form of hairs, glands, or thorns; venation is unicostate reticulate, or multicostate divergent; in *Phyllanthus niruri* leaves are arranged in 2 rows so that each branch resembles with a compound leaf.

Inflorescence Highly variable from a raceme (*Croton*), a spike (*Acalypha*), a dichasium (*Jatropha*), or even the flowers are solitary axillary (*Phyllanthus asperulatus*); but in majority of the cases, as in *Euphorbia*, the inflorescence is a *cyathium*.

Each *cyathium* (Fig. 21.109) contains terminally a single, naked female flower, usually represented by a tricarpellary gynoecium. The female flower is surrounded by a cup-like involucre formed by 4 or 5 connate sepaloid bracts. In the axil of each bract develops a group of stamens in a scorpioid manner. Each stamen represents a naked male flower because it is a jointed structure which indicates that its upper portion is the filament bearing the anther and its lower portion represents the pedicel of the male flower bearing stamen. On the rim of the cup-like involucre are present nectar-secreting glands. Glands are oval or crescent-shaped and often brightly coloured. A cyathium appears like a single flower but actually it is an inflorescence because (i) naked male flowers (stamens) start ripening centrifugally i.e. from the centre towards periphery, (ii) they are jointed, of which the lower portion represents pedicel which bears the male flower (stamen), (iii) development of male flower (stamen) is in a scorpioid manner, and (iv) in the centre is present a single naked female flower.

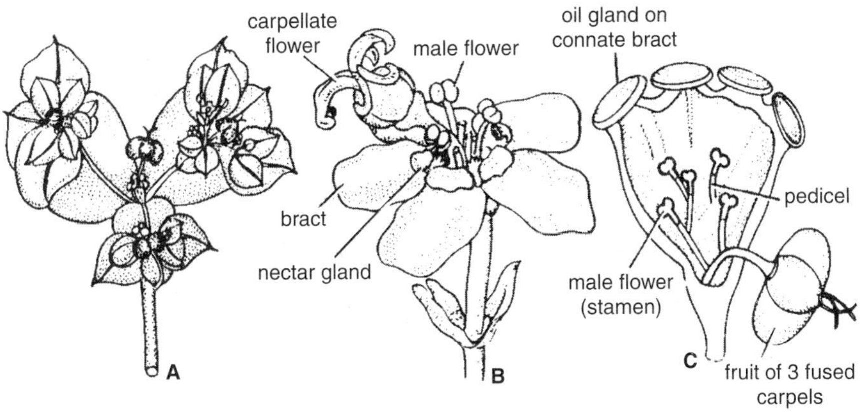

Fig. 21.109 A: Cyathia of *Euphorbia helioscopia*; B: A cyathium of *E. corollata,* and C: L.S. cyathium of *E. helioscopia.*

Flower Bracteate, usually bracteolate; generally unisexual, monoecious (*Baliospermum*) or dioecious (*Bridelia*), actinomorphic, hypogynous, rarely perigynous (*Bridelia*).

Floral structures are highly variable, and some aspects are undermentioned:

Perianth Consisting of both calyx and corolla, or only of corolla, or sometimes both are absent; valvate or imbricate; calyx, when present, consists of 5 parts, usually distinct or united; corolla of 0 to 5 petals, free or united in various ways as under:

(i) In *Chrozophora* sepals are 5, united and valvate, and petals are 5, free and twisted; (ii) in *Croton*, the flowers are often apetalous, and the tepals are distinct in male flowers and usually absent in female flowers; (iii) in *Euphorbia* the perianth parts of individual flowers are absent (Fig. 21.110); (iv) in *Jatropha* same as in *Chrozophora*; (v) in *Mercurialis* sepals are only 3 and united while petals, if present, are free, (vi) in *Phyllanthus niruri* 6 tepals are present in 2 whorls of 3 each, (vii) in *Ricinus communis* male flowers have 5 fused, valvate tepals while the female flowers have 3 to 5 fused, valvate tepals (Fig. 21.111).

Androecium Stamens 1 to 100 or more but usually as many or twice as many as tepals, free or monadelphous; anthers bilocular, rarely 3 to 4 locular; longitudinal or transversely dehiscent. Stamens are highly variable in number, position and adhesion, varying from (i) only 1 stamen in each male flower in *Euphorbia*, consisting of a pedicel, joint, filament and anther lobes (Fig. 21.110); (ii) 3 stamens in *Breynia*; (iii) a whorl of 3 monadelphous stamens in *Phyllanthus niruri*; (iv) 5 stamens with their profusely branched, tree-like filaments having dithecous, introrse anther lobes at their apex in *Ricinus communis* (Fig. 21.111); (v) 8 stamens in *Acalypha*, (vi) 10 stamens in 2 whorls of 5 each in *Jatropha*, (vii) 15 stamens in 3 whorls of 5 each in *Chrozophora*, (viii) 15 to 30 stamens in *Codiaeum*, and (ix) 80 to 100 or more stamens in *Croton*. Stamens are either free or more or less monadelphous; forming a staminal column by the fusion of the filaments as in *Phyllanthus niruri*; in male flowers the disc is usually present in the form of intrastaminal or extrastaminal glands; sometimes staminodes are present in pistillate flowers.

Gynoecium Usually tricarpellary, rarely bicarpellary (*Bridelia*, *Mercurialis*) or pentacarpellary (*Wielandia*); syncarpous; superior, rarely semi-inferior (*Bridelia*); trilocular, one (*Ricinus*, *Euphorbia*)

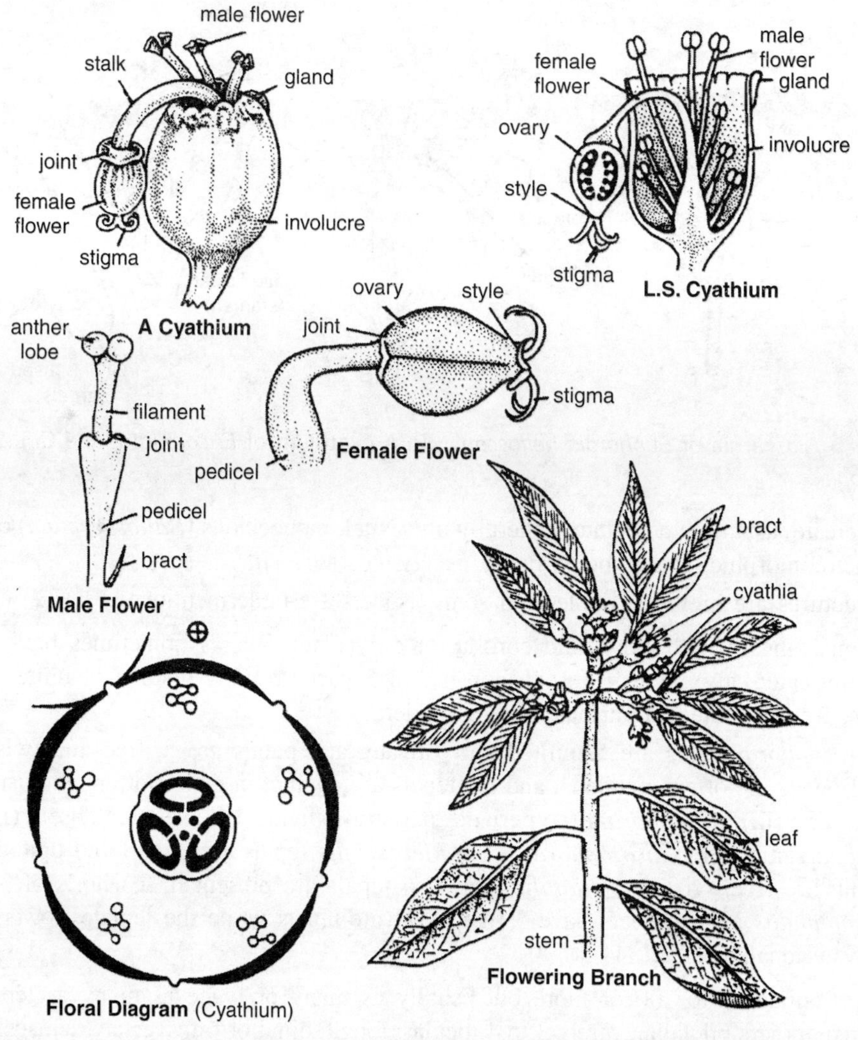

Fig. 21.110 *Euphorbia pulcherrima* Willd. syn. *Poinsettia pulcherrima* R. Grah.

or two (*Phyllanthus*) anatropous ovules in each locule; axile placentation; styles 3, each bifurcating apically into two feathery stigmas; a nectariferous disc is present at the base of the ovary; in staminate flowers, the gynoecium is sometimes present as a pistillode.

Fruit and Seeds Fruit usually a capsule, but sometimes a drupe. Seeds endospermic, oily, with a knob-like caruncle (*Ricinus communis*); embryo straight.

Pollination Cross pollination is necessary because of unisexual flowers. Certain characters (coloured bracts, well-developed nectaries and glands) also favour insect pollination in some plants.

General Floral Formulae

(a) *Male Flower:* Br, ⊕, ♂, $P_{3+3 \text{ or } (3-5) \text{ or } 0}$, $A_{1-\infty \text{ or } (3-\infty)}$, $G_{0 \text{ or pistillode}}$.

(b) *Female Flower:* Br, ⊕, ♀, $P_{3+3 \text{ or } (3-5) \text{ or } (5)}$, $A_{0 \text{ or staminodes}}$, $G_{(\underline{3})}$.

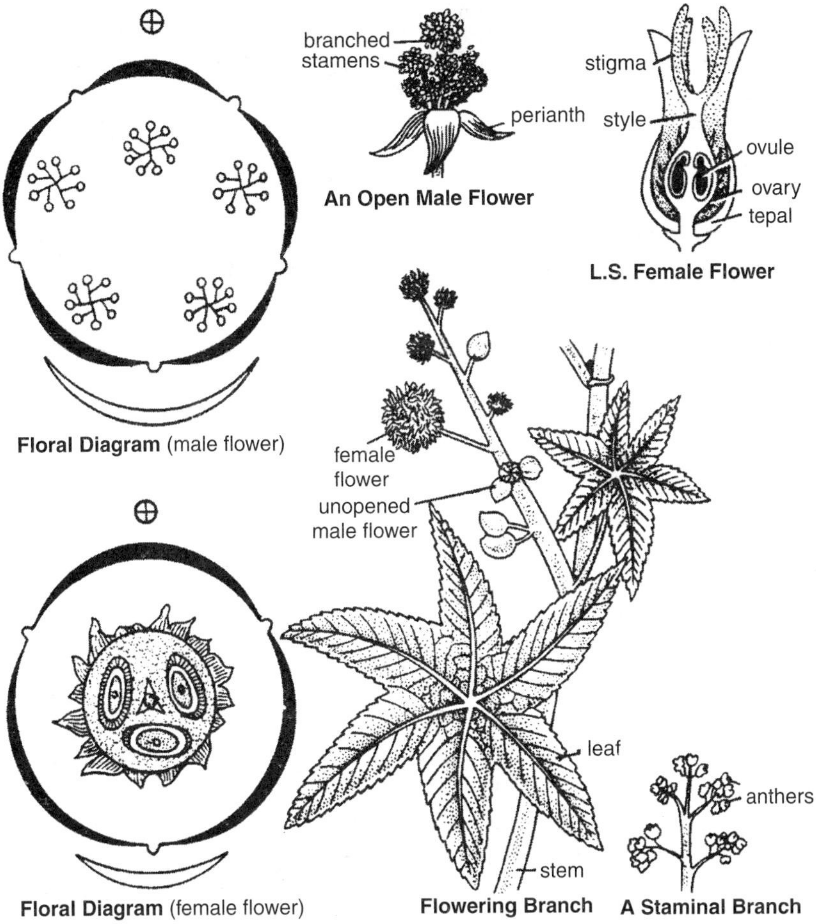

Fig. 21.111 *Ricinus communis* L.

21.94.6 Economic Importance

Several plants of high economic importance belong to Euphorbiaceae, such as *Hevea brasiliensis* (the main commercial source of rubber), *Ricinus communis* (the source of castor oil), *Manihot esculenta* (the source of cassava or tapioca), *Emblica officinalis* (the source of Amla), *Euphorbia pulcherrima* (the famous ornamental Poinsettia), and *Codiaeum variegatum* (the popular ornamental plant with variegated leaves, often incorrectly called "Croton"). However, several Euphorbiaceae are poisonous, causing sickness or death if ingested, or dermatitis if juice contacts the skin. Even the rainwater dripping from certain plants is enough to cause dermatitis, and therefore should be handled with utmost care. Some of their specific uses are mentioned below:

- *Ornamental Plants:* These include *Acalypha ciliata*, *A. hispida*, *A. wilkesiana*, *Codiaeum variegatum* (plants with variegated leaves), *Croton tiglium* (crotons), *Euphorbia antiquorum* (Tidhara Sehund), *E. neriifolia* (Sehund), *E. pulcherrima* syn. *Poinsettia pulcherrima*

(Poinsettia, with red, pink or white floral leaves), *E. splendens* (Crown-of-thorns, with bright red bracts), *E. tirucalli* (Milk-bush), *Jatropha hastata*, *J. gossypifolia*, *J. podagrica*, *J. panduraefolia*, and *Trewia nudiflora* (False white teak).

- *Oil Plants:* (i) Croton oil, obtained from the seeds of *Croton tiglium* (Jamalghota) is used as a powerful purgative, (ii) Castor oil, obtained from the seeds of *Ricinus communis* (Arandi) is mainly used as a vegetable oil and also as a mild laxative, as a lubricant and in paint, varnish and plastic industries, (iii) Jatropha oil, obtained from the seeds of *Jatropha curcas*, is used as a purgative and also in skin diseases, in rheumatism, and also in the manufacture of soaps, lubricants, candles, etc., (iv) Tung oil, obtained from the seeds of *Aleurites fordii*, *A. moluccana* and *A. montana*, is used in preparing paints, varnishes, linoleum, India ink and also in waterproofing the paper, wood, etc.
- *Rubber:* (i) Over 98% of the total natural rubber, produced in the world, is obtained from the coagulated latex of *Hevea brasiliensis*, (ii) *Manihot glaziovii* (Manicoba rubber) is the another rubber-yielding plant of this family.
- *Cassava or Manioc:* Tuberous roots of *Manihot esculenta* (Cassava) are rich in starch (Arrowroot) and used for preparing bread, biscuits and other foodstuffs. Bitter cassava contains hydrocyanic acid (HCN) and must be used with due care.
- *Medicinal Plants:* (i) *Emblica officinalis* syn. *Phyllanthus emblica* (Amla) fruits are rich source of vitamin C, used in treating scurvy, in preparing shampoos, and also in making the hair dyes (bark and leaves of this tree are used for tanning); (ii) Castor (*Ricinus communis*) oil is a valued purgative known; (iii) *Jatropha gossypifolia* leaves are used in eczema, and roots in leprosy and snake bites; (iv) Entire plant of *Synadenium grantii* is used as a stimulant of central nervous system; (v) *Croton cascarilla* and *C. elateria* bark is used as a tonic, (vi) The drug, euphorbium, obtained from the latex of *Euphorbia resinifera*, is used as a purgative, (vii) *Mallotus philippinensis* (Kamela tree) fruits are used as anthelmintic.
- *Dyes:* (i) Kamela dye, obtained from the fruits of *Mallotus philippinensis*, is used for dyeing wool and silk, (ii) Blue dye, obtained from the bark of *Jatropha curcas*, is used for dyeing fishing nets, (iii) Purple dye, obtained from the bark of *Chrozophora tinctoria* is used in textile industry, (iv) Red dye is obtained from the roots of *Kirganelia reticulata*.
- *Timber Plants:* Valuable timber, used for packing cases, tea boxes, veneers, plywood, match industry and several other similar purposes, is obtained from *Aporosa dioica*, *Bischofia javanica*, *Drypetes roxburghii*, *Gelonium multiflorum*, *Hemicyclia andamanica*, *H. elata*, *Hura crepitans* (sandbox tree) and *Trewia nudiflora* (False white teak).

21.94.7 Systematics and Phylogeny (Affinities)

Treated under Unisexuales of Monochlamydeae by Bentham and Hooker, Euphorbiaceae was placed under Geraniales by Engler, and under Euphorbiales by Hutchinson (1959), Takhtajan (1969), Cronquist (1981) and Thorne (1983). Hickey and King (1988) divided Euphorbiaceae into 4 subfamilies as under:

1. *Phyllanthoideae* (Cotyledons broader than radicle; ovules 2 per locule; no latex).
2. *Euphorbioideae* (Cotyledons broader than radicle; ovules 1 per locule; latex present).

3. *Porantheroideae* (Cotyledons as wide as radicle; ovules 2 per locule; no latex).
4. *Ricinocarpoideae* (Cotyledons as wide as radicle; ovules 1 per locule; latex present).

Euphorbiaceae resembles Geraniaceae and allied families in the characters of fruit. Characters such as monadelphous or polyadelphous stamens and carunculate seeds bring it close to Sterculiaceae. Cronquist opined that Euphorbiales are closer to Sapindales as well as to Geraniales, Linales and Polygalales, and all these have been derived from Rosaceous-stock (Cronquist, 1981).

21.95 URTICACEAE (NETTLE FAMILY)

21.95.1 Systematic Position

Monochlamydeae or Apetalae, Unisexuales.

21.95.2 Field Recognition

Usually herbs with stinging hairs and watery sap; inflorescence cymose; gynoecium monocarpellary with single ovule; style one.

21.95.3 Selected Indian Genera

Laportea, Boehmeria, Pouzolzia, Sarchochlamys, Urtica, Girardinia, Elatostema.

21.95.4 Size, Distribution and General Information

A family of about 45 genera and 1060 species (Hickey and King, 1988), Urticaceae are distributed in both tropical and temperate regions. About 12 genera and over 100 species have been reported from India, mainly distributing in Assam, West Bengal, Kashmir and sub-Himalayan regions. Some of the largely represented genera along with the number of their approximately reported species are *Pilea* (400), *Elatostema* (200), *Boehmeria* (100), *Cecropia* (100), *Urtica* (50) and *Laportea* (23). *Urtica* (stinging nittle) has sharp, stinging hairs which penetrate in the skin of the animals coming in their contact and cause sharp burning pain for hours.

21.95.5 Description of the Family

General Habit Mostly fibrous herbs (*Urtica*, Fig. 21.112), infrequently shrubs (*Maoutia, Boehmeria*), or rarely small trees (*Laportea crenulata*); without latex but with watery sap; a few are climbers; stinging hairs, present in several genera (*Urtica, Urera*), are long hollow cells with silicified tips which penetrate the skin like a fine needle; stinging hairs usually absent in several genera (*Pilea, Elatostema, Boehmeria, Maoutia, Parietaria, Cecropia*); epidermal cells usually contain cystoliths.

Leaf Simple, alternate or opposite; stipulate but rarely exstipulate (*Parietaria*); often with stinging hairs.

Inflorescence Usually cymose with condensed flowers; sometimes catkin-like or heads; or flowers solitary.

Flower Bracteate or ebracteate, ebracteolate; usually incomplete, unisexual, actinomorphic; rarely bisexual (*Parietaria*); hypogynous; usually very small or greatly reduced.

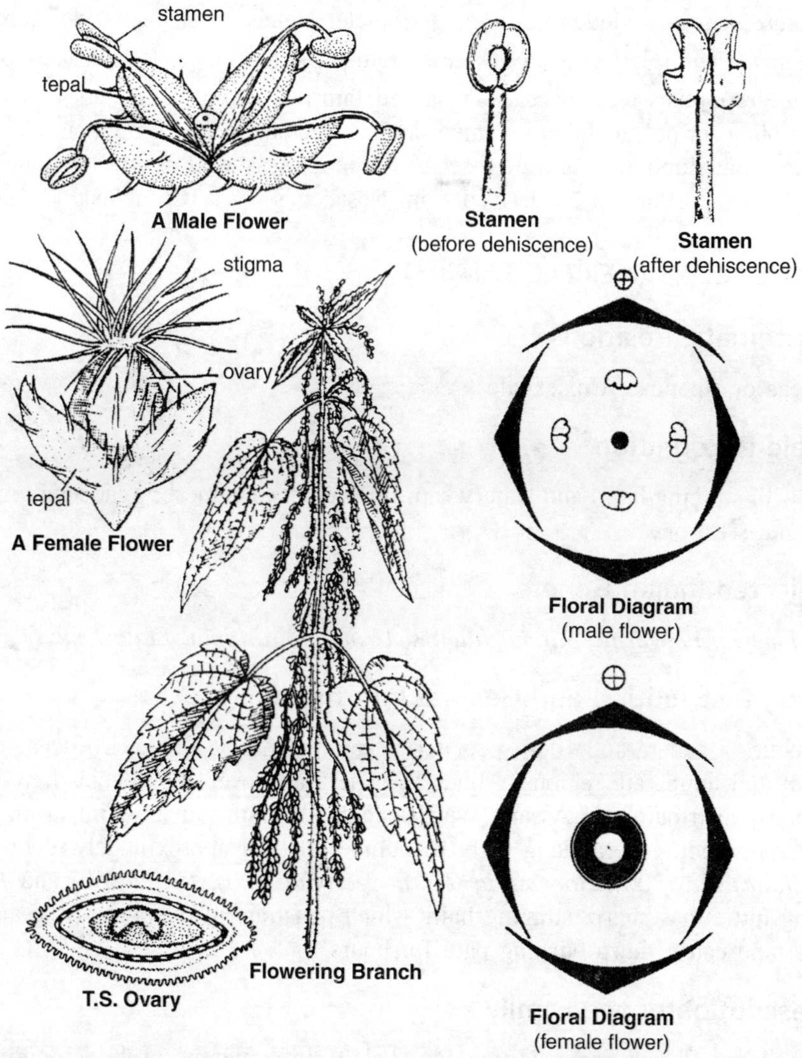

Fig. 21.112 *Urtica dioica* L.

Perianth 4 or 5, free or united tepals; sometimes even tepals are absent (*Forskohlea*); mostly green or sepaloid; petals absent.

Androecium In staminate flowers the stamens are mostly 4, rarely 3–5; present opposite the perianth lobes; straight (*Cercopia*, *Poikilospermum*) or bent down inwards in bud and exploding when ripe; anthers dithecous, dehiscing longitudinally; scale-like staminodes are present in pistillate flowers; only 1 stamen is present in each male flower in *Forskohlea*.

Gynoecium In pistillate flowers the gynoecium is monocarpellary; superior; unilocular with 1 basal, erect, orthotropous ovule; style 1; stigma 1 and often with a brushlike tuft; a rudimentary pistillode is present in staminate flowers.

Fruit and Seeds Fruit usually an achene or fleshy drupe, enclosed by persistent perianth lobes. Seeds endospermic, with small and straight embryo.

Pollination *Urtica dioica* flowers are wind pollinated. The anemophilous flowers of *Parietaria diffusa* are well-adapted to ensure cross pollination.

General Floral Formulae

(a) *Male Flowers:* Br or Ebrl, ⊕, ♂, $P_{4-5 \text{ or } (4-5)}$, A_4, $G_{0 \text{ or pistillode}}$.

(b) *Female Flowers:* Br or Ebrl, ⊕, ♀, $P_{4-5 \text{ or } (4-5)}$, $A_{0 \text{ or staminode}}$, $G_{\underline{1}}$.

21.95.6 Economic Importance

The family is of little economic importance as under:

- Fibre (ramie), obtained from the inner bark of *Boehmeria nivea*, is perhaps the longest and most silky of all vegetable fibres. It is used for cordage, ropes, etc. Fibres obtained from *Maoutia puya* and *Gigardinia zeylanica* are also used for similar purposes.
- Bank notes are manufactured from the paper pulp made from the tough and silky fibres of *Boehmeria nivea.*
- Ornamental plants belong to several species of *Pilea*, *Pellionia*, *Soleirolia*, and *Urera.*
- Young tops of *Urtica* are cooked and eaten like spinach.
- Fruits of *Debregeasia hypoleuca* are edible.

21.95.7 Systematics and Phylogeny (Affinities)

Family Urticaceae of Unisexuales (Bentham and Hooker) was placed under order Urticales by most of the latter taxonomists (Engler and prantl, Hutchinson, Cronquist, Takhtajan, and Thorne). Hickey and King (1988) divided Urticaceae into 6 tribes (Urticeae, Procrideae, Boehmerieae, Parietarieae, Forskohleeae and Conocephaleae).

Urticaceae is closely related to Moraceae, but differs from the latter in having erect ovule and a straight embryo. From the other families of Urticales, Urticaceae are considered advanced for having herbaceous plants and monocarpellary ovary.

21.96 MORACEAE (MULBERRY FAMILY)

21.96.1 Systematic Position

Monochlamydeae or Apetalae, Unisexuales.[1]

21.96.2 Field Recognition

Shrubs or trees and only rarely herbs; milky sap present; leaves stipulate; flowers unisexual; tepals usually 4; stigmas 2.

[1] Bentham and Hooker did not recognise Moraceae as an independent family but discussed its members under two tribes (Moreae and Artocarpeae) belonging to family Urticaceae of Unisexuales. However, most of the other workers recognised Moraceae as an independent family of order Urticales.

21.96.3 Selected Indian Genera

Artocarpus, Broussonetia, Ficus, Morus.

21.96.4 Size, Distribution and General Information

A family of 53 genera and over 1400 species (Hickey and King, 1988), Moraceae are distributed mainly in tropical and subtropical regions of the world, and a few are also found in temperate regions. Only 15 genera and about 150 species have been reported from India. Major genera along with the approximate number of their worldover reported species and common names of some of them are *Ficus* (800, fig or rubber plant), *Dorstenia* (170), *Artocarpus* (47, jackfruit), *Maclura* (12, hedge apple), *Morus* (10, mulberry), and *Broussonetia* (8, paper mulberry).

21.96.5 Description of the Family

General Habit Mostly trees (*Ficus benghalensis, F. religiosa, Morus alba*), or shrubs with milky juice, rarely herbs (*Dorstenia*) or climbers (*Ficus benjamina*); aerial prop roots grow vertically downward, enter the soil and provide support to the plant in several species of *Ficus* (*F. benghalensis*).

Stem Woody, contains milky, sticky latex in long sacs, specially in secondary cortex and phloem (*Ficus elastica*, rubber plant).

Leaf Simple, palmately or pinnately veined; alternate; usually with 2 cuducous stipules; petiolate; leathery; margin entire (*Ficus*), serrate (*Morus*) or lobed; on falling, the caducous stipules leave a semicircular scar in *Ficus*.

Inflorescence Basic plan is cymose, but the flowers are grouped in variously shaped clusters of racemes, spikes, umbels, heads, or catkins, or even on the inside of a hollow receptacle i.e. syconium or hypanthodium (Fig. 21.113). Plants may be monoecious or dioecious. In *Morus*, female flowers are arranged in pseudospikes while male flowers in catkins; in *Broussonetia* the female flowers are arranged in pseudoheads while male flowers in pseudoracemes; in *Artocarpus* female flowers in pseudoheads while male flowers in pseudocatkins; in the syconium of *Ficus*, the male flowers are arranged towards the ostiole in the hollow cavity while rest of the cavity is occupied by female flowers, as in *Ficus carica* (Fig. 21.113).

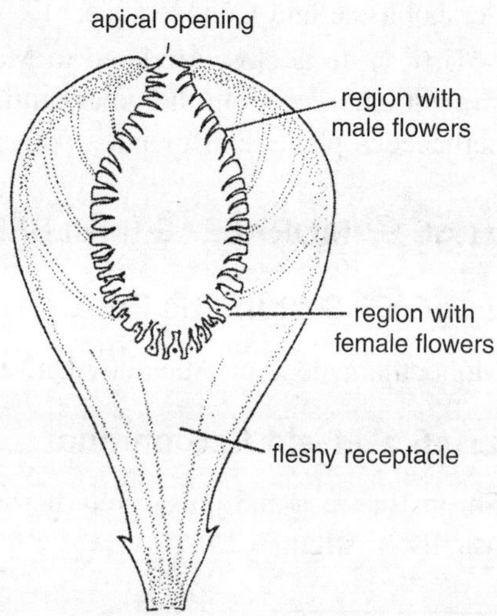

Fig. 21.113 L.S. syconium of *Ficus carica*.

Flower Unisexual, plants monoecious or dioecious, incomplete, actinomorphic; pistillate flowers hypogynous; tetramerous, small and inconspicuous.

Perianth Usually 4 tepals, free and in 2 whorls of 2 each (*Morus*), or more or less united, persistent, imbricate or valvate; tepals usually ovate and hairy; sepaloid; petals absent.

Androecium Usually 4 stamens, present only in male or staminate flowers, opposite the tepals (*Morus*), polyandrous, dithecous, basifixed or versatile; introrse, dehiscence longitudinal; anthers not exploding; only 1–2 stamens in *Ficus*, *Artocarpus*; sometimes rudimentary stamens present in female flowers.

Gynoecium Present only in female or pistillate flowers; bicarpellary, syncarpous, ovary superior (*Morus*) to inferior (*Dorstenia*); out of the 2 carpels usually only 1 develops; unilocular, with 1 usually pendulous ovule; style short and bifurcating into 2 coiled stigmas, sometimes rudimentary pistil is present in male flowers.

Fruit and Seeds Fruit an achene or drupe-like, but usually a multiple fruit developing from the union of fruits of several different flowers; several achenes develop inside the fleshy receptacle in *Ficus*; in *Morus* several fruits are enclosed in individual fleshy perianth and collectively called a *sorosis*. Seeds are with or without endosperm, and with usually curved embryo.

Pollination In *Morus*, the anthers of the male inflorescence dehisce, and their pollens are carried by the wind to the prominent paired stigmas of the female flowers. Pollination in *Ficus carica* is effected by a female insect (*Blastophaga psenes*).

General Floral Formulae

(a) *Male Flower:* Ebr, ⊕, ♂, P_{2+2}, A_4, G_0.

(b) *Female Flower:* Ebr, ⊕, ♀, P_{2+2}, A_0, $G_{(\underline{2})}$.

21.96.6 Economic Importance

Moraceae are important from the economic point of view in providing several fruit plants, ornamentals, rubber plant and medicinal plants, as under:

- Fruit plants belonging to Moraceae include *Morus nigra* (black mulberry or Shahtoot), *M. alba* (white mulberry or Tunt, Fig. 21.114), *M. serrata* (Himalayan mulberry), *Ficus carica* (fig or Anjeer), *F. palmata* (Anjiri), *Artocarpus heterophyllus* syn. *A. integrifolia* (jack fruit or Kathal), *A. incisia* (bread fruit) and *A. lakoocha* (monkey fruit or Barhal).
- Certain sacred trees used in religious ceremonies and also for shade include *Ficus benghalensis* (banyan tree, or Bargad), *F. religiosa* (Peepal), *F. glomerata* (Goolar), and *F. virens* syn. *F. infectoria* (Pilkhan).
- Famous ornamental plants include *Ficus elastica* (Indian rubber plant), *Ficus krishnae* (Krishan's butter cup), *Morus nigra*, *M. alba*, *Maclura pomifera* (Osage orange), and *Dorstenia cordifolia*.
- Paper is obtained from the fiber present in the inner bark of *Broussonetia papyrifera* (paper mulberry).
- Rubber is obtained from the latex of *Ficus elastica* (Indian rubber plant) and *Castilla elastica* (Panama rubber).
- Leaves of *Morus alba* and *M. nigra* are used to feed silkworms.
- Useful timber, utilized in the manufacture of sport goods and several other articles, is obtained from *Morus alba*, *M. serrata*, *Ficus sycomorus*, *Artocarpus altilis*, *A. chaplasha*, and *A. hirsuta*.

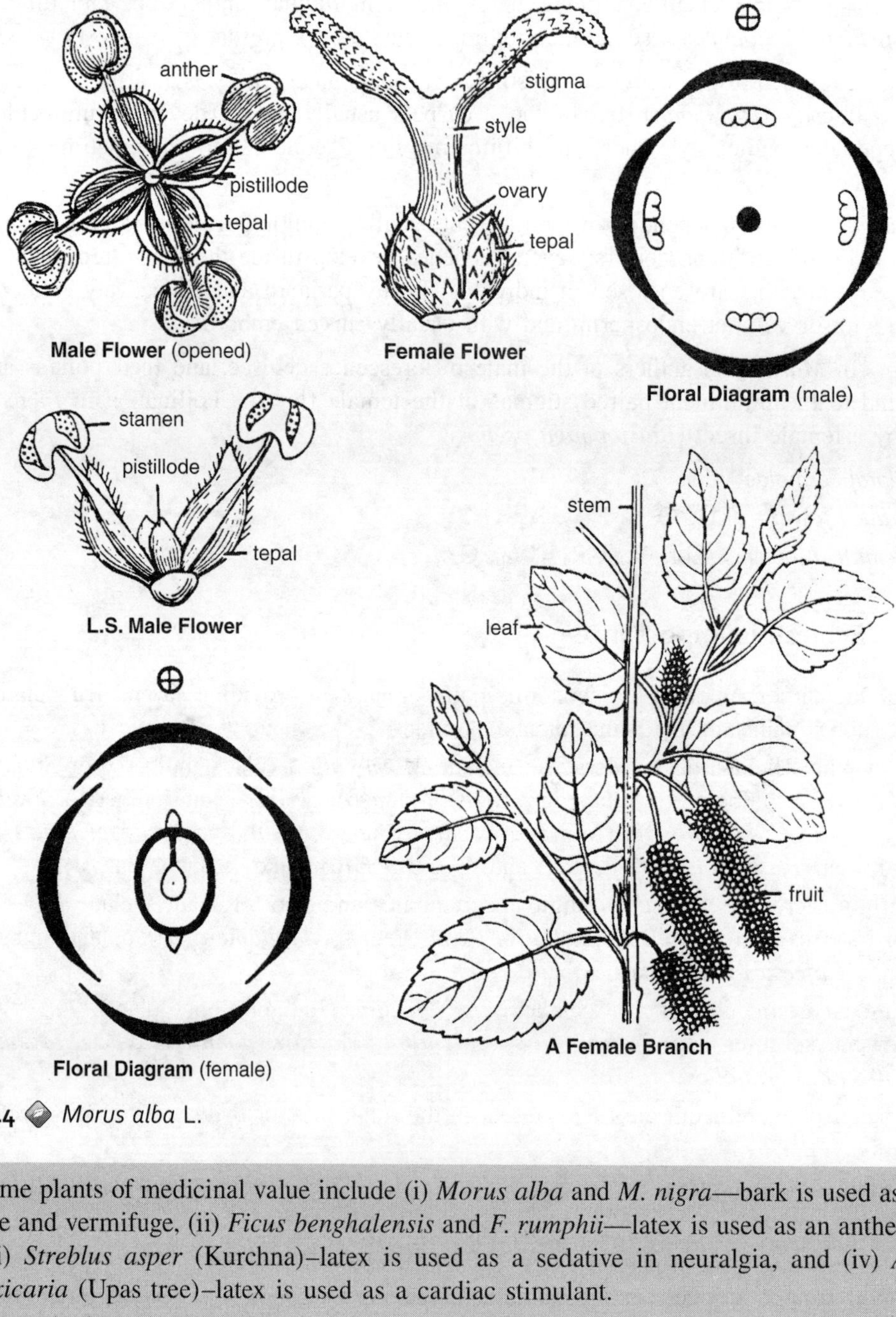

Fig. 21.114 *Morus alba* L.

- Some plants of medicinal value include (i) *Morus alba* and *M. nigra*—bark is used as purgative and vermifuge, (ii) *Ficus benghalensis* and *F. rumphii*—latex is used as an anthelmintic, (iii) *Streblus asper* (Kurchna)–latex is used as a sedative in neuralgia, and (iv) *Antiaris toxicaria* (Upas tree)–latex is used as a cardiac stimulant.
- Fibre obtained from the bark of *Ficus articulata* and *F. semicordata* is used for making ropes.

21.96.7 Systematics and Phylogeny (Affinities)

Moraceae is not recognized as an independent family by Bentham and Hooker. They treated its members in tribes Moreae and Artocarpeae under family Urticaceae of Unisexuales. However, almost all other workers treated Moraceae as an independent family under order Urticales (Engler and Prantl, 1931; Hutchinson, 1959; Takhtajan, 1969; and Cronquist, 1981). Several workers regard Moraceae as an advanced family of dicotyledons.

21.97 CANNABINACEAE (HEMP FAMILY)

21.97.1 Systematic Position

Monochlamydeae or Apetalae, Unisexuales[1].

21.97.2 Field Recognition

Aromatic herbs with watery juice; *Cannabis* is erect while *Humulus* is climbing; leaves opposite, simple in *Humulus* while palmately compound in *Cannabis*; flowers unisexual, mostly dioecious while few are monoecious (*Humulus japonica*); calyx persistent; gynoecium unilocular; fruit achene.

21.97.3 Common Indian Genera

Cannabis, *Humulus*.

21.97.4 Size, Distribution and General Information

A very small family of only 2 genera (*Cannabis* and *Humulus*) and 5 species, Cannabinaceae are mainly distributed in north temperate regions. Sometimes placed under Moraceae, the Cannabinaceae differ from Moraceae in possessing 5 floral parts instead of 4. Both the genera of hemp family grow in India. *Cannabis sativa* (Fig. 21.115), the only species of this genus, grows wild all over India. Of the 4 species of *Humulus*, *H. lupulus* grows wild in Himalayas and other parts of our country, *H. americanus* grows in USA, *H. japonica* in Japan and *H. scandens* in China as well as Japan.

21.97.5 Description of the Family

General Habit Erect (*Cannabis sativa*) or climbing (*Humulus lupulus*) aromatic herbs with watery juice; female plants of *Cannabis sativa* are usually larger than the male plants; latex absent.

Leaves Alternate or often opposite leaves, which are simple or palmately compound and palmately veined; stipulate, stipules persistent; petiolate; margins coarsely toothed; venation reticulate multicostate.

[1] Sometimes included in Urticaceae, Lawrence discussed all members of Cannabinaceae under subfamily Cannaboideae of family Moraceae while Rendle as well as Hutchinson treated all members of Cannaboideae under a separate family Cannabinaceae. Bentham and Hooker treated all genera of Urticaceae, Moraceae and Cannabinaceae under a single family Urticaceae of order Urticales. Recently, Cronquist (1981) and Jones and Luchsinger (1988) discussed Cannabinaceae (or Cannabaceae) under order Urticales of subclass Hamamelidae of class Magnoliopsida (Dicots).

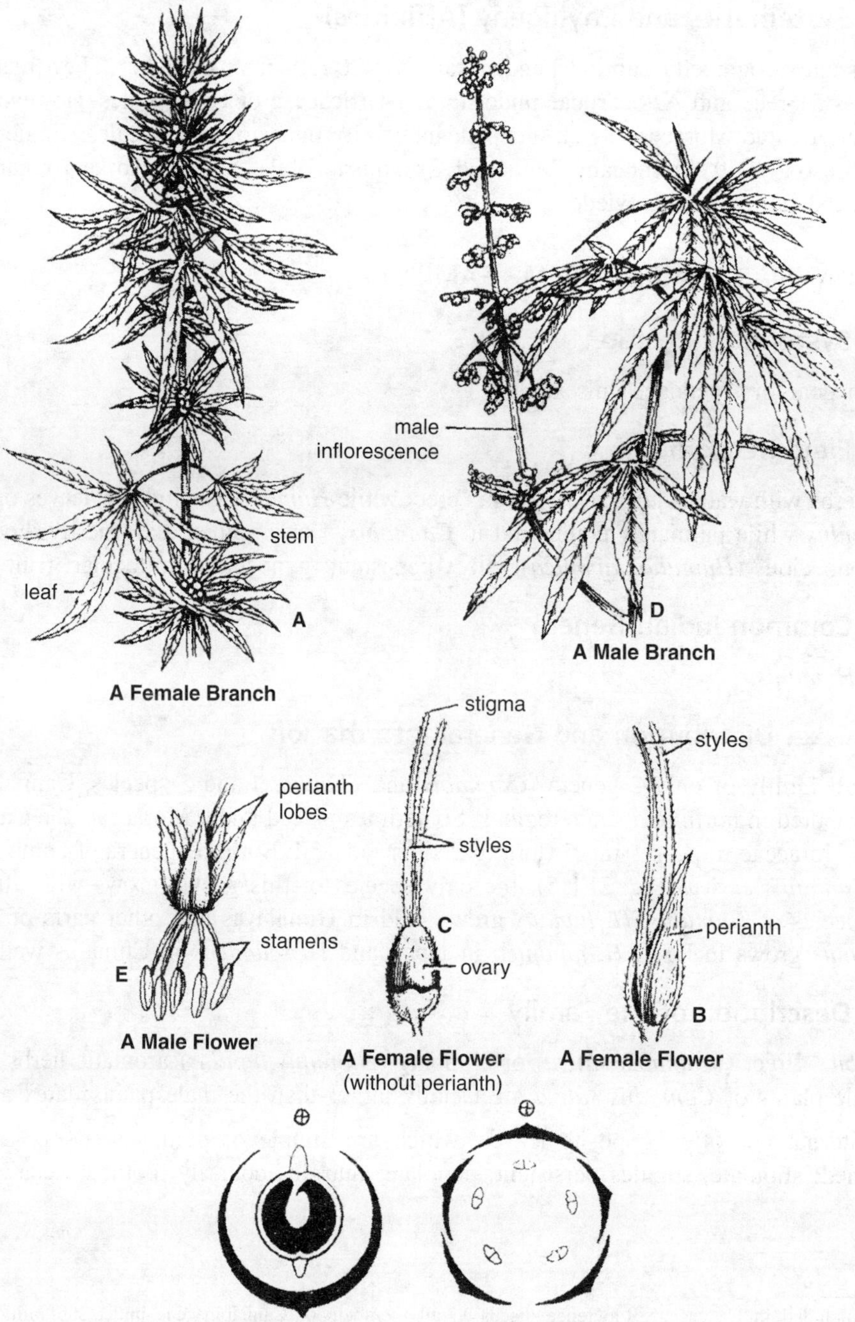

Fig. 21.115 *Cannabis sativa* L.

Inflorescence In *Cannabis*, the male plants bear flowers arranged in axillary paniculate cymes whereas female plants bear flowers arranged in cone-like axillary spikes; in *Humulus*, the male plants have mixed panicles of cymes and racemes whereas female plants bear biparous cone-like flowers.

Flowers Flowers unisexual and plants are usually dioecious; usually apetalous and axillary; actinomorphic; small and inconspicuous; bracteate in *Cannabis sativa* and bracts are covered by glandular hairs on their outer surface; the glands on these hairs remain filled with a secretion containing tetra-hydrocannabinol.

Male Flower

Perianth Staminate flowers arranged in panicles; each flower with a usually 5-parted perianth; lobes valvate or imbricate, free (*Cannabis*) or connate at base (*Humulus*); green.

Androecium Stamens 5, free, opposite the perinath lobes, dithecous, dorsifixed.

Gynoecium Absent.

Female Flower

Perianth Usually absent; if present, then only one perianth lobe enveloping the ovary is present; colourless.

Androecium Absent.

Gynoecium Monocarpellary, superior, unilocular, one-ovuled ovary; ovule pendulous and curved; placentation basal; style one with two long arms; stigmas two, minute and simple.

Fruit An achene; usually glandular and surrounded by persistent perianth lobe.

Seeds Endospermic, flattened with curved embryo.

General Floral Formulae

(a) *Male Flower:* ⊕, ♂, $P_{5 \text{ or } (5)}$, A_5, G_0.

(b) *Female Flower:* ⊕, ♀, P_1, A_0, $G_{\underline{1}}$.

21.97.6 Economic Importance

(a) Cannabis *Cannabis sativa* (hemp) plants provide a valuable fiber of great strength and durability. The fiber is widely used for making ropes, twine, carpets, sacks, bags, etc. The finer grades of this fiber are woven into a cloth that looks like coarse linen. The famous "Jeans" are prepared from this fiber. The "oil of hemp", obtained from the seeds of *C. sativa*, is used for making varnishes and soft soap. Narcotics (e.g. *bhang, ganja* and *charas*) are obtained from the flowering tops and leaves of *C. sativa*. The drug *hashish* or *charas*, a resinuous substance containing several powerful alkaloids, is obtained from this plant. *Bhang* or *marijuana* is actually the resinuous substance obtained from the upper leaves, bracts, flowers of the female inflorescences of *C. sativa*.

(b) Humulus The inflorescences of *Humulus lupulus* (Hops) contain a bitter aromatic narcotic substance called lupulin. Hops are highly used in medicine for their sedative and soporofic properties. Their principal use, however, is in brewing industry. They are added to beer to prevent bacterial action and consequent decomposition, and also to improve flavour and to impart the characteristic bitter taste to the beverage.

21.97.7 Systematics and Phylogeny (Affinities)

As already mentioned in the footnote of Article 21.97.1, the family Cannabinaceae has been treated variously by different taxonomists i.e. either under Moraceae, or under Urticaceae, or also as an independent family under order Urticales. But due to its characters like (i) herbaceous or vines with watery sap, (ii) aromatic plants without stinging hairs, (iii) five floral parts instead of four, (iv) stamens short and straight, (v) ovule apical and anatropous, and (vi) achene type of fruit, Cannabinaceae should be treated as an independent family.

21.98 CASUARINACEAE (CASUARINA FAMILY)

21.98.1 Systematic Position

Monochlamydeae or Apetalae, Unisexuales.

21.98.2 Field Recognition

Equisetum-like jointed stem; minute, whorled, scaly leaves; unisexual flowers; fruit woody, cone-like.

21.98.3 Size, Distribution and General Information

It is a monotypic or unigeneric family represented by only one genus (*Casuarina*) with about 65 species distributed mainly in north-east Australia and also in Fiji, New Caledonia, Malaysia and Mascarene Islands. Only one species (*Casuarina equisetifolia*; Fig. 21.116) occurs in India, and this is the most common species of the world.

21.98.4 Description of the Family

Habit Xerophytic, evergreen, shrubs or trees.

Stem Woody, much branched, with jointed, whorled or striate branches having prominent nodes and internodes; internodes with longitudinal ridges and grooves.

Anatomically, the stem contains (i) ridges and grooves, (ii) thick cuticle, (iii) sunken stomata in grooves protected by long hairs, (iv) sclerenchyma and green palisade parenchyma in the ridges, (v) vascular bundles in two alternate rings, (vi) wood contains vessels, tracheids, fibres and parenchyma.

Leaf Scale-like, 4–16 on each node; whorled; minute, basally connate, forming a sheath around the node; usually linear to lanceolate; leaf tips appear only as minute teeth; leaves resemble with that of *Ephedra* and *Equisetum*.

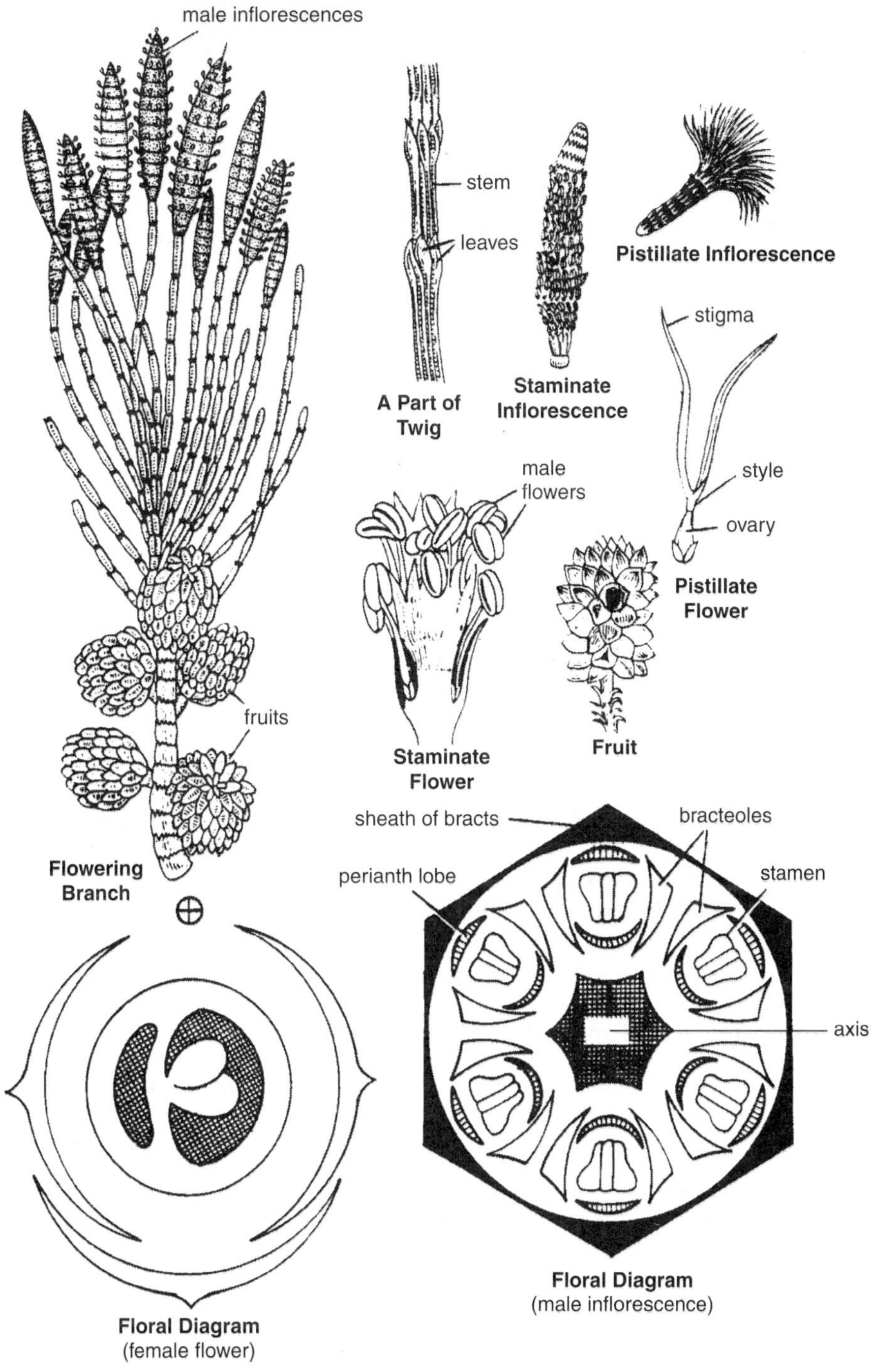

Fig. 21.116 *Casuarina equisetifolia* Forst.

Inflorescence Male flowers arranged in catkin-like erect spikes while the female flowers usually in spherical heads (Fig. 21.116); in both male and female branches the internodes are short; plants are monoecious or dioecious.

Flower Bracteate, bracteolate, bracteoles usually 2; incomplete, unisexual; perianth generally absent or rudimentary.

Male Flower Several male flowers develop at each node of the inflorescence axis; these flowers at each node remain protected by a sheath formed by the combination of bracts; each male flower is surrounded by 2 bracteoles, two small tepals or perianth lobes, and a central stamen which usually hangs out over the edge of the sheath of bracts; anthers dithecous, basifixed, dehiscing by vertical slits.

Female Flower Several female flowers remain crowded in spherical heads at the ends of the branches; each female flower is surrounded by a bract, a pair of bracteoles and a centrally located gynoecium; perianth lobes are absent; gynoecium is bicarpellary, syncarpous; ovary superior, originally bilocular, but becomes unilocular by the suppression of posterior locule; ovules 2, of which 1 usually aborts; parietal placentation; style 1, very short; stigmas 2, long, linear and feathery.

Fruit and Seeds Fruit usually a 1-seeded, small-winged samara, surrounded by 2 woody bracteoles and a bract; several such fruits remain arranged together to form cone-like dry multiple fruit. Seeds are non-endospermic and the embryo is straight.

Floral Formulae

(a) *Male Flower:* Br, Brl, ⊕, ♂, P_2, A_1, G_0.

(b) *Female Flower:* Br, Brl, ⊕, ♀, P_0, A_0, $G_{(\underline{2})}$.

21.98.5 Economic Importance

Casuarina is important mainly as a timber tree. The wood of *C. equisetifolia*, *C. cunninghamia* and *C. stricta* is used for furniture, for fuel and also for basket making. The bark of *C. suberosa* and *C. stricta* is utilised for tanning. *C. equisetifolia* is usually grown as an ornamental and also in order to reclaim sandy dunes in India and several countries.

21.98.6 Systematics and Phylogeny (Affinities)

Treated under Unisexuales by Bentham and Hooker, Casuarinaceae was included under Casuarinales by Hutchinson (1959), Takhtajan (1969), Cronquist (1981) and Thorne (1983). Several workers believe Casuarinaceae to be derived from the Hamamelidales (Takhtajan, Thorne, etc.). In the systems of Engler and Wettstein, Casuarinaceae is placed at the beginning of the dicotyledons. According to Kuznetsov (1936), *Casuarina* is one of the simpler types of angiosperm and is "unconnected with any other family", and the Casuarinales consist of a single family Casuarinaceae with a single genus (*Casuarina*).

On the other hand, Takhtajan (1969) mentioned that because of its special types of leaves and reproductive structures, *Casuarina* looks "more like *Ephedra* than an angiospermous plant". Takhtajan listed several characters of *Casuarina* (such as unisexual anemophilous flowers, unistaminate male flowers, absence of perianth, reduction of posterior locule in the ovary, etc.) which all "are signs of high evolutionary development" of this taxon.

In several morphological features Casuarinaceae approaches families such as Urticaceae, Betulaceae, Corylaceae, Myricaceae and also Hamamelidaceae. "In many features of floral construction, *Casuarina* undoubtedly comes near to the Urticales" (Takhtajan, 1969). However, the pollen grains of *Casuarina* are very similar to those of Betulaceae and Myricaceae (Erdtman, 1952). Takhtajan (1969) finally stated that *Casuarina* comes more closer to Hamamelidaceae because of similarity in several of their characters, such as 'unisexual anemophilous flowers, structure of the infloresence, reduction of perianth, gynoecia of two median carpels and reduction in the number of ovules".

21.99 SALICACEAE (WILLOW FAMILY)

21.99.1 Systematic Position

Dicotyledons, Monochlamydeae, Ordines Anomali

21.99.2 Field Recognition

Deciduous dioecious trees, shrubs or sub-shrubs; flowers arranged in catkins; each flower subtended by a bract or scale with a cup-like disc or a gland; seeds with tufts of hairs.

21.99.3 Selected Indian Genera

Populus alba (Poplar), *Salix tetrasperma* (Willow).

21.99.4 Size, Distribution and General Information

Family is represented by only 3 genera (*Populus*, *Salix* and *Chosenia*) and about 530 species, distributed widely in north-temperate regions of the world. *Salix*, with about 500 species and *Populus* with about 35 species are distributed mainly in north-temperate regions while the single known species of *Chosenia* (*C. arbutifolia*) is found in temperate and sub-arctic regions of north-east Asia.

(i) *Salix:* Bud remains covered by a single scale or bract; nectaries 1–4; stamens generally 2; leaves usually short-stalked with narrow blades.

(ii) *Populus:* Bud remains covered by several imbricate scales or bracts; nectaries absent; stamens generally 4 to many; leaves usually long-stalked with broad blades.

(iii) *Chosenia:* Bud in this monotypic genus remains covered by a single scale or bract; nectary absent; stamens 5.

21.99.5 Description of the Family

General Habit Woody deciduous trees or shrubs and sub-shrubs; mostly dioecious and only rarely monoecious (e.g. *Salix medemil*); many species are prostrate shrubs containing creeping stem sometimes, half-burried in the soil (e.g. *Salix divergens*, *S. furcata* and *S. flabellaris*). *S. nigra* and *S. fragilisi* are tall trees reaching upto 100 feet or more. Poplars are lofty trees attaining a great height.

Stem Aerial, erect or sometimes prostrate, solid, cylindrical, glaucous or glabrous.

Leaves Simple, stipulate; usually alternate or rarely subopposite (e.g. *Salix purpurea*), deciduous; petiolate, petiole short (*Salix*) or long (*Populus*); ovate to obovate (*Salix caprea*, Fig. 20.117).

Inflorescence Plants dioecious; flowers arranged in dense erect or pendulous catkins; small bracts present at the base of the catkins (*Salix caprea*, Fig. 21.117).

Flower Flowers unisexual, naked, each subtended by a fringed or hairy bract or scale; actinomorphic; hypogynous, small, cyclic and inconspicuous; bracts entire (*Salix*) or toothed (*Populus*); in *Salix caprea,* the unisexual flowers are solitary in the axil of bract and arranged spirally round a central axis and collectively form catkin-type of inflorescence (Fig. 21.117).

Perianth Absent or vestigial and represented by a cupular disc or small nectary.

Staminate Flowers With 1 or 2 nectariferous glands and 2 to 30 free or basally connate stamens; in *Salix sitchensis* the stamens are wholly connate; anthers dithecous, dehiscing vertically.

Pistillate Flowers Bicarpellary to tetracarpellary (*Salix tetrasperma*), syncarpous, superior, unilocular; more or less flask-shaped, numerous anatropous ovules on 2 to 4 parietal or basal placentae; 1 or 2 nectariferous glands are present in *Salix* while absent in *Populus*; style 1 with 2 to 4 stigmas; staminodes absent.

Fruit 2 to 4–valved capsule.

Seeds With silky hairs developing from the funicle; hairs not scattered over the entire seed; embryo straight, endosperm little or absent.

General Floral Formulae

(a) *Staminate Flower:* Br, ⊕, ♂, P_0, A_{2-30}, G_0.

(b) *Pistillate Flowers:* Br, ⊕, ♀, P_0, A_0, $G_{\underline{(2)}}$.

21.99.6 Economic Importance

Several species of Salicaceae are grown as ornamentals and also used in basket making and for pulpwood and floor-boards.

- **Salix** (Willow): Salicylic acid, the root compound for aspirin, is named for *Salix*. The soft inner bark of several species of this plant was widely used as a cure for headache in ancient times. *Aspirin*, its chemical derivative, is still used in headache throughout the world. Cricket bats (*S. alba*) and polo balls are traditionally made from the wood of willow (*Salix*). Several species of *Salix* are source of wood made into charcoal and twigs used in making baskets. Some forms of *Salix* are especially popular as decorative trees in damp situations.
- **Populus** (Poplar): Besides being well-known for its ornamental value, the bark of some species of *Populus* is also of medicinal value. *P. nigra* (Lombardy Poplar) has been a familiar avenue tree for over two centuries. Wood of *P. tremula* is used for making match-boxes and match-sticks.

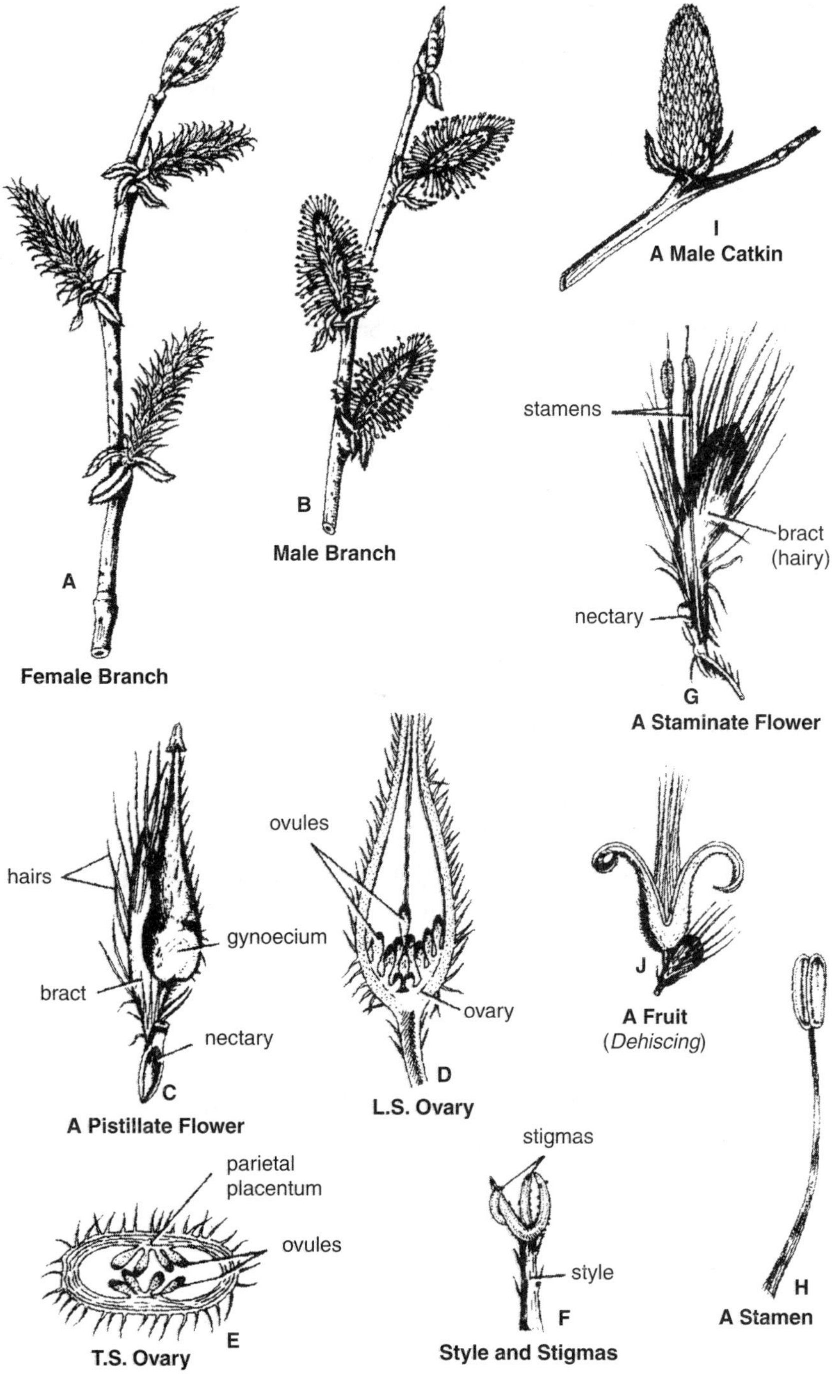

Fig. 21.117 *Salix caprea* L.

21.99.7 Systematics and Phylogeny (Affinities)

According to Lawrence, there is much evidence that Salicaceae "has been derived from advanced ancestral stocks but data are yet needed before the phyletic position can be indicated". Hutchinson (1926) retained Salicaceae with other amentiferous families of presumed hamamelidaceous affinities and treated it as the most primitive of them. Lawrence mentioned that "there is evidence to indicate the Salicaceae to be a more highly specialized group than most of the other amentiferous families" and the only one of an order (Salicales) that stands well apart from other orders of dicotyledons.

Bentham and Hooker included 4 families (Salicaceae, Lacistemaceae, Empetraceae and Ceratophyllaceae)under the series Ordines Anomali of Monochlamydeae and regarded all of them as advanced families. Benson (1957), however, included Salicaceae under order Salicales of Amentiferae and regarded it as an advanced family. Workers like Hutchinson (1973), Takhtajan (1980) and Cronquist (1981) treated Salicaceae under order Salicales while Thorne (1983) treated it under order Violales.

Salicaceae is related to Juglandaceae, Myricaceae and Fagaceae in possessing unisexual flowers arranged in catkins and their perianth is represented simply by glands or scales. It, however, differs from these families in possessing capsular fruits with numerous hairy seeds and not the drupe or nut borne by these families.

Test Your Understanding

1. Bentham and Hooker divided Dicotyledons into three subclasses. What are these subclasses? How can these subclasses be differentiated?
2. Write the characteristic features of the androceium and gynoceium of Magnoliaceae and Annonaceae.
3. Describe family Ranunculaceae or Papaveraceae in semitechnical language.
4. Write the name of the family of following genera:

1. *Iberis*	2. *Coronopus*	3. *Cleome*	4. *Fumaria*
5. *Argemone*	6. *Eschscholzia*	7. *Nelumbo*	8. *Delphinium*
9. *Nigella*	10. *Clematis*	11. *Annona*	12. *Michelia*

5. Describe the economic importance of family Malvaceae or Tiliaceae.
6. *Citrus* fruits belong to which family? Describe the *Citrus* family in semitechnical language. Also write a note on its economic importance.
7. Name any three families of your choice belonging to Disciflorae. Describe details of any one of them in semitechnical language.
8. What are the three subfamilies of family Leguminosae. Make a key to differentiate between these three subfamilies.
9. Describe in detail the economic importance of Papilionoideae.
10. "Rosaceae show great diversity in types of fruit". Justify this statement. Also write a note on the edible fruits of this family.

11. Following genera belong to series Calyciflorae of Polypetalae. Name the family they belong to:

1. *Daucus*	2. *Foeniculum*	3. *Mammillaria*	4. *Passiflora*
5. *Begonia*	6. *Luffa*	7. *Momordica*	8. *Lagenaria*
9. *Lawsonia*	10. *Eucalyptus*	11. *Callistemon*	12. *Terminalia*
13. *Pyrus*	14. *Prunus*	15. *Dalbergia*	16. *Bauhinia*

12. Give the floral description of Cucurbitaceae or Umbelliferae (Apiaceae).
13. Name all the three series of Gamopetalae proposed by Bentham and Hooker. Describe the floral structure of any of the families of your course belonging to Gamopetalae.
14. Describe in detail the inflorescence and detailed floral structure of Asteraceae.
15. Compositae (Asteraceae) occupy highest position among angiosperms. Comment on this statement.
16. Describe family Asclepiadaceae in semitechnical language.
17. Write the botanical name and the family of following plants:

1. Sarpgandha	2. Karonda	3. Chiku	4. Coffee
5. Coriander	6. Ajwain	7. Heeng	8. Zeera or cumin
9. Saunf	10. Kheera		

18. Describe family Convolvulaceae or Solanaceae using semitechnical language.
19. Compare the floral structures of Acanthaceae and Lamiaceae.
20. Give characteristic features of all the eight series of Monochlamydeae. Give at least one example of family of each of these series.
21. Describe the floral structure of family Euphorbiaceae.

Suggested Reading

Bentham, G. and J.D. Hooker, 1862–1883, *Genera Plantarum*, 3 Vols., Reeve & Co., London.

Bessey, C.E., 1915, Phylogenetic taxonomy of flowering plants, *Ann. Mo. Bot. Gard.* **2**: 109–164.

Britton, N.L. and J.N. Rose, 1919–1923, *The Cactaceae*, Vols. 1–4, The Carnegie Institution, Washington.

Cronquist, A., 1981, *An Integrated System of Classification of Flowering Plants*, Columbia Univ. Press, New York.

Engler, A.W. and K. Prantl, 1887–1915, *Die Naturlichen Pflanzenfamilien*, 23 Vols., Leipzig.

Heywood, V.H., 1978, *Flowering Plants of the World*, Mayflower Books, New York.

Hickey, M. and C. King, 1988, *100 Families of Flowering Plants*, (2nd ed.), Cambridge University Press, Cambridge.

Holmes, S., 1983, *Outline of Plant Classification*, Longman, London.

Hutchinson, J., 1969, *Evolution and Phylogeny of Flowering Plants*, Academic Press, London.

_______ 1973, *The Families of Flowering Plants*, (3rd ed.), Clarendon Press, Oxford.

Jones, S.B. Jr. and A.E. Luchsinger, 1987, *Plant Systematics*, (2nd ed.), McGraw-Hill Book Co., New York.

Lawrence, G.H.M., 1951, *Taxonomy of Vascular Plants*, Macmillan Co., New York.

Radford, A.E., 1986, *Fundamentals of Plant Systematics*, Harper & Row Publishers, Inc., New York.

Rendle, A.B., 1925, *The Classification of Flowering Plants* Vol. II (Dicotyledons), Cambridge Univ. Press, London.

Takhtajan, A., 1969, *Flowering Plants: Origin and Dispersal*, Oliver & Boyd Ltd., Edinburgh.

_______ 1980, Outline of the classification of flowering plants (Magnoliophyta), *Bot. Rev.* **46**: 225–359.

Thorne, R.F., 1983, Proposed new realignments in the angiosperms, *Nordic J. Bot.* **3**: 85–117.

Tutin, T.G., 1964, *Flora Europaea* Vol. I, Cambridge Univ. Press, London.

Willis, J.C., 1973, *A Dictionary of the Flowering Plants and Ferns*, (8th. ed., revised by H.K. Airy Shaw), Cambridge Univ. Press, London.

SELECTED FAMILIES OF MONOCOTYLEDONS

22.1 MONOCOTYLEDONS AND THEIR CLASSIFICATION

Monocotyledons (usually characterised by the presence of one cotyledon, fibrous and adventitious roots arising from the base of the stem, narrow leaves with parallel veins, herbaceous stem, irregular distribution of vascular bundles in the stem, trimerous flowers, and perianth often not clearly divisible into calyx and corolla) were also divided into varying number of families by different workers, similar to dicotyledons. Bentham and Hooker (1862–1883) recognised 34 families under Monocotyledons while Takhtajan (1969) divided class Liliatae (or Monocotyledons) into 69 families. Cronquist (1981) included 65 families under class Liliopsida (Monocots) whereas Thorne (1983) treated Monocotyledoneae (= Liliidae) as a subclass of class Angiospermae (= Annonopsida) and discussed 53 families under this subclass. Only 16 monocot families are discussed in the present chapter. The criteria of selecting the discussed families, compilation of their characters, method of family description, and their arrangement in this chapter, are almost the same as followed in Chapter 21 (Selected Families of Dicotyledons).

Bentham and Hooker divided Monocotyledons into 7 series as under:

Series 1. Microspermae, characterised by the presence of epigynous flowers, inferior ovary, parietal placentation and very small and numerous non-endospermic seeds. It includes 3 families, namely Hydrocharitaceae, Orchidaceae[1] and Burmanniaceae.

Series 2. Epigynae, characterised by epigynous flowers, inferior ovary, and large endospermic seeds. It includes 7 families, namely Iridaceae, Amaryllidaceae[1], Haemodoraceae, Taccaceae, Dioscoreaceae, Bromeliaceae[1] and Scitamineae[1].

Series 3. Coronarieae, characterised by coloured or petaloid perianth, superior ovary and endospermic seeds. It includes 8 families, namely Liliaceae[1], Commelinaceae[1], Pontederiaceae, Rapatiaceae, Xyridaceae, Roxburghiaceae, Philyderaceae and Mayaceae.

Series 4. Calycineae, characterised by green or sepaloid perianth, superior ovary and endospermic seeds. It includes 3 families, namely Juncaceae[1], Palmae[1], and Flagellariaceae.

---

[1] Discussed in the present text.

Series 5. Nudiflorae, characterised by perianth mostly lacking or present in the form of scales or bristles, superior ovary and seeds endospermic. It includes 5 families, namely Typhaceae[1], Araceae[1], Pandanaceae, Lemnaceae and Cyclanthaceae.

Series 6. Apocarpeae, characterised by free carpels, superior ovary and non-endospermic seeds. It includes 3 families, viz. Alismaceae[1], Najadaceae, Triuridaceae.

Series 7. Glumaceae, characterised by small, scale-like or chaffy perianth or no perianth, large scaly bracts, flowers in spikelets or heads, ovary unilocular with one ovule in locule, and seeds with abundant and starchy endosperm. It includes 5 families namely Cyperaceae[1], Eriocaulaceae, Restionaceae, Centrolepidaceae and Gramineae[1].

22.2 ORCHIDACEAE (ORCHID FAMILY)

22.2.1 Systematic Position

Monocotyledons, Microspermae, Orchidaceae.

22.2.2 Field Recognition

Perennial herbs; perianth inconspicuous; leaves 2-ranked; flowers zygomorphic, usually trimerous; usually labellum, pollinia and gynostegium present; ovary inferior.

22.2.3 Selected Indian Genera

Cattleya, Cymbidium, Cypripedium, Dendrobium, Habenaria, Orchis, Vanda, Zeuxine.

22.2.4 Size, Distribution and General Information

Orchidaceae is one of the largest families of the flowering plants, represented by about 1000 genera and 20,000 species which are cosmopolitan but primarily distributed in tropical areas. Its members are rare in arctic regions. In Indian flora this is the second largest family represented by about 130 genera and over 880 species, distributed mainly in Eastern Himalayas, Western Ghats and Khasi hills.

This is the family of greatest variety of flowers among flowering plants. Orchid flowers are known for their strange shape, longevity and beautiful look. Some of the largely represented genera along with the number of their approximately reported species are *Dendrobium* (1500), *Habenaria* (600), *Epipendrum* (400), *Oncidium* (350), *Odontoglossum* (200), *Cattleya* (60), *Vanda* (60), *Cypripedium* (50) and *Cymbidium* (40).

22.2.5 Description of the Family

General Habit Perennial herbs; either terrestrial (*Orchis*), epiphytic (*Cattleya, Cypripedium*) or saprophytic (*Neottia*); orchids in temperate regions are mainly terrestrial while those in tropical regions are mainly epiphytic; sometimes climbers (*Bulbophyllum, Vanilla*); often with fibrous or tuberous roots or rhizomatous; epiphytic orchids usually contain fleshy pseudobulbs consisting of one or more

[1] Discussed in the present text.

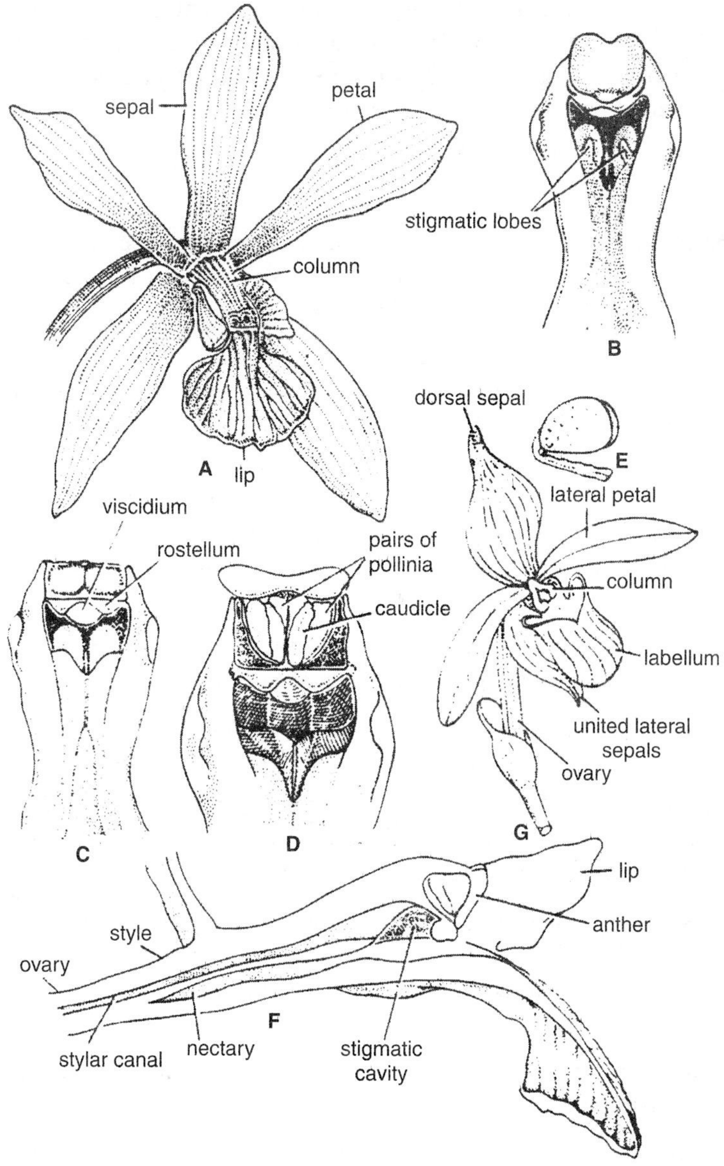

Fig. 22.1 A–F. Floral details of *Epidendrum tampene*–A: A flower; B: Column; C: Column with anther removed; D: Top of column with anther lifted; E: A pair of pollinia; F: L.S. flower; G: Flower of *Cypripedium*.

thickened internodes of stem; aerial roots often contain velamen; mycorrhiza often present; stems leafy or scapose, may be sympodial (*Dendrobium*) or monopodial (*Vanda*); saprophytic forms usually devoid of chlorophyll (*Corallorhiza*, *Neottia*).

Leaf Simple, alternate, often distichous, rarely opposite; sometimes reduced to achlorophyllous scales (e.g. saprophytic species); usually fleshy and linear to ovate in shape, often encircling the stem with a sheathing base; foliage leaves absent in some species of *Dendrophylax* and *Epipogon*; only a single leaf per shoot is present in *Pleurothallis.*

Inflorescence Variable from racemose, paniculate (*Oncidium*), spicate, or flowers solitary (*Cypripedium*).

Flower Usually bracteate; bisexual or rarely unisexual; zygomorphic; epigynous; showy, beautifully coloured; and extraordinarily diverse in shape, size and colour; mostly resupinate i.e. twisted 180° or upside down.

Perianth Usually 6 tepals in 2 whorls of 3 each; the outer whorl (representing calyx) of 3 tepals is green or coloured; the inner whorl (representing corolla) of 3 tepals is petaloid, of these the posterior tepal is highly modified, often projected basally into a spur and called *labellum* or *lip.*

Labellum is strictly the uppermost petal but looks as if located on the lower side of the flower in most orchids. This is caused either by the twisting of the ovary through 180° (as in many-flowered orchids), or by the pedicel bending back over the apex of the stem (as in single-flowered orchids). This phenomenon is called *resupination.* It enables the labellum to work as a landing-place for pollinating insects.

Androecium Represented by 2 fused lateral stamens or 1 terminal stamen; anthers dithecous, introrse; pollen grains granular or bound together by viscin threads into masses called *pollinia.*

Stigmas, style, and stamens are adnate to form a single, highly complex structure called *column* or *gynandrium* (Fig. 22.1), which represents the most characteristic part of the orchid flower.

In several orchids (*Zeuxine*, Fig. 22.2) a pair of pollinia remain connected to the *glandula* or *corpusculum* with the help of a stalk called *caudicle.* A connection between the ovary and stamen is made by *rostellum.* Sometimes staminodes are also present.

Gynoecium Tricarpellary, syncarpous, inferior, unilocular, many ovules in the locule; parietal placentation; rarely the ovary is trilocular and the placentation is axile (*Apostasia*); style 1 (part of the column); stigmas 3, of which 2 lateral are often fertile and the third one is sterile forming a beak called *rostellum.*

Fruit and Seeds Fruit usually a capsule, containing a very large number of very small, non-endospermic seeds, which are distributed easily by wind.

Pollination Flowers of Orchidaceae are well-adapted for insect pollination because of several characters, such as large-sized and gaudy perianth, pleasant smell, and presence of nectar secreted in nectary or sac-like rostellum. The flowers are protandrous. When an insect enters the flower, the pollinia get attached on its body by their caudicles. When this insect visits another flower, it may deposit the pollen grains on its stigmatic surface below the rostellum, thus completing the process of pollination.

General Floral Formula Br, ⊕, ⚥, P_{3+3}, $A_{(2) \text{ or } 1}$, $G_{\overline{(3)}}$.

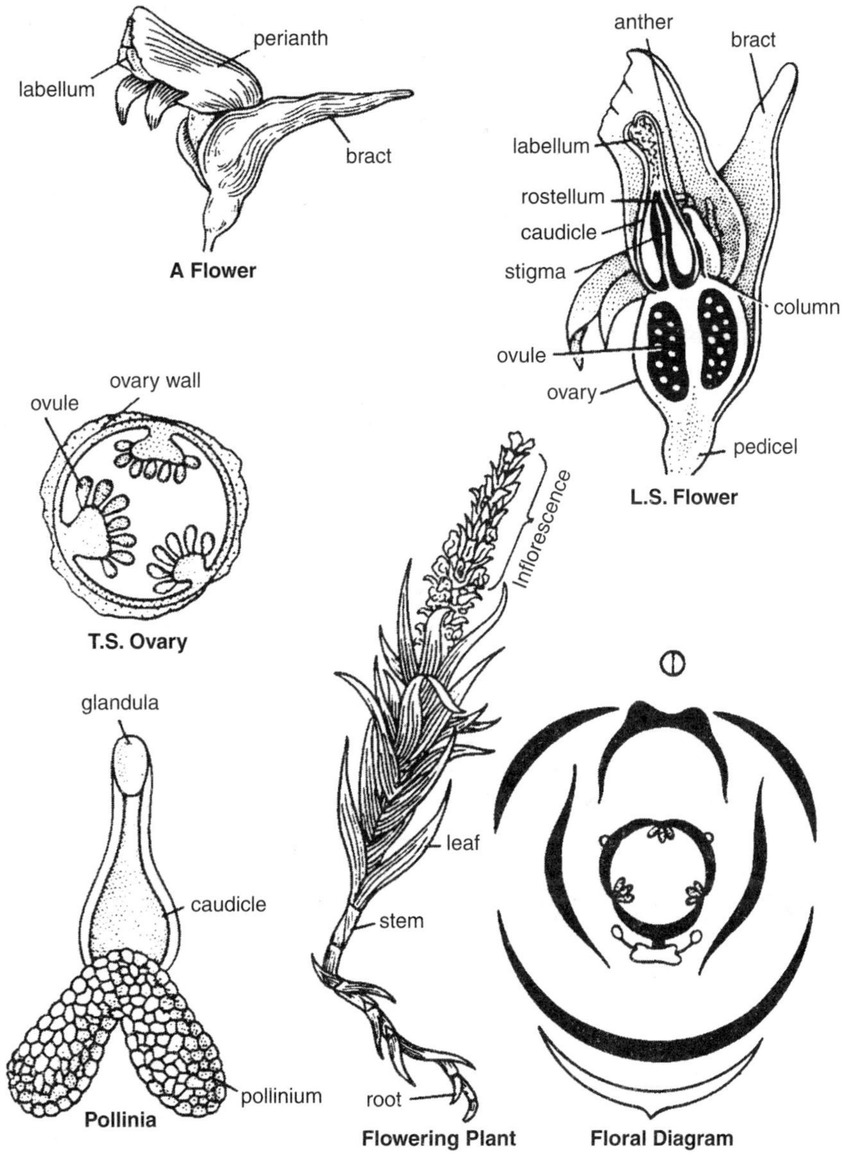

Fig. 22.2 *Zeuxine strateumatica* L.

22.2.6 Economic Importance

- The fruits of the climbing genus *Vanilla fragrans* syn. *V. planifolia* are the source of highly fragrant extract used in perfumery and confectionery.
- Tuberous roots of *Habenaria susannae* and *Orchis latifolia* are edible and eaten during scarcity.

- Starchy tubers of some genera (*Cymbidium*, *Dactylorhiza*, *Eulophia*, *Orchis*, etc.) are dried to form "Salep" of commerce, used for culinary and medicinal purposes.
- Orchids are known throughout the world for their beautiful flowers of various forms, shapes and highly attractive colours. Some orchid genera commonly grown in glasshouses are *Cattleya* (florist's orchid), *Cymbidium*, *Cypripedium* (lady's slipper), *Dendrobium*, *Epidendrum* (green-fly orchid), *Habenaria* (fringe orchid), *Miltonia*, *Odontoglossum* (baby orchid), *Oncidium* (butterfly orchid), *Orchis* (showy orchid), *Paphiopedalum*, *Phalaenopsis*, *Vanda*, *Vanilla* and *Zeuxine*.

22.2.7 Systematics and Phylogeny (Affinities)

Treated as a family under series Microspermae by Bentham and Hooker, Orchidaceae was placed under a separate order Orchidales by majority of the later workers including Hutchinson (1959), Takhtajan (1969) and Cronquist (1981). Thorne (1983), however, placed Orchidaceae under the suborder Orchidineae of the order Liliales. Willis (1973) divides Orchidaceae into 3 subfamilies (Apostasioideae, Cypripedioideae and Orchidoideae) divisible further into 6 tribes (Apostasieae, Cypripedieae, Orchideae, Neottieae, Epidendreae and Vandeae).

Several taxonomists consider Orchidaceae to be the most advanced and highest evolved among monocotyledons. The characters which support this view include (i) reduction in number of stamens, (ii) resupinate epigynous ovary, (iii) presence of rostellum, (iv) non-endospermic seeds, (v) herbaceous habit, and (vi) presence of several epiphytes.

According to the majority of taxonomists, Orchidaceae originated from a Liliaceous stock. Hutchinson opined that Orchidaceae originated from Liliaceae through Hypoxidaceae and Apostaciaceae. Some, however, also trace another line of parallel evolution of Orchidaceae from Musaceae.

22.3 IRIDACEAE (IRIS FAMILY)

22.3.1 Systematic Position

Monocotyledons, Epigynae, Iridaceae.

22.3.2 Field Recognition

Perennial herbs often with bulbs, corms, or rhizome; contain equitant leaves; flowers trimerous with 3 petaloid sepals and 3 petals; stamens 3; ovary inferior.

22.3.3 Selected Indian Genera

Belamcanda, *Crocus*, *Gladiolus* and *Iris*.

22.3.4 Size, Distribution and General Information

A family of about 70 genera and over 1500 species, Iridaceae are distributed in both tropical and temperate regions with the chief centres of distribution in tropical America and South Africa. Chief areas of distribution in India are Kashmir and Kumaon hills. Common species cultivated in Indian gardens include *Belamcanda chinensis*, *Crocus sativus*, *C. biflorus*, *C. speciosa*, *Gladiolus primulinus*, *G. colvillei*, *Freesia refracta*, *Iris kumaonensis*, *I. germanica* and *I. foetidissima*. Iridaceae are cultivated throughout the world for their beautiful flowers.

22.3.5 Description of the Family

General Habit Perennial herbs with roots produced from corm (*Crocus*, Fig. 22.3A, B), bulb (*Gladiolus*), or rhizome (*Iris*, Fig. 22.3C; *Belamcanda*); very rarely subshrubs; stems solitary or several, or plants scapose.

Leaf Usually simple, exstipulate, sessile, linear to ensiform, equitant in 2 ranks; mostly basal, numerous and crowded at the base of stem; with sheathing leaf base; parallel venation.

Inflorescence Racemose raceme or paniculate, or flowers solitary (*Sisyrinchium*, *Romulea*); invariably terminal; in several genera the flowers develop on distinct aerial flowering axis or scape (*Iris*).

Flower Bracteate, complete; actinomorphic or zygomorphic (*Gladiolus*, *Tritonia*); bisexual, epigynous, trimerous; showy, large and beautifully coloured; usually subtended individually or in groups by 2 spathe-like bracts.

Perianth Usually 6 tepals arranged in 2 whorls of 3 each; of these the outer whorl of 3 tepals represents calyx and often petaloid; and inner whorl of 3 tepals represents corolla; tepals of both the whorls are often distinguishable from each other by colour, size, or texture; free or united, but usually basally connate into a long tube (*Crocus*).

Androecium Usually 3 distinct stamens situated opposite the tepals of the outer whorl; mostly epiphyllous; anthers dithecous, basifixed, extrorse, dehiscing by vertical slits.

Gynoecium Tricarpellary, syncarpous; ovary inferior, trilocular, with few to many anatropous ovules in each locule, axile placentation; in *Hermodactylus* the ovary is unilocular with parietal placentation; style 1, often 3-branched, long and coloured; stigmas 3, filiform to subulate or fimbriate; ovary rarely superior (*Isophysis*); style is very short in *Eleuthrine*.

Fruit and Seeds Fruit a loculicidal capsule dehiscing by 3 valves; seeds with small embryo and copious endosperm, sometimes arillate.

Pollination Usually entomophilous, effected by bees and butterflies.

General Floral Formula $\text{Br}, \oplus \text{ or } \%, ⚥, \overline{P_{(3+3) \text{ or } (3+3)}, A_3}, G_{(\bar{3})}$.

22.3.6 Economic Importance

- Dried styles of *Crocus sativus* are used to make '*saffron*', an orange-yellow dye used chiefly for flavouring and colouring dishes. Saffron is also used to give colour to the cooked rice. "Around 100,000 styles are required to produce 1 kilogram of saffron" (Jones and Luchsinger, 1987).
- Orris root, used in perfumes and in dentifrices, is obtained from the rhizomes of *Iris germanica* var. *florentina*.

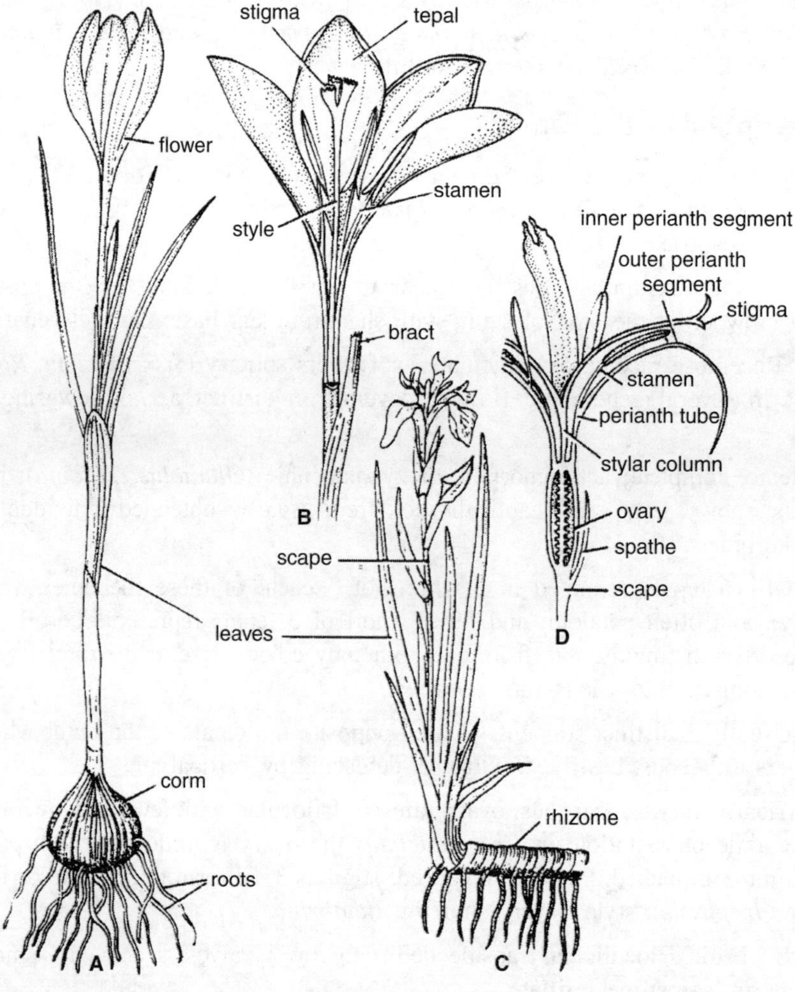

Fig. 22.3 Entire plant (A) and L.S. of the upper part of flower (B) of *Crocus tomasinianus*; Entire plant (C) and L.S. flower (D) of *Iris pseudacorus*.

- Roots of *Belamcanda chinensis* are an antidote to snake poison. A drug obtained from its rhizomes is used in tonsilitis.
- Roots of *Iris germanica* are diuretic and used in diseases of gall bladder.
- Iridaceae are highly-prized decorative plants of gardens and parks. Some of the ornamental genera are *Belamcanada* (blackberry lily), *Crocosmia*, *Crocus*, *Dierama*, *Freesia*, *Gladiolus*, *Iris*, *Ixia*, *Romulea*, *Sisyrinchium* (blue-eyed grass), *Tigridia* (tiger flower) and *Tritonia*.

22.3.7 Systematics and Phylogeny (Affinities)

Hutchinson (1959) and Takhtajan (1969) placed Iridaceae under order Iridales while Cronquist (1981) and Thorne (1983) treated it under order Liliales. Engler and Prantl included Iridaceae under Liliiflorae and divided it into 3 tribes viz. Sisyrinchieae, Ixieae and Irideae, and the same classification has been followed by Hickey and King (1988).

Iridaceae is considered to be close to Liliaceae and Amaryllidaceae. Hutchinson traced its origin from Liliales while Takhtajan opined that Iridales "probably derived directly from the family Liliaceae, most likely through subfamily Melanthioideae".

22.4 AMARYLLIDACEAE (DAFFODIL FAMILY)

22.4.1 Systematic Position

Monocotyledons, Epigynae, Amaryllidaceae.

22.4.2 Field Recognition

Bulbous or rhizomatous, perennial herbs; inflorescence usually a leafless scape or umbel subtended by spathe-like bracts; perianth gamophyllous; stamens 6, antiphyllous and epiphyllous; ovary inferior.

22.4.3 Selected Indian Genera

Agave, Crinum, Narcissus, Zephyranthes.

22.4.4 Size, Distribution and General Information

A family of about 85 genera and over 1100 species (Hickey and King, 1988), Amaryllidaceae are usually distributed in tropical and subtropical regions. About 9 genera and over 60 species have been reported from India. Some of the larger genera along with the number of their worldover reported species are *Crinum* (110), *Hippeastrum* (75), *Narcissus* (60), *Hymenocallis* (50), *Haemanthus* (50), *Cyrtanthus* (47), *Zephyranthes* (40), *Galanthus* (20), and *Pancratium* (15).

22.4.5 Description of the Family

General Habit Usually perennial, bulbous (*Crinum*) or rhizomatous (*Hypoxis*), xerophytic herbs leafing only in the rainy or spring season; some are arborescent (*Agave*, *Furcraea*).

Leaf Mostly basal, simple, alternate, often distichous, exstipulate, sessile with sheathing leaf base (*Crinum*); very large, more or less linear, strap-shaped or ensiform; entire; parallel venation.

Inflorescence Usually on a long leafless scape, with one or more spathes or bracts, cymose, but generally condensed in the form of an umbel or head; sometimes solitary or paired flowers are present (*Narcissus*, Fig. 22.4; *Zephyranthes*); inflorescence appears after several years in *Agave*.

Flower Bracteate, complete, bisexual; actinomorphic (*Crinum*, Fig. 22.5) or zygomorphic (*Amaryllis*); trimerous, epigynous; large, showy and usually fragrant.

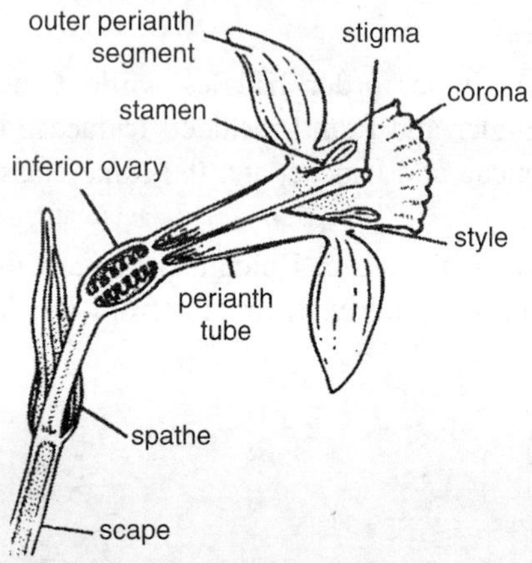

Fig. 22.4 L.S. flower of *Narcissus*.

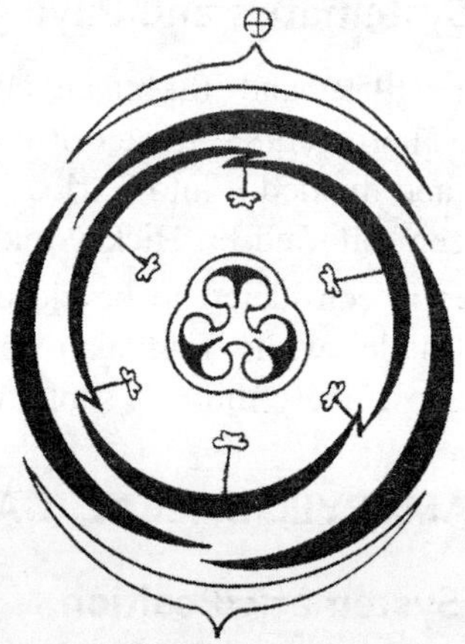

Fig. 22.5 Floral diagram of *Crinum asiaticum*.

Perianth Tepals 6, arranged in 2 whorls of 3 each, not differentiated into calyx and corolla; petaloid; free or often united basally to form a long tube; salver-shaped or funnel-shaped (*Crinum*); large and showy; valvate or twisted or imbricate; sometimes corona present (*Narcissus*) and looks like an extra whorl of perianth-segments in between the normal whorl of perianth and stamens.

Androecium Stamens 6, arranged in 2 whorls of 3 each, epiphyllous; present opposite the tepals; inserted at the throat of perianth tube; filament usually long and coloured (*Crinum*); anthers dithecous, basifixed or versatile (*Crinum*); introrse; dehiscence by longitudinal slits or rarely by apical pore (*Galanthus*); sometimes some stamens become staminodial.

Gynoecium Tricarpellary, syncarpous; ovary inferior, rarely half-inferior; trilocular or rarely unilocular (*Alstroemeria*); numerous anatropous ovules in each locule; axile placentation; style 1; stigma capitate or trilobed.

Fruit and Seeds Fruit a loculicidal capsule or berry (*Haemanthus*). Seeds with small and straight embryo, and endospermic.

Pollination It is entomophyllous. Insects are attracted because of gaudy-coloured, often scented perianth and presence of nectar. Nectar is secreted in the inner tepals in *Galanthus* and in corona in *Narcissus*, *Hippeastrum*, etc.

General Floral Formula Br, $\oplus$ or ⊕, ⚥, $\overbrace{P_{3+3}, A_{3+3}}$, $G_{(\overline{3})}$.

22.4.6 Economic Importance

- Common ornamental plants of Amaryllidaceae are *Agave americana* (century plant), *A. mexicana*, *A. vera-cruz* (Kuwarbuti), *Alstroemeria aurantiaca*, *Amaryllis belladona* (Amaryllis), *Chlidanthus fragrans*, *Clivia miniata* (Kafir-lily), *Crinum asiaticum* (Sukhdarsan), *C. defixum*, *C. latifolium*, *Curculigo capitulata*, *C. latifolia*, *Eucharis grandiflora* (Amazon lily),

Galanthus nivalis, *Haemanthus coccineus* (blood lily), *Hippeastrum vittatum*, *Hymenocallis americana* (spider lily), *Hypoxis hirsuta* (stargrass), *Leucojum vernum* (Snowflake), *Narcissus pseudo-narcissus* (Daffodil or Nargis), *N. bulbocodium*, *N. tazetta*, *Nerine filifolia*, *Pancratium maritinum*, *Polianthes tuberosa* (Tuberose), *Zephyranthes grandiflora* (Zephyr lily), *Z. candida* and *Z. tubispatha*.

- The oil obtained from the flowers of several species of *Narcissus* and *Polianthes* are used in high grade perfumery.
- The fibres obtained from *Agave cantala*, *A. sisalana* (Sisal hemp), *A. vera-cruz* and *Furcraea foetida* (Mauritius hemp) are used for making cordage, mats, socks, shoe soles, ropes, twines, etc. The fibres obtained from the leaves of *Curculigo latifolia* are used for making fishing nets.
- Fruits of *Curculigo latifolia* and roots of *C. orchioides* and *Alstroemeria* are ground and eaten as flour.
- Tubers of *Curculigo orchioides* are used in the treatment of asthma and jaundice.

22.4.7 Systematics and Phylogeny (Affinities)

Treated under series Epigynae by Bentham and Hooker along with other families with inferior ovary, Amaryllidaceae is included variously by different workers under Liliiflorae (Engler and Prantl), Amaryllidales (Hutchinson) and Liliales (Takhtajan).

Most authors treat Amaryllidaceae very close to Liliaceae. Hutchinson (1973) opined that the character of inferior verses superior ovary has been overemphasized in the monocotyledons, although he treated Amaryllidaceae and Liliaceae as separate families. Cronquist (1981), however, considers the separation of Amaryllidaceae and Liliaceae on the basis of ovary position to be unreasonable and merges these two families together under Liliaceae, and the same treatment is followed by Jones and Luchsinger (1987). Takhtajan (1969) stated that "Amaryllidaceae probably have a common origin with Agavaceae from Liliaceae".

22.5 BROMELIACEAE (PINEAPPLE FAMILY)

22.5.1 Systematic Position

Monocotyledons, Epigynae, Bromeliaceae

22.5.2 Field Recognition

Short-stemmed, herbaceous, epiphytic plants, often showing xerophytic habit; stem reduced; leaves stiff, often brightly coloured towards the base; presence of multicellular or stellate scales or hairs; flowers trimerous and floral bracts usually coloured; calyx herbaceous; anthers versatile.

22.5.3 Selected Indian Genera

Ananas, *Bromelia*, *Billbergia*, *Tillandsia*.

22.5.4 Size, Distribution and General Information

Pineapple family is represented by about 45 genera and over 2000 species, and distributed chiefly in tropical America and West Indies. The chief genera along with their approximately recorded species are *Tillandsia* (500), *Pitcairnia* (250), *Vriesea* (190), *Aechmea* (150), *Puya* (120), *Dyckia* (80), *Navia* (60), *Billbergia* (50), *Bromelia* (40) and *Ananas* (5). The delicious fruit of "pineapple" comes from this family. Its bracts, pedicels, ovaries and stalk develop and fuse into a sweet, juicy mass. A familiar epiphyte of this family is *Tillandsia usneodes* (Spanish moss).

22.5.5 Description of the Family

General Habit Mostly epiphytes (*Billbergia nutans*, Fig. 22.6), often xerophytic in habit, short-stemmed.

Leaves Reduced stem contains a basal rosette of linear and spinulose fleshy leaves showing a close fitting at the base, thus providing the whole plant the shape like that of a funnel; minute scaly hairs cover the base of the leaves; usually stiff and often spiny; brightly coloured towards the base.

Inflorescence A terminal raceme, panicle, spike or head with flowers developing in the axils of often brightly-coloured bracts.

Flowers Bracteate; actinomorphic or weakly zygomorphic; usually bisexual, rarely functionally unisexual; epigynous to hypogynous.

Perianth 6 tepals in 2 series, of which outer 3 calyx-like free parts and inner 3 corolla-like free or variously connate parts; often brightly coloured; scales or prominent nectaries are present in some members within perianth lobes.

Androceium Stamens 6, often epipetalous and develop mostly at the base of perianth lobes; free or partially fused to them; anthers dithecous, linear, usually versatile (*Billbergia nutans*, Fig. 22.6); introrse.

Gynoceium Tricarpellary, syncarpous; ovary inferior, half-inferior, or superior; trilocular; numerous anatropous ovules in each locule on axile placentation; style 1; stigmas 3, spirally twisted (Fig. 22.6).

Fruit A berry, capsule, multiple type (e.g., sorosis in pineapple); often more or less enveloped by the persistant perianth.

Seeds Very light, sometimes winged; with small embryo and mealy endosperm.

General Floral Formula Br, $\oplus$, ⚥, $\overparen{P_{3+3 \text{ or } (3+3)}, A_6}$, $G_{(\bar{3})}$.

22.5.6 Economic Importance

- Bromeliaceae is known throughout the world for being the family of famous and delicious fruit "pineapple" (*Ananas comosus*). Its leaves are used for making "pina-cloth".

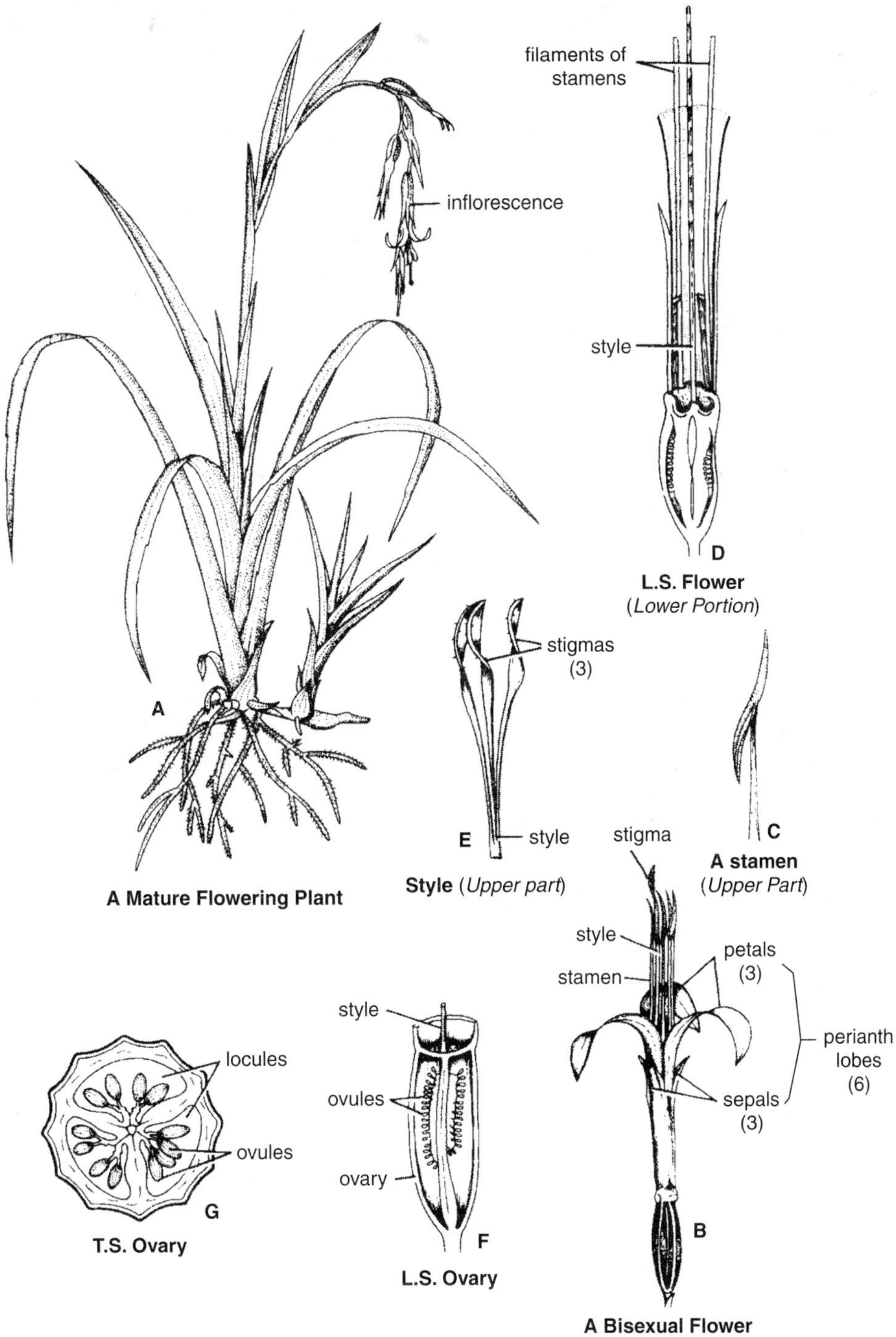

Fig. 22.6 *Billbergia nutans* H. Wendl.

- Several species of a number of genera of this family are grown as indoor plants for ornamental purposes. Some of these genera include *Billbergia*, *Cryptanthus*, *Nidularium*, *Tillandsia* and *Vriesea*.
- "Spanish moss" (*Tillandsia usneoides*) is grown in greenhouses as a curiosity because it appears like that of a lichen and grows pendent.
- Some species, including *T. usneoides*, are used as packaging material and also in upholstery.
- Several species of Bromeliaceae are sources of cordage and fiber for fabrics.

22.5.7 Systematics and Phylogeny (Affinities)

Bromeliaceae has been included under series Epigynae along with six more families, namely Iridaceae, Amaryllidaceae, Haemodoraceae, Taccaceae, Dioscoreaceae and Scitamineae. Engler and Prantl included Bromeliaceae under order Farinosae along with 12 more families. Modern workers like Cronquist (1981) treated Bromeliaceae under a separate order Bromeliales of subclass Zingiberidae of class Liliopsida while Dahlgren (1983) discussed it under order Bromeliales of superorder Bromelliflorae of subclass Liliidae (Monocotyledoneae).

Following Engler and Prantl, Hickey and King (1988) classified Bromeliaceae under 4 subfamilies, including (i) Navioideae (with spinulose-dentate leaves, superior ovary, capsule fruit and seeds without wings, e.g. *Navia*), (ii) Pitcairnioideae (with entire leaves, superior or half-inferior ovary, and appendaged seeds, e.g. *Pitcairnia*), (iii) Tillandsioideae (with entire leaves, superior ovary and hairy seeds, e.g., *Tillandsia*), and (iv) Bromelioideae (with spinulose leaves, inferior ovary, fruit berry and naked seeds, e.g. *Ananas*, *Bromelia*). From the phylogeny view point, Bromeliaceae represents "the climax of a line of descent wherein the calyx and corolla have remained distinct or fairly distinct from each other, a feature retained from the Dicotyledonous stock" (Hutchinson, 1934). Hutchinson treated Bromeliaceae as related to Commelinales.

22.6 CANNACEAE (CANNA FAMILY)

22.6.1 Systematic Position

Monocotyledons, Epigynae (Treated by Bentham and Hooker under Scitamineae).

22.6.2 Field Recognition

Large, coarse, perennial herbs with a tuberous rootstock; leaves large, foliaceous, spirally arranged, with its petiole sheathing the stem; inflorescence raceme or panicle or 2-flowered cincinni; flowers large zygomorphic; usually only 1 functional stamen and remaining 4 staminodes; inferior ovary; ligule absent.

22.6.3 Common Indian Genus

Canna.

22.6.4 Size, Distribution and General Information

Cannaceae is a family of the single genus, *Canna* (Fig. 22.7), with about 55 species. This famous, large-flowered ornamental is distributed widely in tropical and subtropical regions of America, Africa and Asia, including India.

Cannaceae can be distinguished from Marantaceae by the absence of a pulvinus and from the Zingiberaceae by the absence of a ligule.

22.6.5 Description of the Family

Habit *Canna* is a perennial herb perennating by means of rootstock which is a tuberous rhizome.

Leaves Cauline, large and foliaceous, oblong to broadly elliptical in shape; pinnately veined with a well-developed midrib; petiolate, petiole sheathing the stem; ligule absent.

Inflorescence Racemose raceme or panicle, made up of 2-flowered cincinni.

Flowers Large, showy, bracteate, bisexual, zygomorphic, shortly-pedicellate or sessile, epigynous, beautifully coloured.

Calyx Sepals 3, polysepalous, herbaceous or more or less green to purple coloured; persistent, sepals persist even in fruit.

Corolla Petals 3, gamopetalous or only basally connate, long one of the petals is usually smaller than the others; imbricate.

Androecium Highly modified and form the showy part of the flower; stamens usually 2–5, or sometimes 6 or even 4, all briefly basally connate and petaloid; only one stamen is fertile and bears a single fertile anther cell on its edge, the other stamens are represented by infertile staminodes; one of the staminodes is reflexed and is called *labellum* or *lip*; the other 2 staminodes, if present, are called *wings*; 4th staminode, if present, is located behind the fertile stamen; in some species, labellum is the only staminode present.

Gynoecium Tricarpellary, syncarpous, trilocular; many anatropous ovules in each locule arranged in 2 rows; axile placentation; ovary inferior, green and warty; style one, long and petaloid; stigma one, flat.

Fruit Capsule, warty, usually crowned by persistent sepals (Fig. 22.7); dehiscing by collapse of the warty pericarp.

Seeds Small, numerous, subglobose, contain very hard endosperm and straight embryo.

General Floral Fromula Br, ⊕, ⚥, K_3, $C_{(3)}$, $A_{1+4 \text{ staminodes}}$, $G_{(\bar{3})}$.

22.6.6 Economic Importance

- Cannaceae is of economic importance because of ornamental value of several species of *Canna* and number of their hybrids, e.g. *C. indica*, *C. orchioides*, *C. generalis*, *C. orientalis*, etc.
- Edible starch is obtained from the tubers of Queensland Arrowroot (*Canna edulis*).

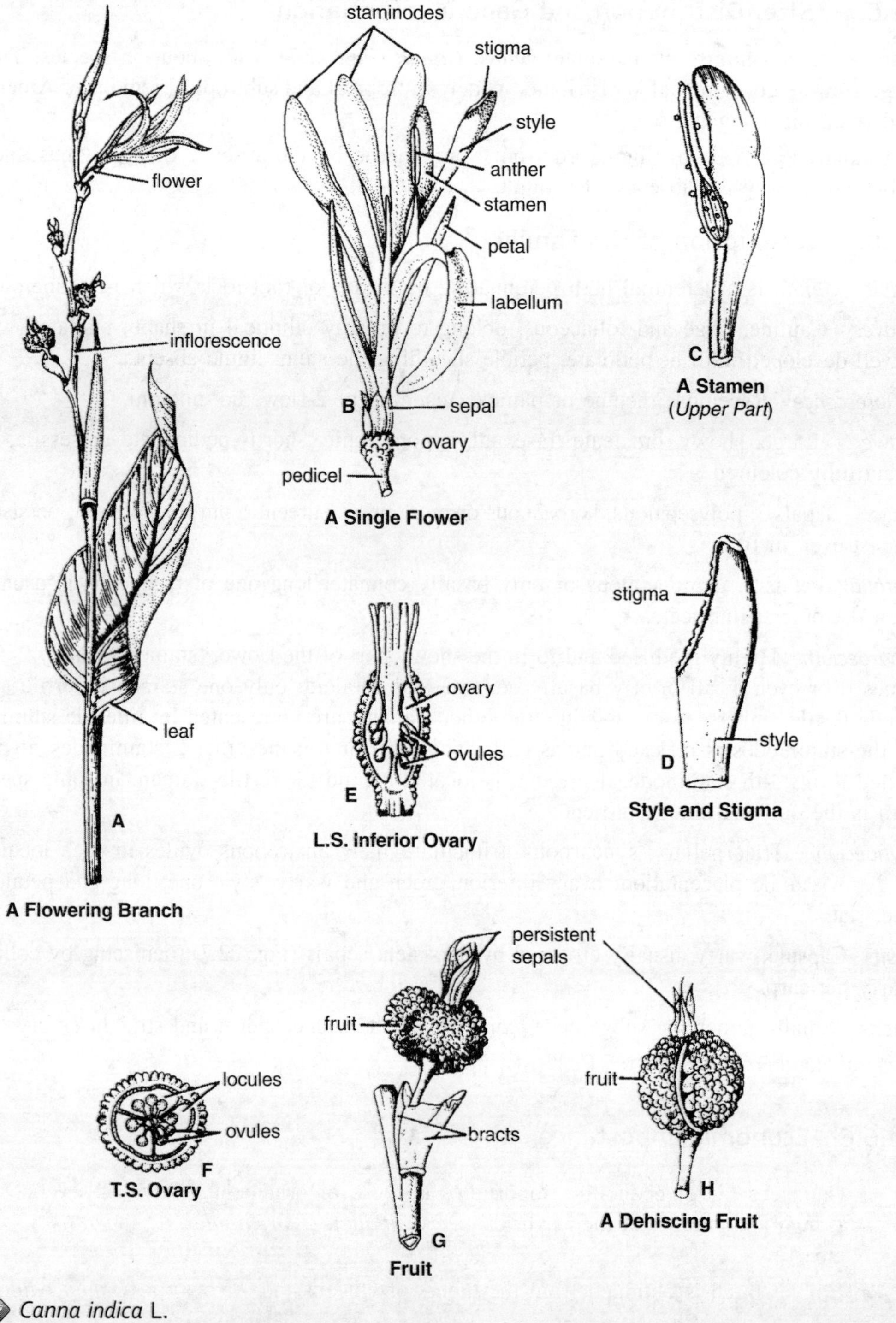

Fig. 22.7 *Canna indica* L.

- The rhizomes of *C. bidentata* are sometimes considered as emergency foods.
- The roots of *C. indica* are diphoretic, diuretic and are given in fevers and dropsy.

22.6.7 Systematics and Phylogeny (Affinities)

Bentham and Hooker treated *Canna* and its species under Scitamineae of series Epigynae of Monocotyledons. Engler and Prantl, however, split Scitamineae into 4 independent families, viz. Musaceae, Cannaceae, Zingiberaceae and Marantaceae. Lymon Benson, however, placed all these 4 families under order Musales, whereas Hutchinson placed all these 4 under order Zingiberales. Taxonomists treated all these 4 families together because they are all characterised by characters like (i) presence of 1 or 5 functional stamens, (ii) zygomorphic flowers, (iii) leaves distichous or in spirals, (iv) inferior ovary, and (v) endospermic seeds.

22.7 MUSACEAE (BANANA FAMILY)

22.7.1 Systematic Position

Monocotyledons, Epigynae, Musaceae (syn. Scitamineae of Bentham and Hooker).

22.7.2 Field Recognition

Perennial giant herbs appearing like trees; leaves large, forming crown at the apex of stout unbranched stem; flowers zygomorphic; inflorescence spadix covered by a spathe; ovary inferior; fruit berry.

22.7.3 Size, Distribution and General Information

A small family of 6 genera (*Musa*, *Ensete*, *Orchidantha*, *Ravenala*, *Heliconia* and *Strelitzia*) and 130 species distributed only in tropical regions. Professor A.D.J. Meeuse (H. de Vries Laboratories, Amsterdam), while commenting on author's "*A Manual of Practical Botany Vol. II*" in 1975, stated about the number of genera in Musaceae that "... some phanerogamists recognize only *Musa* and *Ensete* (or only *Musa* if *Ensete* is not segregated from it) whereas other botanists include the Heliconiaceae, Lowiaceae and Strelitziaceae in the Musaceae which accounts for their mentioning 5 to 6 genera"[1] in this family. Banana (*Musa paradisiaca* subsp. *sapientum*) is the most common plant of the family, found also in India.

22.7.4 Description of the Family

General Habit Large perennial herbs, often tree-like in appearance; containing unbranched, tall, aerial *pseudo-stems* formed by the leaf sheaths; plants sometimes attain a height of 10 to 15 metres; plants persist by underground rhizome.

Leaf Very large (up to 2 metres in length), alternate, sometimes distichous or spirally arranged; oval or oblong with a stout midrib; entire; numerous parallel veins extending up to the margins; rolled in bud.

[1] Personal communication with the author in 1975.

Inflorescence Flowers arranged in racemes and subtended by coloured bracts or spathes; usually single terminal inflorescence develops from the rhizome and comes out at the top of pseudo-stem (Fig. 22.8).

Flower Mostly unisexual, plants monoecious having male flowers within the upper bracts of inflorescence and the female flowers within the lower bracts; zygomorphic; trimerous and epigynous; flowers sometimes bisexual; in *Heliconia* and *Ravenala* the bracts are 2-ranked; in *Orchidantha*, the flowers are orchid like.

Perianth Tepals 6, arranged in 2 whorls of 3 each; petaloid; unequal-sized; one posterior tepal of inner whorl is free and boat shaped; remaining 5 tepals (i.e. 2 of inner whorl and 3 tepals of outer whorl) are united to form a tubular body.

Androecium Stamens 6, arranged in 2 whorls of 3 each; of these 6 stamens 5 are fertile and antiphilous while 1 posterior stamen of the inner whorl is represented by a staminode (Fig. 22.9); in *Ravenala* and sometimes in *Ensete* all the 6 stamens are fertile; filaments long; anthers dithecous, basifixed or adnate; introrse; dehiscing by vertical slits; present only in male or bisexual flowers.

Gynoecium Tricarpellary, syncarpous; ovary inferior, trilocular; with one to numerous anatropous ovules in each locule; axile placentation; style filiform; stigma capitate.

Fruit and Seeds Fruit an elongated berry (*Musa*), or trilocular capsule (*Ravenala, Strelitzia*), or schizocarp splitting into mericarps (*Heliconia*). Seeds often arillate, with straight embryo in mealy perisperm.

General Floral Formula (Bisexual Flower) Br, ⊕, ⚥, $P_{(5)+1}$, $A_{3+2+1 \text{ staminode}}$, $G_{(\overline{3})}$.

22.7.5 Economic Importance

- *Musa paradisiaca* L. sub sp. *sapientum* Schum. (banana or Kela) is one of the most famous tropical fruits, eaten throughout the world. Over 14 species of *Musa* occur in India. Green bananas are used as vegetable, and dried banana powder is a good baby food used in the manufacture of biscuits, chocolate, etc. Starch is prepared from banana stem. Fibre obtained from the sheathing leaf bases of *M. chinensis* is used for cordage. *M. textilis* also provides a useful fibre known as Manila hemp.
- *Ravenala madagascariensis* (Traveller's Tree), *Strelitzia reginae* and species of *Heliconia* and *Orchidantha* are most favoured for decorative purposes.
- *Ensete ventricosa* inflorescences are used as vegetable.

22.7.6 Systematics and Phylogeny (Affinities)

Some taxonomists include only *Musa* and *Ensete* in Musaceae, *Heliconia* in Heliconiaceae, *Orchidantha* in Lowiaceae, and *Strelitzia* and *Ravenala* in Strelitziaceae. Other phanerogamists divide Musaceae into 3 subfamilies viz. Lowioideae (*Orchidantha*), Musoideae (*Musa*) and Strelitzioideae

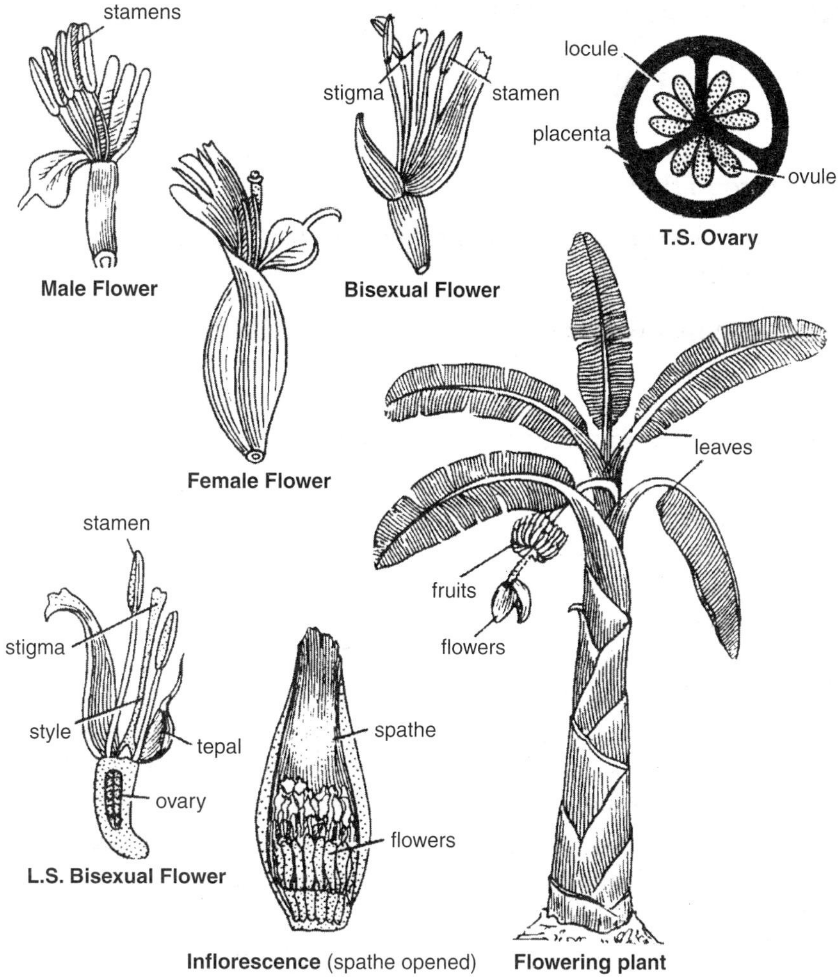

Fig. 22.8 ◈ *Musa paradisiaca* L.

Fig. 22.9 ◈ Floral diagram of bisexual flower of *Musa paradisiaca*.

(*Heliconia*, *Ravenala* and *Strelitzia*). Bentham and Hooker included only *Musa* and *Ensete* in tribe Museae of family Scitamineae.

Musaceae is closely related to other families of Scitaminales, and originated from Liliaceous stock. Engler believed Musaceae to be the ancestral stock of Orchidaceae. Takhtajan (1969) treated Musaceae, Heliconiaceae, Lowiaceae and Strelitziaceae under order Zingiberales along with 4 other families and believed them to be derived from Liliales.

22.8 ZINGIBERACEAE (GINGER FAMILY)

22.8.1 Systematic Position

Monocotyledons, Epigynae, Zingiberaceae.

22.8.2 Field Recognition

Aromatic herbs; ligule present at the top of leaf sheath; perianth differentiated into calyx and corolla; only single fertile stamen; staminodes petaloid.

22.8.3 Selected Indian Genera

Amomum, *Curcuma*, *Elettaria*, *Zingiber.*

22.8.4 Size, Distribution and General Information

A family of 49 genera and over 1000 species, Zingiberaceae are distributed in tropical regions, chiefly in Indomalaysia. About 17 genera and over 115 species have been reported from India, chiefly in Western Ghats and Eastern Himalayas. Some largely represented genera along with the number of their worldover reported species are *Alpinia* (250), *Amomum* (150), *Costus* (150), *Zingiber* (90), *Kaempferia* (70), *Curcuma* (60), *Hedychium* (50), *Globba* (50) and *Elettaria* (7).

22.8.5 Description of the Family

General Habit Perennial, often aromatic herbs; usually with sympodial tuberous (*Curcuma*) or horizontal (*Zingiber*) rhizomes; rarely with fibrous roots (*Cautleya*); aerial stem, if present, is short; sometimes an apparent stem is formed by the rolled-up leaf-sheaths.

Leaf Alternate, sessile or petiolate, distichous or spirally arranged, with sheathing leaf bases; ligulate, ligule present at the junction of blade with petiole or sheath; leaf blade large, linear to elliptical, with closely parallel-pinnate venation.

Inflorescence Variable from a compact spike (*Alpinia*, *Zingiber*), raceme (*Curcuma*), panicle (*Globba*), or flowers solitary; cymose in some species of *Globba*; each flower or group of flowers subtended by a large bract; in several genera the inflorescence develops on leafless scapes.

Flower Bracteate, pedicellate or sessile, bisexual, zygomorphic, only rarely actinomorphic, trimerous, epigynous, large, brightly coloured, usually aromatic or fragrant; perianth differentiated into calyx and corolla.

Calyx 3 united sepals forming a tubular or spathiform, green or herbaceous calyx; the odd sepal is anterior; calyx tube is very short in *Globba* and splitted in *Roscoea.*

Corolla 3 unequal-lobed petals, free or more or less united into a tubular corolla with the posterior lobe usually the largest; delicate and usually showy.

Androecium Basically 6 stamens, in 2 whorls of 3 each; 2 lateral stamens of the outer whorl modify into staminodes while the 3rd anterior stamen of this whorl is usually absent or highly suppressed; 1 median or posterior stamen of the inner whorl is fertile and 2 lateral stamens of this whorl unite to

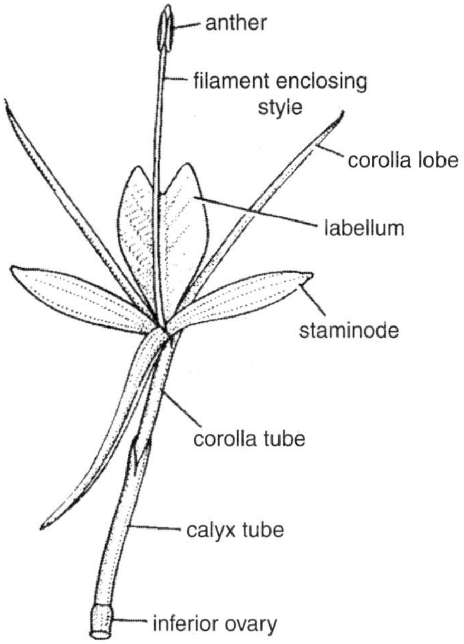

Fig. 22.10 A flower of *Hedychium*.

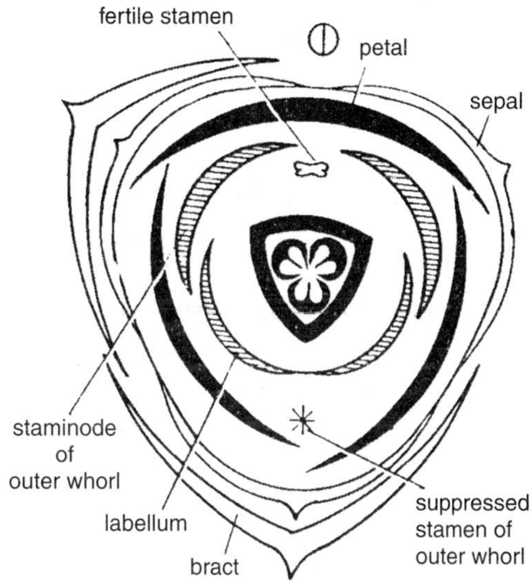

Fig. 22.11 Floral diagram of *Curcuma longa*.

form a labellum (Figs. 22.10, 22.11); fertile stamen is 2-celled, usually epipetalous, with its filament usually slender and deeply grooved.

Gynoecium Tricarpellary, syncarpous; ovary inferior, usually trilocular with numerous anatropous or semianatropous ovules on axile placentation, or unilocular with 3 parietal placentae (*Globba*); style usually filiform and more or less enveloped in the groove of filament of fertile stamen; stigma protruding beyond the anther lobes; usually epigynous nectar-secreting gland present.

Fruit and Seeds Fruit usually a 3-valved, loculicidal capsule, or rarely fleshy, indehiscent, berry-like. Seeds often arillate, and each with straight embryo and abundant endosperm.

General Floral Formula Br, $\oplus$, ⚥, $K_{(3)}$, $C_{(3)\text{ or }3}$, A_1, $G_{(\bar{3})}$.

22.8.6 Economic Importance

Economically, the family is important as a source of some spices and condiments, fragrant oils, and ornamentals.

- *Spices and Condiments* Seeds of *Amomum cardamon* (cardamon or Choti Elayachi) and *Elettaria cardamomum* (true cardamon or Elayachi), roots of *Alpinia officinarum* (gangal root), and rhizomes of *A. gangala* (Siamese ginger), *Curcuma longa* (turmeric or Haldi) and *Zingiber officinale* (ginger or Adrak) are used as spices, condiments and flavouring agents. Ginger oil, obtained from the rhizome of *Zingiber officinale*, is used in perfumery and for medicinal purposes. Turmeric is also used for medicinal purposes.

- *Abir* Abir, the famous scented powder, is prepared from *Hedychium spicatum* (Kafur-Kachri).
- *Zedoary* The product 'Zedoary', used as a tonic and in perfumery, is obtained from the tubers of *Curcuma zedoaria.*
- *Cultivated Ornamentals* Some of the ornamental genera cultivated in greenhouses and gardens for decorative purposes are *Alpinia* (shell ginger), *Brachychilum, Cardamon, Cautleya, Costus, Globba, Hedychium, Kaempferia* and *Roscoea.*

22.8.7 Systematics and Phylogeny (Affinities)

Treated only as a tribe Zingiberae of the family Scitamineae by Bentham and Hooker, Zingiberaceae has been treated as an independent family of the order Zingiberales by Hutchinson (1959); Takhtajan (1969), Cronquist (1981) and Thorne (1983). It is divided into 2 subfamilies viz. Zingiberoideae (plants aromatic, with oil cells, leaves 2-ranked) and Costoideae (plants without oil cells, leaves spirally arranged). Willis treated *Costus* and a few related genera in an independent family Costaceae.

Zingiberaceae is closely related to Musaceae in habit, zygomorphic flowers and inferior ovary. However, due to the reduction in androecium, Zingiberaceae has been considered to be more advanced.

22.9 LILIACEAE (LILY FAMILY)

22.9.1 Systematic Position

Monocotyledons, Coronarieae, Liliaceae.

22.9.2 Field Recognition

Perennial herbs often with bulbs, corms, or rhizomes; leaves usually radical; perianth showy, usually of 6 tepals, arranged in 2 whorls of 3 each; stamens 6; ovary superior.

22.9.3 Selected Indian Genera

Allium, Aloe, Asparagus, Asphodelus, Colchicum, Dracaena, Fritillaria, Lilium, Ruscus, Sansevieria, Smilax and *Yucca.*

22.9.4 Size, Distribution and General Information

A family of about 250 genera and 3700 species showing cosmopolitan distribution. About 35 genera and over 195 species have been reported from India. Some of the largely represented genera along with the number of their worldover reported species are *Allium* (450), *Smilax* (350), *Aloe* (300), *Asparagus* (300), *Dracaena* (150), *Ornithogalum* (150), *Haworthia* (150), *Tulipa* (100), *Fritillaria* (85), *Lilium* (80), *Colchicum* (65), *Sansevieria* (60), *Yucca* (40), *Asphodelus* (12) and *Ruscus* (7).

22.9.5 Description of the Family

General Habit Mostly perennial herbs with sympodial bulbs (*Allium cepa*) , rhizomes (*Paris quadrifolia*), corms (*Colchicum*), or bulbils (*Lilium*); some are woody shrubs (*Dracaena, Yucca*), or trees

(*Xanthorrhoea*); a few are succulents (*Aloe*, *Gasteria*), and climbers (*Gloriosa*, *Smilax*); in some, the ultimate branches are modified into phylloclades (*Ruscus*) or cladodes (*Asparagus*) and the leaves are reduced to scales.

Leaf Basal or cauline, simple; exstipulate but stipulate in *Smilax* where the stipules modify into tendrils; usually alternate, less commonly whorled (*Paris*), and rarely opposite (*Gloriosa*, *Scolyopus*); sometimes fleshy or spiny margined (*Aloe*), acicular, long, fleshy and hollow in *Asphodelus* (Fig. 22.12); reduced to scales in *Ruscus* and *Asparagus*; usually parallel venation but in *Smilax* and *Paris* the venation is reticulate.

Inflorescence Usually racemose raceme (*Asphodelus*), sometimes in spikes (*Aloe*), terminal panicle (*Yucca*), or monochasial cyme in *Hemerocallis*, or cymose in apparent umbels in *Allium*, *Agapanthus*; flowers solitary axillary (*Gloriosa*), or solitary terminal (*Tulipa*).

Flower Bracteate, usually ebracteolate except *Dianella* and *Lilium*; bisexual, actinomorphic, trimerous, hypogynous; rarely unisexual (*Smilax zeylanica*, Fig. 22.13 A, B; *Ruscus*) and such species are usually dioecious; rarely tetramerous (*Maianthemum*, *Paris*); slightly zygomorphic (*Lilium*, *Haworthia*).

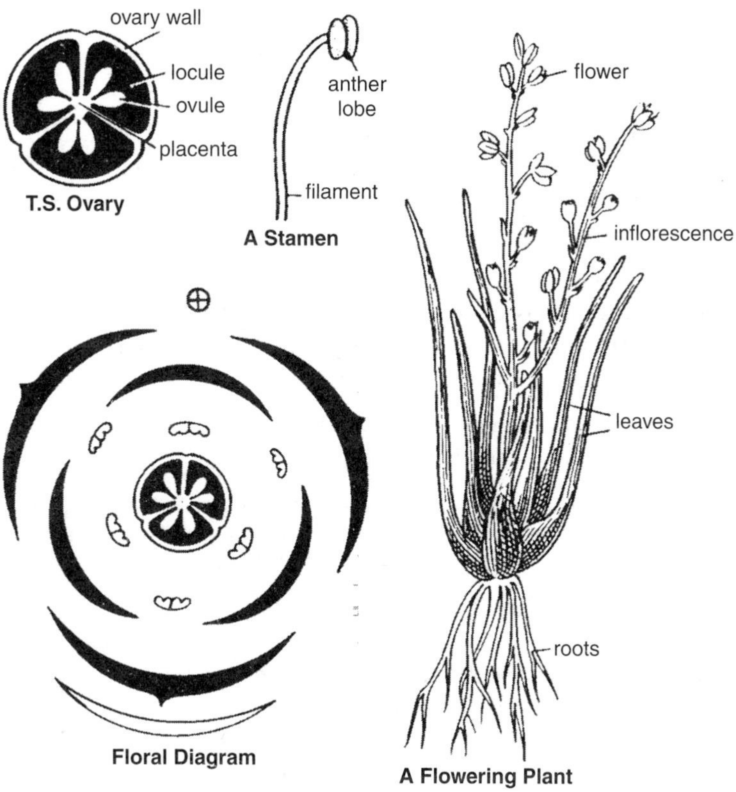

Fig. 22.12 *Asphodelus tenuifolius* L.

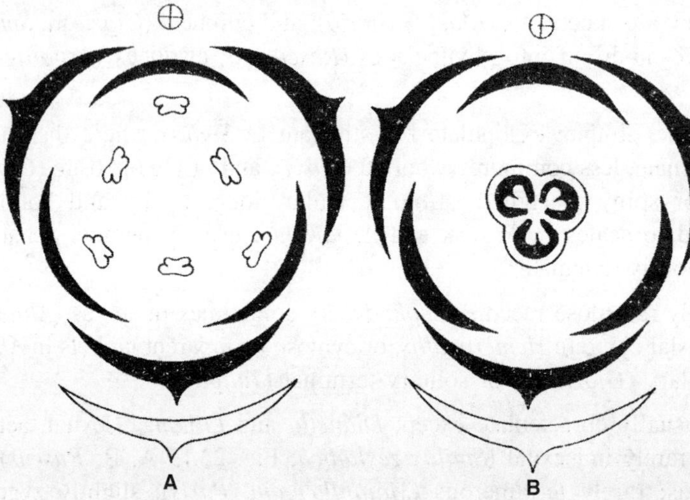

Fig. 22.13A, B *Smilax zeylanica:* Floral diagrams—A: Male flower; B: Female flower.

Perianth 6 tepals arranged in 2 whorls of 3 each, free or rarely united (*Aloe*); usually petaloid or sometimes sepaloid; odd tepal of the outer whorl is anterior in position; valvate or imbricate; more than 6 tepals in *Paris quadrifolia*.

Androecium Usually 6 stamens arranged in 2 whorls of 3 each; rarely the stamens are 3 (*Ruscus*), 4 (*Maianthemum*) or up to 12; polyandrous; opposite the tepals; sometimes epiphyllous; filaments distinct or connate; anthers dithecous, basifixed or versatile, extrorse or introrse, dehiscing usually by vertical slits and sometimes by terminal pores; rarely synandrous (*Ruscus*).

Gynoecium Tricarpellary, syncarpous, the odd carpel usually anterior; ovary superior, trilocular, with 2 rows of numerous anatropous ovules, axile placentation; rarely unilocular with parietal placentation; style usually 1; stigmas 1 or 3; rarely the ovary is inferior (*Haemodorum*); usually the nectar-secreting septal glands are present in the ovary.

Fruit and Seeds Fruit usually a septicidal or loculicidal capsule or a berry *Asparagus*, *Smilax*). Seeds with curved or straight embryo and abundant endosperm.

Pollination Flowers are usually insect-pollinated. Insects are attracted usually because of gaudy and scented perianth lobes and the nectar secreted in the septal glands. In *Paris*, the insects are attracted because of the foetid smell and dark purple colour of the floral parts. Bright colours of tepals in *Tulipa* also attract insects. Snails which come to eat the fleshy tepals of *Rhodea* bring about the cross pollination.

In *Yucca* the cross-pollination is carried out by a special moth, *Pronuba yuccasella*. Fully expanded flowers emit perfumes and are visited by the female moth, especially during nights. She collects a lot of pollen grains from one flower and visits another flower. Life-history of this moth is intimately associated with the pollination mechanism in *Yucca*.

General Floral Formula Br, Ebrl, ⊕, ⚥, P_{3+3}, A_{3+3}, $G_{(\underline{3})}$.

22.9.6 Economic Importance

Family is important from economic point of view in several ways, of which some are undermentioned:

- *Edible Plants* (i) Bulbs of *Allium cepa* (onion, Piyaz), used throughout the world as vegetable, possess stimulative, diuretic, expectorant and bactericidal properties, (ii) Bulbs of *Allium sativum* (garlic, Lahsun), used as condiment like onion, are also good for heart, (iii) Fleshy shoots of *Asparagus officinalis* and tuberous roots of *A. racemosus* are used as vegetable.
- *Medicinal Plants* (i) *Aloe barbadense* (Ghikanwar) leaves are the source of resinuous drug, used as a purgative, (ii) *Aloe vera* (Aloe) leaves provide mucilaginous liquid, used in piles and inflammations, (iii) Medicated oil prepared from the roots of *Asparagus racemosus* is used for nervous and rheumatic complaints and also in skin diseases, (iv) *Colchicum luteum* roots are used in the treatment of gout and rheumatism, (v) *Fritillaria roylei* bulbs are used as expectorant and antipyretic, (vi) *Gloriosa superba* tubers prove helpful in promoting labour pains in women, (vii) *Hemerocallis fulva* flowers are blood purifier and given to women during child birth, (viii) *Iphigenia indica* corms are used in stomach pains and headache, (ix) *Paris polyphylla* rhizomes are used as anthelmintic, (x) *Scilla hyacinthiana* bulbs are used as heart stimulant, (xi) *Smilax glabra* and *S. ovatifolia* roots are used in the treatment of venereal diseases, (xii) *Urginea indica* bulbs are heart stimulant and also used in skin diseases and rheumatism, (xiii) Drug obtained from the roots of *Veratrum viridae* is used in hypertension.
- *Resins* Resins are obtained from the stems of *Dracaena* and *Xanthorrhoea.*
- *Fibre Plants* Excellent fibres, used for cordage, fishing nets, mattings, twines etc., are obtained from *Phormium tenax* (New Zealand Flax), *Sansevieria roxburghiana*, *Smilax glabra* and *Yucca filamentosa.*
- *Insecticides and Raticides* Bulbs of *Urginea indica* are used for killing rats while that of *Veratrum album* are used as insecticides.
- *Polyploidy* Colchicine, an alkaloid obtained from *Colchicum luteum*, is used to induce polyploidy.
- *Ornamental Plants* Some of the well known garden ornamentals of Liliaceae are *Agapanthus africanus* (African lily), *Asparagus plumosus*, *Dracaena*, *Fritillaria tenella*, *Gloriosa superba* (Malabar Glory Lily), *Hemerocallis fulva* (Orange Day Lily), *Lilium candidum* (Lily), *L. giganteum*, *Ruscus aculeatus* (Butchers broom), *Tulipa suaveolens* (Tulip), *Yucca aloifolia* and *Y. gloriosa.*

22.9.7 Systematics and Phylogeny (Affinities)

Discussed under the series Coronarieae by Bentham and Hooker, Liliaceae was treated under Liliiflorae by Engler and Prantl (1931), and Liliales by Hutchinson (1959), Takhtajan (1969), Cronquist (1981) and Thorne (1983). Cronquist merged Amaryllidaceae under Liliaceae. Some taxonomists include certain genera of Liliaceae under independent families, e.g. *Aloe* under Aloeaceae and *Smilax* under Smilacaceae. Bentham and Hooker divided Liliaceae into 20 tribes whereas Engler and Prantl divided it into 12 subfamilies.

Family Liliaceae represents most primitive basic monocotyledonous stock from which the other families of monocots have originated. According to Takhtajan (1969), order Liliales, together with orders Alismales and Triuridales "have more probably a common origin from a hypothetical extinct group with endospermous seeds and 2-celled pollen grains, as in the Liliales", and subfamily Melanthioideae "of the family Liliaceae is nearest to the ancestral type".

22.9.8 Description of a Common Plant

Asphodelus tenuifolius Cav. (vern. Piazi, Fig. 22.12)

Habit: An annual weed with fibrous roots, and condensed, underground and reduced stem. *Leaf:* Radical, simple, exstipulate, sessile, with sheathing leaf bases; long, acicular, acute, fleshy, hollow. *Inflorescence:* Racemose raceme with flowers on long scape or peduncle. *Flower:* Bracteate, pedicellate, bisexual, actinomorphic, trimerous. *Perianth:* 6 tepals arranged in 2 whorls of 3 each, free, petaloid, valvate. *Androecium:* 6 stamens arranged in 2 whorls of 3 each, polyandrous, epiphyllous, present opposite the tepals. *Gynoecium:* Tricarpellary, syncarpous, superior, trilocular, 2 ovules in each locule; axile placentation; style long; stigma trifid. *Fruit:* Capsule.

Floral Formula: Br, ⊕, ⚥, $\widehat{P_{3+3},\ A_{3+3}}$, $G_{\underline{(3)}}$.

22.10 COMMELINACEAE (SPIDERWORT FAMILY)

22.10.1 Systematic Position

Monocotyledons, Coronarieae, Commelinaceae.

22.10.2 Field Recognition

Herbs with succulent stem, and leaves with tubular sheath; floral parts in 3s; perianth differentiated into calyx and corolla; ovary superior; fruit capsule.

22.10.3 Selected Indian Genera

Aneilema, *Commelina*, *Murdannia*, *Streptolirion* and *Tradescantia*.

22.10.4 Size, Distribution and General Information

A family of about 50 genera and 700 species, Commelinaceae are mostly tropical and subtropical in distribution. About 11 genera and over 75 species have been reported from India, chiefly from eastern Himalayas and parts of southern and western India. Some larger genera along with the total number of their reported species and common names of some of them are *Commelina* (200, dayflower), *Aneilema* (100), *Tradescantia* (65, spiderwort), *Cyanotis* (30), *Dichorisandra* (30), *Zebrina* (4, wandering Jew) and *Rhoeo* (1, Moses-in-the bulrushes).

22.10.5 Description of the Family

General Habit Annual or perennial herbs with fibrous or tuber-like roots, jointed and more or less succulent stems; occasionally twining (*Streptolirion*); stem either with purplish tinge (*Commelina*) or deep violet (*Tradescantia*).

Leaf Simple, alternate, sheathing at the base, flat or trough-like, linear or ovate or lanceolate; entire margins; parallel-veined.

Inflorescence Usually an umbellate cyme subtended by two large spathaceous bracts (*Tradescantia*), thyrsoid panicles (*Floscopa*), or flowers solitary.

Flower Pedicellate; usually actinomorphic but zygomorphic in *Commelina*; bisexual; usually subtended by boat-shaped spathe or foliaceous bract; trimerous, hypogynous, commonly blue; in *Commelina* some flowers are also subterranean and cleistogamous.

Perianth Usually biseriate i.e. differentiated into calyx and corolla; calyx is represented by an outer whorl of 3 sepal-like green tepals which are usually persistent, free and imbricate, and only rarely connate; corolla is represented by an inner whorl of 3 petal-like, usually ephemeral, coloured, equal or unequal-sized tepals which are also usually free and imbricate, and only rarely united (*Cyanotis*).

Androecium Typically 6 stamens in 3+3 arrangement, distinct or rarely fused, frequently some reduced to staminodes; there are 2 fertile stamens and 4 staminodes in *Aneilema*, 3 fertile stamens and 3 staminodes in *Commelina* (Fig. 22.14), 5 fertile stamens and 1 staminode in *Floscopa*, and all the 6 fertile stamens and no staminode in *Tradescantia* (Fig. 22.15); filaments often hairy or bearded; anthers 2-celled (parallel or divergent), dehiscence longitudinal or poricidal (*Dichorisandra*).

Gynoecium Tricarpellary, syncarpous; ovary superior, trilocular, with 1 to few orthotropous ovules in each locule; axile placentation; style 1; stigma 1, capitate or trifid.

Fruit and Seeds Fruit a loculicidal capsule, rarely fleshy and indehiscent. Seeds often arillate, 1 to few in each locule, with small embryo and copious endosperm; seed tissue contains calcium oxalate.

Pollination and Dispersal Flowers in *Tradescantia* and several other genera are protandrous and adapted for insect pollination. As the flowers fade, the petals become pulpy and their surface gets covered with a layer of liquid which attracts insects. Fruits are dispersed by animals and birds.

General Floral Formula Br, ⊕ or %, ⚥, K_3, C_3, $A_{3+3 \text{ or } 3+3 \text{ staminodes or } 2+4 \text{ staminodes or } 5+1 \text{ staminode}}$, $G_{\underline{(3)}}$.

22.10.6 Economic Importance

- Except for some ornamental plants, the family is of little importance. The ornamental species include *Commelina suffruticosa, Cyanotis cristata, Dichorisandra thyrsiflora, Rhoeo discolor, Tradescantia virginiana* (spiderwort) and *Zebrina pendula* (wandering Jew).
- Some minor aspects of utility include: (i) the roots of *Aneilema scapiflorum* are used in piles and asthma, (ii) the rhizomes of *Commelina benghalensis* are used as vegetable, (iii) the roots of *C. obliqua* are used as an antidote to snake poison, (iv) the stem juice of *Floscopa scandens* is put in sore eyes, (*v*) plants of several species of *Aneilema, Murdannia* and *Commelina* are used in leprosy and leucoderma, (vi) roots of *Cyanotis* are used to expel worms in cattles.

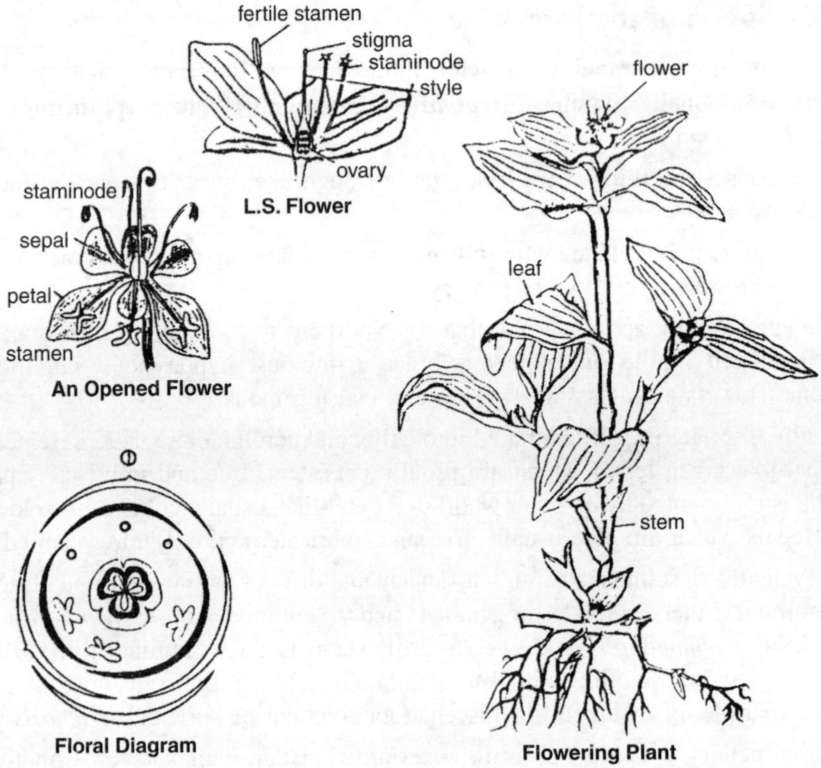

Fig. 22.14 *Commelina benghalensis* L.

22.10.7 Systematics and Phylogeny (Affinities)

The family is included under series Coronarieae by Bentham and Hooker, order Farinosae by Engler, and under order Commelinales by most of the other workers including Hutchinson, Takhtajan, Cronquist, Thorne, etc. Commelinaceae are usually divided into 2 tribes viz. Tradescantieae (flowers usually actinomorphic) and Commelineae (flowers usually zygomorphic).

Hutchinson (1959) derived order Commelinales from 2 most primitive orders viz. Butomales and Alismatales, i.e. from Helobiae. However, Takhtajan (1969) opined that Commelinales "probably has a common origin with the Bromeliales" from Liliales.

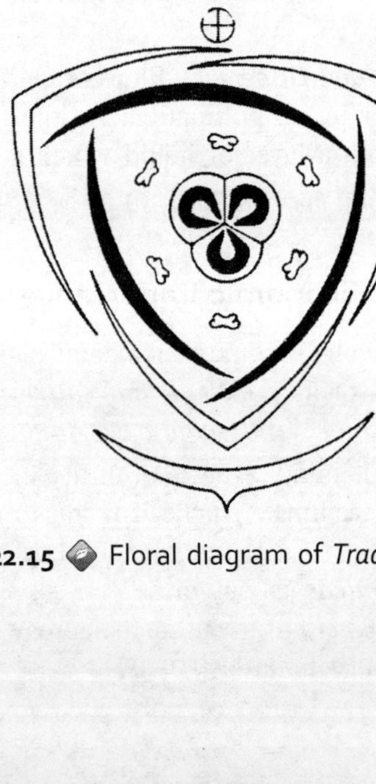

Fig. 22.15 Floral diagram of *Tradescantia*.

22.10.8 Description of a Common Plant

Commelina benghalensis L. (Fig. 22.14)

Habit: Annual weed with semi-erect or creeping stem having nodes and internodes. *Leaf:* Simple, exstipulate, sessile with sheathing leaf base, ovate to oblong-lanceolate, multicostate parallel venation; greenish with purplish tinge. *Inflorescence:* Cymes surrounded by 1–3 spathes or bracts. *Flower:* Bracteate, pedicellate, bisexual, zygomorphic, hypogynous, trimerous, blue coloured; some flowers subterranean and cleistogamous. *Perianth:* Biseriate; outer whorl of 3 tepals represents calyx, polysepalous, valvate or imbricate; inner whorl of 3 represents corolla, polypetalous, 1 petal larger; imbricate. *Androecium:* 6 stamens of which 3 are fertile and 3 are sterile staminodes; antiphyllous, light blue filaments, anthers dithecous, discrete and divaricate. *Gynoecium:* Typical as discussed for the family, except that 1 locule contains 1 ovule and the other 2 locules contain 2 ovules each.

Floral Formula: Br, ⊕, ⚥, P_{3+3}, $A_{3+3 \text{ staminodes}}$, $G_{\underline{(3)}}$.

22.11 JUNCACEAE (RUSH FAMILY)

22.11.1 Systematic Position

Monocotyledons, Calycineae, Juncaceae.

22.11.2 Field Recognition

Grass-like or rush-like, annual or perennial, tufted herbs; leaves basal, tufted, linear, grass-like, sheathing basally or reduced to basal sheath; perianth usually sepalloid, segments 6, in 2 whorls of 3 each; stamens 6, generally in 2 whorls of 3 each; ovary tricarpellary, syncarpous, superior; fruit capsule.

22.11.3 Selected Indian Genera

Juncus, *Luzula*.

22.11.4 Size, Distribution and General Information

Juncaceae includes about 400 species of only 8 genera (*Juncus, Luzula, Rostkovia, Marsippospermum, Andesia, Oxychloe, Distichia* and *Prionium*), of which only 2 (*Juncus* and *Luzula*) are found in India. The two Indian genera are cosmopolitan in their distribution. Juncaceae members are distributed mainly in temperate and arctic regions and also in tropical mountains, specially in moist cool places. The chief genera are *Juncus* (300, rush; Fig. 22.16) and *Luzula* (80, wood rush). Rushes often provide stepping stones for crossing a wet meadow.

22.11.5 Description of the Family

General Habit Annual or perennial tufted herbs, only rarely shrub-like (*Prionium*), with their stems usually leafy only at the base; these plants of damp and cool places usually contain a creeping sympodial rhizome, of which one joint usually appear above ground each year and develop into a leafy shoot.

Leaves Basal, narrow, flat or cylindrical; grass like, sessile; linear or filiform; sheathing at the base or reduced to only a sheath; tufted; exstipulate; mostly hairy and parallel-veined.

Inflorescence Flowers usually a crowded mass of cymes arranged in a panicle, corymb or head; rarely flowers solitary.

Flowers Usually bracteate, pedicellate, sessile, bisexual, actinomorphic, trimerous, small and wind-pollinated; if unisexual, then the plants are dioecious (e.g. *Oxychloe*); hypogynous.

Perianth Segments 6, arranged in 2 whorls of 3 each; polyphyllous; usually sepalloid or green, rarely coloured; scale-like or glumaceous or coriaceous.

Androecium Stamens 6, arranged in 2 whorls of 3 each, or inner whorl absent; antiphyllous (opposite the perianth lobes); polyandrous; anthers dithecous, basifixed, introrse; dehiscing by vertical slits; pollen in tetrads.

Gynoecium Tricarpellary, syncarpous, superior; unilocular, with 3 parietal placentae (e.g. *Luzula*), or trilocular with one to three or many anatropous ovules in each locule showing axile placentation; styles 1–3, simple; stigmas 3, brush-like.

Fruit A 1–3 loculicidal capsule.

Seeds Small, with straight embryo and starchy endosperm.

General Floral Formula Br, ⊕, ⚥, P_{3+3}, A_{3+3}, $G_{\underline{(3)}}$.

22.11.6 Economic Importance

Juncaceae are of little economic importance. In some countries, various species of *Juncus* (e.g. *J. effusus*) are woven into mats, hats, chair seats, baskets, etc. Pith of some of the rushes is used for candlewicks. Few rushes are grown for ornamental purposes in locations adjacent to aquatic habitats, and also used horticulturally for waterside planting. *Juncus squarrosus* is used as fodder for sheeps. Some species of *Luzula* (e.g. *L. pilosa*, *L. campestris*) are used in kidney trouble. Paper is made from fibres obtained from some species of *Luzula*.

22.11.7 Systematics and Phylogeny (Affinities)

Juncaceae has been included under series Calycineae by Bentham and Hooker along with two more families, Palmaceae and Flagellariaceae. Engler and Prantl, however, discussed Juncaceae under order Liliales. Hutchinson included Juncaceae under order Juncales of division Glumiflorae.

Due to the presence of typical trimerous flowers with superior ovary, both Junaceae and Liliaceae are said to be quite close to each other, and Juncaceae is considered to be derived from Liliaceae by reduction. Anatomical studies of two families, however, do not support this view. Due to the similarities in the distribution of vessels and type of stomata, Juncaceae resemble with families like Cyperaceae, Restionaceae and Gramineae. Morphology and pollen development of Juncaceae and Cyperaceae are quite similar and these bring the two families together.

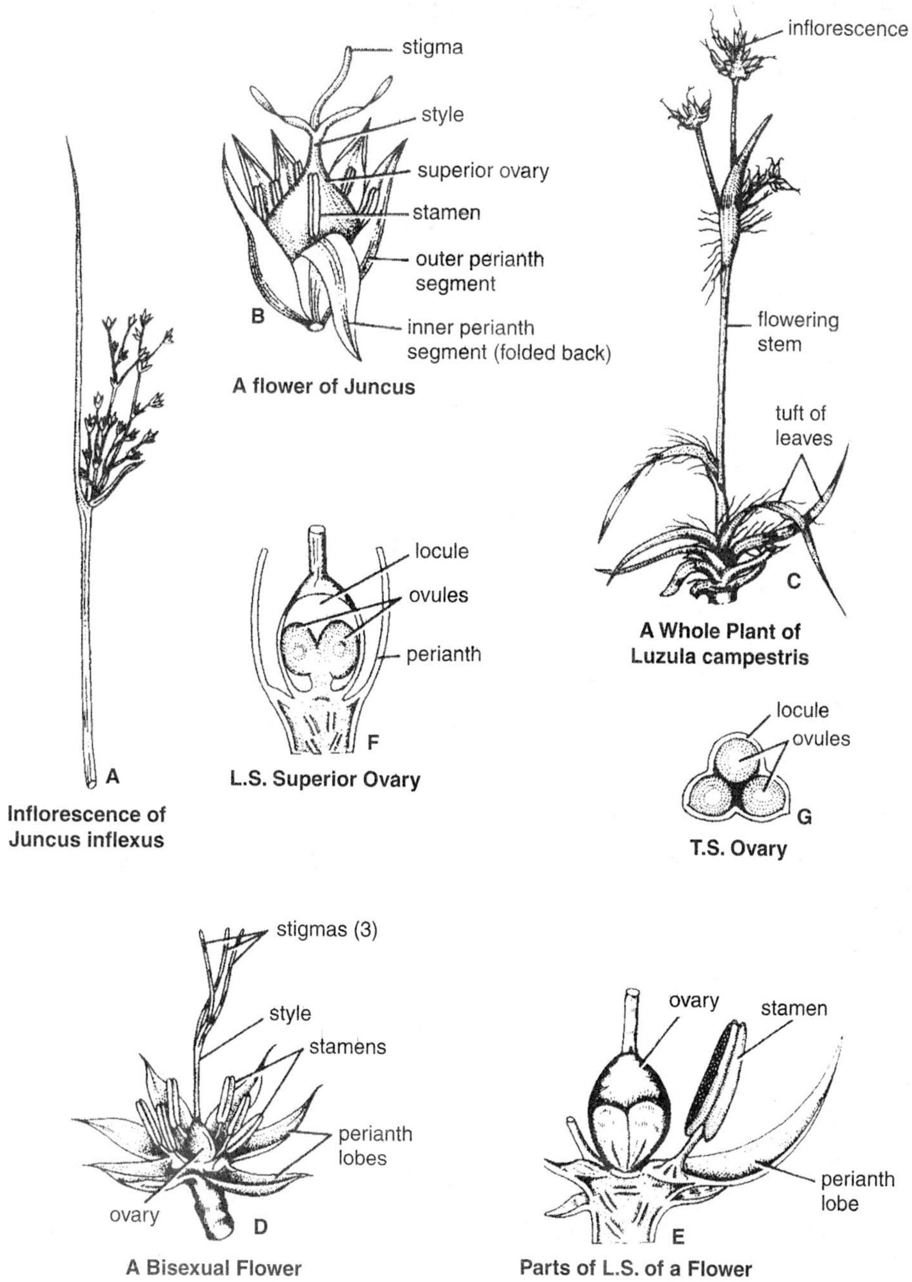

Fig. 22.16 A, B, ***Juncus inflexus***; C–G, Various floral parts of *Luzula campestris*.

22.12 PALMAE OR ARECACEAE (PALM FAMILY)

22.12.1 Systematic Position

Monocotyledons, Calycinae, Palmae or Arecaceae.

22.12.2 Field Recognition

Shrubs or trees with arborescent stem having prominent scars of leaf bases and crown of large fan-shaped or pinnately compound leaves; inflorescence spadix or paniculate, often with spathes; flowers-small, unisexual or bisexual, actinomorphic, trimerous.

22.12.3 Selected Indian Genera

Areca, Borassus, Calamus, Cocos, Nypa, Phoenix.

22.12.4 Size, Distribution and General Information

A family of 217 genera and over 3000 species distributed widely in tropical and subtropical regions of the world, and a few also in warm temperate regions. About 28 genera and over 95 species have been reported from India. Chief genera along with the number of their total reported species and the common names of some of them are *Calamus* (300, cane palm), *Chamaedorea* (130), *Areca* (54, betel-nut palm), *Coccothrinax* (50, biscayne palm), *Copernicia* (30, wax palm), *Raphia* (30, wine palm), *Sabal* (25, cabbage palm), *Phoenix* (17, date palm), *Roystonea* (17, royal palm), *Metroxylon* (15, sago palm), *Phytelephas* (15, ivory-nut palm), *Caryota* (12, fish-tail palm), *Elaeis* (2, oil palm) and *Cocos* (1, coconut palm).

22.12.5 Description of the Family

General Habit Woody shrubs, trees, or sometimes vines (*Calamus*, *Desmoncus*); with usually unbranched, slender to stout stem; short to over 30 metres tall; and with leaves often forming a terminal cluster in the arborescent species.

Root Fibrous and adventitious roots arising from the base of stem; in *Iriartea* prop roots develop.

Stem Variable in different forms, such as (i) very short with leaves appear to arise from the ground (e.g. *Nypa*), (ii) thin and slender with long internodes (e.g. *Calamus*), or (ii) tall, stout, pillar like, covered by persistent leaf bases and containing a terminal cluster of leaves (e.g. *Cocos*, *Phoenix*); stem is rarely branched.

Leaf Leaves are palmate (fan palms) or pinnate (feather palms), and rarely simple; usually large with petiole base often sheathing the stem; usually in a terminal cluster; usually alternate in climbing species (*Calamus*); petiole is large, strong, smooth or spiny; ochreate stipule present in *Calamus* while leaves are ligulate in *Thrinax*; parallel venation.

Inflorescence Large, much branched, paniculate (*Daemonorops*) or spadix (*Phoenix*), often reaching up to 1 metre, and covered by one or more boat-shaped woody spathes or bracts; interfoliar i.e. arise amongst the leaves(e.g., *Borassus*) or intrafoliar i.e. arise below the leaves (e.g. *Areca*); plants monoecious or dioecious or flowers bisexual.

Flower Ebracteate, sessile, actinomorphic; usually unisexual and monoecious but sometimes dioecious (*Phoenix*), rarely bisexual (*Livistona*); trimerous, hypogynous, in *Borassus*, the male flowers are smaller than female flowers.

Perianth 6 free or united tepals in 2 whorls of 3 each; tough, leathery and usually persistent; imbricate, valvate or twisted in the bud; in *Phoenix* the tepals of outer whorl are united and valvate while that of inner whorl are free and twisted (Fig. 22.17); usually the tepals of outer whorl are smaller than of inner whorl.

Androecium Present only in staminate or bisexual flowers; usually 6 stamens arranged in 2 whorls of 3 each (*Phoenix*); positioned against tepals; filaments short, distinct; anthers dithecous, basifixed

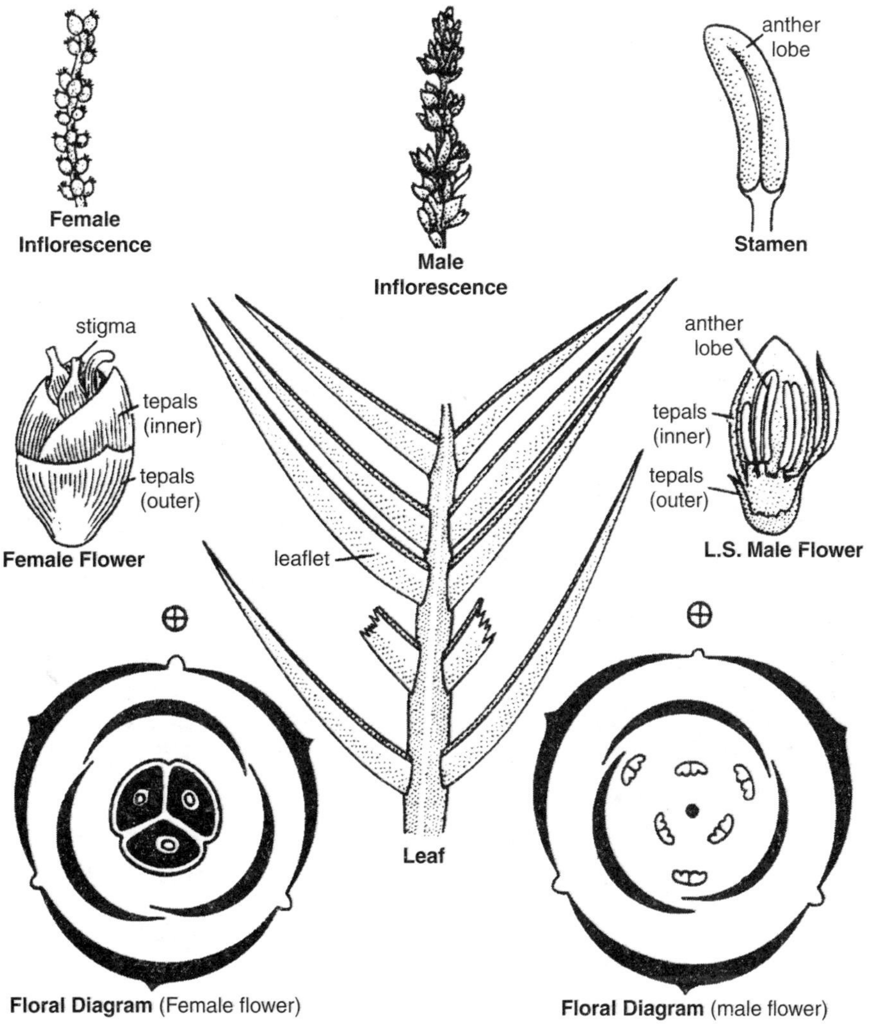

Fig. 22.17 Phoenix *sylvestris* (L.) Roxb.

or dorsifixed, introrse, dehiscence by vertical slits; stamens are only 3 in *Nypa* and numerous in *Caryota* and *Phytelephas*.

Staminodes and Pistillodes A pistillode is usually present in the centre of each male flower. Staminodes may (*Phytelephas*) or may not be present in the female flowers.

Gynoecium Present only in pistillate or bisexual flowers; tricarpellary, syncarpous; or carpels are partly (*Nypa*) or completely (*Phoenix*) free; ovary superior, 3-locular with a single anatropous ovule in each locule; axile placentation; in *Phoenix* the placentation is basal; sometimes the ovary is unilocular and the placentation is parietal; style extremely short or absent; stigmas 3, sessile.

Fruit and Seeds Fruit berry with fleshy exocarp (*Phoenix*) or drupe with fibrous epicarp (*Cocos*). Seeds with small embryo and abundant endosperm.

Pollination and Dispersal Pollination in palms is usually anemophilous, and the flowers are protandrous. Some palms (*Sabal*), however, are insect pollinated. Dispersal of fruit is either by animals or by water (*Cocos*).

General Floral Formulae

(a) *Male flower:* Ebr, ⊕, ♂, $P_{(3)+3 \text{ or } 3+3}$, A_{3+3}, $G_{0 \text{ or pistillode}}$.

(b) *Female Flower:* Ebr, ⊕, ♀, $P_{(3)+3 \text{ or } 3+3}$, $A_{0 \text{ or staminodes}}$, $G_{\underline{(3)} \text{ or } \underline{3}}$.

(c) *Bisexual Flower:* Ebr, ⊕, ⚥, $P_{(3)+3 \text{ or } 3+3}$, A_{3+3}, $G_{\underline{(3)} \text{ or } \underline{3}}$.

22.12.6 Economic Importance

Palm family, supposed to be second in importance to Poaceae, is used for food, oils, shelter, coconuts, copra, dates, clothing, etc. As many as 801 uses of only *Borassus flabellifer* have been mentioned in one ancient Tamil song. *Cocos nucifera* (coconut palm or Nariyal) is another such plant of this family whose almost every part of the entire plant body is utilized in some ways or other, i.e. its (i) *root* decoction is used as mouthwash, (ii) *stem* is used for pillars and as a fuel, (iii) *leaves* are utilized for making mats, fans, baskets, etc., (iv) toddy, obtained from the young *inflorescences*, provides an intoxicating alcoholic drink on fermentation, (v) *endosperm of young fruits* is edible and their milky fluid is a sweet and refreshing drink; (vi) vegetable fat, obtained from the *dried endosperm of mature fruits*, is used for cooking and several toilet preparations, (vii) dried fibrous *mesocarp of fruits* is used for brushes, ropes, mats, floor coverings, etc. (viii) woody *endocarp of fruits* is prepared into several articles of decoration, and (ix) famous coconut oil is used for anointing the body.

Some of the other similar generalized uses of Palmae are undermentioned:

- *Ornamental Palms:* Palms are universally known for their majestic look and handsome appearance. Some of them are *Adonidia merilli* (Manila palm), *Arenga saccharifolia* (Gomuti palm), *Caryota urens* (wine palm or fish-tail palm), *Chamaerops humilus* (European fan palm), *Livistona chinensis* (Fountain palm or Chinese fan palm), *Phoenix rupicola*, *Pritchardia pacifica*, *Roystonea elata* (Floridean royal palm), *R. obracea*, *R. regia* (Royal palm or bottle palm), *Sabal minor* (Bush palmetto) and *S. umbraculifera* (Cabbage palm).

- *Oils:* Coconut oil is obtained from the dried ripe endosperm of *Cocos nucifera* while the palm oil is obtained from the fleshy mesocarp of the fruits of *Elaeis guineensis* (oil palm).
- *Edible Products:* (i) Seeds of *Areca catechu* (Supari or betel nut palm) are sliced and chewed along with "Paan" (*Piper betel*) by millions of the people, (ii) kernel of fruits and tender leaves of *Arenga pinnata* is edible, (iii) young seedlings of *Borassus flabellifer* are used as vegetable and its fruits are eaten after roasting, (iv) fruits of *Copernicia ceribera* and *Hyphaene thebaica* are edible, (v) Sago, used as an article of diet, is obtained from the pith of the stem of *Metroxylon rumphii* (sago palm or Sabudana), (vi) stem buds, young peduncles and immature seeds of *Nypa fruticans* are eaten raw or cooked as vegetable, (vii) ripe fruits of *Phoenix dactylifera* (date palm or "Pind Khajoor") and *P. sylvestris* (wild date palm or "Jangli Khajoor") are eaten throughout the world for their high food value, (viii) fruits of *Sabal palmetto* (cabbage palm) are edible, and honey is prepared from its flowers.
- *Toddy:* A sap or sugary solution, obtained by cutting or tapping the stems, young peduncles or inflorescence of several palms, is used for manufacturing jaggery, and by fermentation it yields an intoxicating drink or beverage called "toddy". This sap is also used for manufacturing number of commercially useful products such as 'Nira', palm sugar, alcohol, vinegar, etc. Palms, commonly used for this purpose are *Arenga pinnata*, *A. saccharifera*, *Borassus flabellifer*, *Caryota urens*, *Metroxylon vinifera*, *Nypa fruticans*, *Phoenix dactylifera* and *P. sylvestris*. The sap of *Raphia hookeri* is the source of famous 'Bourdon wine'.
- *Hats, Mats and Baskets:* These and many other similar articles such as umbrella handles, walking sticks, hand fans, etc, are prepared from *Borassus flabellifer*, *Nypa fruticans*, *Phoenix dactylifera*, *P. sylvestris*, etc.
- *Brushes:* Brushes are prepared from fibres obtained from the leaf stalks of *Arenga pinnata*, *Borassus flabellifer* and *Sabal palmetto*.
- *Timber:* Dense stem wood of several palms (*Borassus*, *Caryota*, *Cocos*, *Hyphaene* and *Phoenix*) is used as timber.
- *Cane:* Common 'cane' or 'rattan' of commerce, used for matting chairs, baskets, furniture frames, walking sticks etc. is obtained from several species of *Calamus* (Cane or Rattan palms), such as *C. extensus*, *C. latifolius*, *C. ovatus*, *C. rotundus* and *C. tenuis*. *Daemonorops adspersus*, *D. jenkinsianus* and *Korthalsia horrida* are also utilized for similar purposes.
- *Vegetable Ivory:* Stony endosperm of *Phytelephas macrocarpa* is used for making beads of necklaces and other similar articles as cheap substitute of ivory.
- *Wax:* Wax, utilized for manufacturing candles, boot polishes, etc. is obtained from *Ceroxylon andicola*, *Copernicia cerifera*, etc.
- *Sacred Writing Material:* Narrow strips of leaves of *Corypha umbraculifera* have been used for several sacred writings in olden days. Leaves of *Borassus flabellifer* were also used for several such writings by ancient Hindus.
- *Resin:* Dragon's blood, a resinuous exudate obtained from the fruits of *Daemonorops ruber*, is used in colouring marbels, varnishes, etc.

22.12.7 Systematics and Phylogeny (Affinities)

Palmae or Arecaceae was variously placed under series Calycinae by Bentham and Hooker, order Principes by Engler and Prantl, order Palmales by Hutchinson, and order Arecales by Takhtajan, Cronquist, and Thorne.

Palmae is closely related to Araceae, and it was discussed by Rendle under Spadiciflorae along with Araceae and Lemnaceae. Hutchinson, however, opined that Palmae is allied to Pandanaceae on one hand and Agavaceae on the other. He believed that Palmae probably originated from Agavaceae. Workers like Takhtajan and Cronquist opined that because of the presence of unisexual inconspicuous flowers developing in spadix in both Palmae and Araceae, the two families are closely related.

22.13 TYPHACEAE (CATTAIL FAMILY)

22.13.1 Systematic Position

Monocotyledons, Nudiflorae, Typhaceae.

22.13.2 Field Recognition

Monoecious, large perennial herbs of fresh to brackish open marshes; leaves alternate, erect, long, linear, in 2-ranks, sessile and with parallel venation; inflorescence a dense spike, fuzzy-brown at maturity, usually divided into upper part containing male flowers and lower part bearing female flowers; fruit achene.

22.13.3 Indian Genus

Typha.

22.13.4 Size, Distribution and General Information

Typhaceae is a unigeneric family comprising only 1 genus (*Typha*, Fig. 22.18) and about 20 species distributed in temperate and tropical regions throughout the world. Due to the dense spike-like inflorescence and long grass-like leaves, this is commonly called *water grass* or *cattail family. Typha angustata*, *T. elephantina* and *T. laxamanni* are some common Indian species.

22.13.5 Description of the Family

General Habit Erect perennial herbs; rootstocks rhizomatous, rhizome thick and creeping; fibrous adventitious roots, arise from the nodes of rhizome; the aerial shoot, developing from the axil of the scaly leaves, terminate into an inflorescence.

Leaves Alternate, long, linear, sessile, mostly basal, developing in 2 ranks; sheathing at the base; venation parallel; leaves often overtop the inflorescence.

Inflorescence Thick, dense cylindrical spikes, usually divided into two parts: the upper, usually yellowish part bearing male flowers, and the lower, usually brownish part bearing female flowers; the two parts usually contiguous.

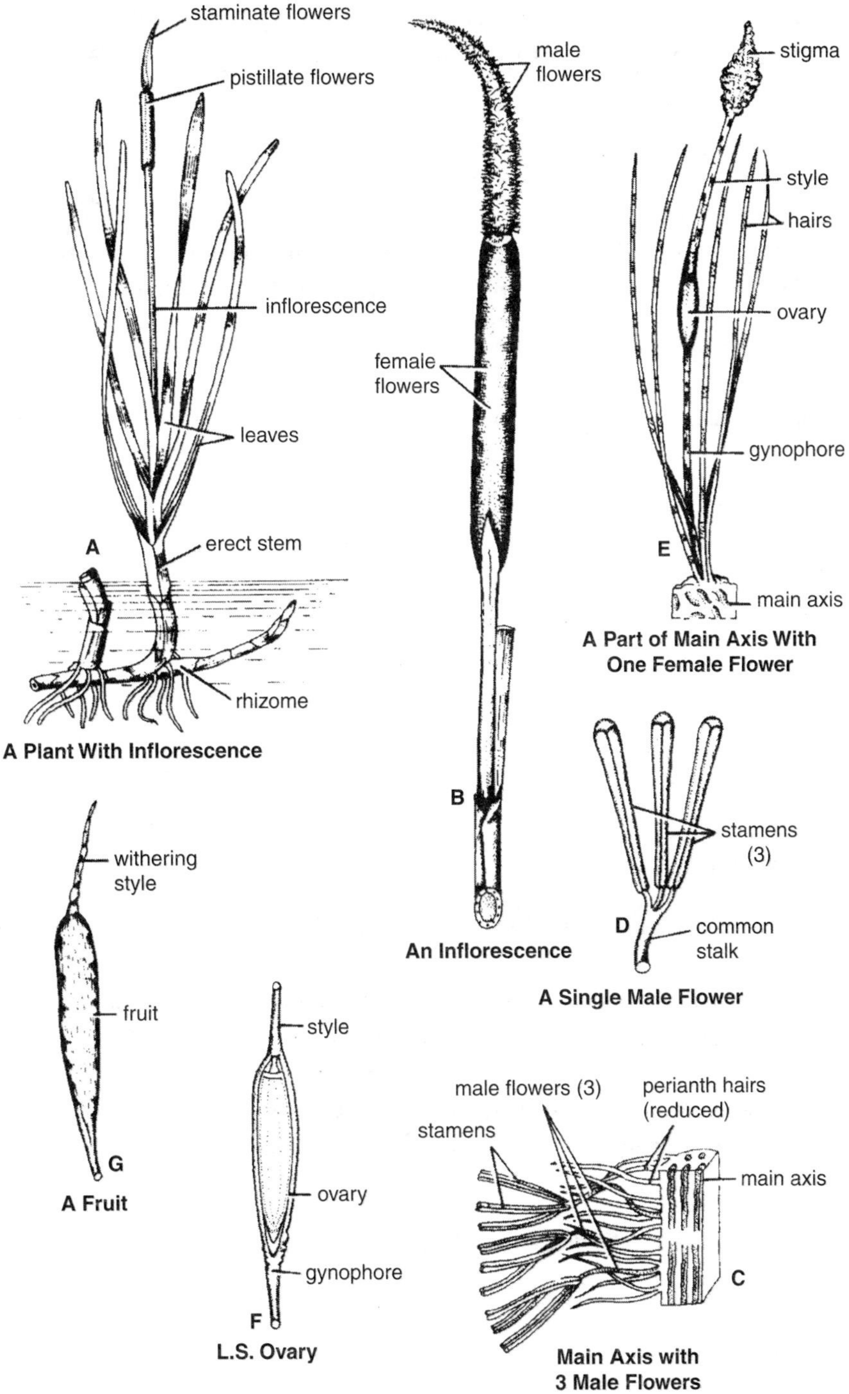

Fig. 22.18 *Typha latifolia* L.

Flowers Unisexual, surrounded by slender hairs or more or less forked scales, representing possibly the reduced perianth; flowers borne on a cylindrical spadix; bracteate, shortly pedicellate, actinomorphic, hypogynous.

Male Flowers

Perianth Perianth segments in the form of numerous slender threads or elongated spoon-like scales.

Androecium Male flowers with 2–5 stamens; variously monadelphous; filaments free or connate; connate filaments bear long silky hairs; anthers basifixed, dithecous; connective projecting beyond the anthers.

Gynoecium Absent.

Female Flowers

Perianth Same as that of male flowers.

Androecium Absent.

Gynoecium Monocarpellary, superior, unilocular, 1 seeded; present on a stipe bearing slender hairs and containing 1 pendant ovule; style 1, long; stigma 1, linear to spathulate or rhomboidal.

Fruit An achene bearing the persistent style and usually surrounded by persistent scales.

Seeds With mealy endosperm; embryo long or straight and narrow.

Floral Formulae

(a) *Male Flower:* Br, ⊕, ♂, P_{∞}, $A_{(2-5)}$, G_0.

(b) *Female Flower:* Br, ⊕, ♀, P_{∞}, $\overline{A_0, G}_{\underline{1}}$.

22.13.6 Economic Importance

- Dried leaves of *Typha* are used as weaving materials in making mats, chair bottoms, baskets, etc.
- Dried, dense cylindrical spikes of inflorescences are used in floral arrangements.
- Occasionally, some species of *Typha* are grown as ornamentals in ponds and river margins.
- Leaf bases and rhizomes are eaten by mammals including humans in some parts of India. Bur, a yellow cake prepared from the flowers of *T. angustata*, is eaten by natives.
- Rootstocks of *Typha angustata* are astringent and diuretic.

22.13.7 Systematics and Phylogeny (Affinities)

Typhaceae has been included under series Nudiflorae by Bentham and Hooker along with 4 more families, viz. Araceae, Pandanaceae, Lemnaceae and Cyclanthaceae. Typhaceae has however been treated by Engler and Prantl under order Pandanales along with one more family, Pandanaceae. Benson also treated Typhaceae under Pandanales as treated already by Engler and Prantl. Hutchinson, however, discussed Typhaceae under order Typhales along with one more family, namely Sparganiaceae.

Modern taxonomists like Hutchison (1964), Takhtajan (1980) and Cronquist (1981) consider Typhaceae as an advanced family of monocotyledons because of several characters, including (i) exstipulate leaves with sheathing bases, (ii) flowers arranged in dense spikes, (iii) unisexual flowers, (iv) reduced perianth in the form of hair-like outgrowth's, and (v) monocarpellary gynoecium.

22.14 ARACEAE (ARUM FAMILY)

22.14.1 Systematic Position

Monocotyledons, Nudiflorae, Araceae.

22.14.2 Field Recognition

Rhizomatous or tuberous herbs, found in both aquatic and terrestrial habitats; inflorescence spadix subtended or enveloped by a single spathe; flowers small and often bad smelling; fruit a berry.

22.14.3 Selected Indian Genera

Acorus, Alocasia, Amorphophallus, Colocasia, Pistia, Pothos.

22.14.4 Size, Distribution and General Information

A family of 115 genera and over 2000 species, Araceae are distributed mostly in tropical and some also in temperate regions of the world. About 25 genera and over 140 species have been reported from India, chiefly from western and southern parts. Some larger genera along with the number of their worldover reported species are *Anthurium* (500), *Philodendron* (275), *Arisaema* (150), *Amorphophallus* (100), *Rhaphidophora* (100), *Pothos* (75), *Alocasia* (70), *Monstera* (50), *Arum* (15), *Colocasia* (8), *Acorus* (2) and *Pistia* (1). Colocasia (*Colocasia antiquorum*) and 'money plant' (*Pothos aureus*) are two famous plants of Araceae.

22.14.5 Description of the Family

General Habit Highly variable; usually perennial, rhizomatous or tuberous herbs (*Arum*), climbers (*Pothos*), or tree-like (*Dracontium, Philodendron*); rarely free-floating aquatic herbs (*Pistia*); some are epiphytes (*Anthurium*), or occur in marshy conditions (*Lasia*); often raphides or calcium oxalate crystals present in the sap of the plants.

Root Fibrous, adventitious, and usually of two types i.e. (i) climbing or clasping roots, which cling to the support, and (ii) absorbing roots which enter into the soil and absorb nutrients; velamen develops in the aerial roots of epiphytic species.

Stem Subterranean or underground in the form of tubers (*Arum*), corms (*Amorphophallus*), rhizome (*Acorus*) or root stocks (*Colocasia*); or aerial (*Pothos*) showing monopodial branching or sympodial; accessory buds often develop in leaf-axils.

Leaf Variable in shape and types; single (*Arisaema*) or a few; basal (*Arisaema*), or cauline and alternate (*Pothos*); often hastate (*Typhonium*), sagittate or round; pinnately or palmately divided; petiolate (*Philodendron*) or sessile (*Pistia*); venation is usually parallel but rarely reticulate (*Arum*).

Inflorescence Flowers usually grouped together to form a more or less cylindrical spadix subtended by a large coloured spathe; the upper portion of spadix is usually naked and the lower portion bears whorls of small unisexual flowers, the male flowers above and the female below (Fig. 22.19); usually some sterile male and female flowers are present close to fertile whorls of male and female flowers; male and female flowers develop on different spadix in some genera, e.g. *Arisaema*.

Flower Ebracteate, ebracteolate, usually unisexual but sometimes bisexual (*Acorus*, *Pothos*), actinomorphic, hypogynous or epigynous, dimerous or trimerous, small, often bad-smelling.

Perianth Usually absent in unisexual flowers (*Arum*, *Pistia*) and present in bisexual flowers (*Acorus*, Fig. 22.20); if present, the perianth lobes are small, scale-like, 4 to 6 in number, present generally in 2 whorls of 2 + 2 or 3 + 3, free or rarely connate (*Spathiphyllum*).

Androecium Usually 6 or lesser stamens present in one or 2 whorls, located opposite the perianth lobes; free or united into a synandrium (*Colocasia*, *Spathicarpa*); anthers dithecous, introrse, dehiscing by slits or pores; female flowers usually bear staminodes.

Gynoecium Varied in structure but often reduced to a single carpel; sometimes 2 to 9 united carpels; superior or rarely inferior and embedded in the spadix; one to many locules and one to numerous ovules in each locule; placentation is basal (*Arisaema*, *Typhonium*), axile (*Pothos*) or parietal (*Arum*); style various, sometimes greatly reduced; stigma variable.

Fruit and Seeds Fruit usually a berry, clustered closely on the spadix and appearing as a multiple fruit. Seeds with or without endosperm.

Pollination Flowers are protogynous and insect-pollinated. Insects are attracted because of the coloured spathe and characteristic bad smell emitted by inflorescences. Insects come into contact with the fertile female flowers and transfer to them any pollen they may be carrying from the male flowers of another plant. Some species of *Arisaema* show self-pollination.

General Floral Formulae

(a) *Male Flowers:* Ebr, ⊕, ♂, P_0, $A_{3+3 \text{ or } 2+2}$, G_0.

(b) *Female Flowers:* Ebr, ⊕, ♀, P_0, A_0, $G_{1 \text{ or } \underline{(2-9)}}$.

(c) *Bisexual Flowers:* Ebr, ⊕, ⚥, $P_{3+3 \text{ or } 2+2}$, $A_{3+3 \text{ or } 2+2}$, $G_{\underline{1} \text{ or } \underline{(2-9)}}$.

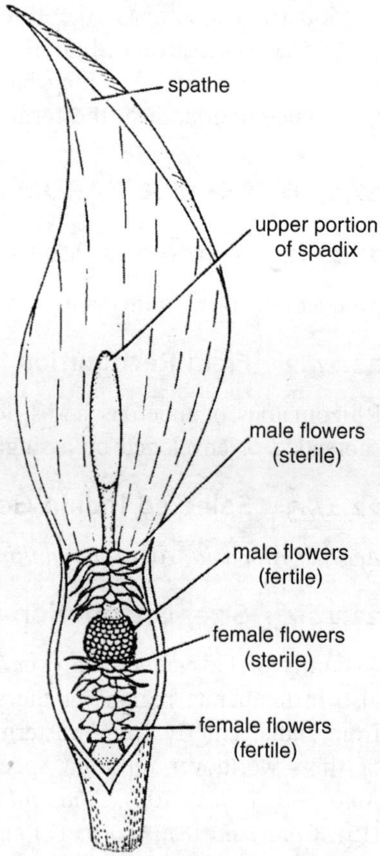

Fig. 22.19 L.S. inflorescence of *Arum maculatum*.

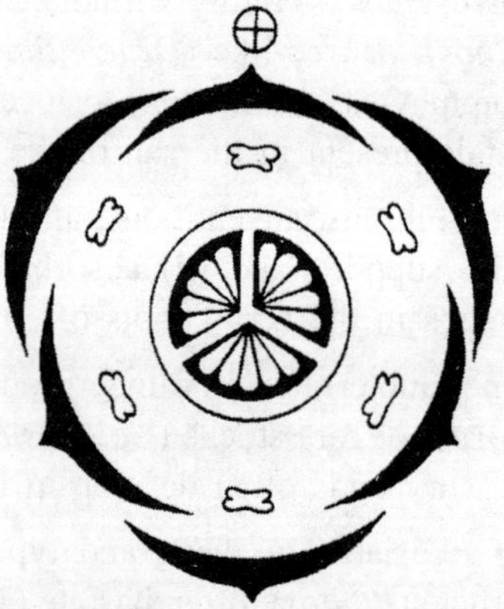

Fig. 22.20 Floral diagram of *Acorus*.

22.14.6 Economic Importance

- Widely used vegetables are obtained from (i) the stem and root-stocks of *Alocasia indica* ('Mankanda'), (ii) tuberous corms and leaves of *Amorphophallus campanulatus* ('Zimikand' or 'elephant foot'), (iii) tubers of *Colocasia esculenta* syn *C. antiquorum* ('Arvi', 'Ghuiyan', 'Kachalu' or Colocasia), and (iv) corms and leaves of *Remusatia vivipara* and *Xanthosoma niger*.
- Industrial alcohol and starchy baby foods are also prepared from the tubers of *Colocasia esculenta*.
- Certain plants are of medicinal value. These include (i) *Acorus calamus*, of which the rhizomes are beneficial in chronic diarrhoea and dyspepsia, (ii) *Alocasia macrorrhiza*, of which the stem juice is used to relieve pain from scorpion bite, (iii) *Amorphophallus campanulatus*, of which the corms are used in treating piles and dysentery, and (iv) *Scindapsus officinalis*, of which the dried fruits have anthelmintic properties.
- Plants of ornamental value, grown commonly in gardens for their often variegated and strikingly handsome leaves are *Acorus calamus* var. *variegatus*, *Alocasia indica* var. *metallica*, *Caladium bicolor*, *C. picturatum*, *Colocasia antiquorum*, *Dieffenbachia picta*, *Monstera deliciosa*, *Pothos aureus* (Money plant), *Scindapsus officinalis* and *S. pictus*.
- Ornamental plants cultivated mainly for their flowers and spathes include several species of *Anthurium*, *Arisaema* (jack-in-the-pulpit or Indian turnip), *Dracunculus* (dragon arum), *Lysichiton* (western skunk cabbage), *Sauromatum* (*Voondoo lily*) and *Spathiphyllum*.
- *Pistia stratioites* is cultivated for being used in aquaria.

22.14.7 Systematics and Phylogeny (Affinities)

Placed under series Nudiflorae by Bentham and Hooker, family Araceae was included in order Spathiflorae in Engler's system, and was placed under order Arales by most of the other taxonomists including Hutchinson, Takhtajan, Cronquist, Thorne, etc. In Engler's system, the family was divided into eight subfamilies *viz*. Pothoideae, Monsteroideae, Calloideae, Lasioideae, Philodendroideae, Colocasioideae, Aroideae and Pistioideae.

Takhtajan (1969) includes only two families in Arales *viz*. Araceae and Lemnaceae. He opined that Arales "most likely has a common origin with the Arecales and Cyclanthales from intermediate ancestors of the Liliales."

In affinities, Araceae appears to be closely related to Palmae, because both families have same type of inflorescence, and small, hypogynous, actinomorphic flowers.

22.15 ALISMATACEAE OR ALISMACEAE (WATER PLAINTAIN FAMILY)

22.15.1 Systematic Position

Monocotyledons, Apocarpae, Alismataceae.

22.15.2 Field Recognition

Aquatic or marshy herbs with perennating rhizomes; bisexual flowers; perianth biseriate with 3 free sepals and 3 free petals; many free carpels.

22.15.3 Selected Indian Genera

Alisma, Limnophyton, Sagittaria.

22.15.4 Size, Distribution and General Information

Represented by about 14 genera and over 90 species, Alismataceae are cosmopolitan in distribution. About 7 genera and over 10 species have been reported from India. The chief genera along with the number of their reported species and common names of some are *Echinodorus* (25, burhead), *Sagittaria* (20, arrowhead), *Alisma* (10, water plaintain), and *Damasonium* (5).

22.15.5 Description of the Family

General Habit Annual or perennial, aquatic or marshy herbs with a stout perennating rhizome; roots are fibrous, and stem is thin, erect or with floating leaves; in *Sagittaria* the runners terminate into tubers; usually the latex is present in stem and leaves.

Leaf Radical or clustered at nodes, long-petioled, leaf base sheathing; erect, floating or submerged; blades ovate or linear to lanceolate with bases sometimes sagittate or hastate; parallel venation; usually small scales develop in axils.

Inflorescence Usually much branched or whorled; primary branches often in racemose racemes or panicles, the secondary branches often cymose.

Flower Bracteate (*Butomus*) or ebracteate, pedicellate; bisexual (*Alisma*, Fig. 22.21; *Butomus*) or unisexual (*Sagittaria*); actinomorphic, trimerous, hypogynous with a flat, convex or dome-shaped receptacle.

Perianth Biseriate, consisting of 6 perianth lobes arranged usually in 2 whorls of 3 lobes; the outer whorl, representing the calyx, consists of 3 free, green sepals which are twisted (*Alisma*, Fig. 22.22), valvate (*Sagittaria*, Fig. 22.23) or imbricate (*Butomus*); the inner whorl, representing the corolla, consists of 3 free, coloured petals which are also twisted, valvate or imbricate.

Androecium Stamens are usually 6 (*Alisma*) but variable between 3 (*Wisneria*) to numerous (*Sagittaria*) in different genera; 9 stamens in *Butomus* are arranged in an outer whorl of 6 and inner whorl of 3 stamens; filaments free and anthers are dithecous, basifixed and extrorse in most of the genera.

Gynoecium Carpels are usually 6 and apocarpous but variable from 3 (*Wisneria*) to numerous (*Alisma*) free carpels; ovary of each carpel is superior, unilocular, with a single ovule, basal placentation; rarely the ovules are 2 or more in each locule; style one, persistent, bear stigmatic papillae (*Alisma*, Fig. 22.21).

Fruit and Seeds Fruit a group of achenes arranged spirally or in ring. Seeds with a horse-shoe shaped or curved embryo and without endosperm.

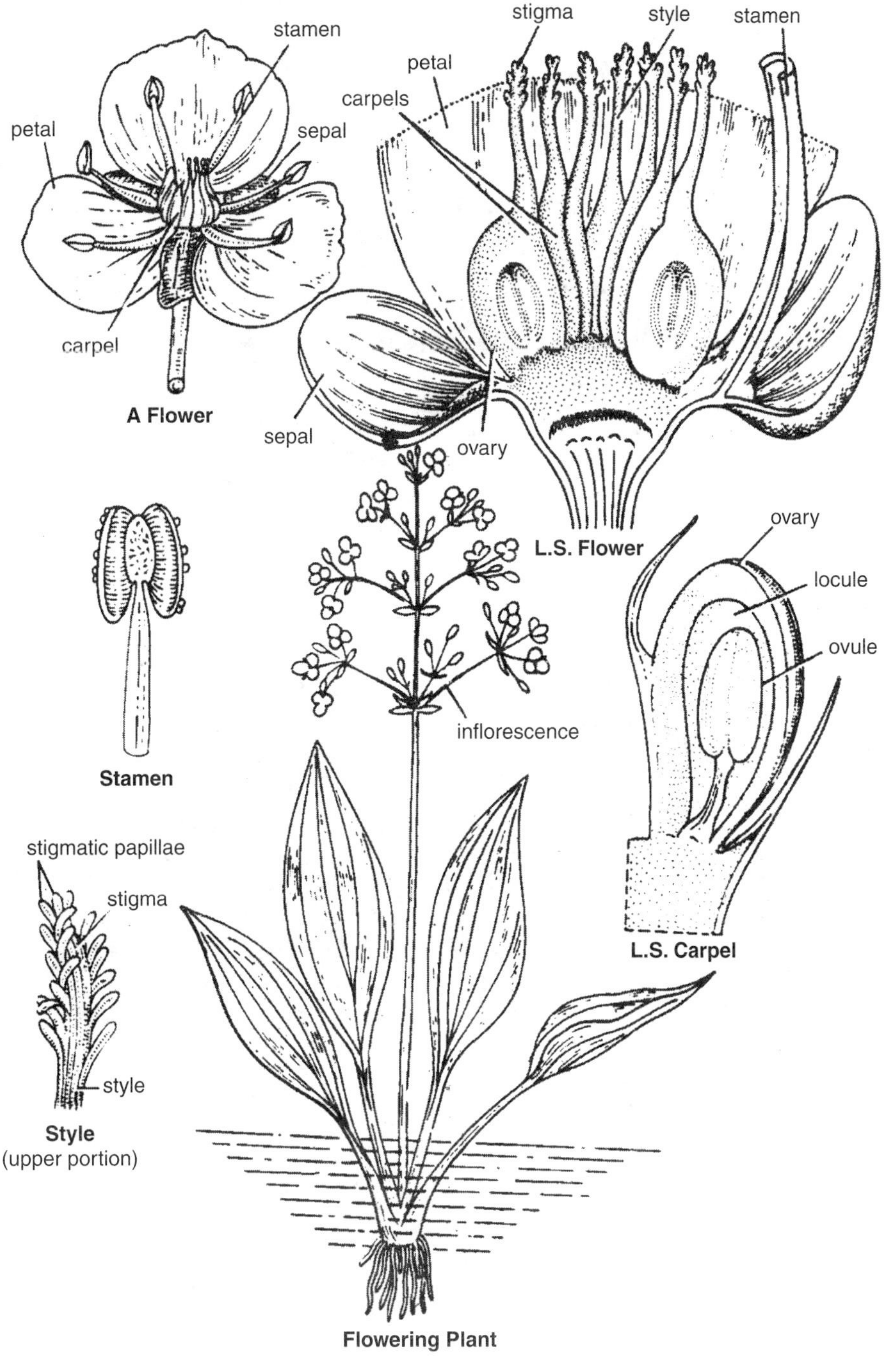

Fig. 22.21 *Alisma plantago-aquatica* L.

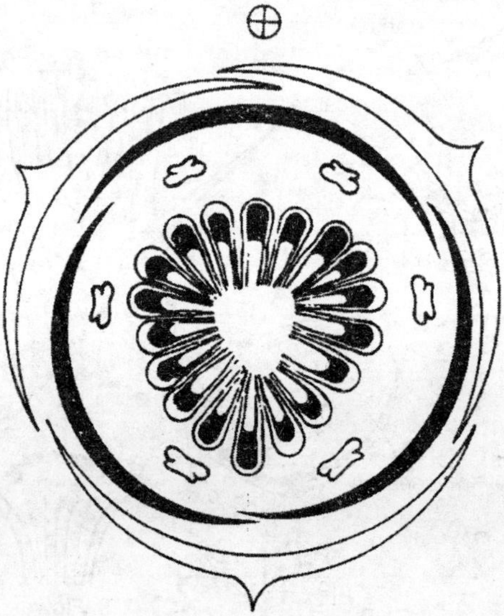

Fig. 22.22 Floral diagram of *Alisma*.

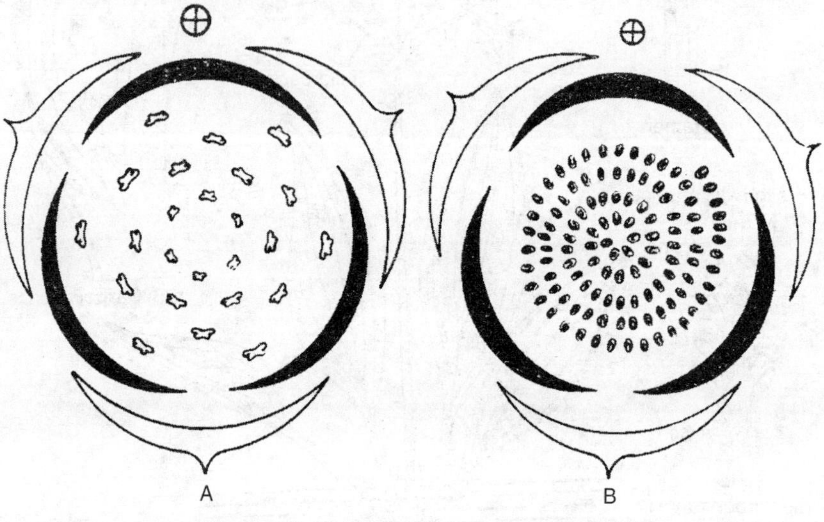

Fig. 22.23 Floral diagrams of *Sagittaria guayanensis*—A: Male flower; B: Female flower.

Pollination and Dispersal Flowers are entomophilous, and get pollinated by flies, short-tongued bees and other similar insects which visit them in search of nectar. Dispersal of seeds takes place mainly by water and partly by air.

Floral Formula (Bisexual Flower) Br or Ebr, ⊕, ⚥, K_3, C_3, A_6, $G_{\underline{6}}$ or $_{\underline{3-\infty}}$.

22.15.6 Economic Importance

Economically, the family is not of any special significance. *Sagittaria* rhizomes are sometimes grown for their food value and used as food under the name "swamp potato". Tubers of *Alisma plantago-aquatica* and corms of *Sagittaria sagittifolia* are edible. Some species of *Sagittaria* and *Alisma* are used as aquarium plants. Some species of *Alisma*, *Echinodorus* and *Sagittaria* are planted for decoration besides ponds and streams.

22.15.7 Systematics and Phylogeny (Affinities)

Family Alismataceae was included under series Apocarpae by Bentham and Hooker, order Helobiae by Engler, order Alismales by Takhtajan, and under order Alismatales by Cronquist, and also by Thorne. Usually, the family is divided into two subfamilies *viz. Alismatoideae* (ovules solitary in each carpel) and *Butomoideae* (ovules many in each carpel).

Alismataceae is closely related to Ranunculaceae of dicotyledons in possessing hypogynous flowers, biseriate perianth differentiated into calyx and corolla, free sepals and petals, and often indefinite number of free stamens and carpels. Because of these similarities, certain theories of origin of monocots from dicots suggest Alismataceae as the connecting link between the two.

Brown (1875) believed that most primitive flowers were those of Magnoliaceae and Alismaceae. However, Takhtajan (1969) stated that Alismales "exhibits definite links with the Nymphaeales among the dicots", and the flowers of both are primitive. However, the seeds are endospermic in Nymphaeales and without endosperm in Alismales. Because of several close similarities between Alismatales and dicotyledons, Cronquist (1981) discussed order Alismatales in the very beginning of Liliopsida (Monocots).

22.16 CYPERACEAE (SEDGE FAMILY)

22.16.1 Systematic Position

Monocotyledons, Glumaceae, Cyperaceae.

22.16.2 Field Recognition

Herbs with triangular stem; leaves 3-ranked, ligule absent; flowers in spikelets; bracts glumaceous; lodicules absent; fruit an achene or nut.

22.16.3 Selected Indian Genera

Bulbostylis, Carex, Cyperus, Eleocharis, Fimbristylis, Scirpus.

22.16.4 Size, Distribution and General Information

A family of approximately 90 genera and over 4000 species, showing worldwide distribution in damp and wet habitats. About 23 genera and over 400 species have been reported from India. Larger genera of the family, along with the number of their approximately reported species from the world and the common names of some of them are *Carex* (1100, Carex or sedge), *Cyperus* (700, sedge), *Fimbristylis* (300), *Scirpus* (250, bulrush), *Rhynchospora* (225, beak rush), *Eleocharis* (200, spike rush), *Scleria* (200, nut rush), *Schoenus* (100) and *Uncinia* (35). *Cyperus papyrus*, used for paper-making by Egyptians even as early as 2400 B.C., and *C. rotundus*, a serious weed of lawns and fields, are two important plants of the family.

22.16.5 Description of the Family

General Habit Annual or perennial, grass-like herbs found in damp or wet places; persisting usually by a creeping, underground, sympodial rhizome.

Roots Fibrous, developing from the underground rhizome or tuberous stem.

Stem Often called *'culm'*, the aerial stem is often triangular, solid or rarely hollow, leafless and generally unbranched below the inflorescence.

Leaf Grass-like, often develop in a crowded tuft on the lower basal part of the culm, three-ranked (in 1/3 phyllotaxy), with a closed sheath; leaf blades long, narrow, with a thick midrib; ligule absent.

Inflorescence Basic unit of the inflorescence is a spikelet (Fig. 22.24); the spikelets are usually 1 to many-flowered and variously arranged in racemes, panicles, spikes, or umbels, and usually subtended by well-developed leaf-like bracts; large bracts are distichously or spirally arranged.

Flower Develop singly in the axil of glumes; unisexual (*Carex*, Fig. 22.25) or bisexual (*Cyperus*, Fig. 22.26); trimerous, hypogynous; small or inconspicuous; unisexual species may be monoecious or dioecious; the female flower in some genera (*Carex*) is enclosed by a modified glume called utricle or *perigynium* (Fig. 22.25B).

Perianth Absent (*Cyperus*) or represented by usually 6 hypogynous bristles (*Scirpus*), scales, or hairs (Fig. 22.27A).

Androecium Stamens are generally 3 (*Cyperus rotundus*), less commonly 2 (*C. alulatus*, *C. iria*) or 1 (*C. squarrosus*), and rarely 6 (*Gahnia*); filaments free; anthers basifixed, dithecous, introrse; dehiscence longitudinal.

Gynoecium Bi- to tricarpellary, syncarpous; ovary superior, unilocular, with 1 anatropous ovule; basal placentation; ovary in *Carex* is subtended and enclosed in a perigynium (Fig. 22.25B); styles 2 to 3 or deeply divided into as many branches as the number of carpels.

Fruit and Seeds Fruit a triangular or biconvex achene or nutlet. Seed one per fruit, with small embryo, embedded in fleshy or mealy endosperm; seed coat free from the pericarp.

Pollination Flowers are usually wind-pollinated, although pollen-eating beetles may also transport pollen grains in some species of *Carex*. Several *Carex* species are protandrous.

General Floral Formulae

(a) *Male Flowers:* $\oplus$, ♂, P_0 or bristles, A_{1-3}, G_0.

(b) *Female Flowers:* $\oplus$, ♀, P_0 or bristles, A_0, $G_{\underline{(2-3)}}$.

(c) *Bisexual Flowers:* $\oplus$, ⚥, P_0 or bristles, A_{1-3}, $G_{\underline{(2-3)}}$.

Fig. 22.24 Inflorescence of a few species of *Cyperus*. A: *C. compressus*; B: *C. globosus* All. syn. *Pycreus globosus* Reich; C: *C. metzii* (Hochst) Mattf. syn. *Kyllinga squamulata* Than; D: *C.* triceps (Rottb.) Endl. syn. *Kyllinga triceps* Rottb; E: *C. triceps* showing five spikes; F: *C. paniceus* (Rottb.) Boeck. var. *roxburghianus* syn. *Mariscus paniceus* Vahl. var. *roxburghianus* Clarke (C, Culm; LB, Leafy bract; S, Spikelet; SP, Spike) (Photographs from author's Ph. D. Thesis).

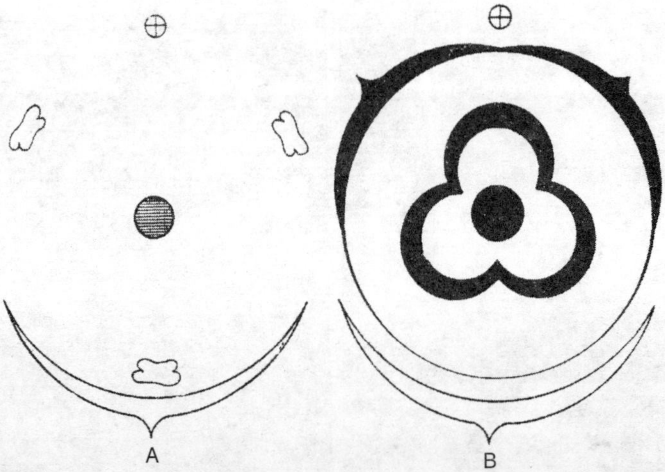

Fig. 22.25 Floral diagrams of *Carex cruciata*—A: Male flower; B: Female flower (Note the perigynium in 'B').

Fig. 22.26 Floral diagram of *Cyperus rotundus*.

22.16.6 Economic Importance

Sedges are not of much significant importance. However, some are grown as ornamentals, few are used in medicines, and some are even edible. Several of them are noxious weeds (*Cyperus rotundus*) of lawns and cultivated fields. Some aspects of their economic importance are undermentioned:

- Starchy tubers of some members are edible, e.g. *Cyperus esculentus* ('Chufas' or Tiger-nut), *C. bulbosus*, and *Eleocharis dulcis* (Chinese water-chestnut).

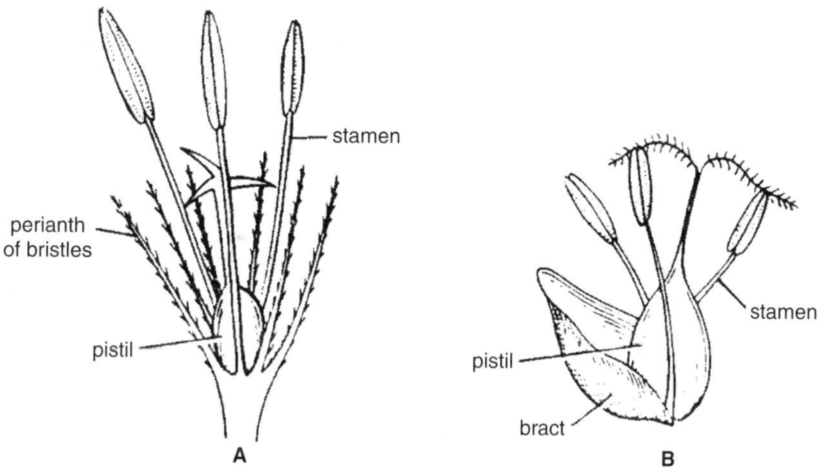

Fig. 22.27 Vertical view of flowers of *Scirpus* (A) and *Cyperus* (B).

- Tubers of *Cyperus rotundus* are used in stomach disorders, while that of *C. articulatus* are used as tonic, and that of *C. stoloniferous* are considered as stimulant for the heart. Tubers of *C. scariosus* are used in treating stone trouble of kidney and urinary bladder.
- Chufa-oil, obtained from the tubers of *Cyperus esculentus*, is used in soap industry.
- Tubers of *Cyperus rotundus* and *C. scariosus* are used in the preparation of fragrant sticks called "Agarbattis" in India.
- *Cyperus papyrus* (Papyrus plant), formerly used in paper-making, is nowadays grown as an ornamental plant. Other plants of ornamental value are *C. flabelliformis* syn. *C. alternifolius* (umbrella plant), *C. fertilis* and *C. gracilis*.
- Culms of *Cladium mariscus*, *Cyperus corymbosus*, *C. elatus*, *C. exaltatus*, *C. iria*, etc. are used for making mats, baskets, etc.
- Several species of *Cyperus*, *Carex*, *Scirpus*, *Rhynchospora*, etc. are used as fodder, though of inferior quality.

22.16.7 Systematics and Phylogeny (Affinities)

Taxonomists generally treat Cyperaceae to be closely allied to Gramineae, and that is why Bentham and Hooker treated both of them under the same series Glumaceae, and the two families have been included under the same order Glumiflorae by Engler as well as Rendle. Cronquist (1981) also included both of them under order Cyperales, and the two families have been included under one and the same order Commelinales by Thorne (1983). Hutchinson (1959), however, put them under 2 monotypic orders, i.e. Cyperaceae under Cyperales and Gramineae under Graminales, and according to him both the families have been derived from Juncales but from different stocks on parallel lines. Takhtajan (1969) opined that Cyperaceae, the only family of order Cyperales, "evidently derived directly from the most primitive Juncaceae".

Hickey and King (1988) divided Cyperaceae into 3 subfamilies viz. (1) *Cyperoideae* (flowers bisexual, spikelets many-flowered, rarely flowers unisexual; e.g. *Cyperus*, *Scirpus*), (ii) *Rhynchosporoideae* (flowers unisexual or bisexual, with or without perianth; in few-flowered, spike-like cymes grouped into heads or spikes; e.g. *Rhynchospora*), (iii) *Caricoideae* (unisexual flowers, without perianth; usually in many-flowered spikes; perigynium present; e.g. *Carex*, *Uncinia*).

Radford (1986) mentioned 8 tribes under Cyperaceae, viz. 1. Cypereae (*Cyperus*), 2. Dulichieae (*Dulichium*), 3. Rhynchosporeae (*Cladium*), 4. Scirpeae (*Scirpus*) 5. Hypolytreae (*Mapania*), 6. Sclerieae (*Scleria*), 7. Cryptangieae (*Cryptangium*), and 8. Cariceae (*Carex*).

22.17 GRAMINEAE OR POACEAE (GRASS FAMILY)

22.17.1 Systematic Position

Monocotyledons, Glumaceae, Gramineae.

22.17.2 Field Recognition

Herbs with round stem (culm) having usually hollow internodes; leaves mostly flat, 2-ranked and usually with open leaf sheath; ligule usually present; bracts of glumes, lemma and palea; fruit caryopsis.

22.17.3 Selected Indian Genera

Avena, *Bambusa*, *Cynodon*, *Hordeum*, *Oryza*, *Poa*, *Saccharum*, *Secale*, *Sorghum*, *Triticum*, *Zea*.

22.17.4 Size, Distribution and General Information

Gramineae, the largest and most important family of angiosperms from the economic point of view, is represented by about 620 genera and over 10,000 species, and distributed widely in almost all regions of the world and in almost every type of habitat. About 240 genera and over 1200 species have been reported from India. Some of the largely represented genera along with the total number of their species and common names of some of them are *Panicum* (500, panic grass), *Digitaria* (380, crab grass), *Aristida* (330, wire grass), *Poa* (300, blue grass), *Eragrostis* (300, love grass), *Stipa* (300, needle grass), *Paspalum* (250, Bahia grass), *Agrostis* (150–200, bent grass), *Arundinaria* (150, giant cane), *Pennisetum* (150, Bajra), *Setaria* (140, Italian millet), *Agropyron* (100–150, wheat grass), *Muhlenbergia* (100, bull grass), *Avena* (70, oats), *Sorghum* (60, sorghum), *Oryza* (25, rice), *Triticum* (20, wheat), *Hordeum* (20, barley), *Saccharum* (12, sugarcane), *Cynodon* (10, Doob grass), *Zea* (1, maize).

22.17.5 Description of the Family

General Habit Usually annual or perennial herbs; rarely woody shrubs or tree-like, reaching up to a height of 30 metre (*Bambusa*) or even more.

Roots Fibrous and adventitious; primary roots are often short-lived; some adventitious roots (stilt roots) in *Zea mays* develop from the basal nodes of stem for providing extra support.

Stem The stems ('*culms*') are erect, ascending, prostrate or creeping; simple or commonly branched at the base; usually round, jointed, and bear hollow internodes, and solid and swollen nodes; terminates usually by inflorescence; many species have a perennial, creeping rhizome or stolon formed by the lower internodes of stem.

Leaf Simple, alternate, in two rows on opposite sides of the stem (2-ranked), originating at the nodes, often crowded at the base, and consisting of sheath, blade and ligule; *sheath* with margins free and overlapping or encircling the culm to form a tube; *blade* or *lamina* usually long, narrow, flat, linear to lanceolate, rarely with a constricted petiole-like base (*Bambusa*); veins parallel; *ligule*, present at the junction of blade and sheath on the adaxial surface, is membranous, or reduced to a ring of hairs, or even absent (*Echinochloa*).

Inflorescence Basic unit of inflorescence is a *spikelet* (Fig. 22.28); each spikelet consists of one or more flowers (or florets) and their subtending bracts arranged on an inflorescence axis called *rachilla*; spikelets are either sessile or with pedicel, and arranged in spikes, racemes or panicles forming compound inflorescences; at the base of each spikelet are generally present two *sterile glumes*, of which the lower one is called *first glume* and the upper one is called *second glume*; in some genera, either the first glume or both the first and second glumes are absent; up to 6 sterile glumes are present in some genera; the flowers are arranged on the rachilla just above the second glume.

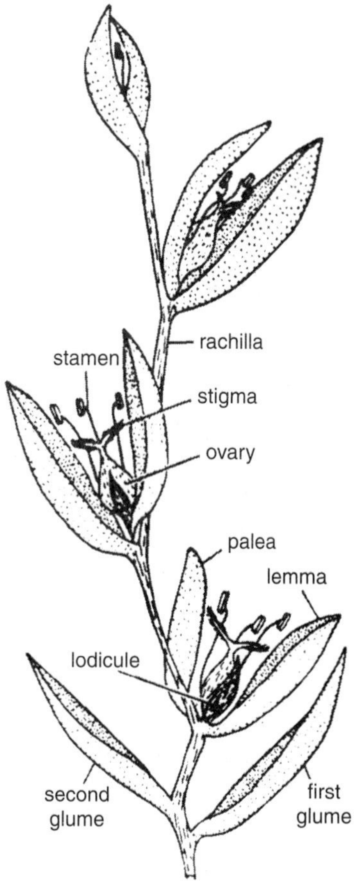

Fig. 22.28 Diagrammatic representation of a spikelet of *Triticum aestivum*.

Flower or Floret Flowers, also called *florets* (Fig. 22.29), are small, inconspicuous, bisexual or unisexual (*Zea*, Fig. 22.30), zygomorphic, hypogynous; develop on the rachilla above the glumes; and each floret is subtended by two bracts called lemma and palea; *lemma*, also called fertile or flowering glume, is greenish, keeled, nerved or awned, and often resembles with sterile glumes; *palea*, present in between rachilla and lemma, is thin, membranous, 2-nerved or 2-keeled, and often partially enclosed by the lemma.

Palea, morphologically representing a bracteole, bears a flower proper in its axil.

Perianth Absent (*Dendrocalamus*) or reduced to usually 2 or rarely 3 minute scales called *lodicules*; lodicules are fleshy or hyaline and present antero-laterally, if 2 in number.

Androecium Stamens 1 to 6 or rarely more, but usually 3 in each floret (Fig. 22.31); the odd stamen is always anterior; only 1 anterior stamen is present in *Uniola*; filaments free; anthers dithecous, basifixed or versatile, introrse, and usually opening by a longitudinal slit.

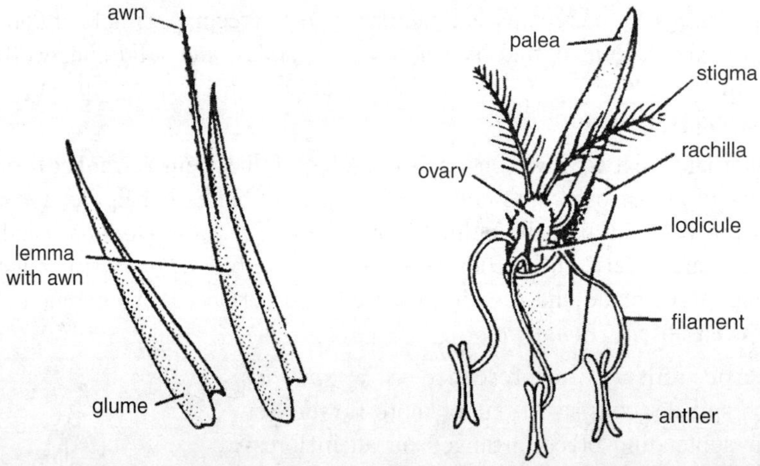

Fig. 22.29 Parts of a bisexual flower of Gramineae.

Gynoecium Bi- or tricarpellary, syncarpous; ovary superior, unilocular, containing one anatropous ovule, basal placentation; styles 1 (*Nardus*) to 3 (*Bambusa*), commonly 2; stigmas typically 2, often plumose or feather-like.

Fruit and Seeds Fruit usually a caryopsis or rarely a nut, utricle (*Sporobolus*) or berry (some bamboos). Seed 1 per fruit, with abundant and starchy endosperm; embryo variable in size.

Pollination Plants, such as *Triticum*, are self-pollinated whereas majority of grasses are wind-pollinated. Flowers are protogynous. Characters, such as small and inconspicuous flowers, and feather-like stigma, are the adaptations for anemophily. Flowers produce loose and powdery pollen grains in large quantity.

General Floral Formula ⊕ or ⦶, ⚥ or ♂ or ♀, $P_{2 \text{ or } 3 \text{ or absent}}$, $A_{3 \text{ or } 1-6}$, $G_{\underline{(2-3)}}$.

22.17.6 Economic Importance

Gramineae are of greater importance than any other family of flowering plants from the economic point of view. Plants of this family provide food for man (wheat, oat, barley, sugarcane, etc.), fodder for our domestic animals, grasses and ornamental plants for our lawns and gardens, and several other products including starch, ethyl alcohol, paper, fibres, edible oils, adhesives, thatching and building materials, wildlife food, etc.

- *Cereals*[1]
  - (i) *Avena sativa* (Oats, Jai): This important cereal and fodder crop is highly nutritious.
  - (ii) *Hordeum vulgare* (Barley or Jaun): Used as a feed grain and also in manufacturing several alcoholic beverages including beer, barley water, etc.

[1] Plants arranged alphabetically.

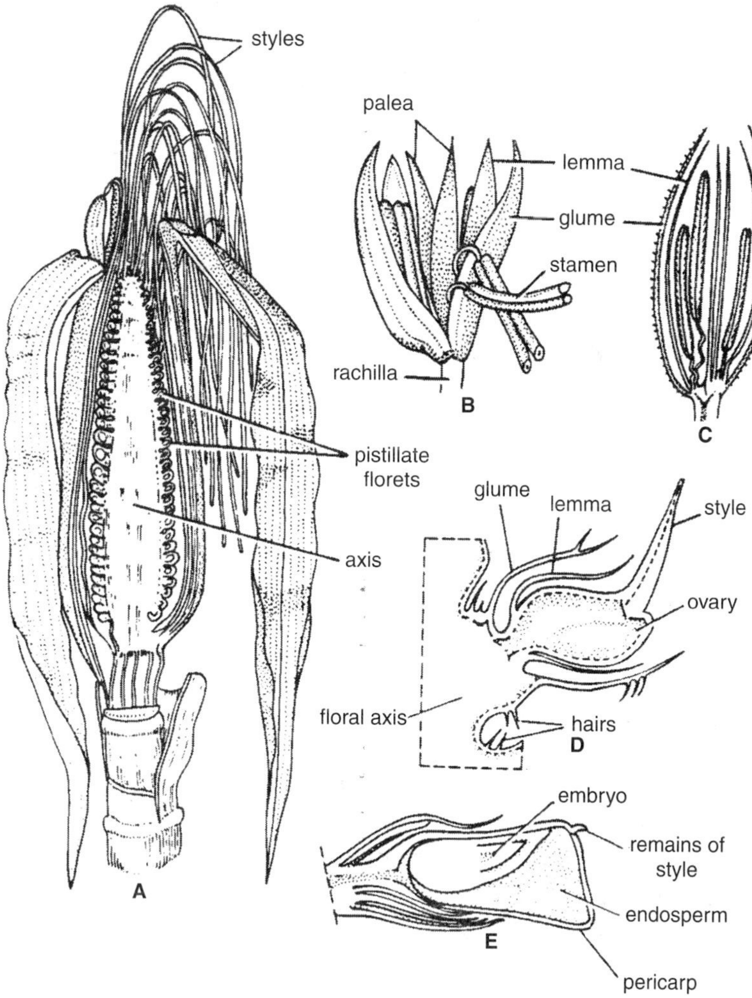

Fig. 22.30 *Zea mays:* A: L.S. female inflorescence; B: A spikelet of 2 male florets, of which one is at anthesis; C: L.S. of two male florets (only nearest stamens shown); D: L.S. female floret; E: L.S. caryopsis surrounded by other floral parts.

(iii) *Oryza sativa* (Rice, Chawal or Paddy): Used as a common staple food by millions of people throughout the world, rice is also used as a livestock and for manufacturing paper, alcohol, starch etc. Its husk is used as fuel and its bran is the source of fatty oil.

(iv) *Secale cereale* (Rye): Besides its limited use as a cereal crop, rye is used for making alcoholic beverages, pickles, sambar, etc.

(v) *Triticum aestivum* syn. *T. vulgare* (Wheat, Gehun): Wheat is used universally as a staple food for man. Its straw is used as stockfeed for cattles, for packing goods, and for

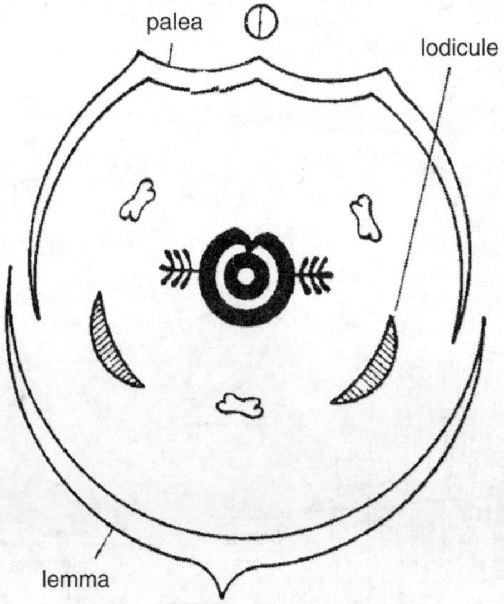

Fig. 22.31 Floral diagram of *Triticum aestivum*.

manufacturing paper. Grains are also used for manufacturing starch, industrial alcohol, etc.

(vi) *Zea mays* (Maize, Makka): Used as a common food by man, maize is also used as a livestock feed for cattles and also for manufacturing several industrial products such as explosives, adhesives, soaps, starch, plastics, asbestos, linoleum, etc. Paper is manufactured from its leaf stalks, and the corn oil is used in preparing paints, varnishes, etc. Several alcoholic beverages are also manufactured from maize.

- *Millets:* Millets[1], used as coarse grains by man and common fodder for animals, belong to Gramineae. Some of the important millets include *Eleusine coracana* (Ragi or finger millet), *Panicum miliaceum* (sama or common millet), *Pennisetum typhoides* (Bajra or pearl millet), *Setaria italica* (Italian millet), and *Sorghum vulgare* (Jowar or sorghum). Jowar and Bajra are chiefly used for poultry and cattles, and also for preparing alcoholic beverages and diastase.
- *Sugar:* It is obtained from the stem juice of *Saccharum officinarum* (sugarcane or Ganna). Baggase is used for manufacturing paper while the molasses is utilized for manufacturing industrial alcohol and alcoholic beverages such as rum.
- *Fodder:* The whole aerial green part of several members of Gramineae is cut and used as green fodder. Several species of some commonly used genera for this purpose belong to *Avena, Eleusine, Hordeum, Pennisetum, Phalaris, Setaria* and *Sorghum.*

[1] Plants arranged alphabetically.

Almost all wild grasses are browsed by the cattles, of which some commonly browsed species are *Andropogon pertusus*, *Apluda mutica*, *Cenchrus ciliaris*, *Cynodon dactylon*, *Dichanthium annulatum*, *Iseilema laxum* and *Panicum maximum*.

Straw of various species of *Oryza* and *Triticum* is also used as a common fodder for cattles.

Some other fodder grasses are *Dactylis glomerata*, *Heteropogon contortus*, *Lolium multiflorum*, *Paspalum scrobiculatum*, *Poa annua*, *Sporobolus diander* and *Stipa orientalis*.

- *Lawn Grasses:* Many grasses are used in lawns and other turfed areas, e.g. *Agrostis*, *Cynosurus*, *Festuca*, *Lolium*, and *Poa*.
- *Paper Manufacturing:* Pulp, prepared from the straw of several species of this family, is used in manufacturing paper of almost all coarse and fine qualities, strawboard, artificial rayon, etc. Some commonly used genera for the purpose are *Bambusa*, *Cephalostachyum*, *Dendrocalamus*, *Erianthus* (*E. arundinaceus*, *E. munja*, *E. ravennae*), *Gigantochloa*, *Melocanna*, *Narenga*, *Ochlandra*, *Oryza*, *Saccharum* (*S. benghalense*, *S. officinarum*, *S. procerum*, *S. spontaneum*), *Triticum aestivum*, *Vetiveria zizanioides* and *Zea mays*.
- *Bamboos:* 'Bamboos', a common term used for some tall tree-like woody grasses, are universally used for papermaking. They are also used for house building, as substitute of timber, and also for preparing several other articles such as walking sticks, beds, tent poles, umbrella handles, kites, musical instruments, furnitures, toys, mats, etc. Some bamboos are harvested for cattle fodder, and several others are grown for checking soil erosion, and also for ornamental purposes. Some common bamboos are *Arundinaria*, *Bambusa*, *Cephalostachyum*, *Dendrocalamus*, *Melocanna*, *Ochlandra* and *Phyllostachys*.
- *Khus-Khus:* Roots of *Vetiveria zizanioides* are used for making curtains and screens (Khus-Khus), which, when wetted, bring about cooling and scenting of rooms.
- *Essential Oils:* Volatile oil, obtained from the leaves of several grasses, is used in perfumery, medicines, mosquito-repellent creams; soaps, etc. Some grasses utilised for the purpose are *Cymbopogon caesius* (ginger grass oil), *C. citratus* (lemon grass oil), *C. martinii* (palmarosa oil), *C. nardus* (Citronella oil) and *Vetiveria zizanioides* (khus oil or oil of vetiver).
- *Ornamental Plants:* Grasses, grown for ornamental purposes in the gardens, or dried and used for floral decoration include *Axonopus affinis* (carpet grass), *Briza maxima* (quaking grass), *Coix lacryma-jobi* (job's tears), *Cortaderia selloana* (Pampas Grass), *Cynodon dactylon* (Doob grass or Bahama grass), *Lagurus ovatus* (Hare's-tail grass), *Lamarckia aurea* (golden top grass), *Phalaris arundinacea* (ribbon grass), *P. canariensis* (canary grass), *Setaria italica* (Foxtail millet) and *Stipa pinnata* (feather grass).
- *Brooms:* Panicles of *Thysanolaena maxima* (Broom grass) are used as soft brooms.
- *Ropes, Mats, Baskets:* Ropes are made from the fibres obtained from the leaves of *Erianthus munja* syn. *Saccharum munja* (Munj) and *Saccharum spontaneum* (Kaans) while mats, baskets, musical instruments, etc. are prepared from the stems of *Arundo donax* (Nal or reed), *Arundinaria falcata*, *A. racemosa* and *Phragmites karka* (Bansi or common reed).

22.17.7 Systematics and Phylogeny (Affinities)

Family Gramineae or Poaceae has been variously placed under series Glumaceae by Bentham and Hooker (1862–1883), order Glumiflorae by Engler and Prantl (1931), order Graminales by Hutchinson (1959), order Poales by Takhtajan (1969), order Cyperales by Cronquist (1981), and order Commelinales by Thorne (1983).

Family Gramineae is usually divided into 6 subfamilies, divisible further into 25 tribes as under:

Subfamily 1. *Festucoideae:* Tribe 1. Festuceae (*Poa*); Tribe 2. Aveneae (*Avena*); Tribe 3. Triticeae (*Triticum*); Tribe 4. Meliceae (*Melica*); Tribe 5. Stipeae (*Stipa*); Tribe 6. Brachyelytreae (*Brachyelytrum*); Tribe 7. Diarrheneae (*Diarrhena*); Tribe 8. Nardeae (*Nardus*); Tribe 9. Monermeae (*Monerma*).

Subfamily 2. *Panicoideae:* Tribe 10. Paniceae (*Pennisetum*); Tribe 11. Andropogoneae (*Saccharum*).

Subfamily 3. *Eragrostoideae:* Tribe 12. Eragrosteae (*Eragrostis*); Tribe 13. Chlorideae (*Cynodon*); Tribe 14. Zoysieae (*Zoysia*); Tribe 15. Aleuropodeae (*Allolepis*); Tribe 16. Unioleae (*Uniola*); Tribe 17. Pappophoreae (*Cottea*); Tribe 18. Orcuttieae (*Orcuttia*); Tribe 19. Aristideae (*Aristida*).

Subfamily 4. *Bambusoideae:* Tribe 20. Bambuseae (*Bambusa*); Tribe 21. Phareae (*Pharus*).

Subfamily 5. *Oryzoideae:* Tribe 22. Oryzeae (*Oryza*).

Subfamily 6. *Arundinoideae:* Tribe 23. Arundineae (*Arundo*); Tribe 24. Danthonieae (*Danthonia*); Tribe 25. Centotheceae (*Chasmanthium*).

Superficially, Gramineae resemble Cyperaceae. In both these families, the flowers are small, inconspicuous, subtended by bracts, and present in the form of spikelets. However, Gramineae differ from Cyperaceae by their usually terete, hollow, jointed stems, distichous leaves, and structure of spikelet. Hutchinson opined that the Gramineae have been derived from Restionaceae of Juncales which again have been derived from Liliaceous stock. Some are also of the opinion that Gramineae may be distantly related to the Commelinaceae.

Test Your Understanding

1. Following genera belong to which family of monocotyledons?

1. *Zeuxine*	2. *Vanda*	3. *Gladiolus*	4. *Crocus*
5. *Agave*	6. *Ananas*	7. *Canna*	8. *Ravenala*
9. *Curcuma*	10. *Elettaria*	11. *Aloe*	12. *Asparagus*
13. *Ruscus*	14. *Tradescantia*	15. *Phoenix*	16. *Colocasia*
17. *Pothos*			

2. Describe floral structure of family Arecaceae (Palmae) in semitechnical language.
3. Describe the basic structure of the spikelet of Cyperaceae.
4. Discuss the inflorescence and floret details of Poaceae.
5. "Poaceae (Gramineae) are of greater economic importance than any other family of flowering plants". Comment.

Suggested Reading

Bentham, G. and J.D. Hooker, 1862–1883, *Genera Plantarum*, 3 Vols., Reeve & Co., London.

Bessey, C.E., 1915, Phylogenetic taxonomy of flowering plants, *Ann. Mo. Bot. Gard.* **2**: 109–164.

Cronquist, A., 1981, *An Integrated System of Classification of Flowering Plants*, Columbia Univ. Press, New York.

Engler, A.W. and K. Prantl, 1887–1915, *Die Naturlichen Pflanzenfamilien*, 23 Vols., Leipzig.

Heywood, V.H., 1978, *Flowering Plants of the World*, Mayflower Books, New York.

Hickey, M. and C. King, 1988, *100 Families of Flowering Plants*, (2nd ed.), Cambridge University Press, Cambridge.

Holmes, S., 1983, *Outline of Plant Classification*, Longman, London.

Hutchinson, J., 1969, *Evolution and Phylogeny of Flowering Plants*, Academic Press, London.

_______ 1973, *The Families of Flowering Plants*, (3rd ed.), Clarendon Press, Oxford.

Jones, S.B. Jr. and A.E. Luchsinger, 1987, *Plant Systematics*, (2nd ed.), McGraw-Hill Book Co., New York.

Lawrence, G.H.M., 1951, *Taxonomy of Vascular Plants*, Macmillan Co., New York.

Radford, A.E., 1986, *Fundamentals of Plant Systematics*, Harper & Row Publishers, Inc., New York.

Rendle, A.B., 1925, The *Classification of Flowering Plants* Vol. II (Dicotyledons), Cambridge Univ. Press, London.

Takhtajan, A., 1969, *Flowering Plants: Origin and Disperal*, Oliver & Boyd Ltd., Edinburgh.

_______ 1980, Outline of the classification of flowering plants (Magnoliophyta), *Bot. Rev.* **46**: 225–359.

Thorne, R.F., 1983, Proposed new realignments in the angiosperms, *Nordic J. Bot.* **3**: 85–117.

Willis, J.C., 1973, *A Dictionary of the Flowering Plants and Ferns*, (8th ed., revised by H.K. Airy Shaw), Cambridge Univ. Press, London.

APPENDIX 1

EXAMINATION TOOLS—COMPARATIVE TABLES OF SELECTED FAMILIES

Table A.1 Magnoliaceae and Annonaceae

Magnoliaceae	Annonaceae
1. Leaves usually stipulate.	1. Leaves usually exstipulate.
2. Usually the sepals and petals are undifferentiated.	2. Sepals and petals are clearly differentiated.
3. Anthers are usually introrse.	3. Anthers are usually extrorse.
4. Style and stigma absent in some species.	4. Style and stigma usually present.
5. Sarcotesta present in the seeds.	5. Sarcotesta absent.
6. Wood is not usually aromatic.	6. Wood usually is aromatic.

Table A.2 Ranunculaceae and Nymphaeaceae

Ranunculaceae	Nymphaeaceae
1. Usually terrestrial, and only rarely aquatic.	1. Mostly aquatic.
2. Plants do not possess latex.	2. Plants usually possess colourless or coloured latex.
3. Stem is usually aerial, erect or climbing; only few possess rhizome.	3. Stem is mostly a rhizome, usually burried in mud.
4. Leaves are small, simple or compound.	4. Leaves are very large, simple; with sometimes petiole as large as the depth of water.
5. The inflorescence may be cymose clusters, racemose, or rarely solitary terminal.	5. The inflorescence is solitary terminal or solitary axillary.
6. Flowers usually small, actinomorphic or zygomorphic.	6. Flowers usually large, long-peduncled, beautifully coloured, actinomorphic.
7. Sepals and petals vary in number from 3 to numerous; usually 5.	7. Sepals usually 3 and petals usually many.

(*Contd.*)

8. There is no transition of petals into stamens.	8. Petals (usually inner ones) frequently transform into staminodes and stamens.
9. Several genera possess spurred sepals and petals.	9. Sepals and petals are not spurred.
10. Gynoecium usually polycarpellary and apocarpous.	10. Carpels usually 3 and sometimes many, apocarpous or syncarpous.
11. Placentation is basal or marginal.	11. Placentation is marginal or parietal.
12. Fruit an etaerio of achenes, or etaerio of follicles, or berry, or capsule.	12. Fruit berry or follicle.

Table A.3 Ranunculaceae and Cruciferae (Brassicaceae)

Ranunculaceae	Cruciferae (Brassicaceae)
1. Usually herbs, without a smelling watery juice.	1. Herbs with a smelling watery juice.
2. Leaves simple or compound.	2. Leaves simple and not compound.
3. Inflorescence may be cymose clusters, racemose, or flowers solitary terminal.	3. Inflorescence usually corymbose raceme.
4. Sepals and petals may or may not be differentiated.	4. Sepals and petals always differentiated.
5. Sepals and petals vary in number from 3 to numerous, usually 5.	5. Sepals and petals usually 4.
6. Sepals and petals are spurred in some genera (e.g. *Delphinium*).	6. Sepals and petals never spurred.
7. Stamens 5 to many and never tetradynamous.	7. Stamens usually 6 and tetradynamous.
8. Gynoecium usually polycarpellary and apocarpous.	8. Gynoecium usually bicarpellary and syncarpous.
9. Ovary unilocular.	9. Ovary unilocular but becomes bilocular due to the development of a false septum or replum.
10. Placentation is basal or marginal.	10. Placentation is parietal.
11. Stigma linear.	11. Stigma usually bifid.
12. Fruit an etaerio of achenes, or etaerio of follicles, or berry, or capsule.	12. Fruit siliqua or silicula.
13. Seeds usually endospermic.	13. Seeds usually non-endospermic.

Table A.4 Ranunculaceae and Rosaceae

Ranunculaceae	Rosaceae
1. Leaves compound, alternate but sometimes opposite.	1. Leaves compound, sometimes simple, alternate.
2. Leaf bases mostly sheathing.	2. Stipulate.
3. Flowers usually regular, rarely irregular (*Delphinium*), entomophilous.	3. Flowers regular, entomophilous but rarely anemophilous (*Poterium*).
4. Sepals variable in number and often petaloid.	4. Sepals usually 5.
5. Epicalyx absent.	5. Epicalyx often present.
6. Petals variable in number, often absent (*Caltha*).	6. Petals usually 5, sometimes absent (*Alchemilla*).

(*Contd.*)

7. Stamens numerous, spiral or cyclic.	7. Stamens numerous but only 4 in *Alchemilla*, whorled.
8. Gynoecium of 3 to many carpels, separate.	8. Gynoecium of highly variable carpel number, ovary free or adnate to hypanthium.
9. Fruit follicle, achene, and rarely berry (*Actaea*) or capsule (*Nigella*).	9. Fruit achene, aggregate, drupe (*Prunus*), pome (*Malus*), follicle (*Spiraea*).

Table A.5 Papaveraceae and Fumariaceae

Papaveraceae	Fumariaceae
1. Herbs usually with latex (*Papaver*, *Argemone*).	1. Herbs with watery juice but latex absent (*Fumaria*).
2. Usually the inflorescence is solitary terminal.	2. Usually the inflorescence is racemose.
3. Flowers actinomorphic.	3. Flowers zygomorphic.
4. Thalamus is always convex.	4. Thalamus is not always convex.
5. Petals rolled or crumpled in the bud.	5. Petals not always rolled or crumpled in the bud.
6. Usually the petals are unspurred.	6. Petals are usually spurred.
7. Stamens numerous.	7. Stamens are 6 and arranged in 2 groups.
8. Gynoecium bi- to polycarpellary.	8. Gynoecium is always bicarpellary.

Table A.6 Papaveraceae and Cruciferae (Brassicaceae)

Papaveraceae	Cruciferae (Brassicaceae)
1. Plants mostly herbs with latex.	1. Plants mostly herbs which usually contain a pungent watery sap.
2. Inflorescence solitary terminal.	2. Inflorescence typically a raceme or corymb.
3. Flowers usually bi- to trimerous.	3. Flowers usually tetramerous.
4. Corolla not cruciform.	4. Corolla cruciform (i.e. petals arranged in a cross).
5. Stamens numerous.	5. Stamens usually 6, arranged in two whorls.
6. Stamens do not show tetradynamous condition.	6. Stamens show tetradynamous condition.
7. Gynoecium bi-to polycarpellary, syncarpous.	7. Gynoecium always bicarpellary, syncarpous.
8. Locule does not bear a false septum or replum.	8. Usually the locule develops a false septum or replum.
9. Fruit usually a capsule.	9. Fruit usually a siliqua or silicula.

Table A.7 Capparidaceae and Cruciferae (Brassicaceae)

Capparidaceae	Cruciferae (Brassicaceae)
1. Includes herbs, shrubs and trees.	1. Plants are mostly herbs, and only rarely small shrubs.
2. Plants usually devoid of smelling watery juice.	2. Plants with a smelling watery juice.
3. Leaves usually stipulate.	3. Leaves usually exstipulate.
4. Leaves simple or palmately compound.	4. Leaves simple or pinnately compound.
5. Flowers bracteate.	5. Bracts are absent.

(*Contd.*)

6. Androgynophore or only gynophore present in some genera.	6. Androgynophore or gynophore absent.
7. Corolla is never cruciform.	7. Corolla is usually cruciform.
8. Stamens 4 to numerous but never tetradynamous.	8. Stamens usually 6 and tetradynamous.
9. False septum or replum absent.	9. False septum or replum present.
10. Fruit berry, capsule, drupe or nut.	10. Fruit usually siliqua or silicula.

Table A.8 Capparidaceae and Caryophyllaceae

Capparidaceae	Caryophyllaceae
1. Includes herbs, shrubs, small or tall trees.	1. Plants are mostly annual or perennial herbs.
2. Inflorescence usually raceme or simple umbel.	2. Inflorescence usually dichasial cyme.
3. Flowers generally tetramerous.	3. Flowers generally pentamerous.
4. Sepals usually 4, and rarely 5.	4. Sepals usually 5, and rarely 4.
5. Petals usually 4 and free.	5. Petals usually 5 and free.
6. Stamens are not obdiplostemonous.	6. Stamens are obdiplostemonous.
7. Androphore and gynophore present in several genera.	7. Androphore and gynophore absent.

Table A.9 Malvaceae and Tiliaceae

Malvaceae	Tiliaceae
1. Herbs or shrubs, rarely trees.	1. Usually trees or shrubs, rarely herbs.
2. Flowers always bisexual.	2. Flowers bisexual as well as unisexual.
3. Epicalyx usually present.	3. Epicalyx is not often present.
4. Sepals 5, free or basally connate, usually persistent.	4. Sepals 3–5, free or basally connate, caducous.
5. Stamens numerous, usually monadelphous.	5. Stamens numerous, usually polyadelphous.
6. Anthers monothecous, reniform, with spiny pollen grains.	6. Anthers dithecous, pollen grains smooth.
7. Filaments of anthers fuse to form a staminal column around the ovary.	7. Filaments do not form the staminal column around the ovary.
8. Seeds with curved embryo and scanty or no endosperm.	8. Seeds with straight embryo, and scanty or copious endosperm.
9. Phloem fibres are not very strong.	9. Phloem fibres are very strong.

Table A.10 Oxalidaceae and Geraniaceae

Oxalidaceae	Geraniaceae
1. Predominantly herbs, often shrubs or trees.	1. Predominantly herbs, rarely shrubs.
2. Leaves usually palmately compound, or pinnately compound, exstipulate.	2. Leaves usually simple or sometimes compound, stipulate.
3. Leaves with a sour taste due to the presence of oxalic acid.	3. Oxalic acid absent.

(Contd.)

4. Flowers actinomorphic.	4. Flowers actinomorphic but some are zygomorphic.
5. Sepals never spurred.	5. Sepals spurred in zygomorphic flowers.
6. Styles usually free.	6. Styles usually united.
7. Capsular fruit not dehiscing into 3–5 mericarps.	7. Capsular fruit dehiscing into 3–5 mericarps.
8. Seeds with straight embryo and fleshy endosperm.	8. Seeds usually with curved embryo and scanty or no endosperm.

Table A.11 Rutaceae and Meliaceae

Rutaceae	Meliaceae
1. Leaves are gland-dotted or contain translucent, very clear dots.	1. Leaves are not gland-dotted.
2. Stamens obdiplostemonous.	2. Stamens not obdiplostemonous.
3. Filaments of anthers usually do not unite to form a columnar tube.	3. Filaments unite to form a columnar tube.
4. Some genera (*Citrus*) show polyadelphous condition.	4. Polyadelphous condition is very rare or absent; monadelphous condition is present.
5. A nectariferous disc present below the ovary.	5. Nectariferous disc usually present between androecium and gynoecium.
6. Fruit usually a hesperidium.	6. Fruit is never hesperidium.
7. Seeds are not winged.	7. Seeds are often winged.
8. Polyembryony is very common.	8. Polyembryony is not of common occurrence.

Table A.12 Rutaceae and Anacardiaceae

Rutaceae	Anacardiaceae
1. Leaves are gland-dotted or contain translucent, very clear-dots; aromatic and strong-smelling.	1. Leaves are not gland-dotted and not even aromatic and strong-smelling.
2. Inflorescence cymose.	2. Inflorescence axillary or terminal panicles.
3. Flowers are hypogynous, mostly pentamerous.	3. Flowers hypogynous, rarely perigynous or epigynous; pentamerous.
4. Cup-shaped nectariferous disc present between ovary and stamens.	4. Cup-shaped nectariferous disc is extrastaminal or intrastaminal.
5. Sepals and petals mostly 4 or 5.	5. Sepals and petals mostly 5, rarely 3 to 7.
6. Stamens 8 to 10, obdiplostemonous; in some members show polyadelphous condition (e.g. *Citrus*).	6. Stamens usually 10, sometimes only 1 stamen and 4 (*Mangifera*) or 6–9 staminodes (*Anacardium*).
7. Gynoecium of 4 or 5 carpels, syncarpous; ovary tetra- to pentalocular,	7. Gynoecium of 1 to 5 fused carpels to form a unilocular or multilocular ovary.
8. Fruit usually a hesperidium.	8. Fruit usually a drupe.
9. Polyembryony is very common.	9. Polyembryony is not very common.

Table A.13 Sapindaceae and Anacardiaceae

Sapindaceae	Anacardiaceae
1. Trees, shrubs or climbers, without any irritant or acrid juice.	1. Trees, shrubs or climbers having irritant or acrid juice.
2. Resin ducts absent.	2. Resin ducts present.
3. Leaves usually pinnately compound.	3. Leaves simple or pinnately compound.
4. Flowers actinomorphic as well as zygomorphic.	4. Flowers mostly actinomorphic.
5. Sepals 4–5, poly- or gamosepalous.	5. Sepals 3–7 but usually 5, free or basally connate.
6. An annular disc is present, usually extrastaminal.	6. The cupular nectariferous disc is extrastaminal or intrastaminal.
7. Gynoecium tricarpellary, syncarpous, trilocular.	7. Gynoecium of 1 to 5 fused carpels, unilocular or multilocular.
8. Fruit drupe, berry, nut, capsule or samara.	8. Fruit usually a drupe.

Table A.14 Mimosoideae, Caesalpinioideae and Papilionoideae

Mimosoideae	Caesalpinioideae	Papilionoideae
1. Predominantly trees and shrubs, rarely herbs.	1. Mostly trees and shrubs, sometimes herbs.	1. Predominantly herbs, but shrubs, trees and climbers are also common.
2. Leaves mainly bipinnately compound, and never simple.	2. Leaves paripinnate, and never simple.	2. Leaves imparipinnate, sometimes simple.
3. Flowers actinomorphic.	3. Flowers zygomorphic.	3. Flowers zygomorphic.
4. Corolla aestivation valvate or rarely imbricate.	4. Corolla aestivation asscending-imbricate; posterior petal innermost.	4. Corolla aestivation descending-imbricate; posterior petal outermost.
5. Petals 5, fused or free.	5. Petals 5, polypetalous.	5. Petals 5, unequal, 2 anterior petals basally fused.
6. Stamens 10 or more.	6. Stamens usually 10, free or monadelphous, often a few stamens reduce to staminodes.	6. Stamens usually 10 and diadelphous i.e. (9) + 1, or monadelphous i.e. (10).

Table A.15 Rosaceae and Myrtaceae

Rosaceae	Myrtaceae
1. Perennial herbs, shrubs or trees.	1. Usually large trees or shrubs.
2. Leaves stipulate, alternate or rarely opposite.	2. Leaves exstipulate, usually opposite.
3. Inflorescence highly variable from solitary flowers to racemose and cymose clusters.	3. Inflorescence usually cymose.
4. Often epicalyx is present in several genera (e.g. *Fragaria*).	4. Epicalyx is absent.
5. Sepals 5, free or basally connate.	5. Sepals 4–5, free or united.

(Contd.)

6. Petals 5, usually attached to the rim of hypanthium; sometimes numerous (e.g. *Rosa*).	6. Petals 4–5, free or united.
7. Stamens numerous, usually in multiple of 5.	7. Stamens numerous with their anther connectives gland-tipped.
8. Ovary superior to inferior with intermediate stages.	8. Ovary inferior or semi-inferior.
9. Placentation is marginal or axile.	9. Placentation is axile or rarely parietal.

Table A.16 Rosaceae and Cucurbitaceae

Rosaceae	Cucurbitaceae
1. Perennial herbs, shrubs or trees.	1. Mostly climbing or prostrate, tendril-bearing herbs; shrubs and trees very rare.
2. Leaves stipulate.	2. Leaves exstipulate.
3. Flowers usually bisexual and only rarely unisexual (e.g. *Pygeum*).	3. Flowers usually unisexual, and only rarely bisexual (e.g *Schizopepon*).
4. Epicalyx often present in several genera (e.g. *Fragaria*, *Geum*).	4. Epicalyx absent.
5. Androecium usually made of numerous distinct stamens in whorls of 5.	5. Androecium usually made of 5 stamens, usually more or less united into a column.
6. Usually a honey-secreting disc is present in between stamens and carpels.	6. Disc is usually not present.
7. A transition between petals and stamens is usually seen as in *Rosa*.	7. Such a transition is not seen.
8. Gynoecium usually varies from monocarpellary to polycarpellary.	8. Gynoecium usually tricarpellary.
9. Ovary superior to inferior with intermediate stages.	9. Usually the ovary is inferior.
10. Fruit of various types varying from drupe, pome, or follicle.	10. Fruit usually pepo or capsule.

Table A.17 Rosaceae and Saxifragaceae

Rosaceae	Saxifragaceae
1. Mostly trees or shrubs, and only rarely herbs.	1. Mostly herbs or shrubs, and only sometimes small trees.
2. Leaves stipulate and not deciduous.	2. Leaves exstipulate and usually deciduous.
3. Inflorescence highly variable from solitary flowers to racemose and cymose.	3. Inflorescence usually cymose, and only sometimes racemose to paniculate.
4. Epicalyx often present.	4. Epicalyx absent.
5. Sepals usually 5, free or united.	5. Sepals 4 or 5, often gamosepalous.
6. Petals usually 5 or even absent (e.g. *Alchemilla*).	6. Petals 4 or 5, sometimes connate or absent.
7. Stamens numerous, or 2-, 3-, or 4-times as many as sepals, and whorled.	7. Stamens usually twice as many as petals, in 2 whorls, obdiplostemonous.

(*Contd.*)

8. Ovary superior, sometimes inferior (*Malus*).	8. Ovary superior to inferior.
9. Carpels 1 to many, usually free.	9. Carpels usually 2, basally connate.
10. Fruit of various types from drupe (*Prunus*), pome (*Pyrus*), follicle (*Spiraea*) to etaerio of achenes (e.g. *Rosa*).	10. Fruit capsule or berry.

Table A.18 Umbelliferae and Rubiaceae

Umbelliferae (Apiaceae)	Rubiaceae
1. Mostly herbs, and only a few are shrubs.	1. Mostly trees, or shrubs, and only some are herbs.
2. Mostly aromatic and contain essential oil or oleoresin.	2. Mostly do not contain essential oil or oleoresin.
3. Leaves generally pinnately compound or decompound, and much divided; rarely simple.	3. Leaves mostly simple.
4. Leaves exstipulate.	4. Leaves stipulate.
5. Inflorescence a simple or compound umbel.	5. Inflorescence basically a dichasial cyme.
6. Sepals 5, free or united, very small.	6. Sepals 4 or 5, free or basally united.
7. Petals 5, polypetalous.	7. Petals 4 or 5, gamopetalous.
8. Stamens 5, free and alternipetalous.	8. Stamens 4 or 5, epipetalous.
9. Styles often with a thickened or swollen base called stylopodium.	9. Stylopodium absent; an epigynous disc often present.
10. Fruit a dry schizocarp called cremocarp, splitting into two mericarps.	10. Fruit a capsule (*Cinchona*), or berry (*Mussaenda*) or schizocarp (*Galium*).

Table A.19 Umbelliferae (Apiaceae) and Compositae (Asteraceae)

Umbelliferae (Apiaceae)	Compositae (Asteraceae)
1. Mostly aromatic herbs containing the essential oil or oleoresin in all organs.	1. Herbs, shrubs, vines or rarely trees but not aromatic.
2. Leaves generally pinnately compound or decompound, and much divided, rarely simple.	2. Leaves mostly simple, and only rarely truly compound.
3. Inflorescence simple or compound umble.	3. Inflorescence is a head or capitulum.
4. Disc florets and ray florets absent.	4. Disc florets and ray florets present.
5. Flowers are mostly actinomorphic.	5. Ray florets are usually zygomorphic.
6. Sepals are 5, free or fused, and never represented by pappus.	6. Calyx represented by pappus.
7. Corolla of 5, free petals.	7. Corolla of 5, fused petals.
8. Stamens are 5, free, and never syngenesious.	8. Stamens 5, epipetalous, and syngenesious.
9. Ovary bilocular, and placentation axile.	9. Ovary unilocular, and placentation basal.
10. Stylopodium present.	10. Stylopodium absent.
11. Fruit is a cremocarp.	11. Fruit is a cypsela.
12. Seed with a minute embryo and copious endosperm.	12. Seed with a large embryo and no endosperm.

Table A.20 Rubiaceae and Compositae (Asteraceae)

Rubiaceae	Compositae (Asteraceae)
1. Mostly trees or shrubs, and only some are herbs.	1. Mostly herbs, only some are shrubs and only rarely trees.
2. Leaves simple and stipulate.	2. Leaves simple and exstipulate.
3. Inflorescence basically a dichasial cyme.	3. Inflorescence basically a head or capitulum.
4. Disc florets and ray florets absent.	4. Disc florets and ray florets present.
5. Calyx made up of 4 or 5 sepals, polysepalous, well-developed.	5. Calyx represented by pappus of bristles, awns or scales, or even absent.
6. Petals 4 or 5, gamopetalous.	6. Petals 5, gamopetalous; represented by 3 basic types viz. (i) 5-lobed, tubular, (ii) ligulate, and (iii) bilabiate.
7. Stamens 4 or 5, epipetalous and never syngenesious.	7. Stamens 5, epipetalous and syngenesious.
8. Ovary inferior, bilocular.	8. Ovary inferior, unilocular.
9. Placentation axile.	9. Placentation basal.
10. Fruit capsule, berry or schizocarp.	10. Fruit a cypsela.

Table A.21 Asclepiadaceae and Apocynaceae

Asclepiadaceae	Apocynaceae
1. Mostly perennial herbs, sometimes shrubs, and rarely trees.	1. Mostly trees and shrubs, only some are herbs.
2. Stamens free or united to form a tube.	2. Stamens usually free.
3. Staminal corona present.	3. Corona absent.
4. Pollinia present.	4. Pollinia absent.
5. Translators and corpusculum present.	5. Translators and corpusculum absent.
6. Gynostegium or gynandrium present.	6. Gynostegium or gynandrium absent.
7. Nectariferous disc is absent.	7. Nectariferous disc is often present around or at the base of the gynoecium.
8. Fruit a pair of follicles.	8. Fruit usually follicle, but sometimes a berry, drupe, or capsule.

Table A.22 Convolvulaceae and Solanaceae

Convolvulaceae	Solanaceae
1. Herbs, shrubs, or climbing or twining vines; with milky sap.	1. Herbs or often shrubs but rarely vines; without milky sap.
2. Bicollateral bundles in stem absent.	2. Vascular bundles are often bicollateral in stem.
3. Calyx of 5, usually free, persistent sepals.	3. Calyx of 5, usually fused, persistent sepals.
4. Bracts often large and showy, sometimes forming involucre.	4. Bracts, if present, are not usually showy.

(*Contd.*)

5. Corolla often infundibuliform, tubular, salverform or campanulate; induplicately valvate or twisted.	5. Corolla tubular to rotate, infundibuliform or bell-shaped; rarely bilipped and zygomorphic (*Schizanthus*).
6. Anthers not connivent.	6. Anthers usually connivent.
7. Gynoecium bicarpellary syncarpous, but carpels not obliquely placed.	7. Gynoecium bicarpellary, syncarpous, but carpels obliquely placed in the flower.
8. Placenta not swollen.	8. Placenta swollen.

Table A.23 Labiatae, Boraginaceae and Scrophulariaceae

Labiatae	Boraginaceae	Scrophulariaceae
1. Leaves opposite or whorled.	1. Leaves alternate.	1. Leaves alternate, sometimes opposite.
2. Flowers zygomorphic, bilabiate.	2. Flowers actinomorphic.	2. Flowers zygomorphic.
3. Calyx of 5 united sepals, sometimes 2-lipped.	3. Calyx of 5 united sepals, sometimes only basally fused (*Lithospermum*).	3. Calyx of 5 united sepals.
4. Stamens usually 4, di-dynamous, epipetalous, sometimes only 2 stamens.	4. Stamens 5, epipetalous.	4. Stamens usually 4, didynamous, sometimes 2, rarely 5, epipetalous; staminodes present.
5. Gynoecium with gynobasic style, 4-lobed ovary, bicarpellary.	5. Gynoecium with 4-lobed ovary, bicarpellary.	5. Gynoecium 2-carpellate.
6. Fruit 4 achenes (nutlets) separating at maturity, sometimes a drupe.	6. Fruit same as in Labiatae.	6. Fruit usually a capsule, sometimes a drupe or berry.

Table A.24 Acanthaceae, Labiatae and Verbenaceae

Acanthaceae	Labiatae	Verbenaceae
1. Mostly herbs or shrubs, rarely trees, not aromatic.	1. Aromatic herbs or shrubs; with square or quadrangular stem.	1. Mostly shrubs or trees; stem often quadrangular; plants not usually aromatic.
2. Leaves simple, opposite decussate, exstipulate.	2. Leaves simple, opposite or whorled, exstipulate; rarely compound.	2. Leaves simple, opposite or whorled, exstipulate; sometimes pinnately (*Peronema*) or palmately (*Vitex*) compound.
3. Inflorescence usually a dichasial cyme.	3. Inflorescence usually a verticillaster.	3. Inflorescence racemose raceme, umbel, or spike.
4. Calyx of 5 united or free sepals.	4. Calyx of 5 united sepals, usually bilipped (1/4 in *Ocimum* and 3/2 in *Salvia*).	4. Calyx of 5 united, persistent sepals.

(*Contd.*)

5. Corolla of 5 united petals, bilipped (2/3) or bilabiate personate.	5. Corolla of 5 united petals, bilipped (4/1 in *Ocimum* and 2/3 in *Salvia, Leucas*).	5. Corolla of 5 united petals, often bilipped.
6. Stamens usually 4, didynamous, epipetalous, rarely only 2 or 5 stamens.	6. Stamen usually 4, didynamous, epipetalous, sometimes only 2 stamens.	6. Stamens 4, didynamous, epipetalous.
7. Style narrow and long, ovary bilobed.	7. Style gynobasic, ovary 4-lobed.	7. Style single, terminal, ovary not lobed.
8. Fruit usually a capsule, rarely drupe.	8. Fruit schizocarp of 1–4 nutlets, rarely drupe.	8. Fruit generally a drupe; rarely capsule, berry or schizocarp.
9. Seeds with jaculators.	9. Seeds without any jaculator.	9. Seeds without any jaculator.

Table A.25 Chenopodiaceae and Amaranthaceae

Chenopodiaceae	Amaranthaceae
1. Predominantly halophytic or xerophytic annual or perennial herbs; only a few shrubs and rarely small trees.	1. Mostly annual or perennial, non-halophytic weedy herbs; rarely shrubs.
2. Leaves and stem mealy to touch.	2. Leaves and stem non-mealy to touch.
3. Inflorescence is not very dense or congested.	3. Inflorescence is usually very dense or congested.
4. Flowers monochlamydous, small and greenish.	4. Flowers monochlamydous, small with dry bracts.
5. Scarious bracts absent.	5. Usually scarious bracts present.
6. Tepals 5, usually persistent in fruit.	6. Tepals 5, dry or membranous, usually not persistent in fruit.
7. Stamens 5, usually fee and not connate.	7. Stamens 5; usually connate filaments, at least at the base.
8. Usually staminodes are not found.	8. Often with equal number of staminodes alternating with the anthers (e.g. *Celosia*, *Achyranthes*).

Table A.26 Amaranthaceae and Polygonaceae

Amaranthaceae	Polygonaceae
1. Mostly weedy herbs, rarely shrubs or climbers (*Deeringia amaranthoides*).	1. Mostly weedy herbs with swollen node, a few climbers and only rarely trees (*Triplaris*).
2. Leaves exstipulate.	2. Leaves stipulate, and stipules are ochreate.
3. Inflorescence dense or congested, axillary or terminal spikes, heads, or racemes, or flowers solitary.	3. Inflorescence primarily racemose but sometimes cymose, or clusters of panicled raceme (*Rumex*) or cymose umbels (*Eriogonum*).
4. Perianth lobes usually 5, free or basally connate.	4. Perianth lobes usually 3–6, free or basally connate.
5. Androecium usually of 5 stamens, opposite tepals; often monadelphous (*Achyranthes*), often with equal number of staminodes.	5. Stamens usually 6–9 in two series.

(*Contd.*)

6. Gynoecium bi- to tricarpellary, syncarpous; ovary superior, unilocular.	6. Gynoecium usually tricarpellary, syncarpous; ovary superior.
7. Ovule pendulous or campylotropous.	7. Ovule orthotropous.
8. Styles 1 to 3; stigma 1–3, capitate.	8. Style 1; stigma 2–4.
9. Fruit usually a utricle (*Achyranthes*) or nutlet (*Digera*) or capsule (*Celosia*).	9. Fruit usually a triangular achene or nut (*Polygonum*).

Table A.27 Euphorbiaceae and Moraceae

Euphorbiaceae	Moraceae
1. Herbs, shrubs or trees with milky latex; some are xerophytes or cactus-like.	1. Mostly trees or shrubs, rarely herbs; contain milky juice.
2. Stem and leaves usually lacking stinging hairs.	2. Stinging hairs usually present.
3. Leaves stipulate, stipules sometimes in the form of hairs, glands, or thorns.	3. Leaves often with 2 caducous stipules.
4. Several genera bear caducous or highly reduced leaves.	4. Leaves are well-developed in all members.
5. Leaves of several genera contain extrafloral nectaries.	5. Extrafloral nectaries absent.
6. Characteristic type of inflorescence is cyathium.	6. Syconium or hypanthodium is the characteristic inflorescence of several genera.
7. Male flowers have one or more (sometimes up to 100, e.g. *Croton*) stamens.	7. Usually the male flowers have only 4 stamens.
8. Gynoecium is usually tricarpellary, syncarpous, superior, trilocular.	8. Gynoecium is usually bicarpellary, syncarpous, superior to inferior and unilocular.
9. Placentation is usually axile.	9. Usual type of placentation is basal.
10. Fruit is usually a capsule or regma.	10. Usually a multiple fruit develops from the union of fruits of several different flowers.

Table A.28 Musaceae and Zingiberaceae

Musaceae	Zingiberaceae
1. Plants herbaceous but some appear like trees.	1. Plants herbaceous and none appears like tree.
2. Leaves large with long petiole and prominent sheath.	2. Leaves large or small; petiole and sheath not very long; leaves may also be sessile.
3. Spathe usually conspicuous.	3. Bracteate.
4. Inflorescence usually a spadix or cincinnus.	4. Inflorescence usually a raceme or spike.
5. Perianth not distinguishable into calyx and corolla.	5. Calyx and corolla are distinct.
6. Perianth lobes may be free or united.	6. Perianth lobes are usually fused.
7. Fertile stamens are 6 or 5.	7. Fertile stamen is only 1, and all others are modified into staminodes.

Table A.29 Liliaceae and Iridaceae

Liliaceae	Iridaceae
1. Leaves radical and not equitant.	1. Leaves equitant.
2. Inflorescence usually racemose, sometimes umbellate (*Allium*) or solitary flowers (*Tulipa*).	2. Usually cymose inflorescence, sometimes flowers solitary (e.g. *Sisyrinchium*).
3. Flowers actinomorphic.	3. Flowers actinomorphic or zygomorphic (*Gladiolus*).
4. Perianth segments 6, often petaloid or sometimes sepaloid.	4. Perianth segments 6, often petaloid.
5. Stamens 3+3, anthers usually introrse.	5. Stamens 3 + 3, anthers extrorse.
6. Ovary tricarpellary, syncarpous, superior.	6. Ovary tricarpellary, syncarpous, inferior.
7. Fruit loculicidal or septicidal capsule or berry.	7. Fruit loculicidal capsule.

Table A.30 Liliaceae and Amaryllidaceae

Liliaceae	Amaryllidaceae
1. Storage organ usually rhizome, sometimes bulb, or roots are fleshy.	1. Storage organ usually a bulb, and sometimes a rhizome.
2. Inflorescence usually racemose.	2. Inflorescence usually cymose.
3. Flowers usually actinomorphic.	3. Flowers actinomorphic or zygomorphic.
4. Tepals 3 + 3, free or united, petaloid or sometimes sepaloid.	4. Tepals 3 + 3, free or united, petaloid.
5. Stamens usually 3 + 3.	5. Stamens usually 3 + 3 but sometimes staminodes present.
6. Ovary superior.	6. Ovary inferior.
7. Fruit a loculicidal or septicidal capsule, or a berry.	7. Fruit usually a loculicidal capsule, rarely a berry.

Table A.31 Juncaceae and Cyperaceae

Juncaceae	Cyperaceae
1. Leaves usually in basal tufts, grass-like and linear.	1. Leaves usually in basal tufts, grass like and 3-ranked.
2. Stem usually cylindrical, pith often present but not solid.	2. Stem usually triangular and with solid pith.
3. Inflorescence cymose panicle or head.	3. Inflorescence is 1-many-flowered spikelets, arranged variously in racemes, panicles, spikes, or umbels.
4. Perianth segments 6, in 2 whorls of 3 each; usually sepaloid.	4. Perianth segments often in the form of 6 bristles or scales, or absent.
5. Stamens usually 6, in 2 whorls of 3 each, or inner whorl absent.	5. Stamens usually 1–3; only rarely 6 (*Gahnia*).
6. Style simple, often with 3 brush-like stigmas.	6. Styles 2 or 3, or deeply divided into as many branches as the number of carpels.
7. Fruit loculicidal capsule.	7. Fruit a triangular or biconvex achene or nutlet.
8. Seeds 3 to many in each fruit.	8. Seed 1 per fruit.

Table A.32 Gramineae (Poaceae) and Juncaceae

Gramineae (Poaceae)	Juncaceae
1. Stem jointed, usually cylindrical.	1. Stem usually without joints, cylindrical.
2. Pith is usually present only at the nodes.	2. Pith often present throughout.
3. Leaves basal, and, if cauline, 2-ranked, usually linear to lanceolate.	3. Leaves mostly in basal tuft, usually linear, or terete.
4. Ligule usually present.	4. Ligule usually absent.
5. Basic unit of inflorescence is spikelet.	5. Inflorescence is usually cymose panicle or head.
6. Perianth is usually reduced into 2 or rarely 3 lodicules.	6. Perianth segments are generally 6, in 2 whorls of 3 each, usually sepaloid.
7. Stamens usually 3 in each floret.	7. Stamens usually 6, in 2 whorls of 3 each, or inner whorl is absent.
8. Styles usually 2, rarely 1 (*Nardus*) to 3 (*Bambusa*).	8. Style simple with 3 brush-like stigmas.
9. Fruit usually a caryopsis or rarely a nut, utricle (*Sporobolus*) or berry (some bamboos).	9. Fruit a loculicidal capsule.
10. Seed usually 1 per fruit.	10. Seeds usually 3 to many in each fruit.

Table A.33 Cyperaceae and Gramineae (Poaceae)

Cyperaceae	Gramineae (Poaceae)
1. Stem not jointed, usually triangular, with solid pith throughout.	1. Stem jointed, usually circular, with hollow internodes; pith often present only at nodes.
2. Leaves 3-ranked or tristichous, leaf sheath usually closed.	2. Leaves 2-ranked or distichous, leaf sheath usually open.
3. Leaves without ligule.	3. Ligule often present.
4. Bract, immediately subtending the flower, is usually with odd number of nerves.	4. Bract, immediately subtending the flower, is usually 2-nerved.
5. Inflorescence various, from a simple spike to a well-branched panicle.	5. Inflorescence spike, raceme or panicle.
6. Perianth represented often by 6 scales or bristles or absent, lodicules never present.	6. Perianth usually in the form of 2 or 3 lodicules.
7. Stamens usually 1 to 3.	7. Stamens usually 3.
8. On reduction division, one microspore mother cell gives rise to 1 functional and 3 non-functional microspores.	8. On reduction division, one microspore mother cell gives rise to a tetrad of 4 functional microspores.
9. Style 1, often deeply divided into 2 or 3 branches.	9 Styles usually 2.
10. Fruit usually trigonous or biconvex achene or nut, and never a caryopsis.	10. Fruit caryopsis.
11. Embryo usually embedded in the endosperm.	11. Embryo usually lateral to the endosperm.

EXAMINATION TOOLS—MAJOR CHARACTERS OF DISCUSSED FAMILIES

Discussed Families of DICOTYLEDONS	Leaves					Flowers				Sepals		Petals		Stamens		Ovary		Carpels		Major Distinguishing Character/Characters
	Alternate	Opposite	Simple	Compound	Stipule	Actinomorphic	Zygomorphic	Bisexual	Unisexual	Free	United	Free	United	Many	Few	Superior	Inferior	Free	United	
1. Magnoliaceae	*		*		*	*		*	+	*		*		*		*		*	+	Deciduous stipules; perianth lobes petaloid.
2. Annonaceae	*		*		+	*		*		*		*		*		*		*		Leaves exstipulate, 2-ranked; perianth 3 + 3 + 3, gynoecium polycarpellary.
3. Ranunculaceae	*	+	+	*	+	*	+	*	+	*		*		*		*		*	+	Stamens numerous, spirally arranged; carpels numerous.
4. Nymphaeaceae	*		*			*		*		*		*		*		*		+	+	Aquatic; leaves with long petiole.
5. Papaveraceae	*		*			*		*		*	+	*		*		*		+	*	Sepals deciduous; parietal placentation.
6. Fumariaceae	*		*				*	*		*		*			*	*			*	Zygomorphic; spurred petals.
7. Capparidaceae	*		*	+	*	*		*		*		*			*	*			*	Androphore and gynophore present.
8. Brassicaceae (Cruciferae)	*		*			*	+	*		*		*			*	*			*	Corolla cruciform; stamens tetradynamous; fruit siliqua or silicula.
9. Violaceae	*	+	*		*	+	+	*		*		*			*	*			*	Corolla spurred; one stamen spurred.
10. Caryophyllaceae	+	*	*		+	*		*	+	+	+	*			*	*			*	Flowers obdiplostemonous; placentation free-central.
11. Portulacaceae	+	+	*		*	*		*		*		*		+	*	*	+		*	Stipules scarious; sepals 2, persistent.

(*Contd.*)

12. Malvaceae	*		*		*	*		*		+	+	*		*	+	*		*	Epicalyx present; stamens monadelphous; petals twisted.
13. Sterculiaceae	*		*		*	*		*		*	+	*			*	*		*	Stamens in 2 whorls; gynoecium pentacarpellary.
14. Tiliaceae	*		*		*	*		*		+	+	*		*	+	*		*	Mostly trees or shrubs; stamens many, polyadelphous.
15. Bombacaceae	*		*	+	*	*		*			*	*		+	*	*		*	Large-sized trees; stipules caducous; flowers large-sized.
16. Geraniaceae	+	+	+	+	*	*	+	*		*	+	*			*	*		*	Stamens with filaments united at the base; fruit beaked or lobed.
17. Oxalidaceae	*		+	*		*		*		*	+	*	+		*	*		*	Leaves usually palmately compound; sour due to presence of oxalic acid.
18. Rutaceae	+	+	+	*		*	+	*	+	+	+	*	+	+	*	*	+	*	Fruit hesperidium; translucent dots on the leaves; cymose inflorescence.
19. Meliaceae	*			*		*		*	+	+	+	*	+		*	*		*	Leaves exstipulate; monadelphous stamens; disc present between androecium and gynoecium.
20. Rhamnaceae	*	+	*		*	*		*	+	*	+	*			*	*	+	*	Shrubs or trees, armed with spines.
21. Vitaceae	*	+	+	+	*	*		+	+		*	*	+		*	*		*	Climbing herbs; tendrils present.
22. Sapindaceae	*			*		*	+	*	+	+	+	*			*	*		*	Petals glandular; leaves pinnately-compound.
23. Anacardiaceae	*	+	+	*		*		*	+	+	*	*			*	*		*	Intrastaminal disc present; fruit drupe.
24. Leguminosae	*		+	*	*	+	*	*		+	*	+	+	+	+	*		*	Fruit a legume.
25. Subfamily Mimoseae	*			*	*	*		*			*	+	+	+	+	*	+		Aestivation valvate or imbricate; flowers actinomorphic.

(*Contd.*)

26. Subfamily Caesalpineae	*			*	*		*	*		+	+	*		+	*	*		+		Flowers zygomorphic; corolla aestivation ascending imbricate.
27. Subfamily Papilionaceae	*			*	*		*	*			*	*			*	*		+		Aestivation descending imbricate.
28. Rosaceae	*	+	+	+	*	*	+	*	+	+	+	*		*	+	*	+	*	+	Stamens numerous; fruits of various types.
29. Saxifragaceae	*		*			*		*		+	+	*		+	*	+	+		*	Leaves deciduous, exstipulate; ovary superior to inferior.
30. Combretaceae	*	+	*			*	+	*	+		*	*			*		*		*	Often lianous; hypanthium present.
31. Myrtaceae	+	*	*			*		*		+	+	+	+	*	+		*		*	Trees or shrubs; anther connectives gland-tipped; ovary inferior.
32. Lythraceae		*	*		+	*	+	*			*	*		+	*	*			*	Corolla crumpled; ovary superior; epicalyx present.
33. Cucurbitaceae	*		*			*		+	*	+	*	+	*		*		*		*	Fruit pepo; tendril-bearing herbs.
34. Begoniaceae	*		*		*	+	*		*	*	+	*	+	*			*		*	Unisexual flowers; inferior ovary.
35. Passifloraceae	*		*	+	*	*		*	+	+	+	+	+		*	*			*	Climbing habit; uniflowered peduncles; androgynophores present.
36. Cactaceae	*		*			+	+	*			*		*	*			*		*	Mostly prickly, spiny, fleshy, succulents.
37. Umbelliferae (Apiaceae)	*		+	*		*		*	+	+	+	*			*		*		*	Inflorescence simple or compound umbel; inferior ovary.
38. Araliaceae	*	+		*	*	*		+	+	+	+	*			*	+	*		*	Leaves alternate; flowers actinomorphic.
39. Rubiaceae		*	*		*	*	+	*		*			*		*		*		*	Stamens alternipetalous; inferior ovary.

(*Contd.*)

40. Caprifoliaceae		*	*	+	+	+	+	*		+	*		*		*		*		*	Calyx fused to the ovary; multicarpellary.
41. Compositae (Asteraceae)	*	+	*	+		+	+	+	+				*		*		*		*	Calyx usually a pappus; head or capitulum.
42. Campanulaceae	*	+	*			+	+	*			*		*		*	+	*		*	Stamens epipetalous and show connation of anthers or filaments.
43. Ericaceae	*	+	*			*	+	*	+	+	*	+	*		*	*	+		*	Flowers urceolate; ovary tetra- to multilocular.
44. Primulaceae	+	*	*			*	+	*			*		*		*	*			*	Stamens opposite the petals; free-central placentation.
45. Plumbaginaceae	*		*			*		*			*	+	*		*	*			*	5-styled pistil; fruit dry, 1-seeded.
46. Sapotaceae	*		*			*		*		+	*		*	*		*			*	Plants with milky latex; stamens in 2-3 whorls.
47. Oleaceae	+	*	+	+		*		*	+		*		*		*	*			*	Trees or shrubs with opposite leaves; stamens 2.
48. Asclepiadaceae		*	*		+	*	+	*		+	+		*		*	*		*	+	Herbs with milky sap; corona, pollinia, translators and corpusculum present.
49. Apocynaceae	+	*	*		+	*		*			*		*		*	*			*	Stamens epipetalous; corona, pollinia and corpusculum absent.
50. Loganiaceae		*	*		*	*		*			*		*		*	*			*	Simple, opposite, stipulate leaves; internal phloem present.
51. Boraginaceae	*	+	*			*	+	*		+	+		*		*	*			*	Hispid herbs; inflorescence helicoid cyme; style gynobasic.
52. Convolvulaceae	*		*			*		*		*	+		*		*	*			*	Corolla tubular, infundibuliform of salverform.
53. Solanaceae	*	+	*			*	+	*			*		*		*	*			*	Carpels obliquely placed.

(*Contd.*)

54. Polemoniaceae	+ +	+ +		* +	*	*	*	*	*	*	Stamens attached at different lengths on corolla tube; ovary tricarpellary.
55. Scrophulariaceae	+ +	* +		+ *	*	*	*	*	*	*	Flowers zygomorphic; corolla bilipped.
56. Bignoniaceae	*	*		*	*	*	*	*	*	*	Inflorescence cymose, flowers zygomorphic.
57. Pedaliaceae	+ *	*		*	*	*	*	*	*	*	Mostly herbs; stamens 4, didynamous.
58. Acanthaceae	*	*		*	*	+ *	*	*	*	*	Flowers zygomorphic; corolla bilipped (2/3); seeds with jaculators.
59. Lamiaceae (Labiatae)	*	*		*	*	*	*	*	*	*	Inflorescence verticillaster; coralla bilipped (4/1 or 2/3); style gynobasic.
60. Verbenaceae	+ *	* +		*	*	*	*	*	*	*	Stem quadrangular; ovary not lobed; style terminal.
61. Plantaginaceae	* +	*		*	*	*	*	*	*	*	Leaves with somewhat parallel venation; petals membranous.
62. Chenopodiaceae	* +	*		*	* +	+ *	——	*	*	*	Perianth sepaloid; halophytic herbs; fruit utricle.
63. Amaranthaceae	+ +	*		*	* +	+ +	——	*	*	*	Mostly herbs; bracts dry, scarious; stamens connate; perianth sepaloid.
64. Polygonaceae	*	*	*	*	* +	——	+ +	*	*	*	Tepals petaloid; ochreate stipules.
65. Aristolochiaceae	*	*		+ +	*	*	——	+ +	*	*	Sepals petaloid; petals absent.
66. Piperaceae	* +	*	+	*	* +	——	——	*	*	*	Perianth absent; stamens 1-10; embryo very minute.
67. Loranthaceae	+ *	*		* +	+ +	+ +	+ +	*	*	*	Aerial, parasitic habit; cup-shaped receptacle.

(*Contd.*)

68. Euphorbiaceae	*	+	*	+	*	*			*	*	+	*	+	+	*	*			*	Milky latex present; flowers unisexual; fruit capsule or regma; inflorescence cyathium.
69. Urticaceae	+	+	*		+	*			*	+	+	——			*	*			*	Herbs with stinging hairs and watery sap; mostly sepaloid.
70. Moraceae	*		*		*	*			*	+	+	+	+		*	+	+		*	Leaves stipulate; flowers uni-sexual; tepals usually 4.
71. Cannabinaceae	+	+	+	+	*	*			*	+	+	+	+		*	*		*		Monocarpellary, calyx persistent; unisexual.
72. Casuarinaceae		+	+			+	+		*	——		——			*	*			*	*Equisetum*-like jointed stem; minute scaly leaves; unisexual flowers.
73. Salicaceae	*	+	*		*	*			*	——		——		+	*	*			*	Deciduous dioecious trees or shrubs; flowers in catkins; a cup-like disc or gland present in each flower.
MONOCOTYLEDONS																				
1. Orchidaceae	*		*				*	*	+	*	+	*	+		*		*		*	Leaves 2-ranked; labellum, pollinia and gynostegium present.
2. Iridaceae	——		*			+	+	*		+	+	+	+		*		*		*	Leaves mostly basal and equitant; herbs with bulbs, corms or rhizome; trimerous.
3. Amaryllidaceae	*		*			*	+	*		+	+	+	+		*		*		*	Inflorescence a long leafless scape; bulbous or rhizomatous herbs; stamens antiphyllous.
4. Bromeliaceae	+		*			*	+	*	+	*		+	+		*	+	*		*	Short-stemmed herbs; leaves stiff; anthers versatile.
5. Cannaceae	*		*				*	*		*			*		*		*		*	Leaves very large with petiole sheathing their base; inflorescence a 2-flowered cincinii.

(*Contd.*)

6. Musaceae	*		*				*	+	*	+	+	+	+		*		*		*	Perennial giant herbs; leaves very large; inflorescence spadix covered by a spathe.
7. Zingiberaceae	*		*				*	*			*	+	*		*		*		*	Aromatic herbs; only one fertile stamen; ligule present at the top of leaf sheath.
8. Liliaceae	*	+	*		+	*		*	+	+	+	+	+		*	*			*	Herbs with bulbs, corms or rhizome; perianth showy.
9. Commelinaceae	*		*			*	+	*		*		*	+		*	*			*	Herbs with succulent stem; leaves with tubular sheath; ovary superior.
10. Juncaceae				*		*		*	+	*		*			*	*			*	Leaves mostly basal; perianth sepaloid.
11. Palmae (Arecaceae)	+		+	*		*		+	*	+	+	+	+		*	*		+	*	Leaves usually in terminal clusters and leave prominent scars of leaf bases.
12. Typhaceae	*		*			*			*	—		—			*	*		*		Perianth often of hairs or scales; fruit achene; monocarpellary.
13. Araceae	*		+	+		*		+	*	+	+	+	+		*	*	+		*	Inflorescence spadix enveloped by a spathe; aquatic or terrestrial habit.
14. Alismataceae	—		*			*		+	+	*		*		+	*	*		*		Marshy herbs with perennating rhizomes and basal leaves.
15. Cyperaceae	*		*			*	+	*	+	—		—			*	*			*	Herbs with triangular stem; leaves 3-ranked, bract glumaceous; fruit achene or nut.
16. Gramineae (Poaceae)	*		*				*	*	+	—		—			*	*			*	Perianth represented by lodicules; ligule usually present; leaves 2-ranked; fruit caryopsis.

*, present; +, present only sometimes; —, absent or represented by bristles or minute scales. In the families where symbols appear in both columns for contrasting characters (+ * and * +); * is the typical condition but sometimes + is found; + +, both characters equally common. Columns left blank or without any symbol means that this character is not found in the particular family.

APPENDIX 2

EXAMINATION TOOLS— SELECTED MEDICINAL PLANTS AND THEIR UTILITY: AT A GLANCE

Name of the Plant*	Family	Parts Used in Treatment of
1. *Michelia champaca*	Magnoliaceae	Flowers and fruits (kidney trouble, gonorrhoea).
2. *Illicium verum*	Magnoliaceae	Seeds (colic disorders).
3. *Uraria narum*	Annonaceae	Roots (urinary problems).
4. *Aconitum napellus*	Ranunculaceae	Roots (nerve sedative, rheumatism, heart palpitation, gastritis).
5. *Actaea spicata*	Ranunculaceae	Roots (nerve sadative).
6. *Anemone pulsatilla*	Ranunculaceae	Plant extract (nervous disorders in women during menstrual periods).
7. *Coptis teeta*	Ranunculaceae	Roots (brain and liver disorders).
8. *Delphinium zatil*	Ranunculaceae	Seeds, leaves and roots (jaundice, enlarged spleen).
9. *Delphinium vestitum*	Ranunculaceae	Seeds (cardiac disorders).
10. *Helleborus niger*	Ranunculaceae	Flowers (madness).
11. *Nigella sativa*	Ranunculaceae	Seeds (cough, asthma, fever).
12. *Paeonia officinalis*	Ranunculaceae	Tubers (colic and uterus disorders).
13. *Ranunculus aquatilis*	Ranunculaceae	Flowers (asthama and rheumatism).
14. *Ranunculus ficaria*	Ranunculaceae	Flowers and seeds, (piles).
15. *Nelumbo nucifera*	Nymphaeaceae	Flowers (cardiac tonic and liver disorders), rhizome (piles).
16. *Nymphaea nouchali*	Nymphaeaceae	Rhizome (dysentery and diarrhoea).
17. *Nymphaea stellata*	Nymphaeaceae	Flowers juice (cardiac pain).

(*Contd.*)

*Plants arranged according to families discussed in the book.

18. *Papaver somniferum*	Papaveraceae	Milky latex of unripe fruits produce "opium" (known for its sedative properties).
19. *Argemone mexicana*	Papaveraceae	Seeds (skin infections).
20. *Sanguinaria canadensis*	Papaveraceae	Rhizome and roots (emetic and dyspepsia i.e. indigestion).
21. *Hypecoum procumbens*	Fumariaceae	Alkaloid protopine (high blood pressure).
22. *Corydalis govaniana*	Fumariaceae	Roots (syphilis).
23. *Fumaria indica*	Fumariaceae	Dried plants and seeds (fever and also as blood purifier).
24. *Cleome gynandra*	Capparidaceae	Leaves and seeds (expelling roundworms).
25. *Cleome religiosa*	Capparidaceae	Bark (increase appetite; reduce bile secretion).
26. *Brassica oleracea* var. *botrytis* (Cauliflower)	Brassicaceae	Inflorescence (possess anticancer properties).
27. *Viola odorata* (Banafsha)	Violaceae	Dried flowers (cough, influenza).
28. *Hybanthus enneaspermus*	Violaceae	Roots (urinary tract infections, gonorrhoea).
29. *Dianthus sinensis*	Caryophyllaceae	Flowers and seeds (gonorrhoea).
30. *Spergula arvensis*	Caryophyllaceae	Seeds (lungs tuberculosis).
31. *Saponaria vaccaria*	Caryophyllaceae	Plant extract (urinary bladder diseases).
32. *Stellaria media*	Caryophyllaceae	Leaves (inflammations of digestive and respiratory tracts).
33. *Stellaria semivestita*	Caryophyllaceae	Entire plant (anticancerous).
34. *Portulaca oleracea*	Portulacaceae	Entire plant extract (ailments of kidney and urinary bladder).
35. *Hibiscus rosa-sinensis*	Malvaceae	Roots (cough and cold).
36. *Althaea rosea*	Malvaceae	Roots (dysentry).
37. *Gossypium sp.*	Malvaceae	Bark (stops haemorrage after child birth in ladies).
38. *Malva verticillata*	Malvaceae	Roots (whooping cough).
39. *Urena repanda*	Malvaceae	Roots and bark (hydrophobia).
40. *Helicteres isora*	Sterculiaceae	Juice of roots (diabetes).
41. *Corchorus capsularis* and *C. olitorius*	Tiliaceae	Dried leaves and roots decoction (dysentry and diarrhoea).
42. *Grewia asiatica*	Tiliaceae	Root bark (rheumatism).
43. *Grewia tiliaefolia*	Tiliaceae	Wood (emetic and as antidote of opium poisoning).
44. *Triumfetta bartramia*	Tiliaceae	Leaves, fruits and flowers (gonorrhoea).
45. *Durio zibethinus*	Bombacaceae	Root decoction (fever).
46. *Geranium nepalense*	Geraniaceae	Root and leaf decoction (kidney troubles).
47. *Acranychia laurifolia*	Rutaceae	Bark poultice (ulcers and sores).
48. *Aegle marmelos*	Rutaceae	Fruit pulp (mild laxative), roasted fruits (diarrhoea), bark (intermittent fever).
49. *Atalantia monophylla*	Rutaceae	Oil from fruits (paralysis and rheumatism).
50. *Barosma betulina*	Rutaceae	Leaves (kidney diseases).

(Contd.)

51. *Cusparia febrifuga*	Rutaceae	Bark (malaria).
52. *Dictamnus albus*	Rutaceae	Bark (nervous disorders and intermittent fever).
53. *Feronia limonia*	Rutaceae	Ripe fruits (cardiac tonic).
54. *Peganum harmata*	Rutaceae	Seeds (asthma, neuralgia and rheumatism).
55. *Pilocarpus pinnatifolius*	Rutaceae	Leaves (kidney troubles).
56. *Toddalia asiatica*	Rutaceae	Root bark (antimalarial).
57. *Zanthoxylum alatum*	Rutaceae	Seeds (cholera).
58. *Azadirachta indica*	Meliaceae	Seeds (skin diseases), bark (malaria), twigs (pyorrhea).
59. *Melia azedarach*	Meliaceae	Leaf juice (anthelmintic), seeds (rheumatism).
60. *Aglaia odorata*	Meliaceae	Fruits (leprosy).
61. *Dysoxylum malabaricum*	Meliaceae	Wood oil (eye and ear diseases).
62. *Walsura piscidia*	Meliaceae	Bark (stimulant, expectorant and skin diseases).
63. *Rhamnus purshiana*	Rhamnaceae	Bark (purgative).
64. *Cayratia carnosa*	Vitaceae	Leaves and roots (high fever).
65. *Leea aequata*	Vitaceae	Oil from the plant (tuberculosis).
66. *Leea macrophylla*	Vitaceae	Tubers (ringworms).
67. *Acer indica*	Sapindaceae	Fruits and seeds (rheumatism).
68. *Caesalpinia bunduc*	Caesalpinioideae	Seeds (diarrhoea, rheumatism).
69. *Cassia fistula*	Caesalpinioideae	Fruit pulp (purgative).
70. *Cassia occidentalis*	Caesalpinioideae	Leaves and seeds (skin infections).
71. *Cassia sophera*	Caesalpinioideae	Leaf decoction (acute bronchitis).
72. *Crotalaria albida*	Papilionoideae	Roots (purgative).
73. *Glycyrrhiza glabra*	Papilionoideae	Roots (cough and sore throat).
74. *Krameria triandra*	Papilionoideae	Roots (chronic diarrhoea).
75. *Moghania strobilifera*	Papilionoideae	Roots (induce sleep).
76. *Psoralea corylifolia*	Papilionoideae	Seeds (leucoderma and leprosy).
77. *Teramnus labialis*	Papilionoideae	Entire plant (tuberculosis).
78. *Rosa* sp.	Rosaceae	Petals yield "Gulkand" (tonsilitis, cough, cold; tonic).
79. *Hagenia abyssinica*	Rosaceae	Female flowers (source of "cusso", an antihelmintic drug).
80. *Potentilla anserina*	Rosaceae	Leaf decoction (arthritis and kidney stones).
81. *Terminalia arjuna*	Combretaceae	Bark (cardiac tonic).
82. *Terminalia bellirica*	Combretaceae	Fruits ("Bahera", one of the three constituents of Triphala, along with Hararh and "Amla").
83. *Terminalia chebula*	Combretaceae	Fruits ("Hararh", one of the three constituents of Triphala).
84. *Combretum roxburghii*	Combretaceae	Leaves (malaria fever).
85. *Syzygium aromaticum*	Myrtaceae	Clove oil (indigestion, antiseptic, antispasmodic).
86. *Eugenia jambolana*	Myrtaceae	"Jamun" fruits (diabetes).
87. *Ferula assafoetida* and *F. narthax*	Umbelliferae	Dried latex from roots ("Heeng", stimulates respiratory, nervous and intestinal systems).

(Contd.)

88. *Centella asiatica*	Umbelliferae	"Brahmi Booti" (brain tonic, used in madness and leprosy).
89. *Ferula sumbule*	Umbelliferae	Plant decoction (hysteria).
90. *Panax ginseng*	Araliaceae	Ginseng roots (stimulant and aphrodisiac).
91. *Cinchona* sp. (*C. calisaya*, *C. officinalis*)	Rubiaceae	Bark (yield quinine, used in malaria).
92. *Cephaelis ipecacuanha*	Rubiaceae	Roots (amoebic dysentery).
93. *Anthemis mobilis*	Compositae	Dried capitula (dyspesia).
94. *Artemisia cina*	Compositae	Flower heads (drug "santonin" used in intestinal worms).
95. *Blumea balsamifera*	Compositae	Leaves (curie excitement and insomnia).
96. *Grindelia camporum*	Compositae	Floral heads (bronchitis, whooping cough).
97. *Inula helium*	Compositae	Leaves (tuberculosis).
98. *Sphaeranthus indicus*	Compositae	Floral heads (stomach ache, piles).
99. *Tanacetum vulgare*	Compositae	Leaves (chronic ulcers, rheumatism).
100. *Lobelia inflata*	Campanulaceae	Dried leaves and tops (drug "lobelia" is used as expectorant and emetic, and in bronchitis and asthma).
101. *Anagallis arvensis*	Primulaceae	Plant decoction (leprosy, gout, hydrophobia).
102. *Achras sapota*	Sapotaceae	Chicle gum (dental surgery).
103. *Mimusops elengi*	Sapotaceae	Seeds and dried fruits (piles).
104. *Madhuca indica*	Sapotaceae	Oil from seeds (skin diseases, rheumatism).
105. *Nyctanthes arbortristis*	Oleaceae	Leaves (fever, rheumatism).
106. *Calotropis procera*	Asclepiadaceae	Roots (cough).
107. *Tylophora indica*	Asclepiadaceae	Roots (bronchitis, asthma, whooping cough).
108. *Rauvolfia serpentina*	Apocynaceae	Dried roots ("Sarpagandha" is used in hypertension, mental disorders and schizophrenia).
109. *Thevetia peruviana*	Apocynaceae	Seeds (rheumatism).
110. *Alstonia scholaris*	Apocynaceae	Bark (malaria and dysentery).
111. *Catharanthus roseus*	Apocynaceae	Plant extract (leukaemia)
112. *Strychnos nux-vomica*	Loganiaceae	Seeds (nervous disorders, paralysis).
113. *Strychnos toxifera*	Loganiaceae	Seeds (shock therapy, muscle relaxant, tetanus).
114. *Lithospermum officinale*	Boraginaceae	Seeds (urinary bladder diseases).
115. *Exogonium purga*	Convolvulaceae	Tuberous roots (purgative).
116. *Nicotiana tabacum*	Solanaceae	Dried leaves (sedative. antispasmodic).
117. *Atropa belladona*	Solanaceae	Roots (drug at tropine relieves spain; dilates pupil of eyes).
118. *Datura stramonium*	Solanaceae	Leaves and flowers (drug stramonium used in asthma).
119. *Mandragora autumnalis*	Solanaceae	Roots (sedative and hypnotic).
120. *Withania coagulans*	Solanaceae	Fruits (asthma and liver troubles).
121. *Withania somnifera*	Solanaceae	Roots (the drug "asgandha" cures cough, rheumatism, sexual weakness; promotes urination).

(*Contd.*)

122. *Bacopa monniera*	Scrophulariaceae	Decoction of mature plants (epilepsy, insanity).
123. *Digitalis purpurea*	Scrophulariaceae	Dried leaves (congested heart failure).
124. *Adhatoda vasica*	Acanthaceae	Leaves (cough, asthma, bronchitis).
125. *Hygrophila spinosa*	Acanthaceae	Leaves and roots (jaundice, rheumatism).
126. *Ocimum sanctum*	Labiatae	Leaves (cough, cold, fever).
127. *Mentha spicata*	Labiatae	Leaves (indigestion, rheumatism).
128. *Thymus vulgaris*	Labiatae	Leaves (hookworm infection).
129. *Chenopodium ambrosioides*	Chenopodiaceae	Decoction of plant (hookworm infection).
130. *Achyranthes aspera*	Amaranthaceae	Inflorescence (antidote against snake and scorpion bites).
131. *Polygonum aviculare*	Polygonaceae	Leaves and seeds (diabetes, rheumatism).
132. *Rheum emodi* and *R. officinale*	Polygonaceae	Roots and rhizomes (drug "rhubarb" is used as laxative, purgative and stomach tonic).
133. *Rumex acetosella*	Polygonaceae	Leaves (tumours and cancer).
134. *Polygonum glabrum*	Polygonaceae	Seeds (leucorrhoea).
135. *Emblica officinalis*	Euphorbiaceae	Fruits ("Aamla", rich in vitamin C; used in scurvy; one of the three constituents of "Triphala").
136. *Ricinus communis*	Euphorbiaceae	Seeds (purgative).
137. *Jatropha gossypifolia*	Euphorbiaceae	Leaves (eczema), root (leprosy).
138. *Synadenium grantin*	Euphorbiaceae	Entire plant (stimulates central nervous system).
139. *Morus alba* and *M. nigra*	Moraceae	Bark (purgative and vermifuge).
140. *Antiaris toxicaria*	Moraceae	Latex (cardiac stimulant).
141. *Salix* sp.	Salicaceae	Soft inner bark and root ("aspirin" is used for headache).
142. *Belamcanda chinensis*	Iridaceae	Rhizome (tonsilitis).
143. *Iris germanica*	Iridaceae	Roots (gall bladder diseases).
144. *Curculigo orchioides*	Amaryllidaceae	Tubers (asthma, jaundice).
145. *Curcuma longa*	Zingiberaceae	Rhizome ("turmeric" is used as stomachic, antiseptic, and blood purifier).
146. *Zingiber officinale*	Zingiberaceae	Rhizome (carminative, digestive stimulant).
147. *Aloe barbadense* and *A. vera*	Liliaceae	Leaves (purgative, piles, inflammations).
148. *Asparagus racemosus*	Liliaceae	Roots (nervous and rheumatic complaints; skin diseases).
149. *Colchicum luteum*	Liliaceae	Roots (gout, rheumatism).
150. *Gloriosa superba*	Liliaceae	Tubers (promoting labour pains in women).
151. *Smilax glabra* and *S. ovatifolia*	Liliaceae	Roots (veneral diseases).
152. *Veratrum viridae*	Liliaceae	Roots (hypertension).
153. *Cyperus scariosus*	Cyperaceae	Tubers (stone trouble of kidney and urinary bladder).
154. *Cyperus stoloniferous*	Cyperaceae	Tubers (heart stimulant).

INDEX